ADVANCES IN URBAN ENGINEERING AND MANAGEMENT SCIENCE VOLUME 1

Advances in Urban Engineering and Management Science contains the selected papers resulting from the 2022 3rd International Conference on Urban Engineering and Management Science (ICUEMS 2022). Covering a wide range of topics, the Proceedings of ICUEMS 2022 presents the latest developments in: (i) Architecture and Urban Planning (Architectural design and its theory, Urban planning and design, Building technology science, Urban protection and regeneration, Urban development strategy, Ecological construction and intelligent control, Sustainable infrastructure); (ii) Logistics and supply chain management (Warehousing and distribution, Logistics outsourcing, Logistics automation, Production and material flow, Supply chain management technology, Supply chain risk management, Global service supply chain management, Supply Chain Planning and Inventory Management, Coordination and collaboration of supply chain networks, Governance and regulatory aspects affecting supply chain management); (iii) Urban traffic management (Smart grid management, Belt and Road Development, Intelligent traffic analysis and planning management, Big data and transportation management).

The Proceedings of ICUEMS 2022 will be useful to professionals, academics, and Ph.D. students interested in the above-mentioned fields. Emphasis was put on basic methodologies, scientific development and engineering applications. ICUEMS 2022 is to provide a platform for experts, scholars, engineers and technical researchers engaged in the related fields of urban engineering management to share scientific research achievements and cutting-edge technologies, understand academic development trends, broaden research ideas, strengthen academic research and discussion, and promote the industrialization cooperation of academic achievements. Experts, scholars, business people and other relevant personnel from universities and research institutions at home and abroad are cordially invited to attend and exchange.

PROCEEDINGS OF THE 3RD INTERNATIONAL CONFERENCE ON URBAN ENGINEERING AND MANAGEMENT SCIENCE (ICUEMS 2022), WUHAN, CHINA, 21–23 JANUARY 2022

Advances in Urban Engineering and Management Science

Volume 1

Edited by

Rashwan Khalil
Sichuan International Studies University, China

Jun Yang
Liaoning Normal University, China

CRC Press
Taylor & Francis Group
Boca Raton London New York Leiden

CRC Press is an imprint of the
Taylor & Francis Group, an **informa** business

A BALKEMA BOOK

CRC Press/Balkema is an imprint of the Taylor & Francis Group, an informa business

Typeset in Times New Roman by MPS Limited, Chennai, India

Library of Congress Cataloging-in-Publication Data

A catalog record has been requested for this book

First published 2023
Published by: CRC Press/Taylor & Francis Group
4 Park Square, Milton Park / Abingdon, Oxon OX14 4RN / UK
e-mail: enquiries@taylorandfrancis.com
www.routledge.com – www.taylorandfrancis.com

ISBN: 978-1-032-38495-5 (SET Hbk)
ISBN: 978-1-032-38497-9 (SET Pbk)

ISBN Volume 1: 978-1-032-30426-7 (Hbk)
ISBN Volume 1: 978-1-032-30427-4 (Pbk)
ISBN Volume 1: 978-1-003-30502-6 (eBook)
DOI: 10.1201/9781003305026

ISBN Volume 2: 978-1-032-38483-2 (Hbk)
ISBN Volume 2: 978-1-032-38492-4 (Pbk)
ISBN Volume 2: 978-1-003-34532-9 (eBook)
DOI: 10.1201/9781003345329

Advances in Urban Engineering and Management Science – Khalil & Yang (Eds)
© 2023 The Editor(s), ISBN 978-1-032-30426-7

Table of contents

Preface xiii
Committee members xv

VOLUME 1

Urban construction engineering and facility planning and design

BIM-based construction cost control research 3
Y. Chen & W. Zhao

Feasibility analysis and practice of engineering construction above strip goaf 9
X. Li

Preliminary design of the water stabilization pond system in Puerto Ayora, Ecuador 15
X. Yang, T. He, H. Chan, M. Li & J. Jiang

Regional impact analysis of the application of artificial water body on the change of urban underlying surface 32
W.Y. Lin, B. Tang & H. Zhou

Study on standard application guide for municipal sewage treatment 37
B. Xu, X. Zhao & J. Huang

Research on integrated solution for municipal sewage treatment 42
B. Xu, J. Huang & X. Zhao

5G MIMO antenna design and AI study 46
Z. Sun

Research on train adaptive speed tracking based on Lyapunov design method 52
C. Ma, B. Liu & Z. Qi

Rural landscape planning and design based on spatio-temporal big data 57
G. Yang, Y. Zhong & Y. Cheng

Antenna design of 5G millimeter wave at 26 GHz 66
F. Chen, Y. Chen & P. Li

Innovative technology and application of the delicacy management of port safety production 75
F. Liao, Y. Hu & J. Lv

The application of BIM in building project safety management 81
P. Li, C. Qi, C. Wei & J. Jiang

Design and application of digital seawall management platform 86
H. Qian, L. Xu & C. Zhou

Study on optimization of urban interchange design scheme 91
J. Chen

Comprehensive evaluation and decision-making model for university innovation base construction modes based on Matlab and AHP 99
Q. Wen, X. Han, J. Liu & T. Huangfu

Suitability evaluation of sports tourism development in ethnic villages based on entropy weight matter-element extension mode 108
S. Zhang & L. Tang

Research on comprehensive evaluation system of new campus pre-planning work of colleges and universities based on the analytic hierarchy process 116
Q. Wen, J. Liu & X. Han

Dynamic detection method of subgrade and pavement compactness for small span highway bridge maintenance 125
J.W. Zhang

Analysis on renovation application of existing buildings an example of urban reading space design of "Haoxi Book Garden" 132
W. Huang & H. Tong

A dynamic emergency network planning model with data-driven distributionally robust probabilistic constraints 139
Q. Li, L. Qin & K. Liu

Research on improvement design scheme of power distribution system in university building 149
L. Yang, F. Hu & L. Chen

Research on schedule optimization of the construction project based on resource supply 157
C.-M. Hu, J.-Y. Gong, S. Wu & X.-F. Fan

Research on design and practice of the structural health monitoring system of highway tunnel based on BIM 165
L.X. Bao, Q.L. Wang, N. Ji, Y. Jiang & Y. Zhang

Analysis on compression deformation of surrounding rock at the top arch of deep underground powerhouse 171
S. Zhang, H. Duan & C. Chen

Leakage analysis and treatment scheme design of dam foundation of Mengjiagou reservoir in Gaomi city 185
Z. Wei & L. Liu

Modal analysis of typical masonry structure of Anhui rural residence in Tan Lu fault zone 192
T. Wu & S. Xu

Study on height selection and operation mode of surge tank in long-distance water supply project 200
T.C. Zhou, G.H. Li & W.J. Danzhen

Research on research methods and protective measures of rockfall disasters on slope 206
C. Ma, T. Zhao, J. Yang, L. Cheng, Z. Zeng & Y. Huang

Research on the flood control safety design strategies of the riverside community based on the BIM analysis 216
M. Qiu, Y. Lei, Y. Lv & Y. Cui

Research on the joint upgrading scheme of series shallow tunnels of rainwater system in built-up areas 226
C. Liu, N. Luo & H. Liu

Research on design strategy and application of prefabricated tourism residential building – Taking Chongqing agricultural tourism complex project "Heshan four seasons" as an example 237
Y. Du, Z. Zhang, Z. Yin & Y. Li

A study on the correlation of vertical greening design on the outside facade of coastal city office building 243
Z. Wu & K. Chul-Soo

Construction technology and civil engineering management

HFACS-based analysis of accidents related to lifting operations in construction 251
X. Lv, Y. Liu & Z. Song

Sensing strategy of autonomous emergency braking based on V2X in non-line-of-sight scenarios 258
X. Guo & C. Piao

Study on progress management based on building information modeling and plan-do-check-action cycle 263
M.F. Khan & C. Nie

The control and improvement of anchorage construction quality under complex and deep soil filling conditions 269
Z. Li, N. Jia, J.X. Tong, J.F. Li & B.B. Ma

Civil engineering construction in the concrete structure construction technology research 276
L. Wei

Analysis of the structure safety of block 0 with tower crane 281
X. Li, J. Zhou, H. Ma & P. Xie

Quality control of emery wear-resisting curing construction for underground garage 288
M. Li

Multi-factor monitoring method for construction of fabricated continuous rigid-frame bridge based on BIM 294
J. Zhang

Study on bearing capacity of reinforced sleeve grouting connected concrete shear wall after fire 300
J. Zheng, J. Li & J. Chen

Construction technology of large span double-deck cast-in-situ beam in the deep water area 312
W. Qiang, J. Zhou, Y. Shi, T. Bo & P. Xie

One-off tensioning construction control technology for tied-arch bridge 319
P. Xie, J. Zhou, X. Liu, X. Zhao & J. Sun

Research on influencing factors of urban viaduct collapse accident based on DEMATEL 326
Z. Tian

Study on horizontal spacing of H-shaped steel piles for HU combined sheet pile support system 332
J. Chen, J. Liu & Q. Zhen

Simulation study on vibration frequency of monolayer cable net glass curtain wall 338
G. Xu & G. Chen

Research on influencing factors of prefabricated construction cost based on DEMATEL 344
Y. Zhao, D. Mei & M. Cheng

Study on the optimization method of multi-component construction scheme based on the empirical mode decomposition method 350
D. Xie, Z. Wang, K. Zhang, L. Chen & D. Zhang

The study of the relationship between natural forests and modern wooden dwellings 357
R. Chen & C. Dong

Research on the application of the parametric design method in special-shaped buildings 363
B. Shi & K. Hao

Risk assessment of building damage caused by undercrossing shield metro construction using Bayesian network 369
X. Lu, C. Xu, X. Du, J. Zhu & Y. Wang

Research on the engineering technical optimization of Village planning at Zhejiang J village based on the collaborative concept 376
Y. Zhong, L. Tong, F. Su & X. Zhu

Research on influencing factors of prefabricated construction supply chain management based on ISM-AHP 382
Z. Tian

Analysis of the optimization model of the renovation project management in colleges and universities based on time constraint theory 389
Q. Wen, J. Liu & J. Ren

Research on the control of the settlement of buildings by automatic groundwater recharge device 398
J.J. Yan, Y.M. Yao, Y.W. Zang, C.B. Liu, H.W. Ying & Z.Y. Luo

Urban distributions of producer services in building industries in China based on information entropy models: A case of top 100 construction supervision enterprises 405
Q. Zhang & Y.-J. Zhang

Assessment of prefabricated concrete buildings construction safety based on objective and subjective weighting combined with fuzzy comprehensive evaluation 411
G. Zhang, X. Fan & H. Zhu

The implication of using modular construction projects on the building sustainability: A critical literature review 422
Z. Zhou, D. Syamsunur & X. Wang

Research on the influencing factors of prefabricated building quality based on AHP 428
Z. Song & S. Chuan

Research on influencing factors of construction economy development based on principal component analysis 433
Z. Song & S. Chuan

Application of muddy lime soil flushing and backfill method in water conservancy and civil engineering 438
Y. Zhu, L. Xiong & X. Ji

The key points of technical management in the construction of high-rise shear wall buildings 447
H. Jia

Study on technology of gangue backfilling and reclamation for construction land 453
G. Sun

Public infrastructure and urban traffic management

Research on the needs of public space for old people sojourning in the background "Internet Plus"—Take Southern Hunan as an example 463
D. Liu, B. Tang, K. Zhao, L. Chen, C. Qiu & P. Fu

Application and practice of intelligent infrastructure in urban transportation management 473
Y. Wang & L. Pan

Analysis of spatial characteristics of intercity traffic travel network based on mobile phone signaling datum—A case study of Fujian Province 482
Y. Liu, J. Ma, T. Liu & D. Wu

Short-term passenger flow prediction of urban rail transit based on SSA-GRU 490
E.Y. Chen, T. Zhang, Y.X. He & Q. Luo

Infrastructure construction of elderly community based on digital network intelligence 499
Y. Hou

A dynamic scheduling method of satellites communication in an uncertain environment 507
Y. Liu & P. Yang

Multi-lanes detection in complex scenes based on Res2Net 515
F. You, Z. Xiao, Z. Wu, J. Yang & J. Yang

Optimization study of train stopping scheme of Guangzhou-Zhuhai intercity railway 528
S.Q. Chen, W. Li, Q. Luo & J.T. Lu

Evaluation of forward collision warning systems on driver behavior and EEG in heavy fog conditions: Based on a driving simulator 539
R. Liu & X. Yan

Study on the influencing factors and implementation path of China's freight structure adjustment under the background of "Carbon Peak" and "Carbon Neutrality" 546
Y. Gao, J. Cheng, L. Bai, G. Zhu, S. Liu & X. Yang

Quantitative risk analysis of major accidents in dangerous cargo consolidation areas at port 553
P. Shao & M. Lan

Comprehensive importance evaluation of urban rail transit network based on complex network 562
S. Han, J. Chen, J. Zhu, F. Tang & X. Yan

Comprehensive evaluation of highway development based on DEA and gray correlation 572
J. Yu, D. Wu & S. Qin

A framework for intelligent network architecture-based transportation infrastructure resilience assessment index system 577
X.H. Yu, G.X. Chang, W. Wei & Z.Q. Zeng

Research and application of safety performance model in port enterprises based on fuzzy theory 584
C. Huang, Y. Bai, L. Lu & H. Jin

Research on urban road safety early warning system 591
S. Ying

A review and insight of vosviewer-based China community renewal-related studies 600
S. Lou

Modeling cascading failures in wireless mesh network technology-based communication networks 608
X. Wang, W. Wei, M. Shi & Z. Zeng

Analysis of freeway traffic crash severity considering time differences 615
L. Jiang, C. Duan, Z. Li, X. Wu, Y. Jiao & Y. Chen

Exploring the factors affecting cyclists' detour decisions in the curbside bus station area 622
J. Liang & L. Qiu

Verification of improved social force model based on vehicle trajectory data 631
C. Lin, Y. Huang, G. Su, X. Yan, C. Long & Z. Xie

One-way traffic organization optimization method of microcirculation road network in Handan City 637
S. Li

The impact of the opening of high-speed railways on corporate financing constraints based on the spatial effect 644
Z. Tian

PM2.5 spatial distribution characteristics analysis in subway station 649
L. Liu, H. Liu, P. Xu & Y. Ma

Information systems architecture development of smart expressway for large scale road network across China 656
K. Chen & J. Ding

Research on the coupling and coordination of transportation and socio-economic system in southwestern minority areas 664
J. Liu, Y. Zheng, X. Han & X. Zhou

Evaluation of community grid management effect based on gray scale correlation model takes Jiang'an District of Wuhan, Hubei Province as an example 671
Y. Wang & F. Guo

An improvement of car retrieve process and system configuration of compact robotic automated parking system 678
Y. Ma, J. Fan & L. Ren

Research and practice of digital protection of historic buildings relying on information models—Take the historical buildings of Taizhou Prefectural City as an example 691
Y. Xin & Y. Xu

The analysis of high-speed railway opening and enterprise performance—Empirical test based on difference-in-difference method 698
L. Sun

An evaluation index system based on the supply-demand of urban parks from the perspective of "Park City" 703
J. Shi & Q. Guo

Research on BIPV design strategy and technology of green public buildings based on data analysis—Taking Xiong'an high-speed railway station as an example 711
B. Xu & Y. He

Study on multi-level analysis method of highway slope maintenance decision during operation period 719
Z. Cheng, Y. Wang, K. Zhang & Q. Li

Study on the construction of China's international logistics system under the international public health crisis 728
H. Hu & L. Yan

The feasibility research of post-occupancy evaluation of mega-event sports venue 737
X. Wu, H. Liu, C. Liu & X. Wan

Study on landscape renewal of old residential area based on microclimatic improvement 744
L.S. Cao

Obstacle detection based on semantic segmentation for autonomous vehicles 751
Y. Liu, Q.X. Guan, X. Yuan & Y. Shi

Author index 757

Advances in Urban Engineering and Management Science – Khalil & Yang (Eds)

Preface

The 2022 3rd International Conference on Urban Engineering and Management Science (ICUEMS 2022) was held on January 21-23, 2022 as a virtual conference due to the growing concerns over the coronavirus outbreak (COVID-19), and in order to protect the well-being of our attendees, partners, and staff as our number one priority. The meeting arounds the network communication and recent research achievements in the field of information security, from institutions of higher learning, scientific research institutes, enterprises and institutions at home and abroad experts, professors, scholars, engineers and so on to provide a shared professional experience, expand the professional network, face to face to new ideas and show the research results of the international platform, Exploring key challenges and research directions in this field, with a view to promoting the development and application of theories and technologies in this field in universities and enterprises, as well as establishing business or research connections for attendees and seeking global partners for future ventures.

This scientific event brings together more than 150 national and international researchers in urban engineering and management science. On top of the local participants coming from different national universities, international participants are also registered from different countries, namely Malaysia, India, Singapore and Australia. During the conference, the conference model was divided into three sessions, including oral presentations, keynote speeches, and online Q&A discussion. In the first part, some scholars, whose submissions were selected as the excellent papers, were given about 5-10 minutes to perform their oral presentations one by one. Then in the second part, keynote speakers were each allocated 30-45 minutes to hold their speeches.

In the second part, we invited four professors as our keynote speakers. Prof. Run Wang, School of Resources and Environmental Science, Hubei University, performed a speech: *Challenges for achieving carbon reduction target in Metropolitan Wuhan*. And then we had Assoc. Prof. Shouliang Zhao, Huazhong University of Science & Technology, China. He presented an insightful speech: *Brilliantness and Limitation of Rationality: A 60-year Review of Brasilia's Plan*. Assoc. Prof. Mohd Johari Mohd Yusof, Department of Landscape Architecture, Universiti Putra Malaysia, delivered a speech: *Monitoring Urban Green Space Changes Using High Resolution Aerial Imagery: A Case Study of Kuala Lumpur, Malaysia*. Assoc. Prof. Alessandro Bianchi, our finale keynote speaker, Politecnico di Milano, Italy. He presented an insightful speech: *Disaster, recovery, design: Proposal for scarred landscapes*. Their insightful speeches had triggered heated discussion in the third session of the conference. Every participant praised this conference for disseminating useful and insightful knowledge.

The proceedings are a compilation of the accepted papers and represent an interesting outcome of the conference. Topics include but are not limited to the following areas: Architecture and Urban Planning, Logistics and supply chain management, Urban traffic management and more related topics. All the papers have been through rigorous review and process to meet the requirements of International publication standard.

We would like to acknowledge all of those who supported ICUEMS 2022. The help and contribution of each individual and institution was instrumental in the success of the conference. In particular, we would like to thank the organizing committee for its valuable inputs in shaping the conference program and reviewing the submitted papers.

The Committee of ICUEMS 2022

Advances in Urban Engineering and Management Science – Khalil & Yang (Eds)
© 2023 The Editor(s), ISBN 978-1-032-30426-7

Committee members

Conference Chairman
Prof. Rashwan Khalil, Sichuan International Studies University, China

Publication Chairs
Prof. Jun Yang, Northeastern University, China
Senior Assoc. Prof. Christiane M. Herr, Xi'an Jiaotong-Liverpool University, China
Distinguished Pro. Shih-Wen Hsiao, National Cheng Kung University, Taiwan, China

Technical Program Committee Chairs
Prof. Zakiah Ahmad, University Teknologi Mara, Malaysia
Assoc. Prof. Juan Carlos Dall'Asta, Xi'an Jiaotong-Liverpool University, China
Dr. Isha Suwalka, Geetanjali Institute of Technical Studies, India

Local Organizing Chairs
Assoc. Prof. Xiuhua Li, Stat Grid Technology College, China
Siu Ming Fung Francis, The Hong Kong Polytechnic University, China

Technical Program Committee
Prof. Li Bing, Shenyang Jianzhu University, China
Prof. Zhigang Zhang, Airforce Engineering University, China
Prof. Ir. Dr. Hj. Ramli Nazir, Faculty of Engineering, University Teknology Malaysia, Malaysia
Prof. Dr. Muhd Zaimi Bin Abd Majid, Universiti Teknologi Malaysia, Malaysia
Prof. Qin Xiaosheng, Nanyang Technological University, Singapore
Assoc. Prof. Mamoun Alazab, Darwin University, Australia
A. Prof. Weijun Cen, Hohai University, China
A. Prof. Mohammad Arif Kamal, Aligarh Muslim University, India
A. Prof. Shah Kwok Wei, National University of Singapore, Singapore
A. Prof. Bon-Gang Hwang, Department of Building, Singapore/National University of Singapore, Singapore
Dr. Nur Mardhiyah Aziz, University of Malaya, Malaysia
Dr. Mohd Rosli Mohd Hasan, School of Civil Engineering, University Sains Malaysia, Malaysia
Dr. Kim Hung Mo, Faculty of Engineering, University of Malaya, Malaysia
Dr. Yuen Choon Wah, Centre for Transportation Research, University of Malaya, Malaysia
Dr. Suhana Koting, Faculty of Engineering, University of Malaya, Malaysia
Dr. Sharifah Akmam Syed Zakaria, University Sains Malaysia, Malaysia
Dr. Derek MA, University of Warwick, UK

Urban construction engineering and facility planning and design

Advances in Urban Engineering and Management Science – Khalil & Yang (Eds)

BIM-based construction cost control research

Yujie Chen & Wanhua Zhao*
ECivil Engineering and Architectural Institute, Wuhan Polytechnic University, Wuhan, Hubei

ABSTRACT: With the maturity and popularization of BIM in China, assembly buildings have also begun to develop rapidly. In 2020, the construction area of newly started assembly across the country increased by nearly 50% over the previous year. However, the market share of assembly projects is still low. Research shows that construction costs are still the biggest obstacle to the development of assembly buildings. With the rapid development of construction information technologies, this report discusses the use of BIM, illustrates how to control the cost of assembly buildings and their related supports, and explores the feasibility of combining all aspects of related technologies, combined with the development wave of domestic assembly buildings. Our aim is to provide a reference for reducing China's assembly construction costs.

1 INTRODUCTION

Despite the fact that China proposed significant concepts in the mid-twentieth century, China's building industrialization development level is not behind that of developing countries due to objective factors such as material economics and national policies. Although the construction industry is booming as a result of reform and opening up, China's construction industry had high energy consumption and pollution status until the beginning of the twenty-first century (energy consumption accounted for 25% of China's energy consumption) (Mao & Yu 2014). Under national macro-policy, the adoption of sophisticated foreign expertise and technology has become a key trend, aligned with the concept of new building industrialization and combined with BIM assembly building applications to become the focal point of construction industry development.

2 THE COMPOSITION OF THE COST OF AN ASSEMBLY CONSTRUCTION PROJECT

2.1 *The current state of assembly construction costs*

According to the relevant research, the construction cost of assembly construction projects is 200–400 yuan/m2 higher than that of traditional current-watered buildings, and the high cost of assembly construction projects is the main reason for limiting the development of assembly buildings in China (Mao et al. 2016). Compared with traditional buildings, the diversified use of assembly buildings in China still has more shortcomings, and authoritative, standardized industry standard indicators are still lacking. The experience of managers in assembly building projects is also scarcer, and the synergy with other emerging technologies is not yet mature. At present, China's prefabricated component assembly plants have not formed a cluster effect, and the relevant supporting facilities

*Corresponding Author: 374936999@qq.com

DOI 10.1201/9781003305026-1

have not been mature. Moreover, most factories do not have the machines to produce large components. The degree of automation and productivity are insufficient, which requires a lot of human and material resources and uneven completion quality. In general, assembly building technology is consistent with other new technologies, such as 3D printing building technology, and faces the main reasons for such as lack of experience in integrated management mode, high efficiency ratio of output of prefabricated components, low degree of design standardization, small economic scale, and insufficient high-end technology in some industries.

2.2 *The components of an assembly building project*

In order to study the link where BIM can be applied to reduce costs, it is essential to analyze its constituent elements. According to the relevant research, the construction cost of an assembly construction project is analyzed, and it is pointed out that the production cost of prefabricated components accounts for 70%–80% of the total cost of an assembly construction project. For assembly construction projects, if the profit is divided according to the cost composition of the whole life cycle of the construction project proposed by EL-Haram. Construction costs, operating and maintenance costs, disposal costs, operating costs, management costs, and financing costs are all composed of the project's full life cycle costs (Figure 1). Throughout the whole life cycle of the project, cost control based on BIM technology can be used to maximize effectiveness (Xue 2020).

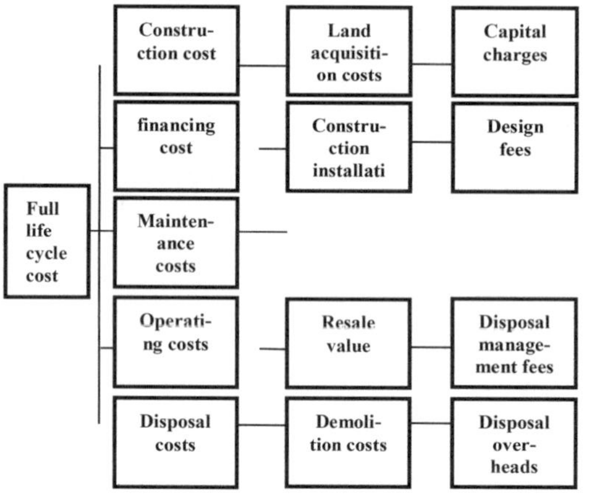

Figure 1. Components of full-life cycle cost.

Obviously, the difference between the construction cost of an assembly building and the cost of a current-poured building is the most important part of the entire assembly building project. Among them, the difference in construction installation costs is obvious. The survey and design fee in other costs of engineering construction is also one of the differences between the cost composition of assembly buildings and currently poured buildings (Table 1). Cast-in-place components are generally constructed on the spot and there is a standardized, low degree of scale, which leads to serious environmental pollution, the quality of the finished product being uneven, etc., and also leads to an increase in the cost of measures. Most assembly components are built in factories far from the city, and they show the characteristics of large clusters, the same standards but more environmentally friendly.

Analyses above show that the cost composition difference between prefabricated building projects and cast-in-place buildings mainly lies in the related cost and management level of component manufacturing. With the improvement of relevant supporting facilities in the industry, the characteristics of BIM technology can be used to effectively control the cost of each link.

Table 1. The difference in the cost composition of an assembly building with a current-poured building.

Construction Installation Costs (partial)	Current-Poured Building	Assembly Building
Part of the project costs	Structural engineering costs Floor engineering costs Other engineering costs	Structural engineering costs Floor engineering costs Engineering costs for prefabricated components Other engineering costs
Measures for project fees	Construction technical measures Other measures	Construction technical measures PC technical measures fee Other measures

3 COMPREHENSIVE APPLICATION

3.1 *BIM-based assembly building project cost control feasibility*

The cost control of the project can be classified through the sub-items of the project, the progress and duration of each sub-item at the macro level, and the comprehensive analysis of the project cost is realized by factories, such as the duration of the project, the value of the completed project, the resource consumption index, etc. At the same time, the value of the deviation range between the cost and progress of the project can be controlled by the comprehensive deviation analysis by the earned value analysis index. ACWP, or Actual Cost for Work Performed, is primarily an indicator of actual consumption of project execution (Han 2019). BCWP, or Budgeted Cost for Work Performed, is the amount of work (or expenses) that is calculated at a stage in the implementation of a project based on actual work done and budget quotas, i.e., Earned Value. Its theoretical formula is shown as below.

$$CV = BCWP - ACWP \tag{1}$$

$$CPI = BCWP / ACWP \tag{2}$$

In each cost element, the comparative analysis of divisional engineering costs and total engineering costs can use the progress, cost deviation, and deviation rate indicators, which can reflect the construction process, so as to regulate the project cost through deviation analysis. In a construction project, it is found that on the current date, ACWP > BCWS > BCWP, CV < 0, reflecting the characteristics of low efficiency, slow progress, and lack of investment, and then the development of Figure 2 (a). If we replace inefficient personnel with efficient personnel on the current date, more strict management of material waste and other measures, then the subsequent development

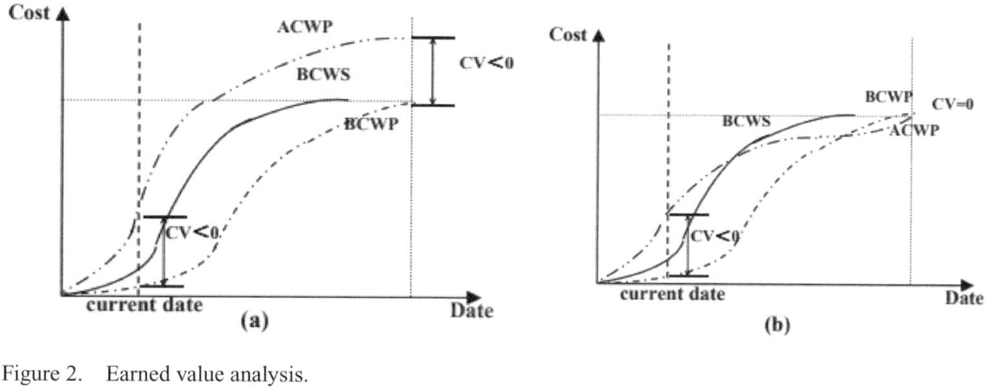

Figure 2. Earned value analysis.

of Figure 2 (b) shows. At this point, it can be found that the BCWP, ACWP, and CV-0 projects are completed according to the plan, providing the decision maker with the opportunity to make a decision.

The above model analysis can be seen to provide project managers with construction dynamic information is very important. BIM technology functions are very consistent with the characteristics of cost control, and through the combination of Revit, Naviworks, and other software 3D models can achieve project dynamic, simulation, visualization, and performance management, so that assembly building projects can achieve real-time dynamic task tracking, optimize the various construction links, as far as possible, to reduce project costs and control the project construction progress.

3.2 *Cost control based on BIM information synergy*

Prefabricated component engineering and PC measure increase fees are closely related to BIM technology. Use BIM data exchange to better standardize industry guidelines and let future users learn from them. In addition, information sharing using BIM is applied to the automatic production of factories, which is coordinated with the RFID platform. When designing prefabricated components, it is generally necessary to build a BIM information model to carry out basic applications of BIM technology such as pipeline synthesis, construction simulation, performance analysis, visual design, etc. according to line requirements. And the use of BIM standards, libraries, and other data effectively shortens the design time. During the factory production phase, positions such as PC sampler and model technician can more intuitively understand the technical parameters of the component through the BIM building information model, avoiding errors such as rework that could lead to increased project costs. In the construction phase of the site, BIM simulation project construction is used to optimize the assembly-type construction design construction tools, design, and construction work condition map, and develop a new assembly-type construction method. Intuitive animation demonstrations can accelerate the training of on-site construction workers to ensure the safe and efficient completion of project construction tasks during the total construction period. Through BIM to contact the information and data exchange of the main body of each link, a three-party common information cooperation platform for design institutes, factories, and construction sites to improve the efficiency of information and data exchange and reduce subjective consciousness expression errors. Through the platform of BIM, we will connect the next generation of information technology, such as block chain, cloud data, big data, and other hot technologies to promote the improvement of assembly building project integrated management.

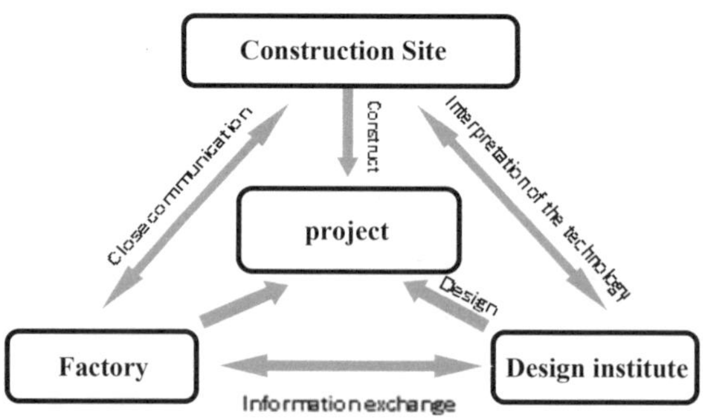

Figure 3. Multi-party collaboration.

3.3 *Cost analysis of BIM-based assembly building operation and maintenance*

Later operation and maintenance can make use of the pre-information data in BIM to facilitate the inspection and repair. In assembly buildings using BIM technology, the clear and detailed component information can be extracted from the library when components need to be maintained and replaced. Through the 3D model, it also helps the maintenance personnel better understand the composition and structure of the repair components when combined with existing technology, such as RFID, to achieve operational maintenance management information. If the components are equipped with self-test devices, their own information can be transmitted to managers in real time, greatly reducing the safety hazards of building-related accidents. When evaluating the cost-benefit indicators of the operation and maintenance of assembly buildings and traditional buildings using BIM, OWA operators can be used for comprehensive evaluation. OWA operators are often used in multi-attribute decision-making methods. Managers can sort N benefit indicators $(A_1, A_2, A_3, \ldots A_n)$ based on importance and then reevaluate to get the criteria weights $(B_1, B_2, B_3 \ldots B_n)^T$. Managers need the form of a two-type triangular fuzzy number to give the decision value of the scheme and generally need computer mathematical modeling at this time. Part of the value stored in BIM can be directly used. Where the decision matrix is made $R = (r_{ij})$. The cost formula is used here (Wang & Han 2013).

$$a_{ij} = \left\langle \left[a_{ij}^1, a_{ij}^2, a_{ij}^3\right]; \left[\mu_{ij}^{\alpha}, \mu_{ij}^{\beta}, \mu_{ij}^{\gamma}\right] \right\rangle \tag{3}$$

$$a_{ij}^k = \frac{\max_j \mu_{ij}^3 - \mu_{ij}^j}{\max_j \mu_{ij}^3 - \min_j \mu_{ij}^1} \tag{4}$$

$$\partial_{a_{ij}}^s = \partial_{\mu_{ij}}^s, k = 1, 2, 3, S = \alpha, \beta, \gamma \tag{5}$$

After obtaining the above values, calculate the confidence level matrix and confidence level $\varepsilon = \mu_i^{\alpha} + 2\mu_i^{\beta} + \mu_i^{\gamma}$. Then the two-type triangular induced OWA operator assembles the standard value of the scheme to obtain the comprehensive benefit index of the scheme and changes the order of cost-benefit indicators, with the possibility of two comparisons, to obtain the optimal solution. At the same time, the cost component of using BIM assembly buildings has also increased, compared with traditional buildings (Hua et al. 2019). The assembly building using BIM contains the first acquisition hardware fee, the system maintenance fee during operation, a software update fee, an equipment hardware replacement fee, etc. Thus, the total cost of the new technology related to hardware and software is zero. The operation of BIM assembly buildings requires professionals who can use BIM and related technologies. The sum of the additional expenses caused by the higher salaries of relevant technicians than ordinary employees and the training costs used to train employees in new technologies is V. Therefore, the total additional costs are

$$W_1 = V + U \tag{6}$$

At the same time, we need to calculate the economic benefits of BIM technology. Assembly buildings using BIM save quantifiable resources such as water, electricity, and gas compared to conventional buildings and have high energy efficiency. So, we get the sum of direct economic benefits P. Dynamic management to reduce damage to buildings and obtain required maintenance components directly from factories is part of the simple economic benefits of using BIM assembly buildings, while BIM assembly buildings are less carbon-based and more environmentally friendly. In the near future, the government will also provide relevant tax breaks for prefabricated construction as well as direct economic subsidies and other related policies. The sum of these more common indirect economic benefits is given as O. In addition, we have some indirect economic benefits that are completely unquantifiable in daily life, such as improving the mood of residents, benefiting human health, and other complex multivariate issues, but perhaps the use of BIM assembly buildings has a positive impact on these other indirect economic benefits. The total economic benefit is

$$W_2 = P + O + Q \tag{7}$$

Overall, the BIM-based assembly building operation and maintenance compared to the cost savings (overspending), W (the traditional building), is equal to the total additional cost (W_1) minus the total economic benefits (W_2).

$$W = W_1 - W_2 \tag{8}$$

4 CONCLUSIONS

In order to solve the problem of high construction costs for assembly construction projects, this report explores the realization path of standardization and economies of scale for assembly construction projects based on the development status and definition of BIM and assembly buildings, and starts from the cost components of traditional cast-in-place construction projects and assembly construction projects. Through the earned value analysis method, it is proved that the cost control of BIM assembly construction projects is proven. In all links of the whole life cycle, the characteristics of BIM technology and new technologies are used to reduce the cost of assembly buildings, and the operation of assembly projects combined with BIM technology is analyzed. The advantages of maintenance cost control and the comprehensive study of BIM's assembly building cost control in many aspects. The research results show that stakeholders of assembly construction projects can apply BIM technology to the formation of collaborative mechanisms, the establishment of platforms, operation governance mechanisms, and construction cost performance feedback, which is conducive to helping stakeholders make decisions, control project progress, and reduce the construction costs of assembly construction projects.

REFERENCES

Han H.L. The application of BIM technology in the cost control of assembly building. Hebei University of Technology, 2: 19–21 (2019).
Hua Y.N., et al. "Study on BIM Application Value Index System in Green Building Operation Phase." Value Engineering 038.021: 216–218 (2019).
Mao C, Xie F, Hou L, et al. Cost Analysis for Sustainable Off-Site Construction Based on a Multiple-Case Study in China [J]. Habitat International, 57: 215–222 (2016).
Mao Z.B., Yu Z.P. Reflections on a number of issues to promote the development of green construction in China. Construction Technology, 43 (01): 14–16 (2014).
Wang Q, and Han Z.Q. "Multi-criterion decision-making method based on type 2 triangular induction OWA operators." Control and Decision 000.007: 1037–1040 (2013).
Xue H. Research on stakeholder collaboration mechanism of assembly building projects under the guidance of construction cost. Diss. Harbin University of Technology, 2: 23–25 (2020).

Advances in Urban Engineering and Management Science – Khalil & Yang (Eds)
© 2023 The Author(s), ISBN 978-1-032-30426-7

Feasibility analysis and practice of engineering construction above strip goaf

Xueliang Li*
China Coal Science and Technology, Ecological Environment Technology Co., Ltd., Beijing, China
China Coal Research Institute, Beijing, China

ABSTRACT: With the increasing shortage of urban construction land, the land resources above the surrounding goaf have to be used. In order to ensure the safety of buildings (structures) during construction above the mined-out area, a feasibility analysis of the construction of the involved area must be carried out, and corresponding foundation treatment and anti-deformation measures should be taken when necessary. At present, the evaluation method for the foundation stability of longwall goaf is relatively mature. Because strip mining may have many problems such as design flaws, coal pillar softening, groundwater erosion and weathering, etc., the stability judgment has not yet formed a unified evaluation standard. Based on the analysis of the characteristics of strip mining and the evaluation method of coal pillar stability, the article studies the feasibility of engineering construction above the strip goaf. It is introduced and analyzed based on engineering examples to provide references for the design and construction of related fields.

1 INTRODUCTION

In the mining cities of central and eastern my country, the per capita land occupation is small, and land resources are very precious. When carrying out engineering construction, many places have to consider using the land above the coal mining subsidence area. In order to protect the buildings (structures) above coal resources in the early days, strip mining was adopted in many areas. Reduce the damage of underground mining to the overlying rock layer, thereby reducing the movement and deformation of the ground surface, so as to achieve the purpose of protecting ground structures (structures). However, the overlying rock and ground surface movement and deformation caused by this method are not sufficient, and there are still hidden dangers of further movement and deformation, especially whether the remaining coal pillars can maintain long-term stability requires further research (W. J et al. 2015; W. Q. Z et al. 2016; Y. Y et al. 2017). Therefore, it is particularly important to seek a reasonable method to explore the feasibility of the construction of strip goaf projects.

Over the years, a large number of experts and scholars have done a lot of research on the theory, technology, and practice of goaf engineering construction, and their results have played a certain guiding role in the development and utilization of the land above the strip goaf. The article comprehensively considers various factors that may affect the construction of strip goaf projects and conducts a more detailed analysis and verification, respectively, to provide ideas for the construction of similar complex old goaf projects.

2 STRIP-MINING THEORY

The strip-mining method was first widely used in Poland, the United Kingdom, the former Soviet Union, the United States, South Africa and other countries, followed by Australia, India, and

*Corresponding Author: 371986152@163.com

DOI 10.1201/9781003305026-2

Germany. Since 1967, my country began to apply the strip-mining method for coal mining under buildings, under water bodies, and under railways. At present, it has been popularized and applied in more than 10 provinces and 100 strip working faces across the country, and a large amount of measured data has been obtained.

2.1 *Strip mining characteristics*

Strip mining refers to the division of strips along a certain direction in the mining block. The mining and retention strips are arranged alternately, and the reserved strip coal pillars are used to support the load of the overlying rock to control the movement of the surrounding rock. The area is quite large. It is a coal mining method that is based on the total area of the mining block, so that the surface movement and deformation are kept within the allowable range (see Figure 1).

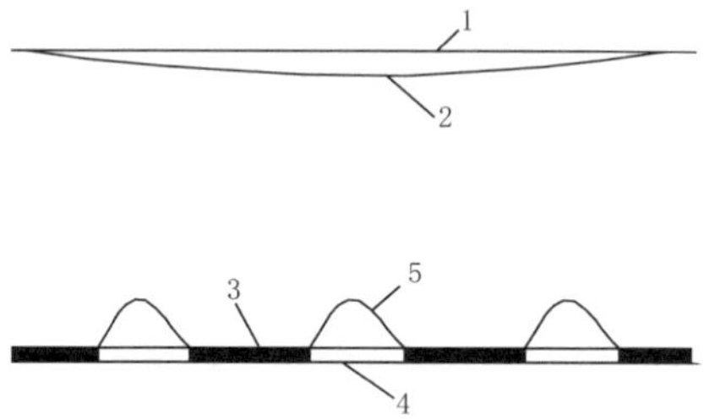

1: Original surface; 2: Surface after sinking; 3: Coal pillar; 4: Goaf; 5: Natural balance arch.

Figure 1. Schematic diagram of strip mining.

Compared with the longwall full collapse method to manage the roof, the surface movement law of strip mining has the following characteristics:

(1) The surface subsidence is small. The surface subsidence coefficient of domestic strip mining is 4.8%~26.8% of the subsidence coefficient of all caving mining methods. The variation range of the subsidence coefficient is 0.024~0.206.
(2) The surface movement tends to be gentle, and the main influence angle tangent Tgβ is small, generally 1.0~2.0.
(3) The active period of strip mining and the moving duration are relatively short.
(4) The amount of horizontal movement and the coefficient of horizontal movement are small. The variation range of the horizontal movement coefficient of strip mining is 0.2~0.3.

2.2 *Design principles*

Reasonable determination of strip-mining size is the key to the success of strip mining. When designing strip mining dimensions, the following principles should be observed:

(1) The maximum mining width of the strip is not more than 1/4 of the thickness of the minimum bedrock overlying the coal seam to ensure that the overlying rock above the mining strip forms a stable, self-supporting arch and avoids wavy subsidence on the surface;
(2) The width of the strip coal pillars should not be less than 1/10 of the maximum mining depth at the lower boundary of the strip-mining area to ensure the long-term support and stability of the strip coal pillars;
(3) The ratio of the width of the strip coal pillar to the thickness of the coal should generally be greater than or equal to 5;
(4) According to domestic and foreign experience, in order to ensure the long-term stability of coal pillars, the safety factor of coal pillar bearing is required to be greater than 1.

3 STABILITY ANALYSIS OF COAL PILLARS

3.1 *Influencing factors of coal pillar stability*

In strip mining, the stability of coal pillars is one of the key factors that determine the success or failure of strip mining (L. X. et al. 2014; S. J. C et al. 2020; Y. B. Z et al. 2013; Y. T et al. 2019). Practice has proved that the factors affecting the stability of coal pillars in strip mining mainly include the following aspects:

Geological factors mainly refer to the initial ground stress, the bulk density of the overburden, the mining depth, the geological structure, the inclination of the coal seam, the occurrence conditions of the roof and floor of the coal seam, and the influence of groundwater.

Mining factors mainly refer to the mining width of the coal pillar, the remaining width, the recovery rate, the height of the strip coal pillar, the treatment method of the goaf, the mining method, the technology, etc.

The mechanical properties of the coal pillar itself mainly include the uniaxial compressive strength, tensile strength, shear strength, elastic modulus, cohesion, internal friction angle, and internal structure of the coal pillar, weak surface or coal pillar cohesion, and internal friction angle of the top and bottom plate interface.

3.2 *Instability mechanism of coal pillars*

After the underground mining is over, despite the long-term natural compaction, the underground cavities, separation layers, cracks and collapse zones caused by the mining are still under-compacted and saturated with water for a long time. Building buildings above the goaf, seismic activity, mining in adjacent areas, mining in multiple coal seams, and forced drainage of groundwater, etc., may break the original stress balance and "activate" the goaf. Research shows that there are big differences in the foundation stability of the mined-out area with different roof management methods; at the same time, the rock mass structure at different positions in the mined-out area and its overlying rock is quite different, and this difference in the structure of the secondary rock mass caused by mining. It has an important influence on the "activation" law of the overburden in the goaf, and its existence will determine the basic characteristics of the "activation" of the overburden in the goaf and directly affect the stability of the foundation above the goaf.

In the mining process of strip mining, with the continuous increase of the mining space, the huge pressure of the overlying rock mass on the roof can only be borne by the coal pillar, which gradually creeps under the increasing pressure. The roof of the overlying rock mass can collapse due to the loss of support caused by yield. It is not difficult to see that the time effect of coal pillar rock mass strength cannot be ignored. To ensure the long-term stability of coal pillars, the rheological properties of coal rock mass must be considered. It is an important research topic for safe mining technology under thick coal seam buildings.

4 FEASIBILITY ANALYSIS OF ENGINEERING CONSTRUCTION

4.1 *The impact of new building loads on the goaf*

Different loads, mining depths, mining thicknesses, lithology, etc., will have a certain impact on the movement and deformation of the building foundation (J. X. F. et al. 2016; L. W. R. et al. 2018; Y. F. Z. & H. B. C. 2014).

Under different load conditions at the same position above the goaf, the maximum value of various movements and deformations of the foundation increases linearly with the load of the building, which shows that the load weight of the building is the main factor that causes the activation of the foundation of the goaf. As the load increases, the amount of ground subsidence above the goaf increases to varying degrees, and the impact of the load on the overlying rock masses on the goaf gradually decreases from top to bottom.

The relative position of the building and the goaf is also one of the main factors affecting the foundation deformation of the goaf. Under the same load acting on different positions above the mined-out area, the amount of surface deformation and subsidence caused is different. When the load is located above the middle of the mined-out area, the amount of deformation and subsidence is the largest, which means that the rock mass in the middle of the mined-out area is the most severely damaged by mining, the mechanical properties of the rock are reduced, and the compressive capacity is the least. The most serious case: when the building load is located above the boundary of the goaf, the rock layer near the center of the goaf is more damaged than the edge. The edge rock layer in the goaf has lower mechanical properties and weaker resistance to deformation. As a result, the building above it sinks unevenly, and the tilting deformation is more serious. The uneven sinking causes the most serious damage to the building, and it is easy to cause cracks in the wall and instability of the foundation.

When the load acting position is the same, the residual subsidence value gradually decreases with the increase of mining depth. With the increase of the additional load, the movement and residual deformation of the ground surface gradually increase, and the maximum deformation of the ground surface occurs directly below the load acting position in the goaf or close to the positions on both sides of the load. Under the same load, the surface subsidence and residual deformation are the largest in the middle area of the goaf, the inner edge area is larger, and the outer edge area is the smallest. When the load size and acting position are the same, the residual subsidence and deformation of the ground surface will decrease with the increase of mining depth.

4.2 *Evaluation method of foundation stability in old goaf*

The strip goaf is different from the longwall goaf. Its stability is mainly reflected in the stability of the coal pillar under the interference of external factors. If the coal pillar loses its stability, it will cause the instability of the entire goaf and its overlying rock structure. The key issue for buildings above the old goaf is the evaluation of the stability of the foundation of the old goaf (C. Y. et al. 2017; J. Y. Z 2009; L. W. et al. 2010). At present, the methods for evaluating the stability of the building foundation in the old goaf are shown in Table 1.

Table 1. Methods for evaluating the stability of building foundations in old goafs.

Method	Principles
Load influence depth method	The main evaluation basis is whether there is a protective layer between the depth of the impact of the building load and the top interface of the collapsed fracture zone.
Critical mining depth mining thickness ratio method	The critical mining depth and thickness ratio are used as the criterion to evaluate the stability of the building foundation in the mined-out area, and a method for determining the critical mining depth and thickness is given.
Additional stress analysis method	According to the critical depth calculation formula provided in the regulations, it is mainly suitable for small coal mine goaf sites with shallow buried depth and simple geological conditions.
Numerical simulation method	Based on the theory of elastoplasticity, the stratum of the old goaf is generalized into a continuous medium model, and the influence of different overburden combinations and different building loads on the stability of the building foundation in the goaf is analyzed by the finite element or finite difference method.
Fuzzy comprehensive evaluation method	Transform geological and mining information into quantitative fuzzy language, comprehensively identify and evaluate the stability of coal mining subsidence areas, and comprehensively consider the influence and interaction of various factors.

In actual work, the load influence depth method and the additional stress analysis method are generally used to analyze the influence of the new construction on the stability of the mined-out area. The load influence depth method is suitable for the quantitative evaluation of the degree of influence of engineering construction on the stability of the goaf of types such as tunnels, rooms with pillars, and single roadways. The additional stress analysis method is suitable for the quantitative evaluation of the influence degree of engineering construction on the site stability of shallow and medium-deep mined-out areas where the collapsed fault zone is developed and the degree of compactness is poor. Strip mining is a short-wall cave mining method and does not belong to the types of tunneling, room-pillar, and single roadway, so the additional stress analysis method is generally used.

5 ENGINEERING PRACTICE

Only No.3 coal seam is mined in and around the evaluation area. It mainly involves working faces in the 3 mining areas. The mining depth is 250~470 m, the mining thickness is 3.2~5.0 m, the coal seam dip is mostly 3°~10°, and the mining time is 2000–2014. The underground area in the eastern part of the evaluation area is the strip-mining face under thc village, and the other areas are the longwall fully mechanized mining face.

Strip mining is designed for safety reasons and to avoid village relocation. The final mining and retention ratio are set at 35 m and 55 m. According to the data of strip-mining face size, coal pillar size, mining thickness, and overburden depth, the strength safety factor of a coal pillar can be calculated. It can be seen from Table 2 that the safety factor of coal pillars in the strip-mining area is 2.29, which meets the requirement of being greater than 1.5, and the aspect ratio of coal pillars is 13.8, which also meets the requirement of being greater than 5. The size of the strip coal pillars under the village meets the strip design principles and stability requirements. Therefore, when the strip coal pillars are stable, the ground surface generally does not deform discontinuously.

Table 2. Calculation table of coal pillar strength for strip mining in the evaluation area.

Average mining Depth(h/m)	Mining thickness (n/m)	Strip size(m)			Safety factor
		Stay wide (a)	Mining width (b)	a/b	
295	4.2	55	35	1.6	2.29

The overlying rock lithology in this area is mainly sandstone, mudstone, claystone, etc., which are considered medium-hard rocks. After calculation, the maximum mining thickness in the evaluation area is 5.0 m, and the height of the water-conducting fracture zone is 55.0 m.

It is planned to build a high-rise residence here. The high-rise building is equipped with a 2-story basement, the standard floor height is 3 m, the raft foundation, or box foundation, is 6 m deep, and the pile length is 30 m. The width of a unit is calculated as 15 m, and the length is calculated as 25 m. The plane load of each building is considered to be 20 KPa. The location of additional pressure on the foundation caused by the building load is taken at the end of the pile. The residence is composed of different numbers of units, and the calculation results of the depth of influence of the different floors of the building are shown in Table 3.

Table 3. Load influence depth of high-rise residential buildings (width B = 15 m).

width(m)	length(m)	unit number	Load influence depth			
			15 floors	20 floors	25 floors	30 floors
15	25	1	56	59	61	64
15	50	2	60	64	67	70
15	75	3	61	66	69	73

The minimum burial depth of the goaf is 250 m, and the height of the water-conducting fracture zone is 55 m, so the minimum burial depth of the collapsed fracture zone is 195 m. Since the maximum impact depth of additional stress is 73 m, the minimum burial depth/maximum impact depth of additional stress is 2.7, so the evaluation result of the impact of engineering construction on the stability of the goaf is small according to the additional stress analysis method. It is feasible to carry out engineering construction above this.

At present, the newly built structures above the strip goaf are operating well, and no obvious movement, deformation and cracks have been seen.

6 CONCLUSIONS

(1) The "activation" of the goaf is easily affected by various factors inside and outside, especially in the strip goaf. The coal pillars are subject to weathering, water erosion, etc., which will easily cause the surface layer to peel off and reduce their size. Its strength and stability are difficult to grasp. When it comes to engineering construction under complex mining conditions such as multiple mining methods, large changes in mining depth (especially shallow coal seams), and unstable surface settlement, various factors must be comprehensively considered and special foundation stability evaluation and analysis must be carried out. Key areas are supplemented by engineering surveys.

(2) The stability of coal pillars in strip mining is related to the safety of the goaf and the surface. The stability of coal pillars refers to the fact that the coal pillars only deform due to the redistribution of stress within a certain period of time and under a certain load. It does not produce destructive collapse and sliding. Research on the stability of coal pillars in strip mining requires consideration of many factors, and there is still a lot of work to be done at present.

(3) The allowable values specified in the specifications in the text include the total limit values for the inclination and compression deformation of the foundation caused by the conventional and residual deformation of the mined-out area. When planning the design of the building structure, the conventional inclination and settlement of the foundation should be calculated according to the geotechnical survey data, foundation type, and superstructure load, and then the residual inclination and settlement caused by the goaf should be superimposed. The value should not exceed the corresponding limit. At the same time, anti-deformation measures are taken.

ACKNOWLEDGMENTS

We would like to thank the editors and the anonymous reviewers for their insightful and helpful comments. Research on this study was supported by Beijing science and technology project (Z181100005118012) and China coal science and industry group science and technology innovation venture capital special key project (2018-2-ZD007).

REFERENCES

C. Y, S. Y. L, Y. F. D. Rock and Soil Mechanics, 38, 875 (2017).
J. X. F, W. D. S, Y. Y. T. Chinese Journal of Rock Mechanics and Engineering. 35, 217 (2016).
J. Y. Z. Journal of University of Science and Technology Beijing, 31, 1368 (2009).
L. W, G. L. G, X. N. Z. Journal of Mining & Safety Engineering, 27, 57 (2010).
L. W. R, G. L. Z, Z. L. D. Rock and Soil Mechanics, 39, 2922 (2018).
L. X, Z. T. W, W. G. H. Journal of Mining & Safety Engineering, 31, 447 (2014).
S. J. C, D. W. Y, B. N. H. Journal of Mining & Safety Engineering, 37, 110 (2020).
W. J. G, H. L. W, Z. P. L. Journal of Mining & Safety Engineering, 32, 369 (2015).
W. Q. Z, H. L. L, K. Z. Journal of Mining & Safety Engineering, 33, 1065 (2016).
Y. B. Z, Y. F. Z, D. H. L. Journal of China University of Mining & Technology, 42, 567 (2013).
Y. F. Z, H. B. C. Journal of China Coal Society, 39, 1473 (2014).
Y. T, W. B. G, E. H. B. Journal of China Coal Society, 44, 1003 (2019).
Y. Y, K. Z. D, H. D. F. Journal of China Coal Society, 42, 3089 (2017).

Advances in Urban Engineering and Management Science – Khalil & Yang (Eds)
© 2023 The Author(s), ISBN 978-1-032-30426-7

Preliminary design of the water stabilization pond system in Puerto Ayora, Ecuador

Xiao Yang*
School of Environmental Science & Engineering, Huazhong, University of Science & Technology, Wuhan, Hubei, China

Tianhao He
Shanghai Weiyu International School, Shanghai, China

Heiyu Chan
Hong Kong International School, Hong Kong, China

Muzi Li
The Madeira School, Virginia, USA

Jing Jiang
Shenzhen College of International Education, Shenzhen, Guangdong, China

ABSTRACT: This is a paper about the design of a waste stabilization pond for the Galapagos Islands in Ecuador, where Charles Darwin proposed his theory of evolution a hundred years ago. It is an ideal tourist attraction for people to enjoy the charm of nature. However, one of the problems they must face is the lack of a waste water treatment system. Dealing with waste water is significant because the sea is all around us and it causes environmental issues if people discharge polluted wastewater directly into the ocean. In order to improve water quality and work out wastewater problems effectively, designing a wastewater treatment system is warranted. Nonetheless, the construction is to take place in a low-income country, Ecuador. Traditional methods of treatment will be unrealistic because of the constructional and operational costs that will put tremendous pressure on the government budgets. Moreover, such treatment reactors will require a high level of control and automation, meaning that higher upfront investments and a crew with relevant knowledge and skills are irreplaceable. Thus, in light of the constant high ambient temperature of the island, a simple, low cost, and efficient biological treatment method is the ideal solution to the problem. This paper combines a particular solution which contains an anaerobic pond, a facultative pond, and an ultraviolet disinfection chamber based on researching various data about actual values. At the beginning, there is a brief introduction. Step-by-step measurements occupy a large proportion of the paper. The conclusion, which analyzes the future strengths and weaknesses of the establishment of this waste stabilization pond, is at the end.

1 INTRODUCTION

Water is one of the most needed and precious resources for any settlement or society. Therefore, ensuring water usage safety is essential for the welfare of Puerto Ayora's citizens and their livelihoods. Puerto Ayora is an island city in Ecuador's Galapagos Islands, and it has around 15,000 residents. A usable supply in Puerto Ayora is the best way to improve public health, industry, and the environment. However, since the local government may not be able to afford the construction and

*Corresponding Author: 1178601193@qq.com

DOI 10.1201/9781003305026-3

operation costs of the conventional treatment methods, Furthermore, the infrastructure on the island may lack reliable energy sources and experienced technicians to operate the system. Therefore, a solution that can utilize the local environment while compensating for the fore-mentioned shortcomings is warranted. Water Stabilization Pond Systems (WSPs), which are well-known for their cost-effectiveness, dependability, and low automation requirements, are ideal for treating domestic wastewater in tropical climates. Also, WSPs can be run with the least amount of energy input and a minimum crew, resulting in lower operational costs. Hence, adopting WSPs in this case can well utilize the advantages of the tropical climate and bypass the disadvantages of high costs. This report will introduce a two-pond wastewater stabilization system for Puerto Ayora, its design and methodology, maintenance, its positive influence on the environment, and how it compares to other designs.

2 BASIC DESCRIPTION

2.1 *Problems and aims*

The main issue with Puerto Ayora's water supply is that citizens receive largely untreated groundwater. Since Puerto Ayora is surrounded by unusable sea water, its source of water comes from underground aquifers under porous bedrock (CNHTours 2020). This means that all forms of harmful substances from above ground can seep into the water source through cracks in the rocks, including human waste and microorganisms. The current water treatment system provides no aid for this problem. Consisting of septic tanks, it is capable of doing basic water filtering and some BOD reduction, but not sludge treatment and other essential processes. As a result, the water sent to homes and establishments is a combination of rainwater, seawater, minerals, and human waste, which is a significant failure in terms of protecting citizens' public health and living standards. Citizens cannot drink or use the current water supply without facing risks of intestinal infection and skin allergies, and are thus forced to purchase bottled water (CNHTours 2020). An unclean water supply affects not only the inhabitants' welfare, but also tourism, one of the island's biggest industries. Hotels and other establishments are unable to provide adequate tap water, leading to inconveniences that decrease the island's appeal to tourists.

2.2 *Maintenance and costs*

Waste stabilization ponds are the optimal system for Puerto Ayora due to its usage of natural energy sources and large land area, both of which are readily available on the island. The system generates high levels of reduction of BOD, solids, pathogens, and nutrients via natural processes such as sunlight, wind mixing, and microorganisms (Dorothee Spuhler & EAWAG 2020). As such, it requires few daily workers, reducing the cost of maintenance. Cleaning the ponds is essential every five or ten years to remove the sludge that accumulates. On the other hand, the UV disinfection chamber needs to be regularly cleaned to avoid fouling of tubes and checked that the level of suspended solids is low enough to maintain UV disinfection's effectiveness (EPA 1999). In addition, building materials for the ponds are very common if repair work is needed. For all these reasons, waste stabilization ponds are suited to making treatment systems for small local communities. However, initial construction costs are quite large, estimated at 73179 US dollars. It should be noted that this design assumed general soil quality and the cost of construction materials, since data local to South America was inaccessible, so the estimation can be reduced. To optimize this design, this work referenced other case studies of the construction of wastewater stabilization ponds. Through analyzing these materials and incorporating local conditions, this work optimized the structure and measurements of the ponds, pipes, inlets, and outlets as well as the UV disinfection chamber.

3 DESIGNING PAST

3.1 *Estimation of water flows and pollutant concentrations*

As for the designing part, first the work will have to start by analyzing the perimeters of the wastewater, where it comes from and where it ends up. The major source of the wastewater will be domestic waste from the residential area of Puerto Ayora. According to the characterization of the legal limits for wastewater discharged into the sewerage system, the TULSMA Standard (Juan Francisco Webster Moscoso 2020) and the legal discharge limits for marine waters (Environmental quality and effluent discharge standard: water environment), as shown in Table 1, the preferred treatment measure can be determined. After being treated by the facility, the water will be discharged directly into the marine waterbody.

Table 1. The comparison of the legal limits of pollutant levels and the actual levels in the wastewater, according to TULSMA standards.

Item	Legal limits for wastewater discharged into the sewerage system, TULSMA Standard (mg/L)	Legal discharge limits for marine waters (Conservatory Water) (mg/L)	The aim for treated effluent's levels (mg/L)
BOD5	250	200	≤ 30
Total Nitrogen	50	40	≤ 40
Total phosphorus	15	not mentioned	≤ 10
Total suspended solids	220	250	30
pH	6 to 9	6 to 9	≥ 7
Faecal coliform	6 × 10^7 /100 ml	2000/100 ml	$\leq$ 2000/100 ml

The requirement for BOD and nutrient removal is relatively looser, so a system consisting of an anaerobic pond and a facultative pond will do the job. Considering the relatively longer hydraulic retention time and shallower depth, building such a structure will occupy a considerable amount of land, which on an island will be of significant economic and utility value.

Because in the system the latter maturation pond is omitted, to eliminate the pathogens and toxic organisms in the wastewater, a process of disinfection is necessary. In Table 1, this design will adopt the concentration of faecal coliform to represent the biological objective of the treatment.

Although the pH value of the wastewater is within the legal limits, the system needs to keep it above 7 in the facultative pond to ensure qualifiable nutrient removal to reach the effluent specification. The mean water flows should be carefully estimated since they have direct effects on the size of the ponds and the construction costs. After comparing multiple rules of thumb applied around the world (Abdullahi 2014; Kayombo *et al.* 2020; Liu *et al.* 2017; Long *et al.* 2017; United State Environmental Protection Agency 1983), a suitable design is 85% of the in-house water consumption. The design here assumes the water consumption per capita as 140 L per day (according to an approximation of British standards (James Brockett). And the estimated local population of the Galapagos Islands is approximately 15,000, which necessitates the construction of a wastewater treatment plant. The daily flow should be

$$Q = 15000 * 140 * 85\% = 1785\ m^3 \text{ (Kayombo } et\ al. \text{ 2020)} \quad (1)$$

and that is rounded to the hundredth digit to be 1,800 m^3. The BOD may be measured if wastewater exists, based on 24-hour flow weighted data. Alternatively, the BOD will be estimated from the legal limitation of the BOD level in wastewater discharged into the sewer network in Ecuador, which is 250 mg/L (Juan Francisco & Webster Moscoso 2020). And, since the industrial development on the island is minor and the source of waste is mainly domestic waste, such an assumption is satisfactory.

The usual range of faecal coliform in domestic wastewater is 10^7–10^8 faecal coliform/100 ml, with a suitable design value being 6 × 10^7/100 ml (United State Environmental Protection Agency 1983).

3.2 *Design of anaerobic pond*

Rules of thumb may vary considerably according to regional differences. (Long *et al.* 2017) Considering the norms applied in China (Liu *et al.* 2017) may fail to represent a construction on a tropical island, this design will not adopt the Oswald Method or the Wehner Method, which are widely used in China.

The anaerobic ponds are designed on the basis of volumetric loading (λv, g/m^3/d) (Kayombo et al. 2020), which is given by the equation:

$$\lambda v = Li * Q/Va \tag{2}$$

Where Li is influent BOD (mg/l), Q is flow rate (m^3/day), and Va is anaerobic pond volume (m^3) (from Mara & Pearson 986; Mara *et al.* 1997).

Since the local temperature is constantly above 20°C according to data in the previous years (World Climate Guide 2020), the equation on the third line will imply. In order to maintain anaerobic conditions, this design will employ a value of 300 g/m^3/d according to Table 2. The volume of the pond can then be calculated after the organic loading is chosen and compared to empirical values from similar facilities (Long *et al.* 2017).The hydraulic retention time is then calculated as follows:

$$Va = Li * Q/\lambda v = 250 * 1785/300 = 1487.5\,m^3, \quad t_{an} = Va/Q = tan = 0.833 \text{ day}. \tag{3}$$

Table 2. Design value of permissible volumetric BOD loadings on, and percentage BOD removal in, anaerobic ponds at various temperatures. (Kayombo *et al.* 2020)

Temperature(°C)	Volumetric loading(g/m^2.day)	BOD removal (%)
<10	100	40
10~20	20T−100	2T+20
20~25	10T+100	2T+20
>25	350	70

A retention time of less than one day should not be used for anaerobic ponds; if it occurs, however, a retention time of one day should be used, and the volume of the pond should be recalculated. (Kayombo *et al.* 2020) Therefore, the revised Va should be 1,785 m^3, which is rounded to the hundredth digit to be 1,800 m^3. The BOD removal rate, according to the forementioned equation presented by Mara and Pearson, is calculated as follows:

$$Se/S0 = 2 * T + 20 = 60\% \tag{4}$$

The hydraulic characteristics of rectangular and trapezoidal cross-section ponds (as shown in Figure 1) have been found to be superior to those with the geometry of a square, a circle, or with

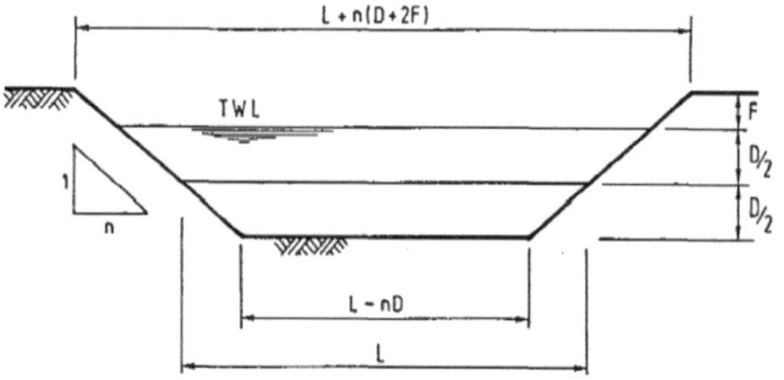

Figure 1. The cross-sectional view of the anaerobic pond (Abdullahi *et al.* 2014).

irregular geometry (Kayombo et al. 2020), and have been widely adopted in WSP designs around the world. A length to breath ratio of 2:1 is adopted for stability (Cui *et al.* 2017; Kayombo *et al.* 2020). The ponds are to be trapezoidal in cross-section and rectangular in plain.

The embankment slope of 1:3 should be satisfactory for its stability and relative economics in most soil conditions. Yet, since local land resources are scarce, a steeper slope must be adopted to reduce the occupation of land. Moreover, if the design adopts steeper slopes as we planned, their stability should be reinforced by standard soil mechanics procedures like concrete hardening. The in-situ concrete works (like a concrete base and embankment) can effectively stop vegetation growth along the banks and so prevent the breeding of mosquitoes, which can reduce the cost of maintenance and daily operation. Moreover, the designed shape of the cross-section can be stable even though random disruptions brought by the desludging process exist.

As for the size of the anaerobic pond (Liu *et al.* 2017), the surface area is estimated given that the available depth of the pond is set to 4 m, plus 0.5 m for sludge accumulation. Then the average surface area of the pond is given by:

$$A = Q/h = 1800/4 = 450\,m^2 \tag{5}$$

Applying the length to width ratio of 2:1, 2W2 = 450, and the ratio of slope of 2.5.

$$W = 15\,m,\ L = 30\,m.$$

Assuming at the depth of 1/2 effective depth, the surface area is as mentioned.
Then
$L_{surface} = 30 + 2 * 2.5 * 2 = 40$ m, $W_{surface} = 15 + 2 * 2.5 * 2 = 25$ m
$L_{bottom} = 30 - 2 * 2.5 * 2 = 20$ m, $W_{bottom} = 15 - 2 * 2.5 * 2 = 5$ m
The effective volume of the pond is

$$V' = \frac{d}{2} * (LsBs + 4LmBm + LbBb) \tag{6}$$

$$V = 4/6 * (40 * 25 + 4 * 30 * 15 + 20 * 5) = 1933\,m^3.$$

The result is quite different from the theoretical value. Hence, recalculation is required.

$$1800 = 4/6 * ((x + 10)(2x + 10) + 2x2 + (x - 10)(2x - 10)) \tag{7}$$

$$X = 14.43\,m.$$

After the recalculation, $W_m = 14.5$ m.

$$\text{Then}: L_{surface} = 29 + 2 * 2.5 * 2 = 39\,m$$

$$W_{surface} = 14.5 + 2 * 2.5 * 2 = 24.5\,m.$$

The actual area occupied by the pond, given 0.5 m of reserved depth:

$$A = (L_{surface} + 0.5 * 2.5 * 2) * (W_{surface} + 0.5 * 2.5 * 2) = 1120.5\,m^2. \tag{8}$$

The effective volume of the pond:

$$V = 4/6 * (39 * 24.5 + 4 * 29 * 14.5 + 19 * 4.5) = 1815\,m^3. \tag{9}$$

And in this design, we must make the pond an additional 0.5 m deeper (according to industry practice in China) to reserve room for sludge accumulation and 0.5 m above the effective depth for any impact load of wastewater and the normal fluctuation of the flow per day, which is determined according to the present standards in relevant business in China. (Standards for the Design of

Outdoor Wastewater Engineering 2021) Otherwise, the additional wastewater will flow over the embankment.

The total volume of the pond, plus reserved depth of 0.5 m and 0.5 m for sludge accumulation:

$$V = 5/6 * (41.5 * 27 + 4 * (41.5 - 2.5 * 4.5) * (27 - 2.5 * 4.5) + 19 * 4.5) = 2593.5\,m^3. \quad (10)$$

The bottom of the pond should be impermeable (this design will adopt a concrete base). According to previous experience of similar facility (Abdullahi *et al.* 2014), the sludge layer is expected to serve as sealant for the micro pores in the soil. Nonetheless, since there hasn't been any actual measurement of the local soil seepage rate, such effectiveness is not wholly guaranteed. Thus, sealing the base with concrete is necessary to prevent possible groundwater pollution. To summarize, this design will use concrete reinforcement on the base, sides, and top of the embankment to protect it from erosion.

3.3 *Design of facultative pond*

The wastewater flow from the anaerobic pond is 1,800 m^3/day with a BOD5 of 100 mg/L. According to experience value presented by Mara and similar facility designs in China, the goal of Se (BOD5 of the outflow) is 30 mg/L.

There are several prevalent ways used in the design of facultative ponds. The two most commonly used methods are based on first-order kinetics and maximum permissible surface loading (Kayombo *et al.* 2020). The former is not currently recommended for this design since there is too little reliable data on which to base design values for the various kinetic constants, and there hasn't been reliable data on such values in Ecuador or regions of similar climate. Pond design procedures based on BOD surface loading are empirical and thus more reliable, and pond performance in similar tropical climates can be used to establish a recommended design value (Long *et al.* 2017).

And, according to the design manual published by Mara (Mara 1998), the following equation is satisfactory when the temperature is up to 20°C:

$$\begin{aligned} \lambda_s &= 50(1.072)^T \\ &= 50 * (1.072)\hat{}20 \\ &= 200.8472\,kgBOD_5.ha^{-1}.d^{-1}, \end{aligned} \quad (11)$$

It's within the recommended SLR range in tropical and subtropical climate regions by von Sperling.

The total mass of the BOD_5 is

$$1800 * 0.1 = 180(kg/d) \quad (12)$$

The effective surface area is

$$A = BOD_{5total}/\lambda_s = 180/200.8472 * 10\hat{}4 = 8996\,m^2, \quad (13)$$

which is approximately equal to 9,000 m^2.

And since the retention time with pond depth (H) is set to be 1.5 m according to empirical value (Standards for Design of Outdoor Wastewater Engineering 2021), the calculation is a little bit different from the anaerobic pond. Because unlike the anaerobic pond, which is smaller and covered with a lid to collect gas from anaerobic digestion, the surface area of the facultative pond is relatively bigger and the HRT is also relatively longer. Then the surface evaporation and seepage at the bottom are not negligible. And in this case, the bottom of the pond is reinforced with a concrete base, so the seepage is minor enough to be ignored. Therefore, the HRT of the facultative pond must be calibrated with the following equation (Kayombo *et al.* 2020):

$$HRT = A_f * H/Q_m \quad (14)$$

Where H is the pond depth, and Q_m is the mean flow (m^3/day).

The mean flow is the mean of the influent and effluent flows (Q_i and Q_e). If seepage is negligible, Q_e is given by:

$$Q_e = Q_i - o.001A_{fe}, \quad (15)$$

where e is net evaporation rate (mm/day), which is set to be 4.5 mm per day.

Thus, the forementioned equation can be transformed into

$$\begin{aligned} HRT &= A_f * H/(2Q_i - 0.001A_{fe}) \quad (16) \\ &= 2 * 9000 * 1.5/(2 * 1800 - 0.001 * 9000 * 4.5) \\ &= 7.58d. \end{aligned}$$

For temperatures below 20°C, a retention time of 5 days should be used, and for temperatures above 20°C, a retention time of 4 days should be used (Mara 1998).This is for the purpose of minimizing hydraulic short-circuiting, which can prevent wastewater from leaving the pond without proper treatment. and to give algae sufficient time to multiply to serve as a supplier of dissolved oxygen (i.e., to prevent algal washout).

And the flow of the pond will be set to be perpendicular to the prevalent wind direction locally to utilize wind's mixing effect.

As for the design of the facultative pond's geometry, base and embankment, this design will continue to adopt those used in the design of the facultative pond (Liu *et al.* 2017). Applying the length to width ratio of 2:1, 2W2=9000, and the ratio of slope of 2.5.

$$W = 67m, L = 134m.$$

Assuming at the depth of 1/2 effective depth, the surface area is as mentioned.

Then $L_{surface} = 134 + 2 * 2.5 * 0.75 = 137.75$ m, $W_{surface} = 67 + 2 * 2.5 * 0.75 = 70.75$ m

$L_{bottom} = 134 - 2 * 2.5 * 0.75 = 130.25$ m, $W_{bottom} = 67 - 2 * 2.5 * 0.75 = 63.25$ m

The effective volume of the pond is

$$\begin{aligned} V' &= \frac{d}{2} * (LsBs + 4LmBm + LbBb) \quad (17) \\ V &= 1.5/6 * (137.75 * 70.75 + 4 * 134 * 67 + 130.25 * 63.25) \\ &= 13474.03\, m^3. \end{aligned}$$

This calculation is the satisfactory representation of the effective volume of the pond.

The total volume of the pond, plus reserved depth of 0.5 m and 0.5 m for sludge accumulation:

$$V = 2.5/6 * (138.75 * 71.75 + 4 * 134 * 67 + 129.25 * 62.25) = 22463.8\, m^3. \quad (18)$$

3.4 *Sludge volume and desludging period*

The total Q is 1,800 m^3/d, with a drop of 220 mg/L in BOD_5 and 190 mg/L in SS.

$$Q_{sludge} = 365QXi/C\rho \quad (19)$$

$$X_i = X_a + X_b + 0.23YS_0 \quad (20)$$

Where Q stands for the flow of the wastewater.

Xi stands for inert matter level (mg/L).

C stands for the mass fraction of the dry solids in the sludge, which is set to be 0.1.

P stands for the density of the sludge, which is set to be 1.03 g/ml.

X_a stands for the non-biodegradable VSS level (mg/L).

X_b stands for nom-volatile SS level (mg/L).
Y stands for the growth rate, which is set to be 0.6.

$$X_a = 0.4 * 0.75 * 220 = 66\,mg/L, \tag{21}$$

$$X_b = 0.25 * 190 = 47.5\,mg/L, \tag{22}$$

$X_i = 66 + 47.55 + 0.23 * 0.6 * 250 = 148.05$ mg/L,
$Q_{sludge} = 365 * 1800 * 148/(0.1 * 1.03 * 10\hat{}6) = 944.03\,m^3/a$
In the situation of pond in series, the first pond is estimated to produce 30% to 50% of the total sludge, according to the empiric experience of China. in this design, it will be set to be 40%.
Then Q_{sludge} of the anaerobic pond is $944.03 * 40\% = 377.62\,m^3/a$,
Time period for using pumps to extract the sludge from the anaerobic pond:

$$V_{sludge}/Q_{sludgeF} = 259.16/377.62 = 0.68\text{ a}. \tag{23}$$

Time period for using pumps to extract the sludge from the facultative pond:

$$V_{sludge}/Q_{sludgeF} = 4497.012/(944.03 - 377.62) = 7.93\text{ a}. \tag{24}$$

3.5 *Nutrient removal*

The Total Kjeldahl Nitrogen level of the inflow is 50 mg/L.
Total nitrogen removal in the individual facultative and maturation ponds was presented by Reed (1995) (Kayombo *et al.* 2020), as follows:

$$Ce = Ci * e^{\{-[0.0064*1.039^{T-20}\}[\Theta+60.6*pH-6.6]]} \tag{25}$$

Where *Ce* and *Ci* is the total nitrogen concentration of the inflow and outflow, respectively(mg/L), T represents the temperature (°C), and Θ refers to the retention time (days).
In this case, the pH value is between 6 and 9.
Then the removal rate (Kayombo *et al.* 2020) is

$$1 - Ce/Ci = 2.7182\hat{}(-(0.0064 * 1.039) * (7.58 + 60.6 * (pH - 6.6))) \tag{26}$$

According to the calculation result, the value of the rate is between 64% ($pH = 9$) to 0 ($pH \leq 6.5$).
Hence the inflow of the facultative pond must be controlled above 7 to ensure proper nutrient removal to meet the need of outflow specifications.

3.6 *Inlet and outlet of the structure*

There's been a great variety of designs for inlet and outlet structure, which should be simple and inexpensive as the pond itself. The inlet to anaerobic and primary facultative ponds (shown in Figure 2) should discharge below the constant liquid level so as to minimize short-circuiting (especially in deep anaerobic ponds) and reduce the quantity of scum (which is great in facultative ponds) (Kayombo *et al.* 2020).

Inlets to secondary facultative and maturation ponds (shown in Figure 3) can discharge either above or below the liquid level, although discharge at mid-depth is preferable as it reduces the potential for short circuiting and the disturbance to the sludge layer to a minimum.

As for the outflow structure (shown in Figure 4), the outlet of all ponds should be equipped with a scum guard to prevent the contained scum from being discharged. The take-off level for the effluent, which is controlled by the scum guard depth, is therefore crucial as it can have a significant influence on effluent quality by blocking out sludge particles and scum. In facultative ponds, the scum guard should extend below the maximum depth of the algal band in the waterbody to minimize

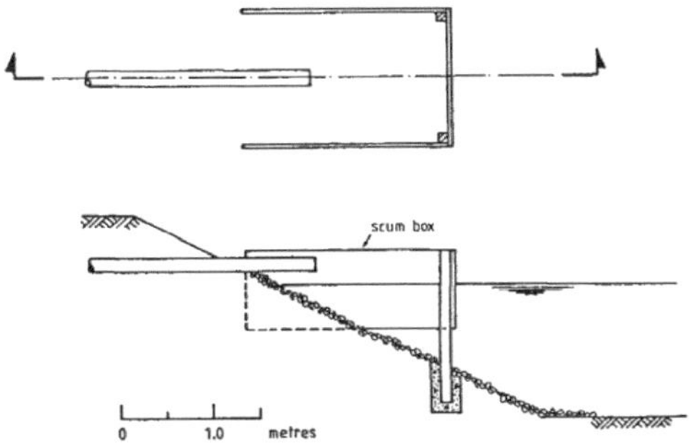

Figure 2. The inlet structure for anaerobic and primary facultative ponds. (Design of Waste Stabilization Pond for Sewage Treatment at Nigerian Defence Academy Staff Quarters, Permanent Site Mando Kaduna. Vol. 1, no. 2, Nov. 2014).

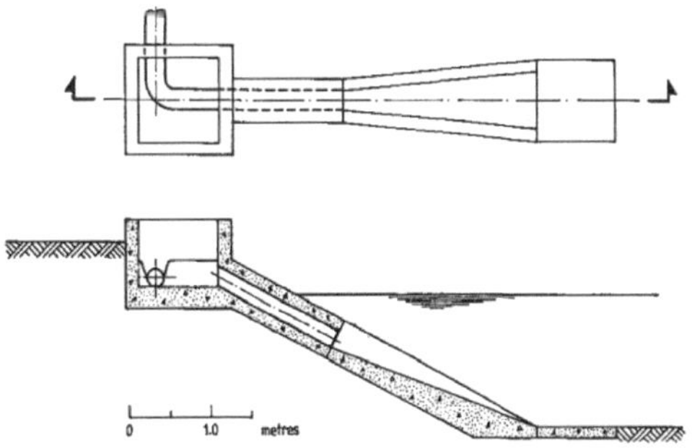

Figure 3. The inlet structure for secondary facultative and maturation ponds. (Design of Waste Stabilization Pond for Sewage Treatment at Nigerian Defence Academy Staff Quarters, Permanent Site Mando Kaduna. Vol. 1, no. 2, Nov. 2014).

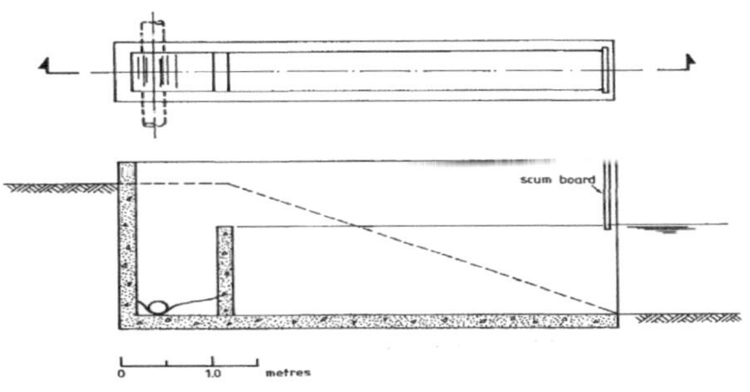

Figure 4. The outlet weir structure. (Design of waste stabilization pond for sewage treatment at Nigerian defence academy staff quarters, permanent site Mando kaduna. Vol. 1, no. 2, Nov. 2014).

the quantity of algae, which can add up to the effluent BOD level and turbidity, leaving the pond. Moreover, the pond must retain a sufficient algal population to provide DO for biodegrading. In anaerobic systems, where algal banding is irrelevant with the presence of a lid, the takeoff should be nearer the surface; in anaerobic ponds, it should be well above the maximum depth of the sludge layer but also below any surface crust to prevent poisonous gas and settled sludge particles from entering the pond. The following effluent take-off levels are recommended (Kayombo *et al.* 2020):

- 30 cm anaerobic ponds;
- 60 cm facultative ponds.

In this design, it will be possible to install a variable height scum guard since it permits the take-off level to be at an optimal level at a changeable water level during the operation of the pond.

The weir need not necessarily be strictly linear, according to similar examples (Kayombo *et al.* 2020). Often, a U-shaped structure is more economical, especially for a pond with high flow rates. (Mara & Pearson 1987).

3.7 *Configuration of the system and pipe network*

To maintain normal production when pipes in the systems need maintenance, this design adopts a twin pipe structure (as shown in Figure 5). Each pipe in one set serves as a backup to the other. And there will be valves on both ends of each pipe that function as isolators to isolate the pipe from the system when maintenance for a certain segment of the pipe is needed.

Also, in this system, it is often advantageous to bypass the anaerobic pond to make way for any maintenance process like desludging. This results in a temporary load surge on the facultative pond, but it's not critical because of the relatively longer detention time of the pond. Plus, since the temperature of the Galapagos Islands stays constantly above 20°C, the increase in SLR will not pose a severe threat to the processing efficiency of the system. The pond layout should be such that the length of by-pass pipework is as short as possible. The designed flowchart of the system is shown in Figure 6.

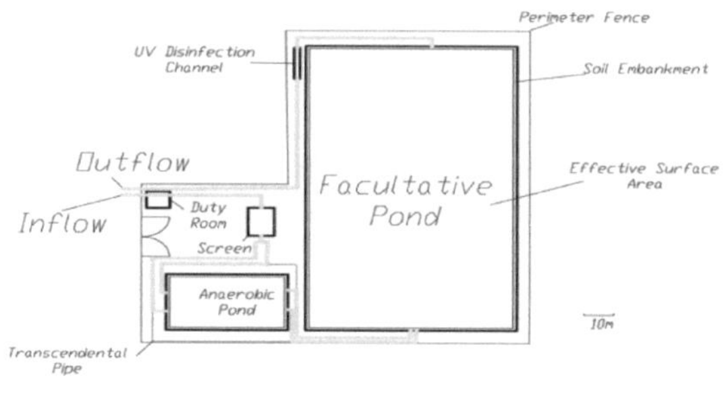

Figure 5. Platform of configuration of the system.

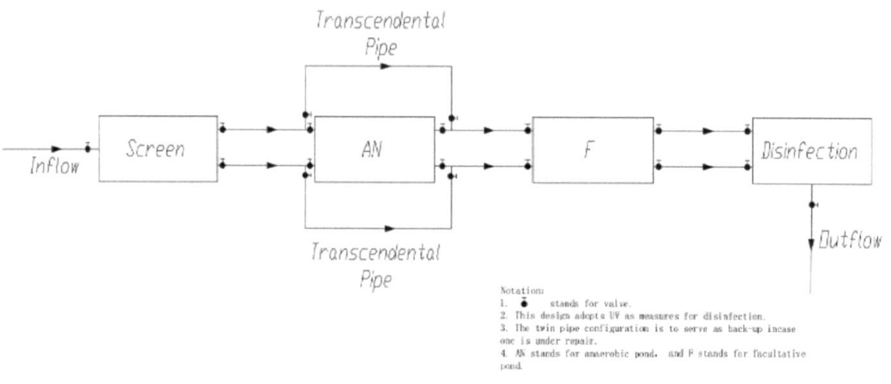

Figure 6. Flow chat of the whole process with pipe network.

3.8 *Design of the disinfection unit*

Disinfection is widely considered to be the primary mechanism for the inactivation of pathogenic organisms to prevent the spread of waterborne diseases to users of the same waterbody. Since the construction takes place on the Galapagos Islands, where land resources must have been scarce, instead of using a maturation pond as a measure to bring down pathogen concentrations, this design will use a UV disinfection reactor to do the job.

An ultraviolet disinfection system can transfer electromagnetic energy from a mercury arc lamp to an organism's genetic material. When UV radiation penetrates the cell wall of an organism, it destroys the cell's ability to reproduce. UV radiation, generated by an electrical discharge, can penetrate the genetic material of microorganisms to retard their ability to reproduce.

Its advantages and disadvantages (Wastewater Technology Fast Sheet 1999) are listed below. Advantages:

1. UV disinfection is effective in inactivating most viruses, spores, and cysts. Such effectiveness is valuable, especially during the pandemic of COVID-19.
2. Since it's a physical instead of a chemical process, it can avoid the necessity to generate, transport, handle, or store hazardous or toxic chemicals like chlorine, chloros, or chlorine dioxide.
3. There's no residual effect that can be harmful to humans or aquatic life, which is crucial because the wastewater will be discharged directly after the disinfection process.
4. It has a relatively shorter contact time (approximately 20 to 30 seconds), which means a smaller contact chamber and less occupied land in comparison with a maturation pond.

Disadvantages:

1. Turbidity and total suspended solids (TTS) in the wastewater will render the UV disinfection ineffective. For low pressure lamps, UV disinfection is not effective for effluent with TTS levels above 30 mg/L, which does not apply to this case because the effluent of the facultative pond has a TTS level of below 30 mg/L.
2. The temperature of the media around the UV modules, which is wastewater itself, influences the effectiveness by altering the energy transmission of the UV. With lower temperatures, the effectiveness will drop accordingly. In this case, in the Galapagos Islands, where the temperature is constantly above 20°C, such a shortcoming is rendered inconsequential.

Hence, adopting UV disinfection in this design is preferred. There are two types of UV disinfection reactor configurations that exist (as shown in Figure 7): contact types and non-contact types. In this case, because the flow is relatively smaller and the need to save costs is necessary, this design will adopt the configuration (a) in the picture. Such a design can allow modules to be easily removed and replaced in the event of malfunction, or even added to suit the increase in daily flow.

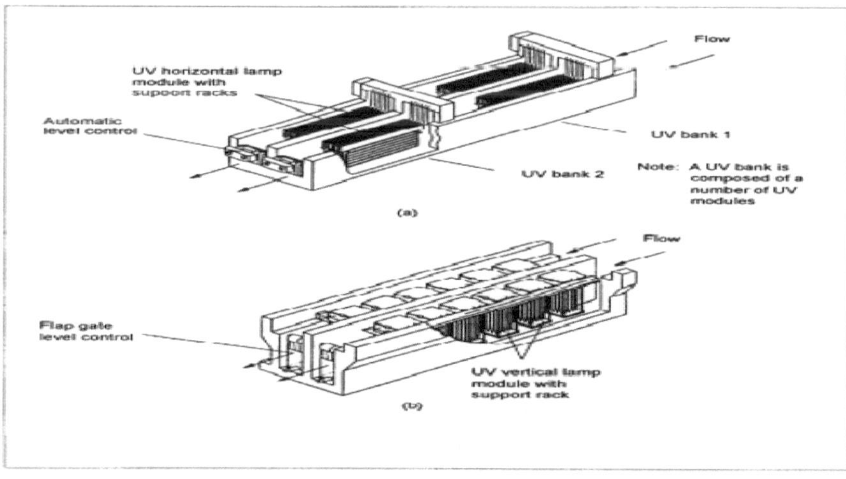

Source: Crites and Tchobanoglous, 1998.
(a) adapted from Trojan Technologies, Inc.
(b) adapted from Infilco Degremont, Inc.

Figure 7. Layouts of two types of UV disinfection (Wastewater Technology Fast Sheet 1999).

Table 3. Main parameters of UV disinfection system produced by Canadian TROJAN company (Liu *et al.* 2017). Specific calculations (shown in Table 3) and relevant calculations are as follows (Liu *et al.* 2017).

System	UV3000PTP	UV3000PLUS	UV3002PTP
Treatment Capacity (mean value)	40–2900 m^3/d	2900–38000 m^3/d	38000 m^3/d and above
Performance	28 lamps for 3800 m^3/d of inflow	14 lamps for 3800 m^3/d of inflow	2.5 lamps for 3800 m^3/d of inflow
Inflow Parameters	TSS: 10–30 mg/L UVT≥45%	TSS: 10–30 mg/L UVT=45 to 70%	TSS: 10–100 mg/L UVT>15%
Number of Lambs on each Module	2 or 4	2, 4 or 8	6 to 24
Power of one Lamp	44W	250W	280W
Cleaning Methods	manual	mechanical and chemical cleaning (automatic)	mechanical and chemical cleaning (automatic)

In this design, a UV disinfection system from Trojan Technologies, Inc. will be adopted. This design will select the UV300PTP disinfection equipment, which needs 28 lamps for 3,800 m^3/d.

Number of the lamp: $1800/3800 * 28 = 13.26$, which is rounded to 14. This design plans to select 4 lamps as one module.

Design of the disinfection channel: According to the specifications of the equipment, the depth of the channel will be 0.35 m, with flow rate of 0.3 m/s.

The flow cross-section is

$$A = Q/v = 1800/0.3/24/3600 = 0.070 \text{ m}^2 \tag{27}$$

The width of the channel is

$$W = A/H = 0.07/0.35 = 0.2 \text{ m}. \tag{28}$$

The length of every module will be 2.46 m. The distance between two modules will be 1.0 m. and there will be 1.5 m form the last module to the weir for automatic level control. The length of the channel is

$$L = 4 * 2.46 + 2 * 1.0 + 1.5 = 13.34 \text{ m} \tag{29}$$

Recalculate the exposure time is

$$t = 4 * 2.46/0.3 = 32.8 \text{ s}. \tag{30}$$

Hence, the design meets the requirements. Here is the configuration of the disinfection channel (Figure 8), the actual design the length of the channel can be extended to suit the potential increase of wastewater flow:

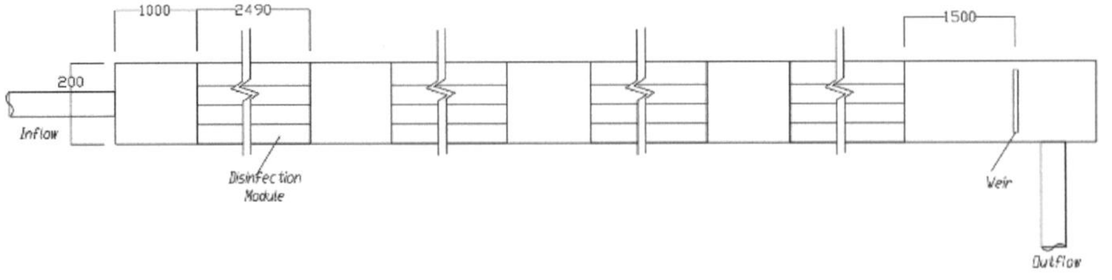

Figure 8. Configuration of the disinfection channel.

4 ESTIMATED COSTS

And the last part is the budget estimation. Since there's little local example of a similar facility in central America, this estimation is made based upon similar WSP budgets in the USA (Estimating Costs and Manpower Requirements for Conventional Wastewater Treatment Facilities 1971), China (Abdullahi 2014; Liu *et al.* 2017; Long *et al.* 2017; World Climate Guide 2020) (Standards for Design of Outdoor Wastewater Engineering 2021), Greece, France (SAMCO 2019) and Nigeria (Abdullahi *et al.* 2014), and a similar treatment facility in Ecuador. Also, since the composition of the soil and the geotechnical condition of the island are unknown, the part concerning excavation and relocation of the soil is all hypothetical. And the concrete work is limited to the range of the pond, including inlets and outlets, which means actual construction could cost more if hardening the entire facility's ground or additional traffic infrastructure is necessary. The rudimentary estimation of the cost of the construction is shown as follows in Tables 4–8.

Table 4. Budget for construction of the anaerobic pond.

Budget for construction of the anaerobic pond					
Item	description	unit	quantity	rate	amount in dollars
1.1	Clearing site of all bush, shrub grasses and all trees, roots, anthills, and cart to spoil with an average depth if 0.5m.	Square meter	1120.5	0.097251	108.9697231
1.2	Excavating any material in the pond pit to form trapezoidal shape.	Cubic meter	1815	0.850946	1544.467126
1.3	Back-filling of excavated material (for re-use as embankment filling materials)	Cubic meter	1000	0.607819	607.818625
1.4	Removing surplus excavated mate	Cubic meter	815	0.413317	336.853082
1.5	Concrete works	Cubic meter	494.2	36.46912	18023.03787
	Sum				20621.14642

Table 5. Budget for construction of the facultative pond.

Budget for construction of the facultative pond					
Item	description	unit	quantity	rate	amount in dollars
2.1	Clearing site of all bush, shrub grasses and all trees, roots, anthills, and cart to spoil with an average depth if 0.5m.	Square meter	9955.313	0.097251	968.1638968
2.2	Excavating any material in the pond pit to form trapezoidal shape.	Cubic meter	13474.03	0.850946	11465.67294
2.3	Back-filling of excavated material (for re-use as embankment filling materials)	Cubic meter	5000	0.607819	3039.093125
2.4	Removing surplus excavated mate	Cubic meter	8474.03	0.413317	3502.457819
2.5	Concrete works	Cubic meter	4300.325	36.46912	156829.0577
	Sum				175804.4455

Table 6. Budget for construction of the fence and pipe works.

Budget for construction of the fence and pipework					
Item	description	unit	quantity	rate	amount in dollars
3.1	Fences with height off street level of 2m.	Meter	566	1.458765	825.6608202
3.2	Back-filling of excavated material Around fence foundation	Cubic meter	7	0.364691	2.552838225
3.3	Removing surplus excavated materials away from site	Cubic meter	130	0.413317	53.73116645
3.4	Concrete works	Cubic meter	33.96	36.46912	1238.49123
3.5	Bricks for the wall	Cubic meter	30	5.348804	160.464117
3.6	Gate of the facility (19m)	None	1	3000	3000
3.7	Pipes within the facility (150mm, PVC) (Supply and installment)	Meter	423	5.288022	2236.833322
3.8	Inlets and outlets (200mm, PVC) (Supply and installment)	Meter	221	7.026383	1552.83071
					9070.564204

Table 7. Budget for construction of the screen and UV disinfection modules.

Budget for construction of the Screen and UV disinfection modules					
4.1	Concrete works	Cubic meter	142	36.46912	5178.614685
4.2	Disinfection modules (Trojan Technologies)	none	4	16000	64000
4.3	Accessories	none	none	3000	3000
4.4	Screen	none	1	1000	1000
					73178.61469

Table 8. Total cost for the WSP system in Puerto Ayora.

Item	descriptions	amount in dollars
1	Budget for construction of the anaerobic pond	20621.15
2	Budget for construction of the facultative pond	175804.4
3	Budget for construction of the fence and pipework	9070.564
4	Budget for construction of the Screen and UV disinfection modules	73178.61
5	15% as Contingencies	41801.22
	5% as VAT	13933.74
	Salaries for workers	1000
Total		335409.7

5 DISCUSSION

Based on the design of the anerobic pond, facultative pond, and ultraviolet disinfection chamber, the waste water problem that existed in Puerto Ayora will be solved. Additionally, since the waste stabilization ponds are established, it will have some advantages for the local communities but will encounter some building barriers at the same time.

Firstly, the waste water stabilization ponds can improve the water quality of the waste water.

Due to the report of the WHO (World Health Organization) on Puerto Ayora's drinkable water, it shows that there are some security risks in the drinkable water in Puerto Ayora. The main reason for the problem is that the water still contains some pollutants for the local people's tap water supply. (Liu, Jessie & d'Ozouville, Noémi. 2013) Local people can't use the water directly since the awful water quality may hurt people's skins. People in Puerto Ayora need to buy bottles of water from other places in order to satisfy the daily demand. The design of two waste stabilization ponds could purify the water again and reduce the BOD (Biochemical Oxygen Demand) when it leaves the sewage treatment plant. The less BOD in the water, the purer the water people use. The advancement of water quality will change the patterns that people have developed in the past.

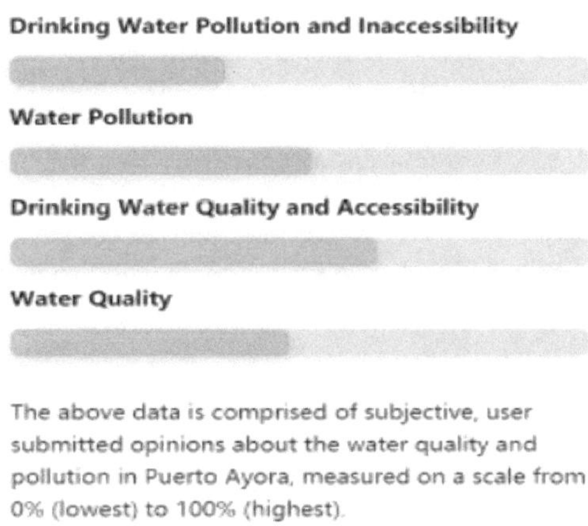

Figure 9. User submitted ratings for Puerto Ayora tap water.

People will not necessarily import any bottles of water from other places, which will stimulate the local economy in the market. There will be more firms investing more energy and money in the tertiary industry because the daily needs of local residents for water have been satisfied. Citizens will improve their quality of life by spending more money on services. Moreover, due to the improvement in water quality, citizens will stay away from the disease to a certain extent when there are fewer pollutants carried by the water. The local people will be healthier and have a longer life span than before, which offers the local people chances to contribute more to the world with their talent and wisdom. The population in Puerto Ayora will increase in the following decades, which also benefits the education and economy in society a lot.

Secondly, the design of the ponds is very suitable for the Puerto Ayora climate. The design of the ponds manages to reduce the concentration of BOD effectively at high temperatures. When the temperature is less than 10 degrees Celsius, the removal of BOD is just 40%, which is ineffective. The volumetric loading can only reach 100 grams per day, which isn't able to treat the whole water in the local area. However, in Puerto Ayora, the temperature varies between 18 and 29°C (64 and 82°F), which offers pleasant weather. The climate in Puerto Ayora matches this design of two ponds so that the ponds are always efficient throughout the whole year. Therefore, it will cost a small amount of money during the operation time. It will decrease government spending due to the high efficiency of the machines, which improves the feasibility of the establishment.

Thirdly, the ponds will protect the marine animals living near Puerto Ayora. In Puerto Ayora, marine animals are not only a food source for the local people but also a sustainable income in tourism. There is a huge number of people who come to visit Puerto Ayora for joy. People can have close contact with marine animals, and even sleeping seals can be seen on the street. However, with the development of tourism, both the local people and the tourists pollute the sea to a great extent. Even though the local residents do some treatment before the sewage directly enters into the sea, the waste water still contains some organic solids inside. These solids will hurt the health of marine animals when the animals eat the solids accidentally. The marine animal population is declining currently due to the crisis in the health and safety of the animals in Puerto Ayora. Without these animals, it is hard for the local people to hunt for food resources due to the decreasing population of marine animals. According to the food chain, those chemicals will eventually accumulate in humans' stomachs, which causes serious health problems. Additionally, it will make the people poorer since fewer tourists will visit Puerto Ayora. Puerto Ayora is losing a major tourist attraction, which is irreversible.

The design of the ultraviolet disinfection chamber can solve the problem effectively. The machines create a harmonious relationship between animals and human beings, which does not produce any toxic chemicals or chlorine dioxide. The process of ultraviolet disinfection will be less harmful to the human beings who directly discharge the waste water after disinfection. Additionally, the protection of the marine community will sustain the development of local tourism. Tourism will offer the local people some extra income, which will advance employment in society.

However, the design of the waste stabilization ponds also has some disadvantages. To begin with, the construction size of the machines will be a great problem. In Puerto Ayora, it is impossible for the anerobic ponds and the facultative ponds to be constructed because the area can only hold 11,974 people in total. Consequently, the ponds may not be completed during the construction.

6 CONCLUSION

To conclude, this research paper designs a waste stabilization pond design for Puerto Ayora, Galapagos Islands, Ecuador, including introduction, problems and aims, maintenance and costs, a detailed design of an anerobic and facultative pond with an ultraviolet disinfection chamber, and conclusion. The biological treatment process has a unique advantage in tropical climates, in which the organisms responsible for the bio-degradation reach maximum bio-activity. In this case, WSPs excel among other biological treatment reactors for their cost-effectiveness, reliability, adoptability, and lower level of automation. Plus, the comparative lower construction and operational costs, simple structure, and high removal rate will render it the optimal option for low-income countries and territories.

Higher sanitary conditions must be maintained to make this world a better place for people in developing countries to live in. The construction of this design will not only address the problem faced in Puerto Ayora, but also provide a solid example of a cheap, effective, and reliable water treatment system. For similar projects and constructions to be adopted in developing countries of tropical climates, this design serves as an ideal model to keep their people away from the forementioned waterborne diseases with the least expenditure.

REFERENCES

Abdullahi, I Nasiru, A Saminu, L Sagir, E Charity. Design Of Waste Stabilization Pond for Sewage Treatment at Nigerian Defence Academy Staff Quarters, Permanent Site Mando Kaduna. International Journal of Engineering and Applied Sciences (IJEAS) ISSN: 2394–3661, Volume-1, Issue-2, November 2014.

CNHTours, 10 Sept 2020, https://www.cnhtours.com/news/2020/9/10/potable-water-in-puerto-ayora-always-3-years-away/. Accessed 19 Sept 2021.

Design Manual for Waste Stabilization Ponds in Mediterranean Countries. (1998) www.personal.leeds.ac.uk/~cen6ddm/WSPmanualmedcountries.html. "Design Manual Municipal Wastewater Stabilization Ponds." *United* State Environmental Protection Agency, Oct. 1983. Accessed 19 Sept. 2021.

Design of Waste Stabilization Pond for Sewage Treatment at Nigerian Defence Academy Staff Quarters, Permanent Site Mando Kaduna. Vol. 1, no. 2, Nov. 2014. International Journal of Engineering and Applied Sciences (IJEAS), www.ijeas.org/download_data/IJEAS0102005.pdf. Accessed 19 Sept. 2021.

Dorothee Spuhler & EAWAG, 2020, https://sswm.info/factsheet/waste-stabilisation-ponds. Accessed 19 Sept 2021.

Environmental quality and effluent discharge standard: water environment. TULSMA. Table 10. Pp. 25–27, May, 2012

EPA, Sept 1999, https://www3.epa.gov/npdes/pubs/uv.pdf. Accessed 19 Sept 2021.
James Brockett. PR19 Challenge Report #5. Water consumption. WWT-Report-.pdf (waterwise.org.uk). pp 3–5 October, 2019.
Juan Francisco Webster Moscoso.(2020) Design of a wastewater treatment plant for the company Lácteos San Antonio in Cuenca – Ecuador. Department of Chemical Engineering and Analytical Chemistry. University of Barcelona. Pp. 11–13.
Liu, Jessie & d'Ozouville, Noémi. (2013). Water contamination in Puerto Ayora: Applied interdisciplinary research using Escherichia coli as an indicator bacteria. Galapagos Report 2011–2012. 76–83.
Long T. Ho*, Wout Van Echelpoel, Peter L.M. Goethals.(2017) Design of waste stabilization pond systems: A review. Water Research 123(2017)236–248.
Mara, D.D. (1998) Design manual for waste stabilization ponds in Mediterranean countries. Lagoon Technology International. ISBN: 0951986929.
Mara, D.D., and H.W. Pearson. (1987) Water Stabilization Ponds Design Manual for Mediterranean Europe. World Health Organization.
Mara, D.D., and H.W. Pearson. *Water Stabilization Ponds Design Manual for Mediterranean Europe*. World Health Organization, 1987.
Moreno, Lorena, et al. "Integrating water-quality analysis in national household surveys: water and sanitation sector learnings of Ecuador." *npj Clean Water*, doi.org/10.1038/ s41545-020-0070-x.
"News : Potable water in Puerto Ayora: Always 3 years away." *Cnh Tours*, 10 Sept. 2020, www.cnhtours.com/news/2020/9/10/potable-water-in-puerto-ayora-always-3-yearsaway/. Accessed 19 Sept. 2021.
Nesterova, N., Makarieva, O. and Post, D.A. (2019). Understanding hydrological processes at a remote mountainous continuous permafrost watershed in a changing environment. In Elsawah, S. (ed.) MODSIM2019, 23rd International Congress on Modelling and Simulation. Modelling and Simulation Society of Australia and New Zealand, December 2019, pp. 1181–1187. ISBN: 978-0-9758400-9-2. https://doi.org/10.36334/modsim.2019.K28.nesterova
Office of Water Washington, D.C. "Waste Water Technology Fact Sheet Ultraviolet Disinfection." *USEPA*, Sept. 1999, www3.epa.gov/npdes/pubs/uv.pdf. Accessed 19 Sept. 2021.
S. Kayombo, T.S.A. Mbwette, J.H.Y Katima, N. Ladegaard, S.E. Jørgensen. (2020) Water Stabilization Ponds and Constructed Wetlands Design manual. University of Dar es Salaam. Danish University of Pharmaceutical Sciences.
SAMCO. (2019) How Much Do Biological Wastewater Treatment Systems Cost? How Much Do Biological Wastewater Treatment Systems Cost? (samcotech.com)
Shanghai Municipal Commission of Housing and Urban – Rural Construction Management. (2021) Standards for Design of Outdoor Wastewater Engineering. China Planning Press, Beijing.
Tap Safe. 8 Apr. 2021, www.tapsafe.org/tap-water-safety-in-ecuador/. Accessed 19 Sept. 2021. *Water Stabilization Ponds for Waste Water Treatment, Anaerobic Pond*. home.eng.iastate.edu/~tge/ce421-521/Fernando%20J.%20Trevino%20Quiroga.pdf.
United State Environmental Protection Agency. (1971) Estimating Costs and Manpower Requirements for Conventional Wastewater Treatment Facilities. Water Pollution Control. Clean Water Series.
United State Environmental Protection Agency. (1983) Design Manual Municipal Wastewater Stabilization Ponds.
United State Environmental Protection Agency. (1999) Wastewater Technology Fast Sheet. Ultraviolet Disinfection. EPA 832-F-99-064.
World climate guide(2020). Climate – Galapagos. Galapagos climate: average weather, temperature, precipitation, when to go (climatestotravel.com).
Zhengjiang Liu, Yuchuan Cui, Hongping Chen, Yantao Wang. (2017) Design Manual for Municipal Wastewater Treatment Plant. Chemical Industry Press. Beijing.

Advances in Urban Engineering and Management Science – Khalil & Yang (Eds)
© 2023 The Author(s), ISBN 978-1-032-30426-7

Regional impact analysis of the application of artificial water body on the change of urban underlying surface

WenYang Lin*
Zhejiang Tongji Vocational College of Science and Technology, Hangzhou, China

Bin Tang & Huan Zhou
Zhejiang Design Institute of Water Conservancy and Hydroelectric Power, Hangzhou, China

ABSTRACT: With the improvement of people's living standards, the demand for water security, environmental protection, and water ecology is becoming increasingly urgent. Under the background of building "happy rivers and lakes" as proposed by governments at all levels, it is necessary to consider the planning scheme of artificial water bodies from multiple angles and directions. Therefore, in the new era of urban construction, urban artificial water bodies are given new characteristics and functions. Therefore, this paper mainly analyzes the problems of flood control and drainage safety and ecological water use caused by the change in urban underlying surface and discusses how to further improve the role of urban water bodies in water security and water quality by establishing regional hydrodynamic and water quality models to simulate the effect of urban water body application on the above problems.

1 INTRODUCTION

An urban water body is an important part of urban ecological construction and one of the basic conditions for the harmonious coexistence of man and nature. In the process of urbanization, urban construction has changed local climate conditions and created a heat island effect. Thus, the proportion of impervious area increases, the underlying surface conditions change randomly, the runoff coefficient increases, the confluence speed increases greatly, the peak flow increases, and the urban storage capacity is weakened. In some cities, rivers, lakes, and artificial water bodies have been filled into land during construction or artificial obstacles have been set on the water body, resulting in the obstruction of water fluidity. Due to the rapid economic development, they have not been matched with the urban basic drainage and sewage facilities, resulting in the decline of water quality to a certain extent. At present, there are many problems in the urban water ecosystem, such as water pollution and deterioration of the water environment, lagging drainage systems and sewage treatment facilities, insufficient water area, water culture and water landscape development. Among these problems, this paper mainly introduces how to do a good job in an urban artificial water body scheme under the condition of water environment design, and takes solving water safety and water environment as the breakthrough point of the scheme.

2 URBAN ARTIFICIAL WATER BODY

Urban artificial water bodies generally refer to the water bodies used for flood detention, water source reserve, water landscape, etc., formed after human processing and repair within the city. It mainly includes artificially built lakes, rivers, aquariums, fountains, waterfalls, and other forms in parks, squares, living quarters, and so on. Most urban artificial water bodies are closed, shallow

*Corresponding Author: 11669155@qq.com

 DOI 10.1201/9781003305026-4

water bodies. Due to their poor water surface fluidity, high water temperature, low biodiversity and high surrounding environmental pollution load, the anti-pollution ability of water bodies is low (Chu et al. 2008).

In recent years, people have begun to expand the area of urban water bodies by building artificial lakes, restoring the original natural rivers and increasing the fluidity of water bodies, so as to improve urban water security and people's quality of life. Urban artificial lakes play a very important role in the urban ecosystem inhabited by human beings. It has irreplaceable functions in people's lives and urban ecological construction. The specific functions are as follows:

1) Water security function is the basis of building an urban ecological water conservancy system and is an important prerequisite and content of urban ecological construction. Water security functions include urban flood control and drainage security, water supply security, and ecological water use. They make full use of the role of urban artificial water bodies in flood storage and peak shaving, improve urban flood control and drainage capacity, and finally realize a new ecological urban water system combined with the water environment.
2) The quality of life of urban residents depends on the function of the water environment. Therefore, it is necessary to start with the causes of water pollution and carry out treatment and restoration according to the results and objectives of corresponding water function areas and water environment function areas. At the same time, engineering means or biotechnology are used to assist water purification to meet the functional needs of the urban water environment.
3) The function of water culture is the combination of urban water culture construction and urban waterscape view, which reflects the culture with water as the axis in the water landscape. The main contents include historical and cultural excavation of urban relics, historical figures, folk art, and so on, as well as scientific, technological, and cultural contents of modern society.
4) The function of the water economy is to integrate the demand for water resources into its own urban economic development into the urban water market after the construction of water supply and drainage systems. For example, the construction of water environments and water culture improves the living comfort of local residents, raises the prices of surrounding real estate and supporting facilities, and promotes local economic development, which is also the embodiment of indirect water economic function.

3 WATER ENVIRONMENT DESIGN CONDITIONS

Water environment design conditions mainly include urban hydrological conditions, water quality and pollution source conditions.

Urban hydrological conditions are mainly based on urban hydrology, water level, rainfall and other measuring stations to calculate urban design rainstorm, urban runoff and the combination of urban waterlogging and outer river (SEA) tide (water) level.

1) The rainstorm intensity of urban design rainstorms with different duration and return periods is an important parameter for urban municipal drainage and regional drainage design. It directly affects the design of municipal drainage pipelines and the amount of regional drainage work. At present, most cities have general rainstorm formulas. According to the rainstorm formula, the rainstorm intensity for different durations can be obtained. However, in practice, the difference in rainstorm intensity of different durations is rarely considered. There are many cases where the 60-min average rainfall intensity is adopted uniformly, that is, the problem of design rain pattern is not considered. In addition, due to the great difference between the return period specified for municipal drainage and the return period specified for regional waterlogging drainage by the water conservancy department, for example, the standard for municipal drainage is mostly once every 2–5 years, and the standard for regional waterlogging is mostly once every 5–20 years. How to connect the two is worth discussing and studying.
2) Due to the influence of urbanization, the consistency of hydrological data is destroyed in the urban runoff generation and concentration models. The change of underlying surface in urban

areas leads to the difference in runoff generation and concentration between hilly areas and areas with an insignificant impact of human activities. It is difficult to use traditional methods to calculate runoff results. In order to solve the problem of urban flood control and drainage, we must start with the study of hydrological characteristics in urban areas, fully analyze the impact of urbanization on hydrology, and adopt appropriate urban rain and flood simulation techniques and methods. There are many kinds of urban runoff calculation methods and the commonly used runoff calculation models, such as the runoff coefficient method, full storage runoff method, infiltration curve method, index method, SCS curve method, etc. The mathematical models of urban surface confluence calculation mainly include hydrodynamics and hydrology. The commonly used hydrological models include the reasoning formula, linear reservoir, nonlinear reservoir, instantaneous unit hydrograph, isochron method, etc. (Peng et al. 2007).

3) Water quality and pollution source conditions in an urban water body or river water quality boundary are obtained by statistical analysis of measured water quality data. The planned water quality objectives can be used for planning according to the corresponding classification types and target water quality of the water environment and water functional areas issued by the city. Pollution source investigation and analysis is an important factor in improving water quality and the environment. It mainly analyzes pollution sources such as industry and life around urban water bodies, urban non-point source pollution and farmland non-point pollution according to the data provided by local environmental protection and urban construction departments (Shen 2008).

4 APPLICATION EXAMPLES

A standby water source is located in the southwest of an area, between the Qiantang River and the Puyang River, and adjacent to Sanjiangkou (Qiantang River, Fuchun River, and Puyang River). In order to better ensure the safety of urban and rural drinking water supply sources in an area, it can effectively solve the problems of single water source and low water supply guarantee in an area.

4.1 *Hydrological conditions*

Through the analysis and statistics of local hydrological data, the urban rainstorm results are obtained, and the runoff production results of each underlying surface are obtained through the calculation of urban runoff generation and concentration.

4.2 *Water quality and pollution source conditions*

According to the results of the water quality monitoring in recent years, except for some indicators such as total nitrogen and ammonia nitrogen, other indicators of diversion water quality can basically meet the water quality requirements of water function zoning. Referring to the environmental quality standard for surface water, the single evaluation standard index method is used to evaluate the quality of surface water in the Qiantang River and lake area.

4.3 *Establishment of hydrodynamic and water quality model*

The governing equation of the one-dimensional hydrodynamic model is the Saint Venant equations (Geng 2006).

$$\frac{\partial A}{\partial t} + \frac{\partial Q}{\partial x} = q \tag{1}$$

$$\frac{\partial Q}{\partial t} + \frac{\partial}{\partial x}\left(\alpha Q^2/A\right) + Ag\frac{\partial Z}{\partial x} + g\frac{Q|Q|}{C^2RA} = 0 \tag{2}$$

Where x and t are space coordinates (m) and time coordinates (s) respectively. A is the sectional area (M^2); Q is the flow of any section in the river (m^3/s); q is the side inflow flow (m^2/s); R is the hydraulic radius; G is gravity acceleration (M/S^2); α is the momentum correction coefficient; and C is Xie Cai coefficient.

Navier Stokes equation under the assumption of incompressible fluid, shallow water and Boussinesq is adopted. The component of vertical acceleration in the vertical momentum equation relative to the horizontal direction is a small quantity and can be ignored. Therefore, hydrostatic pressure equation is adopted vertically. Considering the small temperature gradient in the calculation area, it can be approximately considered that the influence on the flow field can be ignored. A series of partial differential equations and a series of corresponding definite solution conditions are solved based on finite difference grid. After calculation, the TP concentration distribution under current working conditions is shown in Figure 1.

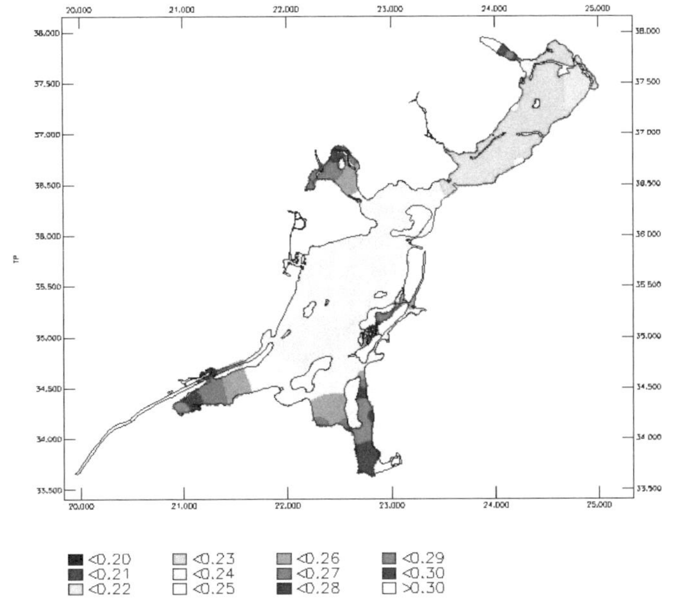

Figure 1. TP concentration distribution under current working conditions.

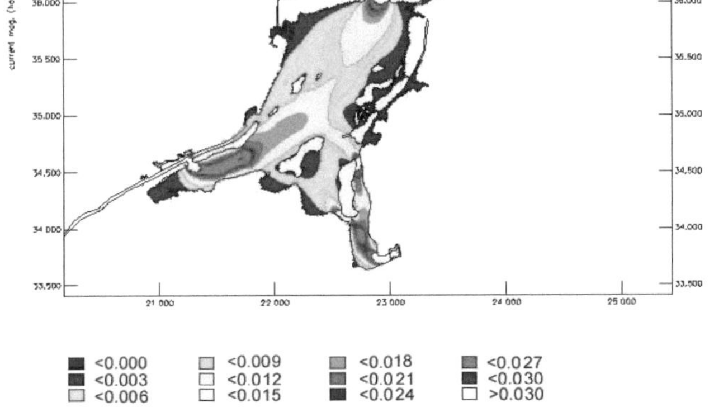

Figure 2. Velocity distribution.

Through the calculation of a hydrodynamic mathematical model, the water storage capacity, water diversion, drainage gate scale, and flood season water level of the city are calculated. The calculation results of the water quality model show that adding a water outlet can greatly reduce the dead water area in a diversion. Under the above conditions, the use of water quality models can simulate the main functional effects of urban water bodies and find solutions accordingly, so as to achieve the harmonious unity of water safety and water environment.

5 CONCLUSION

With the development of urbanization, the underlying surface changes obviously, and the urban water body is the basic condition of urban water safety. Therefore, through the hydrodynamic model, we can put forward the effect of flood detention within the surrounding planning range and how to dispatch and use the water body. At the same time, non-engineering measures can be put forward after the implementation of the project to further improve urban water safety (Peng 2003).

Biological self-purification requires oxygen consumption. If oxygen is not supplemented in time, aerobic microorganisms will be busy and the biological self-purification process will be terminated (Zhu 2005). By diverting water from a lake to enhance the self-purification capacity of the body and controlling water with water, it cannot only increase the water volume and dilute the sewage but also increase the self-purification coefficient of the body, so as to enhance the self-purification capacity of the body. At the same time, the diversion increases the flow rate of the water body, enhances the reoxidation capacity of the water body, increases the concentration of dissolved oxygen in the water body, correspondingly increases the number and species of aquatic microorganisms and plants, and enhances the activity of aquatic organisms. The purpose of purifying water quality is achieved through the metabolism of various organisms, which can effectively.

REFERENCES

Chu Junda, Zhang Yongchun, Hu Mengchun and Zheng Xiaoyu. Ecological effects and protection of urban artificial water environment [M]. Science Press, 2008.

Geng Yanfen. Study on hydrodynamic coupling model of urban rain and flood [D]. Dalian University of technology, 2006.

Peng Kewei. Ecological planning and utilization of urban water environment [D]. Southeast University, 2003.

Peng Zezhou, Yang Tianxing, Liang Xiujuan and Gu Zhaosheng. Mathematical model of water environment and its application [M]. Chemical Industry Press, 2007.

Shen Qingji. Urban ecology and urban environment [M]. Tongji University Press, 2008.

Zhu Chunlong. Research on decision support technology for urban water environment system control [D]. Hohai University, 2005.

Advances in Urban Engineering and Management Science – Khalil & Yang (Eds)
© 2023 The Author(s), ISBN 978-1-032-30426-7

Study on standard application guide for municipal sewage treatment

Bingsheng Xu*, Xuezhi Zhao & Jin Huang
Research Branch of Resource and Environment, China National Institute of Standardization, Beijing, China

ABSTRACT: This study focused on the comprehensive application of standards in the field of municipal sewage. The standard application guide was proposed by strengthening the correlation between standards and process technology after the best technical route and process combination of municipal sewage treatment technology were clarified. Furthermore, the innovative application mode of "formulation of important technical standards—optimization and control of integrated schemes—application practice of typical industries" was realized for better standardization of the municipal sewage treatment industry. This study leveraged support for efficient operation of sewage treatment system facilities in terms of standardization of municipal sewage treatment.

1 INTRODUCTION

On the one hand, municipal sewage treatment can effectively reduce the pressure of sewage discharge on the ecological environment, and on the other hand, the recycling of sewage is also an effective way to alleviate the regional water shortage. Although the technologies and equipment performance for municipal sewage treatment have been improved continuously, there is still a lack of standard application mode for "formulation of important technical standards-optimization of control and integration scheme-typical industry application practice" as well as comprehensive standards. It is necessary to improve the integrated capacity and operational performance, as well as technical, economic, and management elements of systems and facilities. Therefore, it is of great significance to strengthen the integration of standards and technology, continuously standardize the municipal sewage treatment industry, and provide strong support for the efficient operation of municipal sewage treatment systems and facilities (Zhang et al. 2012; Zhou 2001).

The application of technological processes corresponding to standards was then analyzed by improving the economic effect, pollutant removal capacity, management level, and social and environmental effects of facilities. Efforts were also made to promote the integrated application of standards for better municipal sewage engineering and comprehensive sewage treatment, thus providing guidance on process selection and standard application in the municipal sewage industry (Redey et al. 2011; Shao et al. 2016; Sun et al. 2013).

2 METHODS

In this study, efforts were made to systematically sort out the key indicators of municipal sewage treatment, build a standard system for municipal sewage treatment, and analyze the internal logical relationship and endogenous extensibility with relevant standards, thus substantially affecting the in-depth implementation of sewage discharge and relevant standards and offering theoretical support and advanced technological guidance for standardization in this regard.

*Corresponding Author: bingshengxu@cnis.ac.com

DOI 10.1201/9781003305026-5

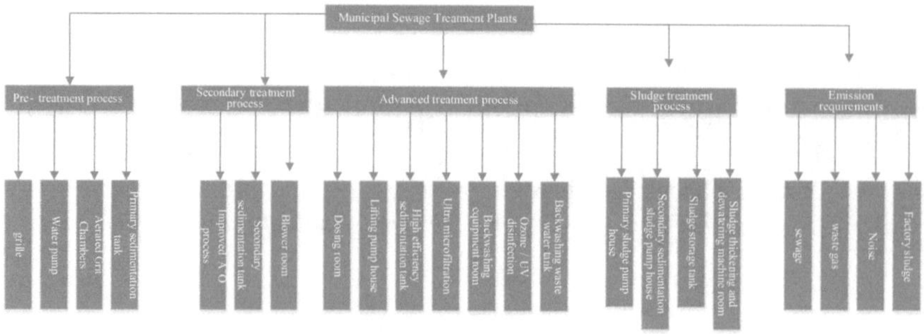

Figure 1. Application composition of sewage treatment facilities.

3 RESULTS AND DISCUSSIONS

3.1 *Scope of application*

This guide is applicable to the process selection, integrated application of equipment and facilities, and improvement of the comprehensive performance of medium- and large-sized municipal sewage treatment plants, which are newly built, expanded, and reconstructed.

3.2 *General requirements*

Municipal sewage treatment enterprises must develop a unified management system and institutional structures. Environmental, quality, energy, and occupational health and safety management systems must adhere to the national standards GB/T 24001, GB/T 19001, GB/T 23331, and GB/T 28001, respectively. The effluent discharge of municipal sewage treatment enterprises shall meet the effluent limit indicator requirements stated in the pollutant discharge standards of urban sewage treatment plants or the requirements of local discharge standards. Municipal sewage treatment plants' main work must be equipped with a complete waste gas cleaning system. The treated exhaust gas shall meet the requirements of GB 16297 or local discharge standards. Effective noise reduction measures, such as sound insulation, noise elimination, and greening, should be considered and taken during the design and construction of municipal sewage treatment plants. The noise and vibration control should meet the requirements of GBJ87 and GB50040. The operation performance of municipal sewage treatment facilities shall be evaluated according to the Technical Requirements for the Evaluation of Operation Performance of Comprehensive Municipal Wastewater Treatment Facilities. Based on the evaluation results, measures for improving the environmental impact, reducing energy and material consumption, and comparing and selecting advanced technologies will be developed.

3.3 *Requirements on influent*

If the primary treatment is adopted according to the national standard of GB/T 31962, the sewage quality shall meet the requirements for Class C specified in 4.2.1 of GB/T 31962-2015. If the secondary treatment is adopted, the sewage quality will meet the requirements for Class B specified in 4.2.1 of GB/T 31962-2015. According to the requirements of GB/T 51347, the influent amount and quality of municipal sewage treatment plants shall be put forward and determined according to local population size, water consumption status, living habits, economic conditions, and regional planning, in combination with the sewage treatment process and regional inflow, and by reference to the influent quality of other sewage treatment plants, with the approval of the environmental protection authority. The sewage treatment plant must treat all sewage within its capacity, according to GB/T 34173.

3.4 *Application of technologies and standards for pre-treatment process*

The reasonable operation mode is adopted to ensure the pollutant removal efficiency of the bar screen and save operation costs. The rotary and grabbing type bar screens can be used to remove the coarse suspended solids in sewage as the first coarse bar screens in municipal sewage treatment plants. The bar screen can be controlled by water level difference, intermittent operation, or manual control. Operators should clean up the pollutants wound on the bar screen in time to ensure its pollutant removal capacity. Pollutants removed from sewage by the bar screen must be recycled.

The energy efficiency indicators of rotary, axial, mixed-flow, and other types of electric pumps with different energy efficiency grades shall meet the requirements specified in Table 1 in GB 32031-2015 respectively. Electric pumps with energy efficiency grade 1 or 2 are recommended. For submersible sewage pumps with energy efficiency grade 1 or 2, their lift shall not be less than 97% of the specified flow rate; the lift of axial-flow pumps shall not be less than 94%. Centrifugal, mixed-flow, and axial-flow pumps must operate economically in accordance with GB/T 13469-2021.The operation of a pump set shall be regulated scientifically and efficiently according to the set flow or liquid level. In the chemical dosing system, the automatic control system is adopted to realize the dynamic regulation of the pump operation and process indicator, thus realizing the fine operation, energy saving, and consumption reduction of the water pump.

The rotary aerator's energy efficiency limit and grade must meet the standards of GB 37483. The energy efficiency ratios of rotary aerators of various energy efficiency grades must not be less than the value specified in Table 1 of GB 37483-2019. It is recommended to use rotary aerators with an energy efficiency rating of 1 or 2. A rotary aerator's motor must have at least an energy efficiency grade of 2 as specified in GB 18613. In the rotary aerator, the matching reducer with a coefficient of performance of 1.75–2 should be used instead of those with a coefficient of performance of 2.4. Furthermore, the reducer's transmission efficiency should be greater than 95%. The hydraulic components of the rotary aerator must be able to absorb air, transmit mass, oxygenate, raise, and stir. Pusher submersible agitators for sewage treatment must fulfill the energy efficiency limit and grade specified in GB 37485-2019. The specific power of such pusher submersible agitators with various energy efficiency classes must not exceed the limits set out in Table 1 of GB 37485-2019. The motor of such pusher submersible agitators must have an energy efficiency grade of at least 2 as defined in GB 18613. A planetary gear reducer, ring gear reducer, less differential gear reducer, or parallel shaft gear-driven reducer with a transmission efficiency of at least 95% shall be used in such pusher submersible agitators. The pusher submersible agitator's hydraulic components must be capable of stirring, mixing, pushing flow, and preventing winding. It is encouraged to select air blowers with an energy efficiency grade of 1 or 2 as specified in GB 28381. The sand eliminator will be used to remove the inorganic particles with a density higher than water in sewage, thus improving the service life of wastewater pumps, valves, and dehydrators.

The design of the primary sedimentation tank shall comply with the provisions specified in Section 6.5 of GB50014-2006. The primary sedimentation tank is provided to remove over 90% of the precipitable solid substances in suspended solids, functioning for wastewater treatment, water quality balance, and hydrolysis (acidification). The chemical coagulation treatment, mechanical filtration, and incomplete biological treatment shall be adopted on the basis of conventional primary treatment (gravity sedimentation) and in accordance with GB 18918, thus improving the treatment effect of primary treatment.

3.5 *Application of technologies and standards for secondary treatment process*

This process is equipped with a pre-anoxic tank (dissolved oxygen concentration of 0.2–0.5 mg/L), an anaerobic tank (dissolved oxygen concentration of less than 0.2 mg/L), a primary AO tank, and a secondary AO tank. The quality of the water shall be determined by reference to the provisions specified in Section 3.4 of GB50014-2006. The sewage entering the biological nitrogen and phosphorus removal systems should meet the requirements that the ratio of BOD5 to TKN in sewage should be greater than 4 during nitrogen removal and that the ratio of BOD5 to TP should be greater than 17 during phosphorus removal. In addition, the process reasonably matches the anaerobic, anoxic, and aerobic microbial flora in different environmental conditions and species, and gives full play to the functions of removing organic matter, nitrogen, and phosphorus. The detection and control system should be set up according to the process operation requirements to realize the automation of operation management. It is required to ensure the influent pH and nutrient level, keep the alkalinity in the reactor at 100 mg/L and the residual alkalinity in the aerobic zone at greater than 70 mg/L. When discharged directly, the effluent from the A2O system must meet national or local discharge standards, and it must also meet the influent requirements when discharged into the next treatment system. The influence of water temperature should be considered during A2O design. Mechanical stirring should be adopted in anoxic and anaerobic zones with a mixing power of 5–8 W/m3.

The horizontal-flow, radial-flow, and vertical-flow sedimentation tanks used in primary sedimentation tanks can be used as secondary sedimentation tanks. The circular radial-flow sedimentation tanks with mechanical wastewater absorption can be adopted by medium and large-sized sewage treatment plants, while the multi-bucket horizontal-flow sedimentation tanks can be adopted by medium-sized sewage treatment plants. The maximum allowable horizontal or ascending velocity should be lower than that of the primary sedimentation tank.

The air blower shall be selected according to air volume and pressure. Single-stage high-speed centrifugal blowers or multi-stage low-speed centrifugal blowers should be selected for large and medium-sized sewage treatment plants. HJ/T 278 applies to single-stage high-speed centrifugal blowers. It is recommended to select blowers with an energy efficiency grade of 1 or 2 as specified in GB 28381. The total air supply capacity of common blowers in the blower room shall meet the requirements of designed air supply capacity (Gs), with a 10% surplus maintained.

3.6 *Application of technologies and standards for advanced treatment process*

When BOD5/TKN entering the reaction tank is less than 4, the carbon source should be added into the anoxic zone. The capacity of the carbon source storage tank should be 7–30 daily dosages of the theoretical dosage; at least 2 sets of dosing systems and a metering pump should be adopted for dosing. When the TP in the effluent cannot meet the discharge requirements, chemical phosphorus removal should be adopted as an auxiliary measure. The best medication type, dosage, and dosing point must be determined in a scientific and reasonable manner. The capacity of the chemical agent storage tank should be 4–7 daily dosages of the theoretical dosage; at least 2 sets of dosing systems and a metering pump should be adopted for dosing.

The design of the sedimentation tank shall comply with the provisions specified in Section 6.4 of GB 50014-2006. The membrane module shall comply with the relevant provisions in GB/T 36137, HY/T 213 and CJ/T 169, and the technical requirements for evaluation of the operation performance of the MBR process system for urban sewage treatment. The relevant equipment in the backwashing equipment room shall comply with the relevant provisions of JB/T 10410.

3.7 *Application of technologies and standards for wastewater treatment process*

The operation of wastewater treatment and disposal facilities must ensure that all wastewater is treated and disposed of within the design capacity. The wastewater and chemical storage yard of the sewage treatment facility must meet the standards of GB18599. According to the treatment methods, the quality of treated wastewater must meet the requirements of GB/T 23484, GB/T 23485, GB/T 23486, GB/T 24600, GB/T 24602, GB/T 25031, and GB/T 4284.The quantity and destination of wastewater discharged from municipal sewage treatment plants must be tracked and recorded if the wastewater is concentrated, dehydrated, and digested. The wastewater must not be discharged or discarded anywhere other than the treatment and disposal sites designated by the competent department. The wastewater shall be stored temporarily for at most 30 days, covered and isolated to prevent secondary pollution.

The wastewater storage tank must be constructed and set up for the purpose of storing wastewater. The aeration or stirring equipment must be designed to prevent the effluent from sinking to the bottom. The wastewater concentration tank can be configured based on the project's current status and following wastewater treatment process.

The wastewater concentration and dehydration machine room are used to constantly concentrate and dehydrate various qualities of wastewater in order to generate high-solids-content wastewater cakes. In the concentration zone, wastewater with a solid content of around 2% is concentrated to a solid content of over 5% (appropriate for pressure filtering) and subsequently dehydrated to a solid content of over 20% in the pressure filtration zone.

3.8 *Discharge requirements*

The standards for sewage disposal must be in accordance with GB 18918. Class I (A) criteria should be met by municipal sewage treatment plant effluent used as recycle water. The Class I (B) standards must be followed when municipal sewage treatment plant effluent is discharged into Class III functional waters of surface water as defined in GB 3838, Class II functional waters of seawater as defined in GB3097, and closed or semi-closed water areas such as lakes and reservoirs. The secondary standard must be applied when municipal sewage treatment plant effluent is released into Classes IV and V functional waters of surface water as defined in GB 3838, or Classes III and IV functional waters of seawater as defined in GB 3097. If the quality of sewage discharged into urban sewers exceeds the design specifications but does not exceed GB/T 31962, the effluent quality must meet the operation contract's norms. If the sewage quality exceeds GB/T 31962, prompt treatment is necessary, as well as timely reporting to the responsible department for municipal drainage and other relevant authorities. The appropriate units will be notified by the competent department to perform a specific investigation and treatment. The released effluent shall meet the GB/T 24188 quality criteria. The exhaust emission must meet the requirements of GB 18918 or GB 16927. The noise level in the environment must meet the requirements of GB 12348. If a piece of equipment's noise level during operation exceeds GBJ 87, it must be replaced.

4 CONCLUSION

The municipal sewage treatment process technology and standard application guide were developed using technology integration, standard analysis, and case application mode. The "essential technical standards development – optimization control integration scheme – typical industry application practice" application mode was the first to emerge, providing standardized assistance for the efficient operation of municipal industrial sewage treatment systems.

ACKNOWLEDGMENT

This work was supported by the National Key R&D Program of China (Grant No. 2018YFF0213203).

REFERENCES

Redey, A., Somogyi, V., Anyos, J., Domokos, E., Thury, P. (2011) Simulation of the influence of industrial wastewater on a municipal sewage treatment plant-a case study. Environmental Science & Pollution Research.

Shao, S., Mu H., Yang, F., Yun, Z., Li, J. (2016) Application of Emergy Analysis to the Sustainability Evaluation of Municipal Wastewater Treatment Plants. Sustainability. pp. 8.

Sun, F., Wang, X., Li, X. (2013) An innovative membrane bioreactor (MBR) system for simultaneous nitrogen and phosphorus removal. Elsevier, pp.11

Zhang, S.R., Zhang, T.J., Liu, J.L., Li, Y. (2012) Study on A^2O Method for Co-Treatment of Landfill Leachate and Municipal Sewage. Advanced Materials Research. pp. 2908–2913.

Zhou, L. (2001) Technical Keys to Municipal Sewage Treatment with High-efficiency and Low-cost.

Advances in Urban Engineering and Management Science – Khalil & Yang (Eds)
© 2023 The Author(s), ISBN 978-1-032-30426-7

Research on integrated solution for municipal sewage treatment

Bingsheng Xu*, Jin Huang & Xuezhi Zhao
Research Branch of Resource and Environment, Beijing, China
National Institute of Standardization, Beijing, China

ABSTRACT: This study focused on the research of technology integration and the comprehensive application of standards in the field of municipal sewage. By analyzing the correlation between key process indicators and standards, the technological processes and key technologies for municipal sewage treatment were sorted out. Based on evaluations of influent parameters, process selection, and investment and operating costs, typical municipal wastewater treatment procedures were chosen and used to construct an integrated municipal wastewater treatment scheme.

1 INTRODUCTION

Under the guidance of China's national policy on energy conservation and emission reduction, as well as proactive fiscal policies, municipal sewage treatment technologies have witnessed rapid development, and remarkable achievements have been made in urban water environment treatment in recent years. China's overall water quality has met policy-mandated standards with a steadily decreasing proportion of IV-inferior class V water. It was put forward at the "14th Five-Year Plan" working meeting that the water quality should satisfy such a standard. The water in a river can raise aquatic plants and fish, and can even satisfy the standard for swimming. Therefore, more attention will be paid to ecological protection and restoration while improving the water environment. In addition, upgrading and renovation are still the key tasks of the industry.

According to China's 13th Five-Year Plan for the Construction of Urban Sewage Treatment and Recycling Facilities, the treatment rate of sewage in cities and counties must be 95% and no less than 85% by the end of 2020, with the Beijing-Tianjin-Hebei, Yangtze River Delta, and Pearl River Delta regions expected to meet the target ahead of schedule. The capacity of sewage treatment facilities should be expanded from 217 million m^3/day by the end of 2015 to 267 million m3/day by the end of 2020, including the upgraded and reconstructed capacity of 42.2 million m3/day. As indicated in the China Urban-Rural Construction Statistical Yearbook, the municipal sewage discharge in 2018 was up to 62.052 billion m^3. The total sewage discharge in 2017 was 69.97 billion tons, including 51.78 billion tons of municipal domestic sewage (or 74.0% of the total discharge). As the proportion increases, municipal domestic sewage has become the main sewage source. By 2018, the city and county sewage treatment rates in China were 95.49% and 91.16%, respectively, exceeding the goal set in the 13th Five-Year Plan. However, the sewage treatment rate in towns was only 53.18%, far behind the goal of 70% in 2020.

As the urban resident population keeps increasing and the requirements for effluent quality of urban sewage treatment plants have become stricter in recent years, many urban sewage treatment plants in China have been operated at full capacity. In the meantime, most sewage treatment plants were constructed many years ago, so the quality of effluent from these plants is still at Class I (B) as stated in the Discharge Standard for Pollutants from Municipal Wastewater

*Corresponding Author: 356403520@qq.com

 DOI 10.1201/9781003305026-6

Treatment Plants (GB 18918-2002), which means they can no longer meet the effluent quality requirements.

The optimal control of technology, equipment, and systems for municipal sewage treatment in China can provide efficient guidance for the planning, construction, operation, maintenance, and management of municipal sewage treatment facilities. The reasonable treatment processes shall be selected from such dimensions as influent characteristics, process selection, investment, and operation cost, in accordance with existing national and industrial policies and standards. The following elements of municipal sewage treatment shall be clearly defined: the best technical route, combination of technology and process, process integration scheme, project construction, as well as operation and maintenance of facilities. All these efforts are attributed to the scientific and reasonable selection of treatment processes in actual investment, construction, upgrading, and renovation.

2 METHODS

As current discharge standards (especially local standards) raise high requirements on COD, nitrogen, phosphorus, and other indicators, stricter requirements will be proposed for nitrogen and phosphorus removal efficiency based on the environment-technology-economy-based multi-objective optimization analysis. The conventional secondary treatment may not meet the requirements for nitrogen and phosphorus removal rates. Therefore, it is necessary to strengthen the nitrogen and phosphorus removal from the secondary sewage, increase the advanced treatment processes, and optimize the existing processes of sewage treatment plants.

3 RESULTS AND DISCUSSIONS

There will be various problems if the nitrogen and phosphorus removal processes are completed in the single wastewater system of the traditional A2/O process, such as the contradiction of wastewater age, the competition of carbon sources, and the residual interference of nitrate and dissolved oxygen (DO). With the development of a process, the oxidation ditch process may be used on the basis of the A2/O process, i.e., the A2/O oxidation ditch process. Compared with the traditional A2/O completely mixed reactor, the A2/O oxidation ditch process has a larger treatment load and higher resistance to impact load. In addition, it can better design the air distribution system, save electricity costs, and achieve better nitrogen and phosphorus removal through the sequential arrangement of anaerobic, anoxic, and aerobic zones in the ditch. For large water treatment plants, the nitrogen and phosphorus removed can be recycled many times with better effect. The oxidation ditch process should be adopted for the treatment project with a large designed discharge.

The modified A2/O process improves the traditional A2/O nitrogen and phosphorus removal processes by setting the pre-anoxic zone to remove NO3-N from the returned wastewater, which provides a good anaerobic environment for the phosphorus-accumulating bacteria in the anaerobic tank to obtain more carbon sources and enhances the phosphorus removal capacity of the system. The modified A2/O process is characterized by the small land occupation of biochemical tanks, compact structure layout, fewer pipeline crossings, stable and efficient effluent quality, and high purification potential. The capacity expansion project is equipped with external carbon source facilities, which can selectively add carbon sources in anaerobic and anoxic zones as per the operation's needs, thus stabilizing and enhancing the biological nitrogen removal capacity of the system.

In conclusion, medium and large-sized municipal sewage treatment plants are advised to use the tertiary water treatment process (including water regeneration and recycling) outlined below:

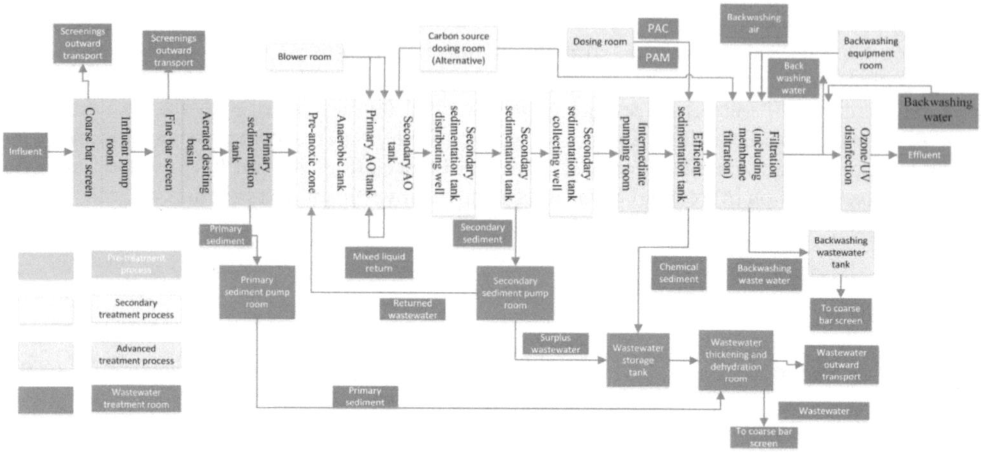

Figure 1. Tertiary water treatment process for medium and large-sized municipal sewage treatment plants.

The municipal sewage treatment process route consists of three stages, i.e., pre-treatment, secondary biochemical treatment (modified A2/O process, A-A/A/O process, multi-inlet reversed A/A/O process, UCT process, and MUCT process), and advanced treatment (including membrane filtration), which can meet the increasingly stringent municipal sewage discharge requirements. To ensure stable and up-to-standard effluent, the widely accepted and applied coagulation, sedimentation, and (membrane) filtration advanced treatment process is recommended to be used in the advanced treatment stage. In addition, the treated municipal sewage may be recycled, further improving the utilization rate of water resources.

This technology is applicable to municipal sewage treatment and recycling, upgrading and reconstruction of original sewage plants, and advanced treatment and recycling of up-to-standard effluent from original secondary sewage plants.

4 CONCLUSION

Based on the results and discussions presented above, the conclusions are as below:

(1) The related standard internal logic relationships were analyzed, as well as the correlation between the endogenous ductility. This research can give theoretical support as well as sophisticated technology for the application of wastewater discharge standards.

(2) The three-stage process technology of system integration optimization, modularization, and high automation of pretreatment-bio enhancement-(membrane) advanced treatment and reuse was clearly proposed through the multi-objective optimization analysis based on environment-technology-economy and the comparison and selection of municipal sewage treatment processes based on the improvement of standard requirements. The findings may help to improve municipal sewage treatment levels through the use of standards that are applied in an integrated manner.

ACKNOWLEDGMENTS

This work was supported by the National Key R&D Program of China (Grant No. 2018YFF0213203).

REFERENCES

Bode, H., Grünebaum, T. (2000) The cost of municipal sewage treatment–structure, origin, minimization-methods of fair cost comparison and allocation. EBSCO.

Shao, S., Mu H., Yang, F., Yun, Z., Li, J. (2016) Application of Emergy Analysis to the Sustainability Evaluation of Municipal Wastewater Treatment Plants. Sustainability. pp. 8.

Zhang, S.R., Zhang, T.J., Liu, J.L., Li, Y. (2012) Study on A^2O Method for Co-Treatment of Landfill Leachate and Municipal Sewage. Advanced Materials Research. pp. 2908–2913.

Zhang, X.H., Deng, S.H., Wu, J., Jiang, W. (2010) A sustainability analysis of a municipal sewage treatment ecosystem based on energy. Ecological Engineering. pp. 685–696

Advances in Urban Engineering and Management Science – Khalil & Yang (Eds)
© 2023 The Author(s), ISBN 978-1-032-30426-7

5G MIMO antenna design and AI study

Zhe Sun*
School of Electrical Engineering, The University of Sydney, Sydney, Australia

ABSTRACT: The Fifth-Generation New Radio (5G NR) communications framework provides an entirely new approach to cellular communications. With its invention, many new technological innovations are supported by it, such as scalable waveforms, multiple access schemes, and service multiplex across wide bandwidths. It also supports existing services while being forward-compatible with future requirements. Multiple-input, multiple-output, or MIMO, is also an important technology that is being used in 5G. It can reduce bit errors, improve signal range, and lower power consumption. Although 5G wireless requires more complex signal processing and the ability to handle a much higher data rate than previous protocols, one of the keys to its success is antenna design. This paper reviews two kinds of 5G MIMO antenna designs and one type of AI design that can make designing 5G antennas easier.

1 INTRODUCTION

In 2021 and in the future, mobile communication technology and the mobile communication industry will enter the development phase of the fifth generation of mobile communication (5G). 5G can satisfy people's demand for super high internet speeds and ultra-high mobility to provide users with high-definition video, virtual reality, augmented reality, cloud desktop, online games, and other extreme business. It will penetrate into the Internet of Things and other fields. Along with in-depth integration of industrial facilities, medical instruments, and transportation tools, the "Internet of Everything" will be fully realized and effectively meet the industry's information services for medical care, transportation, and other vertical industries, such as the requirements. The 5G antenna is one of the essential system components essential to achieving these visions.

To achieve the perfect performance of 5G, Massive MIMO technology should be used. Multiple-input, multiple-out, which is also called MIMO technology, is an established wireless communication technique and has been around for decades (Tracy 2016). MIMO techniques play a prominent role in Wi-Fi communications, as well as 3G, 4G, and 4G LTE networks. Massive MIMO (also known as Large-Scale Antenna Systems, Very Large MIMO, Hyper MIMO, Full-Dimension MIMO, and ARGOS) makes a clean break with current practice through the use of a very large number of service antennas (e.g., hundreds or thousands) that are operated fully coherently and adaptively (Maxwell 1892). Additional antennas help to concentrate the transmission and reception of signal energy into smaller and increasingly smaller spatial regions. Throughput and energy efficiency are substantially improved by the additional antennas, especially when the simultaneous scheduling of a large number of user terminals is combined (e.g., dozens or hundreds).

As wireless technology evolves from 1G to 5G, the frequency step length mainly determines whether the evolving standards do not require large technical changes or discontinuity, and so do the changes required in antenna design (Jacobs 1963). However, the 5G implementation provides up to more than tenfold frequency increases for certain applications, which is a huge change from previous technology. It will bring major challenges and new opportunities for society. There are more and more elements being put into the antenna, which will cause the electromagnetic interference (EMI) (Ling 2009) or larger size requirement of the board. Also, in the past 20

*Corresponding Author: zsun0442@uni.sydney.edu.au

 DOI 10.1201/9781003305026-7

years, mobile terminal devices and other mobile terminal devices have developed rapidly, and the development trend is concentrated on "high screen ratio" and "ultra-thinning." Especially with the popularity of comprehensive-screen mobile phones, the design size of antennas in mobile terminal equipment is becoming more and more stringent. How to realize the miniaturization of the antenna is the most important problem with the antenna research and design of mobile communication terminal equipment.

This paper introduces the research situation of 5G antennas and shows the two designs of MIMO antennas and one AI system.

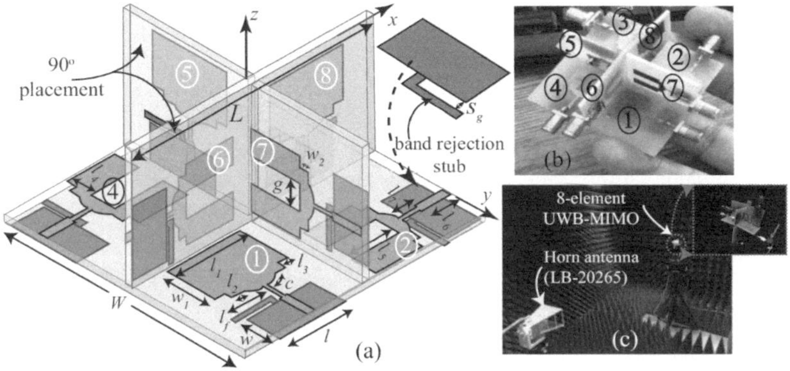

Figure 1. (a) Perspective view of the proposed UWB-MIMO antenna with dimension in mm. L = 50, W = 50, wl = 10, l1 = 15, l2 = 2.25, l3 = 2, l4 = 5, l5 = 10, g = 5, w2 = 1.5, l6 = 6, l7 = 3.82, c = 3.1, l = 13.5, w = 7, Lf = 7.25, and Sg = 0.5, (b) perspective view of fabricated prototype with dimensions shown in Figure 1, and (c) photo during pattern measurement in fully calibrated anechoic chamber.

2 EASE OF USE

2.1 *An eight element UWB-MIMO/diversity antenna*

This antenna is an eight-element UWB-MIMO/diversity antenna with WLAN band rejection capabilities. Many antenna designs have tradeoffs between design complexity, size, number of ports, and bandwidth. They partly increase their size and are not able to reject the band selected. Therefore, an eight-element, compact Ultra-Wideband Multiple Input Multiple Output (UWB-MIMO) antenna capable of providing high data rates for future Fifth Generation (5G) terminal equipment along with the provision of necessary bandwidth for Third Generation (3G) and Fourth Generation (4G) communications that accomplishes band rejection from 4.85 to 6.35 GHz by deploying an Inductor Capacitor (LC) stub on the ground plane is presented (Yorozu 1982).

This antenna design has many aspects to be mentioned. The first element is that it has a band rejection stub. It is being placed in the center of the ground plane and beneath the transmission line because this is the most effective place for the stub connection to draw the current from the stub. The second method is that the second monopole is placed orthogonally at 6.15 mm edge distances to exploit the polarization diversity. At this distance, it can be inserted into a U-shaped slot (Satam et al. 2018), which decreases the power level of antenna element 2 and, therefore, can lower the polarization of element 2. The third method is that four more monopoles (labeled as 5, 6, 7, and 8 in Figure 1(a)) were employed perpendicularly. In this design, it can preserve the compactness of the design, achieve polarization diversity and wideband impedance matching, low mutual coupling, and band rejection capabilities. The last method is that to suppress the surface current and mitigate the near-field coupling, the separated ground plane is used.

There are three main advantages of this antenna compared to other antennas, which are shown in the following paragraph:

- It has a stub to reject the WLAN band in Figure 1(a). The stub was placed on the backside of a monopole and connected to the ground. By changing the length of the stub, the rejected bands can be varied, and the length is calculated by $L_f = \lambda_g/4 = c/4f_o\sqrt{\epsilon_r}$ where $f_o \approx 5GHz$ and ϵ_r is the relative permittivity of the substrate. With the stub increasing, the rejected band also increases. Another way to change the rejected bandwidth is to change them by changing the gap between the stub and the ground plane. This is because the changing gap will affect the capacitance between them and the resultant rejection bandwidth change.
- It has a smaller board size compared to the design presented in (Palaniswamy 2016) is 70 × 70 mm^2, whereas the board size is 90 × 90 mm^2 in (Alsath 2018). With the smaller board size, it still has 86% efficiency and both simulated and measured mutual couplings are not exceeding the −17 dB level.
- It is a disconnected ground plane MIMO/diversity antenna, and it has perpendicular placement of antenna elements. Monopoles 58 are placed orthogonally to the planar board (in between the monopoles 1–4) as shown in Figure 1. Compared to another connected ground plane, it can be used in specific applications. LTE/WiMAX mobile terminals (Zhai 2014), communication in vehicular networks, and radar imaging, all of which require multiple antennas in a confined space.

2.2 *Miniaturization MIMO for 5G mobile phone antenna design(Zhang 2021)*

The rapid development of 5G has increased the demand for 5G communication on mobile phones. The sales volume of 5G mobile phones and 5G-related applications is experiencing explosive growth. In wireless communication, the mutual conversion function of telecom signals and electromagnetic waves cannot be achieved without the antenna. Therefore, an antenna suitable for the 5G communication needs of the mobile phone will be discussed, which can meet the smaller size of the mobile phone on the premise that it can also meet the communication performance needs.

This antenna uses FR-4 material as a substrate, and 10 antennas with ideal conductors are placed around the substrate whose relative dielectric constant is 4.4 and loss tangent is 0.02. To simulate the size of the mobile phone with this material, the medium plate design was grown to 80 mm wide, 150 mm wide, and 0.8 mm high. The shape is similar to a mobile phone. The specific size is given in Figure 2.

Ten 5G antenna modules were placed symmetrically on the media plate, with three antennas placed beside 150mm, two in the same direction and the other in reverse at 180°. Two antennas are placed on each side of 80mm, and two rectangular medium slot spaces are placed below each antenna to improve the external radiation of the antenna. Specific size is given in Figure 3, and they are 16.5 mm × 2.5 mm and 0.5 m × 0.5 mm. In Figure 2, L1 is 10 mm and L2 is 5.5 mm. The distance between the antenna on the 150 mm side of the other side is 28.5 mm,70.8 mm, and

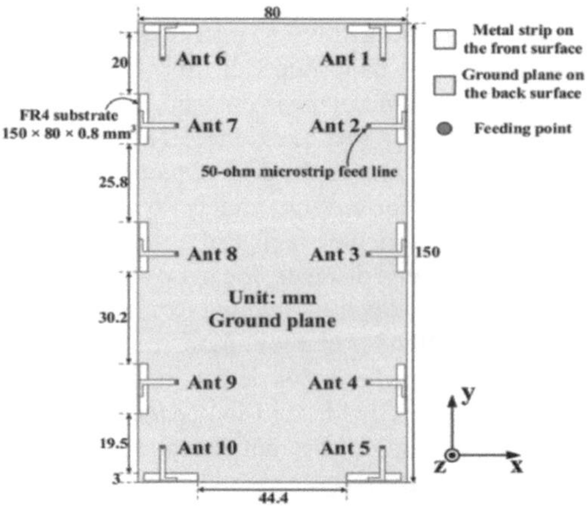

Figure 2. Overall antenna size diagram.

116 mm. And Ant9 and Ant4 and Ant8 are the mirror relationship of Ant2, Ant3, Ant 8 on the Y axis, Ant1 and Ant5, and the mirror relationship of Ant6 and Ant10 on the X-axis, this design can improve the radiation range of the antenna.

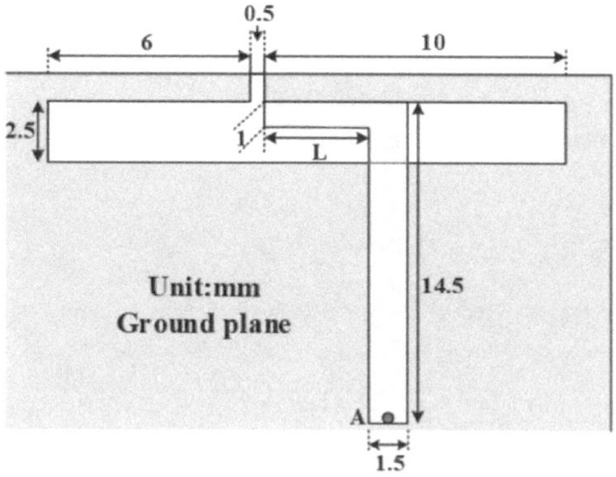

Figure 3. Rectangular medium slot spaces.

The antenna is all applied to a 50 Ω coaxial feeding mode. To facilitate the feeding of the 50 Ω SMA connector from the board, the fitting surface of the ten antenna units is perforated with the feeder in the middle of the medium board, which can ensure the stability of the transmission. This antenna is a single-pole antenna of type L. The antenna position determines the isolation degree. After simulation optimization, the L length is 3.5m, and the width of ending A is 1.5mm. The goal of the antenna is designed to achieve the communication requirements of 5G, with the target coverage of frequency bands of 3300–3400 MHz and 4800–5000 MHz.

This antenna mainly uses MIMO technology to realize the performance of multi-band and high bandwidth, reduces the impact of the mutual coupling between cells, and increases the isolation through simulation optimization. A good antenna scheme is provided in 5G mobile phones.

Considering the compactness of the internal space of the phone, miniaturization the antenna while maintaining good gains, which has high application values.

3 AI

AI is a technology that mimics human intelligence, allowing computer applications to learn from the experience of iterative processing and algorithmic training. AI systems get smarter with each successful round of data processing since each interaction allows the system to test and measure solutions and develop expertise in the task they've been set to accomplish (CSU 2021). Since this can be completed rapidly, much faster than the rate a human being would be able to perform similar work, AI systems can become experts far more quickly than humans, making them incredibly effective options for any process requiring intelligent decision-making. With the availability of large datasets and ever-increasing computing power, there has been a growing use of data-driven artificial intelligence systems, which have shown their potential for successful application in diverse areas (Lecci).

3.1 *Machine learning-aided design of thinned antenna arrays for optimized network level performance*

In this AI system, it wants to solve the question of whether Massive Uniform Planar Arrays (UPAs) operating in the mmWave systems will be adopted in the 5th generation of mobile networks (5G) as the key enablers to meet the challenging requirements of the new standard. Many ambitious targets have been listed and are waiting to be solved, such as: decreasing the side-lobe level or maximizing the directivity. More global, network-oriented requirements need to be taken into account. However, designing and optimizing the antenna arrays can fix the questions about requiring network topology and hence dramatically increase the complexity of the optimization problem. It likely moves from the bare electromagnetic to the network domain. Optimization based on heuristic simulations is generally not feasible, as such detailed emulators are time-consuming and computationally expensive. The optimization algorithm takes a large amount of time to simulate the iterations. Therefore, a machine learning (ML) framework that can simulate a given simulator and allow us to achieve any network optimization goal in a reasonable time is proposed.

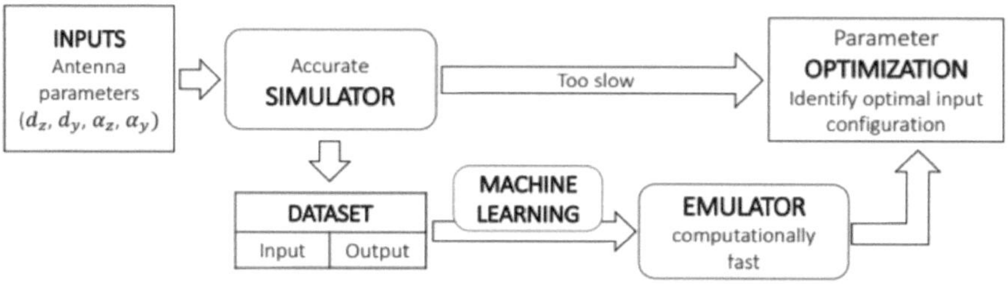

Figure 4. Workflow of the proposed framework.

In Figure 4, the highlight shows the diagram of how the parameter optimization is achieved using an ML-based emulator and machine learning. plane arrays. Returning an optimized antenna family rather than a specific configuration successfully reduces the search space for possible configurations, making it possible to further refine it through more precise simulations.

It uses Monte Carlo iterations and random forests for the emulator. To assess performance, the normalized Root Mean Square Error (nRMSE) metric is used. Due to the limitation of Monte Carlo iterations, there are only 500 data points, but if they give the best results shown in Figure 5.

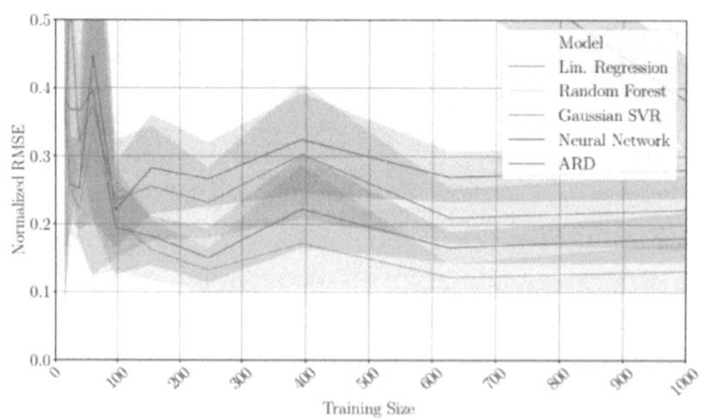

Figure 5. Plot showing cross-validation scores on the nRMSE metric with increasing training size.

Figure 6 shows a performance comparison between different classes of antennas. The chosen 300 antennas generated from non-optimized input configurations from the input space described in Section II-B are shown in grey dots. Red dots, instead, show 30 antennas, all generated with the optimal input configuration. Blue and red triangles represent the baseline and the optimal antenna found in Level Performance 2020, respectively.

This ml-based optimization framework can successfully optimize the antenna design very efficiently. Because the antenna parameterization is chosen from this study; the framework can explore more complex configurations than regular-spaced plane arrays. Returning an optimized antenna family rather than a specific configuration successfully reduces the search space for possible configurations, making it possible to further refine it through more precise simulations.

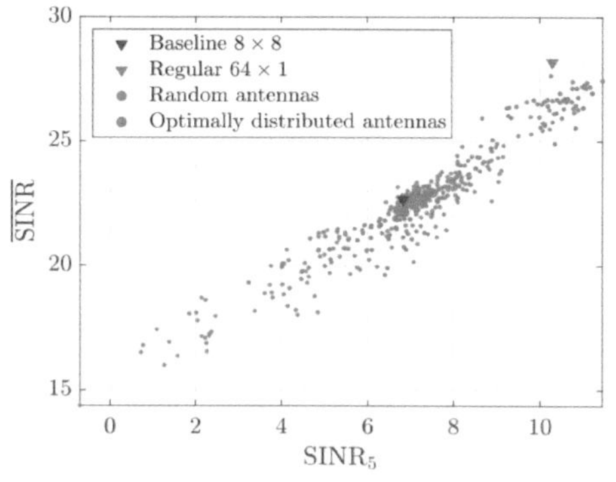

Figure 6. Performance.

4 CONCLUSION

In this paper, two types of MIMO antenna design with their advantages and one AI system which can be used to optimize the antennas are discussed. From the antennas, it clearly shows that with a 5G network and quality of life developing, the antenna should be designed to adapt to more specific situations. Hence, it can meet the needs of human beings. To solve this requirement for different designs quickly, the AI technique is a good option to optimize the antenna from the previous version, which can save a lot of time and money.

REFERENCES

CSU Global, July 5th, 2021 retreieved from "https://csuglobal.edu/blog/why-is-artificial-intelligence-so-important", in 20/11/2021

G. Li, H. Zhai, Z. Ma, C. Liang, R. Yu, and S. Liu, "Isolation-improved dual-band mimo antenna array for lte/wimax mobile terminals," IEEE Antennas and wireless propagation letters, vol. 13, pp. 1128–1131, 2014

I. S. Jacobs and C. P. Bean, "Fine particles, thin films and exchange anisotropy," in Magnetism, vol. III, G T. Rado and H. Suhl, Eds. New York: Academic, 1963, pp. 271–350.

J. Clerk Maxwell, Massive MIMO systems, 3rd ed., vol. 2. Oxford: Clarendon, 1892, pp.68–73.

J. Z. Zhang, B. L. You, X. H. Wang, J. Li, "A 5G Mobile Phone Antenna Design for the Miniaturized MIMO Technology", ELECTRONICS WORLD, vol. 20, pp150–151, 2021

Level Performance, "14th European Conference on Antennas and Propagation (EuCAP 2020)", Copenhagen, Mar. 2020

M. G. N. Alsath, H. Arun, Y. P. Selvam, M. Kanagasabai, S. Kingsly, S. Subbaraj, R. Sivasamy, S. K. Palaniswamy, and R. Natarajan, "An integrated tri-band/uwb polarization diversity antenna for vehicular networks," IEEE Transactions on Vehicular Technology, vol. 67, no. 7, pp. 5613–5620, 2018.

M. Lecci, P. Testolina, M. Rebato, A. Testolin, and M. Zorzi, "Machine Learning-aided Design of Thinned Antenna Arrays for Optimized Network

P. Tracy, "Understanding massive MIMO and what it means for 5G" Phil. Trans. Roy. Soc. London, vol. A247, pp. 529–551, August, 2016.

S. K. Palaniswamy, Y. P. Selvam, M. G. N. Alsath, M. Kanagasabai, S. Kingsly, and S. Subbaraj, "3-d eight-port ultrawideband antenna array for diversity applications," IEEE Antennas and Wireless Propagation Letters, vol. 16, pp. 569–572, 2016.

S.-Y. Lin and H.-R. Huang, "Ultra-wideband mimo antenna with enhanced isolation," Microwave and Optical Technology Letters, vol. 51, no. 2, pp. 570–573, 2009.

V. Satam, S. Nema, and S. S. Thakur, "Spanner shape monopole mimo antenna with high gain for uwb applications," in Proceedings of International Conference on Wireless Communication. Springer, 2018, pp. 129–138.

Y. Yorozu, M. Hirano, K. Oka, and Y. Tagawa, "Electron spectroscopy studies on magneto-optical media and plastic substrate interface," IEEE Transl. J. Magn. Japan, vol. 2, pp. 740–741, August 1987 [Digests 9th Annual Conf. Magnetics Japan, p. 301, 1982].

Advances in Urban Engineering and Management Science – Khalil & Yang (Eds)
© 2023 The Author(s), ISBN 978-1-032-30426-7

Research on train adaptive speed tracking based on Lyapunov design method

Chenlong Ma*, Bohong Liu* & Zhengsheng Qi
School of Automation and Electrical Engineering, Lanzhou Jiao tong University, Lanzhou, China

ABSTRACT: During the operation of high-speed trains, in order to achieve the goal of punctuality and efficiency, the speed of the train is usually monitored in real time, and real-time speed adjustments are made, so as to meet the requirements of ideal speed tracking. In this paper, a basic model of train force is established for the change of resistance parameters of trains on different lines, and the control law of reference adaptive control theory is used to achieve the effect of coping with the effect of parameter changes, that is, to adjust speed to reduce or eliminate uncertain resistance factors. The train adaptive control strategy in the article is based on the Lyapunov design method and has a faster convergence. Through the analysis of the simulation results of the train operation process, it is proved that the uncertainty problem caused by the complicated operation line conditions of the high-speed train can be effectively dealt with, and the tracking purpose within a small error can be achieved, thus ensuring the safety and reliability of the high-speed train operation. It also shows the feasibility of the reference model adaptive control in the application of train speed tracking.

1 INTRODUCTION

In the field of train control research, the main research is the study of train ATO (Automatic Train Operation, ATO) control strategies. Its main function is to replace the driver in order to "manipulate" the train to a certain extent. At the same time, train operation control systems are key technologies to ensure train safety, and they play an important role in the efficiency and safety of train operation. Therefore, train operation speed control has always been the research content that researchers have always favored. currently used in train operation control methods, including intelligent algorithms, classical control theory, and deep learning, etc. Therefore, the widespread use of these control theories enables the train to achieve high-precision speed control, which in turn achieves the effects of comfort, energy-saving, and high efficiency. Train operation control is mainly realized by the automatic train operation system, so the entry point for its research on speed control is to start with the study of train ATO speed control strategy. From the perspective of ATO system research, institutions including Hitachi in Japan, Siemens in Germany, Westinghouse in the United Kingdom, etc., have also invested in research and have achieved some considerable gains. Later, countries, including the Soviet Union and Denmark, joined one after another, so the research investment of various countries provided a certain knowledge reserve for subsequent ATO-related research. In addition, in the research of ATO control algorithms, the ones that currently receive more attention from researchers include classic control algorithms, adaptive control algorithms, and intelligent control algorithms. Regardless of the algorithm, it contains its own control characteristics and advantages. Therefore, to get a good application, it is necessary to dig out the point of convergence between its control object and control theory.

ATO control research provides more valuable exploration for train operation control. In this paper, the reference model for adaptive control is used as a support. The main function is to reduce the

*Corresponding Authors: 2289768638@qq.com and 2497741476@qq.com

 DOI 10.1201/9781003305026-8

dimension of the changing parameters and then continue to adjust the control output to achieve the adjustment purpose of the strain parameter changes, thereby improving the stability and reliability of the train. The article uses the adaptive control law of the Lyapunov design method, which is one of the model adaptations, to design the controller, and then changes the train control strategy to complete the speed optimization adjustment.

2 ANALYSIS OF TRAIN MATHEMATICAL MODEL

According to the second law of Newton's mechanics, we can regard the train as the research object of a single particle, and analyze the force on it as:

$$0 = F - F_b - D(t) - Ma \tag{1}$$

Where M is the total mass of the train; a is the acceleration of the train; F is the control force acting on the train; F_b is the basic resistance encountered during the train operation. The resistance is the basic resistance. The basic resistance F_b can be expressed as

$$F_b = a_0 + a_1 v + a_2 v^2 \tag{2}$$

Among them, a_0, a_1 and a_2 are all constants greater than 0. According to the "Training Regulations", it is known that their value is related to the type of train and the operating environment, and the formula shows that the basic resistance of the train will be one yuan two as the speed increases. The linear quantitative relationship to the power.

$D(t)$ is the random resistance of the tunnel and the additional resistance of the train, including the random resistance of the ramp.

$$D(t) = L + C + R \tag{3}$$

According to the "Training Regulations", under restricted conditions, the ramp resistance is expressed as

$$L = 0.0001lv \tag{4}$$

l Indicates the length of the tunnel; the curve resistance is expressed as

$$C = \frac{600}{R} \tag{5}$$

And 600 represents the empirical constant of the curve; R is the radius of the line curve.
Thus, the ramp resistance is expressed as

$$R = 1000 \sin \alpha \tag{6}$$

Expressed as the ramp angle; because the ramp angle of the following cars is usually small, it is expressed in thousandths, and it is obtained according to the characteristics of the sine function

$$R = 1000\alpha \tag{7}$$

3 DESIGN OF ADAPTIVE SPEED CONTROLLER

The adaptive speed controller of the train is mainly used for the control of the train when the quality is relatively certain, but the basic resistance parameters and random disturbance are unknown. In this paper, the influence of traction and braking force on speed in reference is designed, and its effectiveness is verified. It realizes the tracking control of the train with the ideal speed as the tracking target under the bounded disturbance of the outside world, so that the train The speed curve can approach the ideal speed curve in real time, thus realizing effective control of the train speed.

Under normal circumstances, if the line environment parameters are relatively determined, the traction and braking force of the control train will be relatively determined, and the actual output speed will have an error value with the ideal speed. It is difficult to determine a_0, a_1, a_2 and D in the actual line. Therefore, the purpose of the design is to obtain a traction or braking force that can calculate the train model parameters under the line environment changes. The model reference adaptive control design idea structure is as shown in the figure below.

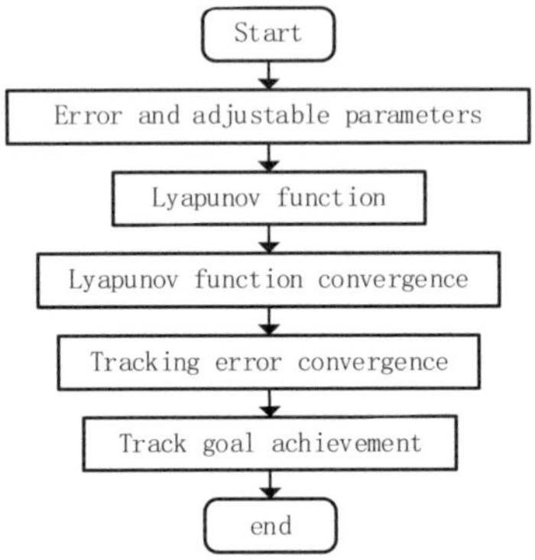

Figure 1. Model reference adaptive control design drawing.

The function of the traction or braking force of the train is to output the corresponding speed so that the error e between the actual speed and the ideal speed is close to 0 as the time t increases, that is, the speed error e is expressed as follows

$$e = v - v_s \tag{8}$$

In the formula, v is the real-time speed; v_s is the ideal speed. The control force F is relatively certain in the quality of the train, and the output under unknown conditions such as the basic resistance parameters of the line and random disturbance is as follows

$$\mathrm{F} = \hat{\mathrm{a}}^{\mathrm{T}}\phi(\cdot) - \mathrm{Mk_0 e} \tag{9}$$

Where k_0 is a positive parameter; $\hat{a}^T$ is $a = [a_0, a_1, a_2, M, D]$ of estimated value; $\varphi(\cdot) = [1, v, v^2, \dot{v}, 1]$. Therefore, the estimated value of a in the complex and changeable line section is obtained, and a matrix is formed

$$\begin{pmatrix} a_{11} & a_{12} & \cdots & a_{1n-1} & a_{1n} \\ a_{21} & a_{22} & \cdots & a_{2n-1} & a_{2n} \\ a_{31} & a_{32} & \cdots & a_{3n-1} & a_{3n} \\ M_{41} & M_{42} & \cdots & M_{4n-1} & M_{4n} \\ D_{51} & D_{52} & \cdots & D_{5n-1} & D_{5n} \end{pmatrix}$$

In order to facilitate the convergent control of the speed error on the line, we need to find a general adaptive law that changes with the line to indirectly control the train speed. Therefore, the equation of the basic force model of the train can be further expressed as

$$0 = F - a^T\phi(\cdot) - Mk_1\dot{e} \tag{10}$$

Where k_1 is a positive compensation parameter.

Then

$$\dot{e} = \frac{1}{Mk_1}\tilde{a}^T\phi(\cdot) - \frac{k_0}{k_1}e \tag{11}$$

Where $\tilde{a} = \hat{a}(t) - a$.

According to the composition characteristics of the Lyapunov function, the selected Lyapunov function is

$$V = \lambda_0^{-1} e^2 + \lambda_1^{-1} \tilde{a}^2 \tag{12}$$

Where $\lambda_0 > 0$ and $\lambda_1 > 0$ are constant. And no matter or $e \neq 0$ or $\tilde{a} \neq 0$, they are positive definite. Therefore, the derivative of V with respect to time is

$$\dot{V} = 2(\lambda_0^{-1} e\dot{e} + \lambda_1^{-1} \tilde{a}\dot{\tilde{a}}) \tag{13}$$

With the help of $\dot{e}$ and $\dot{\tilde{a}}$, we can obtain

$$\dot{V} = \frac{2}{Mk_1\lambda_0} e\tilde{a}^T \phi(\cdot) - \frac{2k_0}{k_1\lambda_0} e^2 + \frac{1}{\lambda_1} \tilde{a}\dot{\tilde{a}} \tag{14}$$

When the adaptive law $\dot{\tilde{a}} = -\frac{2\lambda_1}{Mk_1\lambda_0} e\phi(\ )$, then

$$\dot{V} = -\frac{2k_0}{k_1\lambda_0} e^2 \tag{15}$$

Therefore, V is a monotonically decreasing function.

$$V[e(t), \tilde{a}(t)] < V[e(0), \tilde{a}(0)] \tag{16}$$

V is an ellipse with $e(t)/\sqrt{\lambda_0}$ and $\tilde{a}(t)/\sqrt{\lambda_1}$ as the axis, indicating that the velocity error is bounded. Therefore, the integral of the uniform continuous function produced by e^2 in this period is bounded. According to Barbalat's theorem, we know that $\lim_{t \to \infty} e^2(t) = 0$, which is the controller It can manage the actual output speed v(t) of F under unknown conditions such as the basic resistance parameters of the train and random disturbance $V(t)$ to approach $v_s(t)$ to achieve the purpose of speed tracking.

4 RESULTS AND DISCUSSION

The simulation was carried out in order to verify the feasibility and effect of the above-mentioned speed control strategy method. The new generation of high-speed trains operated by the Beijing-Shanghai high-speed railway was used as the object. The high-speed train in this model was running between two stations in various types of traction during the simulation. Idling, constant speed, braking and other working conditions in cooperation with each other. The mass of the train is $M = 890$ t, and the goal of the controller is to have the train track the ideal speed curve when the basic resistance coefficient and the additional resistance are unknown, that is, to achieve speed tracking. As shown in Figure 2, the selected ideal speed curve is as follows:

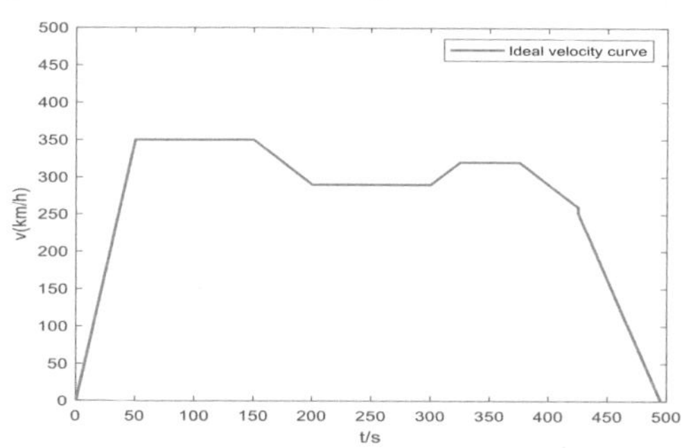

Figure 2. Ideal speed curve.

The train speed is tracked in real time through the controller. Under the interference of the uncertain environmental resistance during the train operation, the tracking effect achieved is shown in Figure 3

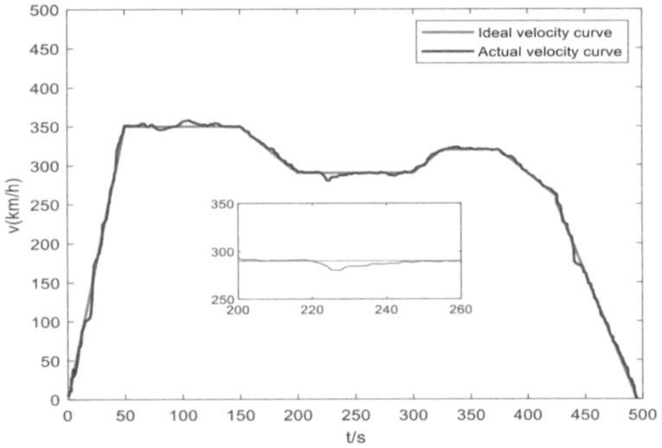

Figure 3. The effect diagram of controller speed tracking.

Under the influence of random resistance, the actual running speed curve of the train fluctuates around the ideal speed curve, that is, the train can pass the control method of this article, and the deviation caused by the speed can achieve rapid convergence when the train is subjected to random external disturbances. It can achieve the purpose of fast tracking of speed. At the same time, by analyzing the change rate of the actual speed curve, it can be obtained that the speed change rate error of the train is low, that is, the comfort of the train is also guaranteed. The train achieves fast speed tracking and accurate tracking performance, as shown in Figure 3.

5 CONCLUSIONS

The control method used in the article can make the train achieve a fast-tracking effect with a large convergence speed under the random line, so the continuity and rapidity of the basic resistance parameter change in actual operation can be effectively controlled. It is mainly related to the appearance of the car, the structure of the car, the quality of marshalling and traction, the route conditions and the weather factors. Without relying on system parameters, the train can respond to changes in the environment in real time, so it also verifies the feasibility and effectiveness of the designed method, and provides a certain reference value for subsequent related research.

REFERENCES

Jiangtao Zhang, Xiaochun Wu.Multi-objective optimization of high-speed railway ATO operation and manipulation based on improved MH algorithm[J].Journal of Railway Science and Engineering,2021,18(02):334342.DOI:10.19713/j.cnki.43-1423/u .T20200341.

Lin Huang,Ying Yang.,Zhongkui Li,Several questions about intelligent control[J].Science in China:Information Science,2018,48(08):1112–1120.

Qi Song,Qing Gu,Feng Liu,Yongduan Song.Adaptive speed and position control of high-speed trains[J].Control Engineering,2010,17(S1):35–37.DOI:10.14107/j.cnki.kzgc.2010.s1 .024.

Qujiang Dong,Yingying Nie, Yanhong Guo.ATO Speed Curve Research Based on Improved Genetic Algorithm[J].Railway Communication and Signal Engineering Technology,2020,17(09):63–68.

Ri Liu,Tong Sun,Yiwei Zhang,Mengjiao Li,Junying Chu,Sheng Zhang.ATO controller for high-speed train based on model-free adaptive control[J].Electric Drive for Locomotives,2021(04):119–125.

Yiliang Zhu. Research on optimization of high-speed train ATO control strategy[D]. Lanzhou Jiaotong University,2021.DOI:10.27205/d.cnki.gltec.2021.000264.

Yu Huan,Hong Luo,Xu Ze. Adaptive Controller Designed for High-Speed Train ATO System with Nonlinear and Uncertain Resistance[J]. Applied Mechanics and Materials,2013,2307(300–301):

Zhiyu He,Ning Xu.Automatic train driving algorithm based on adaptive iterative learning control[J].Transportation System Engineering an Information,2020,20(02):69–75.

Advances in Urban Engineering and Management Science – Khalil & Yang (Eds)
© 2023 The Author(s), ISBN 978-1-032-30426-7

Rural landscape planning and design based on spatio-temporal big data

Guangpei Yang & Yue Zhong
Guangzhou Nanyang Polytechnic, Guangdong, China

Yunli Cheng*
Software Engineering Institute of Guangzhou, Guangdong, China

ABSTRACT: The 19th big conference put forward the strategy of rural revitalization at national level. It embodies the party and country-to-country attaches great importance to the construction work. The country's revitalization strategy is based on the prosperous industry, ecological livable, local custom civilization, effective governance, and life-rich multi-dimensional plots that make up the overall layout for rural development. A gradient optimization method for landscape spatial pattern was proposed based on big data analysis, and the gradient pattern of the landscape was analyzed by the landscape pattern index. The ecological contribution value of the landscape spatial pattern gradient was calculated according to the analysis results, and the evaluation index of impact factors was obtained by the single factor evaluation method. The comprehensive measurement of the landscape spatial pattern gradient was calculated according to the evaluation index. The gradient of landscape flow was obtained by the exponential method, and the gradient of landscape spatial pattern was optimized by combining the contribution value of the ecological environment and comprehensive force measurement. The current gradient optimization method for landscape spatial pattern can't effectively optimize the gradient of a landscape spatial pattern due to the lack of plant species and uneven distribution. The optimization results show that the gradient pattern optimized by the proposed method has many plant species and a uniform distribution, which can effectively optimize the gradient of a landscape spatial pattern.

1 INTRODUCTION

The first put forward to implement the strategy of rejuvenating the country was the party's 19th, 2018. The central files in the prosperous industry, ecological livable, local custom civilization, effective governance, and life-rich second cross policy, ecological livable is key. Countries continue to increase the intensity of ecological protection, strengthen rural infrastructure construction, and make rural homes beautiful places to live and work in peace and contentment. Building a beautiful and ecologically livable countryside is not only in line with people's demands for a better living environment, but also an integral part of rural revitalization and the building of a new socialist countryside. Accelerate rural construction, promote the integrated development of primary, secondary, and tertiary industries in rural areas, improve infrastructure construction, and improve the rural governance system. By 2050, rural revitalization should be realized in an all-round way. It is not only the sporadic appearance of beautiful rural models, but the blooming of beautiful villages everywhere that reflects their distinctive development.

The spatial pattern of a landscape includes landscape configuration, type, spatial distribution, and unit number. It is an important factor in landscape dynamic and functional change and also a

*Corresponding Author: 767220031@qq.com

DOI 10.1201/9781003305026-9

concrete embodiment of landscape spatial pattern structure. The main research content of landscape spatial ecology is landscape pattern. Quantitative analysis of the spatial distribution characteristics of a landscape is the basis of an in-depth study of landscape dynamics and functions. The landscape pattern is getting more and more attention in recent years. The landscape spatial pattern used in said model obtained the rapid development of the related staff and academics in the temporal dimension to the dynamic evolution of the landscape pattern is studied, found in the forests, wetlands, and farmland landscape dynamic and landscape evolution law, The influence of urbanization on landscape patterns was studied, but less attention was paid to landscape gradient patterns. When the current landscape optimization method is used to optimize the gradient pattern of a landscape space, the obtained optimized gradient pattern has few plant species and an uneven distribution, which means it cannot effectively optimize the gradient pattern. Therefore, it is necessary to study the planning and design method of the rural landscape with big data.

2 RELEVANT THEORETICAL BASIS

2.1 *Rural landscape*

At the 19th National Congress of the CPC, rural revitalization was proposed as a strategy at the national level, which reflects the great importance the Party and the country attach to rural construction. The rural revitalization strategy is planned around multiple dimensions of industrial prosperity, ecological livability, civilized local customs, effective governance, and affluent life, and it creates an overall layout for rural development. In general, time migration causes the evolution of rural landscapes, primarily from the original landscape, traditional rural landscapes, to modern rural landscapes, and other levels of evolution. The development of China's rural landscape is mainly faced by the natural landscape to modern landscape transformation process. Through the reasonable transformation of traditional rural landscape planning, can achieve the effective allocation of environmental resources, as well as protect the rural ecological environment, develop agricultural tourism potential, create a favorable rural living environment, and also protect the integrity of the country's context inheritance. To promote rural economic development, it is of great significance to realize rural livable and suitable for business and tourism.

Compared with the urban landscape, the rural landscape, as a complex "organic product", has different research categories under different perspectives and has different connotations for different disciplines. From the perspective of geography, a rural landscape is a kind of natural landscape that is represented by one or more rural settlements. Compared with cities, the rural landscape is characterized by inefficient land use, large scale, small population density, and a rural lifestyle. From the perspective of landscape ecology, the rural landscape is composed of different land units, which are not only affected by natural environmental conditions but also restricted by the size, shape, structure, and other factors of these mosaics, with social, ecological, and aesthetic values. The so-called "rural landscape planning and design" refers to a region's development plan over a certain period of time, usually dominated by the local government's rural landscape construction, the main use of rural landscape in the village's economic development, and finally to improve the rural ecological environment, to realize rural society, economy, and culture through the coordinated development of ecological level. It can be said that rural landscape planning is an important basis for rural construction and rural management.

2.2 *Feature analysis of spatio-temporal big data of landscape*

Big data is realizing great changes in human work, life, and thinking. Its "power" has also had a strong impact on education and is becoming a subversive force to promote innovation and reform of the education system. The 2013NMC horizon report (higher education edition) is very prescient that "big data and learning analytics" will become mainstream technology in the next 2–3 years. Under the impact of big data concepts and technologies, a "quiet revolution" is taking place in the field of education, leading to the transformation and transformation of teaching, learning, research, service, management, evaluation, and other aspects.

Compared with ordinary big data, spatio-temporal big data has the characteristics of space, time, multi-dimension, mass, and complexity. Landscape spatio-temporal big data not only has the basic 6V characteristics of big data, that is, it is characterized by large volume, variety, velocity, veracity, valence, and value, but it is also characterized by spatio-temporal big data. The characteristics of spatio-temporal big data in the landscape are described by the following six aspects.

1) The objectivity of landscape spatio-temporal big data;
2) Multi-sources of spatio-temporal big data in landscape;
3) The dynamics of spatio-temporal big data;
4) The fineness of landscape spatio-temporal big data;
5) The present situation of landscape spatio-temporal big data;
6) The human nature of landscape spatio-temporal big data.

2.3 *Principle of spatial pattern gradient optimization*

When the current pattern gradient optimization method is used to optimize the gradient pattern of a landscape space, the gradient pattern of the landscape is analyzed by patch density, edge density, average shape index, diversity index, dominance index, and distance index. The basic service value model of the landscape spatial gradient pattern was constructed according to the analysis structure, and the gradient optimization of the landscape spatial pattern was completed according to the total value of ecosystem services obtained by different gradient patterns.

PD is set as patch density, which can reflect the degree of fragmentation of landscape spatial components, and the calculation formula is as follows:

$$\mathrm{PD} = \mathrm{N}/\mathrm{A} \tag{1}$$

where N represents the total number of patches in the landscape, and A represents the total area of landscape space. Let ED represent the edge density, and the calculation formula of ED is:

$$\mathrm{ED} = \mathrm{E}/\mathrm{A} \tag{2}$$

where E represents the total length of all patch boundaries in the landscape, The fragmentation degree of the landscape is reflected by edge density, which has a direct impact on species composition and edge effect in the landscape. *MSI* represents the average shape index, and its calculation formula is as follows:

$$MSI = \sum (0.25p_{ij}/\sqrt{a_{ij}})/N \tag{3}$$

Based on the above landscape indices, the gradient pattern of the landscape pattern is analyzed, and the basic service value model of the landscape spatial gradient pattern is constructed according to the analysis results. Assume that *Vi* represents the service value generated by a single item in the spatial gradient pattern of landscape architecture, and its calculation formula is as follows:

$$V_i = \sum q_{ij} \cdot A_j \tag{4}$$

In the formula, A_j represents the distribution area of landscape type i in landscape architecture, and Q_{ij} represents the unit price of ecological function of gradient pattern type i in site j. According to Formula (4), the total value of ecosystem services of landscape spatial gradient pattern is obtained. In Formula (5), m and n are value coefficients.

$$V = \sum^{m} \sum^{n} A_j \cdot q_{ij} \tag{5}$$

3 FEATURE ANALYSIS OF BIG DATA IN LANDSCAPE PLANNING, DESIGN AND CONSTRUCTION

Urban and rural landscape planning and design are based on the cognition of the natural and cultural value of the site. Through the analysis of the natural and cultural elements of the site, they determine the development plan of protection and utilization, construction and management in a certain period of the future. Landscape planning and design have entered the time and space big data era, thanks to the rapid development and widespread application of smart cities, smart scenic spots, mobile Internet, and intelligent terminals.

First, the objectivity of big data in landscape hierarchical planning and construction. In the process of spatial hierarchical planning and construction of urban plant landscapes, it is necessary to use big data technology to truly reflect considerable things or phenomena. Because the information obtained by using technology does not go through any interference, it has strong authenticity and objectivity.

At present, the big data collection technology mainly includes remote sensing image data, video control data, environmental parameter sensing data, etc. During the information collection process, human intervention needs to be reduced. If necessary, the flow of people and traffic can be recorded for staff reference and analysis.

Secondly, the multi-source of big data in landscape hierarchical planning and construction. During the use of big data technology, due to the planning and design of plant landscape spatial hierarchy, during the actual operation, need to use the types of information technology, the information is also more source sex, mainly including the landscape vegetation data, humidity, temperature data, the botanical garden birds, and so on, also including the garden near the noise data monitoring, etc. Through the human environment, garden plants, and other factors, a comprehensive analysis of garden plant landscape planning is conducted.

Finally, the dynamics of big data in landscape hierarchical planning and construction. During the planning and design of the spatial hierarchy of plant landscape, the amount of garden vegetation is limited, and there is great variability in the utilization of garden space. Therefore, the sense of hierarchy of the vegetation landscape needs to be considered comprehensively according to the local environmental dynamics as well as the vegetation situation and vegetation distribution of the garden. As species and quantities of plants are generally relatively fixed, the variation range is small, but for the citizens, tourists, the environment of gardens, ecological benefits, and other factors, it is very variable, so the data obtained has a certain dynamic. In addition, with the application of big data technology, the space design of plant landscape can be refined through data, thus promoting the harmony of garden plant landscape.

4 RESEARCH PROGRESS IN GRADIENT PLANNING OF LANDSCAPE SPATIAL PATTERN BASED ON BIG DATA

4.1 *Research on big data based on mobile communication technology*

Mobile communication devices and means have become quite popular and have become an integral part of residents' lifestyles. Mobile phone signaling data can well reflect the spatial and temporal activities of residents or tourists, providing an interactive relationship between people and the site for planning and design. Niu Xinyi et al. (2014) took the central city of Shanghai as an example and proposed a method to identify urban spatial structure by using mobile phone positioning data. Through the use of mobile communication base station location data and mobile phone signal data, using the method of kernel density generated at different times for mobile phone users' working days and rest days density figures, further identifying the type, level, and function of the urban public center and identifying employment, leisure, and residential areas with a mixed degree of auxiliary decision significance for fine urban planning design. From the large sample data of mobile phone signaling, Fang Jia et al. (2018) extracted the data information of departure, arrival, departure time, and residence coordinates of tourists in 32 large parks in Shanghai from the large sample data of mobile phone signaling, analyzed the spatio-temporal behavior of tourists in different parks, and divided the parks into 8 types according to the distribution characteristics of tourist source space and visitor quantity over time. Respectively, through 8 class park space and time indexes (group space area, the core gathered habitat area, visitors to the park's per capita OD, peak start and duration, etc.), and quantitatively describe the characteristics of each type of park, to parse out the cause of the excavation and the reason of the park positioning dislocation improvement, the present situation of the planning and management of the park has a guiding significance.

4.2 *Research on big data based on positioning and navigation*

With the development of information technology, the function of information technology is increasing and providing convenience for people's lives. After the application of positioning and navigation technology, location information and trajectory information are obtained by information data, which provides convenience for landscape design. During the garden planning and construction, there was a lot of vegetation and trees in the garden. In the process of implementing spatial hierarchical planning and construction of the plant landscape, it is necessary to determine the specific location of plants of different shapes and then locate the trees after establishing the garden landscape model. The hierarchical planning and construction of plant landscape space can be implemented in landscape gardens, urban ecological gardens, and other environments. Through big data technology and the application of positioning and navigation technology, the hierarchical design of a garden plant landscape can be more accurate, and the plants in the garden can be precisely managed.

Location navigation data is a very typical location service (LBS) data that contains location information and trajectory information, which can well describe the spatio-temporal behavior pattern of crowds, reveal the relationship between people and the site, as well as the spatio-temporal characteristics of crowds and the site. Li Yuan et al. (2018) conducted a quantitative analysis of tourist behavior patterns in Gulangyu Scenic Spot based on the trajectory of satellite positioning and navigation data and the types of scenic spots visited by tourists as clustering elements. The research showed that tourists in the Gulangyu scenic area have four kinds of spatial behavior models, namely the history and culture-island beach, a foodie-culture-food, and a food shopping-history and culture-island beach hybrid, in which food shopping in the typical behavior mode occupies a larger proportion. This is to better understand the Gulangyu island tourism spatial structure and space optimization organization is of great significance.

4.3 *Research on big data based on environment awareness*

Environmental perception big data includes remote sensing big data and perception big data. The obtained data can reveal the natural and cultural environment or resource status of the site, or users' feedback and response to the planning and design, which is helpful to improve or optimize the landscape planning and design and reflect the people-oriented. The research of Xue Fei et al. (2018) shows that urban vegetation cover is an important basis for measuring urban ecological status and promoting urban ecological landscape planning. Through comparative analysis of multi-temporal space remote sensing data, the vegetation cover and its spatial distribution changes within the sixth ring road of Beijing from 1984 to 2014 were obtained. It provides a quantitative basis and new ideas for the construction of an urban ecosystem in Beijing by means of spatial planning. Liu Song et al. (2018) studied a kind of wearable biosensor method to evaluate real-time environments of real emotions, relying on experimental methods with ECG sensors, electrical sensors, corrugator muscle electricity sensors, skin temperature sensors, galvanic skin sensors, and respiratory sensors, to record people's emotional experiences in real-time environments while walking. Then data fusion is made with GPS spatial location to generate sentiment evaluation trajectory with spatial attributes, which can better reflect the emotional response affected by space and improve the positive effect of landscape environment planning and design on users' psychology and emotions.

4.4 *Research based on social network big data*

Social network data contains a large amount of spatio-temporal information, semantic information, emotional information, association information, etc. Analysis and mining of it can help us to understand landscape planning and design sites in many aspects. In recent years, with the use of information technology, more and more social platforms have emerged for information transmission. Therefore, garden plant landscape space information hierarchical design can undertake reaction through social networks of big data and obtain more information data for the perfection of the plant landscape space. The deficiencies and advantages of urban landscape planning can be understood through the number of visitors, time, composition of tourists, and emotion detection in landscape garden areas through social networks, and the deficiencies can be adjusted and improved to implement fine management. Through the analysis of the original hierarchical design of plant landscape space, it provides a valuable reference for the subsequent hierarchical planning and design of plant landscape space.

Branch of Li (2018) in Huangshan Scenic Area: As an example, the study to build the scenic area based on the social networking (SNS) data traffic research model, through access to the scenic guests during the National Day of Sina Weibo social network data, analyzes the Huangshan Scenic Area's passenger flow of tourists, time characteristics, spatial distribution, and sentiment detection. Based on this, suggestions are put forward for smart planning, fine management, and accurate service of the Huangshan Scenic Spot. Shao Juan (2018), such as using the day-trip travels of honeycomb and Ctrip net website big data mining analysis, reveals the Huashan scenic tourist group composition, spatial behavior, preferences and satisfaction evaluation of the scenic spot, etc., further analyzing the correlation between Huashan and related tourism node strength, provides support for the scenic area planning and design.

4.5 *Research based on numerical simulation of big data*

Based on the characteristics of landscape planning and design factors and the understanding of the environmental benefits, can set different types of landscape planning and design, and numerical simulation analysis can further combine fluid dynamics equations for temperature field, humidity field, wind field, and so on, carrying on the three-dimensional simulation, and the quantitative evaluation of the results is also the direction of digital landscape planning and design. Zhang Li et al. (2018), based on the architectural layout of typical residential areas in Wuhan, selected eight kinds of local representative trees for modeling and quantified the vegetation layout by using aspect ratio of trees (ART). The environmental benefits (cooling and ventilation effects) of eight plants in three different vegetation layouts in summer were simulated and analyzed. The results showed that tree layout and plant physical attributes (leaf area index, canopy width, and tree height) all affected the cooling and ventilation effects of green space. Furthermore, tall trees with large leaf area indexes and large crowns have strong temperature regulation ability and can actively improve outdoor thermal comfort. When arbors were arranged in ART$\geq$2, the average temperature of the residential community was effectively reduced and outdoor thermal comfort was significantly improved. When trees are arranged with ART$\geq$2, trees have the least resistance to wind speed in residential areas.

5 FUTURE DEVELOPMENT TREND ANALYSIS

5.1 *The ecological environment of big data application is gradually constructed*

Driven by the development and application of information technology in China, information technology is becoming more and more advanced. At present, big data technology has been integrated into various industries in society and has promoted the development of social economies. In order to ensure the standardized application of big data information technology, China has issued relevant action programs for big data information technology in order to understand the application direction of big data information technology. Big data technology can't only reflect consumers' preferences and concepts of consumption, but also reflect the quality of urban landscape construction and tourists' preferences for urban landscape construction. Therefore, the application of big data is becoming more and more pervasive, permeating all aspects of people's lives. Therefore, the application of big data information technology in urban construction, combined with the construction of landscape architecture, improves the quality of spatial hierarchical planning and design of garden plant landscapes so as to provide people with a good urban ecological environment. At the same time, we should create an ecological environment for big data to assure future advancements in big data information technology.

5.2 *The technical system of big data application is gradually constructed*

In the process of China's economic development, information technology plays a driving role. It not only improves the efficiency of China's industrial production, but also helps understand the problems existing in China's industrial production process and timely improve it. Based on this, you need to improve the data information technology used in the construction of garden plant landscape planning and establish the technical system in view of the city's landscape planning and hierarchical planning and construction of urban garden plant landscape space, so as to make the design of urban garden plants combined with the city's economic development, beautify the urban environment as a whole, and provide good entertainment places for urban residents. By analyzing big data information technology, we can better understand the preferences of tourists and continue to improve urban development strategies.

5.3 *The talent force of big data application is gradually expanding*

Talents are the guarantee for promoting urban development in the process of urban development. Therefore, in order to further promote the spatial planning and construction of urban garden plant landscapes, the state needs to increase the training of big data talents. To do this, our country established the cultivation of the talent for big data, especially in view of the garden plant landscape space hierarchical planning and construction aspects of the talent. During the period of landscape professional knowledge teaching, not only for students to popularize knowledge of garden plant landscape design, but also to increase large data information technology related knowledge, promote the student to the big data information technology master degree, and during the actual teaching, combine theory with practice to cultivate comprehensive talents in line with social needs as a whole. Through school education and teaching, we can create a group of compound talents with strong technical ability for China's urban construction, increase the reserve of talents in China, and promote China's urbanization construction, which not only improves the city's appreciation but also improves the ecological environment of the city.

5.4 *Innovation and development of multi-source integration of big data application*

The research work of landscape planning and design based on big data that has been carried out is often an experimental exploration of a certain kind of big data. In fact, the landscape spatio-temporal big data of the two categories, 12 types and more than 30 types listed in Tables 1 and 2, each has its own characteristics and application advantages. Relying solely on a certain kind of data may only explain or depict a certain aspect or some dimensions of landscape planning and design. With the construction of big data application ecological environments and application systems as well as the training of professionals, it is an important development trend to integrate, analyze, and apply various types of spatio-temporal big data.

5.5 *Humanistic characteristics of big data application*

Combined with the above analysis of the spatio-temporal big data humanistic characteristics, in fact, urban and rural landscape planning and design are under the premise of respecting natural and social laws to meet the needs of people's production, life, leisure, recreation, and entertainment, to reflect the people-oriented. Therefore, landscape planning and design based on spatio-temporal big data must always embody humanistic characteristics, quantitatively depict multi-dimensional characteristics of human nature, society, culture, and emotion through big data analysis, and then satisfy multi-dimensional characteristics of humans in space, time, facilities, and environment through planning and design techniques.

6 CONCLUSION

During urban economic development, the urban population increases, and the size of the city increases. Many cities have increased the construction of urban gardens as part of their efforts to provide good recreational and entertainment opportunities for the city's residents. Plant types have been introduced to enhance the beauty of the ornamental gardens. While planning landscapes, it is necessary not only to achieve diversity in urban landscapes, but also to ensure that ornamental plants are not strewn about at random, but to implement hierarchical planning of garden spaces, allowing the plants to be strewn about less, providing people with more place for leisure, entertainment, ornamental, and into local history and culture. The adoption of big data information technology can enhance the quality of spatial hierarchical planning and design of plant landscapes as a whole, thus contributing to the sustainability of urban construction.

ACKNOWLEDGMENT

This paper is supported by three projects, including the study on construction technology and quality process control of PHC pipe pile (Grant No.: NY-2020KYYB-04), 2021 Key Scientific Research Project of Guangdong Provincial Education Department (Grant No.: 2021ZDZX4093), and 2019 Guangzhou Nanyang Polytechnic College Teaching Reform Project (Grant No.: NY-2019CQ2KC-22).

REFERENCES

Bian Fuling, DU Jiangyi, Meng Xiaoliang. Demand, application and challenge of spatio-temporal big data processing [J]. Geoinformation of Surveying and Mapping, 2020(4): 1–4.

Dang A R, Jian X, Biao T, et al. Research Progress of the Application of Big Data in China's Urban Planning [J]. China City Planning Review, 2017, 24(1): 24–30.

Geng Qing, Li Bing, Zhan Wei. Design of big data platform for spatio-temporal information data [J]. Geospatial Information, 2019(10): 52–54; 56.

Gao Qiang, Zhang Fengli, Wang Ruijin, et al. Trajectory Big Data: A Review of Key Technologies in Data Processing [J]. Journal of software, 2017, 28(4): 959–992.

jeje. Spatial and temporal big data analysis to support business decision making [J]. Beijing Planning and Construction, 2017(6): 21–25.

Li Deren, Ma Jun, Shao Zhenfeng. On space-time big data and its application [J]. Satellite Application, 2021(9): 7–11.

Liu Yu, Kang Chaogui, Wang Fahui. Geomatics and information science of Wuhan university, 2014, 39(6): 660–666.

Luo Jiancheng, Hu Xiaodong, Wu Wei, et al. Journal of geo-information science, 2016, 18(5): 590–598.

Mao Mingrui. Application of big data in urban planning: Reflections and practice from Beijing Institute of Urban Planning and Design [J]. Urban planning international, 2017, 29(6): 51–57.

Song Weiwei. Research on spatial big data management and high-performance computing based on spatio-temporal information cloud platform [D]. Kunming: Kunming University of Science and Technology, 2019.

Song W J, Wang L Z, Xiang Y, An analysis of geographic big data correlation via The Hilbert – Huang Transformation [J]. Journal of Geographical Science Computer and System Sciences, 2017: 89.

Tan Mengqian, Shao Xiongkai, Liu Chun. Location big data Mining based on spatio-temporal analysis [J]. Journal of Hubei University of Technology, 2018(2): 53–57.
Wang Jiayao, Wu Fang, Guo Jianzhong, et al. Science of Surveying and Mapping, 201, 42(7): 1–7.
Xiao Jianhua, Wang Houzhi, Peng Qingshan, et al. Geospatial-temporal big data management and application cloud platform construction [J]. Bulletin of Surveying and Mapping, 2019(4): 38–42.
Xiang Hongmei, Guo Mingwu. Chinese Journal of Surveying and Mapping, 2018(11): 91–95. (In Chinese)
Zhu Qing, Fu Xiao. Acta Geodaetica et Cartographica Sinica, 2017, 46(10): 1672–1677.

Advances in Urban Engineering and Management Science – Khalil & Yang (Eds)

Antenna design of 5G millimeter wave at 26 GHz

Feifan Chen
College of Electrical Engineering, Sichuan University, Chengdu, Sichuan, China

Yuhao Chen
Department of Intelligent Manufacturing, Wuyi University, Jiangmen, Guangdong, China

Peiyu Li*
Communication Engineering College, Xidian University & Heriot-Watt University, Xi'an, Shaanxi, China

ABSTRACT: The fifth-generation mobile communication system (5G) is penetrating into all industries and fields of the economy and society, which needs miniaturized, wide-band, tunable antennas. In this research, a 26 GHz rectangular microstrip patch antenna is designed and simulated, with a mini size of 8.3 mm × 10 mm × 0.254 mm. And the parameters of the patch are 5.7 mm × 3.8 mm × 0.035 mm. This antenna resonates at 26 GHz with a bandwidth of 720 MHz, a gain of 4.662 dBi and an efficiency of 89%. The inset feed transmission line is adopted to match the patch and 50 microstrip feedline, with an insert depth of 0.9 mm. As for the material, a Roger 5,880 substrate was selected, whose dielectric constant is 2.2. The parameters of the antenna were calculated and the simulated results have been displayed and analyzed using the Computer Simulation Technology Microwave Studio.

1 INTRODUCTION

1.1 *5G mobile communication network*

As demand for wireless traffic increases, which is expected to grow exponentially over the next decade, there are growing concerns that 4G wireless cellular systems and their associated mobile antenna configurations cannot meet the requirements (Hong 2017). Therefore, a global promotion development for 5G happened worldwide, which has enormous potential with wider bandwidth (Fuentes et al. 2020), higher frequency, and ultra-high reliability (Amini & Baidas 2020). Meanwhile, the transmission rate of 5G is 100 times higher than 4G, achieving 10Gps. In terms of storage capacity, 5G is 1000 times bigger than 4G (Gai et al. 2021). With 5G, a brand-new mobile communication network, users will not only experience augmented reality, virtual reality, 3D video and other more immersive business experiences, but also mobile medical, networking, intelligent homes, industrial controls, environmental monitoring, and other IoT-related applications.

Nowadays, the 5G spectrum is divided into sub-6GHz and millimeter-wave bands, and most operators choose to adopt the millimeter-wave spectrum. For example, the U.S. Federal Communications Commission has recommended 28 and 38 GHz bands for 5G standards, while the Office of Communications, U.K., has developed 5G test beds on 26 GHz (Jilani et al. 2019). This means that high-quality antennas need to be developed.

*Corresponding Author: Perilla@stu.xidian.edu.cn

 DOI 10.1201/9781003305026-10

1.2 *Microstrip patch antenna*

Single-story patch antennas are suitable for covering single-story small offices, small stores, and other indoor locations where access points cannot be centrally located. Antenna patches provide hemispheric coverage and propagate from the installation point at a range of 30 degrees to 180 degrees.

Microstrip antennas are divided into three categories: microstrip patch antennas, microstrip slot antennas, and microstrip traveling wave antennas. Here, the microstrip patch antenna refers to the resonant microstrip patch antenna. The biggest characteristic of this antenna is its high efficiency, but the impedance band is narrow.

Microstrip patch antennas have been widely used because of their high relative efficiency and mature analytical methods, but they are limited in their application due to their narrow bandwidth. In addition to military applications, microstrip patch antennas can also be used in mobile communication systems, satellite communication systems, global positioning systems, and remote sensing systems. If applied in a mobile communication system, it can be used as a mobile phone antenna and the same type of mobile phone body, so as to solve the problems of whip antenna low power, not being easy to carry, and electromagnetic radiation on the human brain (Chen & Long 1999).

In this research, a micro-strip patch antenna is adopted, with small volume, easy fabrication, compactness, and low cost. In spite of the narrow bandwidth (Kaur & Malhotra 2016), the main shortage of this type of antenna, we can change parameters on the feed-line to widen the bandwidth.

2 MATERIAL AND DESIGN

2.1 *Material*

Printed circuit boards play an important role in electronic products, so we need to choose the right materials for PCB manufacturing. There are many common and widely used PCB boards in life. PCB materials, such as FR-4 material, HDI material, and Rogers PCB material, are all different types of materials that can be adopted. Among them, FR-4 materials and Roger materials are more important and common than other PCB materials, and they are also the two candidate materials we want to adopt in this study.

2.1.1 *FR-4*

Typically, standard FR-4 materials are used for printed circuit boards, often to help customers reduce product costs. It has the following characteristics as well, so this is a good choice because FR-4 materials are affordable and effective in many applications.

Fr-4 is a composite of glass-reinforced epoxy laminates made of fiberglass braided cloth and fire-retardant epoxy binder. For electronics engineers and PCB designers, FR-4 glass epoxy is a popular and versatile brand of high-pressure thermosetting plastic laminate with a good strength-to-weight ratio.

Fr-4 has near zero water absorption and is most commonly used as an electrical insulator with considerable mechanical strength, thus maintaining its high mechanical value and electrical insulation properties under both dry and wet conditions.

These properties, together with good manufacturing characteristics, make the class available for a wide range of electrical and mechanical applications. Fr-4 therefore performs well in most environmental conditions.

2.1.2 *Rogers 5880*

We decided to choose Rogers 5880 as the substrate. The advantages of this material are as follows:

- Fr-4 material has a higher Df or dissipation factor than Roger material, resulting in greater signal loss.

- It is easy to be cut into the required shape, which can be perfectly adapted to the shape requirements of various cavity structures in high-frequency applications.
- In terms of impedance stability, Rogers' material has a wider Dk value range than FR-4 material.
- In terms of temperature management, compared with FR-4 material, Roger material has less change.

The FR-4 material provides the basic standard for PCB substrates, maintaining a wide and effective balance between cost, durability, performance, manufacturability, and electrical characteristics. However, Roger has the advantages of low electrical and dielectric loss, better thermal management, and improved impedance control. Performance and electrical characteristics play a major role in our design, and hence Roger materials were selected.

2.2 *Design*

Depending on Microstrip antenna, the rectangular shaped microstrip patch antenna operating at 26 GHz for 5G applications is shown in the figure below:

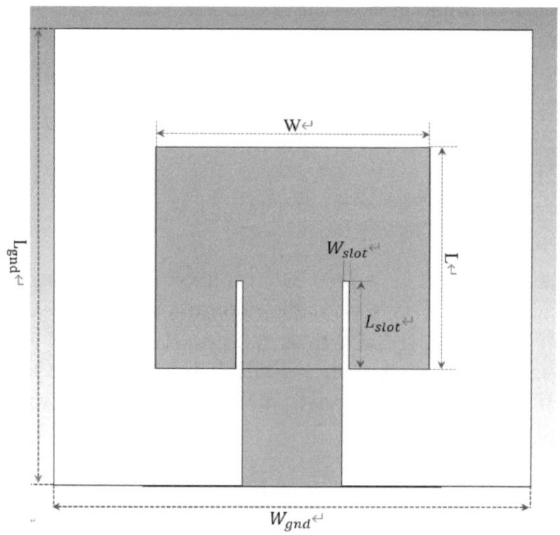

Figure 1. The shape of microstrip antenna.

Table 1. The meaning of parameters and their symbolism.

Parameter	Symbol
Patch Length	L
Patch Width	W
Patch Thickness	th
Substrate Thickness	h
Slot Length	L_{slot}
Slot Width	W_{slot}
Microstrip Line Width	W_{mic}
Ground Length	L_{gnd}
Ground Width	W_{gnd}

Therefore, what needs to be done is to calculate the theoretical values for each parameter.

3 DESIGN PARAMETERS SETTINGS

After choosing the operating frequency (26 GHz) and dielectric constant of the substrate (Rogers 5880), the main parameters are shown below.

3.1 *The dielectric constant (ε_r)*

Through searching for material information, we can get:

$$\varepsilon_r = 2.2 \tag{1}$$

3.2 *For the TM01 mode*

$$k_{mn}^2 = \left(\frac{m\pi}{W}\right)^2 + \left(\frac{n\pi}{L}\right)^2 \tag{2}$$

$$f_{mn} = \frac{k_{mn}c}{2\pi\sqrt{\varepsilon_r}} \tag{3}$$

$$k_{01} = \frac{\pi}{L} \tag{4}$$

$$f_{01} = \frac{c}{2L\sqrt{\varepsilon_r}} \tag{5}$$

3.3 *The patch length (L)*

$$L = \frac{c}{2f_{01}\sqrt{\varepsilon_r}} = 3.8 \text{ mm} \tag{6}$$

3.4 *The patch width (W)*

The cross-polarization level is dependent on W : L.

For a patch fed at the edge operating at mode, the cross-polarization is smallest when W : L = 1.5 : 1 (i.e., 5.7 mm)

3.5 *The substrate thickness (h)*

$$BW = \rho \cdot \frac{h}{\lambda_0} \cdot \sqrt{\frac{W}{L}} \tag{7}$$

When $\rho = 180$, $\frac{h}{\lambda_0\sqrt{\varepsilon_r}} \leq 0.045$ or $\frac{h}{\lambda_0\sqrt{\varepsilon_r}} \geq 0.075$

When ρ=200, &0.045 $\leq \frac{h}{\lambda_0\sqrt{\varepsilon_r}} \leq 0.075$

Now, the length and width of the patch are determined. According to the standard thickness of Rogers 5880 and the bandwidth of the antenna is 459 MHz (h = 0.254 mm).

3.6 *The microstrip line width (W_{mic})*

When h equals 0.254 mm and Z_0 equals 50 Ω, then

$$Z_0 = \frac{120\pi}{\sqrt{\varepsilon_{eff}} \times [\frac{W_{mic}}{h} + 1.393 + \frac{2}{3}\ln(\frac{W_{mic}}{h} + 1.44)]} \tag{8}$$

$$\varepsilon_{eff} = \frac{\varepsilon_r + 1}{2} + \frac{\varepsilon_r - 1}{2}\left[\frac{1}{\sqrt{1 + 12\left(\frac{h}{W_{mic}}\right)}}\right] \tag{9}$$

Thus, we can get $W_{mic} = 0.775$ mm.
Finally, determine the parameters' values after several calculations and comparisons.

Table 2. Dimension of the proposed antenna.

Symbol	Value(mm)	Symbol	Value(mm)
L	3.8	L_{slot}	1
		W_{slot}	0.3
W	5.7	W_{mic}	0.775
th	0.035	L_{gnd}	10
h	0.254	W_{gnd}	8.3

The simulation environment of the proposed rectangular microstrip patch antenna is shown in Figure 2 below.

Figure 2. The microstrip patch antenna simulation model in CST.

4 RESULTS AND DISCUSSION

Through the above calculation process, we can roughly determine the size of the antenna we need to design, and then we need to complete the modeling of the 26 GHz microwave millimeter patch antenna. We use Computer Simulation Technology Microwave Studio to achieve this purpose. Once the model is built, we need to adjust the parameters slightly to make sure that the final model parameters meet our expectations. To ensure that the data fits the ideal model, it is constantly adjusted based on the simulation results.

4.1 *Frequency and Lpat*

By slightly adjusting the calculated value of Lpat, we can get the simulation result (Figure 3). As can be seen from the figure, with the increase in patch length, the working frequency presents a downward trend. When Lpat = 3.65 mm, the operating frequency is closest to 26 GHz, which is the final value we selected. This is not far from the ideal value calculated earlier.

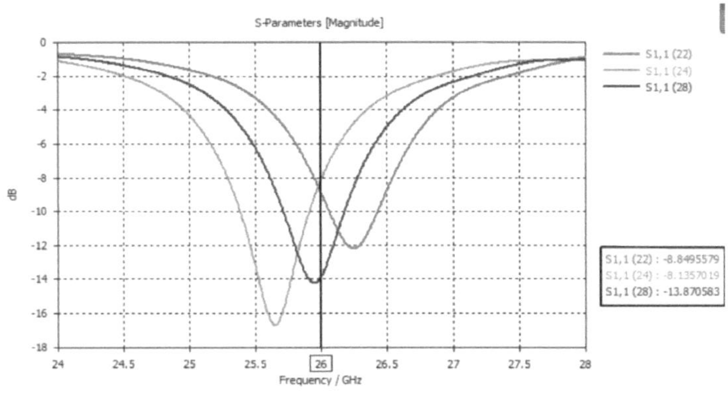

Figure 3. The frequency varies with the length of patch.

4.2 *S11 and Bandwidth*

Absolute bandwidth is the frequency range in which the antenna operates, usually less than −10dB, because the return loss value of −10dB represents a reflection factor of 10%, which is acceptable. As can be seen from Figure 4, when S11<10dB, the working bandwidth of the antenna is: 26.36 GHz − 25.64 GHz = 0.72 GHz. The S11 is 23.87 dB at 26 GHz.

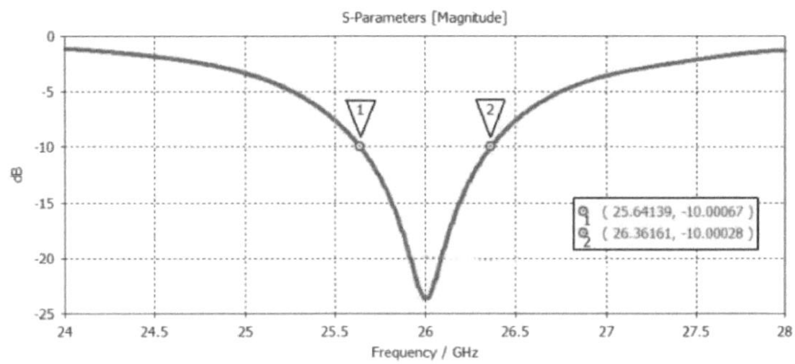

Figure 4. Bandwidth of the patch antenna operating at 26 GHz.

4.3 *VSWR*

Ideally, the patch antenna should have a standing wave ratio of 1. Figure 5 shows that the ratio of voltage standing wave to frequency is minimum at 26 GHz, which is 1.14 (very close to 1). This confirms that our model meets our expectations. In addition, we know that the smaller the voltage standing wave ratio, the smaller the reflection, the better the match, and vice versa. The value of VSWR in this model is very small and close to the ideal value, indicating that the model we designed has reached the expected value.

4.4 *Radiation pattern and gain*

4.4.1 *Radiation pattern*

In order to observe the directivity and gain of the antenna, we use the radiation model to describe the radiative transmission process quantitatively. The following figures are 3D plots and 2D plots of the radiation pattern, respectively.

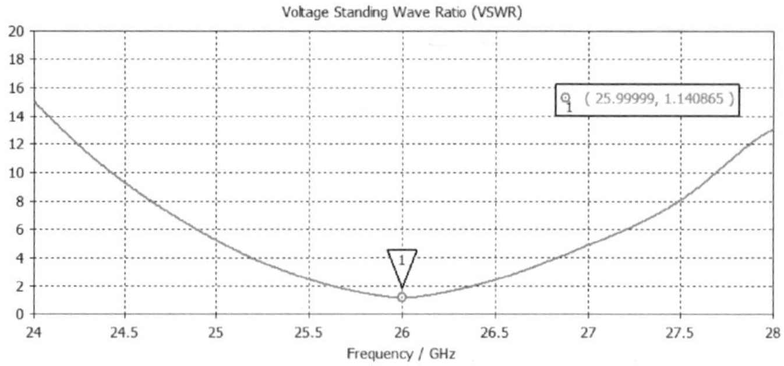

Figure 5. VSWR of the patch antenna.

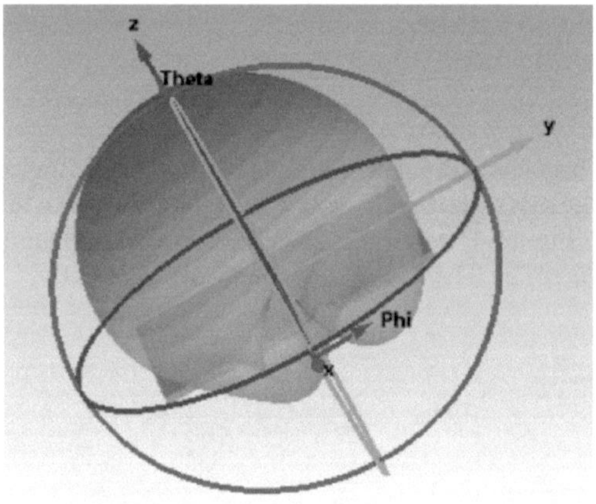

Figure 6. Radiation pattern in 3D and the color scale of gain.

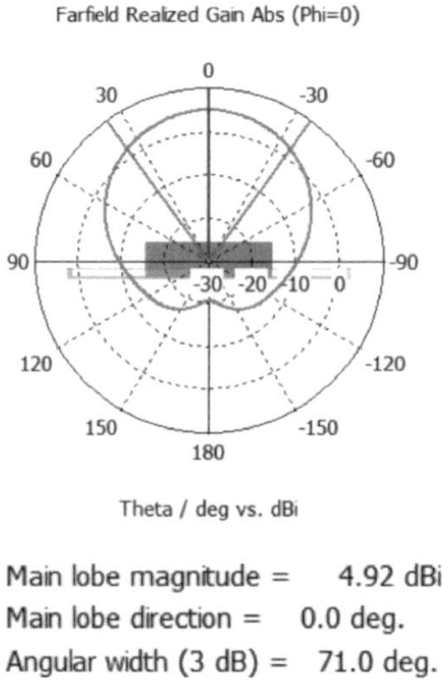

Figure 7. Radiation pattern of the antenna on H-plane in 2D.

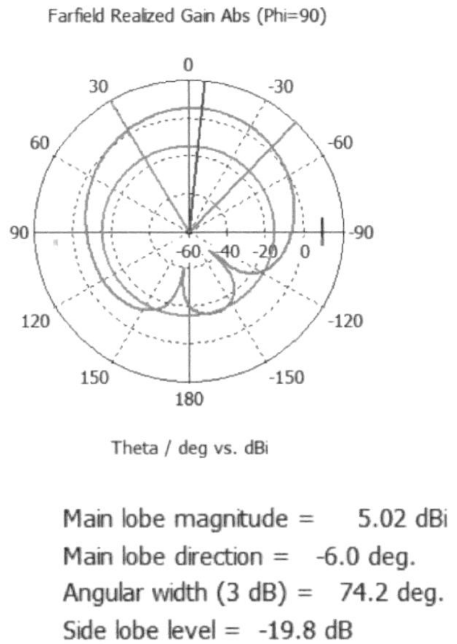

Figure 8. Radiation pattern of the antenna on E-plane in 2D.

The figures above (Figures 7 and 8), They show the radiation pattern and directivity of the antenna. From the CST simulation result, we can get the angular width of 71.0 degrees on the H-plane and 74.2 degrees on the E-plane. And the main lobe magnitude on the H-plane and on the E-plane is 4.92 dBi and 5.02 dBi, respectively.

4.4.2 *Gain*

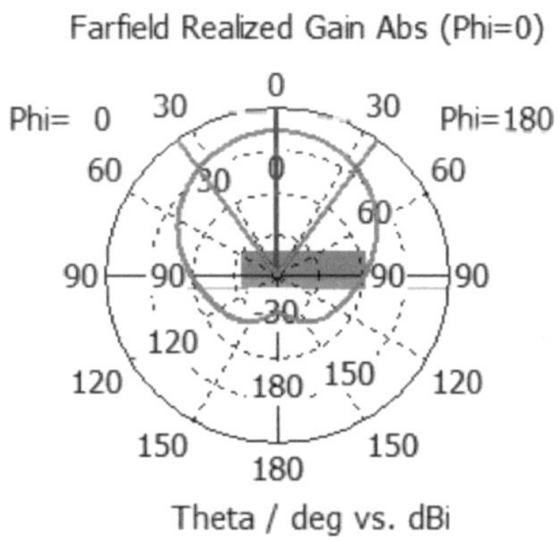

Figure 9. Far-field realized gain in 2d.

Antenna gain is used to measure the ability of an antenna to send and receive signals in a particular direction. Under the same conditions, the higher the gain, the farther the radio wave travels. The simulation results in CST show that the realized gain of antennas is 4.622 dBi, which is relatively high.

The above data are summarized in Table 3.

Table 3. Simulation results about several parameters.

Antenna Parameter	Simulation Results
S11	−23.87 dB
Bandwidth	720 MHz
VSWR	1.14
Gain	4.662 dBi
Efficiency	89%

5 CONCLUSION

This paper mainly studies the design and simulation of microwave patch antennas in the 5G range. A 5G millimeter wave microstrip patch antenna has been proposed for 5G applications. From the selection of materials to the calculation and adjustment of parameters, and then to the output results, we continue to try and error, finally designing the ideal antenna model. It is a good candidate for 5G.

The antenna resonates at 26 GHz with a return loss of −23.87 dB. The proposed antenna shows a radiation efficiency of 89% and a gain of 4.662 dBi. The results also show that the bandwidth is 720 MHz and the proposed antenna can be a good candidate for 5G mobile communication.

REFERENCES

Chen Yajuan, Long Yunliang. Research Progress of broadband microstrip patch Antenna [J]. Chinese Journal of Radio Science, 1999, 14(3): 357–361.

M. Fuentes et al., "5G New Radio Evaluation Against IMT-2020 Key Performance Indicators," in IEEE Access, vol. 8, pp. 110880–110896, 2020, doi: 10.1109/ACCESS.2020.3001641.

M. R. Amini and M. W. Baidas, "Random-Access NOMA in URLL Energy-Harvesting IoT Networks with Short Packet and Diversity Transmissions," in IEEE Access, vol. 8, pp. 220734–220754, 2020, doi: 10.1109/ACCESS.2020.3042744.

N. Kaur and S. Malhotra, "A review on significance of design parameters of microstrip patch antennas," 2016 5th International Conference on Wireless Networks and Embedded Systems (WECON), 2016, pp. 1–6, doi: 10.1109/WECON.2016.7993491

R. Gai, X. Du, S. Ma, N. Chen and S. Gao, "A Summary of 5G applications and prosprcts of 5G in the Internet of Things," 2021 IEEE 2nd International Conference on Big Data, Artificial Intelligence and Internet of Things Engineering (ICBAIE), 2021, pp. 858–863, doi: 10.1109/ICBAIE52039.2021.9389985.

S. F. Jilani, M. O. Munoz, Q. H. Abbasi and A. Alomainy, "Millimeter-Wave Liquid Crystal Polymer Based Conformal Antenna Array for 5G Applications," in IEEE Antennas and Wireless Propagation Letters, vol. 18, no. 1, pp. 84–88, Jan. 2019, doi: 10.1109/LAWP.2018.2881303.

W. Hong, "Solving the 5G Mobile Antenna Puzzle: Assessing Future Directions for the 5G Mobile Antenna Paradigm Shift," in IEEE Microwave Magazine, vol. 18, no. 7, pp. 86–102, Nov.–Dec. 2017, doi: 10.1109/MMM.2017.2740538.

Advances in Urban Engineering and Management Science – Khalil & Yang (Eds)
© 2023 The Author(s), ISBN 978-1-032-30426-7

Innovative technology and application of the delicacy management of port safety production

Fei Liao
Tianjin Jingang Huian Technology Co., Ltd, Tianjin, China

Yanhua Hu* & Jianyu Lv
Tianjin Research Institute for Water Transport Engineering, M.O.T., Tianjin, China

ABSTRACT: With the goal of promoting the construction of a strong transportation country and the intrinsic safety of the world's first-class ports, this paper is committed to closely focusing on the demand for high-quality development of safety in the industry, deepening the structural reform on the supply side of safety professional scientific research, and promoting the in-depth integration and collaborative innovation of "safety + traditional professional". To this end, the innovative technology system of port safety production refinement management, covering four fields including human-oriented safety, essential safety, operation safety, and management innovation, and nine types of new technologies, has been formed to effectively improve the level of delicacy in the management of safety production in port enterprises. The intelligent management of port production will be improved comprehensively, and the management service level and safety production level will be improved accurately, and provide technical support to achieve the industry's development goal of "technology consolidates safety and innovation enhances safety."

1 INTRODUCTION

In September 2019, the Central Committee of the Communist Party of China and the State Council issued the Outline for Building a Powerful Transportation Country, proposing to achieve the development goal of "perfect and reliable safety guarantees and rapid response" by improving the level of essential safety, enhancing the transportation safety production system, and strengthening the transportation emergency rescue capability. In November 2019, the Ministry of Transport, together with the Development and Reform Commission, the Ministry of Finance, and other nine departments, issued the Guidance on Building World-Class Ports, which requires focusing on strengthening essential safety, promoting the construction of dual prevention mechanisms, and strengthening safety and emergency response capabilities as a way to accelerate the development goal of building safe ports. In February 2020, the General Office of the Central Committee of the Communist Party of China and the General Office of the State Council issued Opinions on Comprehensively Strengthening the Work Safety of Hazardous Chemicals. Priority should be placed on strengthening the safe production of hazardous chemicals, vigorously preventing and resolving systemic safety risks, significantly reducing serious accidents, and protecting property and lives effectively. In April 2020, the Work Safety Committee of the State Council issued the Three-Year Action Plan for Special Remediation of Work Safety in the Country, with requirements to compact responsibilities at all levels, strengthen risk prevention and control, fundamentally eliminate potential accidents, and effectively safeguard people's lives and property.

---

*Corresponding Author: huyanhua@tk-aq.com

DOI 10.1201/9781003305026-11

At present, port enterprises face many pressures and challenges in safe production. Firstly, the requirements for safe production at the national level are getting higher and higher, and the assessment is getting more and more stringent; secondly, the incremental risks and port stock risks brought by new technologies, new equipment, new processes, and new materials are interwoven and superposed, and all kinds of unstable and uncertain factors are increasing; thirdly, there are many types of operations, large volumes of operations, various berthing ships, and serious aging of personnel; fourth, there are more dynamic three-dimensional crossing operations in ports, the safety foundation is relatively weak, and dangerous omen events occur from time to time; Fifth, the rough management model still exists, and the concept of emphasis on production while light on safety is deep-rooted; Sixth, the thinking and awareness of employees cannot keep up with the development needs of the safe port and smart port, and safety management effectiveness needs to be further enhanced (Chen 2019, (2020) Three-year Action Plan for Special Rectification of National Production Safety (2021) Opinions of the Ministry of Transport on Intensifying the Prevention and Resolution of Major Risks in Production Safety, Shi 2019). Therefore, it is very necessary to establish a delicacy management and innovative technology system for port safety production and to improve the level of delicacy management of port enterprise safety production.

In view of this, this paper proposes a delicate and innovative technology system for port safety production, with a focus on introducing the content of the system and providing a reference for the development of a delicate and innovative port safety production. The rest of this paper is structured as follows. Section 2 provides an overview of the research on port safety and production management. Section 3 introduces the content of the delicate and innovative technology system of port safety production. Section 4 summarizes the findings of this paper and analyzes the innovation and limitations of the research and future research directions.

2 DEMAND ANALYSIS

The main purpose of safety management is to prevent accidents and focus on solving the four aspects of unsafe behavior of people, unsafe conditions of objects, unsafe factors of the environment, and management deficiencies (Chen 2018; Liu 2016; Wang, 2019). Therefore, the research involves relevant technologies in four major fields, including human safety, operational safety, intrinsic safety, and management innovation, with an effort to accomplish behavior pre-determination, risk pre-control, and hidden danger pre-screening through the coupling and integration of innovation systems, thereby achieving the purpose of accident prevention.

In his research on safety management and control at home and abroad, Taylor proposed the training theory for the first time. He believes that the safety training theory needs to be based on technical standard training (2004: "Notice on Doing a Good Job in Port Safety Production and Safety Management"). According to Wang, safety training needs to start with the training needs of corporate training, emphasis is put on the power of teachers, and Internet technology is used for safety training (Wang 2007).

Safety management must aim at improving management efficiency (CCH: (2005); Glayton 2009; Lind 2008; Luo 2010; Peter 2004; Wu 2014; Zhou 2008; Zheng 2015). Firstly, it is committed to executing system implementation, responsibility implementation, risk management, and target management, as well as an automated assessment of the safety performance of all staff and an automated assessment of the health status of equipment. Secondly, with the use of visualization and control, the complex safety management details are presented intuitively, clearly, and transparently to practitioners. Thirdly, with efficient real-time and intelligent control, it is possible to ensure that key work nodes can advance smoothly while meeting safety and quality requirements by using risk warnings, big data analysis, Internet of Things technology, and information technology (Bai 2016; Dhillon 2006; George 2006; Heinrich & Eterson 1980; Liu 2013; Zhao & Zhang 2017). In this regard, a new pattern of safe development is established, which can better serve the construction of smart ports and safe ports, create first-class facilities, first-class technology, first-class management, and produce first-class services in order to foster a safe and high-quality development. The purpose of this paper is to elaborate on the construction content of an innovative technology system that optimizes port safety production and safety management in conjunction with the essential requirements of safety management.

3 SYSTEM CONSTRUCTION

3.1 *Port enterprise employees' post safety competency assessment system*

For the first time, this assessment system systematically created the concept of "post safety competency," which is the sum of safety competencies required for employees to take up the task role of a particular post. In general, the higher the overall safety quality of an employee matching the post, the stronger the safety competency of the position.

Firstly, the elements of job safety competency are determined through method and system research, and a generic job safety competency model for port operators is established through data analysis and calculation. Secondly, using theoretical methods such as self-adaption and BP neural network, a non-linear generic evaluation index system for job personnel safety competency is constructed, and evaluation criteria are established in the way of self-evaluation and other evaluation. Finally, system dynamics entropy weight theory is used to quantitatively evaluate the job safety quality and job matching degree of operators, and obtain the job safety level and job safety competency index of key job operators

3.2 *Safety operation analysis and tolerance assessment of port equipment and facilities*

The safety operation analysis and tolerance capacity assessment of port equipment and facilities mainly include the following aspects. With the application of structural load analysis and numerical simulation to port equipment and facilities based on nonlinear elasto-plastic fracture analysis, we explore the influence law of different defect types and sizes on structural load carrying capacity and specify tolerance limits and defect characterization thresholds for its safe operation.

The economic-safety double-layer analysis model is constructed by evaluating the capacity of port equipment and facility structures. Based on the principle of the reasonable use of the structure and risk alarm, the safety margin and risk index of the structure under different operating cargo, defect type, and size conditions are clarified to provide a quantitative basis for its safe operation.

Based on the results of the numerical simulation, model construction, and theoretical analysis, an accident risk analysis of port equipment and facilities is performed, and control measures and suggestions for safe operation are offered to provide technical assistance for its on-site safety management.

3.3 *The wind resistance performance evaluation and wind speed monitoring and alarm information system of large-scale port cranes*

The construction of the wind resistance performance evaluation and wind speed monitoring and alarm information system of large-scale port cranes First, the wind resistance performance (anti-skid and anti-overturning) of large-scale port cranes (gantry cranes, quay cranes, ship loaders) in working conditions is evaluated and calculated to determine the influence of the law of average wind speed, instantaneous wind speed, wind level, and wind pressure on its wind resistance performance, and a windproof standard for field operations based on redundant safety margin analysis is proposed.

Second, based on the numerical simulation and calculation analysis of the wind field in the port area, the probability distribution model of key parameters of the wind field and the probability distribution model of wind field extremes are established; based on the results of wind field simulation research, a cluster analysis model is established to optimize the installation points of anemometer and linked alarm; through the study of anemometer alarm calibration and its correction criteria, a gradient model of wind height coefficient is established; combined with the built-in algorithm of PLC to convert the wind height coefficient, the wind speed monitoring and alarm information system is built.

3.4 *The classification, monitoring, and alarm system for the storage, loading, and unloading of hazardous goods in the port*

The construction of the classification, monitoring, and alarm systems for the storage, loading, and unloading of hazardous goods in the port is based on the type and mechanism of port storage tank accidents. A storage tank safety risk assessment index system and a classification early warning model are constructed to obtain real-time monitoring operating parameters of large-scale storage tanks in the port, including the temperature, pressure, liquid level, foundation settlement, and grounding resistance. Dynamical prediction of the risk level of the storage tank can ensure the safe operation of the storage tank.

Through the research of the automatic control system for the real-time detection and monitoring of the temperature of the hazardous cargo container yard in the port, the technical indicators of the wireless temperature measurement sensor, the wireless temperature measurement relay, and the operating system are determined, and key modules such as the temperature measurement sensor and the relay structure are formed, thus realizing the full-time monitoring, identification, and early warning of the temperature control dynamics of the hazardous goods container yard.

3.5 *Safety risk assessment of hazardous cargo transshipment in ports*

The risk assessment of port hazardous cargo transshipment security is to summarize the main causes of accidents involving hazardous cargo in waterway transportation and port storage through statistical analysis of port hazardous cargo accident cases; a systematic study of the hazardous cargo port transshipment process is conducted to comprehensively analyze hazardous cargo security risks, including inbound transport security risks, outbound transport security risks, hazardous cargo in-port transshipment security risks, etc. According to the selected representative hazardous goods and accident research areas, quantitative calculation methods are used to predict the impact of the accident, analyze the consequences of the accident, and consequent safety management countermeasures are proposed to promote the safety of hazardous cargo operations at the port.

3.6 *Port safety production data management*

The evaluation index system and database of port enterprises' safety production are constructed. There are 10 primary indicators, 52 secondary indicators, and 95 tertiary indicators in the index system, covering ten categories of data such as personnel management, emergency management, equipment and facilities management, safety culture construction, occupational health management, hidden danger management, risk management, hazardous operation management, related party management, and monitoring and surveillance.

A safety production data analysis model and evaluation criteria for each index are established to scientifically and quantitatively analyze the influence law of each basic index on enterprise safety production management so that the direction and suggested measures for improving safety production work are targeted according to the evaluation results.

3.7 *Visual management of port safety production*

In the context of the construction of the "double prevention mechanism", the visual management of hazards and their control measures in the operating area is achieved using safety colors, safety labels, signs, and placards, etc.; for the specific management objects such as the behavior of on-site personnel, equipment and facility management, material management, area management, etc., the system is implemented by using colors, labels, signs, and placards. Safety Production Visualization Management Technical Standards is formulated in accordance with the actual operation of port enterprises. In order to achieve the normal operation of the technical standards, the analysis of business sectors such as container terminals and yards, ro-ro terminals and yards, bulk cargo terminals and yards, as well as storage and logistics enterprises, is utilized to solidify the support means for visualization management.

3.8 *Port safety culture construction and operation evaluation system*

In view of the actual production, operation characteristics, and management status of the enterprise, the unified definition, scope of application, and implementation norms of safety culture shall be clarified. The technical index system for the management and evaluation of safety culture in the port industry includes 7 primary core indexes, 24 secondary indexes, and 65 tertiary indexes, and a series of technical methods such as pressure-state-response evaluation model, intelligent behavior diagnosis, risk perception pre-control, and hidden danger pre-screening, which are researched by using collaborative theory, scenario design, self-adaption, and BP neural network. A calculation analysis software and system platform have been developed based on the port industry's enterprise safety culture evaluation system.

3.9 *Management and control system for port hazardous goods*

The construction of a risk management and control system for port hazardous cargo development mainly includes two aspects: First, in order to ensure the efficient operation of the information system for port risk management, a management model of "1+3+1" has been developed, where "1" construction standard refers to Implementation Opinions on the Construction of Double Prevention Mechanism for Port Hazardous Cargo Operation Enterprises, which clarifies the main responsibilities, main contents, and construction procedures of the system; "3" guidance manuals indicate the operation manual for the construction of the "double prevention" mechanism, information reporting guidelines, and guidelines for operational supervision and inspection; and "1" technical guarantee represents the provincial and municipal port production safety supervision of hazardous goods list of powers and responsibilities, with a focus on clearing powers, responsibilities, supervision content and tasks. Second, in the process of risk control construction, the responsibilities and needs of continuously hierarchical management departments and the content and methods of supervision are combined to build a list of responsibilities for them.

4 CONCLUSION

With regard to the essence of safety management, this paper establishes an innovative technology system for port production delicacy management, which is an effective embodiment and concrete practice of the concept of "safety is the priority with prevention as the principle" for port enterprises, based on the actual production, operational characteristics, and management status of port enterprises. It has boosted the level of delicacy management of safety production, prompted port enterprises to change from "I wanted to be safe" to "I want to be safe," further guaranteed the safe production of port enterprises, and provided strong support for building a new pattern of delicacy management of safety production for port enterprises. Therefore, the level of port intelligent management will be improved, as well as the intelligent management of port safety.

In order to better promote the delicacy development of port safety production, the research will continue from the following two aspects: First, the innovative technology and application results of delicacy are coupled to integrate risk warning and control, standardization, fire and emergency, occupational health and energy conservation and environmental protection and other management modules, thus forming an integrated, full-coverage control mechanism of intelligent safety production; second, it is important to advocate the concept of "no safety, no production' to realize real-time risk control, intelligent early warning analysis and standardized job operation of production operations, ensuring that production is always in a risk-controlled state.

ACKNOWLEDGMENTS

This research is supported by the special fund for the basic research business of the central public welfare research institutes (TKS 20200320, TKS 20210306).

REFERENCES

Bai, Z. D. (2016) Analysis of the Status Quo of the Enterprise Safety Emergency Management System and Suggestions for Improvement, pp.11–15.

CCH: (2005) Planning Occupational Health and Safety. 6th edn. Sydney: Australia.

Chen, L. Q. (2019) Analysis of Port Safety Production Management Issues and Countermeasures, City of Industry (first half of the month), pp. 1–2.

Chen, Q. (2018) Principle and Implementation of "Occupational Health and Safety Management System Requirements and Guidelines". Beijing: China Quality Inspection Press, pp. 277–279.

Dhillon, B.S. (2006) Human error in maintenance: A review, Journal of quality in maintenance, pp. 28–35.

George, D. H. (2006) Introduction to Emergency Management, Burlington: Butterworth Heinemann Press.

Glayton, G. (2009) Field Behavioral Based SHE Process (CORE) &Personal SHE Charter, Singapore.

Heinrich, E. H., Eterson, D. (1980) Industrial accident prevention. New York: McGraw-Hill Book Company, pp. 37–51.

Lind, S. M. (2008) Theory safety risks assessment in industrial maintenance, Journal of Quality in Maintenance Engineering, pp. 217–220.

Liu, J. Q. (2016) Promoting the Safe Development of Port Enterprises by Implementing Safety Production Standardization, Port economy, pp. 47–48.

Liu, K. (2013) Research on new mode of port safety production early warning management, China Water Transportation, pp.73–74.

Luo, Y. (2010) Modern Security Management, Beijing: Chemical Industry Press, pp. 6–7.

Ministry of Communications. (2004) "Notice on Doing a Good Job in Port Safety Production and Safety Management".

Ministry of transport of the people's Republic of China. (2021) Opinions of the Ministry of Transport on Intensifying the Prevention and Resolution of Major Risks in Production Safety.

Peter, M. (2004) Reducing the harms associated with risk assessments, Environmental Impact Assessment Review, pp. 733–748.

Safety Committee of the State Council. (2020) Three-year Action Plan for Special Rectification of National Production Safety.

Shi, Z. P. (2019) Analysis of Weak Links in Port Safety Management and Countermeasures, Consumer Guide, pp.132–133.

The Central Committee of the Communist Party of China and the State Council. (2021) The Fourteenth Five-Year Plan for the National Economic and Social Development of the People's Republic of China and the Outline of Long-Term Goals for 2035.

The Central Committee of the Communist Party of China and the State Council. (2019) The Outline for Building a Powerful Transportation Country.

The Central Committee of the Communist Party of China, the State Council. (2020) General Office of the CPC Central Committee General Office of the State Council. Opinions on comprehensively strengthening the safety production of hazardous chemicals.

Wang, L. (2007) Research on Safety Training, Beijing, Beijing Jiaotong University.

Wang, S. K. (2019) Safety Management Issues and Measures for Port and Waterway Engineering Construction, Sichuan Building Materials, pp. 222–223.

Wu, N. (2014) On-site Operation Permit Management System Review Meeting, Now Occupational Safety, pp. 46–47.

Zhao, K. G., Zhang, Rui. (2017) Research and Application of Emergency Capability Evaluation Index System for Port Enterprises, Shenhua Technology, pp.78–80.

Zheng, J. (2015) Eight Types of Special Operations Safety Operations Regulated by the National Standard, Labor protection, pp. 98–99.

Zhou, J. (2008) Research on Theoretical Exploration and Operation Improvement of Production Training System, Zhejiang, Zhejiang University of Technology.

Advances in Urban Engineering and Management Science – Khalil & Yang (Eds)
© 2023 The Author(s), ISBN 978-1-032-30426-7

The application of BIM in building project safety management

Peiqi Li, Chunming Qi*, Chunchang Wei & Jinhu Jiang
Department of Civil Engineering, University of South China, Hengyang, China

ABSTRACT: Since entering the information age, all walks of life have been affected to make changes in line with the trend of the times. The construction industry is no exception. The country is vigorously promoting BIM technology to be applied to the construction industry, but the current application is still not common. Based on this, this paper will compare the traditional construction safety construction management methods and the application of BIM technology to construction safety management methods for comparative analysis, and then put forward some suggestions for BIM technology applied to construction safety management.

1 INTRODUCTION

In recent years, the development of China's construction industry has been accelerating with the progress of science and technology. In the process of the rapid development of the construction industry, a major problem restricting its development is the safety problem. There are various risks in the construction process. We should firmly adhere to the people-oriented concept and give priority to personnel safety on the construction site. Traditional construction safety management methods that have many deficiencies cannot keep up with the pace of modern scientific and technological development and cannot find and solve problems faster and more comprehensively. With the increasing development of China's economy, the current construction projects tend to be large and complex projects. The working face of these large buildings is bound to increase. The inspection coverage and workload faced by safety management will increase exponentially, which makes it difficult for managers to control each risk point from a macro perspective. It is necessary to find more efficient and convenient management methods to improve management efficiency. The introduction of BIM technology can remediate many of the shortcomings of traditional construction safety management methods, increase the level of safety management, decrease the number of safety accidents, and provide new and effective safety management methods.

2 THE SHORTCOMINGS OF TRADITIONAL SECURITY MANAGEMENT METHODS

2.1 *Management inefficiency*

Traditional construction safety management mainly relies on the safety management personnel to conduct safety inspections by patrolling the site and submitting paper documents to describe the safety problems they find. Moreover, most managers judge the safety problems of on-site construction based on previous experience (Lang & Wu 2017), which is subjective. The location where the safety problem occurs and the potential safety hazard exists is generally communicated through text description, which often makes the inspection results and standards unable to be updated in

*Corresponding Author: qcm108100@sina.com

DOI 10.1201/9781003305026-12

time. This will result in the construction information being unable to be grasped systematically and comprehensively, and the hazard sources being unable to be predicted and warned about. A collision inspection of the design drawings must be performed prior to construction to identify existing problems and avoid rework during construction. According to the traditional inspection method, not only is the efficiency low, but also the accuracy is not high (Gai et al. 2020).

2.2 *The construction technology is not advanced enough*

Engineering complexity increases with progressive science and technology. The traditional safety management method is no longer suitable for the updated and improved construction methods, which makes it easy to leave potential safety hazards in the construction process and increase the project cost (Kang & Wang 2021). In the field of construction engineering, many units and individuals still use relatively old technology and lack advanced equipment and facilities. And the main personnel in construction engineering are migrant workers, whose cultural knowledge level and safety protection awareness are poor. They often can't come up with effective solutions and make reasonable responses the first time they encounter sudden technical accidents. Therefore, it's essential for them to train in comprehensive and detailed construction technology (Liu 2020). Construction technology has a great impact on the quality and safety level of the project. Backward construction technology cannot guarantee the quality of the construction and may leave many potential safety hazards (Wang & Li 2020).

2.3 *The hazard identification is not accurate enough*

In traditional construction safety management, when identifying the hazard sources before construction, there is no engineering entity and the hazard sources cannot be visualized, which makes the presentation of hazard sources not intuitive enough. People's spatial imagination is limited, which is easy to lead to misunderstanding and omission (Hou & Zhou 2015). Moreover, the identification of hazard sources mainly depends on site inspection. In the inspection process, professional managers mostly rely on personal experience, which is easily affected by subjective factors. The factors considered are often not comprehensive, and it is easy to ignore that the activities on the construction site change in real time and dynamically. In different construction stages, the structural form and engineering parts differ, and there are great differences in the use of materials, machinery, and equipment, and the risk factors in each construction stage are also different (Zhang 2019), which makes the identified hazard sources not accurate enough by the traditional construction safety management methods. It is necessary to introduce BIM technology to ameliorate the management method and improve the accuracy of identifying hazard sources.

3 ADVANTAGES OF APPLYING BIM TECHNOLOGY TO SAFETY MANAGEMENT

3.1 *BIM models improve efficiency of management*

At present, the application of BIM technology in construction engineering is mainly reflected in the establishment of BIM models. The BIM model takes the project as the object and parameterizes various physical and functional characteristics contained in the project. Establish a BIM information database to store all relevant information about the project by BIM technology, so as to effectively integrate the information. The modifications of quantities, costs, progress information, and other contents during construction will also be updated in the database, which can effectively avoid the lag of information transmission and realize barrier-free connection, communication, and management between different stages and different participating units (as shown in Figure 1). The application of BIM technology can directly compare the information in the information database with the established BIM model and the construction site to find out the hazard sources existing in the construction process, which is more intuitive and reliable than the traditional model.

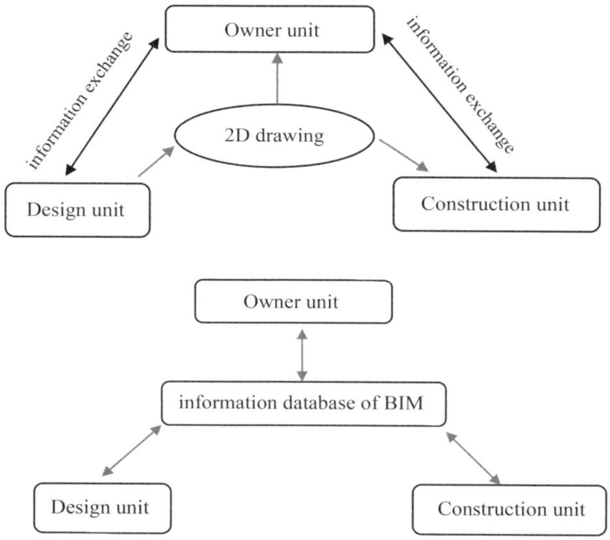

Figure 1. Traditional information exchange and information exchange based on BIM Technology.

In the architectural design stage, construction enterprises can use BIM technology to establish an information model, and then analyze whether the design scheme is scientific and safe based on this model; whether the deduced construction scheme is reasonable; and whether the setting of safety guidance signs on the construction site is reasonable. Simulate how to evacuate relevant personnel in case of an emergency and analyze whether the location of safe passage and disaster avoidance area is reasonable (Liu 2020). Safety management using BIM technology can better achieve pre-control and control the hazard sources before an accident than traditional safety management, which greatly reduces the probability of the accident.

3.2 *Visual application of BIM technology*

In traditional safety management, the information source for the project is based on two-dimensional drawings. It is easy to lose and lag information in the process of information disclosure, which may cause safety accidents due to the construction personnel's incorrect understanding or the missing and untimely transmission of information. However, the information for safety management based on BIM technology mainly comes from 3D building models. The visualization of BIM makes the transmission of information more intuitive, comprehensive, and not easy to lose (Liu & Wang 2021). BIM technology is applied to construction engineering to simulate construction before construction, which can not only show the 3D presentation of the building but also show the site appearance, so as to better plan the construction scheme and carry out the construction organization design. In addition, the visualization of BIM technology can also be used to carry out safety education for workers. Moreover, BIM technology can also provide some geometric and nongeometric information, which can help the progress of complex construction projects.

The Time Liner tool in NavisWorks software can integrate the time and space information of the project, visualize the construction schedule of the project, plan the construction time more scientifically, find and solve problems before construction, and avoid unnecessary engineering changes and rework. When the design of the actual project changes, the BIM software is used to change the construction progress. In addition, the schedule will also be updated simultaneously and reflected in the BIM model, thereby reducing construction delays to a minimum.

3.3 *Coordinated application of BIM technology*

BIM software has the function of conflict inspection. Construction projects are constantly changing in the process of construction. For emergencies on the construction site, managers need to adjust the construction plan and scheme according to the actual situation. The coordination of BIM technology can coordinate the work of participating units as soon as possible to guarantee that all units can communicate quickly and the timeliness and consistency of information, which can save time and money and avoid delays in the project's progress. Under the BIM management platform, project personnel located in different places can work at the same time across the barriers of space. All participants in the project can work together to improve the efficiency of management.

3.4 *BIM technology identifies hazards*

The virtual construction technology of BIM software is used to carry out the simulated construction before the construction, basically classify and analyze the construction hazard sources, and divide the hazard areas and mark different colors in the model, so as to provide the idea of informationized managing construction hazard sources based on BIM Technology (Lin 2021). BIM technology can assist managers to monitor the changes in the construction process in real time and inspect the quality and safety of the project. Managers can find problems and take countermeasures in time, so that the loss degree is always within an acceptable range, reducing the probability of safety accidents and improving the benefits of construction hazard source management.

4 RECOMMENDATIONS FOR BIM APPLICATIONS

4.1 *Improve relevant policy standards*

Nowadays, the application of BIM is not mature enough and there are no perfect government laws and regulations in the development of the construction industry, but the application advantages of BIM in construction engineering are obvious. The wide application of BIM is the general trend. Its application result is to improve the effectiveness of project management, reduce the occurrence of safety accidents, which is conducive to the standardization and informatization of the project, and develop in a safer direction (Fan 2017). In recent years, the state has issued many policies to support the promotion of BIM, but there are limitations. The policies are implemented in just some regions and cannot be carried out nationwide. At present, BIM national standards have been issued, which are very meaningful for information classification and coding, but they are not perfect. BIM delivery standards and storage standards have not been issued, and BIM implementation standards are not perfect (Ran & Zhang 2018). The advantages of BIM technology in building construction are obvious to all. It can improve efficiency, save costs, predict construction difficulties and on-site hazards in advance, so as to formulate preventive measures and management plans in advance and strengthen on-site safety management. Therefore, it is necessary to improve relevant policies, unified application standards, and implementation guidelines for BIM technology. Let all enterprise units implement relevant policies of BIM and you will not cause disputes between enterprises due to different standards.

4.2 *Train BIM managers*

At present, there is still a problem with the application of Bim in safety management. BIM technicians do not know much about safety management, safety managers are not proficient in BIM technology, and there is a lack of training in the use of BIM technology, which greatly reduces the value of BIM technology in the use of safety management. The training of BIM technology is difficult, and the requirements for technicians are high. To create a complex BIM building model, the modelers need not only professional architectural knowledge and skills, but also a good mathematical foundation and certain programming skills, which greatly improves the difficulty of operation and requires the company to invest in relevant training for technicians. In order to better promote and implement the use of BIM technology, BIM courses should be set up in relevant majors in colleges and universities to cultivate professional and technical talents who understand the professional knowledge of construction engineering and can skillfully use BIM. Enterprises should also strengthen the education of managers, provide a platform of learning and education, and a suitable construction environment for practice, to effectively improve the quality level of managers, ensure that safety managers can use BIM technology scientifically, and effectively ensure construction safety (Wang 2020).

4.3 *Increase visibility of BIM*

At first, BIM technology was applied to construction safety management, but the role of BIM was not very obvious (Hu 2020). The importance of BIM should be increased. Set up a professional BIM team that is given a specific role and corresponding responsibilities in the construction enterprise, unify the mentality, thinking mode, and working habits of the employees of the BIM team, and implement the application of BIM in safety management, so as to improve the work efficiency and effectiveness of using BIM for safety management.

5 CONCLUSION

The application of BIM technology in safety management is not only the replacement of tool software and hardware, but also the change of information transmission mode, cooperation mode, and thinking mode. Compared with traditional safety management, the use of BIM technology has promoted safety management in many aspects. It is more comprehensive and convenient to identify hazard sources. The transmission of engineering information and technical disclosure are more efficient and intuitive. A BIM-based safety management system is of great importance to the improvement of safety management and to the construction industry.

REFERENCES

Fan Jiamao. (2017) BIM based construction safety management model. J. Journal of Shandong Agricultural University (NATURAL SCIENCE EDITION), 48 (03): 464–466.

Gai Tongtong, Yu dehu, Sun Baodi, Yang Shujuan. (2020) Research progress of fusion application of BIM and artificial intelligence. J. Architectural Science, 36 (06): 119–126.

Hou Yu, Zhou Jianliang. (2015) Research on hazard source management of construction project based on BIM and SCL technology. J. Construction technology, 44 (12): 84–89.

Hu Guanhua. Research on the application of BIM Technology in construction safety management [C]. Proceedings of 2020 Wanzhi Science Development Forum (smart engineering II), 2020:39–49

Kang Enquan, Wang Jun. (2021) Research on safety technology of smart mine engineering based on BIM. J. Metal mine, 04: 172–177.

Lang Haochuan, Wu Tianhua. (2017) Thoughts on applying BIM to safety management of construction site. J. Construction supervision, 10: 55–57.

Lin Lei. (2021) On the development status and application of BIM Technology. J. Fujian construction technology, 02: 96–98.

Liu Xunmei, Wang Liuyan. (2021) Research on the application of BIM Technology in project management system. J. Construction economy, 42 (S1): 232–235.

Liu Zhao. (2020) Deeply explore the effective application of BIM Technology in building safety management. J. Ceramics, 08: 147–148.

Liu Zhenlong. (2020) Analysis of problems and Countermeasures in construction management of construction engineering. J. Engineering construction and design, 18: 218–219.

Ran Longbin, Zhang Chao. (2018) Data transfer and sharing in BIM application. J. Chongqing Architecture, 17 (07): 33–36.

Wang Wei, Li Guofeng. (2020) Application of BIM Technology in expressway bridge construction safety management. J. Highway, 65 (06): 176–178.

Wang Xiaobing. (2020) Application of BIM Technology in construction safety management. J. Building materials technology and application, 05: 29–31.

Zhang Binbin. (2019) Application of BIM Technology in hazard source management of construction site. J. Construction technology, 48 (06): 6–10.

Advances in Urban Engineering and Management Science – Khalil & Yang (Eds)
© 2023 The Author(s), ISBN 978-1-032-30426-7

Design and application of digital seawall management platform

Hongsheng Qian*, Liming Xu & Chi Zhou
Hangzhou Ding Chuan Information Technology Co., Ltd, Hangzhou, Zhejiang, China

ABSTRACT: Intelligent water conservation is an overall need of the Ministry of Water Resources' 14th Five-Year Plan, as well as an essential component of Zhejiang Provincial government's digital reform. Based on the Provincial "four horizontal, four vertical, and two ends" general framework, this paper constructs the general framework of a digital seawall management platform, designs a specific functional structure of the platform to achieve "accurate monitoring, dynamic research and judgment, automatic warning" of seawall, providing an important support for seawall management departments to prevent and resolve major risks and hidden dangers, and improve monitoring.

1 INTRODUCTION

At the Third Meeting of the Financial and Economic Commission of the CPC Central Committee in October 2018, General Secretary Xi Jinping pointed out that in key areas and weak links, we will promote the construction of a number of key projects, carry out the protection and restoration of coastal zones and build ecological seawalls to enhance our ability to withstand typhoons, storm surges and other marine disasters. In March 2021, the Zhejiang Provincial Party Committee issued the Overall Plan of Digital Reform in Zhejiang Province (2021) Notice on printing and distributing the Overall Plan of Digital Reform in Zhejiang Province by the Commission for Deepening Overall Reform of Zhejiang Provincial Party Committee), requiring the overall usage of digital technology, digital thinking, and digital cognition to promote the process reengineering, digital empowerment, efficient coordination, and overall intelligent governance of various departments, and comprehensively launch the digital reform in Zhejiang.

In order to thoroughly implement the spirit of General Secretary Xi Jinping's important instructions on the construction of an ecological seawall and implement the overall requirements of the Zhejiang Provincial Party Committee on the construction of "digital Zhejiang", the Zhejiang Provincial Department of Water Resources issued the Construction Plan of the Seawall Anlan 100-Billion Project in Zhejiang Province (2021) (Zhejiang Anlan 100-Billion Project Construction Planning), which made decisions and plans to implement the "Seawall Anlan 100-Billion Project" in Zhejiang Province (2021) Zhejiang Anlan 100-Billion Project), to prevent major risks as the starting point, adhering to the safety bottom line, upholding the ecological concept, systematically applying multiple measures and synergistically promoting seawall safety standards, ecological quality, integration, management and conservation, to achieve the "safety +" comprehensive function of the seawall coastal zone. The digital management platform of the seawall, as an important part of the "Seawall Anlan project" construction, is an important guarantee of project management and maintenance efficiency improvement.

2 CONSTRUCTION OBJECTIVE

The digital seawall management platform focuses on tide resistance ability, safety monitoring, operation management, and other business contents of a seawall project. It gathers water level,

*Corresponding Author: qian.hongsheng@163.com

 DOI 10.1201/9781003305026-13

rainfall, tidal level, video, safety monitoring, typhoon, inspection, maintenance, and other data and uses GIS, AI, BIM, 5G, IoT, and mathematical analysis models and other advanced technologies to build a digital seawall intelligent application system integrating seawall inspection and maintenance, safety monitoring, dynamic research and judgment, automatic control, etc., and to realize the dynamic research and judgment system of the digital seawall tide resistance ability with full perception of seawall information, full digital engineering, full operation process, and full intelligence research and judgment, so as to provide support for accurate decision-making, and maximize economic, social, and safety benefit.

3 DESIGN AND IMPLEMENTATION OF THE PLATFORM

3.1 *Overall architecture*

This paper designs and constructs the digital seawall management platform in accordance with the unified Provincial "four horizontal, four vertical and two ends" overall architecture, the unified construction frame system of water conservancy digital reform "1161", the overall layout of "One warehouse, one map, one center platform" and "five unified" (2021) Notice of Zhejiang Provincial Department of Water Resources on Printing and Distributing Implementation Plan of Zhejiang Provincial Water Resources Digital Reform, Yu, Jiang, Jin, Huang, etc.. 2020) unified user, unified portal, unified data, unified map, unified security construction standards. The overall architecture is shown in Figure 1.

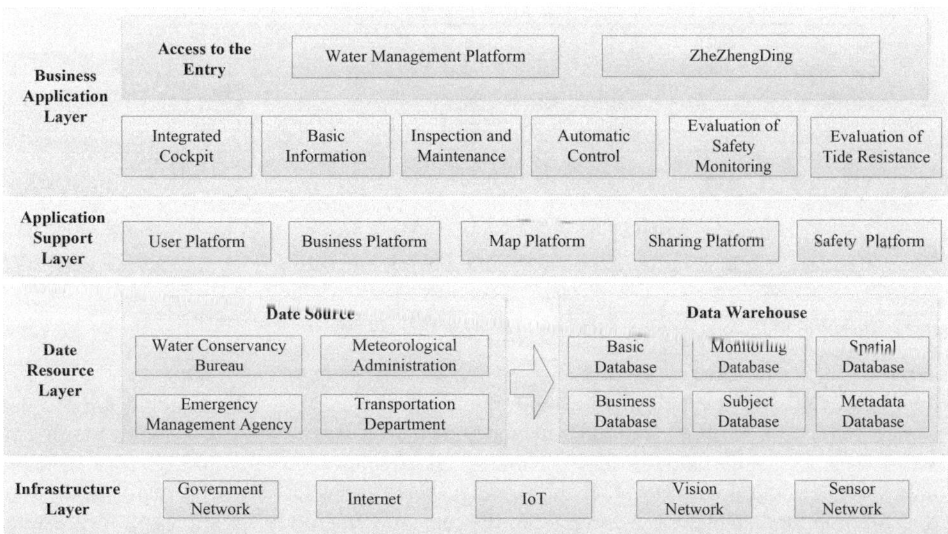

Figure 1. Overall Architecture of Digital Seawall Management Platform.

(1) The Infrastructure Layer is composed of the Government Network, Internet, IoT, Vision Network, Sensor Network, etc., which provides software and hardware infrastructure support for the Data Resource Layer.
(2) The data sources are represented by the Date Resource Layer mainly come from the Water Conservancy Bureau, Meteorological Administration, Emergency Management Agency, Transportation Department, Housing and Urban-Rural Development, Bureau of Culture, Radio, Film, Tourism, and Sports, etc. This data is collected into the data warehouse through extraction, cleaning, and transformation. The Data Warehouse mainly includes basic databases, monitoring

databases, spatial databases, business databases, subject databases, metadata databases, etc., as well as data services formed on this basis.

(3) Application Support Layer: Service platforms such as User Platform, Business Platform, Map Platform, Sharing Platform, and Security Platform shall be built mainly in accordance with the construction specifications of "five unified" to provide application support for specific business applications.

(4) Business Application Layer: The specific business application modules of Digital Seawall Management Platform include Integrated Cockpit, Basic Information, Inspection and Maintenance, Automatic Control, Safety Monitoring and Evaluation, Research and Evaluation of Tide Resistance Ability, etc., which are accessed through Water Management Platform, ZhezhengDing and other entrances.

3.2 *Design of platform functional structure*

According to the overall structure of the platform, the functional design of the platform mainly includes modules such as integrated cockpit, basic information, inspection and maintenance, safety monitoring, and tide resistance ability, which is shown in Figure 2.

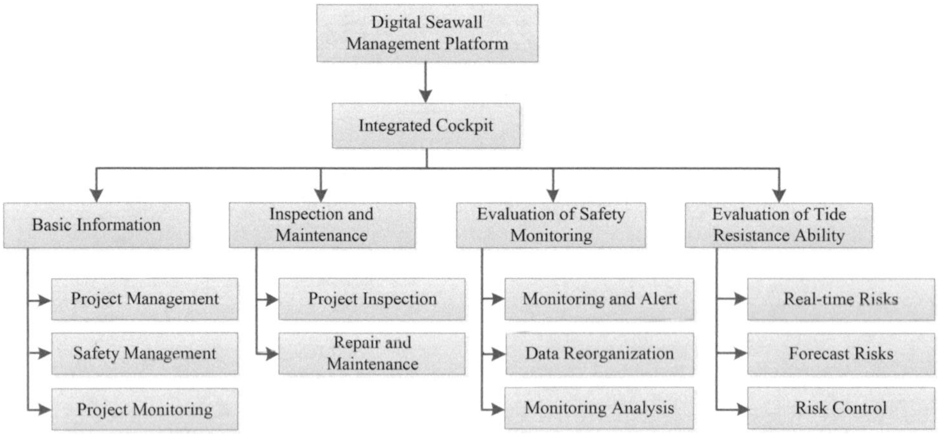

Figure 2. Platform functional structure diagram.

(1) Integrated Cockpit: It displays all factors related to the seawall, which includes basic data (sluice, pump station, sluice station, protected object, protected area, etc.), real-time monitoring data (water level, rainfall, tide level, wind speed, video, safety monitoring, etc.), and tide forecast data (hydrological weather monitoring, storm surge forecast, risk analysis, warning transmission, and statistics). According to the real-time operation situation and forecast and early warning information of the seawall, combined with the visual effect of GIS, the key indicators are dynamically tracked and the monitoring data and tide resistance ability of the seawall are intuitively displayed, which provides decision support for emergency decision-making and engineering safety analysis of the seawall.

(2) 1) Project Management, which includes project characteristics, building management situations, management protecting situations, etc., mainly displays the basic attributes, construction situations, management situations, and construction information, as well as the scope of management and protection, the information of right confirmation, and other information in various forms such as text, graphics, forms, attachments, etc.. 2) Safety Management, which includes safety appraisal, safety reinforcement, emergency plan, safety responsibility system, etc., through a series of safety management activities of the seawall project, lets the management personnel know the safety status of the seawall in real time. 3) Project Monitoring, which

includes engineering photos, video images, video monitoring, etc., shows the main features of the seawall project in various forms, such as photos, videos, videos, and so on, to achieve the combination of dynamic and static monitoring of the seawall.

(3) Inspection and Maintenance: 1) Project Inspection It includes inspection statistics and early warnings. Inspectors carry out inspections through mobile terminals and report inspection data to the platform in real time. The platform makes statistical analysis and views of the inspection situation and timely gives warnings and reminders to those who fail to perform inspections on time. 2) Repair and upkeep. Through maintenance project tracking and maintenance process recording, the "closed-loop" management of engineering maintenance project tracking, recording, and completion can be realized, and the professional and standardized management level of engineering maintenance management can be improved.

(4) Safety Monitoring, Research and Judgment: 1) Monitoring and Early Warning. Through automatic monitoring of all kinds of monitoring data, when the threshold is reached, the system will automatically send early warning information to the relevant responsible person at the seawall, and the responsible person will take the corresponding early warning disposal after receiving the information. 2) Data Reorganization: According to the surface displacement monitoring, deep horizontal displacement monitoring, infiltration line monitoring, and other types of monitoring points, the real-time monitoring values and historical monitoring values of the seawall monitoring points are counted in the form of a list to form a monitoring data report. 3) Monitoring Evaluation: On the basis of data regrouping, the graph analysis of the data of various monitoring types is carried out, and the evaluation coefficient of the overall stability of the seawall is generated by the model algorithm, so as to judge the stability of the seawall.

(5) Tide Resistance Capability Research and Evaluation:1) Risk in real time. According to the judgment standard for seawall storm surge, the warning level and guaranteed level were used as risk discrimination indexes for real-time risk judgment. The real-time risks of the seawall are displayed in three colors: normal (green), super alert (yellow) and super guaranteed (red). 2) Estimate Risk. When a typhoon is coming, based on the typhoon surge seawall risk dynamic warning model, which combines the typhoon's path in China with the current path baseline time, and typhoon, wind speed, tide level, and other monitoring data are used as input indexes, the seawall risk in the next 3 days is analyzed and predicted. 3) Risk management. The system triggers the warning automatically according to the seawall's real-time and forecast risks and sends early warning information to the relevant person in charge through Zhezheng Ding, SMS, and other ways. After receiving the early warning information, the responsible person shall carry out offline early warning disposal and feedback in the background system after completion, after which they shall close the early warning after review, forming a closed loop.

3.3 *Implementation and application of the platform*

The platform has been developed and applied in Haiyan County and other counties (cities, districts), which greatly improves the tide resistance ability and operation management level of seawall projects and provides important support for accurate decision-making in seawall safety management.

4 CONCLUSION

This paper constructs the overall framework of the digital seawall management platform and designs and implements the specific functional modules of the platform according to the overall framework of "four horizontal, four vertical, and two ends" of digital reform in Zhejiang Province and the overall framework of "1161" of digital reform in the Water Resources Department of Zhejiang Province. The platform has been developed and put into use in many places. It provides important support for seawall management departments to prevent and resolve major risks and hidden dangers and improve monitoring, forecasting, and early warning capabilities.

REFERENCES

Bao Zhiyan, Jiang Xiaojun, Huang Kang, etc.. (2020) Research on the Overall Framework and Key Technologies of Zhejiang Water Resources Digital Transformation. Water Conservancy Informatization, 2: 1–8.

Commission for Deepening Overall Reform of Zhejiang Provincial Party Committee. (2021) Notice on printing and distributing the Overall Plan of Digital Reform in Zhejiang Province by Commission for Deepening Overall Reform of Zhejiang Provincial Party Committee. Zhejiang Provincial Committee for Deepening Overall Reform, 1: 1–165.

Yu Kaisen, Jiang Xiaojun, Jin Xuancheng. (2018) Idea on Smart Water Resources in Zhejiang Province under Government Digital Transformation. Water Conservancy Informatization, 6: 11–15.

Zhejiang Provincial Department of Water Resources. (2021) Notice of Zhejiang Provincial Department of Water Resources on Printing and Distributing Implementation Plan of Zhejiang Provincial Water Resources Digital Reform. Zhejiang Provincial Department of Water Resources, 1:1–51.

Zhejiang Provincial Development and Reform Commission, Zhejiang Provincial Department of Water Resources. (2021) Zhejiang Anlan 100-Billion Seawall Project Construction Planning. Zhejiang Provincial Development and Reform Commission, Zhejiang Provincial Department of Water Resources. 1: 1–104.

Advances in Urban Engineering and Management Science – Khalil & Yang (Eds)
© 2023 The Author(s), ISBN 978-1-032-30426-7

Study on optimization of urban interchange design scheme

Junxiang Chen*
Guangxi Transportation Science and Technology Group Co., Ltd., Nanning, China

ABSTRACT: Urban interchange is a systematic project, and the alignment and structural design of the interchange projects are the core issues, which determine the advantages and disadvantages of the design scheme. Based on the design of an interchange project in Nanning, this paper discusses a variety of technical problems in the overall scheme design of interchange under complex environment, puts forward the optimization methods of road alignment and bridge structure design of interchange, and improves the optimization methods in combination with building materials, construction technology, and landscape environment. Then, considering various factors, such as technology and economy, the design schemes of urban interchange are compared and recommended, which provide a reasonable and effective reference for the optimization design of urban interchange.

1 INTRODUCTION

With rapid increase in urban population and economic development, the number of urban vehicles is increasing day by day. To solve traffic congestion and ensure smooth traffic, interchange projects are more widely used in urban road constructions.

Urban interchange is a way of connecting roads or railways at different elevations by using cross-line structures to cross each other. It is a common facility for efficient traffic at important intersections of urban traffic. Therefore, the overall optimization of the urban interchange scheme must be achieved through the comprehensive optimization of various factors, which not only ensures that the substructure of the bridge meets the construction of a variety of crossing conditions but also the alignment of the ramp can meet the requirements of traffic safety.

2 RELYING ON THE CURRENT CONDITIONS OF THE PROJECT (DESIGN 2010)

This article is based on an overpass project in Anning Industrial Park, Xixiangtang District, Nanning City. The interchange center is 1.8 km from Anji Interchange Center to the west, 1.3 km from Lianfu Interchange Center to the north, and 2.05 km from Beihu Interchange Center to the south. The construction planning area is centered on the north-south Anji Avenue, Beibei Road, and Yongwu Road extension lines, and combines the east-west Xiuxiang Avenue, AnWu Avenue, Keli Avenue, and the newly planned urban arterial roads on the north side of the Shuinan Expressway. As shown in Figure 1, a "three-vertical and three-horizontal" main road network structure in the planning area has been formed.

AnWu Interchange also has different requirements for crossing lines, rivers, and subways. Among them, the interchange crosses the underground structure of Beibei Road Station on Metro Line 3. The North Road Station is located at the intersection and is laid diagonally across the intersection. It is an underground two-story island platform station. The layout of bridge piers and abutments meets the requirements of rail transit by optimizing the plane alignment of the interchange.

*Corresponding Author: jychxb@sina.com

DOI 10.1201/9781003305026-14

3 OPTIMIZED DESIGN OF INTERCHANGE ROUTE PLAN IN COMPLEX ENVIRONMENT

3.1 *Optimized design of the route plan of the interchange project*

The technical focus of the urban overpass lies in the determination of the road route plan. Its horizontal alignment includes the main line and ramp, which should meet the traffic requirements in all directions, the vertical section alignment should meet the clearance requirements under the bridge and the bridge deck, and the cross-section elements should meet the traffic volume requirements. Only by optimizing the overall level, vertical and horizontal routes can a reasonable route design plan for interchange road projects in complex environments, such as crossing routes, subways, and water systems, at the same time be obtained (CJJ37-2012).

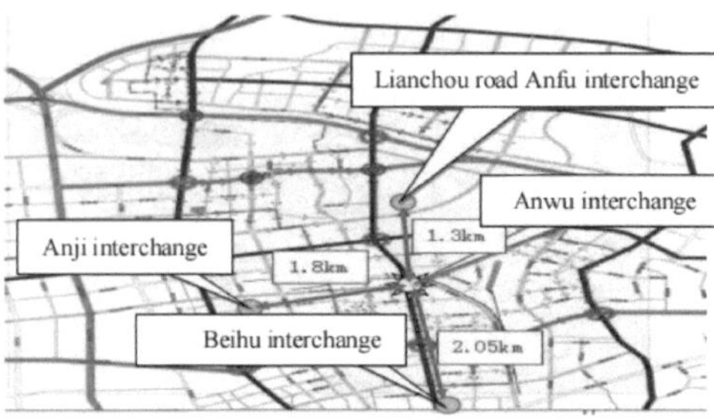

Figure 1. AnWu interchange and nearby interchange positions.

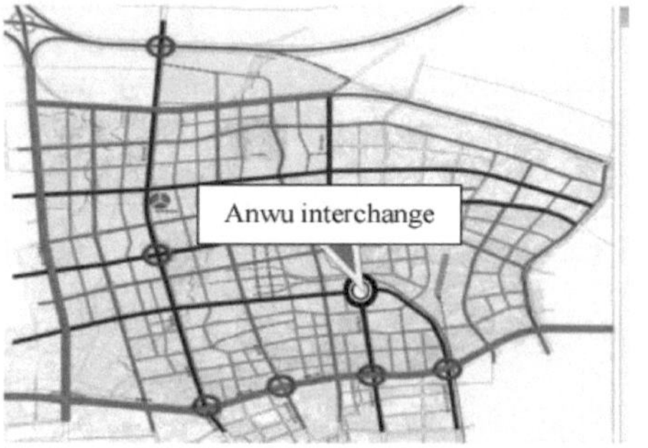

Figure 2. Scope of study planning road network map.

3.1.1 *Route plane design*

The interchange route is formed by the intersection of east-west and north-south lines, with an intersection angle of 90 degrees. As shown in Figure 2, the route scheme design should consider not only the connection of subway station entrances and exits but also the restrictions of Chaoyang Creek land use planning. Therefore, the road centerline of the interchange main line is arranged along the current road centerline, and the ground widening and channelization design shall be carried out within the land scope of the planned road red line as far as possible.

The design speed of the route is 60 km/h for AnWu main line, 50 km/h for Beihu main line, and 40 km/h for the ramp and ground system. Combined with the engineering practice, on the premise of meeting the crossing capacity and limiting conditions, try to adopt a larger radius of curvature and avoid the specified minimum radius of curvature. The design parameters of plane alignment meet the requirements of relevant specifications, which not only ensure the smoothness and simplicity of plane alignment but also avoid large-scale building demolition, to realize the technical and economic optimization of road alignment. A general layout is shown in Figure 2.

3.1.2 *Vertical section design of the route (Zhao 2006)*

The maximum longitudinal slope of the overpass bridge set in this project is controlled by 5%, and the clearance between the bridge deck and under the bridge is designed to be no less than 5.0 m. The elevation control of the ground system is based on the principle of facilitating passage and drainage.

According to the technical code for safety protection of urban rail transit structures and considering the bearing capacity of subway stations, do not load and reduce a large amount of load above the stations, and maintain the existing ground elevation.

Around the intersection, there are five subway passage pedestrian entrances and exits, and the elevation of the five subway entrances and exits is controlled to be 45 cm higher than the current ground. After quantitative comparison, the index values of the road vertical control elements are shown in Table 1.

Table 1. Road vertical control elements table.

Serial number	Road name	Pile no	Designed elevation	Remark
1	AnWu road	K0+320	79.444	Starting point, connected to design road elevation.
2		K0+457.89	80.163	The elevated system is connected to the existing road, and the overpass begins to rise.
3		K0+792.58	89.701	AnWu Avenue crosses the highest point of Beihu North Road on the ground.
4		K1+034.87	85.746	The viaduct system is connected with the current road, and the overpass falls to the ground.
5		K1+140	88.794	End point, connecting the status quo road signs high.
6	Beihu North Road	K0+000	84.695	Starting point, connecting the status quo road signs high.
7		K0+088.28	83.724	The viaduct system connects to the existing road and the overpass begins to rise.
8		K0+592.87	99.24	Beihu North Road crosses the highest point of AnWu Avenue
9		K1+060	79.579	The elevated system is connected to the existing road and the overpass is grounded.
10		K1+240	78.539	Termination point, connected to current road elevation.

3.1.3 *Cross-section design of the route (CJJ11-2011)*

The cross-section design of the interchange route shall be carried out within the width of the planned red line (the red line width of AnWu Avenue is 50 m, and the red line width of Beihu North Road is 40 m), including road grade, road division, lane division, design speed, design traffic volume, traffic organization, traffic facilities, pipeline equipment, greening landscape, upgrading and lighting terrain, and geology and other factors shall be comprehensively optimized to ensure that the cross-section of the road meets the design traffic capacity and ensures safety of vehicles and pedestrians.

The number of overpass lanes is two-way four lanes of the main line, the ground system and ramp are one-way two lanes, and the clearance height of the motor vehicle lane is $\geq$5.0 m. The optimization design of road cross-section is carried out in combination with the current conditions of interchange.

3.2 *Landscape design of interchange project*

In recent years, urban road networks are becoming more complex. Urban interchange is an important part of urban regional landscape design. Urban interchange landscape consists of interchange body, interchange green space, surrounding facilities, and buildings. Urban interchange landscape design can considerably enhance its influence as a city sign, and it is an indispensable part of scheme design. The landscape engineering of the interchange attaches great significance to the horizontal and vertical visual continuity and three-dimensional coordination. The components are scattered and form a strong sense of hierarchy.

In the process of designing an interchange landscape, it is necessary to comprehensively consider urban construction requirements, cost, environmental protection, and other factors, follow the principles of environmental protection, beauty, the dynamic and static combination of landscape, and ensure that the route linear design is coordinated with the landscape environment, have traffic functions, ecological environment, and aesthetic characteristics, highlight the characteristics of comfort, safety, and modernization, and obtain the best landscape effect.

3.3 *Comparison and selection of interchange engineering design schemes (Li 2019)*

According to the requirements of relevant design specifications for urban roads for route design, and in combination with the terrain and buildings around the interchange, the interchange project carries out the optimization design of route alignment and bridge structure, and puts forward two design schemes, as shown in Figure 3.

Scheme 1 is a two-way up-crossing scheme. Considering the long-term planning of the project location and surrounding areas, based on the current level crossing, open the Duantou road of AnWu Avenue, add a two-way up-crossing bridge, and form a three-layer system in the interchange area: the first layer is the ground layer of Beihu North Road, and the second layer is the layer of AnWu avenue that crosses the ground layer of Beihu North Road, The third floor is the elevated floor of Beihu North Road crossing the elevated floor of AnWu Avenue. In terms of traffic organization, the intersection is controlled by signal lights. The ground system uses the space under the bridge to set a U-turn lane to realize the U-turn function in this direction, and the channelized and widened intersection realizes the straight and left-right turning functions, as shown in Figure 3.

Figure 3. Plan view of the interchange. (Scheme 1).

Figure 4. Plan view of the interchange. (Scheme 2).

Scheme 2 for three layer and 3 alfalfa leaf + semi-directional interchange is based on the scheme 1, increase the left four directional ramp, in northeast, northwest, southwest quadrant ramp setting 3 alfalfa leaf type, southeast quadrant set a half directional ramp, using ramp maximum separation of motor vehicles and non-motor vehicles, completely solve the problem of node go straight, turn left from the traffic, and ensure smooth passage of the ground transportation system as shown in Figure 4.

The above two schemes have been selected through technical and economic comparison. Route scheme 2 has the characteristics of strong traffic capacity, high service level, and good safety, which is conducive to the development of long-term traffic. Therefore, it is recommended for this interchange project.

4 OPTIMIZATION DESIGN OF STRUCTURE SCHEME OF MULTIPLE SPAN INTERCHANGE BRIDGES (JTG D60-2015)

The main technical parameters for structural design of the overpass: the design load is urban-A, the design life is 100 years, the design safety level is class I, the clearance under the bridge is ≥4.5 m, the seismic design method is class A, and the bridge deck width (overpass A1 *section*): 8 m = 0.5 m (crash barrier) + 3.5 m (car lane) + 3.5 m (car lane) + 0.5 m (crash barrier).

The basic seismic intensity in this area is 7 degrees, and the seismic intensity is fortified according to 8 degrees. The bridge is located in the urban trunk road, and the seismic fortification category is class B. The corresponding structural measures are taken in the structural design to improve the overall seismic capacity.

4.1 *Bridge type selection*

The selection of bridge type scheme shall be based on the overall layout of the interchange project, comprehensively considering the terrain, geological conditions, landscape effect, reasonable span, construction difficulty, project cost, and construction period of the bridge area, as well as the surrounding traffic demand and traffic demand during the construction period, and on the premise of ensuring smooth development of river regulation at the bridge site and normal operation of the subway station. Based on the design experience of similar scale overpasses, the safety, applicability, and landscape of bridge schemes are optimized and compared.

As far as this project is concerned, there are bridge spans with larger spans. Considering the rationality of hole splitting and the applicable span of construction method, it is proposed to adopt the integral casting method for the bridge spans that cross the main line of the road and the existing ground underground structures. Beam bridge: the integrally cast beam bridge structure has good integrity, comfortable driving, large spanning ability, and good adaptability to ramp beams and obstacles. It can adopt a variety of main girder cross-section forms; a variety of span combinations and various types of piers that enrich the bridge landscape in a limited space.

4.2 *Optimization design of bridge superstructure (CJJ193-2012)*

4.2.1 *Bridge structure type*

The reasonable selection of bridge superstructure is an important link in the design of the interchange. At present, the superstructure of the urban interchange ramp bridge generally adopts cast-in-situ box girder and steel box girder. The cast-in-situ box girder bridge has the advantages of moderate cost, span meeting the actual needs, simple shape, soft structure, overall performance, landscape effect, and adaptability to variable width. The construction of a steel structure box girder bridge is complex, the deck pavement needs special design, and the project cost and later maintenance costs are high, which makes it unsuitable for large-scale use.

To sum up, the concrete cast-in-situ box girder is recommended for the bridge superstructure of the project; when crossing the main road below, the beam section with high clearance requirements can give priority to the steel box girder scheme.

4.2.2 *Bridge span combination*

The selection of bridge span combination has significant influence on project investment, construction method, traffic organization, landscape environment, and more. Therefore, the basic span of the continuous beam must be accurately analyzed and optimized. According to the bridge construction experience, combined with the general layout of the road scheme and the actual site conditions, the ring road cast-in-place box girder with a single span of 25 m or more can meet the requirements of crossing the tunnel of Metro Line 3.

4.2.3 *Selection of cross-section form of superstructure*

As the main load-bearing structure, the main beam of the urban overpass also puts forward higher requirements for landscape. Therefore, the superstructure of the viaduct is planned to adopt the form of a box section continuous beam, and its simple and stable shape is well in line with the breath of urban modernization. Inclined web box girder combines the characteristics of straight web and fish-bellied box girder section, with good structural safety and economy. It can also reduce the pressure brought by sharp chamfer of inclined web to pedestrians under the bridge by adding arc chamfer to the web, to make the beam shape full of streamlined beauty and enhance the overall landscape effect of the overpass.

4.2.4 *Selection of continuous beam joint length*

If the joint length of each continuous beam is too short, there are many expansion joints that are not conducive to driving comfort, and there are many transition piers to be set, resulting in the lack of unified beauty of the bridge. When the bridge connection length is too large, it is not conducive to the parallel operation of the whole line of bridges and the construction period is prolonged. Considering the heavy traffic in the overpass area, the construction period shall be shortened as far as possible to alleviate the traffic pressure during construction. Therefore, the continuous beam should be connected with 3~5 holes, which is convenient for synchronous construction, reducing the construction period, and ensuring smooth driving.

4.2.5 *Superstructure design scheme (Wang 2019)*

Scheme 1: The bridge across the line of AnWu Avenue has a total of two connections, and the span combination is: 3×30 m prefabricated simple support followed by continuous small box girder + (40+60+40) m constant section cast-in-place continuous box girder. Beihu road bridge has three connections in total, and the span combination adopted is (75+90+90+75) m variable section cast-in-place continuous box girder + 4×30 m prefabricated simple support and then continuous small box girder + 4×30 m prefabricated simple support and then continuous small box girder. The cross-section of the bridge is arranged as 0.5 m (guardrail) + 8.25 m (carriageway) + 0.5 m (central divider) + 8.25 m (carriageway) + 0.5 m (guardrail).

Scheme 2: AnWu Avenue span bridge has a total of four connections, the span combination is 2× (3×30 m) equal section cast-in-place continuous box girder + (40+60+40) m equal section continuous steel box girder +2×30 m equal section cast-in-place continuous box girder.

Beihu North Road span bridge has five connections, the span combination is 4×30 m constant section cast-in-place continuous box girder + (55+75+55) m variable section cast-in-place continuous box girder +3×(3×30 m) constant section cast-in-place continuous box girder.

A ramp bridge has six connections, the span combination is (23+30+23) m continuous box girder + 2×25 m continuous box girder + 4×30 m continuous box girder + (30+25+40+35) m continuous box girder + 3×20 m continuous box girder + (23+30+23) m continuous box girder.

B ramp bridge has four connections, the span combination is (25.67+25) m continuous box girder +2×30 m continuous box girder +2×25 m continuous box girder +1×45.91 m steel box girder.

C ramp bridge has four connections, the span combination is (17.37 m+2×25) m continuous box girder + 2×25 m continuous box girder + (17+32.34) m continuous box girder.

D ramp bridge has four connections, span combination 2×30 m continuous box girder +3×20 m continuous box girder + (35+25) m continuous box girder + (20+18.74) m continuous box girder.

4.2.6 *Optimization design of prestress*

For the special-shaped beam (Figure 5), the prestressed beam is anchored at the end of the beam, and the roof is equipped with a long beam and tensioned at both ends. In the cross-section of the box girder, the steel bundle is generally arranged within the web. Except for special-shaped beams, the ends of the prestressed steel beams in other beam segments (Figure 6) are bent at the top of the beam ends and tensioned at both ends. In the bottom plate and roof with a short beam, using tension at both ends or one end of tension. Such beam distribution can shorten the construction period.

The construction sequence of special-shaped beams should precede other beam segments at both ends. Except for special-shaped beams, the construction sequence of other beam segments is not restricted and can be constructed at the same time. This prestressed beam distribution method effectively solves the construction sequence problem of the box girder, and has obvious advantages during the construction period compared with the conventional construction method of lengthening the hole by hole.

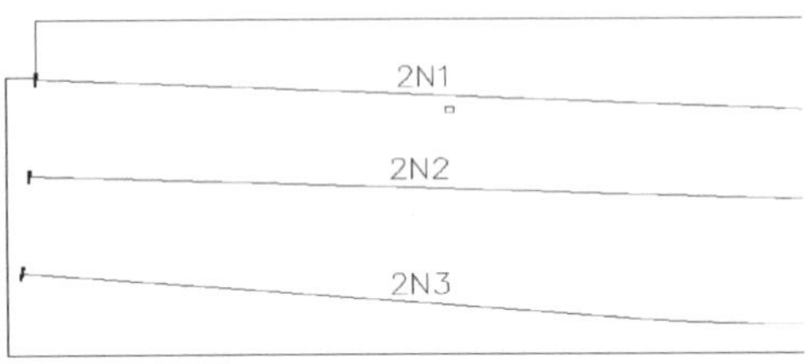

Figure 5. Special-shaped beam end prestressed steel beam configuration diagram.

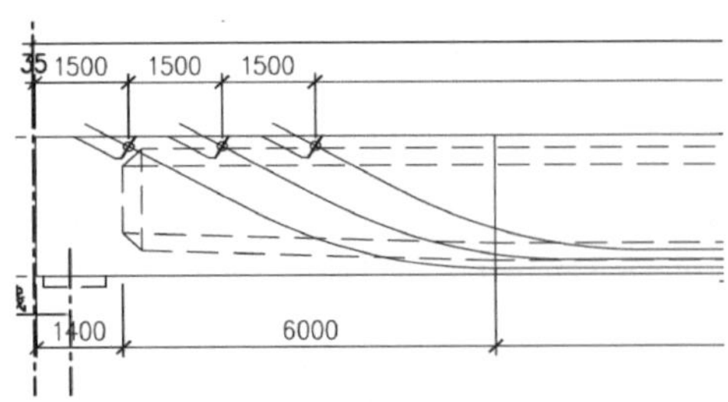

Figure 6. Configuration of prestressed steel bundles at the beam end of general beam sections.

4.3 *Optimization design of bridge substructure*

The bridge pier is an important part of the bridge structure and the most direct visual object for pedestrians and drivers under the bridge. Therefore, the pier form of the viaduct should not only meet the structural stress requirements but also be the focus of urban bridge landscape design. At the same time, the installation of the pier is affected by many factors, such as surface driveway and underground pipeline, so the choice of pier position and pier type is very important.

A folding box girder is recommended for the superstructure. Therefore, the design form of the substructure should be as unified as possible with the shape and style of the superstructure. The two types of pier are analyzed and compared in the substructure design, as shown in Table 2.

Table 2. Pier comparison table.

Serial number	Main bridge piers	Ramp bridge pier	Characteristics
1		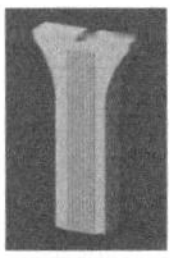	The pier body for architectural decoration processing To increase the vertical groove line, make pier tall and straight for use
2		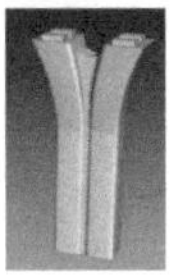	The pier has beautiful lines, a reasonable combination of straight and curve elements, and coordination with the linear shape of the superstructure. This option is recommended.

The selection of pier mainly considers the ground system and landscape effect. After optimization research, the second type of pier (vase pier) is recommended. The viaduct has a wider deck, so double-column vase piers are used. At the junction of the steel box girder and prestressed concrete box girder, the top of the bridge pier must be made into a step, the cross-sectional position of the support is slightly different, and the rest of the structure is the same. The prefabricated pier construction method can be adopted, the prefabricated pier can achieve industrial production, and has good adaptability to BIM technology, can optimize and improve the quality of design results.

Abutment adopts buttress type light abutment. A pile foundation is the main foundation, mainly considering the geological soil conditions. At the same time, the pile foundation has better seismic performance.

4.4 *Influence of materials and process on structural optimization*

The bridge includes a prestressed concrete box girder structure and a steel box girder structure. According to the structure and material characteristics of the girder, the prestressed concrete structure is preferred, and the main material type is concrete. The concrete structure has a low cost, mature construction method, short construction period, and a little influence on traffic. Steel box girder cost is high, prefabricated construction period advantage is not obvious, and the construction is difficult, but can reduce a certain height of the superstructure. Therefore, all spans of the project are prestressed concrete beams, except the steel box girders used in the second bridge of AnWu Avenue.

5 OPTIMIZATION METHOD OF URBAN INTERCHANGE ENGINEERING DESIGN

5.1 *Optimization method of mainline shape of interchange*

The optimization method of the main line alignment in the complex environment based on different crossing requirements is to determine the reasonable line alignment and bridge alignment scheme elements (horizontal, vertical, and horizontal) by analyzing, calculating, and comparing the site conditions so that the alignment has good visual continuity and three-dimensional coordination. Recommend the best scheme after scheme comparison and *selection optimization.*

5.2 *Alignment optimization method of interchange ramp*

The key point of alignment optimization for interconnecting urban interchange is ramp, the A ramp bridge of this interchange is semi-directional trumpet-shaped, linear complex. There are six continuous box beams, which is the longest ramp with large space for optimization in the linear shape. The plane line element of the ramp is mainly the plane curve radius of the ramp and cyclotron parameters, needed to consider the design speed of ramp, cross type, traffic volume, natural conditions, land conditions, bridge layout, cost, and other factors, after analysis, calculation, adjustment, optimization of the selection to ensure the safety and comfort of driving.

5.3 *Structure optimization method of the overpass*

The optimization of ramp bridge interchange design includes structural form, span size, mechanical properties, stability, engineering materials, construction scheme, dynamic landscape, and aesthetics, etc.

Midas series software is the most widely used bridge engineering numerical simulation calculation software at present, the software can form a comparison of calculation results, and effectively verify the accuracy of the calculation results. Based on the numerical simulation analysis and the experience of similar projects, the results of the model analysis are evaluated by using the method of analogy analysis, and the reliability of the simulation results is illustrated.

6 CONCLUSION

Urban interchange is a systematic project, and the main roads of the city mostly adopt the interchange. After in-depth investigation and research, accurate analysis, calculation, and prediction of traffic volume, according to the regional site conditions and the surrounding landscape environment, the intersection road alignment, and bridge structure design scheme. The optimization method of overall scheme design of urban interchange bridge based on different crossing requirements in a harsh and complex environment is explored, which provides a reference for the optimization range of important parameters, such as flat curve radius, bridge span, and joint length, which can be quickly locked in the route selection and scheme formulation stage of urban interchange design. To provide help for technical control in the implementation stage of the design scheme and facilitate construction management and operation and maintenance. The unification of technology, security, and economy of interchange design scheme should be realized as much as possible to obtain good investment benefit and application value and promote the innovation and development of similar projects.

REFERENCES

CJJ11-2011, Code for Urban Bridge Design [S] (2019 Edition). (2011)Industry Standard of the People's Republic of China.

CJJ193-2012, Code for Urban Road Route Design [S]. (2012)Industry Standard of the People's Republic of China.

CJJ37-2012, Code for Urban Road Engineering Design [S]. (2012) Industrial Standard of the People's Republic of China.

Design Code for Urban Road Intersections [S]. (2010)Industry Standard of the People's Republic of China.

JTG D60-2015, General Code for Design of Highway Bridges and Culverts [S]. (2015) Industry Standard of the People's Republic of China.

Li S. Design and Analysis of concrete continuous curve box girder Bridge [J]. Urban Road Bridges and Flood Control, 2019(05):111–114.

Wang Xinlei, LIU Tao. Research on Urban Interchange Scheme Design [J]. Highway Traffic Science and Technology (Application Technology Edition),2019(07):223–225.

Zhao Yueyu, Kang Houjun, Rui et al. Advances in Mechanics, 2006(2):170–186.

Advances in Urban Engineering and Management Science – Khalil & Yang (Eds)
© 2023 The Author(s), ISBN 978-1-032-30426-7

Comprehensive evaluation and decision-making model for university innovation base construction modes based on Matlab and AHP

Qianshan Wen*, Xiaona Han, Jiangtao Liu & Tianxing Huangfu
Beijing Jiaotong University New Campus Construction Office, Beijing, China

ABSTRACT: By examining the development of colleges and universities, this paper analyzes the necessity of building innovation bases in colleges and universities and summarizes four construction modes that are currently used in colleges and universities. In order to provide the basis and decision-making method for colleges and universities to choose the mode of building innovation bases, this paper establishes a comprehensive evaluation and decision-making model for university innovation base construction modes based on the Matlab and AHP methods. The model establishes an evaluation index hierarchy model through AHP and uses the Matlab data analysis software to code, thereby realizing the calculation of the judgment matrix. In the meantime, consideration is given to the weights of the evaluation indexes and the evaluation subjects, and the arithmetic average method and the weighted average method are reasonably combined to obtain the final result. Through case analysis, the rationality and feasibility of the comprehensive evaluation and decision-making model established in this paper are verified.

1 INTRODUCTION

According to the National Statistical Bulletin of Education Development, the total number of students in higher education in China increased by 2.43 times from 2000 to 2020, and the gross enrollment rate of higher education reached 54.4% in 2020.

Table 1. Total number of students in higher education.

Year	2000	2005	2010	2015	2020
Total number of students (10,000 persons)	1220	2300	3105	3647	4183
School building construction area (10 million m2)	2075	5527	7460	8914	10529
Floor area (10 million m2)	4707	13870	17012	19362	20688

As shown in Table 1, from 2000 to 2020, the construction area and floor area of school buildings in colleges and universities across the country have maintained a steady increase trend, among which the increase in the floor area of colleges and universities has decreased year by year since 2015. As higher education has entered the stage of popularization, the large-scale construction in colleges and universities is drawing to a close.

According to the requirements of the Interim Regulations on the Establishment of General Undergraduate Schools formulated by the Ministry of Education in 2006, the floor area per student in a general undergraduate school should reach 60 m2 and the school building construction area per student should reach 30 m2. In 2020, the school building area per student in colleges and universities is 28.77 m2 and the floor area per student is 58.32 m2, both of which do not meet the requirements of the Ministry of Education. In the next 5 to 10 years, many colleges and universities

*Corresponding Author: qswen@bjtu.edu.cn

DOI 10.1201/9781003305026-15

will plan the construction of new campuses to solve the problem of lack of basic school-running land and housing that restricts the development of colleges and universities.

Taking a university in Beijing as an example, according to the Building Area Index of Colleges and Universities (Jian Biao 191-2018), the current shortage of school-running land has reached more than 2,000 mu, and the shortage of basic school-running housing is about 300,000 square meters.

The central government and Beijing Municipality have strengthened the management and control of spatial planning, and the Ministry of Education has strictly controlled colleges and universities to run schools in different areas so that it is difficult for colleges and universities in Beijing to find an expansion and to change the stock. Under such restrictive policies, these schools actively adjust their construction ideas, explore new models of integration of production and education, promote the construction of innovation base projects, and seek to expand their scientific research spaces.

2 UNIVERSITY INNOVATION BASE CONSTRUCTION MODES

University innovation base is the collective name of national laboratories, national key laboratories, national engineering laboratories, national engineering technology research centers, and provincial and ministerial research platforms established by relying on colleges and universities (Duan & Hu 2008), and has been incorporated into the national innovation system construction plan (Wang 2006).

The innovation base is mainly composed of the theoretical basis section, frontier science section, resource support section, evaluation and pilot test section, and social development section, covering the entire technological innovation chain from basic research to applied research to achievement transformation. The theoretical basis section relies on the campus to operate and shares some resources, serving as the basis for the operation of the innovation base. The frontier science section mainly conducts cutting-edge basic research and becomes the innovation center in the relevant basic frontier field. The resource support section utilizes computer hardware resources and focuses on building a comprehensive data platform serving the innovation base. The evaluation and pilot test section is mainly the evaluation center and the pilot test base, which aims to promote the transformation and implementation of scientific research achievements. The social development section provides the society with a platform for cooperation and exchange to promote cross and integrated development. At present, there are mainly four modes of building innovation bases in colleges and universities: simultaneous construction in new campuses, renovation construction in old campuses, independent construction outside campuses, and leasing.

2.1 *Simultaneous construction in new campuses*

The university innovation base is built simultaneously by lying on the new campus. It integrates key scientific research projects and platforms to build innovation highlands in relevant fields and is arranged relatively independently from the campus. The innovation base is arranged on the side close to the city, forming a transition zone between the campus and the city, giving full play to the school's role in talent cultivation, scientific research and discipline construction, and promoting the integration of production and education. The innovation base adopts an open layout, operates relatively independently from the campus, shares scientific research, technology, achievements, platforms, etc. with the society, and provides technical support for regional development.

2.2 *Renovation construction in old campuses*

At present, the distribution of scientific research space in most old campuses of colleges and universities is scattered, and most of the research space is constructed and managed by colleges, lacking top-level design. Under the circumstance that the expansion space outside the campus is limited, colleges and universities should make overall plans and sort out the existing scientific research platforms and facilities in old campuses to optimize the allocation of resources and transform and construct innovation bases. They should utilize the existing scientific research and space conditions, centrally plan scientific platforms and facilities, and gather most of the scientific research resources together for mutual benefit and sharing, which is conducive to the cross-development of disciplines, resource sharing, and space utilization efficiency.

2.3 *Independent construction outside campuses*

If the school does not gain support for the construction of a new campus and the old campus does not have the conditions for renovation, it can give full play to the advantages of the school. It can independently strive for land outside the school to build an innovation base, make overall plans for the existing high-end scientific research platforms, and create a high-level scientific research cluster. There are two main ways to obtain construction land: one is to obtain land for education and scientific research free of charge through state allocation; the other is to obtain commercial land for R&D and design with self-raised funds through agreement.

2.4 *Leasing*

In addition to the self-construction modes, colleges and universities can also cooperate with local governments of advantageous disciplines to jointly build innovation bases, obtain the right to use existing buildings for several years by means of full or partial subsidized rents by local governments, jointly create high-end R&D highlands, and form agglomeration effects with local government industries.

3 HIERARCHICAL STRUCTURE MODEL OF COMPREHENSIVE EVALUATION OF UNIVERSITY INNOVATION BASE CONSTRUCTION MODES

In order to effectively evaluate the pros and cons of the innovation base construction modes and help schools choose the appropriate construction mode correctly, this paper establishes a comprehensive evaluation and decision-making model for university innovation base construction modes. The index system of the comprehensive evaluation includes the target layer, criterion layer, and index layer, as shown in Table 2. The criterion layer, also known as the strategy layer or constraint layer,

Table 2. Hierarchical structure model of a comprehensive evaluation of university innovation base construction modes.

Target layer T	Criterion layer S		Index layer B	
Evaluation index system for planning and building innovation bases in colleges and universities	S1	Scientific research	B1	Achievement transformation
			B2	Scientific research platform
			B3	Major projects
			B4	Integration of industry and education
	S2	Talent cultivation	B5	Discipline construction
			B6	School-enterprise cooperation
			B7	Faculty
			B8	Basic guarantees
	S3	Economic benefits	B9	Construction costs
			B10	Operating costs
			B11	Achievement benefits
			B12	Project funding
	S4	Social impact	B13	Regional development
			B14	Resource sharing
			B15	Human factors
			B16	Ecological environment

mainly conducts a comprehensive evaluation of university innovation base construction modes through the four aspects of talent cultivation, scientific research innovation, economic benefits, and social impact.

3.1 *Scientific research*

The fundamental task of university innovation bases is to give play to the school's disciplinary advantages, concentrate core innovation resources, and create a base for scientific research innovation and talent cultivation. Colleges and universities can conduct an in-depth evaluation of the scientific research and innovation capabilities of university innovation bases through four aspects: integration of production and education; achievement transformation; scientific research platform; and major projects.

3.2 *Talent cultivation*

The fundamental task of colleges and universities is to cultivate talents. The level of talent cultivation must always be taken as the standard to measure the level of colleges and universities. Colleges and universities can effectively evaluate the talent cultivation capacity of their innovation bases through discipline construction, school-enterprise cooperation, teaching staff construction, and basic guarantees.

3.3 *Economic benefits*

Whether the construction of innovation bases in colleges and universities can be successfully completed mainly depends on the economic benefits of the bases. The economic benefits are considered from four aspects: construction and installation costs; operating costs; achievement benefits; and project funding. The financing paths of university innovation bases mainly include state-allocated funds and self-raised funds. State-allocated funds are obtained by applying for the construction of national laboratories, frontier science centers, national collaborative innovation centers, and other R&D institutions. The achievement benefits of university innovation bases are realized through the transformation of scientific and technological achievements, school-run enterprises, etc.

3.4 *Social impact*

The construction of innovation bases in colleges and universities will definitely have a great impact on society, which mainly includes four indicators: regional development, resource sharing, human factors, and ecological environment. The innovation base develops symbiotically with relevant industrial resources through various forms of innovation radiation. With relevant scientific research platforms and projects, the innovation base will attract high-end talent resources in related fields and form an important gathering place for high-end talent innovation and entrepreneurship.

4 ESTABLISHMENT OF THE COMPREHENSIVE EVALUATION AND DECISION-MAKING MODEL FOR UNIVERSITY INNOVATION BASE CONSTRUCTION MODES

AHP is a decision-making method that can transform complex decision-making from qualitative analysis into quantitative analysis according to the hierarchical structure by combining objective mathematical analysis methods with human experience and objective judgments. It is practical, systematic, and indirect (Fan 2012). Combining Matlab and AHP can effectively reduce the difficulty of data processing, reduce analysis time, and improve the accuracy of comprehensive evaluation (Yang et al. 2005).

This paper establishes a hierarchical structure model of evaluation indices and constructs the judgment matrix of each layer of indices; Matlab data analysis software is used to program the algorithm; the judgment matrix is imported into Matlab as a text for the calculation to obtain the combined weight of evaluation indices; in order to distinguish the importance of different evaluation subjects, this study determines the weights of different evaluation subjects; the arithmetic average method and the weighted average method are reasonably combined to process the data of the evaluation results, thus obtaining the final results.

4.1 *Determination of index weights*

The allocation of index weights is a quantitative allocation of the importance of different evaluation indices at each level, and is a differential treatment of each evaluation index in the overall evaluation (Su 2019). The importance of one index u_i relative to another index u_j is described by quantitative relative weights. Assuming that $U = \{u_1, u_2, \ldots, u_n\}$, a pairwise comparison matrix is established with the 1-9 scale value method for the importance between two factors. For example, $u_{ij} = u_{ji} = 1$ means that factor u_i is of equal importance relative to factor u_j, and $u_{ij} = 9$ or $u_{ij} = 1/9$ means that factor u_i is extremely important relative to factor u_j; when $u_{ij} = 2\text{-}8$ or $u_{ij} = 1/2\text{-}1/8$, the importance increases gradually. The relative importance of each index is determined through discussion by experts in relevant fields of the school to obtain the judgment matrix U.

$$U = \begin{bmatrix} u_{11} & u_{12} & \ldots & u_{1m} \\ u_{21} & u_{22} & \ldots & u_{2m} \\ \ldots & \ldots & \ldots & \ldots \\ u_{n1} & u_{n2} & \ldots & u_{nm} \end{bmatrix}$$

4.2 *Consistency check*

The consistency check of the judgment matrix U. The consistency check formula is:

$$\mathrm{CR} = \frac{CI}{RI} \leq 0.1$$

Where $CI = \frac{\lambda_{max} - n}{n-1}$, RI is the average random consistency index. When the CR of the judgment matrix is less than or equal to 0.1, it indicates that the consistency check is passed; when the CR is greater than 0.1, experts should be reorganized to adjust the judgment matrix.

4.3 *Determination of combined weight*

Matlab is used to write the calculation code of the combined weight: the judgment matrix is calculated to obtain the maximum eigenvalue $\lambda_{\max}$ and the corresponding eigenvector; the eigenvector is normalized to obtain the weight vector of each index; the criterion layer weight and the index layer weight are combined to obtain the combined weight vector W.

$$W = \begin{bmatrix} w_1 & w_2 & \ldots & w_{15} & w_{16} \end{bmatrix}$$

4.4 *Determination of the weight of the evaluation subject*

In view of the differences in the professional level and experience of the evaluation subject, in order to obtain more objective and accurate evaluation results, the evaluation weight P of different evaluation subjects is determined through expert discussion, where *e* is the number of evaluation subjects.

$$P = \begin{bmatrix} p_1 & p_2 & \ldots & p_e \end{bmatrix}$$

4.5 *Plan scoring*

The 16 evaluation indices are scored by different evaluation subjects to obtain a score set. The arithmetic mean of scores of an evaluation subject for a single index is taken as the final score $\bar{V}_e$.

$$\bar{V}_e = \left[\bar{v}_{e1}\ \bar{v}_{e2}\ \ldots\ \bar{v}_{e15}\ \bar{v}_{e16}\right]$$

The weighted average V' of the final scores of different evaluation subjects is taken as the final score for a single index.

$$V' = \bar{V}_1 * p_1 + \bar{V}_2 * p_2 + \cdots + \bar{V}_e * p_e$$

The weighted average of the final scores of the 16 indices is taken as the final score for a single plan V''.

$$V'' = V' * W^T$$

5 CASE ANALYSIS

A university in Beijing had two different innovation base construction plans:

In Plan 1, a certain scale of commercial land is obtained with self-raised funds through transfer by agreement, and an independent innovation base is independently built outside the school; the construction, installation, equipment, and other costs of the innovation base are all borne by the school itself; some national-level scientific research platforms are supported by state-allocated funds which are used for platform construction and the purchase of related equipment; the property rights of the land, buildings, equipment, etc. belong to the school.

In Plan 2, the right to use buildings in some areas is obtained through cooperation with the local government for the construction of the innovation base; the local government promises to reduce or exempt building rents and requires the school to produce certain results within a certain period of time; the school bears the secondary decoration and renovation costs and the operating costs.

The judgment matrix of the criterion layer S and the index layer B was determined through research conducted by 16 experts in related fields of the school.

$$S = \begin{bmatrix} 1 & 3 & 2 & 5 \\ 1/3 & 1 & 1/2 & 2 \\ 1/2 & 2 & 1 & 3 \\ 1/5 & 1/2 & 1/3 & 1 \end{bmatrix}$$

$$B_1 = \begin{bmatrix} 1 & 1/5 & 1/4 & 1/2 \\ 5 & 1 & 2 & 4 \\ 4 & 1/2 & 1 & 3 \\ 2 & 1/4 & 1/3 & 1 \end{bmatrix} \quad B_2 = \begin{bmatrix} 1 & 5 & 3 & 7 \\ 1/5 & 1 & 1/2 & 2 \\ 1/3 & 2 & 1 & 3 \\ 1/7 & 1/2 & 1/3 & 1 \end{bmatrix}$$

$$B_3 = \begin{bmatrix} 1 & 6 & 3 & 4 \\ 1/6 & 1 & 1/3 & 1/2 \\ 1/3 & 3 & 1 & 3 \\ 1/4 & 2 & 1/3 & 1 \end{bmatrix} \quad B_4 = \begin{bmatrix} 1 & 3 & 2 & 5 \\ 1/3 & 1 & 1/2 & 3 \\ 1/2 & 2 & 1 & 4 \\ 1/5 & 1/3 & 1/4 & 1 \end{bmatrix}$$

The above judgment matrices were imported with Matlab for calculation. The specific implementation codes were as follows:

```
%1. Input parameters and initialization parameters
S=load(sfile); % Read the judgment matrix of criterion layer
B1=load(b1file); % Read the judgment matrix of sub-index layer B1
B2=load(b2file); % Read the judgment matrix of sub-index layer B2
B3=load(b3file); % Read the judgment matrix of sub-index layer B3
```

```
B4=load(b4file); % Read the judgment matrix of sub-index layer B4
NS=4; % Number of elements in the criterion layer
NB=4; % Number of sub-index elements of the index layer index
NW=NS*NB; % Length of the weight vector
W=zeros(1,NW); % Initialize the combined weight vector
%2. Obtain the eigenvector and eigenvalue matrix of the judgment matrix of criterion layer S
[vector_S,ramda_S]=eig(S);
[row_S colume_S]=find(abs(ramda_S)==max(max(abs(ramda_S)))); % Find the column corresponding to the maximum eigenvalue
vector_S_reg=vector_S(:,colume_S)./sum(vector_S(:,colume_S)); % Eigenvector corresponding to the normalized maximum eigenvalue
%3. Obtain the eigenvectors and eigenvalue matrices of the judgment matrices of sub-index layers B1, B2, B3, and B4
[vector_B1,ramda_B1]=eig(B1);
[row_B1 colume_B1]=find(abs(ramda_B1)==max(max(abs(ramda_B1))));
vector_B1_reg=vector_B1(:,colume_B1)./sum(vector_B1(:,colume_B1));
[vector_B2,ramda_B2]=eig(B2);
[row_B2 colume_B2]=find(abs(ramda_B2)==max(max(abs(ramda_B2))));
vector_B2_reg=vector_B2(:,colume_B2)./sum(vector_B2(:,colume_B2));
[vector_B3,ramda_B3]=eig(B3);
[row_B3 colume_B3]=find(abs(ramda_B3)==max(max(abs(ramda_B3))));
vector_B3_reg=vector_B3(:,colume_B3)./sum(vector_B3(:,colume_B3));
[vector_B4,ramda_B4]=eig(B4);
[row_B4 colume_B4]=find(abs(ramda_B4)==max(max(abs(ramda_B4))));
vector_B4_reg=vector_B4(:,colume_B4)./sum(vector_B4(:,colume_B4));
%4. Store the normalized weight vectors of the judgment matrices of index layers B1, B2, B3, and B4 into the 1st, 2nd, 3rd, and 4th columns of the temporary weight matrix in turn
temp_W(:,1)=vector_B1_reg;
temp_W(:,2)=vector_B2_reg;
temp_W(:,3)=vector_B3_reg;
temp_W(:,4)=vector_B4_reg;
temp_W_vec=reshape(temp_W,1,NW);
%5. Combine the criterion layer weight and the index layer weight to obtain the combined weight vector
for i=1:NS
   for j=1:NB
      W(NB*(i-1)+j)=vector_S_reg(i)*temp_W_vec(NB*(i-1)+j);
   end
end
```

The maximum eigenvalue of each judgment matrix, the corresponding eigenvector, and the weight of each index was calculated with Matlab. The calculation results are shown in Table 3.

Table 3. Analysis of judgment matrices.

Weight vector					λ_{max}	CR
W_S	0.483	0.157	0.272	0.088	4.015	0.006
W_{B1}	0.078	0.491	0.306	0.125	4.048	0.018
W_{B2}	0.586	0.124	0.217	0.073	4.019	0.007
W_{B3}	0.550	0.077	0.252	0.121	4.088	0.033
W_{B4}	0.473	0.170	0.284	0.073	4.051	0.019

From the results in the table, it can be seen that the judgment matrix CR of each factor is less than 0.1, indicating good consistency. The check is passed.

The combined weights of the 16 indices are shown in Table 5.

In the form of a questionnaire to all teachers and students, a comprehensive evaluation of the innovation base construction mode was carried out. After expert review, different weights were given to experts, teachers, and students, as shown in Table 4.

Table 4. Weights of evaluation subjects.

Evaluation subject	Expert	Teacher	Student
Weight	0.40	0.35	0.25

10 experts, 20 teachers, and 20 students from the school were invited to give scores (0-100 points) for the 16 indices of Plan 1 and Plan 2 respectively. After processing and analyzing the scoring data, the scoring results of the 16 indices of the two plans were obtained, as shown in Table 4.

From the combined weight values in Table 4, it can be seen that experts generally believed that scientific research platforms, major projects, and construction costs were the most important considerations for the construction of an innovation base in the university. From the scores of the 16 indices of Plan 1 and Plan 2, it can be seen that the construction with self-raised funds received high scores in scientific research platforms, major projects, discipline construction, and ecological environment, and the construction by leasing in cooperation with government received high scores in the integration of production and education, construction costs, and resource sharing. These results were in line with the reality and were unanimously recognized by the review experts.

To sum up, the final score of Plan 1 was 85.4, and the final score of Plan 2 was 86.1. After a comprehensive analysis of the final results, the school decided to choose Plan 2 as the final plan.

Table 5. Comprehensive evaluation and analysis.

S	Node weight W_S	B	Node weight W_B	Combined weight W	Plan 1	Plan 2
S1	0.483	B1	0.078	0.038	75.6	78.3
		B2	0.491	0.237	92.7	83.4
		B3	0.306	0.148	88.4	84.8
		B4	0.125	0.060	82.1	89.2
S2	0.157	B5	0.586	0.092	93.5	87.7
		B6	0.124	0.019	93.2	94.3
		B7	0.217	0.034	82.3	83.6
		B8	0.073	0.011	87.4	84.3
S3	0.272	B9	0.550	0.150	72.8	94.6
		B10	0.077	0.021	75.7	78.5
		B11	0.252	0.069	82.3	81.7
		B12	0.121	0.033	89.6	88.5
S4	0.088	B13	0.473	0.042	84.4	87.5
		B14	0.170	0.015	75.9	87.4
		B15	0.284	0.025	83.2	79.8
		B16	0.073	0.006	86.8	80.2
Final score					85.4	86.1

6 CONCLUSION

In order to achieve high-quality development, colleges and universities should build innovation bases that aim at the future cutting-edge technologies, gather high-quality scientific and technological innovation elements, improve the school's modern governance capabilities, with an effort to release the school's teaching and research potential, build world-leading innovation platforms, gather first-class talents from all over the world, and promote the in-depth integration of disciplines. Based on the school's advantageous disciplines, the innovation base joins hands with industry backbone enterprises and top scientific research institutes, focuses on key core technologies, concentrates on the transformation of scientific and technological achievements, strengthens platform aggregation and integration, and cultivates top-notch innovative talents.

From the case analysis results of the university in Beijing, it can be seen that the comprehensive evaluation and decision-making model constructed in this paper is in line with reality and has certain reference value for colleges and universities to choose the construction mode of innovation base. It is helpful for colleges and universities to solve the problem of insufficient scientific research space with an ideal plan, promoting meaningful development of the school.

REFERENCES

Duan H R, Hu Y J (2008). Reflections on the Construction of Innovation Bases in Research Universities. J. Chinese and Foreign Entrepreneurs.8, 82–84.

Fa D H, Zhou J (2021). Application of Fuzzy Synthesis Method in Post-construction Evaluation of New Campuses in Colleges and Universities. J. Journal of Changsha University of Science and Technology (Social Science Edition). 27(1), 75–79. DOI: 10.16573/j.cnki.1672–934x.2012.01.024.

Su H H (2019). Research on Quality Evaluation of Post-earthquake Restoration and Reconstruction Projects Based on Matlab AHP .J. China Earthquake Engineering Journal. 41(03), 793–799.

Wang W (2006). Reflections on the Construction of Scientific and Technological Innovation Bases in Colleges and Universities. J. Science-Technology and Management. 3, 152–153+156. DOI:10.16315/j.stm.2006.03.049.

Yang Z Y, Qu J B, Huang C K (2005). Bridge Safety Assessment Based on Fuzzy Comprehensive Evaluation and AHP. J. Journal of Tianjin University.12, 1063–1067.

Advances in Urban Engineering and Management Science – Khalil & Yang (Eds)
© 2023 The Author(s), ISBN 978-1-032-30426-7

Suitability evaluation of sports tourism development in ethnic villages based on entropy weight matter-element extension mode

Sha Zhang & Lei Tang*
College of Physical Education and Health, Chuxiong Normal University, Chuxiong, Yunnan, China

ABSTRACT: This paper evaluates the suitability of the development of traditional ethnic sports tourism in village space and provides new idea for the development of sports tourism in village with ethnic characteristics. Based on the four levels—development basis, development potential, development conditions, and development benefits—the suitability evaluation system of sports tourism industry development in ethnic villages is constructed. Combining the entropy weight method and extension theory, the development suitability evaluation system of the entropy weight matter-element extension model was constructed, the suitability of the sports tourism development of a village with ethnic characteristics in Yunnan Province was evaluated, and the development suitability of each indicator was evaluated, thus verifying the scientific evidence and reasonable applications of the model.

1 INTRODUCTION

In September 2018, the Strategic Plan for Rural Revitalization (2018–2022) issued by the CPC Central Committee and The State Council was released. The plan points out that China's economic development has shifted from high-speed growth to high-quality development, and high-quality integrated rural development has become the connotation of rural development in the new era (Chen et al. 2017). To strengthen the construction of rural culture, intensive efforts to excavate the excellent resources contained in rural traditional culture, and cultivate and inspire villagers' consciousness and confidence in the culture of "agriculture, rural areas, and farmers" is undoubtedly an important prerequisite for realizing the strategic planning goal of rural revitalization. "Yunnan Beautiful livable Countryside Construction Action (2016-2020)" also clearly pointed out that the construction of beautiful livable countryside accelerate the development of rural tourism. In the "14th Five-year Plan for Sports Development" issued by the General Administration of Sport of China, it is emphasized that "new achievements have been made in the development of sports industry and new progress has been made in the construction of sports culture". These achievements need efforts by many parties, and the gathering of model power and the promotion of excellent projects with high quality, which will undoubtedly better drive the development of the industry (Pan & Fu 2007).

The author considers sports tourism development of ethnic villages in Yunnan Province as an example. The study intended to dig the new system of industrial development for ethnic villages to adapt to the new trend of rural development, to gather and integrate rural traditional culture, and let the ethnic villages of sports and tourism industries spawn a new industry system. Combined with entropy matter-element extension model under the village space to produce new system for national traditional sports tourism development suitability analysis, reasoning, the system of tourism and sports industry into a formal problem in the process of model development, demonstration analysis, as to provide a theoretical basis for ethnic village sports tourism development.

*Corresponding Author: tanglei@cxtc.edu.cn

 DOI 10.1201/9781003305026-16

2 THE SUITABILITY EVALUATION SYSTEM OF SPORTS TOURISM DEVELOPMENT IN ETHNIC VILLAGES

To establish a scientific and systematic ethnic village suitability evaluation index system, the development of sports tourism suitability evaluation based on tourism development, ethnic village tourism development of data access, preliminary build multiple factors after the sports tourism development of suitability evaluation index, index selection after three rounds of experts and correction are required. Finally, four first-level indicators are constructed based on development basis, development potential, development conditions, and development benefits, and the four first-level indicators are divided in detail to summarize 15 second-level indicators for the suitability evaluation system of sports tourism development in ethnic villages, as shown in Table 1.

Table 1. Suitability evaluation system of sports tourism development in ethnic villages.

Target	Level Indicators	Secondary Indicators
Suitability evaluation system of sports tourism development in ethnic villages(C)	Development Infrastructure(C_1)	National sports resource reserve degree C_{11}
		Diversity of ethnic sports resources C_{12}
		Value of national sports culture C_{13}
	Development Potential(C_2)	Sports tourism consumption level C_{21}
		Sports tourism market development potential C_{22}
		Attraction of tourist landscape C_{23}
		The government's support for sports tourism development C_{24}
	Development Condition(C_3)	Residents' recognition C_{31}
		National sports heritage C_{32}
		Construction of supporting Settings C_{33}
		Strength of sports tourism industry C_{34}
		Sports tourism employment rate C_{35}
	Development BenefitsC_4	Economic benefits C_{41}
		Social benefits C_{42}
		Inheritance and protection of ethnic sports C_{43}

Development is based on the basic conditions required for sports tourism. Ethnic villages contain many resources, which ensure the important basis for development of national traditional sports tourism, including the national sports resources reserves, national traditional sports resources diversity, national sports resources value, etc. We determine the value of traditional ethnic sports tourism development value and benefits and develop the prerequisites for optimum display of suitability (Yin et al. 2017). Development of potential is how much value exists under the guarantee of basic conditions, and is an important guarantee for the development of the sports tourism industry. Including sports tourism consumption level, sports tourism market development potential degree, tourism landscape attraction, government support for sports tourism development, and other factors, to combine sports and tourism, we need to explore the connection between the two and highlight the important value of sports tourism (Wu et al. 2017). Development conditions are based on foundation and potential, combined with the actual problems for benefits, solve the important problems in the development process of traditional national sports tourism, and transform the theory into practical problems in reality, including The decisive factors are the recognition degree of residents, the scale of national sports inheritance, the construction of supporting facilities, the strength of sports tourism industry institutions, and the employment rate of sports tourism (Liu 2016). Development benefit is the embodiment of the actual value of sports tourism development in ethnic villages under the background of rural revitalization, and the embodiment of industrial value is an important premise of sports tourism development. It includes economic benefits, social benefits, inheritance, and protection of national sports (Li 2020; Liu 2020).

3 CONSTRUCTION OF ENTROPY WEIGHT MATTER-ELEMENT EXTENSION MODEL FOR THE SUITABILITY EVALUATION OF SPORTS TOURISM DEVELOPMENT IN ETHNIC VILLAGES

The paper adopts the extension evaluation method to evaluate the suitability of sports tourism development, obtains the correlation degree of higher index factors through the coincidence operation of single factor correlation degree and its weight, and finally determines the comprehensive multi-level extension evaluation results of the suitability of sports tourism development of ethnic characteristic villages objectively according to the principle of maximum membership degree.

3.1 *Determine the matter element to be evaluated*

There are "n" indexes in the evaluation system of the suitability of sports tourism development, which are respectively $C1$, $C2$, $C3 \ldots Cn$, according to the matter-element theory, the evaluation matter-element model is established as follows:

$$R=(N, C_m, v_m)=\begin{bmatrix} N & C_1 & v_1 \\ & C_2 & v_2 \\ & \vdots & \vdots \\ & C_n & v_n \end{bmatrix} \tag{1}$$

Where, N represents the evaluation level of development suitability evaluation, and V_m represents the amount of N on evaluation index C_m, that is, the specific data collected by each index.

3.2 *Identify the classical and nodal domains*

According to the extension matter-element theory, the evaluation classical domain and nodal domain model are constructed as follows:

$$R_{0i}=(P_{0i}, C, V_i)=\begin{bmatrix} P_{0i} & C_1 & v_{i1} \\ & C_2 & v_{i2} \\ & \vdots & \vdots \\ & C_n & v_{in} \end{bmatrix}=\begin{bmatrix} P_{0i} & C_1 & \langle a_{0i1}, b_{0i1} \rangle \\ & C_2 & \langle a_{0i2}, b_{0i2} \rangle \\ & \vdots & \vdots \\ & C_n & \langle a_{0in}, b_{0in} \rangle \end{bmatrix} \tag{2}$$

The classical domain was determined based on the model, and the evaluation indexes were divided into i grades ($i=1,2\ldots m$), where R_i represents the matter-element model of the suitability evaluation index of sports tourism development of the ith evaluation level, Pi represents the suitability evaluation effect of the ith evaluation level, and Vi is the range of magnitude specified by P_i about C, namely, the data range of the corresponding evaluation index of each level of suitability evaluation of development.

According to the section domain definition, each indicator section domain indicates the allowed value range of each indicator.

$$R_{\mathrm{q}}=(P_q, C, V_q)=\begin{bmatrix} P_q & C_1 & v_{q1} \\ & C_2 & v_{q2} \\ & \vdots & \vdots \\ & C_n & v_{qn} \end{bmatrix}=\begin{bmatrix} P_q & C_1 & \langle a_{q1}, b_{q1} \rangle \\ & C_2 & \langle a_{q2}, b_{q2} \rangle \\ & \vdots & \vdots \\ & C_n & \langle a_{qn}, b_{qn} \rangle \end{bmatrix} \tag{3}$$

Wherein, R_q represents the node domain of the evaluation matter-element model, P_q represents all index levels, $V_q=(a_{qn}, b_{qn})$ represents the allowable range of index C value in P_q.

3.3 *The dimensionless treatment of each index*

Because the statistical data of each indicator are not consistent in units, to facilitate data analysis, it is necessary to normalize each indicator to unify dimensions and facilitate calculation. The smaller the value is, the better it is, and the larger the value is, the normalized processing method for such cases is as follows:

$$C_i^* = \frac{C_i - \min C_i}{\max C_i - \min C_i} \tag{4}$$

In the formula, maxCi and minCi are upper limit and lower limit, respectively, of the corresponding value of each index I.

3.4 *Determine the correlation function value of each evaluation index*

After determining the matter-element to be evaluated for the suitability evaluation of sports tourism development, the correlation function of the index to be evaluated on the evaluation level Ri was established by using the correlation function and calculated as follows:

$$K_n(v_i) = \begin{cases} -\dfrac{\rho(v_m, V_i)}{|V_i|} & v_m \in V_i \\ \dfrac{\rho(v_m, V_i)}{\rho(v_m, V_p) - \rho(v_m, V_i)} & v_m \notin V_i \end{cases} \tag{5}$$

In the formula:

$$\rho(v_m, V_i) = \left| v_m - \frac{a_{0in} + b_{0in}}{2} \right| - \frac{b_{0in} - a_{0in}}{2} \tag{6}$$

$$\rho(v_m, V_p) = \left| v_m - \frac{a_{pn} + b_{pn}}{2} \right| - \frac{b_{pn} - a_{pn}}{2} \tag{7}$$

3.5 *Entropy weight method is used to determine the weight of the evaluation index*

In information theory, entropy is a measure of uncertainty. The greater the amount of information, the smaller the uncertainty, and the smaller the entropy, the smaller the amount of information, the greater the uncertainty, and the greater the entropy (Shao 2008). The specific steps are as follows:

The first step is to build a normalized matrix: $R = (r_{ij})_{m*n}$

The second step is to define the entropy of the evaluation index: $H_i = -k\sum_{j=1}^{m} f_{ji} \ln f_{ji}$

In the formula:

$$k = \frac{1}{\ln(m)} \tag{8}$$

$$f_{ij} = \frac{r_{ji}}{\sum_{j=1}^{m} r_{ji}} \tag{9}$$

The third step is to determine the weight of entropy to evaluate each index:

$$w_i = \frac{(1 - H_i)}{n - \sum_{i=1}^{n} H_i} \tag{10}$$

3.6 *The multistage extension comprehensive correlation degree of the matter element to be evaluated was calculated*

Combined with the weight coefficient *Wn* of the evaluation index and the correlation function value of each evaluation index, the correlation degree of the secondary evaluation index and the whole evaluation with each evaluation grade in the matter-element index system to be evaluated was calculated, respectively.

$$K_n(p) = \sum_{n=1}^{m} w_n K_n(v_i) \tag{11}$$

3.7 *Matter element rating to be evaluated*

Using the principle of maximum membership degree, the comprehensive correlation degree of each grade is compared to evaluate the grade. The larger the grade correlation degree is, it indicates that the degree of conformity of this grade set of development suitability evaluation is the highest, that is, the development suitability evaluation grade is this grade.

$$K_j(p) = \max_{1 \leq j \leq n} K_n(p) \quad (j = 1, 2, \ldots, n) \tag{12}$$

To further accurately reflect the trend degree of the evaluation grade, the characteristic value $J*$ of the grade variable to be evaluated is calculated by the following formula:

$$J^* = \frac{\sum_{\mathrm{j}=1}^{m} \mathrm{j} \cdot \overline{K}_{\mathrm{j}}(p)}{\sum_{\mathrm{j}=1}^{m} \overline{K}_{\mathrm{j}}(p)} \tag{13}$$

In the formula:

$$\overline{K}_j(p) = \frac{k_{\mathrm{j}}(p) - \min_j K_j(p)}{\max_j K_j(p) - \min_j K_j(p)} \tag{14}$$

4 THE EMPIRICAL RESEARCH

Taking a village with ethnic characteristics in Yunnan Province as an example, the villages selected are the five groups of villages with ethnic characteristics published by the state. The development suitability evaluation index system and entropy weight matter-element extension model is used to evaluate the development suitability of ethnic characteristic village sports tourism development.

4.1 *Grading of evaluation*

According to the established evaluation index system shown in Table 1, combined with the relevant requirements, the development level of suitability evaluation on the development suitability evaluation hierarchy is divided into four, namely I (very suitable), II (suitable), III (relatively suitable), IV (inappropriate), because each index in the raw data of the unit is different. Before the evaluation, the evaluation grade should be normalized. After the treatment according to formula (4), the results are I (very suitable) [0.9, 1], II (suitable) [0.8, 0.9), III (relatively suitable) [0.6, 0.8), and IV (unsuitable) [0,0.6).

4.2 *The calculation of the correlation degree of the matter element to be evaluated and the weight of the evaluation index*

If the statistical data of dimension is not consistent, the original data should be normalized for processing using the formula (4), and then according to the article set up the suitability evaluation of the sports tourism development of various indicators of evaluation grades of correlation functions, using the formula (5), (6), (7), (8), (9), and (10) is calculated for each evaluation index weight value and the relative membership degree of evaluation grades. Data is presented in Table 2.

Table 2. The correlation degree and weight calculation results of each evaluation index and evaluation grade.

Level Indicators	Secondary Indicators	Weight	Correlation			
			I	II	III	IV
C_1	C_{11}	0.1524	−0.1545	0.0144	−0.0004	−0.2174
	C_{12}	0.0553	−0.2476	0.0424	−0.0487	−0.2485
	C_{13}	0.1208	−0.2687	0.0843	−0.0824	−0.3114
C_2	C_{21}	0.0345	−0.3358	−0.1432	−0.0042	−0.4009
	C_{22}	0.0848	−0.1457	0.0687	−0.0147	−0.2119
	C_{23}	0.0515	−0.2148	−0.1882	0.0098	−0.3250
	C_{24}	0.0637	−0.3171	0.0957	−0.0630	−0.2972
C_3	C_{31}	0.0741	−0.1489	−0.0097	0.0345	−0.2457
	C_{32}	0.0481	−0.4219	0.0747	0.0002	−0.0735
	C_{33}	0.0582	−0.3741	−0.2481	0.0048	−0.2527
	C_{34}	0.0521	−0.3547	−0.2875	−0.0455	0.0571
	C_{35}	0.0324	−0.3111	−0.1884	−0.0447	0.0488
C_4	C_{41}	0.0518	−0.1884	0.0148	−0.0184	−0.2412
	C_{42}	0.0648	−0.1125	0.0487	−0.0484	−0.2150
	C_{43}	0.0555	−0.3484	0.0611	−0.0312	−0.4188

It can be concluded from Table 2 that the weight values of the four first-level indicators are $C1 = 0.3285$, $C2 = 0.2345$, $C3 = 0.2649$, and $C4 = 0.1721$. Among them, the weight of development basis ($C1$) is the largest $w = 0.3285$, followed by the weight value of development condition ($C3$) $W = 0.2649$, indicating that development basis is the most important factor in the development of sports tourism in ethnic villages of Yunnan Province. In the development of sports tourism, the local resource basis should be fully considered first, followed by development conditions.

4.3 *Comprehensive extension evaluation*

According to the above formula (11), the correlation degree between the first-level index of the suitability evaluation index for sports tourism development of ethnic villages in Yunnan Province and each evaluation grade can be calculated, and the evaluation grade of each first-level index can be determined according to the principle of maximum membership degree. Data is presented in Table 3.

It can be seen from Table 3 that C1 and C4 of the four first-level evaluation indicators are rated as "II", that is, the foundation and benefit of sports tourism development in this ethnic characteristic village reach the "suitable" level. In the first-level guidance, the evaluation grade of C2 and C3 is "III", that is, the development potential and development conditions only reach "relatively suitable" grade.

According to the calculation results from the comprehensive extension evaluation, the evaluation level of the suitability of sports tourism development of this ethnic characteristic village is "III", which is "relatively suitable" for development. To accurately reflect the trend of the evaluation level, Formulae (13) and (14) are used to calculate the characteristic values of the evaluation

Table 3. Results of comprehensive extension evaluation on the suitability of sports tourism development in ethnic villages in Yunnan Province.

	Correlation					
Indicators	I	II	III	IV	Weight	Rating
C_1	−0.0697	0.0022	−0.0001	−0.0331	0.3285	II
C_2	−0.0552	−0.0049	−0.0001	−0.0138	0.2345	III
C_3	−0.0817	−0.0007	0.0026	−0.0182	0.2649	III
C_4	−0.0364	0.0008	−0.0010	−0.0125	0.1721	II
C	−0.0637	−0.0005	0.0005	−0.0211		III

grade variables of the suitability of sports tourism development in the ethnic characteristic villages. Through calculation, $J^* = 2.8755$, indicating that the suitability evaluation level of sports tourism development of the ethnic characteristic's village is between "suitable" and "relatively suitable", mainly biased to "relatively suitable", there is still a certain gap from the "suitable" standard. Combined with the first-level indicators, the main reason is the restriction of development potential and development conditions. Through further in-depth analysis, it is found that the evaluation of the two second-level indicators of "institutional strength of sports tourism industry" and "sports tourism employment rate" in the suitability evaluation of sports tourism development in the ethnic characteristics village is "inappropriate". It shows that sports tourism in this region is underdeveloped, and there is a gap in the development of sports tourism resources, the establishment of institutions, and the employment rate of sports tourism promotion.

5 CONCLUSION

The research combined tourism destination development suitability and the characteristics of national sports tourism covering the four aspects, namely development foundation, development potential, development condition, and development benefit for sports tourism development suitability evaluation index system in the ethnic village in Yunnan Province. It used the entropy weight method and the extension theory to build a common development suitability evaluation of the entropy matter-element extension model. The entropy method is applied to assess the weight of each index value, making the results more objective, using a multilevel extension assessment scheme for the specific case calculation. Thus, it can be concluded that the sports tourism development of ethnic village suitability grades, at the same time, from the secondary indicators and the correlation of evaluation grades, can accurately reflect the level of the evaluation index and provide some help to developers and managers. In addition, the use of variable characteristic value $J^* = 2.8755$ is an accurate reflection of sports tourism development in the region suitability degree of bias, between "appropriate" and "suitable", mainly to "suitable", the results accord with local reality, to validate the use of entropy matter-element extension model of sports tourism in ethnic villages development suitability evaluation of reliability.

ACKNOWLEDGMENT

Thanks to Lei Tang for his hard work in field research and professor Liaokun Ye for his technical guidance in model construction. This work was supported by the Scientific Research Foundation of Yunnan Education Department (No. 2022J0821; No. 2022J0836). The authors state that they have no competing economic interests.

REFERENCES

Chen W, Liu Q, Wang C. (2017) Research on innovation and Development of national sports. Xidian University Press, Xi'an.

Li J R, Chen Y X. (2020) Strategies to Innovate and promote sports tourism industry in Ethnic areas – Based on the investigation and thinking of Enshi Tujia and Miao Autonomous Prefecture.Journal of South-Central University for Nationalities (Humanities and Social Sciences),40:140–144.

Liu G. (2016) Study on rural tourism development. Ocean University of China Press, Qing Dao.

Liu J H. (2020) Evaluation of tourism development potential of ethnic traditional villages. Master thesis of Guangxi Normal University, Guang Xi.

Pan L Y, Fu J X. (2007) Cultural tourism. Zhejiang University Press, Zhe Jiang.

Shao H. (2008) System safety Engineering.Petroleum Industry Press, Bei Jing.

Wu Z J, Dai M, Wu S R. (2017) Cultural change and spatial reconstruction in ethnic minority villages: A case study of the planning and designing of Pingdeng Dong village, a characteristic tourist town. Guangxi Ethnic Studies,135:133–140.

Yin H G, Yao Y G, Xiong L Y. (2017) Research on the integration development of tourism industry and cultural Industry.China Book Publishing House, Bei Jing.

Advances in Urban Engineering and Management Science – Khalil & Yang (Eds)
© 2023 The Author(s), ISBN 978-1-032-30426-7

Research on comprehensive evaluation system of new campus pre-planning work of colleges and universities based on the analytic hierarchy process

Qianshan Wen*, Jiangtao Liu & Xiaona Han
Beijing Jiaotong University New Campus Construction Office, Beijing, China

ABSTRACT: This paper analyzes the content and main features of the pre-planning of the new campus of colleges and universities, sums up 20 work plans, and builds a comprehensive evaluation system for the pre-planning of the new campus in colleges and universities. The evaluation system uses the Analytic Hierarchy Process (AHP) to establish a pre-planning hierarchy model, determine the weight of each work, and obtain the evaluation membership matrix by means of questionnaires. Finally, the final evaluation result is obtained through the matrix operation. This paper takes a new campus of a university in Beijing as the research object and conducts a comprehensive evaluation of its pre-planning work. The resulting conclusions are consistent with the satisfaction of teachers and students, which verifies the reliability of the evaluation system.

1 INTRODUCTION

Due to the early construction time of most domestic colleges and universities, there are many restrictive factors such as constrained school space, backward campus planning, and policy restrictions (Bai 2019). According to the data, there are more than four campuses on average in the national first-class universities already in operation, and the construction of new campuses has become a major strategic decision for colleges and universities to seek development (Wu et al. 2020). At the same time, the planning and construction period of the new campus is relatively short, generally between 3–5 years (Deng 2021), resulting in the lack of a scientific and systematic overall plan for the construction of the new campus.

Good planning and construction work of the new campus must be done through scientific forecasting and planning, to achieve the development goals of the school in a step-by-step and phased manner under the guidance and regulation of the preliminary work. The pre-planning of the new campus of colleges and universities is the key prerequisite for the overall planning and construction of the new campus, serving as an important foundation for the construction and operation of the campus, which requires unified planning, integrated organization, in-depth investigation, and full demonstration.

2 ORGANIZATIONAL PROCESS OF PLANNING AND CONSTRUCTION OF THE NEW CAMPUSES OF COLLEGES AND UNIVERSITIES

The construction of the new campus of colleges and universities is a complex system project. According to the planning and construction sequence and process, it can be divided into four phases: planning phase, overall design phase, monomer design phase, and engineering construction phase; according to the planning and construction conditions, it can be divided into four conditions: External conditions, internal conditions, and support conditions. The specific process is shown in Figure 1.

*Corresponding Author: qswen@bjtu.edu.cn

 DOI 10.1201/9781003305026-17

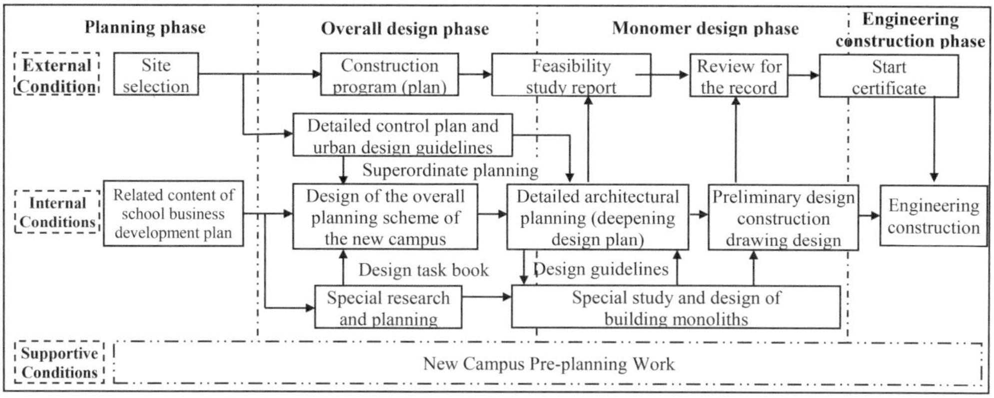

Figure 1. Organizational flow chart of the planning and construction of the new campus.

2.1 *Planning phase*

This phase mainly solves major strategic issues such as the functional positioning of the new campus, subject layout, scale of teachers and students, housing needs and planning concepts, clarifies the investment scale, technical requirements, and phased construction of the new campus, and accurately compiles the design task book, providing design basis and technical standards for the overall design phase.

2.2 *Overall design phase*

As the overall design phase is mainly oriented for specific plots, the upper planning conditions and the requirements of the design task book are the basis to layout various buildings on the campus, allocate building area, coordinate building style, design campus public space, organize road traffic, and plan various types of pipelines, etc., including campus culture and architectural style, ecological environment and landscape features, energy conservation, and emission reduction, green and low carbon, smart campus, municipal infrastructure, etc.

Through the overall design, the results formed in the planning phase are implemented into the spatial layout, to solve the three-dimensional architectural and landscape layout problems, rationally allocate space, improve the efficiency of space use, shape a beautiful campus environment, and form campus planning and design guidelines, providing guidance for the design of architectural monoliths.

2.3 *Monomer design phase*

In the design phase of a monomer project, in accordance with the design guidelines formed in the overall design phase, the monomer design work is carried out and the monomer building feasibility study report is prepared. Part of the design work can be carried out in parallel with the dislocation of the overall design phase. The architectural plan design of key monoliths should start the planning and design work in the overall design phase.

2.4 *Project construction phase*

In accordance with the requirements of the pre-planning work, the construction of the new campus will be started when the construction permit is obtained and the conditions for starting work are met.

3 THE PRE-PLANNING WORK FOR THE CONSTRUCTION OF THE NEW CAMPUS OF COLLEGES AND UNIVERSITIES

According to the organization process of university planning and construction, for the four different phases, the pre-planning work of the new campus construction of the university can be divided into four parts: top-level design, overall planning, key monomers, and construction management.

3.1 *Top-level design*

1. Research on functional positioning and disciplinary layout: Functional positioning is the primary issue in the construction of new campuses. It is necessary to clarify the development positioning, school mode, disciplinary layout, and functional requirements of the new campus through various ways such as sorting out the current situation of the school, investigating the experience of domestic and foreign universities, and analyzing the territorial positioning of the new campus, so as to give full play to the overall advantages of one university and multiple campuses.
2. Research on student programs: It is needed to sort out the scale and composition of students in the existing campus, analyze the future social talent demand and industry development direction, combine with the functional positioning of the new campus, and clarify the scale and composition of students in the new campus, enrollment policies, training plans, and other content.
3. Research on faculty and staff program and related policies: Efforts are made to sort out the scale and composition of faculty and staff in existing campuses, analyze relevant national policies, and clarify the scale and composition of faculty and staff in the new campus, job requirements, staff sources, incentive policies, and guarantee mechanisms.
4. Research on the demand for various types of housing: It is required to sort out the existing school conditions, analyze the school's development needs, clarify the construction scale, spatial layout, site selection requirements and operating conditions of basic school buildings, research and industrial buildings and supporting buildings, thus formulating the principles for the allocation and use of various types of housing on the new campus.
5. Research on the concept of campus planning: Research on the new model of future university space is needed to combine the classic planning cases of innovative campuses at home and abroad. Thus, innovative campus planning concepts adapted to the new campus can be proposed, forming pre-planning principles and suggestions such as overall imagery, functional zoning, spatial structure, ecological landscape, campus transportation, and green low-carbon.

3.2 *Overall plan*

1. Research on campus culture and architectural style: Through research on the campus history and culture, colors, and architectural style of the new campus, the focus will be placed on analyzing the inheritance and development of campus culture and solving the problems such as the formation of a cultural fault line and lack of cultural accumulation in the new campus.
2. Research on ecological environment and landscape style: Special research is conducted on campus ecological environment, the landscape style, and open space of the new campus to build technical guidelines for ecological landscape campus and provide technical support for ecological campus construction.
3. Research on energy conservation, emission reduction, and green low-carbon: Special research is carried out on energy conservation, emission reduction, and green low-carbon for the new campus, to build technical guidelines for green low-carbon campus and provide technical support for green campus construction.
4. Research on smart campus: With the new generation of information technology as the core, special research is performed on the smart campus information system, intelligent equipment and facilities, and smart infrastructure environment of the new campus to provide technical support for the construction of the smart campus.

5. Research on municipal infrastructure: Research and analysis are denoted on the policies and conditions of municipal infrastructure construction in the new campus, to investigate and clarify the layout and planning requirements of municipal facilities such as roads, security, fire-fighting, water, electricity, gas, heating, network communication, and garbage disposal in the new campus, and to propose the construction plan of the comprehensive pipeline corridor.

3.3 *Key monomers*

Special research is dedicated to the key monolithic construction projects of the new campus, so as to investigate and prove the building functions, space layout, equipment, and facilities, to fully consider the future development needs. In this regard, forward-looking design requirements are proposed to provide technical support for the monolithic design, feasibility study submission, and subsequent expansion of preliminary design and construction drawing design. The main key monoliths include the library, data and computing center, integrated gymnasium, and special studies of key research platforms and facilities.

3.4 *Construction management*

1. Research on institutions and mechanisms: Combining with the functional positioning of the new campus, research is conducted on the construction mode and management institutions of the new campus, and schemes of institutions and mechanisms in line with the development of the new campus are proposed.
2. Research on construction bidding and procurement: The characteristics of engineering management in new campus construction projects are analyzed, and a comprehensive, efficient, and low-risk management system is explored to meet the needs of new campus planning and construction.
3. Research on fund raising and management: It is necessary to study the relevant national policies, clarify the fund requirements for the construction of the new campus, sort out the financial situation of the school, propose a practical fundraising plan, and formulate a safe and effective management system to provide guarantee for the smooth progress of the new campus.
4. Research on the whole process of tracking audit: Commitment is made to analyze the characteristics of construction engineering design supervision mode, summarize and generalize the past experience of construction engineering audit, and establish the whole process of design mechanism for the construction management of the new campus.
5. Research on construction supervision and management: Combining with the construction mode of the new campus, institutions and mechanism of supervision and management for the construction of the new campus will be established to clarify the supervision mode and content to ensure that the construction work of the new campus is carried out in an orderly manner.

4 COMPREHENSIVE EVALUATION MODEL FOR THE PRE-PLANNING OF THE NEW CAMPUS

The analytic hierarchy process has strong practicability and effectiveness for multi-objective, multi-criteria, and multi-level complex decision-making problems. It is a combination of qualitative and quantitative analysis methods (Lv & Zhang 2011), without considering the influence of people's subjective judgments, preferences, and other factors (Li & Liu 2009). This paper classifies and groups the pre-construction planning work of the new campus in colleges and universities, establishes a hierarchical structure model using hierarchical analysis, and makes comparisons to construct a pairwise comparison matrix, thereby solving the weight of the judgment matrix. This paper analyzes the pre-planning work of a new campus of a university in Beijing as an example.

4.1 *Build a hierarchical model*

According to the pre-planning work plan for the construction of the new campus of the university, a hierarchical model of the pre-planning work of the new campus can be constructed, as shown in Table 1.

Table 1. Hierarchical structure model of the pre-planning of the new campus.

A: Pre-construction planning for the new campus	B1: Top-level design	C1: Research on functional positioning and disciplinary layout C2: Research on student programs C3: Research on faculty and staff programs and related policies C4: Research on the demand for various types of housing C5: Research on the concept of campus planning
	B2: Overall planning	C6: Research on campus culture and architectural style C7: Research on ecological environment and landscape style C8: Research on energy conservation, emission reduction, and green low-carbon C9: Research on smart campus C10: Research on municipal infrastructure
	B3: Key monomers	C11: Special research on library C12: Special research on data and computing center C13: Special research on museums C14: Special research on the comprehensive gymnasium C15: Research on key research platforms and facilities
	B4: Construction management	C16: Research on institutions and mechanisms C17: Research on construction bidding and procurement C18: Research on fund raising, planning, and management C19: Research on whole process tracking audit C20: Research on construction supervision and management

4.2 *Construct a pairwise comparison matrix*

The set of factors $U = \{u_1, u_2, \ldots, u_n\}$ was set up, and the importance of any two factors in U was compared to form a pairwise comparison matrix according to Table 2 using the 1 to 9 scale value method proposed by T.L. Saaty.

Table 2. Scale value.

Importance relationship	u_{ij}	u_{ji}
Factor u_i is equally important as factor u_j	1	1
Factor u_i is slightly more important than factor u_j	3	1/3
Factor u_i is more important than factor u_j	5	1/5
Factor u_i is far more important than factor u_j	7	1/7
Factor u_i is extremely more important than factor u_j	9	1/9
When $u_{ij} = 2, 4, 6, 8$, it means that the importance is between the levels		

Twenty experts were invited from each college of the university and related departments for comparison, and all experts' opinions were combined to derive the final judgment matrix for factor A as follows:

$$A = \begin{bmatrix} 1 & 3 & 7 & 5 \\ 1/3 & 1 & 3 & 2 \\ 1/7 & 1/3 & 1 & 1/2 \\ 1/5 & 1/2 & 2 & 1 \end{bmatrix}$$

Similarly, the judgment matrix of factors B1, B2, B2, and B4 are determined as:

$$B_1=\begin{bmatrix}1 & 5 & 7 & 3 & 2\\ 1/5 & 1 & 2 & 1/2 & 1/3\\ 1/7 & 1/2 & 1 & 1/3 & 1/4\\ 1/3 & 2 & 3 & 1 & 1/2\\ 1/2 & 3 & 4 & 2 & 1\end{bmatrix} B_2=\begin{bmatrix}1 & 4 & 3 & 2 & 1/2\\ 1/4 & 1 & 1/2 & 1/3 & 1/7\\ 1/3 & 2 & 1 & 1/2 & 1/5\\ 1/2 & 3 & 2 \quad 1 & 1 & /4\\ 2 & 7 & 5 & 4 & 1\end{bmatrix}$$

$$B_3=\begin{bmatrix}1 & 3 & 7 & 6 & 2\\ 1/3 & 1 & 3 & 2 & 1/2\\ 1/7 & 1/3 & 1 & 1/2 & 1/4\\ 1/6 & 1/2 & 2 & 1 & 1/3\\ 1/2 & 2 & 4 & 3 & 1\end{bmatrix} B_4=\begin{bmatrix}1 & 3 & 2 & 7 & 4\\ 1/3 & 1 & 1/2 & 3 & 2\\ 1/2 & 2 & 1 & 4 & 3\\ 1/7 & 1/3 & 1/4 & 1 & 1/2\\ 1/4 & 1/2 & 1/3 & 2 & 1\end{bmatrix}$$

4.3 *Calculate the weight vector and test the consistency*

The consistency test formula is:

$$\mathrm{CR}=\frac{CI}{RI}\leq 0.1$$

Where $\mathrm{CI}=\frac{\lambda_{max}-n}{n-1}$, RI is the average random consistency index, as shown in Table 3:

Table 3. Random consistency indicators.

n	1	2	3	4	5	6	7	8	9
λ	0	0	0.58	0.90	1.12	1.24	1.32	1.41	1.45

Matlab calculation software is used to calculate the judgment matrix A, and the maximum eigenvalue and corresponding eigenvector of A are obtained as:

$$\lambda_{max}=4.019$$

$$\omega=\begin{bmatrix}-0.914 & -0.339 & -0.112 & -0.191\end{bmatrix}$$

The weight vector of the judgment matrix A is obtained after normalizing ϖ:

$$\mathrm{W}_A=\begin{bmatrix}0.587 & 0.218 & 0.072 & 0.123\end{bmatrix}$$

It is calculated that CI = 0.006 and CR = 0.0067 $\leq$ 0.1, which pass the consistency test.

In the same way, the judgment matrices B1, B2, B2, and B4 are calculated, and the results are shown in Table 4. It can be seen from the table that the judgment matrices B1, B2, B2, and B4 have a good consistency.

Table 4. Judgment matrix analysis table.

	Weight vector	λ_{max}	CR
W_1	0.447 0.090 0.056 0.154 0.253	5.034	0.008
W_2	0.247 0.055 0.088 0.143 0.467	5.048	0.011
W_3	0.458 0.152 0.054 0.086 0.250	5.038	0.008
W_4	0.435 0.157 0.257 0.056 0.095	5.04	0.009

From the power vector WA of the judgment matrix A, it is clear that for the new campus planning and construction of colleges and universities, the most important is the study of top-level design, followed by the study of overall planning and construction management, and finally the study of individual design.

From the weight vectors W1, W2, W3, and W4, it can be seen that for the four factors of top-level design, overall planning, key monoliths, and construction management, the largest weights are placed for the research on functional positioning and discipline layout, the research on municipal infrastructure, the special research on libraries, and the research on institutions and mechanisms, and the smallest weights are placed for the research on faculty and staff programs and related policies, the research on ecological environment and landscape style, the special research on museums, and the research on whole process tracking audit.

4.4 *Establish fuzzy relationship matrix*

The evaluation set $P = [p_1 \quad p_2 \quad p_3 \quad p_4 \quad p_5]$, corresponding to excellent, good, moderate, qualified, and unqualified respectively. In this paper, a questionnaire was used to evaluate the pre-planning work of the new campus for 100 teachers and students (including 20 full-time teachers, 20 administrative posts, 20 logistics service posts, 20 graduate students, and 20 undergraduate students), and the degree of affiliation of each factor to the evaluation set was determined, so as to establish the fuzzy relationship matrix R. The survey results are as follows.

Table 5. Questionnaire results.

No.	Pre-planning work	Excellent	Good	Moderate	Qualified	Unqualified
1	Research on functional positioning and disciplinary layout	22	39	24	15	0
2	Research on student programs	32	29	25	14	0
3	Research on faculty and staff programs and related policies	39	26	18	17	0
4	Research on thc dcmand for various types of housing	42	33	12	13	0
5	Research on the concept of campus planning	33	30	21	16	0
6	Research on campus culture and architectural style	18	38	25	19	0
7	Research on ecological environment and landscape style	25	21	33	21	0
8	Research on energy conservation, emission reduction, and green low-carbon	15	24	42	19	0
9	Research on smart campus	20	23	35	22	0
10	Research on municipal infrastructure	24	42	23	11	0
11	Special research on libraries	37	30	19	14	0
12	Special research on data and computing center	19	23	38	20	0
13	Special research on museums	28	32	27	13	0
14	Special research on the comprehensive gymnasium	23	36	21	20	0
15	Research on key research platforms and facilities	16	25	45	14	0
16	Research on institutions and mechanisms	28	31	25	16	0
17	Research on construction bidding and procurement	26	27	24	23	0
18	Research on fund raising and management	19	26	38	17	0
19	Research on whole process tracking audit	23	33	23	21	0
20	Research on construction supervision and management	25	29	27	19	0

The membership matrix of 20 factors to the evaluation set was obtained by collating statistics on the research data.

$$R_1 = \begin{bmatrix} 0.22 & 0.39 & 0.24 & 0.15 & 0 \\ 0.32 & 0.29 & 0.25 & 0.14 & 0 \\ 0.39 & 0.26 & 0.18 & 0.17 & 0 \\ 0.42 & 0.33 & 0.12 & 0.13 & 0 \\ 0.33 & 0.30 & 0.21 & 0.16 & 0 \end{bmatrix} \mathrm{R}_2 = \begin{bmatrix} 0.18 & 0.38 & 0.25 & 0.19 & 0 \\ 0.25 & 0.21 & 0.33 & 0.21 & 0 \\ 0.15 & 0.24 & 0.42 & 0.19 & 0 \\ 0.20 & 0.23 & 0.35 & 0.22 & 0 \\ 0.24 & 0.42 & 0.23 & 0.11 & 0 \end{bmatrix}$$

$$\mathrm{R}_3 = \begin{bmatrix} 0.37 & 0.30 & 0.19 & 0.14 & 0 \\ 0.19 & 0.23 & 0.38 & 0.20 & 0 \\ 0.28 & 0.32 & 0.27 & 0.13 & 0 \\ 0.23 & 0.39 & 0.21 & 0.20 & 0 \\ 0.16 & 0.25 & 0.45 & 0.14 & 0 \end{bmatrix} \mathrm{R}_4 = \begin{bmatrix} 0.28 & 0.31 & 0.25 & 0.16 & 0 \\ 0.26 & 0.27 & 0.24 & 0.23 & 0 \\ 0.19 & 0.26 & 0.38 & 0.17 & 0 \\ 0.23 & 0.33 & 0.23 & 0.21 & 0 \\ 0.25 & 0.29 & 0.27 & 0.19 & 0 \end{bmatrix}$$

A fuzzy comprehensive evaluation is performed on the membership matrix, and the comprehensive evaluation membership vector is obtained by calculation:

$$\mathrm{B}_1 = \mathrm{W}_1 * \mathrm{R}_1 = \begin{bmatrix} 0.297 & 0.342 & 0.211 & 0.150 & 0 \end{bmatrix}$$

$$\mathrm{B}_2 = \mathrm{W}_2 * \mathrm{R}_2 = \begin{bmatrix} 0.212 & 0.356 & 0.274 & 0.158 & 0 \end{bmatrix}$$

$$\mathrm{B}_3 = \mathrm{W}_3 * \mathrm{R}_3 = \begin{bmatrix} 0.273 & 0.286 & 0.290 & 0.154 & 0 \end{bmatrix}$$

$$\mathrm{B}_4 = \mathrm{W}_4 * \mathrm{R}_4 = \begin{bmatrix} 0.248 & 0.290 & 0.283 & 0.179 & 0 \end{bmatrix}$$

To this end, the membership matrix of factor A can be obtained:

$$\mathrm{R} = \begin{bmatrix} B_1 & B_2 & B_3 & B_4 \end{bmatrix}^T$$

A fuzzy comprehensive evaluation is performed on the membership matrix R, and the comprehensive evaluation membership vector of factor A is calculated:

$$\mathrm{B} = \mathrm{W}_A * \mathrm{R} = \begin{bmatrix} 0.271 & 0.335 & 0.239 & 0.156 & 0 \end{bmatrix}$$

4.5 *Result analysis*

According to the final evaluation results, it is decided whether to carry out subsequent planning and construction work in accordance with the results of the pre-planning work. In view of the fact that the construction of the new campus is a complex system project, this paper suggests that if the evaluation result is good or above, the relevant work can be continued; if the evaluation result is moderate or qualified, experts from the university should be invited to review the decision; if the evaluation result is unqualified, it indicates that there are great problems in the pre-planning work, and it is suggested to suspend the study, find out the corresponding problems, and restart the preliminary planning work.

According to the principle of the maximum degree of membership, the overall evaluation level of the pre-planning work for the construction of a new campus of a university in Beijing as an example in this paper is good, indicating that the pre-planning research results can meet the future development needs of the school, and follow-up related work can be continued. The evaluation level of the top-level design is good, the evaluation level of the overall plan is good, the evaluation level of key monoliths is moderate, and the evaluation level of construction management is good.

After investigation and analysis, teachers and students of the school are highly satisfied with the pre-planning of the new campus construction, which is consistent with the conclusions drawn in this paper, proving that the evaluation system constructed in this paper is reliable and has a high reference value for the pre-planning of the new campus.

5 CONCLUSION

The construction of a new campus is a century-long plan for a school, and the pre-planning work occupies an important position, directly affecting the overall planning, cost control, and quality assurance of the new campus. Good pre-planning work can play an important guiding role in the construction of new campuses and play a general role in the overall situation and future development of the university, which is the key prerequisite for the construction of a good new campus. It is proved that the pre-planning scheme and evaluation system of the new campus of colleges and universities constructed in this paper is beneficial for colleges and universities to carry out the pre-planning work smoothly, discover the shortcomings in the pre-planning work, take targeted improvement measures, and lay the foundation for the planning and construction work of new campus.

REFERENCES

Bai Y B (2019). Research on the Planning and Layout Strategy of New University Campuses under the Background of "Double First-Class". J. Asia Pacific Education. 9, 39.

Deng W (2021). Innovative exploration and application of technical guidelines for college campus planning and construction: Taking the Tianfu campus of the Civil Aviation Flight University of China as an example. J. Technology and Industry. 21(6), 189–196.

Li G X, Liu Y Z (2009). Comprehensive evaluation of university laboratories based on analytic hierarchy process. J. Experimental Science and Technology. 7(3), 129–131+134.

Lv P, Zhang W J (2011). Study on Fuzzy Comprehensive Evaluation of University Academic Discipline Construction Based on Analytic Hierarchy Process. J. Journal of Qiqihar University (Natural Science Edition). 27(2), 66–70.

Wu Z L, Wang Y F, Cui S, Cui H Y, Li C M (2020). Analysis of the preliminary work of the construction of the new university campus based on the perspective of organization theory—Taking the construction of the West Coast campus of the Ocean University of China as an example. J. Engineering Economics. 30(3), 38–43.

Advances in Urban Engineering and Management Science – Khalil & Yang (Eds)

Dynamic detection method of subgrade and pavement compactness for small span highway bridge maintenance

JianWei Zhang*
Road & Bridge International Co., Ltd, Beijing, China

ABSTRACT: Aiming at the problem of poor dynamic detection accuracy of subgrade and pavement compactness, a dynamic detection method of subgrade and pavement compactness for small-span highway bridge maintenance is proposed. Combined with subgrade resilience modulus, the stress structure of subgrade and pavement of small-span highway bridge is analyzed, and the evaluation value of subgrade and pavement compactness for long-span highway bridge maintenance is calculated. Finally, the experiment proves that the dynamic detection method of subgrade and pavement compactness for the maintenance of small-span highway bridges has the characteristics of fast, convenient and efficient, and can be better used for the rapid detection and evaluation of subgrade resilience modulus.

1 INTRODUCTION

The importance of highway bridges is self-evident. Their quality can directly affect the development of social economy and the safety of people's travel. Therefore, the quality of subgrade and pavement must be controlled. Among them, it is very important to detect the compactness of subgrade and pavement (Wang et al. 2021). If the compactness of subgrade and pavement does not meet the corresponding standards, highway bridges will have problems. Affect its performance. The following analyzes and discusses the subgrade and pavement compactness detection methods and influencing factors in China, hoping to provide help to the subgrade and pavement compactness detection personnel of highway bridges. Subgrade is the supporting structure of the road (Das et al. 2020). In order to ensure the service performance of the road, the subgrade design and construction should not only meet the overall stability requirements, but also strictly control its deformation, so as to provide solid support for the pavement (Sang et al. 2020). In the design, construction and performance evaluation of asphalt concrete and cement concrete roads in China, subgrade resilience modulus is usually used to reflect the deformation resistance of subgrade. Therefore, the research on subgrade resilient modulus detection technology is of great significance to the quality control and improvement of road construction. In order to realize the synchronous detection of subgrade resilience modulus, shallow layer wave test method and dynamic continuous synchronous detection technology have been gradually applied in road engineering (Hu et al. 2021). Among the above methods, the on-site bearing plate method and BB method are dynamic detection methods, and the measured subgrade resilience modulus is dynamic modulus. By summarizing the application of each on-site detection method in subgrade resilience modulus test, analyze and evaluate the advantages and disadvantages of different test methods, and preliminarily establish the relationship between the measured values, so as to provide reference for subgrade engineering design and construction.

*Corresponding Author: bmw0932@139.com

DOI 10.1201/9781003305026-18

2 DYNAMIC DETECTION OF SUBGRADE AND PAVEMENT COMPACTNESS FOR SMALL SPAN HIGHWAY BRIDGE MAINTENANCE

2.1 *Stress structure of subgrade and pavement of small span highway bridge*

The detection of road compactness has a significant impact on the whole project. During the construction process, the application of on-site compactness detection technology makes the construction personnel have a clear and accurate judgment on the construction quality of road engineering. While detecting the construction quality of previous periods, it provides a reference basis for the construction of subsequent projects (Ye et al. 2019). The high-quality road compactness detection effectively improves its stability and durability, provides a safe and stable traffic environment for people's travel, and helps to improve people's quality of life (Dao & Dong 2020). With the continuous advancement of the compaction process, the air in the material is continuously squeezed out, and the gap between particles is U becoming smaller and smaller. At the end of the compaction process, the moisture in the material is also squeezed out, the contact area between particles is also increasing, the friction resistance between particles is also increasing, and the compaction of the material reaches the limit value $g_1\ (P_L)$. The compacted material also has a certain structural strength and bearing capacity. $g_2\ (A\omega)$ is used to characterize the degree of compaction (paving material), and its relationship with the motion parameters of the vibratory roller.

$$k = g_1\ (P_L) + g_2\ (A\omega) - U \tag{1}$$

Among them, P is the linear load of the steel wheel of the vibratory roller, W is the working amplitude of the vibratory roller, V_s is the working frequency of the vibratory roller, and H_{i} is the working speed of the vibratory roller. According to the formula, the working speed, working frequency, working amplitude of the vibratory roller and the linear load of the vibratory steel wheel all have different effects on the compactness of the pavement paving material. As far as the stone is concerned, its specific gravity is the ratio of the mass of the mineral substance to the mass of water in the same volume, that is, the ratio of the mass of the stone to the volume of the mineral substance, i.e.

$$\delta = PW/V_s - kH_{\text{i}} \tag{2}$$

Where: Wo is the weight of stone; δ is the volume of mineral entity. The apparent specific gravity refers to the ratio of stone weight to mineral entity volume V_{sc} closed pore volume, i.e.

$$\gamma' = Wo - q/(V_{sc} - \delta) \tag{3}$$

Where: z'_{a} is the closed pore volume. However, practical experience tells us that the influence of line load on compaction effect is not very obvious. After the construction process of paving asphalt mixture, the asphalt mixture is in a loose state, leaving many gaps between particles, which are mainly occupied by air and basically do not form strength. Therefore, the asphalt pavement must be compacted by compaction machinery. The better the overall performance of the material, the higher the pavement quality. For asphalt pavement, the degree of compaction is equal to the ratio of the actual dry density after compaction at the construction site to the maximum dry density obtained from the standard compaction test in the laboratory, expressed as a percentage.

$$K = \frac{\rho}{\rho - \gamma z'_{\text{a}}} \times 100\% \tag{4}$$

Where, K is compactness of a measured part of asphalt surface course ρ—The actual density of asphalt mixture measured by the test p the compaction standard density of asphalt mixture. When testing the seismic degree of pavement, the pavement with support as insurance needs to be tested first. When the pier is easy to repair or damaged, the pile foundation. must not be damaged. In order to avoid brittle damage, repair the places prone to damage. The calculation formula of shear capacity is as follows:

$$q_c = \text{f}_c\text{F}_{\text{c}0} \times D_c F_c + \text{K}F_{\text{s}} - \beta \tag{5}$$

In the formula: f_c is the shear capacity; F_{c0} is the bearing capacity of the bearing platform; D_c Is the supporting force of pavement cover; F_c is the bending angle of small span pavement; F_s is the tension between the pavement; β is the tension between each small stage.

2.2 *Evaluation algorithm of subgrade and pavement compactness*

The unit weight index measured by sand filling method on site is divided by the indoor standard unit weight, so that the unit weight compaction degree that can correctly reflect the compaction degree can be obtained. Since all open pores have been included in the indoor test, sand leveling is not required even if the pores on the road surface are large in the field test, otherwise corresponding changes shall be made in the indoor test. The working principle of vibratory roller compaction of asphalt mixture is analyzed from the perspective of force balance, and the two degree of freedom dynamic model of "roller asphalt mixture" is established.

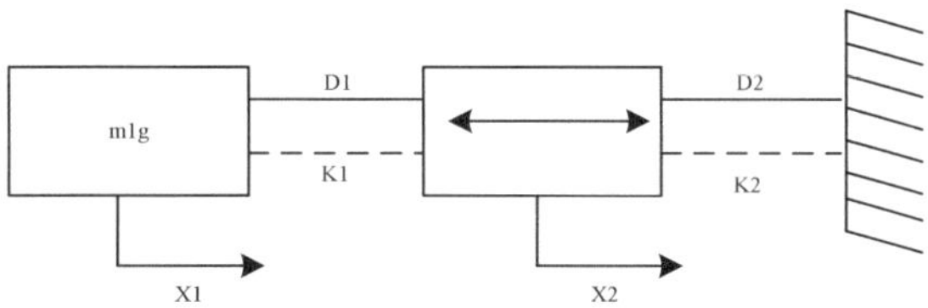

Figure 1. Two degree of freedom dynamic model.

In the same medium, its wave velocity is 0.862 ~ 0.995 times that of shear wave. The elasticity of the pressed material is closely related to the compactness of the material. Rayleigh wave method is divided into two methods to detect compactness.

2.3 *Realization of compactness detection of highway bridge subgrade and pavement*

In order to clarify the problems in compactness detection, it is necessary to analyze which of the above three compactness indexes are obtained in Marshall compactness test. In the Marshall test, the weight η of the test piece can be weighed by the balance, and the volume L of the test piece can be measured by the buoyancy method, that is, the specific gravity of the test piece is taken as:

$$V = LW - q_c + \eta \tag{6}$$

When wax sealing method is adopted:

$$\bar{K} = LE - W_3 - V/Aq_c \tag{7}$$

Where: E is the weight of the specimen in water, A is the weight of the wax sealed specimen in air, W_3 is the weight of the wax sealed specimen in water, and DP is the unit weight of wax. The compactness of the pavement shall be evaluated by 1–3 km, and the on-site compactness sampling inspection shall be carried out according to the detection frequency and method required below. Calculate the average compactness, standard degree and coefficient of variation of each measuring point for an evaluation section:

$$S = y + x\sqrt{\frac{(K_1 - \bar{K})^2 + (K_2 - \bar{K})^2 + \cdots + (K_n - \bar{K})^2}{N - V - 1}} \tag{8}$$

Where, the average compactness of each evaluation section; Standard deviation (%) of the measured value of compactness of evaluation sections; Coefficient of variation of measured value of compactness of each evaluation section; $K_1, K_2, \ldots K_n$ degree of compaction (%) of each measuring point in the evaluation section y the total number of measuring points in the evaluation section, and its degree of freedom is x. The on-site compactness detection method of asphalt concrete layer recommended in the specification is to drill core sampling and measure its compactness (apparent density) ratio indoors, that is, the default definition of compactness in the specification is the apparent density ratio.

3 ANALYSIS OF EXPERIMENTAL RESULTS

In order to carry out the research of this project, five typical filled subgrade sections are selected as the test sections during the construction of an expressway bridge, and comparative test and detection are carried out. Among them, three sections are plain soil subgrade and two sections are lime soil subgrade. The engineering overview of each test section is shown in Table 1.

Table 1. Project overview of test section.

Serial number	Starting and ending chainage	Length (m)	Compaction area	Subgrade filler type	Embankment part
A	S17+120~S17+410	290	Zone 85	Plain soil	Lower embankment
B	S20+800~S20+910	110	Zone 85	Plain soil	Lower embankment
C	S22+200~S22+400	200	Zone 85	Plain soil	Lower embankment
D	S25+850~S26+310	560	Zone 89	5% lime soil	Lower roadbed
E	S28+630~S28+810	170	Zone 89	5% lime soil	Lower roadbed

In the field bearing plate method, the bearing plate is used to load and unload the subgrade step by step, and the rebound deformation value corresponding to each level of load is measured, so as to obtain the subgrade rebound modulus, which can be used in pavement design. This method is one of the standard methods for measuring subgrade resilience modulus, and is widely used in highway and bridge engineering. The table shows the application of on-site bearing plate method in the detection of subgrade resilience modulus of some highway and bridge projects.

Table 2. Example of subgrade resilience modulus tested by bearing plate method.

				Modulus of resilience / MPa		
Section	Measuring point No	Rebound deformation / mm	Number of samples / pieces	Mean value	Standard deviation	Representative value of 90% guarantee rate
Beijing	A	0.3	20	62.5	23.1	36.5
Shijiazhuang	B	0.4	20	59.2	222.	35.5
Expressway	C	0.5	15	58.6	9.98	51.2
	D	1.1	15	56.6	6.58	48.6
Sanling	A	0.3	40	93.2	29.3	58.2
Expressway	B	0.5	39	86.5	26.5	55.2
National	A	0.6	11	68.2	5.35	65.5
Highway 102	B	1.1	11	65.9	5.65	61.3
Yinzhong	A	0.6	12	79.8	12.0	65.2
Expressway	B	1.1	12	75.6	12.5	63.5
Xihu	A	0.6	15	165.2	15.6	152.3
Expressway	B	1.1	16	165.3	26.5	141.6

It can be seen from the table that when the field bearing plate method is used for detection, the measured value of subgrade resilience modulus decreases with the increase of resilience deformation, up to 14.0mpa. The small change of subgrade elastic modulus will have a significant impact on pavement structure design, resulting in design waste or failure to meet safety requirements. Through analysis and conversion, the mechanical parameter r reflecting the subgrade stiffness can be obtained, and then the subgrade resilient modulus can be obtained from the fitting relationship between R and subgrade resilient modulus. The figure shows the relationship between the subgrade resilience modulus obtained by Ye Weimin and others through field test and the back calculated dynamic stiffness R through real-time online detection.

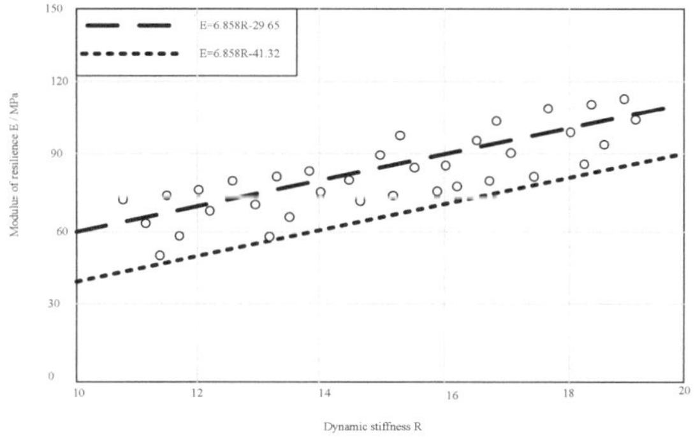

Figure 2. Dynamic stiffness relationship of real-time online detection of resilient modulus.

It can be seen from the figure that the dynamic stiffness r of subgrade is basically linear with the modulus of resilience, and the real-time online detection technology can be used to test the modulus of resilience of subgrade. It should be noted that when using the real-time on-line detection technology to test the subgrade resilient modulus, the parameters should be calibrated first, otherwise the reliability of the measurement results will be greatly reduced. Figure 3 shows the particle grading curve of soil sample.

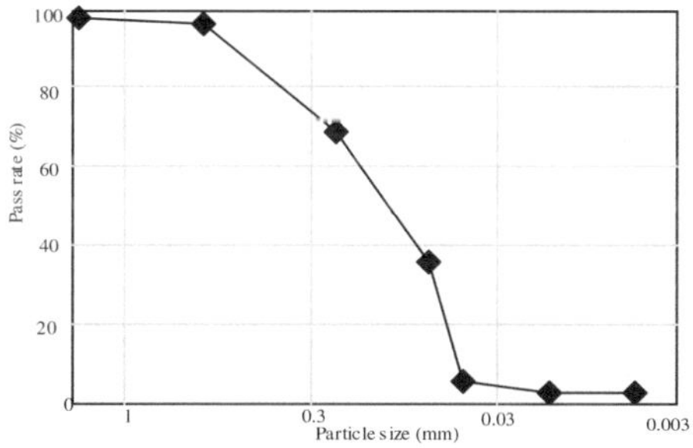

Figure 3. Grading curve of compaction degree of plain soil particles.

According to the screening results, the non-uniformity coefficient Cu of subgrade filler soil sample is 3.71 and the curvature coefficient C ε 1.20: the content of particles <0075 mm is 77.3%, greater than 50%, indicating that the soil sample is fine-grained soil.

A 20s long signal is intercepted through the vibration acceleration signal collected during the pavement construction and compaction of K24 + 910k24 + 930 section. The acceleration signals detected during the first five times of compaction under this section are filtered and piecewise fitted to obtain the vibration acceleration values shown in the table. The acceleration values in the table are not excluded by the confidence interval.

Table 3. Vibration acceleration data before processing (M/S2).

	Rolling times				
Time period (s)	The first time	The second time	third time	Fourth time	Fifth time
0.5	38.965	40.325	40.32	42.652	44.625
0.9	39.15	39.658	41.5	42.025	44.652
1.3	38.251	40.62	41.32	42.052	45.215
1.8	37.652	39.65	41.65	42.325	45.325
2.1	37.625	39.032	42.32	42.635	45.025
2.5	37.999	39.32	41.25	42.322	45.652
2.9	38.65	39.526	41.365	42.989	45.685

The K24 + 910-k24 + 930 rolling section is taken as the test section for field test. During the test, after each pass of rolling, take 5 positions of different pile numbers to enter the hall with a nuclear densimeter to measure their compaction density (two points for each pile number), and then compare with the maximum dry density to obtain the compaction degree of corresponding positions.

Based on this, the bearing load of highway bridges before and after construction is calculated, and the construction quality is compared according to the bearing load. The specific results are shown in the Figure 4 below:

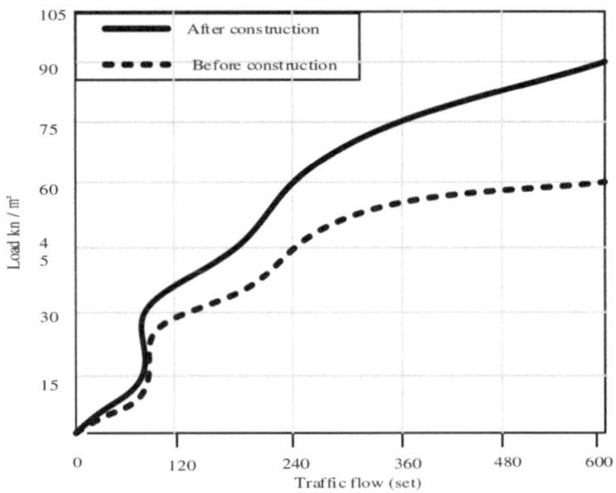

Figure 4. Construction quality inspection results.

The above comparative analysis results reveal that the proposed construction technology for Subgrade and pavement compaction of highway bridges has high bearing capacity in the process of practical application, which can better ensure the construction quality of highway bridges.

4 CONCLUSION

The on-site compactness detection technology has a great impact on the quality control of road engineering. In practice, road constructors only fully realize the importance of compactness detection and continuously optimize the inspection technology on the basis of analyzing its technical application. Only in this way can the application level of on-site compactness detection technology be improved. Only by ensuring the quality of subgrade and pavement of highway bridges can we effectively prolong the service life of highway bridges, promote the development and progress of social economy and improve people's quality of life. In the process of subgrade and pavement compactness detection, relevant personnel must operate in strict accordance with the regulations, and objectively and accurately evaluate the subgrade and pavement of highway bridges, so as to improve the quality of highway bridges and promote the healthy and sustainable development of highway bridges in China.

REFERENCES

A Das, R Beaumont, I Butterworth, I Masters, D Williams. 2020, High rate and temperature-dependent tensile characterization with modelling for gap-bridged remote laser welded (RLW) joint using automotive AA5182 alloy. International Journal of Impact Engineering, 144(2):103672.

Dao N D, Dong C L. 2020, Modulation of Bidirectional AC/DC Converters Based on Half-Bridge Direct-Matrix Structure. IEEE Transactions on Power Electronics, PP (99):1–1.

IP Sang, SH Lee, A Almasi, JH Song. 2020, Extended IFC-based strong form meshfree collocation analysis of a bridge structure. Automation in Construction, 119(3):103364.

J Hu, W Li, Z Su, D Wang. 2021, Analytical design and experiment of radial inductive displacement sensor with full-bridge structure for magnetic bearings. The Review of scientific instruments, 92(6):064703.

Ye X, Zhuo S, Chen B. 2019, Research on modal parameters identification of bridge structure based on adaptive signal de-noising method. Cluster Computing, 22(4):1–11.

Z Wang, N Guo, S Wang, Y Xu. 2021, Prediction of highway asphalt pavement performance based on Markov chain and artificial neural network approach. The Journal of Supercomputing, 77(2):1354–1376.

Advances in Urban Engineering and Management Science – Khalil & Yang (Eds)
© 2023 The Author(s), ISBN 978-1-032-30426-7

Analysis on renovation application of existing buildings an example of urban reading space design of "Haoxi Book Garden"

Wenjuan Huang*
School of Design and Communication, Nantong Institute of Technology, Nantong, China

Haiyu Tong*
School of Civil Engineering, Nantong Institute of Technology, Nantong, China

ABSTRACT: Buildings, one of the physical forms boasting convenience and comfort for people's lives, are of external functional use. Giving full play to the continued application value of existing buildings is important to carry out the concept of green development. This paper analyzes the application of the functional renewal, design planning according to local circumstances and structural transformation technology in the renovation process of existing buildings, and takes the renovation of "Haoxi Book Garden" as a design case to an offer experience for future renovation of existing buildings.

1 INTRODUCTION

In current society, the concept of green development has been deeply rooted in people's minds, and with limited resources on this globe, it is dramatically vital to adhere to the concept of urban renovation and organic renewal, and retrofit and reuse existing buildings (Long 2021). The successful renovation of the old building not only revitalizes the building itself, but also PROMOTES the economic development of the site as well as its surrounding areas.

2 RENOVATION SIGNIFICANCE OF EXISTING BUILDINGS

Cities are the carrier of history and culture, while buildings are the historical and cultural heritage of cities. As they convey profound historical information and nostalgic memories and have precious values and spiritual elements of Chinese culture, hence they become a part of the life of cities (Long 2021). The renovation of existing buildings refers to the structural adjustment of the original building or the conversion of its functions in order to enhance the safety, durability, comfort and environment of the buildings. Ministry of Housing and Urban-Rural Development of the People's Republic of China issued a document entitled *"Notice of the Ministry of Housing and Urban-Rural Development on Further Improving the Preservation, Utilization and Renewal of Existing Buildings in Cities"* in September 2018, which requires close attention paid to the preservation, utilization and renewal of existing buildings in cities; the 2019 Report on the Work of the Government during two sessions of NPC and CPPCC conveys a strong work signal for promoting the preservation and renovation projects of existing buildings with utmost endeavors.

3 MAIN FORMS FOR RENOVATING FUNCTIONS OF EXISTING BUILDINGS

The renovation of existing buildings is a complex and comprehensive system, with the main types of construction to be renovated like old cities, old villages, old plants, old streets and old buildings.

*Corresponding Authors: huangwj@ntit.edu.cn and 2250498220@QQ.com

 DOI 10.1201/9781003305026-19

They are renovated in terms of the renewal of building functions, the update of building culture, reinforcement and renovation of buildings and energy-saving practices for buildings (Xu 2018). To sum up, there are three kinds of common renovation modes of existing buildings for development and protection:

It is important to basically retain the existing buildings, with moderate transformation based on the requirements of modern life, via reinforcing the original structures as a whole, enriching the use of the buildings, replacing the existing building functions, enhancing the openness and publicness of buildings, and presenting the regional features of buildings in the spatial designs. For example, in the case of "Renovation of Library in Yanjiao Campus of Central Academy of Fine Arts", the library and the adjacent art museum were planned and renovated together. The design of the library renovation is conducted from the following terms: adjusting the layout, sorting out the boundaries, building a new entrance for the library, setting up a new spatial center, adopting the Tyson polygonal algorithm (Voronoi diagram) to generate a high grille roof, cascading staircases and "micro-terrace" reading space, etc.

Figure 1. Source: Official account entitled "Design competition".

Combined with the surrounding environmental factors and constraints, the facade renovation is carried out by retaining the exterior walls and roofs to display the iconic image of buildings, while the internal structure would be demolished and rebuilt according to the requirements of commercial space. For example, the case of "the renovation of the Old Sugar Refinery building" was designed by the Original Design Studio in the Architectural Design and Research Institute of Tongji University. The original exterior walls of the south and west sides of this building were maintained as usual, with the facades renovated in the principle of "repairing the old as the old", with the thick reinforced concrete walls of the original warehouse built in the 1930s retained, and with the wooden frames of the doors and windows on the south side and the exhaust fans on the west side with historical characteristics preserved. While as to the north side of the building, a gray cantilevered space was set up after the removal of the original partition wall to display the space section, and then an exterior riverside lookout platform was established; on the east side of the building, a curved retaining wall was built in connection with a grass slope, presenting a positive and compatible gesture of indoor and outdoor space after the physiographic reshaping.

Figure 2. Source: Official account entitled "New micro design".

This mode adopts basic demolition, with buildings re-planned and reconstructed, which only retained partial wall structure and representative characteristics of original buildings. For example, "Riverside Creative Industry Park", whose original buildings had been mostly demolished for historical reasons, have been rebuilt by Chinese Taiwanese architect Deng Kunyan since 2004. His team embarked on the conservation and development of the existing old factory blocks, and they preserved the partial structure and walls of the building, using a lot of recycled materials from the site and the most common building materials to create the most avant-garde spatial environment in the post-industrial era.

Figure 3. Source: http://blog.sina.com.cn/s/blog_65aeee420102wnni.html.

4 RENOVATION DESIGN FOR THE URBAN READING SPACE PROJECT OF "HAOXI BOOK GARDEN"

In the process of development, China's public reading bookstores have been transformed from the single function as bookstores to public reading spaces that integrate culture, aesthetics, creativity and social interaction. According to the survey data of the 2020–2021 China Physical Bookstore Industry Report, about 4,061 new physical bookstores have been opened and 1,573 bookstores closed in China, with 2,488 newly-opened bookstores nationwide. By now, urban public reading spaces, represented by physical bookstores and public libraries, have once again entered the stage and become an important force for urban cultural landmarks and communication.

4.1 *Background of project renovation*

As an important architectural facility in a city, the public library provides the necessary functional space for the development of urban culture. Nantong boasts "the first city of modern China", and Zhang Jian, a famous historical figure in Nantong, founded a series of industrial industries and cultural and educational undertakings, which laid a profound foundation for Nantong. "Haoxi Book Garden" is located at No. 9, Haoxi Road, Chongchuan District, Nantong City, Jiangsu Province, designed by the famous Chinese architect Professor Pan Guxi, with an antique style, exquisite design and clever layout, besides, its east side near the river lies the landmark "Five Pavilions Inviting the Moon", known as one of the "Ten Views of Hao River".

With the promotion of Chinese national policies on reading and the drive of intelligent technology, the building renovation and repositioning of Haoxi Book Garden has become one of the highlights of Nantong urban construction. The renovation project includes two buildings A and B of "Haoxi Book Garden" and the landscape site, with a total area of about 1,200 square meters. A green and convenient 24-hour "new urban reading space" with a beautiful environment, profound cultural heritage, and perfect facilities is constructed via the optimization of the original building structure, re-planning of the space and updating of functions, the increase of intelligent facilities, the refining of building materials and construction and the like.

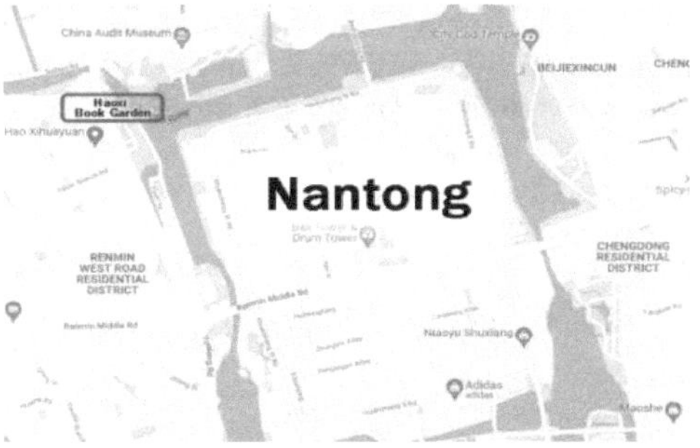

Figure 4. Location map of "Haoxi Book Garden".

Figure 5. Original external image.

4.2 *Project renovation concept*

The life of the integrated building of both traditional regional architectural advantages and project renovation depends on its use values. In the event of time variation, the building can no longer match the contemporary lifestyle, and thus innovation is to endow the existing building with new commercial functions and regional cultural values. Its exterior appearance with the historical memories is retained, and the internal structure is renovated to give the building a new life through new technology, new materials and new structural forms (Wu et al. 2020).

The renovation of "Haoxi Book Garden" adopts the new open urban public reading space as its value target of the new space, focusing on the inheritance and carrying forward of regional culture as well as the development concept of public welfare, providing high-quality reading content and services. It employs modern technology to transform the design of the space planning within the building and the renovation of building facades. On the basis of the existing building and local conditions, it combines the book culture organically with various types of digital cities, incorporating Zhang Qian, one of Nantong's historical celebrities, as well as blue print cloth and other elements with Nantong traditional characteristics. The design shifts from the traditional and conventional antique style of the original scheme to the one mainly featuring a modern shape and a new-Chinese-style core, thus creating an architectural art more in line with the aesthetics of young people nowadays. It absorbs the traditional concept of "Harmony between Humanity and Nature" to be in harmony with the natural environment and in conciseness but not simplicity. Its starting point is to create a green and convenient "urban reading space" with an elegant environment and profound heritage, which strives to make "Haoxi Book Garden" a cultural hall inheriting Nantong culture and enhancing spiritual well-being.

4.3 *Optimization of building structure and site*

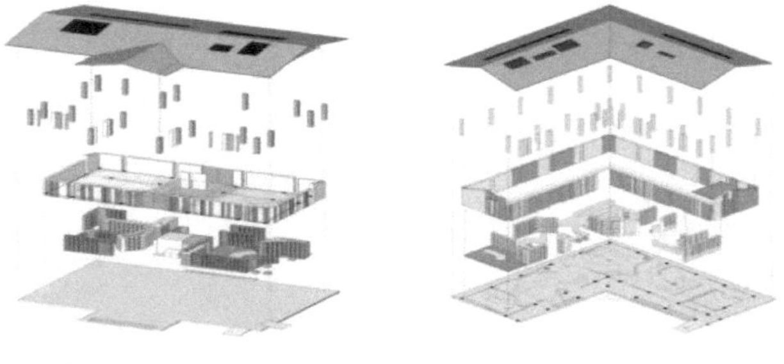

Figure 6. Exploded view of Hall A and Hall B.

When carrying out the renovation of the building facades of "Haoxi Book Garden", the roofs of Hall A and Hall B are extended to solve the problem of partitioning the structure between construction bodies, to render a practical, solid, and unified construction. The combination of light sources and the building leads to the emphasis of the building bodies, and the enhancement of the sense of atmosphere and three-dimensionality of the site. Ground lights are set around the greenery for guidance and directions. At the key spatial nodes of the site outside the building, tall and dense green plants are used to strengthen the spatial transitions, and the trees serve as a strong connection with the rigid geometry and lines of the building via their diverse natural gestures.

Figure 7. Rendering of exterior facade structure & profile map of Hall A and B.

4.4 *Update of space functions and parameters of decoration and construction*

The original symmetrical and dull design has been totally changed during the space renovation within the building, and the functional partitions are linked together with dynamic lines to make the space present a concise, bright, simple and reserved image. In the Chinese space with enormous blank and numerous logs, what visitors feel is not only an elegant, subtle and clean space style, but also a touch of Zen. The elements of Hao River in Nantong are extracted and adopted in the shape of ceiling curves and lamps, echoing with the surrounding environment. Tea sets and Chinese zithers are chosen as furnishings to effectively promote the exchange and dissemination of urban culture, along with the elements of blue print cloth chosen in the design of decorations and cultural creation to explore the material beauty, craft beauty and heritage of traditional regional culture. The main lighting sources are the LED diffuse lights with spot lights, and the hollow fences in the roofs are used to introduce natural light as an aid.

Via the renovation design, the functional areas in Hall A are linked together via the dynamic lines. From the entrance inwards, there appear paper-book reading area, smart borrowing area, and rest area for in-depth reading. The bottom of the bookshelves in the reading area is designed with a sitting board to allow people to rest for a moment while reading, and the half-wrapped high and low bookshelves cater for people's privacy needs when reading. The smart borrowing area in the middle is equipped with some computers and intelligent devices concerning book return system, so that readers can enjoy the public reading space quickly and intelligently. The in-depth reading area is designed with a partially lifted floor, which effectively divides the functional areas. The height of the bookshelves in the library is uneven, which not only breaks the dull space atmosphere while taking into account the privacy, but also achieves the flexibility of space change.

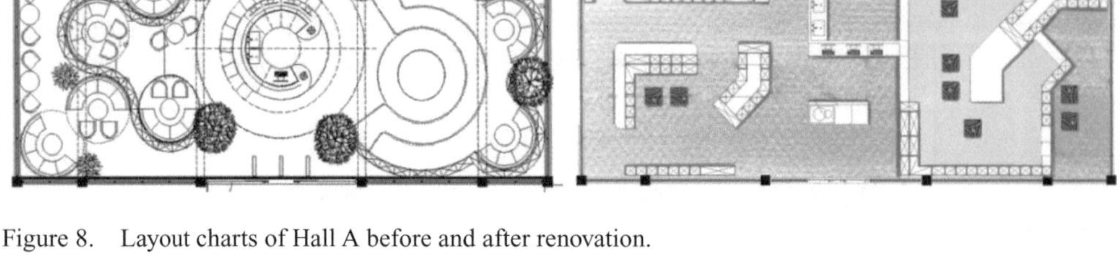

Figure 8. Layout charts of Hall A before and after renovation.

Figure 9. Rendering of Hall A.

Hall B is renovated with the design concerning the necessity of division, and therefore large reading areas and rest areas are set up. Hall B is an L-shaped space, and one can enjoy distinctive cultural exchange space compared to that in Hall A by walking into the nearest entrance to Hall B through a corridor which links Hall A to Hall B. The appropriate interior decorations can beautify the interior environment, boost interior artistic conception, strengthen the interior style and other functions. According to the regional cultural atmosphere, the tea tasting area and the piano playing area are set up in Hall B to enrich people's spiritual and cultural communication. In terms of the interior walls, large floor-to-ceiling windows are employed, which create a hazy atmosphere and Zen beauty when the sunlight pours inside.

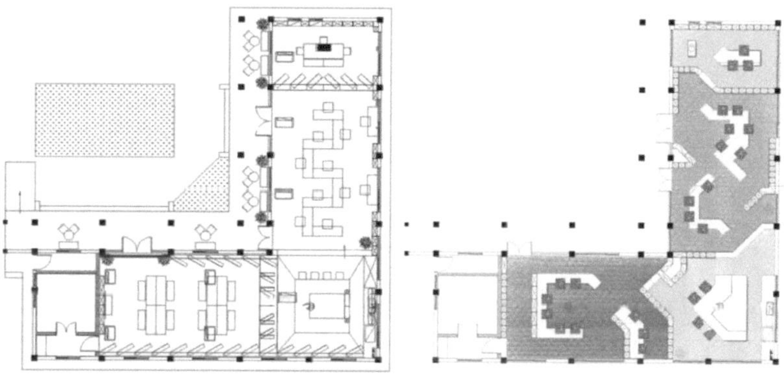

Figure 10. Layout charts of Hall B before and after renovation.

Figure 11. Rendering of Hall B.

In the process of renovating the urban public reading space of "Haoxi Book Garden", the designers analyzed the historical and cultural elements of the building and the development status of architectural resources, and fully investigated the cultural, structural, technical and artistic values of the existing building. They also utilized the main building and the surrounding environment, and carry out space renovation and layout design concepts on the basis of preserving the original building structure according to local conditions. Besides, the renovation updated the social functional needs of the space, and played the essential role of urban public reading, and hence become a new landmark in Nantong.

5 CONCLUSION

Along with the depletion of land supply in urban centers, radical development is becoming less and less. According to relevant data, urban demolition in China has reached 203.3 billion square meters from 2016 to 2020, and the renovation of existing architectures is gradually becoming a new development point. White Paper of the 2018–2019 China Urban Renewal demonstrates that urban renewal and existing building renovation have become a trillion-dollar market, with the total amount of existing buildings being about 61.3 billion square meters, and through the renovation of existing buildings, old buildings can be enabled to meet the needs of production and life in the new era, when the "old" space takes a new lease of life again (Yang & Wang 2021). Meanwhile, resources and manpower consumed in the process of demolition and reconstruction could be saved in this way, with great economic benefits and environmental value.

ACKNOWLEDGMENT

This work was financially supported by the Training Program for Young and Middle-aged Backbone Teachers of Nantong Institute of Technology (ZQNGGJS202110), Provincial Science and Technology Service Platform Cultivation Project of Nantong Institute of Technology (XQPT202102) and Science and Technology Guiding Project of Jiangsu Provincial Department of Housing and Urban Rural Development – Research on Historical Inheritance and Architectural Function Transformation of Existing Buildings (2019ZD001028).

REFERENCES

Long X.F. (2021). Heritage Protection and Cultural Heritage. J. Architecture and Culture. 4(07), 1.

Wu Z.X., Ouyang G.T., Pan X.J., Rao Z.F. & Xiong Y.S. (2020). A Preliminary Study on the Integration of the Advantages of Traditional Regional Architecture into the Transformation of Existing Buildings – Taking Beijing, Shanghai, Guangdong and Hubei as Examples. J. Guangdong Civil Engineering and Architecture. 27 (01), 27–39.

Xu Y.F. (2018). Design Research on Transforming Existing Buildings into Maker Space. D. Shenzhen University. 2018.

Yang J.M. & Wang K.Y. (2021). Study on the Facade Reconstruction Strategy of Urban Street Buildings Under the Concept of Urban Double Repair – Taking the Street Renewal Design of Suzhou as an Example. J. Urban Architecture. 18 (13), 60–63.

Advances in Urban Engineering and Management Science – Khalil & Yang (Eds)
© 2023 The Author(s), ISBN 978-1-032-30426-7

A dynamic emergency network planning model with data-driven distributionally robust probabilistic constraints

Qing Li, Lu Qin & Kanglin Liu*
School of Traffic and Transportation, Beijing Jiaotong University, Beijing, China

ABSTRACT: Efficient post-disaster emergency network design can improve rescue efficiency and service quality. Considering demand uncertainty, the evolution of disasters and supply penalty cost, a dynamic emergency network planning model based on distributionally robust optimization was established by introducing probabilistic constraint based on Wasserstein ambiguity set to guarantee satisfying service quality. This model can be reformulated into a mixed integer programming model and solved by Gurobi efficiently and accurately. Numerical experiments based on the emergency distribution network in some areas of Shanghai during the epidemic were conducted. The results prove that the distributionally robust model has more advantages in demand satisfaction and computational efficiency. The influence of Wasserstein radius on objective value and probabilistic constraint's satisfaction rate are also successfully explored.

1 INTRODUCTION

Emergency logistics service will be urgently needed when natural disasters or epidemics occur. However, it is difficult to guarantee victims' demands due to the uncertainty of sudden emergency events. Thus, constructing a more robust rescue service network has been widely concerned by scholars.

Most of the existing researches on emergency network planning considering demand uncertainty are based on stochastic optimization. Some people established a two-stage stochastic optimization model and used the Sample Average Approximation (SAA) method to take the empirical distribution as the true distribution of random variable (Carmen & Mark 2009). The exact distribution of random variable must be known in this method, while sudden emergencies make it difficult. Even if it can be predicted by exhaustive method (such as SAA), the scenarios may be underrepresented or too many to be solved (Lawrence & Zümbül 2015). Instead, robust optimization does not need to know the exact distribution by assuming uncertain set or distribution form of demand to optimize the worst case (German & Maria 2020). Some scholars proposed a two-stage robust optimization model and designed a column constraint generation algorithm to solve it (German & Maria 2020).

In addition, distributionally robust optimization has attracted much attention in recent years, which is not too conservative compared with robust optimization by using historical data and also significantly improves computational efficiency through its theoretical properties. Some scholars established a distributionally robust model and transformed it using moment-based ambiguity set (Kanglin et al. 2019). However, the moment-based ambiguity set will make it difficult to solve and has no asymptotic consistency. So, some scholars established a multi-period facility location and capacity planning model based on ∞-Wasserstein ambiguity set, which was transformed into a mixed integer second-order cone planning model (Zhuolin et al., press). In contrast, the model based on Wasserstein ambiguity set has low computational difficulty and can approach the real distribution when the sample size tends to infinity (Weijun 2019).

In general, the following aspects remain to be studied: Firstly, most of the above literature rarely include dynamic strategies. Secondly, ∞-Wasserstein ambiguity set is a special case, which cannot

*Corresponding Author: klliu@bjtu.edu.cn

DOI 10.1201/9781003305026-20

be extended to other dimensions. Thirdly, the model involved is more complicated and difficult to solve. Fourthly, the above studies did not consider the actual emergency management in China for case verification.

Therefore, considering the demand uncertainty and dynamic characteristics of emergency, a Distributionally Robust Multi-Period Network Planning (DRMNP) model is established based on the classical Wasserstein ambiguity set in this paper. And the model is transformed into a mixed integer programming model, which is easier to address. Numerical experiments based on the background of Chinese epidemic are carried out to verify and analyze the model.

2 MATHEMATICAL MODEL

2.1 *Problem description and assumptions*

The problem studied in this paper is how to distribute emergency supplies from J alternative distribution centers to I demand sites during T periods. It is assumed that the demand of each demand site is a random variable, and victims will have corresponding penalty costs in the process of waiting for relief (José et al. 2013). Multi-period planning is necessary because the capacity of the distribution center will increase gradually due to resource scheduling, road interruption and so on, and the construction cost will decrease over time on the contrary.

Some assumptions are made: (1) all distribution centers operate normally; (2) demand samples can be estimated through historical records and population data; (3) at the end of the planning cycle, unmet demand will result in high penalty costs.

2.2 *Supply penalty cost calculation*

Supply penalty cost of demand site i consists of two parts: the penalty cost during and after the planning period. Each part is equal to the product of the supply penalty factor and the quantity of supply shown as equation (1).

$$\phi_i = \sum_{t\in[T]} \mu_t q_{it} + \mu_1 \max\{d_i - \sum_{t\in[T]} q_{it}, 0\} \tag{1}$$

In (1), q_{it} is the quantity of goods obtaining at demand point i in period t. d_i is the total demand during planning period. μ_t, μ_1 are the penalty factors during and after the planning period respectively.

Taking water as an example, Jose et al. express the penalty cost of water as the price people are willing to pay (José et al. 2013). Based on this and regional prices, this paper set μ_1 to be 200 and μ_t is calculated according to (2).

$$\mu_t = 0.02 \cdot (e^{1.5031+0.1172\cdot 12\cdot t} - e^{1.5031}) \tag{2}$$

2.3 *Notations*

Sets: $[J]$ is the set of alternative distribution centers indexed by j; $[I]$ is the set of demand sites indexed by i; $[T]$ is the set of time periods indexed by t; $[N]$ is the set of demand samples indexed by n.

Parameters: f_{jt} is the fixed construction cost of distribution center j in period t; β is the unit transportation cost; c_{ij} is the distance from distribution center j to demand site i; α is the demand satisfaction level; $1-\varepsilon$ is the probabilistic level of the emergency network; d_i is the demand of demand site i which is a random variable and $\boldsymbol{d}$ is the corresponding vector; $\hat{d}_{in}$ is the demand of demand site i in sample n and $\hat{\boldsymbol{d}}_n$ is the corresponding vector; M_t is the capacity of distribution center j in period t.

Decision variables: X_{jt} decides whether to build distribution center j in period t; Y_{ijt} refers to the quantity of supplies at demand point i supplied by distribution center j in period t.

2.4 *Model building*

The DRMNP model is shown below:

$$\begin{aligned}\min \mathrm{C} = & \sum_{t\in[T]}\sum_{j\in[J]} f_{jt}X_{jt} + \beta \sum_{t\in[T]}\sum_{i\in[I]}\sum_{j\in[J]} c_{ij}Y_{ijt} + \\ & \sum_{i\in[I]}\sum_{t\in[T]}\sum_{j\in[J]} \mu_t Y_{ijt} + \mu_1 \mathbb{E}_F\{\sum_{i\in[I]} \max(d_i - \sum_{t\in[T]}\sum_{j\in[J]} Y_{ijt}, 0)\}\end{aligned} \tag{3}$$

$$s.t. \mathbb{P}_F\{\alpha d_i \le \sum_{t\in[T]}\sum_{j\in[J]} Y_{ijt}, \forall i \in [I]\} \ge 1 - \varepsilon \tag{4}$$

$$\sum_{i\in[I]} Y_{ijt} \le M_t \sum_{T=1}^{t} X_{jT}, \forall j \in [J], t \in [T] \tag{5}$$

$$\sum_{T=1}^{t} X_{jT} \le 1, \forall j \in [J], t \in [T] \tag{6}$$

$$X_{jt} \in \{0, 1\}, \forall j \in [J], t \in [T] \tag{7}$$

$$Y_{ijt} \ge 0, \forall i \in [I], j \in [J], t \in [T] \tag{8}$$

In this model, objective function (3) is intended to minimize the total operating cost, which includes distribution center fixed construction cost, transportation cost and supply penalty cost. Constraint (4) is a joint probabilistic constraint, which represents that the probability of all demand sites' Class II service level being at least α must not be lower than $1 - \varepsilon$. Constraints (5) ensures that the total amount of supplies transported by each distribution center in each period cannot exceed the corresponding capacity. Constraints (6) ensures that the distribution center will not be built repeatedly. Constraints (7) and (8) are constraints of decision variables.

3 MODEL REFORMULATION

3.1 *Construction of the Wasserstein ambiguity set*

Because equations (3) and (4) contain a random variable d_i, this paper assumes that the true distribution $F$of d_i belongs to a Wasserstein ball, which is centered on the empirical distribution F_N with radius δ_N. It is expressed in (9).

$$\mathcal{F}_N = \{F \in \mathcal{M}(\Xi) : d_W(F_N, F) \le \delta_N\} \tag{9}$$

In (9), $\mathcal{M}(\Xi)$ represent the possible distributions supported by space Ξ. The Wasserstein distance is defined in (10).

$$\begin{aligned} d_W(F_1, F_2) = \inf\{\textstyle\int_{\Xi\times\Xi} d(\xi_1, \xi_2)K(d\xi_1, d\xi_2) : \times \\ \textstyle\int_{\Xi} K(\xi_1, d\xi_2) = F_1(\xi_1), \int_{\Xi} K(d\xi_1, \xi_2) = F_2(\xi_2)\}\end{aligned} \tag{10}$$

(Ξ, d) is a totally bounded Polish space with distance metric shown in (9). $K : \Xi \times \Xi \rightarrow \mathbb{R}_+$ is the joint distribution of $F_1 \in \mathcal{M}(\Xi)$ and $F_2 \in \mathcal{M}(\Xi)$. Besides, $d(\xi_1, \xi_2)$ equals to $\|\xi_1 - \xi_2\|_p$ and $\|\cdot\|_p$ is the $l_p$norm on R^n. In this paper, p can be one, two or infinity, which can come to the same result.

3.2 *Objective function reformulation*

Since (3) contains random variable and is difficult to calculate, this paper linearizes it by introducing auxiliary variable A_i and (12) and (13). The reformulated model is shown as follows:

$$\min\ C = \sum_{t\in[T]}\sum_{j\in[J]} f_{jt}X_{jt} + \beta\sum_{t\in[T]}\sum_{i\in[I]}\sum_{j\in[J]} c_{ij}Y_{ijt} + \sum_{i\in[I]}\sum_{t\in[T]}\sum_{j\in[J]} \mu_t Y_{ijt} + \mu_1\sum_{i\in[I]} A_i \tag{11}$$

$$s.t.\ \mathbb{P}_F\{A_i \geq \hat{d}_i - \sum_{t\in[T]}\sum_{j\in[J]} Y_{ijt}\} \geq 1-\varepsilon_1, \forall i\in[I] \tag{12}$$

$$A_i \geq 0, \forall i\in[I] \tag{13}$$

equation (4) $\sim$ (8)

$1-\varepsilon_1$ is the confidence level of individual chance constraints. When ε_1 is small enough, (12) and (13) can guarantee that (3) and (11) are equivalent.

3.3 *Probabilistic constraints reformulation*

Next, (4) and (12) will be reformulated based on Wasserstein ambiguity set to make them easier to calculate.

Firstly, the left side of (4) refers to the possibility of the worst case. So, it is equivalent to (14).

$$\inf_{F\in\mathcal{F}_N} \mathbb{P}\{\alpha d_i \leq \sum_{t\in[T]}\sum_{j\in[J]} Y_{ijt}, \forall i\in[I]\} \geq 1-\varepsilon \tag{14}$$

And it can be further reformulated into (15).

$$\sup_{F\in\mathcal{F}_N} \mathbb{P}\{\alpha d_i > \sum_{t\in[T]}\sum_{j\in[J]} Y_{ijt}, \exists i\in[I]\} = \sup_{F\in\mathcal{F}_N} \mathbb{E}[\mathbb{I}(\alpha d_i > \sum_{t\in[T]}\sum_{j\in[J]} Y_{ijt}, \exists i\in[I])] \leq \varepsilon \tag{15}$$

In (15), $\mathbb{I}_A$ is an indicator function where it equals to 1 when A is true and 0 otherwise. Because it is a bounded and upper semicontinuous function, we can further get (16) according to theorem 1 (Jose et al. 2019).

$$\min_{\lambda\geq 0}\{\lambda\delta - 1/N\sum_{n\in[N]} \inf_{\boldsymbol{d}}[\lambda||\boldsymbol{d}-\hat{\boldsymbol{d}}_n|| - \mathbb{I}(\alpha d_i > \sum_{t\in[T]}\sum_{j\in[J]} Y_{ijt}, \exists i\in[I])]\} \leq \varepsilon \tag{16}$$

Then, based on Claim 1 (Weijun 2019), when $\lambda \geq 0$ is true, for each sample, we reformulate part of (16) to (17).

$$\inf_{\boldsymbol{d}}[\lambda||\boldsymbol{d}-\hat{\boldsymbol{d}}_n|| - \mathbb{I}(\alpha d_i > \sum_{t\in[T]}\sum_{j\in[J]} Y_{ijt}, \exists i\in[I])] = \min\{\min_{i\in[I]} \inf_{\alpha d_i > \sum_{t\in[T]}\sum_{j\in[J]} Y_{ijt}} [\lambda||\boldsymbol{d}-\hat{\boldsymbol{d}}_n|| - 1], 0\} \tag{17}$$

The Wasserstein distance in (17) is calculated according to (18) (Jose et al. 2019).

$$\inf_{\alpha d_i > \sum_{t\in[T]}\sum_{j\in[J]} Y_{ijt}} ||\boldsymbol{d}-\hat{\boldsymbol{d}}_n|| = \max\{\sum_{t\in[T]}\sum_{j\in[J]} Y_{ijt} - \alpha\hat{d}_{in}, 0\} \Big/ ||\boldsymbol{e}_{\boldsymbol{m+1}}||_* \tag{18}$$

In (18), $\boldsymbol{e}_{\boldsymbol{m+1}} = (0,0,\ldots,0,1)^T_{m+1}, m = J*T$. When p is one, two or infinity, $||\boldsymbol{e}_{\boldsymbol{m+1}}||_*$ equals to one all the time. So (14) can be reformulated into (19).

$$\lambda\delta - \varepsilon \leq 1/N\sum_{n\in[N]} \min\{\min_{i\in[I]}(\lambda\max\{\sum_{t\in[T]}\sum_{j\in[J]} Y_{ijt} - \alpha\hat{d}_{in}, 0\} - 1), 0\}; \lambda \geq 0 \tag{19}$$

Finally, it is assumed that $\delta > 0$, $\gamma = 1/\lambda$ and $\gamma \geq 0$, so (19) can be reformulated into (20). And apparently (19) and (20) are equivalent.

$$\delta - \varepsilon\gamma \leq 1/N \sum_{n \in [N]} \min\{\min_{i \in [I]}(\max\{\sum_{t \in [T]} \sum_{j \in [J]} Y_{ijt} - \alpha\hat{d}_{in}, 0\} - \gamma), 0\}; \gamma \geq 0 \tag{20}$$

However, (20) is still difficult to calculate. In order to linearize it, Z_n is introduced firstly to reformulate (20) into the next several constraints.

$$\delta - \varepsilon\gamma \leq 1/N \sum_{n \in [N]} Z_n \tag{21}$$

$$\begin{aligned} Z_n + \gamma &\leq \min_{i \in [I]} \max\{\sum_{t \in [T]} \sum_{j \in [J]} Y_{ijt} - \alpha\hat{d}_{in}, 0\}, \forall n \in [N] \\ &\leq \max\{\sum_{t \in [T]} \sum_{j \in [J]} Y_{ijt} - \alpha\hat{d}_{in}, 0\}, \forall n \in [N], i \in [I] \end{aligned} \tag{22}$$

$$\gamma \geq 0 \tag{23}$$

$$Z_n \leq 0, \forall n \in [N] \tag{24}$$

And then we introduce an auxiliary variable L_n, a binary variable y_n and a large number M_n related to n to reformulate (22) into the next several constraints.

$$Z_n + \gamma \leq L_n, \forall n \in [N] \tag{25}$$

$$L_n \leq \sum_{t \in [T]} \sum_{j \in [J]} Y_{ijt} - \alpha\hat{d}_{in} + M_n(1 - y_n), \forall i \in [I], \forall n \in [N] \tag{26}$$

$$L_n \leq M_n y_n, \forall n \in [N] \tag{27}$$

$$L_n \geq 0, \forall n \in [N] \tag{28}$$

$$y_n \in \{0, 1\}, \forall n \in [N] \tag{29}$$

In general, the probabilistic constraint (4) can be reformulated into (21), (23), (24) and (25) to (29).

As for the individual probabilistic constraints (12), they can be regarded as the same kind constraint with (4) but only for one demand site. By introducing auxiliary variables t_i, z_{in} and q_{in}, they can be reformulated into the next several constraints.

$$\delta - \varepsilon_1 t_i \leq 1/N \sum_{n \in [N]} z_{in}, \forall i \in [I] \tag{30}$$

$$-\hat{d}_{in} + A_i + \sum_{t \in [T]} \sum_{j \in [J]} Y_{ijt} + Mq_{in} \geq z_{in} + t_i, \forall n \in [N], i \in [I] \tag{31}$$

$$M(1 - q_{in}) \geq z_{in} + t_i, \forall n \in [N], i \in [I] \tag{32}$$

$$t_i \geq 0, \forall i \in [I] \tag{33}$$

$$z_{in} \leq 0, \forall i \in [I], n \in [N] \tag{34}$$

$$q_{in} \in \{0, 1\}, \forall i \in [I], n \in [N] \tag{35}$$

Therefore, based on Wasserstein ambiguity set, DRMNP model is successfully reformulated into a mixed integer model which is much easier to solve.

4 CASE ANALYSIS

Considering the need to distribute supplies in Pudong District of Shanghai during the epidemic, this paper takes 15 existing pharmacies (15 evenly distributed in Pudong) as alternative distribution centers and 36 towns of Pudong as demand sites to analyze the network planning strategy within 5 time periods (delivery every 12 hours). The hardware environment of the experiments was Intel(R) Core (TM) I5-6300HQ CPU @ 2.30ghz, and the memory was 12GB. Gurobi 9.1.1 solver was imported in Python for solving.

4.1 *Parameter values*

Table 1. Parameter values.

Parameter	Values
f_{jt}	U[4+1.5*(T-t),5.5+1.5*(T-t)] *10000 (Yuan/unit)
β	2 (Yuan/KM/unit)
M_t	42+30*t (unit)
α	0.8
ε	0.2
ε_1	0.01
δ	$\{0.001:0.001:0.01;0.02:0.01:0.1;0.2:0.1:1;1.1:0.1:1.5\}$
M, M_n	10000000

Some parameters are given in Table 1. Taking water as an example, it is assumed that about 0.1% to 0.2% of the population needs supplies, and each person consumes 9 liters of water in 60 hours, with 12 bottles per box as the transportation unit. Therefore, demand sample values are given in Table 2, where U represents uniform distribution.

Table 2. Demand sample values.

Demand Site	Demand Values	Demand Site	Demand Values	Demand Site	Demand Values
Beicai Town	U (154,308)	Huinan Town	U (156,312)	Shuyuan Town	U (32,63)
Caolu Town	U (100,200)	Jinqiao Town	U (47,93)	Tang Town	U (88,176)
Chuansha New Town	U (169,337)	Jinyang New Village	U (93,187)	Tangqiao Street	U (40,81)
Datuan Town	U (33,65)	Kangqiao Town	U (154,308)	Wanxiang Town	U (16,31)
Dongming Street	U (64,128)	Laogang Town	U (18,37)	Huaifang New Village	U (53,107)
Gaodong Town	U (53,106)	Lujiazui Street	U (60,120)	Xinchang Town	U (55,109)
Gaoqiao Town	U (97,195)	Nanhui New Town	U (74,147)	Xuanqiao Town	U (40,80)
Gaohang Town	U (84,168)	Nanmatou Street	U (55,109)	Yangjing Street	U (80,161)
Hangtou Town	U (105,210)	Nichen Town	U (53,106)	Zhangjiang Town	U (129,258)
Heqing Town	U (70,140)	Puxing Street	U (86,173)	Zhoujiadu Street	U (67,134)
Hudong New Village	U (53,106)	Snlin Town	U (180,359)	Zhoupu Town	U (136,272)
Huamu Street	U (128,256)	Shanggang New Village	U (47,95)	Zhuqiao Town	U (121,242)

4.2 *Wasserstein radius analysis*

We set N as 30, 60 and 90 and choose the radius from δ. Thirty experiments are conducted independently. Based on the cross-validation method (Peyman et al. 2018), for each experiment, eighty percent of samples are chosen as training set and twenty percent of samples are chosen as test set to calculate the objective value (OBJ), probabilistic constraint satisfaction rate (PROB). The results are shown in Figures 1 to 3.

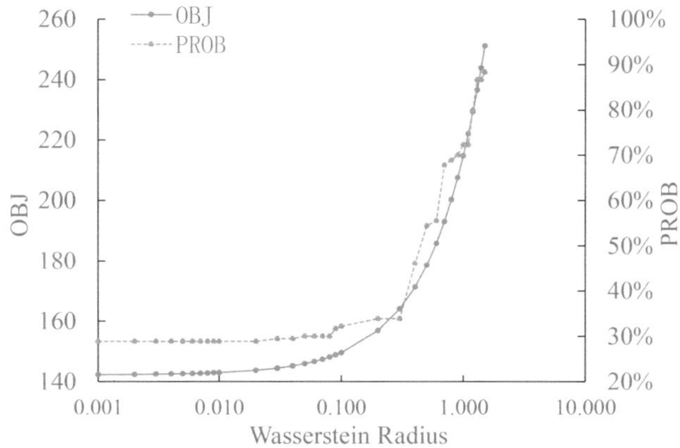

Figure 1. Wasserstein radius effect (N = 30).

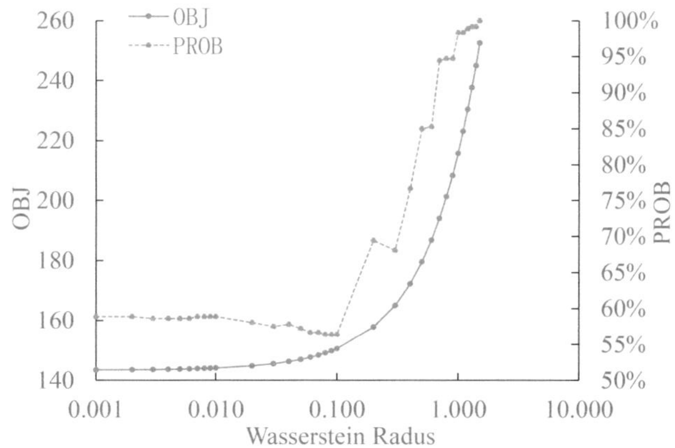

Figure 2. Wasserstein radius effect (N = 60).

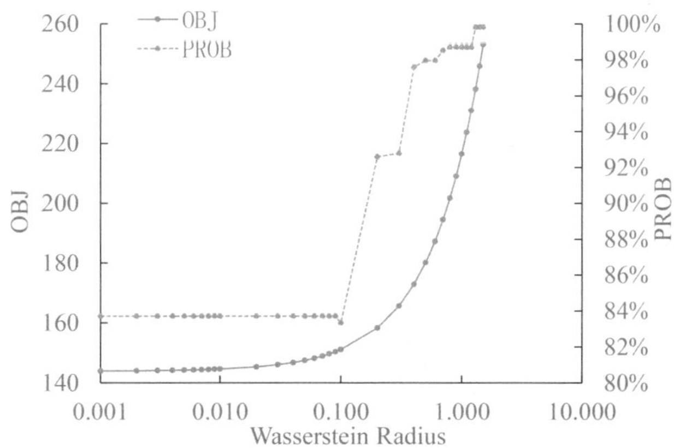

Figure 3. Wasserstein radius effect (N = 90).

It shows that OBJ increases with the Wasserstein radius, and its size is almost not affected by sample size. With the increase of Wasserstein radius, PROB decreased slightly at first and then increased gradually, and increased significantly with sample size.

4.3 *Model analysis*

To verify the DRMNP model, a Stochastic Multi-period Network Planning model (SMNP) is designed to compare with DRMNP. The same experimental method is taken to calculate OBJ, PROB and the average probability of demand satisfaction rate of single demand site (prob). The optimal Wasserstein radius should be selected according to (36) to compare with SMNP effectively, because the optimal OBJ and PROB cannot be guaranteed at the same time.

$$\delta_N \in \arg\max_{\delta_N^i \in \delta} \left\{ 1 - \frac{OBJ(\delta_N^i)}{\sum_{i=1}^{33} OBJ(\delta_N^i)} + \frac{PROB(\delta_N^i)}{\sum_{i=1}^{33} PROB(\delta_N^i)} \right\} \tag{36}$$

In SMNP, it is a two-stage model based on SAA method, so the objective function turns to (37) and three new constraints are added. The SAA method evenly considers the impact of each scenario on the whole.

$$\begin{aligned} \min \sum_{t\in[T]}\sum_{j\in[J]} f_{jt}X_{jt} + 1/N \sum_{n\in[N]} [\beta \sum_{t\in[T]}\sum_{i\in[I]}\sum_{j\in[J]} c_{ij}Y_{ijtn} + \\ \sum_{i\in[I]}\sum_{t\in[T]}\sum_{j\in[J]} \mu_t Y_{ijtn} + \mu_1 \sum_{i\in[I]} \max\left(d_i - \sum_{t\in[T]}\sum_{j\in[J]} Y_{ijtn}, 0\right)] \end{aligned} \tag{37}$$

$$s.t. \frac{1}{N}\sum_{n=1}^{N} \left\{ \alpha \hat{d}_{in} \leq \sum_{t\in[T]}\sum_{j\in[J]} Y_{ijtn}, \forall i \in [I] \right\} \geq 1 - \varepsilon \tag{38}$$

$$\sum_{i\in[I]} Y_{ijtn} \leq M_t \sum_{T=1}^{t} X_{jT}, \forall j \in [J], t \in [T], n \in [N] \tag{39}$$

$$Y_{ijtn} \geq 0, \forall i \in [I], j \in [J], t \in [T], n \in [N] \tag{40}$$

equation (6) $\sim$ (7)

The results are shown in Table 3. It shows that although the costs of DRMNP model are larger than that in SMNP model, PROB and prob both increase greatly, which meets the first principle of emergency logistics to meet the demands of disaster victims. Besides, SMNP model takes a long time to be solved, which is more than 300 seconds, and the optimality gap (GAP) increases with sample size, while the DRMNP model takes less than 10 seconds to be solved. Therefore, the performance of DRMNP model is significantly improved compared with that of SMNP model.

Table 3. Model comparison results.

Sample	SMNP				DRMNP			
Size	OBJ	PROB	prob	GAP (300s)	OBJ	PROB	prob	GAP (10s)
20	1087703.34	0.00%	57.00%	0.02%	2339875.68	73.75%	99.06%	0.00%
40	1066563.71	0.00%	56.68%	0.59%	1767867.16	83.75%	99.43%	0.00%
60	1077354.86	0.00%	59.93%	2.14%	1776114.32	90.42%	99.73%	0.00%
80	1137704.16	0.00%	72.48%	8.02%	1565779.95	92.50%	99.75%	0.00%
100	1154446.12	0.00%	74.89%	10.71%	1566735.70	92.25%	99.78%	0.00%

Figure 4 shows the results of emergency supplies network planning. Eleven distribution centers are gradually built during the first three periods, with distribution flows changing over time. The cross shape represents the selected distribution center. The hollow point represents other candidate facilities, and the solid point represents the location of demand sites.

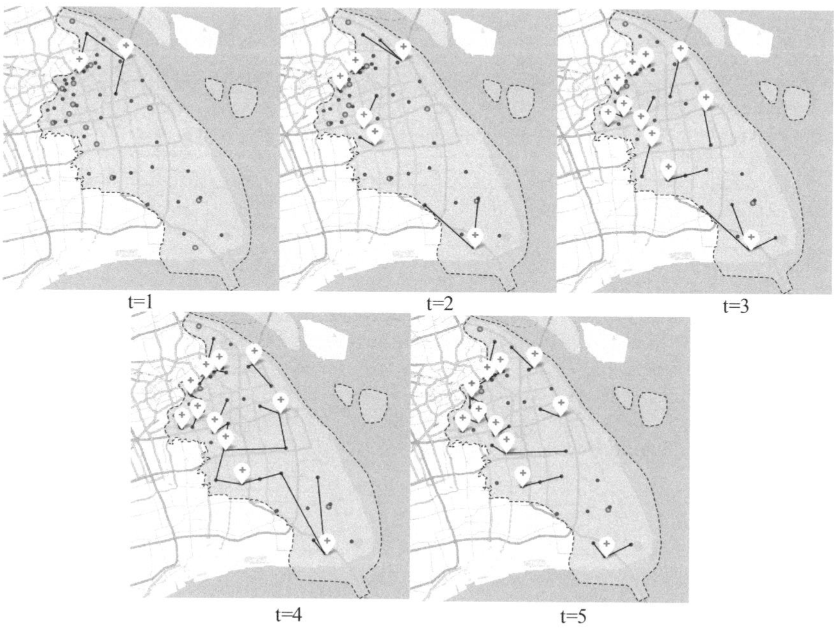

Figure 4. Multi-period emergency network planning.

5 CONCLUSION

The emergency logistics network with strong robustness can better resist various uncertainties after the occurrence of disasters, so as to guarantee people's life safety more effectively. In this paper, a data-driven distributionally robust optimization model is established for dynamic emergency network planning problem considering demand uncertainty, dynamic evolution of emergencies and supply penalty cost. Based on the classical Wasserstein ambiguity set, it is reformulated into a mixed integer programming model. The main conclusions are as follows. With the increase of Wasserstein radius, the satisfaction rate of probabilistic constraint decreases slightly first and then increases gradually, and increases significantly with sample size, but the cost also increases. The optimal Wasserstein radius can be selected by balancing these two aspects. Compared with SMNP model, the DRMNP model based on classical Wasserstein ambiguity set can better meet the requirements of demand sites, and the computational difficulty is lower. The multi-period decision made by the model fully considers the dynamic changes in the rescue process and makes the resource allocation process more reasonable.

ACKNOWLEDGMENT

This work was supported by Fundamental Research Funds for the Central Universities of Ministry of Education of China (2021RC202), National Natural Science Foundation of China (72101021) and China Postdoctoral Science Foundation (2021M690009).

REFERENCES

Carmen, G.R. & Mark, A.T. 2009. Pre-positioning of emergency supplies for disaster response. Transportation Research Part B 44(4): 521–534.

German, A.V. & Maria, E.M. et al. 2020. Prepositioning disaster relief supplies using robust optimization. IISE Transactions 52(10): 1122–1140.
José, H.V. & Noel, P. et al. 2013. On the appropriate objective function for post-disaster humanitarian logistics models. Journal of Operations Management 31(5): 262–280.
Jose, B. & Karthyek, M. 2019. Quantifying Distributional Model Risk via Optimal Transport. Mathematics of Operations Research 44(2): 565–600.
Kanglin, L. & Qiaofeng, L. et al. 2019. Distributionally robust optimization of an emergency medical service station location and sizing problem with joint chance constraints. Transportation Research Part B 119: 79–101.
Lawrence, V.S. & Zümbül, A. et al. 2015. OR/MS models for supply chain disruptions: a review. IIE Transactions 48(2): 89–109.
Peyman, M.E. Daniel, K. 2018. Data-driven distributionally robust optimization using the Wasserstein metric: performance guarantees and tractable reformulations. Mathematical Programming 171(1–2): 115–166.
Weijun, X. 2019. On distributionally robust chance constrained programs with Wasserstein distance. Mathematical Programming 186: 115–155.
Zhuolin, W. & Keyou, Y. et al. 2021. Multi-period facility location and capacity planning under ∞-Wasserstein joint chance constraints in humanitarian logistics. in press.

Advances in Urban Engineering and Management Science – Khalil & Yang (Eds)

Research on improvement design scheme of power distribution system in university building

Li Yang*
Department of Infrastructure and Planning, Beijing Jiaotong University, Beijing, China

Feifei Hu*
The Logistics Group, Beijing Jiaotong University, Beijing, China

Li Chen*
Department of Infrastructure and Planning, Beijing Jiaotong University, Beijing, China

ABSTRACT: In this paper, aiming at the difficulty of insufficient power supply capacity and difficult to implement power capacity increase in the comprehensive building of colleges and universities, a comprehensive building in our school is taken as the research object. Based on the idea of mining power supply capacity from itself, the improvement design scheme of distribution system is studied to alleviate electricity tension, reduce building energy consumption and improve the intelligent level of electricity management.

1 INTRODUCTION

The complex buildings in colleges and universities often have multiple functions such as teaching, office, scientific research and experiment. With the increase of the types of electrical equipment and the total power of electricity, they are faced with the problems of insufficient power supply capacity and low reliability of power supply. Therefore, it is urgent to increase power capacity. However, in the first-tier cities, it is difficult to approve the power capacity increase scheme, and the construction cost is too high to achieve.

Sun Yu adopts intelligent control system to realize active load control based on college students' dormitory, which plays an energy-saving goal of peak shifting and valley filling and stable load operation (Sun & Hou 2019). Han Yuzhuo believes that the distribution trunk line of scientific research buildings should not adopt pre-branch cable or T connection form, and should use intensive bus slot (Han 2020). Wei Jianjun selected a reasonable design scheme according to the existing housing structure, and adopted an intelligent distribution monitoring system to realize the safe, reasonable and economic operation of the power supply system and improve the reliability of the distribution system (Wei 2020). Cheng Wenan proposed the design scheme of fire distribution system in building electrical design (Cheng 2020).

In this paper, the ninth teaching building of our school is taken as the research object. On the premise of keeping the total capacity of high voltage power supply line and transformer unchanged, based on the idea of mining power supply capacity from itself, the composition and characteristics of power load are statistically analyzed, and the distribution system is improved. On the demand side, the technical measures and management measures of saving power and safe power consumption are taken. On the supply side, the load allocation is carried out to reduce the peak-valley difference of load and improve the reliability of power supply. The intelligent monitoring equipment is added to the distribution system at all levels to realize the monitoring of distribution equipment and power load. The corresponding modules of energy supervision platform software are upgraded, hoping

*Corresponding Authors: lyangl@bjtu.edu.cn, ffhu@bjtu.edu.cn and lchen@bjtu.edu.cn

DOI 10.1201/9781003305026-21

to alleviate the tension of school electricity, reduce building energy consumption and improve the intelligent level of power management.

1.1 *Statistical analysis of electricity load*

The ninth teaching building is a complex building, which was put into use in 1991. The building area is 24,000 square meters, 6 floors above ground and some underground layers. The original distribution system: the total distribution room is set in the first layer, two 630KVA transformers, high voltage inlet for a 10kV cable, two low voltage bus segments using single bus connection; The low voltage system in the building adopts the trunk power supply mode, which is supplied by the outlet cabinet of the general distribution room through the trunk cable to each floor. Four distribution boxes (concealed) are evenly arranged on each floor in the corridor, and the distribution box supplies power to each room through the embedded line. With the increase of high-power equipment such as experiments, offices, air conditioners and servers, the capacity of transformers and cables has been unable to meet the growing demand for power load.

1.1.1 *Electricity load statistics*

The rooms in this building are mainly used for classrooms, offices, experiments, scientific research, television stations, as well as security, fire protection, water supply, heating and other infrastructure. See Table 1 for electricity load classification statistics.

Table 1. Electricity load classification and case study (Unit: kW).

Power supply	Lighting	Experimental socket	Air-conditioner	Fire control Security	Emergency Lighting	Installed capacity	Calculated capacity
401 Input line	45	231	190	32	15	1693	543
402 Input line	44	191	230	32	15	1425	542

In the form lighting includes general lighting and professional lighting; The socket includes office equipment, experimental station, multimedia equipment, projection and display equipment, broadcast equipment, server, etc. Air conditioning includes wall hanging machine, cabinet, VRV air conditioning equipment and water heater; Fire control and safety precaution includes fire alarm host, fire broadcast equipment, CRT display equipment, leakage fire alarm, fire door monitoring, fire power monitoring, smoke exhaust fan, fire stabilizer pump, security monitoring equipment, access control system, etc. Elevator includes four elevators and ancillary equipment; Emergency lighting is the emergency lighting equipment in public areas, some computer rooms and laboratories, TV stations, power distribution rooms, water pump rooms, fire control rooms, etc.

1.1.2 *Electricity load analysis*

(1) Load classification: The electric power of single equipment is mainly professional lighting, experimental station, multimedia equipment, server, air conditioning, electric water heater, fire pump, elevator, etc. Electric equipment is divided into continuous power consumption, seasonal power consumption load, fixed period power consumption, repeated short-term power consumption, short-term power consumption load. Among them, the continuous electricity load is mainly the server and special air conditioning in the computer room, and the seasonal electricity load is mainly the general-purpose air conditioning. The electricity load in the fixed period is mainly teaching, experiment, scientific research, electric water heater and television station. The repeated short-term electricity load is mainly elevator, and the short-term electricity load is mainly fire protection.

(2) Electricity peak valley: According to the daily monitoring data of the energy regulatory platform, the peak period of electricity consumption is 11:00–17:00 working period, and the trough

period is 22:00–7:00. The peak daily load curve of power consumption before the improvement of low voltage 402 incoming line in general distribution room is shown in Figure 1.

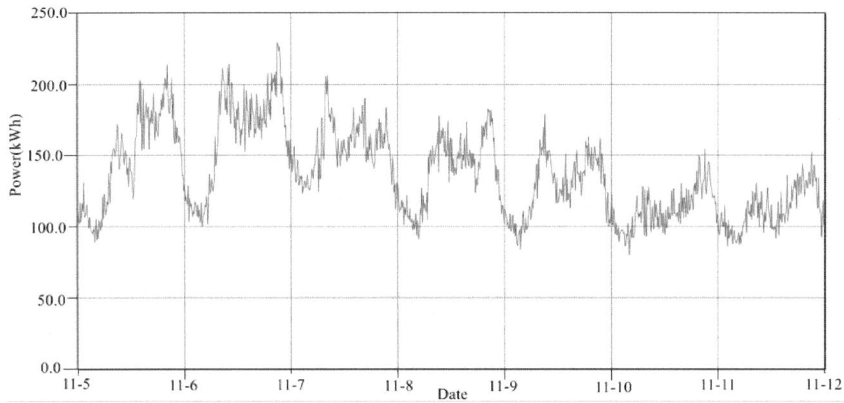

Figure 1. Load curve from NOV 5 to 12, 2018.

2 IMPROVED DESIGN SCHEME OF DISTRIBUTION SYSTEM

2.1 *Power supply and line*

Household electricity management and hierarchical radial distribution are adopted to improve power supply reliability and facilitate fault detection and processing. In order to reduce the influence of construction, the distribution system reconstruction is divided into three stages of construction, and the original strong and weak power system routing is not used.

The inlet cable is changed to double 10KV power supply, the total capacity of high voltage inlet and two transformers remains unchanged.

The low voltage outlet cables of the general distribution room are radial, along the outdoor power tunnel and indoor electrical shaft, and the cable bridge is laid to the south and north distribution rooms of each floor.

One distribution room is added in the middle positions of the south and north sides of each floor, and two distribution cabinets are set in the south and north distribution rooms. The cable bridge is added in the corridor to supply power to the distribution boxes in each room of the floor.

Each room is equipped with an independent distribution box. The incoming line of the power source is taken from the distribution cabinet on the floor, and the outlet line adopts multi-loop power supply according to the load classification in the room.

Plane and vertical figures of improved distribution system as shown in Figures 2 & 4.

2.2 *Load allocation*

According to the statistical analysis results of power load, load allocation is carried out to reduce the peak-valley difference of low voltage incoming line load in general distribution room, avoid transformer overload, reduce line loss and transformer loss, and realize economic operation.

The users with large power and continuous power load, security and fire load are arranged in different low voltage bus sections as average as possible.

The users with seasonal electricity load and fixed period electricity load are arranged in the same low voltage bus section according to the reasonable combination of electricity time. The educational administration department needs to adjust the scheduling time to reduce the peak value of electricity.

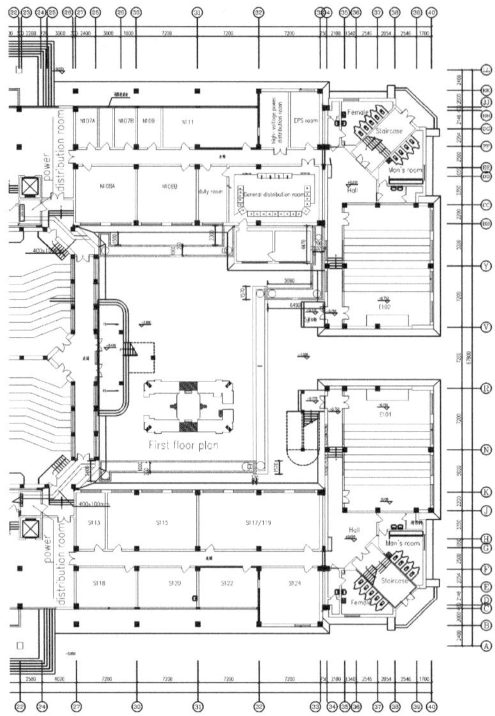

Figure 2. Plane graph of the 1-layer distribution trunk line.

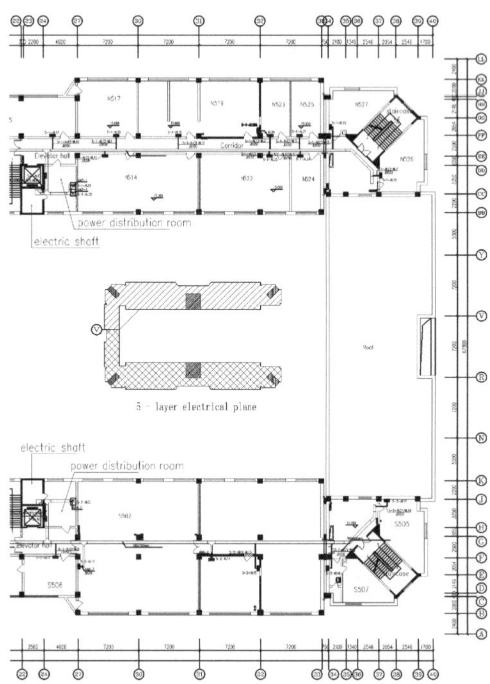

Figure 3. 5-layer electrical plan.

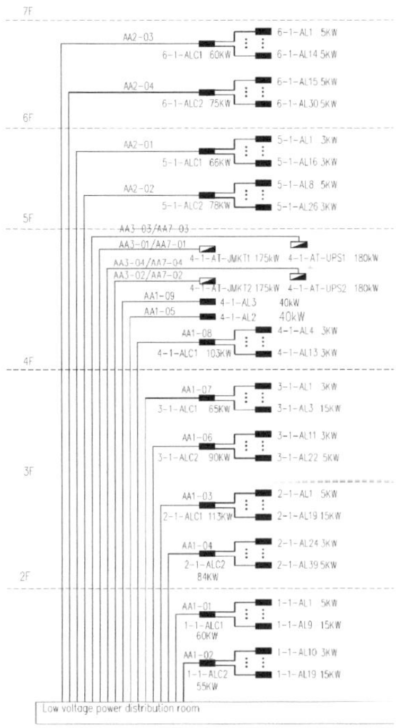

Figure 4. Vertical distribution.

With high-power experimental load and scientific research load as the main user, set up two-way power supply, respectively from two low-voltage bus, according to the load curve automatically switch operation mode in advance to avoid transformer overload.

According to the load allocation situation, re-compiled the low-voltage outgoing sequence of the general distribution room to ensure the reliability of a class of load power supply when one power failure or maintenance.

The improved distribution system is shown in Figure 5.

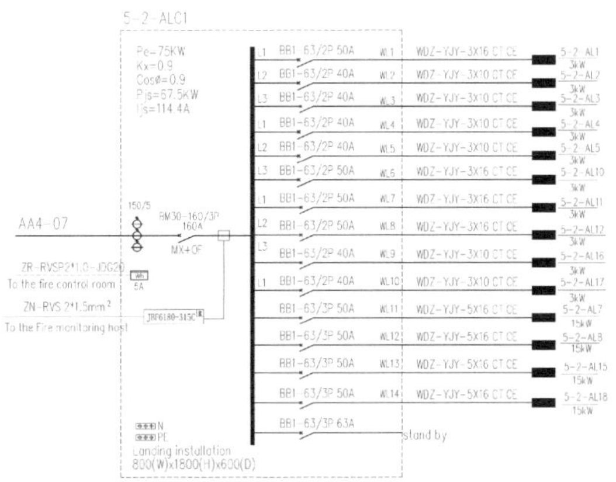

Figure 5. Improved distribution system.

The peak daily load curve of low voltage 402 incoming line in general distribution room is improved as shown in Figure 6. It is expected that the peak daily load of the 402-inlet line after the overall completion of the project will be reduced by 20- 25 %, the peak-valley difference decreases by 25 %.

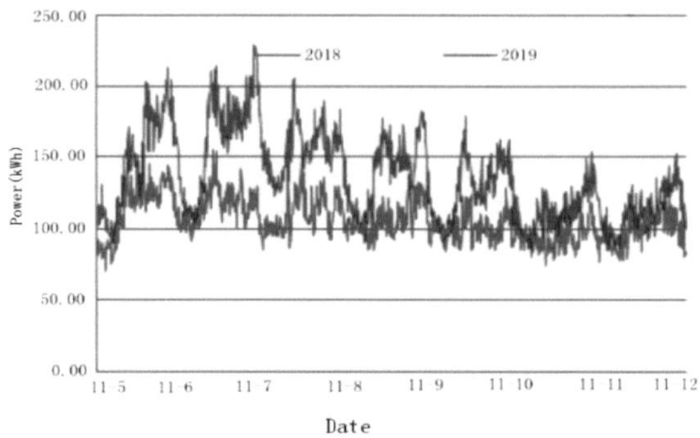

Figure 6. Power curves 5–12 November 2018 and 5–12 November 2019.

2.3 *Adding EPS system*

The original fire protection, emergency lighting, security facilities and other power supply systems in this building are self-contained power lighting devices with dual power switching and decentralized settings. EPS system is added to improve the reliability of power supply.

Set EPS emergency power supply and related equipment (capacity 50kVA) near the general distribution room.

The dual power supply switching distribution boxes are installed in the distribution rooms of each floor, the fire control room, the 6-floor smoke exhaust room and the 7-floor fire water tank to realize the dual power supply switching between the normal municipal power supply and the EPS emergency power supply.

All the emergency lighting, evacuation instructions, fire control host, fire equipment power monitoring host, electrical fire monitoring host, fire smoke exhaust equipment, fire door monitoring and other fire safety and security equipment in this building are powered by dual power mutual switching distribution box. The distribution system is powered by municipal power under normal conditions, and emergency power supply provided by EPS under abnormal conditions.

2.4 *Adding fault detection device*

The automatic inspection of the robot is added to the general distribution room: the general distribution room is unattended, and the guide rail and intelligent robot, high-definition camera are set to realize the automatic inspection of the equipment in the general distribution room without dead angle, and the image is uploaded to the school general supervision room to realize the real-time monitoring of the distribution room.

Pyrolysis Particle Monitoring Host is set up in the general distribution room, Pyrolysis Particle Detector is set up in the distribution cabinet, electrical shaft, cable bridge and cable trench, and communication between the detector and the host is formed through data lines. When the temperature of molded case circuit breaker, connection terminal or cable increases to a set value, the detector perceives the material particles decomposed by heating and starts, and the alarm signal is issued by the pyrolysis particle monitoring host and the school general monitoring room.

Set up an electrical fire monitoring host in the fire control room, set up an electrical fire detector in the outlets of the distribution cabinets in the general distribution room, the incoming lines of the distribution cabinets in the floor distribution room and the incoming lines of the distribution box with dual power supply, the incoming lines of the distribution box in the large computer room and the room of important users, and communicate between the detector and the host through data lines. When the residual current monitored by the detector reaches a set value, the electrical fire monitoring host sends an alarm signal.

Taking the above measures can timely detect electrical equipment failure, effectively prevent the occurrence of electrical fire, improve the reliability and safety of distribution system operation. individually or in bulk, manually or automatically according to preset logic, while the latter can only be operated individually.

3 CIRCUIT RECONSTRUCTION SCHEME IN ROOM

3.1 *Household management*

Each room is equipped with an independent distribution box and multi-functional intelligent monitoring terminal to realize one household, one table, and each household independent measurement, measurement and monitoring.

According to the electrical equipment in the room, the electricity consumption index of each room is formulated in advance, and the basic free electricity consumption quota is provided monthly. The excess part is borne by the users themselves, so as to improve the users' awareness of electricity saving.

Adopting multi-rate time-of-use pricing to encourage users to use electricity at peak load.

3.2 *Approval and loading of high-power electrical equipment*

According to the regulations on the management of laboratory equipment in schools, the high-power electrical equipment for testing, scientific research and teaching needs to be reported to the school electricity management department in advance in the equipment procurement stage, and can be purchased through the school state-owned asset management office after approval. Before the installation of equipment, it is necessary to report to the school's electricity management department again. According to the demand of electricity, the equipment access location and power consumption period are arranged, and the reasonable load allocation is carried out to meet the needs of electricity management.

3.3 *Energy-saving equipment and management strategy*

All lighting fixtures in this building are illuminated by LED light source, and subregional human-induced lighting controllers are used in larger classrooms, computer rooms and public areas to save energy.

The intelligent air conditioning socket is installed in all the air conditioning equipment in this building, and the remote opening and shutdown control of the air conditioning can be realized by the intelligent induction function of the socket. At the same time, the running state of the air conditioning can be automatically adjusted according to the temperature in the room, and the power supply can be automatically alarmed or cut off when the equipment fails.

4 ENERGY REGULATORY PLATFORM UPGRADE

Our school energy supervision platform has realized the campus electricity, water, gas, heat all kinds of total energy classification sub-measurement; The improved design scheme upgrades the related contents of the energy supervision platform and the distribution system of the ninth teaching building.

Multi-functional intelligent monitoring terminals are set up in the main distribution room, the floor distribution room and the room distribution box, which are sent to the collector and the network controller through the data line. Through the communication between the campus network line and the background of the energy supervision platform and the school general monitoring room, the hierarchical monitoring and control of the three-level equipment of the substation, distribution and electricity in the building are realized. Measurement includes voltage, current, frequency and other electrical parameters, transformer, busbar temperature and other non-electrical measurement; Monitor the running state of electrical equipment, realize remote control and automatic switching of switching equipment.

The intelligent monitoring terminal added to the distribution cabinet and room distribution box in the floor distribution room is added to the management software, and the user account is established in the room. For the first time, the distribution system management of our school is upgraded from the secondary management (school general supervision room and building general distribution room) to the four-level management (school general supervision room, building general distribution room, floor distribution room and room), so as to improve the intelligent level of power consumption management and promote the fine management process of energy consumption data.

5 CONCLUSIONS

In this paper, the distribution system of the ninth teaching building of our school is improved. In the supply side, the power supply line is added, the load allocation is implemented, the radiation graded distribution is used, and the fire safety EPS is added to improve the reliability of power supply and reduce the peak-valley difference of load. In the demand side, technical measures such as LED light source lamps and human body induction lighting controller are adopted, and management measures such as formulating electricity consumption indicators and high-power electrical equipment are adopted to save electricity, reduce building energy consumption, smooth load peak and alleviate the pressure of electricity shortage. The automatic inspection device, pyrolysis ion monitoring device, electrical fire monitoring device and intelligent monitoring terminal are added to the distribution system at all levels, and the corresponding modules of the monitoring platform software are upgraded to ensure safe power consumption and improve the intelligent level of power management.

REFERENCES

Cheng WA "Analysis of fire distribution design scheme in building electrical design, Planning and design", 2020,10, pp. 44–45.

Han YC et al, "Distribution system design of a Sino-foreign cooperative university training building", Intelligent building electrical technology, 2020,14 (1), pp. 79–81.

Sun Y, and Hou WJ. "Discussion on intelligent allocation method of transformer load," Electrical technology of intelligent building, 2019,13(6), pp. 58–61

Wei JJ et al, "Power capacity expansion and distribution retrofit schemes and applications of the Institute", Intelligent buildings and smart cities, 2020 (10), pp. 24–26.

Advances in Urban Engineering and Management Science – Khalil & Yang (Eds)
© 2023 The Author(s), ISBN 978-1-032-30426-7

Research on schedule optimization of the construction project based on resource supply

Chang-Ming Hu & Jing-Ya Gong*
School of Civil Engineering, Xi'an University of Architecture and Technology, Xi'an, Shaanxi, China

Shuai Wu & Xiao-Feng Fan
Northwest Branch of China Construction First Building (Group) Co., Ltd., Xi'an, Shaanxi province, China

ABSTRACT: In this paper, construction schedule control is studied based on resource-constrained project scheduling problems. Firstly, the origin and solution of resource-constrained project scheduling problems are introduced through the literature review. Secondly, the optimization model of the construction schedule is established according to the construction characteristics of linear engineering, including the establishment of project data structure, constraint conditions, and the formula of the objective function. Based on this, a research idea is proposed to calculate the construction schedule of a project under the condition of limited resources from the perspective of resource supply. The supplied value of limited resources was innovatively substituted into a genetic algorithm as the gene value of a single-stranded chromosome. Under the constraints of renewable resources and activity priorities, the optimal combination of all kinds of limited resource allocation is selected to calculate the "duration-cost" dual objective optimization scheme of the project. Then, a practical case is used to analyze the calculation experiment and results, which verifies the feasibility of the idea and method.

1 INTRODUCTION

Resource Constraint Project Scheduling Problem (RCPSP) is a polynomial solution to a complex combination Optimization Problem in a system. In construction engineering, it is mainly used in the theoretical calculation of construction project schedule management to describe the management information of each dimension of project activities. Besides calculating project duration and cost by Gantt chart, single code network chart, or double code network chart, scholars from various countries have extended and expanded the content and methods of RCPSP from various angles. For example, activity priority rule, resource constraint type, scheduling optimization goal, activity execution mode, project parallel number, etc., researchers proposed a variety of RCPSP solution models for different combinations of these factors.

We find that most of the mathematical models of resource-constrained scheduling follow the generalized priority relationship between activities and the constraints of renewable resources to solve the objectives of the construction schedule: duration and cost. In other words, we list all the activities of the project, determine the start time of the activities with the overlapping relationship and priority rules, balance the direct cost and management cost of the whole project with the allocation of project resources, and finally reach the best "time - cost" dual objective scheme. However, in the established model, the constraint conditions are about time and cost, and the result

*Corresponding Author: 362425359@qq.com

DOI 10.1201/9781003305026-22

is also about the two-objective optimization of time limit and cost. Most studies just end up with two simple indicators that are better. What we calculated in the mathematical model was not only the two simple indicators of cost and time limit but the construction schedule corresponding to the two optimal indicators, that is, the resource allocation of the project, the start time of various activities, and the entry and exit time of personnel and materials.

However, in the process of construction schedule control, it is more consistent with the actual situation of the project to change the construction schedule due to the influence of uncertain factors and adjust the resource supply. Therefore, this paper will study the construction schedule control based on the scheduling optimization problem of resource-constrained projects. The supplied value of the constrained resource is used as the gene value of a single-stranded chromosome, and the genetic algorithm is used to establish the intelligent optimal scheduling model of the resource-constrained project. Under the constraints of renewable resources and activity priorities, the optimal combination of various limited resource allocations of the project is selected to calculate the project duration and cost target value. Then in the construction process according to the actual situation to further adjust the resource demand, to achieve the purpose of a time limit and cost control.

2 PROBLEM DESCRIPTION

The classic RCPSP problem is described as follows: AON-directed acyclic graph $G=(V,E)$ is used to describe a project network plan consisting of N activities. $V=\{0,1,2,\ldots,n,n+1\}$ represents the set of all activities contained in a project and $E=\{(i,j)\mid i,j\in V; i\rightarrow j\}$ represents the set of logical lap relationships between activities. Let the set of activity duration of projects be $T=\{(S_i,F_i)\mid i\in V\}S_i$ is the start time of activity i, F_i is the finish time, and the duration of activity i is d_i, then $S_i+d_i=F_i$. In this project, all kinds of activities required resource collection for $R=\{(k,r,c)|k\in N\}$, where k means resource types, R said the supply of such resources, c said the unit price of this resource. The demand of activity i for resource K is denoted as r_{ik}. *Pred(j)* represents the set of activities before activity j. The activity j must be set in the *Pred(j)* all can begin after the activity, in which $Pred(j)=\{i|(i,j)\in E\}$. *Succ(i)* represents the set of activities immediately after activity I. Pooled *Succ(i)* all activities can only start after the end of the activity i, namely $Succ(i)=\{j|(i,j)\in E\}$.

According to the above description and notation, the model of dual-objective project scheduling optimization problem studied in this paper under the constraint of variable resource capacity is expressed as follows:

$$min\ maxF_i, i\in pred\ (n+1) \tag{1}$$

$$min\ C=\sum\nolimits_{t=1}^{T}\sum\nolimits_{i=1}^{N}\sum\nolimits_{k=1}^{R} r_k\cdot c_k \tag{2}$$

$$s.t.F_i+FTS_{ij}\leq S_j, j\in Succ(i) \tag{3}$$

$$\sum\nolimits_{i\in V_t} r_{ik}^{\rho}\leq R_k^{\rho}, \forall k,t \tag{4}$$

In the above optimization model, objective (1) is to minimize the project duration, and objective (2) is to minimize the direct cost of the project, including the cost of personnel, materials, and machinery of the sub-project and measure project. Formula (3) is the logical lap relation constraint between activities. Activity J can only start after the completion of the preceding activity I with the interval time. Formula (4) represents the renewable resource constraint, indicating that the demand for resource K of the ongoing activity set V_t in period T does not exceed the overall supply R_k^{ρ} of the project.

Table 1. Project resource allocation and resource requirements of various activities.

					Need							
Num	Resource	Type	Supply	Unit price (RMB)	pile foundation	cushion cap	I-cap	pier stud 1	pier stud 2	Cast-in-place beam	steel box beam	bridge deck system
1	Steel fixed	Manpower	80	350	5	6	6	6	6	30	/	20
2	carpenter	Manpower	70	200	/	6	6	/	/	40	/	10
3	concreter	Manpower	80	300	3	5	5	7	7	25	/	25
4	scaffolder	Manpower	50	350	/	/	/	6	6	35	/	/
5	welder	Manpower	40	400	3	4	4	5	5	20	20	10
6	rebar	Materials	15825t	6310	3t	45t	44t	10t	11t	280t	/	20t
7	commercial concrete	Materials	89148m^3	320	40m^3	55m^3	50m^3	56m^3	70m^3	1200m^3	/	/
8	steel box girder	Materials	19680t	4200	/	/	/	/	/	/	19680t	/
9	bituminous concrete	Materials	166200m^3	380	/	/	/	/	/	/	/	9300m^3
10	wood planking	Revolving material	6000m^2	25	/	80m^2	90m^2	/	/	3000m^2	/	/
11	C steel formwork	Revolving material	3	3000	/	/	/	1	1	/	/	/
12	C1 steel formwork	Revolving material	1	3000	/	/	/	1	1	/	/	/
13	rotary drilling rig	Machine	4	2800	1	/	/	/	/	/	/	/
14	crawler crane	Machine	7	600	/	/	/	/	/	/	2	/
15	Pushing device	Machine	2	50	/	/	/	/	/	/	2	/
16	crane-QY25	Machine	4	840	1	/	/	1	1	2	1	1
17	Self-leveling mixture paver	Machine	2	2500	/	/	/	/	/	/	/	2
18	asphalt distribution truck	Machine	2	200	/	/	/	/	/	/	/	2
19	Intelligent tension system	Machine	6	270	/	/	/	/	/	6	/	/
20	Intelligent vacuum grouting system	Machine	6	320	/	/	/	/	/	6	/	/

3 DATA STRUCTURES AND ALGORITHMS

All kinds of sub-projects of large-scale construction projects are independent and interfere with each other, and have the construction characteristics of parallel lap operation mode and docking operation mode. In this paper, a new calculation idea is designed: first, each construction activity is modeled, that is, the number of human material machines needed to complete the activity and the time and cost needed to complete the activity are listed. Then all the activities are connected in series according to the space-time lapping logic, and the corresponding single code network diagram can be obtained. The resource supply is then calculated, and the number of activities working in parallel in the same period is calculated according to the degree of resource constraint. Plus, the time and resource cost required by the measure project, and then calculate the construction schedule and cost of the system.

3.1 *Data structure unit — Activity execution pattern*

The execution model of activities is to produce structures with the same function for the same project unit, and the production materials required have different combinations. Statistics and measurement of various execution modes of project activities are needed before optimization calculation of construction schedule.

3.2 *Constraints and objectives*

The constraint conditions of construction projects are mainly divided into two categories, one is the logical overlap relationship between activities, the other is the work sequencing relationship between activities due to the limitation of key resources. The former relationship is rigid and generally difficult to change; The latter relationship is flexible and can be adjusted as the total amount of resources changes.

3.2.1 *Basic constraints*

a) Logical lap — generalized priority relation

As far as the construction process of a progress unit of a construction project is concerned, the overlap relation between the work can be simply arranged from the Angle of time and space. For example, the corresponding serial number and working duration of a certain bridge engineering project are shown in Table 2. The logical order of construction of a progress unit of the viaduct is the pile foundation → cap platform → pier column → beam plate → bridge deck system. Therefore, the single code network diagram corresponding to this project is shown in Figure 1.

Table 2. Contents and duration of project activities.

Num	item description	Time/day
0, 19	Start or End Node	
1–5	pile foundation	1 pillar/day
6, 7	cushion cap	1 cap/10days
8–10	I-cap	1cap/10days
12, 14	pier stud 1	1 pillar/12days
11, 13, 15	pier stud 2	1 pillar/12days
16	Cast-in-place beam	1 section/60days
17	steel box beam	1 section/45days
18	bridge deck system	1 section/25days

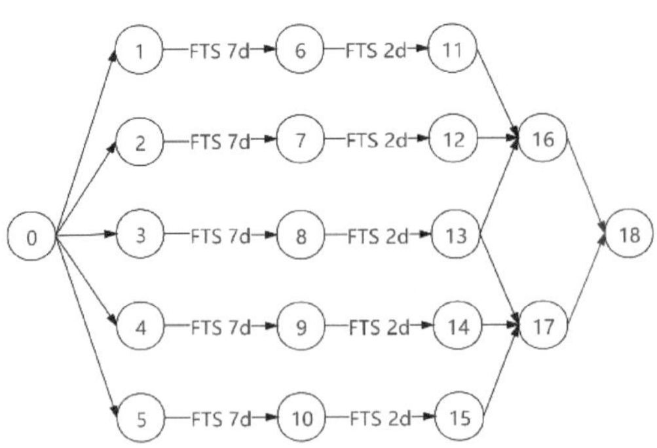

Figure 1. Single code network diagram.

b) Constraints on renewable resources

Renewable resources refer to the resources that can be reused. Its resource capacity is planned by the construction organization and designer in the early stage and can be further adjusted in the construction process. For example, the number of manpower and equipment. In the execution process, the demand of activity i for the renewable resource $k \in R^{\rho}$ is r_{ik}^{ρ}, and the available amount of the kind of k renewable resource at any time is R_k^ρ. The sum of the resources required by the activities under construction at any one time of the project must not exceed the total amount of the resources.

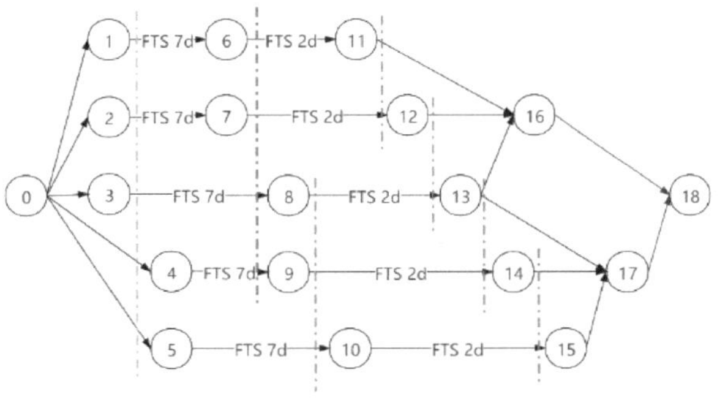

Figure 2. Schematic diagram of single code network with resource constraints.

3.2.2 *Target constraints*

a) Target period

Project duration minimization is the most common objective function of project scheduling. For such linear projects as highways and railways, it is unrealistic to equip all resources at one time with an extremely compressed construction period. However, a certain period target will in turn constrain the amounts of resources invested by the construction site. The model established in this paper also computes the target duration of the scheduling scheme generated by resource constraints. $F_{\mathrm{n}+1}$ is used to represent the completion time of the $n+1$ task, then the objective function can be expressed as:

$$min\, F_{n+1} \tag{5}$$

Among them:

$$F_{n+1} = \max F_i, i \in pred\ (n+1) \tag{6}$$

b) Target cost

The project cost includes resource cost (direct) and management cost (indirect). The most effective way to control project cost is to control total resource cost. The direct cost of the project consists of the sum of the cost of personnel, materials, and machinery and the cost of measures for each sub-project in the project. The principle of resource allocation of general construction units is to minimize the supply of circulating materials. How to control the allocation of resources determines the duration and cost of a project.

$$\min C_{\text{direct}} = C_{\text{project}} + C_{\text{measure}} = C_{manpower} + C_{materials} + C_{machine} = \sum_{t=1}^{T}\sum_{i=1}^{N}\sum_{k=1}^{R} r_k \cdot c_k \tag{7}$$

3.3 *Genetic algorithm*

For a given construction project, the manager should make up at least the following data: (1) the project includes all the activities and work hours, $V = \{(v_i, d_i) | i \in N\}$; (2) activities overlap between logic and time interval, $E = \{FTS_{ij} | \ (i, j)\}$; (3) Resource model required by various activities $R_i = \{k, r\}$; (4) Total limited resources and unit price of the project, $R = \{k, r, c\}$; (5) Schedule risk factors and their impact scope.

This paper uses a genetic algorithm to solve the NP-hard problem. According to the arrangement and analysis of the data structure, the variable data of the project is coded by a single-chain list. Each chromosome represents a group of schemes for the supply of various resources of the project, the gene position of the chromosome represents the number of resources, and the gene value represents the supply of resources. The genetic operators of the algorithm mainly include selection, crossover, and mutation. According to the actual situation, various parameters of the genetic algorithm are considered :(1) Population number; (2) Selection probability; (3) Crossover probability; (4) Mutation probability; (5) Termination criteria.

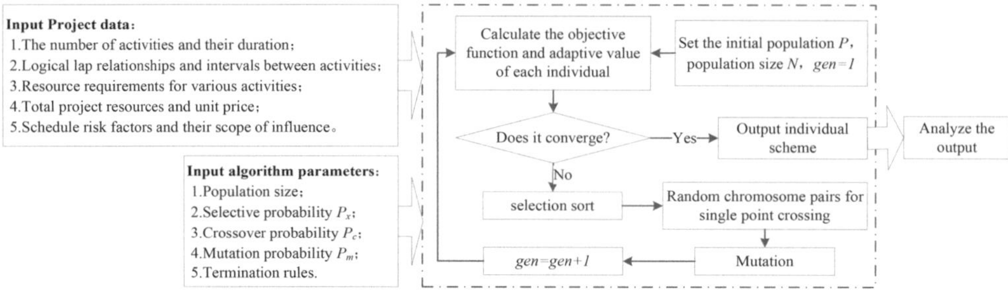

Figure 3. Genetic algorithm flow chart.

4 COMPUTATIONAL EXPERIMENTS

To verify the effectiveness of the model and algorithm in solving practical engineering problems, the scheduling problem of a comprehensive bridge project is analyzed in this paper. The comprehensive bridge project consists of 18 work activities. The main structure is composed of five parts, namely pile foundation, cap, pier column, beam plate, and deck system. Beam and plate parts are divided into the cast-in-place beam and steel box girder construction. Table 4 shows the comparison of resource allocation and schedule of each scheme before and after using the algorithm. By comparing the original scheme with the model optimized scheme, we can know how to control the start time of activities and which part of the resource supply should be increased or decreased to meet the construction needs. The control ability of the construction schedule has obvious directivity.

Table 3. Resource configuration and schedule of the solution.

Items		Content	Time/ Day	Cost/RMB
1	Resource allocation	[30, 40, 35, 35, 40, 3000, 5, 5, 5, 2, 2, 5, 2, 2, 6, 6]	145	10270440
	Activity start time	[0, 1, 1, 1, 1, 1, 12, 12, 12, 12, 12, 24, 24, 24, 24, 24, 36, 36, 96, 146]		
	Activity end time	[0, 4, 4, 4, 4, 4, 21, 21, 21, 21, 21, 35, 35, 35, 35, 35, 95, 80, 145, 146]		
2	Resource allocation	[36, 40, 32, 41, 40, 3000, 1, 1, 3, 2, 2, 3, 2, 2, 6, 6]	178	10237160
	Activity start time	[0, 1, 1, 1, 5, 5, 12, 12, 22, 22, 32, 24, 36, 48, 60, 72, 60, 84, 129, 179]		
	Activity end time	[0, 4, 4, 4, 8, 8, 21, 21, 31, 31, 41, 35, 47, 59, 71, 83, 119, 128, 178, 179]		
3	Resource allocation	[42, 40, 39, 47, 40, 3000, 2, 2, 1, 2, 2, 4, 2, 2, 6, 6]	157	10236400
	Activity start time	[0, 1, 5, 9, 13, 17, 12, 16, 22, 26, 32, 24, 28, 36, 40, 48,48, 60, 108, 158]		
	Activity end time	[0, 4, 8, 12, 16, 20, 21, 25, 31, 35, 41, 35, 39, 47, 51, 59, 107, 104, 157, 158]		

Scheme 1 is a construction layout plan without resource constraints. The same construction units will be constructed at the same time, and the demand for various resources is stable. The construction period of this scheme is extremely compressed, which requires the construction side to invest a lot of turnover resources and lease equipment, resulting in high costs.

Scheme 2 is the construction schedule arranged by the construction project manager according to the existing resources and management experience. Compared to Option 1, Option 2 achieves cost savings by reducing the supply configuration of revolving materials and leased machinery. This is also a principle for most project managers to prepare construction organization design. However, according to the layout results, the construction period of scheme 2 has weak control ability among all construction schedule planning and layout schemes.

Scheme 3 is the result of iterative calculation by the genetic algorithm designed in this paper. Compared with the previous two schemes, this scheme has better control ability in both the time limit and the cost. Compared with Scheme 2, scheme 3 can only better meet the requirement of cost-saving by estimating the resource allocation in the manual compilation of construction organization design according to the principle of reducing turnover resources. However, through the global search calculation of the capacity of resource allocation by the genetic algorithm model in this paper, an appropriate resource allocation plan can be obtained to meet the multi-objective optimization of construction schedule costs for resource-constrained projects.

5 CONCLUSIONS

In this paper, a construction schedule optimization model based on RCPSP is proposed to solve the resource allocation problem of large construction projects with multiple parallel flow construction. It has good guiding significance for actual engineering construction, mainly reflected in the following aspects:

1) A construction schedule optimization model is established to calculate the optimal project duration and cost by mobilizing the capacity of allocated resources. According to the literature review and construction characteristics of the construction project, the data structure is sorted out. It includes the establishment of project data structure unit, constraint conditions, and objective function formula introduction, solving algorithm, and the final result analysis.
2) The supplied value of limited resources was innovatively substituted into the genetic algorithm as the gene value of single-stranded chromosomes to solve the time-limit and cost-double objective optimization scheme of resource-limited projects. By transforming the objective function into the adaptive value, the resource allocation under the optimal solution state of "duration – cost" of the construction project can be sought.
3) Based on the results of computer case experiments, this paper concludes that the scheduling ability of the project manager's plan based on the principle of the construction project schedule is limited, and the model established in this paper can better search the construction schedule optimization plan under the condition of limited resources. The feasibility of the research ideas and methods proposed in this paper is verified.

REFERENCES

Feng Kong, Dong Dou. RCPSP with Combined Precedence Relations and Resource Calendars[J]. Journal of Construction Engineering and Management,2020,146(12):

Heng-Wei Wang, Jia-Rui Lin, Jian-Ping Zhang. Work package-based information modeling for resource-constrained scheduling of construction projects[J]. Automation in Construction,2020,109:

JTG/T 3832-2018. Highway Engineering Budget Quota (Part I, Part II) [S]. Ministry of Transport,2018-12-17

JTG/T 3833-2018. Highway Construction Machinery Station Fee Quota [S]. Ministry of Transport,2018-12-17

LIU Jia. Research on Optimization of Construction Project Scheduling Based on Network Planning [D]. Beijing Jiao-tong University, 2018.

MA Guo-Feng, ZHANG Ling-Zhi. Engineering construction schedule optimization based on improved genetic algorithm [J]. Journal of civil engineering and management, 2019,36(05):1–6.

Ripon K. Chakrabortty, Humayun Fuad Rahman, Michael J. Ryan. Efficient priority rules for project scheduling under dynamic environments: A heuristic approach[J]. Computers & Industrial Engineering, 2020,140:

Shou Yong-Yi. Model and Method of Resource-constrained Multi-Project Scheduling [M]. Hangzhou: Zhejiang University Press, 2010

Wang Jing, Zeng Sha-Jie, Ju Juan. Construction Schedule Management based on Resource-constrained Project Scheduling Model [J]. Journal of Tongji University (natural science), 2017, 45 (10) :1561–1568.

Wang Xue-Yan, Zhang Chao-hui, Hu Chang-Ming, et al. Progress Control of Multi-project Concurrent Construction [J]. Construction Technology, 2013,42(10) :100–104.

Xie Cun-ren, Xu Feng, Ruan Min-Hao. Research on Construction Schedule Optimization based on BIM and Genetic Algorithm [J]. Journal of Engineering Management, 201,35(03):117–122.

Xie Fang, Xu Zhe, Yu Jing. Dual-objective optimization of project scheduling problem with flexible resource constraints [J]. Systems engineering-theory & practice,2016,36(03):674–683.

Advances in Urban Engineering and Management Science – Khalil & Yang (Eds)
© 2023 The Author(s), ISBN 978-1-032-30426-7

Research on design and practice of the structural health monitoring system of highway tunnel based on BIM

L.X. Bao, Q.L. Wang, N. Ji* & Y. Jiang
Shanghai Urban Construction Design and Research Institute (Group) Co. Ltd., Shanghai, China

Yang Zhang
Shanghai Urban-Rural Construction and Transportation Development Research Institute Shanghai, China

ABSTRACT: In recent years, China's highway tunnel construction mileage is growing rapidly, but there are some problems in the tunnel structural health monitoring, such as low degree of information, monitoring is not in place, management level is not high. Aiming at these problems, the building information model is introduced into the structural health monitoring of highway tunnel, and the overall solution of the tunnel structural health monitoring system based on BIM and the establishment of BIM database are proposed. Based on Dalian Road Tunnel in Shanghai, a BIM based highway tunnel structural health monitoring system is established to provide an intuitive and convenient management method for highway tunnel structural health monitoring and improve the information management level of highway tunnel.

1 INTRODUCTION

By the end of 2019, there were 21316 highway tunnels, approximately 2.2 million meters, cross our country, China, including 1394 super-long tunnels and 5541 long tunnels (Ministry of Transport of the People's Republic of China 2020). Compared with the speed of tunnel construction, the structural health monitoring of tunnels in China is relatively backward, which is still in the static management state using inflexible management means, the paper record of pavement diseases and the account of data management (He et al. 2016).

In order to solve these problems, many scholars and engineers have conducted relevant research. Huang et al. (2017) designed and developed a highway tunnel operation and maintenance management system based on BIM. They designed the architecture and planned the tunnel BIM implementation and proposed the coding rule of highway tunnel equipment with BIM. Zhang (2019) researched the early warning and safety evaluation method of structural health monitoring for operating tunnel.

However, there is still necessary to conduct more in-depth and comprehensive research on highway tunnel structural health monitoring, and verify and promote the improvement of relevant technical methods by practice. The structural health monitoring system of highway tunnel based on BIM can provide an intuitive and convenient management tool and improve the information management level of highway tunnel.

2 REQUIREMENT ANALYSIS FOR TUNNEL OPERATION AND MAINTENANCE WITH BIM

At present, most of the existing tunnel management systems are two-dimensional, which can not only show the structural health monitoring of the tunnel various parts, but also difficult to

*Corresponding Author: jinanjy@foxmail.com

DOI 10.1201/9781003305026-23

comprehensively express overall operation state of the tunnel according to correlated multi-source data and characters. One way to overcome above problems is to build a three-dimensional, visual and comprehensive tunnel structural health monitoring system with data analysis, mining and other functions, by which the operation and management of the tunnel can be known, stored, seen, controllable, commanded, decided, shared and serviced.

The full name of BIM is building information modelling, that is one of computer-aided technologies to three-dimensional digital representations of physical and functional characteristics of a facility. The advantages of BIM application are information transmission, storage and sharing, and then modeling and visualization through these information processing. Thus, leading BIM into tunnel structural health monitoring can greatly meet the basic needs of operation and maintenance management in three dimensions. And because of the technical features, BIM can further improve the accuracy of monitoring data and the efficiency of operation and maintenance management. The automatic monitoring information of structural health monitoring sensors and the manual patrol inspection monitoring information and other monitoring information can be integrated in the BIM model. So that the development trend of tunnel structural health can be predicted directly, the weak links can be visualized, and the key monitoring objects can be determined.

3 THE STRUCTURAL HEALTH MONITORING SYSTEM OF HIGHWAY TUNNEL BASED ON BIM

3.1 *The design of system architecture*

Aim to build a safe, smart, efficient-operation tunnel, with data analysis and applications as the core, this system combines with professional structural safety monitoring equipment and modern communication technology, forms a set of abilities including three-dimensional visualization solutions, safety assessment and early warning. In the system architecture, data acquisition, real-time data transmission, central data warehouse construction and applications are demands for providing the integrated service of tunnel structural health monitoring (Duan 2019; He et al. 2016). As shown in Figure 1, the system consists of acquisition layer, communication layer, data layer, analysis layer and application layer.

(1) Acquisition layer. This layer mainly collects the front-end multi-source heterogeneous data. The data acquisition methods include automatic monitoring and manual patrol inspection. The automatic monitoring uses professional sensors to realize automatic environmental monitoring, deformation monitoring, stress and strain monitoring; the records will be written down by manual patrol inspection, that including structural appearance, embedded part status, osmotic water, et al.
(2) Communication layer. This layer builds a safe, stable and efficient communication network to transmit the data collected by the front end to the background center in real time. According to the realization of sensor and data form, the communication methods can adopt serial port, Internet of things, mobile cellular network, LAN network, etc.
(3) Data layer. This layer collects, cleans, integrates and stores data on the basis of data acquisition and communication, and forms a BIM database for decision-making and applications in combined with the construction of BIM model and other information of the tunnel.
(4) Analysis layer. This layer is the key layer for the system to give full play to the value of data application. Through models and algorithms, the data calculates and outputs required results in this layer. Combined with advanced technologies such as big data and artificial intelligence, this layer can better carry out structural health status assessment and risk assessment.
(5) Application layer. This layer provides users with functions such as display, information release, information sharing and report generation, and interfaces with other systems to realize information-based, visualized, integrated, interlinked, and large-scale tunnel management, so as to further improve the management efficiency.

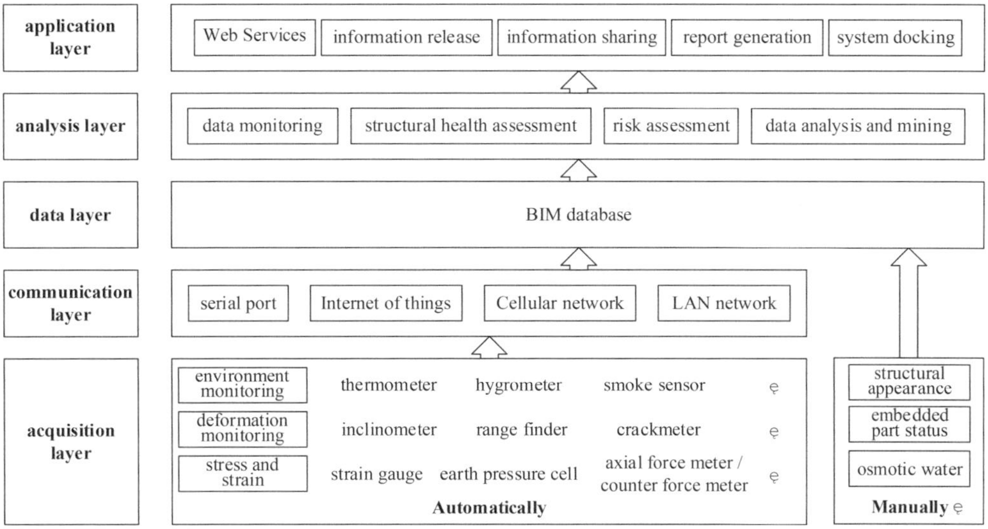

Figure 1. The architecture of the structural health monitoring system of highway tunnel based on BIM.

3.2 *The establishment of BIM database*

As the foundation and core, BIM database is mainly used to store the key information of tunnel structural health monitoring including basic information of tunnel (design data, construction data, completion data, etc.), monitoring data (raw data and processed data), patrol inspection data (constructional element disease description, disease pictures, patrol inspection and maintenance forms), early warning and evaluation data (model parameters, threshold, early warning and evaluation results, etc.), system management information (unit management, personnel management, authority management, etc.). That means that the system can realize effective and whole-process management for all dynamic and static data. Considering the large amount of data and various types of information, the database provides a reliable distributed data exchange and storage platform to share information for the development and using of various applications. The specific technical objectives are as follows:

- Provide tools and places for data storage;
- Provide tools for data security and user management;
- Provide tools to ensure data consistency and synchronization;
- Provide strategies and tools for data backup and recovery.

3.3 *Data backup and access management*

Because BIM database is the core of the system data management and applications, it is significant to ensure its safety. We major consider data backup and access management to ensure safety of the database.

Data backup. Set up the mechanism of database backup. With the help of redundant disk array and the timing function of software, the periodic backup of database can be realized, and the availability of backup need to be tested regularly.

Access management. Multiple user permissions such as user management, user group management and access control are used to manage the secure access to the database.

- User management. Provide the functions such as user list, user query, adding new user, deleting user, modifying user information/ password/ authority and so on.

– User group management. Provide the functions like user group list, user group query, adding new user group, deleting user group, modifying user group and so on.
– Access control. Set the operation permission of each data object, user/ user group and the query permission of operation permission.

4 THE CASE STUDY ON THE STRUCTURAL HEALTH MONITORING SYSTEM OF DALIAN RD TUNNEL, SHANGHAI

Dalian Rd Tunnel, Shanghai is the first shield river crossing tunnel constructed by BOT mode, and it has become a representative case of intelligent transformation of open tunnels (He 2016). Dalian Rd tunnel has a total length of 2.5 km and consists of two two-way four lane tunnels with a diameter of 11 m. The design speed is 40 km/h and the clear height is 4.5 m.

4.1 *The architecture of system function*

The structural health monitoring system of Dalian Rd tunnel is built on the basis of the above architecture. This system, as shown in Figure 2, is composed of user authority and log management, data collection, data monitoring, data analysis, expert evaluation and some extended functions. The system combines BIM technology and some new generation information technologies, such like big data, Internet of things, artificial intelligence and so on. With the aid of these technologies, for each constructional element, modeling and data management can be done intuitively, timely and comprehensively.

On the one hand, these functions focus on providing an operation tool to user. Exactly, it provides a friendly operation and management interface to users to improve their work efficiency. Users can use the interface to maintain database, adjust data acquisition frequency, control some sensors, monitor the full-process of data processing, look into real-time structural monitoring curves and reports, etc. And user can also entry or export the data and reports that they need.

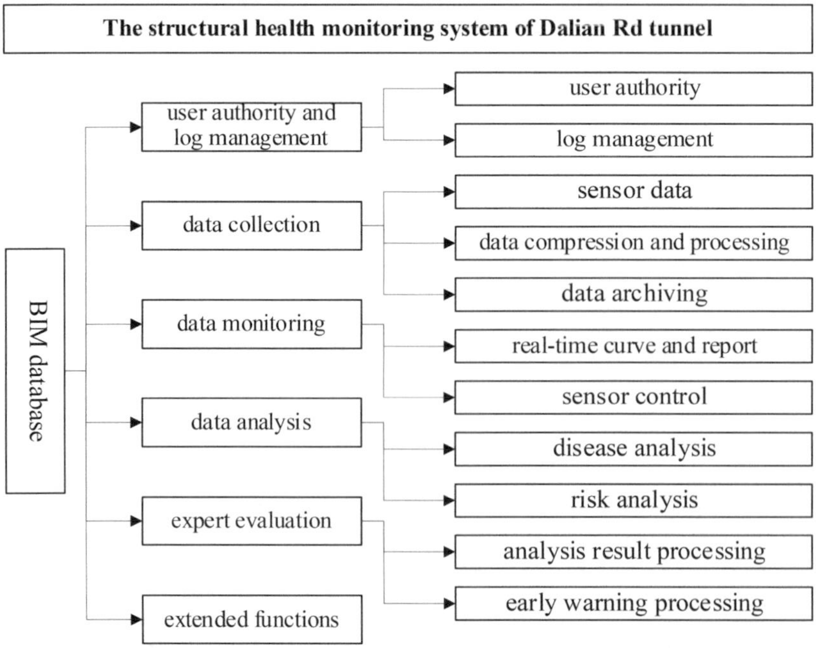

Figure 2. The architecture of system functions.

On the other hand, these functions focus on provide a decision-making tool to user. The tunnel structural diseases are identified, including the main structure of the tunnel, road constructional elements in the tunnel, tunnel ancillary facilities and system equipment. Therefore, the structural health status can be monitored and sound early warning, through which management and maintenance staffs can make right decision and respond rapidly.

4.2 *The system visualization and application based on BIM*

The aims of system visualization include daily maintenance data identification, accurate positioning of disease risk, identifying the diseases with different colors, which are effective to manage each constructional element. The system visualization mainly contains three contains:

- Data import into BIM model. Data sources include tunnel structure monitoring, video monitoring, environmental monitoring, fire early warning, etc. Also, the system accepts manual data entry. Through the interface programs, these data are push to the system and displayed in a comprehensive interface.
- BIM-based function integration and display. The BIM model provides integrated display, like tunnel structure monitoring sensor status, environmental monitoring and fire early warning. These basic information and tunnel traffic information can be queried by clicking the facility and equipment icon.
- The cockpit for tunnel maintenance and management. All diseases and warns are put together in the cockpit. On one cockpit page, the marked in different colors or with flashing constantly remind managers.

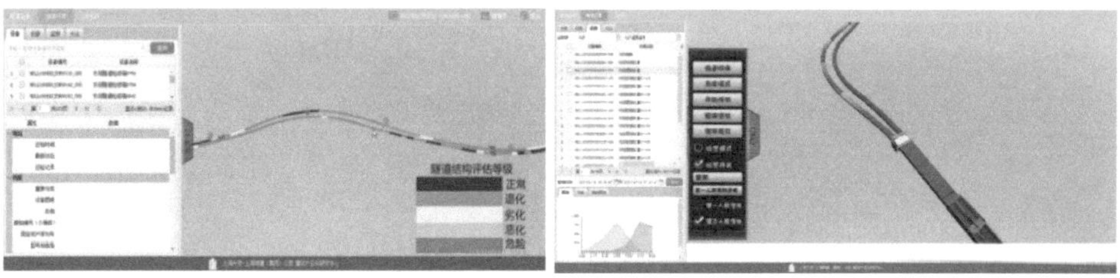

Figure 3. The system visualization and application.

5 CONCLUSION

In this article, we research about the design and the practice of the structural health monitoring system of highway tunnel based on BIM. Practical experience shows that, it has become a kind of reliable tool to help management and maintenance staffs. And with the focus of transportation infrastructure changing from construction to operation and maintenance in China, this informationalized structural health monitoring system based on BIM will play an important role not only on highway tunnel but also bridge to improve the efficiency and reasonableness of operation, management and maintenance.

According to the discussions presented above, the conclusions are obtained as below:

- The architecture of structural health monitoring system of highway tunnel based on BIM is provided. And the BIM database is established to be a vital part of the three- dimensional visual highway tunnel structural health monitoring system.
- In case study, the system has been designed in detail for Dalian Rd tunnel, Shanghai, by which an intuitive and convenient management mode for highway tunnel structural health monitoring is realized. The system further improves the information management level of the highway tunnel.

ACKNOWLEDGMENTS

This work is supported by Research Project of Shanghai Municipal Commission of Housing and Urban-Rural Development "Research on Operation Monitoring and Impact Analysis for Underground Road Network in Shanghai" and Talent Program of Shanghai Municipal Science and Technology Commission (19XD1432500).

REFERENCES

Duan, C.F. (2018). Research and application of whole life cycle health monitoring and evaluation of system for shield method highway tunnel, Building Construction, 40 (08), 1462–1464. (In Chinese)

Ministry of Transport of the People's Republic of China (2020). Statistical bulletin on the development of transportation industry in 2020. https://xxgk.mot.gov.cn/2020/jigou/zhghs/202105/t20210517_3593412.html.

He, Y.J., Zhang, J., Chen, G.Q., Tao, L. (2016) Study of cross river tunnel intelligent operation & maintenance management based on BIM. China Municipal Engineering, 2016 (S1), 81–84+119–120. (In Chinese)

He, X. (2016). A research of structure health monitoring system of long-term operation of Shanghai Dalian Rd. China Municipal Engineering. 2016 (01), 75–77+97. (In Chinese).

Huang, T., Chen, L.J., Shi, P.X., Yu, C.C. (2017). Design and development of BIM-based operation and maintenance system of highway tunnel. Tunnel Construction, 37 (01), 281–304. (In Chinese)

Zhang, K.N. (2019). Research on health monitoring and early warning and safety evaluation of operational tunnel structure. Huazhong University of Science and Technology, China: Wuhan. (In Chinese)

Advances in Urban Engineering and Management Science – Khalil & Yang (Eds)

Analysis on compression deformation of surrounding rock at the top arch of deep underground powerhouse

Shilei Zhang*
China Institute of Water Resources and Hydropower Research, Beijing, China;
Beijing Zhongshui Engineering Group Co., Ltd., Beijing, China

Hang Duan
China Three Gorges Construction Management Co., Ltd., Chengdu, China

Changhua Chen
China Yangtze Power Co., Ltd., Yichang, China

ABSTRACT: To explore the characteristics and mechanism of compression deformation of the top arch of deep-buried caverns, the displacement analysis method is used to analyze the compression deformation characteristics, and the elastoplastic theory is adopted to qualitatively analyze the shrinkage deformation driving mechanism. Furthermore, the characteristics of the prestressing load and their relationship with surrounding rock deformation in the shrinkage deformation zone are discussed. Research results show that the surrounding rock deformation presents significant spatial differentiation characteristics in the top arch. The deep part of the upstream side of the top arch produces shrinkage strain, and the distribution range along the axis of the plant is proportional to the depth. The shrinkage strain is concentrated in the range of 10.0m, and the magnitude decreases with increasing depth. The redistribution of stress caused by the excavation causes the deep layer of the top arch to produce an arch effect zone. The deep surrounding rock of the top arch on the upstream side is located at the "arch abutment", and its force is loaded along the radial direction of the top arch, resulting in shrinkage and deformation; the top arch on the downstream side is an unloading relaxation zone. The shrinkage deformation zone is related to the size of the cavern. As the excavation progresses to the deep part, the shrinkage strain of the relatively shallow surrounding rock weakens and even turns into tensile strain. The anchor cable prestress has significant stage characteristics. The empty bearing period verifies the compression deformation of the surrounding rock in the deep part of the top arch. When the range of the contraction zone exceeds the reinforcement depth of the anchor cable, the prestress is close to the locked value, and the steel strand only bears the pretension load, showing low or empty load.

1 INTRODUCTION

During the excavation process, the stress state of the surrounding rock is restricted by factors such as ground stress and structural planes, and the deformation characteristics of the surrounding rock are complex (JIANG, FENG, LI et al 2019). In recent years, scholars have carried out a lot of research on the deformation characteristics of cavern surrounding rock combined with the Three Gorges Project, Jinping and other major projects.

Peng et al. (PENG, WANG, DENG et al. 2007) analyzed the temporal and spatial laws based on the monitoring data of the displacement of the surrounding rock of the high sidewall, and concluded that the deformation of the surrounding rock of the cavern is mainly controlled by the opening displacement of the structural plane under the conditions of medium and low in-situ stress.

*Corresponding Author: zhangsl@iwhr.com

DOI 10.1201/9781003305026-24

Wei et al. (WEI, DENG, WANG et al. 2010, HUANG, HUANG, DUAN et al. 2011) studied the deformation characteristics of surrounding rock under high-stress conditions and low stress-to-strength ratios, and proposed that the proportion of rock mass unloading deformation under this condition increased. Lu et al. (LU, WANG, DING et al 2010) analyzed the effects of high-stress conditions and structural plane factors on the deformation and failure of surrounding rock based on geological survey and surrounding rock deformation monitoring. Jiang et al. (JIANG, FAN, FENG et al 2017) used intra-hole photography to obtain the progressive destruction of the rock mass inside the surrounding rock at the top arch from shallow to deep, and they believed that the surrounding rock inside the top arch and the surrounding rock of the cave wall have a similar deformation and failure process because they are both subjected to hoop loading. And the radial unloading action produces slabs or fractures; the macroscopic surrounding rock displacement is the expansion and accumulation of fractures. Restricted by the characteristics of the camera technology in the hole, the compression deformation cannot be observed. It is difficult to obtain the essential mechanical response of the deep surrounding rock by only observing the discrete fractures in the axial direction of the borehole. In terms of obtaining internal deformation information of surrounding rock, Yu et al. (YU, Lin, LI 2017) used transparent rock mass technology to establish a physical model of the cavern, and used digital photographic measurement technology to obtain spatial information and evolution process of surrounding rock deformation, which expanded the study area of surrounding rock deformation. On the basis of geological survey and monitoring data, Liu et al. (LIU, ZHU, CAI 2018, MENG, He, Chen et al 2020) elaborated on the control effect of the underground powerhouse tectonic stress field on the deformation and failure of the cavern group, and used a numerical model to analyze the deformation and deformation of the top arch of the Abietin underground powerhouse. The failure mechanism, but the deformation process of the surrounding rock in the deep stress concentration area is equated with the force mechanism of the surrounding rock of the cave wall.

Generally speaking, the existing research focuses on the force characteristics, deformation and failure mechanism of the cave wall or shallow surrounding rock, or simply equates the stress state of the deep surrounding rock with the shallow surrounding rock to explain the discrete fracture failure of the deep surrounding rock. Little research has expounded on the stress state of the surrounding rock in the stress concentration area of the top arch of the cavern in the high ground stress environment and studied the principle of the surrounding rock in the deep shrinkage deformation area of the top arch and the change law of the supporting load.

During the excavation of the underground powerhouse on the left bank of Baihetan Hydropower Station, a widespread shrinkage deformation zone appeared in the deep part of the upstream side top arch, while the downstream side top arch had no shrinkage deformation characteristics. Based on the monitoring data of the surrounding rock deformation and prestress of the top arch of the large underground powerhouse of Baihetan Hydropower Station, this paper intends to analyze the shrinkage deformation and anchor cable prestress characteristics of the upstream side of the cavern. The hard rock cavern with dominant structural stress is discussed. The force principle and deformation characteristics of the deep surrounding rock in the stress concentration area are expected to provide a scientific basis for the optimal design of similar underground projects.

2 PROJECT BACKGROUND

The underground powerhouse of Baihetan Hydropower Station spans 34 m, the largest underground powerhouse built in the world. The main and auxiliary workshops are 453.00 m long and 88.70 m high. The axis direction of the main and auxiliary powerhouses is N20°E, as shown in Figure 1.

During the excavation of the underground powerhouse on the left bank, the planned construction procedures were adjusted according to the monitoring results. The actual excavation sequence is shown in Figure 1 and Table 1. The underground powerhouse on the left bank was excavated in January 2014, and all excavation was completed on June 2, 2018 (Beijing IWHR Technology Co., Ltd. 2019).

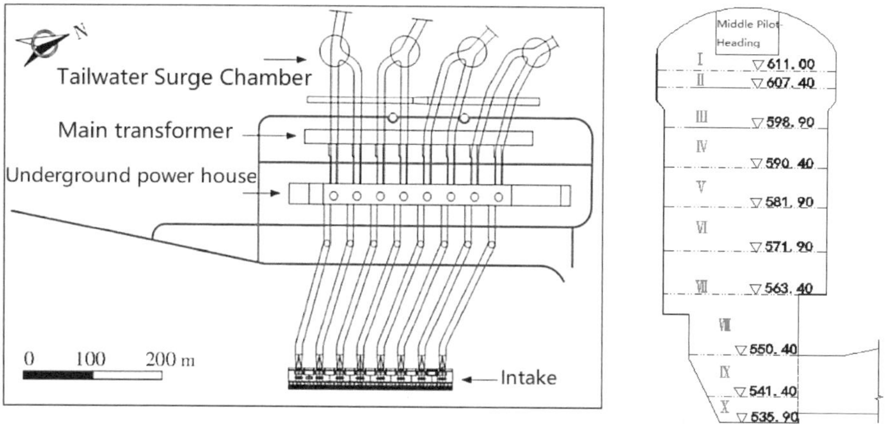

(a)Layout of the underground powerhouse on the left bank (b)Actual excavation sequence plan

Figure 1. Underground caverns on the left bank.

Table 1. Actual excavation information.

Construction procedures	Excavation size (width × height)	Excavation time
I	34m × 12.6m	201401-201412
II	34m × 3.6m	201411-201505
III	34m × 8.5m	201505-201512
IV	31m × 8.5m	201602-201610
V	31m × 8.5m	201609-201612
VI	31m × 10m	201611-201703
VII	31m × 8.5m	201702-201709
VIII	20m × 13m	201708-201711
IX	20m × 9m	201712-201805
X	16m × 5m	201801-201806

3 TEST RESULTS AND DISCUSSIONS

3.1 *Engineering geological conditions*

The dam site of the Baihetan Hydropower Station in the upper reaches of the Jinsha River is composed of Emeishan basalt in the Upper Permian system, with basaltic volcanic agglomerates in the upper part and compact basalt or porphyry basalt in the lower part. The underground powerhouse on the left bank is located in the upper mountain body of the dam abutment, with a horizontal buried depth of 600–1000 m and a vertical buried depth of 260–330 m. The surface is slope No. 2, leaning to the southeast.

The surrounding rock of the powerhouse on the left bank is mainly composed of fresh porphyry basalt, almond basalt, cryptocrystalline basalt, and breccia lava in the $P_2\beta_2^3$ and $P_2\beta_3^1$ layered rock flow layers. The intra-layer dislocation zone LS_{3152} is developed on the top of the $P_2\beta_3^1$ layer, with a width of 2 cm and breccia structure rocks in the zone. In addition, the underground powerhouse area also developed f_{717}, f_{718}, f_{719}, f_{720}, f_{721}, f_{722}, f_{723} and other faults. The fault zone is dominated by jointed structural rocks and structural breccias. The engineering geology section of the plant is shown in Figure 4 (Hangzhou: Power China Huadong Engineering Corporation Limited, 2013, 2007).

The surrounding rock of the roof arch of the powerhouse is dominated by type III_1, accounting for 64%, type II surrounding rock accounting for 17%, type III_2 located at the top arch of unit 7#

and 8#, accounting for 3%. The type IV surrounding rock is divided into a gently inclined structure Near the surface, accounting for 7%. The development of the LS_{3152} intra-layer dislocation zone in the upper part of the top arch is unfavorable to the stability of the top arch.

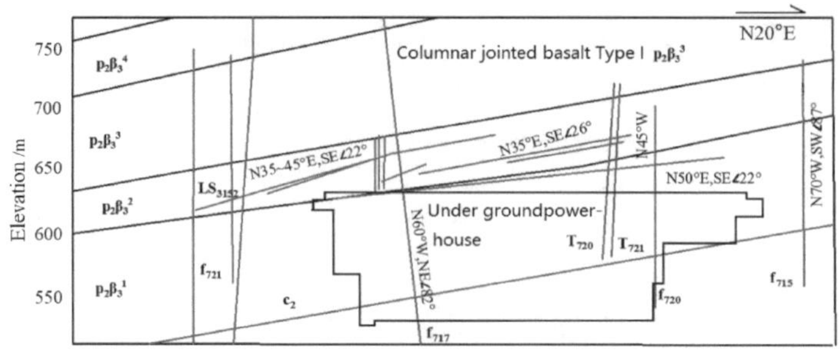

Figure 2. Engineering geological section of the upstream wall of the underground powerhouse on the left bank.

3.2 *Characteristics of the initial stress field*

Baihetan Hydropower Station is located on the north side of the Xiaojiang fault zone and the east side of the Jiaojihe fault. Affected by the Himalayan movement (regional compression in the NW-NWW direction), the stress of the surrounding rock of the main powerhouse on the left bank of the Baihetan Hydropower Station is dominated by tectonic stress.

Excavation unloading redistributes the stress of the surrounding rock and forms a secondary stress field. The mechanical behaviour of the surrounding rock depends on the stress history and the secondary stress field (HUANG, HUANG, DUAN et al. 2011). The distribution of the measured values of the principal stress of the initial stress field is shown in Figure 3, and the relationship between the maximum principal stress and the axis of the powerhouse is shown in Figure 4.

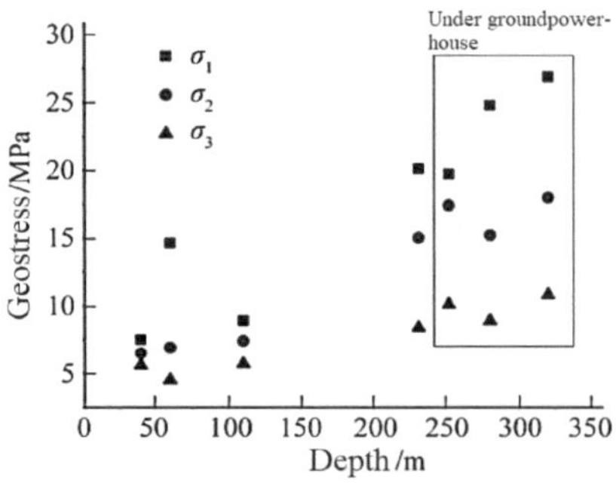

Figure 3. Distribution of the measured values of principal stress in the initial stress field of the underground powerhouse on the left bank.

The horizontal stress in the plant area is greater than the vertical stress. The first and second principal stresses are basically horizontal, and the third principal stress is roughly vertical. The maximum principal stress value of the initial stress field is between 20–30 MPa, locally up to 34 MPa, the azimuth is between N30°–50° W, and the inclination angle is 5°–13°. The direction is nearly horizontal and intersects with the plant axis at a large angle. The minimum principal stress is about 10 MPa, and the direction is nearly vertical. The angle between the direction of the large principal stress and the axis of the plant is relatively large, which is detrimental to the stability of the surrounding rock (Hangzhou: Power China Huadong Engineering Corporation Limited, 2013, 2007). The strength-stress ratio of the rock in the plant area is 3–5. Considering the stress redistribution caused by the surrounding rock of the cavern excavation, the magnitude of the stress concentration is 50–60 MPa, and the strength-stress ratio of the rock in the plant area is revised to be between 2–3 (Hangzhou: Power China Huadong Engineering Corporation Limited, 2019). Generally speaking, it belongs to the high in-situ stress area (Beijing: China Planning Press, 2009).

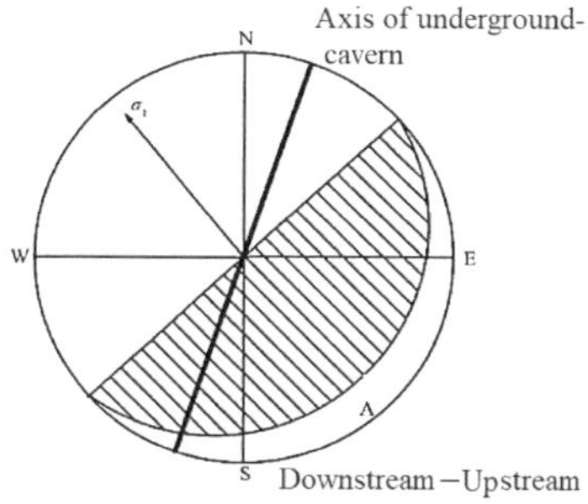

Figure 4. The relationship between the maximum principal stress of the underground powerhouse on the left bank and the axis of the powerhouse.

3.3 *The layout of measuring points of specimens*

The deformation monitoring arrangement of the top arch enclosure eight main monitoring sections are arranged in the underground main powerhouse on the left bank (stake number: left factory 0–051.6, left factory 0–012.9, left factory 0+018.4, left factory 0+076.0, left factory 0+152.0, left factory 0+228.0, left factory0+328.0). The anchor heads of the multi-point displacement meter are respectively 1.5 m, 3.5 m, 6.5 m and 11 m, 17m away from the surface of the cave wall. The plan layout of the main monitoring section and the typical layout of the top arch multi-point displacement meter are shown in Figures 5 and 6 (Beijing IWHR Technology Co., Ltd. 2019, Hangzhou: Power China Huadong Engineering Corporation Limited, 2019).

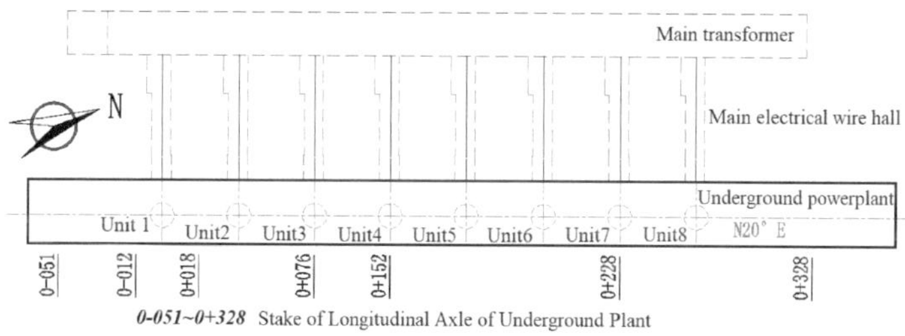

Figure 5. Layout of monitoring section.

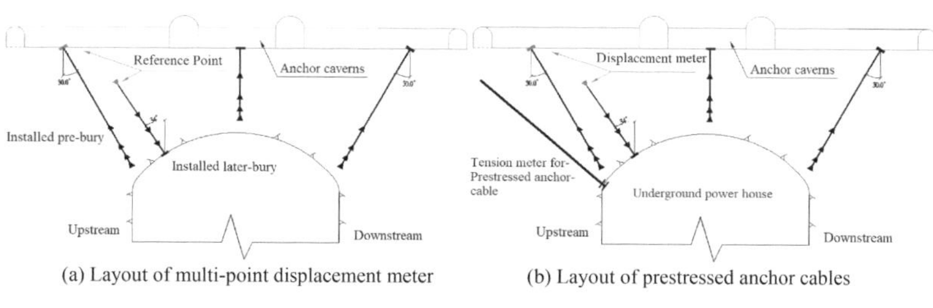

Figure 6. Typical layout of measuring points for the arch.

For the embedded multi-point displacement gauge, the calculation of the deformation value selects the fixed point at the base of the bottom of the plant roof anchor tunnel; for the direct-buried multi-point displacement gauge, the deepest anchor head is selected as the fixed point. The deformation value is positive for stretching and negative for compression.

3.4 *Spatial distribution characteristics of the deformation of the surrounding rock of the top arch*

As shown in Figure 7a, at a depth greater than 3.5m from the upstream side to the cave wall, the surrounding rock deformation has significant spatial differentiation along the axis of the powerhouse. The cumulative deformation of surrounding rock is generally less than 20 mm, and the shrinkage deformation is widely distributed along the axis of the plant, reaching the maximum shrinkage deformation of 3.55 mm in the 0+328 section. The deformation that occurred in the 0+018~0+076 section of the factory is significant, with the maximum magnitude reaching 63.57~63.68 mm, which was 2.5~3.0 times of other sections. The 0+018~0+076 tunnel section of the factory is affected by the dislocation zone LS_{3152} in the soft structure surface with a gentle dip angle, and the rock mass under the dislocation zone is unloaded and deformed significantly at the weak surface.

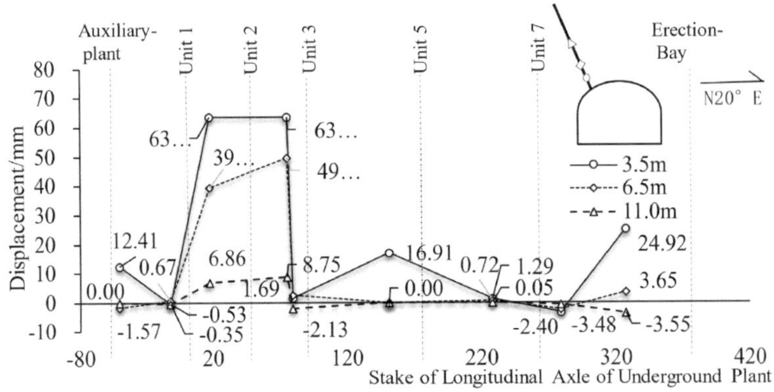

(a) the deformation of the deep surrounding rock on the upstream roof

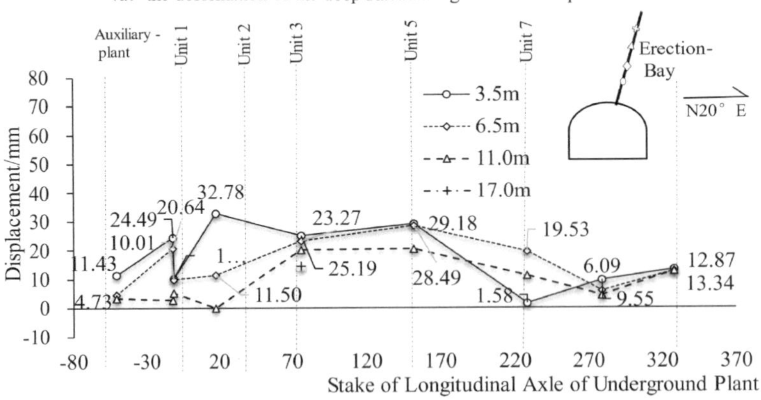

(b) the deformation of the deep surrounding rock on the downstream roof

Figure 7. The displacement distribution characteristics of the upstream and downstream top arches.

As shown in Figure 7b, the surrounding rock at a depth greater than 3.5 m from the cave wall on the downstream side shows tensile deformation, with a cumulative deformation magnitude of 29.18~32.78 mm, and no shrinkage deformation zone. This is obviously different from the cumulative deformation distribution characteristics of the surrounding rock on the upstream side: on the upstream side, except for the 0+018~0+076 section of the factory, the surrounding rock at depths greater than 6.5 m has no opening and deformation characteristics, and "0" displacement or even shrinkage generally appears.

Comparing the upstream and downstream distribution rules, in the III1 surrounding rock tunnel section, the deformation of the deep surrounding rock on the upstream side is relatively low, and the shrinkage deformation zone is widely distributed. The top arch on the downstream side is the unloading relaxation zone, and the surrounding rock deformation is dominated by relatively significant tension. Combined with the geological survey, it is found that the section of the cavity generated by the shrinkage deformation corresponds to the concentrated area. It can be seen that the deep part of the stress concentration area has generated shrinkage deformation. In the cave section with large-scale weak structural planes, structural plane factors become the main reason for controlling surrounding rock deformation.

Figure 8 are the distribution characteristics of the surrounding rock deformation along the axis of the powerhouse between the adjacent measuring points of the multi-point displacement meters on the upstream side top arch after excavating the cavern. In section 1.5–3.5 m away from the cave wall, the shrinkage deformation of the surrounding rock accounts for 1/3, and the opening deformation of the surrounding rock is less than 10mm accounted for 44.4%. In section 3.5–6.5 m away from the cave wall, the proportion of shrinkage deformation of surrounding rock increased to 44.4%, and the magnitude of shrinkage deformation was −1.02∼−1.08 mm. The shrinkage deformation of surrounding rock occurred in the >11m section accounted for more than 50%, and the shrinkage deformation magnitude reached −3.55 mm, which occurred at the section 0+328, as well as the magnitude of the opening deformation less than 10 mm.

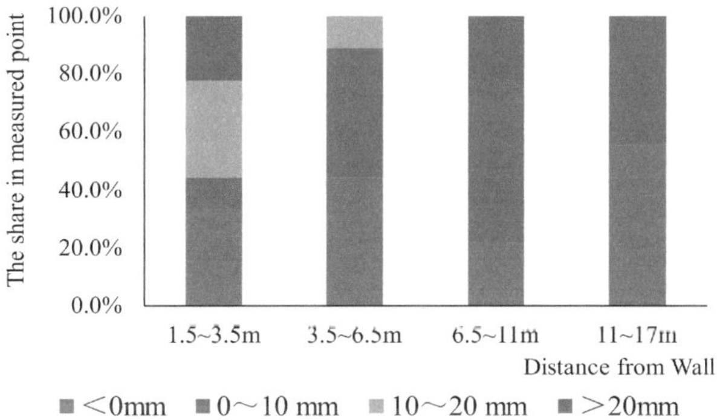

Figure 8. Distribution characteristics of surrounding rock deformation at various depths of the upstream arch.

It can be seen that the deep surrounding rock of the top arch on the upstream side generally undergoes shrinkage deformation along the radial direction of the top arch, which accounts for 22.2% to 55.6% of the distribution along the axis of the powerhouse; The distribution of the shrinkage deformation zone along the longitudinal axis of the powerhouse is basically proportional to the depth of the surrounding rock (distance from the cave wall), that is, the larger the depth, the more extensive the shrinkage deformation zone is distributed.

By introducing the concept of surrounding rock strain ε (PENG, WANG, DENG et al. 2007, LI), the shrinkage deformation caused by the top arch on the upstream side could be analyzed. The strain is calculated from the displacement of each depth anchor head of the multi-point displacement meter:

$$\varepsilon = \frac{\Delta y}{\Delta x} = \frac{y_i - y_{i+1}}{x_i - x_{i+1}} \tag{1}$$

where, i is the anchor head number, $i \geq 0$ represents the anchor head closest to the cave wall, and $i+1$ represents the anchor head farther from the cave wall. y_i represents the displacement of the i anchor head in mm, $y_i - y_{i+1}$ represents the deformation of the surrounding rock between adjacent anchor heads in mm; x_i represents the length of the anchor head rod in mm; is the average dimensionless strain of the surrounding rock between the anchor head i and i+1 anchor head.

The average strain of the surrounding rock between adjacent anchor heads is equivalent to the strain value at the midpoint of the section, and the distribution characteristics of the surrounding rock strain along the radial direction of the crown are obtained. As shown in Figure 9, the horizontal axis represents the depth from the cave wall, and the vertical axis represents the strain of the surrounding rock.

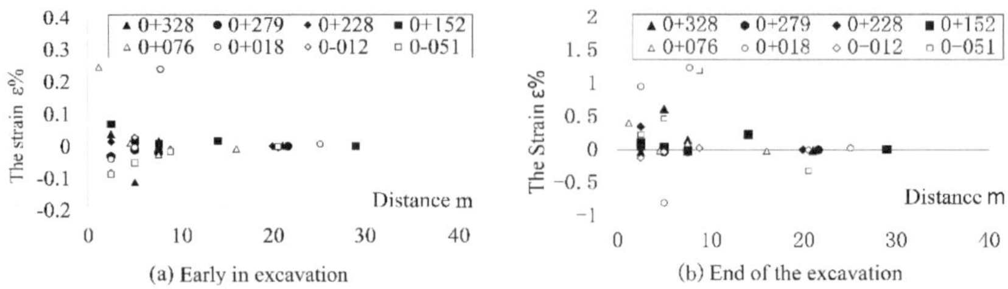

Figure 9. The Strain distribution characteristics of the surrounding rock of the roof on the upstream side.

As shown in Figure 9a, the shrinkage strain and tensile strain of surrounding rock show similar laws regardless of the distribution law along the hole depth or the strain level at the initial stage of excavation. The distribution along the depth of the hole is similar to the relaxation law of the surrounding rock obtained by the acoustic wave test (LI). The shrinkage strain is concentrated in the depth of 2.5~10.0 m from the cave wall value decreases with the increase of the depth. When the hole depth exceeds 15 m, the tension and compressive strain values all tend to "0." The maximum compressive strain is 0.12%, and the maximum tensile strain is about 0.2%. Significant shrinkage strain appears in the 0-051~0+018 tunnel section on the south side of the plant, and the 0+018 section reaches the maximum shrinkage strain, and this part corresponds to the area where the stress concentration occurs at the beginning of the excavation (Hangzhou: Power China Huadong Engineering Corporation Limited, 2007).

As shown in Figure 9b, after the excavation, even if the support is fully put into work, the distribution and magnitude of the shrinkage strain of the surrounding rock along the hole depth have not changed significantly. Compared with the initial stage of excavation, the magnitude of the tensile strain at a depth of only 2.5~10.0 m from the cave wall increased to 1.2%. It can be seen that the stress of the deep surrounding rock of the upper arch on the upstream side has significant characteristics of loading along the radial direction of the upper arch, which is different from the cave wall and surrounding rock of the upper arch on the downstream side (loading along the tangential direction of the upper arch, radial unloading).

3.5 *The development law of strain development in the deep surrounding rock of the top arch*

During the excavation of the cavern, the shrinkage deformation zone produced by the top upstream arch evolves to a deeper part. The compressive strain of the original shrinkage deformation zone gradually weakens and tends to turn into tensile strain. Take the surrounding rock strain process of the upstream side top arch (Mzc0+080-1) of the 0+76 section as an example. Figure 10 shows the surrounding rock strain evolution process, and Table 2 shows the corresponding surrounding rock strain process analysis.

The strain characteristics of Mzc0+080-1 surrounding rock consist of three sub-processes that have profound effects on the strain. "Process 1" lasts for 40 days, during which the third layer is excavated, and the excavation elevation is from 607.4~598.9 m. During "Process 1", the strain of surrounding rock in the 2.5–6.5 m section increased by 0.0055%, while the strain in the deeper section of 6.5–11 m decreased by 0.0076%, and the strain of the surrounding rock in the deepest section greater than 11m remained unchanged. This indicates that, with the excavation of the third

Table 2. Analysis of the strain process of the deep surrounding rock of the upper arch at the 0+76 section.

Date	Surrounding rock strain ε% 2.5~6.5m	6.5~11m	11~21m	Event
2015/6/11	−0.0103	−0.0024	−0.0031	Procedure 1
2015/7/21	−0.0048	−0.0100	−0.0031	Excavation of layer III
Amount of change ε%	0.0055	−0.0076	0.0000	∇607.4 ~ 598.9m Duration: 40d
2016/2/11	−0.0068	−0.0147	−0.0062	Procedure 2
2016/3/2	0.0070	−0.0267	−0.0066	Excavation of layerIV1
Amount of change ε%	0.0138	−0.0120	−0.0004	∇598.9 ~ 595.9m Duration: 22d
2016/8/23	0.0113	−0.0282	−0.0081	Procedure 3
2016/11/30	−0.0200	0.0544	−0.0278	Excavation of layer IV2 and V
Amount of change ε%	−0.0313	0.0826	−0.0197	∇591.9 ~ 583.9m Duration: 99d

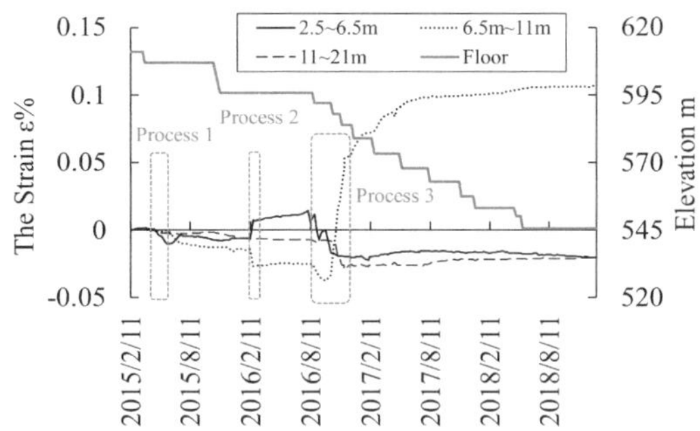

Figure 10. The strain time history curve of the deep surrounding rock of the upper arch of section 0+076.

layer, the load on the surrounding rock in the 2.5–6.5 m section along the radial direction of the crown becomes weaker, while the load on the surrounding rock in the relatively deeper section becomes stronger.

As the plant continued to be excavated ("Process 2"), the surrounding rock at the deepest section greater than 11m also produced loading along the radial direction of the crown, and the shrinkage strain increased by 0.0004%. The shrinkage strain of the relatively shallow surrounding rock is further weakened and has been transformed into tensile strain.

The tendency of the shrinkage deformation zone to evolve to the deep part is confirmed again in "Process 3": In the 6.5–11 m section of the shrinkage state in "Process 1~2", the surrounding rock strain has been transformed into tensile strain, and it continued to grow by 0.0826% in the process. The deepest surrounding rock larger than 11m shows a continuous increase in shrinkage strain.

In short, the shrinkage deformation zone evolves to the deep part during the excavation process of layer III~IV1, and the shrinkage strain of the relatively shallow shrinkage deformation zone weakens and even turns into tensile strain. This reflects the expansion of the plastic loosening zone in the stress concentration zone and the continuous inward movement of the deep compression zone.

3.6 *Analysis of the deformation mechanism of surrounding rocks in the deep top arch*

The shrinkage characteristics and spatial asymmetric distribution characteristics of the deep surrounding rock deformation of the top arch are produced under the stress pattern of the pressure arch effect. When the horizontal ground stress is large principal stress, the surrounding rock stress redistributes after excavation, which will cause stress concentration in the upstream top arch and downstream bottom of the cavern (JIANG, FAN, FENG et al 2017, YU, Lin, LI 2017, LIU, ZHU, CAI 2018, Hangzhou: Power China Huadong Engineering Corporation Limited, 2019, ZHANG, LU, ZHOU et al. 2010). According to the pressure arch theory (Du), in the surrounding rock with sufficient horizontal stress, the main stress direction of the surrounding rock deviates after excavation, and the stress is concentrated to form an arch-shaped stress distribution area. These are the mechanical conditions of the pressure arch effect after the excavation.

The deep surrounding rock of the upper arch on the upstream side corresponds to the pressure arch effect "arch seat" area, and the arch effect area is mainly subjected to compressive stress. The surrounding rock in the "abutment" area produces contraction strain along the radial direction of the crown under compression. The deformation mechanics mechanism of the deep surrounding rock of the top arch is shown in Figure 11. The deep part of the top arch on the downstream side is a typical unloading relaxation zone, and the surrounding rock is subjected to radial unloading and tangential loading along the top arch. This point is consistent with the existing research on the response of surrounding rock under the arch effect (ZHANG, LU, ZHOU et al 2010, Du).

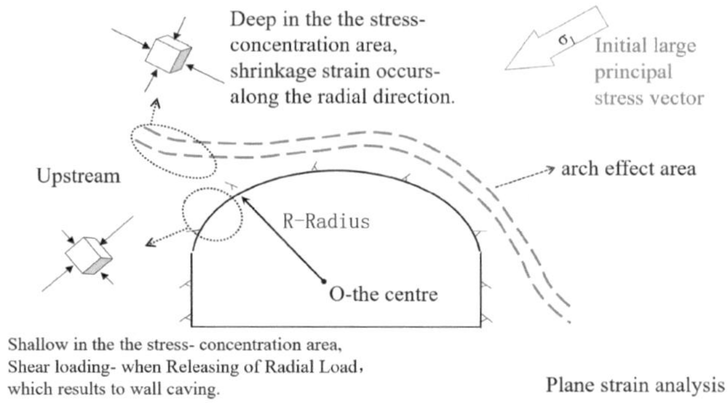

Figure 11. Schematic diagram of the deformation mechanics mechanism of the surrounding rock in the top arch.

The strain of the surrounding rock in the deep part of the top arch is related to the evolution of the arch effect zone. The size and location of the arch effect area are mainly affected by the size of the cavern. With the excavation of the cavern, the arch effect area evolves to the deeper part of the top arch. The stress process of surrounding rock in the corresponding arch effect zone is as follows: the unloading effect of the deep surrounding rock along the radial direction of the top arch increases, and the shrinkage strain decreases correspondingly, and the unloading effect of the shallow surrounding rock along the radial direction of the top arch gradually weakens, and even tensile strain begins to appear.

The relationship between the evolution process of the surrounding rock strain and the size of the cavern is shown in Figure 12. It can be seen that there are two characteristic sizes of caverns: when the cavern is smaller than a certain characteristic size (point a), the surrounding rock of the deeper layer (2.5–6.5 m) increases with the increase of cavern size along the radial contraction strain of the crown; If the size of the cavern is larger than a certain characteristic value (point c), the surrounding rock strain will converge.

3.7 *Analysis of the anchor cable prestress analysis*

The upstream side top arch monitoring anchor cable adopts unbonded prestressed steel strand and pressure dispersion anchor cable construction technology. As shown in Figure 13, the anchor cable bundle is composed of 14 steel strands with a diameter of 15.2 mm and a length of 25 m. The strength standard of a single steel strand is 1860 MPa, which adopts group compensation tensioning, and

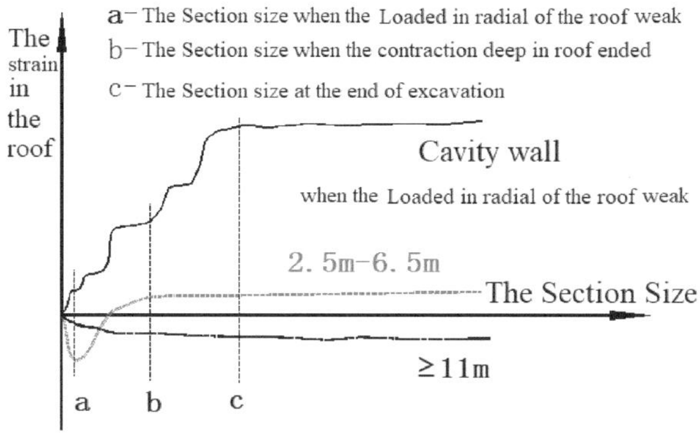

Figure 12. The relationship between the size of excavation and the evolution process of the surrounding rock in the arch on the upstream side.

cyclic tensioning one by one completes the prestress tensioning. The anchor cable dynamometer is installed between the anchor and the anchor backing plate, and is used to measure the total load (unit: kN) transmitted by the steel strand to the anchor. The arrangement of the anchor cable and surrounding rock deformation measurement points for the upstream side top arch monitoring is shown in Figure 6.

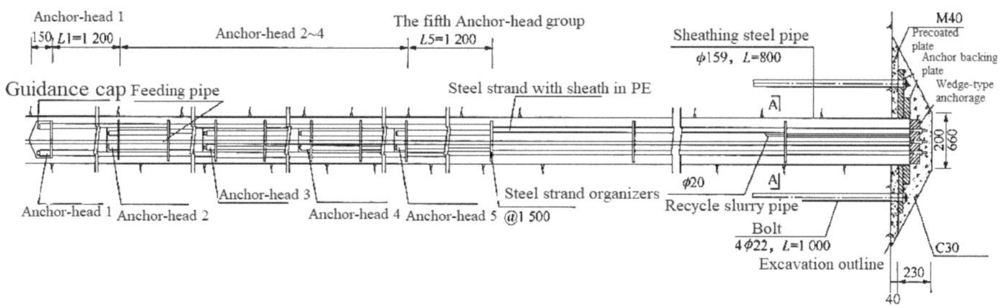

Figure 13. Schematic diagram of pressure dispersive anchor cable structure.

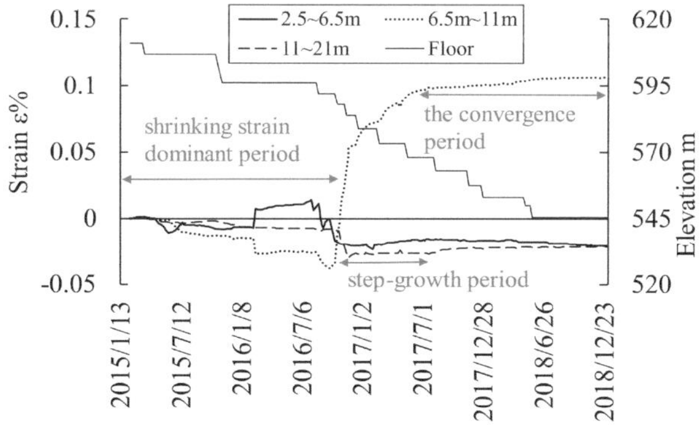

Figure 14. Strain time-history curve of the deep surrounding rock of the upper top arch of section 0+076.

At the initial stage of locking, there is a loss of 5% to 10% of the prestress, which is related to the structure of the anchor plate, the lithology and the group anchor effect (ZHU, SUN, WANG et al 2004, ZHOU, HE, YANG 2006). The characteristics of prestressing and surrounding rock strain of the anchor cable on the upstream side of the 0+076 section are shown in Figures 15 and 16. The change of anchor cable prestress is similar to the evolution law of surrounding rock strain, which is divided into the zero-bearing period or shrinking strain dominant period, the step-growth period and the convergence period.

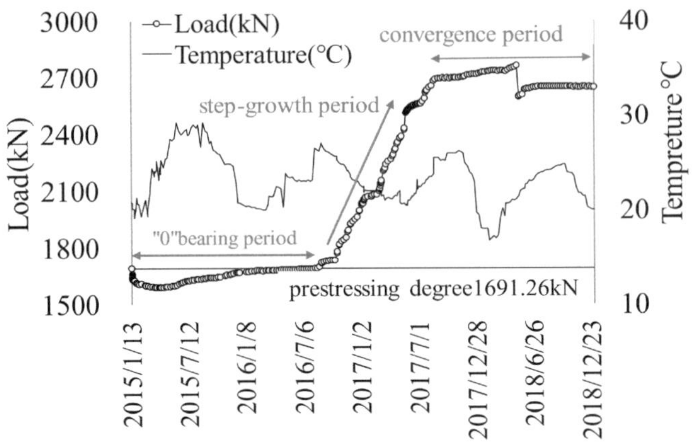

Figure 15. Time-history curve of anchor cable prestress on the upstream top arch of section 0+076.

Before July 2016, the prestress of the anchor cable was close to the locked value, the steel strands only bore the pretension load, and the anchor cable was basically at empty load. As shown in Figure 10, the surrounding rock above 2.5 m all showed contraction strain during this period, and the surrounding rock strain was in the predominant period of contraction strain. The scope of the contraction area exceeds the reinforcement depth of the anchor cable, so the anchor cable steel strand shows the characteristic of zero-load bearing. This load characteristic verifies that in the initial stage of excavation, the pressure arch effect is generated in the cavern, and the top arch on the upstream side produces 4a shrinkage zone along the radial direction of the top arch. In the process of studying cavern layout and top arch support optimization, similar shrinkage deformation zones were also found (Chen, Hua 2018).

When the surrounding rock shows a step increase trend of tensile strain, the growth of prestressing also produces similar growth characteristics. Regression analysis is carried out on the loss rate of prestressing and the tensile strain of surrounding rock, as shown in Figure 16. The relationship between the prestress loss rate P, and the tensile strain of the surrounding rock is as follows:

$$P = -234.11\varepsilon^4 + 174.69\varepsilon^3 - 31.112\varepsilon^2 - 3.087\varepsilon + 0.8839 \quad (2)$$

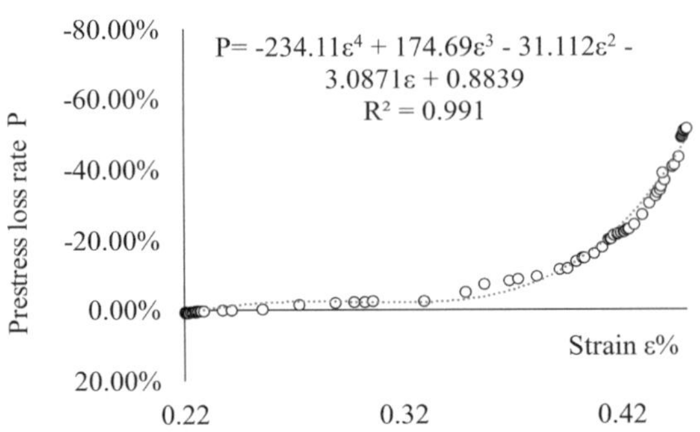

Figure 16. Correlation analysis diagram of prestress loss rate and surrounding rock tensile strain.

It can be seen that the prestress is closely related to the tensile strain of the surrounding rock, and the correlation coefficient of the fourth-order polynomial fitting reaches 0.9955. Since August 23, 2016, as the excavation interval is shortened, the disturbance has increased, and the surrounding rock strain has increased in a "step" shape. And the corresponding anchor cable prestress has also shown a similar increase in characteristics. The excavation of the machine pit was entered in August 2017. After that, the excavation disturbance was significantly reduced. Both the anchor cable prestress and the surrounding rock strain quickly showed convergence characteristics and entered the convergence period. This reflects the brittle deformation characteristics of the surrounding rock to a certain extent.

4 CONCLUSION

Based on the monitoring data of surrounding rock strain and prestress in the whole process, this paper uses strain analysis combined with prestressing analysis to summarize the shrinkage deformation characteristics and supporting load laws of the deep surrounding rock in the stress concentration area under high-stress conditions. The conditions and mechanism of its generation are analyzed qualitatively based on the elastoplastic theory to discuss the mechanical problems of the deformation zones in the deep and shallow parts of the stress concentration area. The main conclusions of this paper are as follows:

(1) The shrinkage strain of the surrounding rock is generated in the deep part of the stress concentration area of the top upstream arch. The distribution range of the shrinkage deformation zone along the axis of the powerhouse is proportional to the depth, while the shallow surrounding rock shows the radial tensile strain corresponding to the lamella. The downstream side is the tensile strain caused by unloading relaxation.
(2) The magnitude and orientation of high ground stress determine the deformation characteristics of the tunnel section in all directions. The redistribution of stress causes the arching effect area in the deep layer of the top arch. The deep surrounding rock of the top arch on the upstream side is located in the "arch abutment," which produces shrinkage strain along the radial direction of the top arch. With the excavation of the cavern, there is a trend of expansion of the plastic-plastic loosening zone in the stress concentration area and the continuous inward movement of the deep compression zone.
(3) The law of anchor cable prestress is divided into three stages: the zero-bearing period, the step-growth period and the convergence period. When the shrinkage area exceeds the anchor cable reinforcement depth, the steel strands only bear the pretension load. The zero-bearing period verifies the shrinkage and deformation characteristics in the deep part of the stress concentration area driven by the arching effect.

The scope of application of the research results in this paper is high tectonic stress, hard rock caverns, and mechanical problems of deep surrounding rock of top arch. The quantitative index of compression zone and sloughing zone needs further discussion. This paper establishes the relationship between prestressing and surrounding rock strain based on the prestress change law throughout the whole process, which can provide a reference for the analysis of the influence of the time-effect deformation of surrounding rock on the prestress of anchor cable during operation and the optimization design of deep support parameters for similar projects.

ACKNOWLEDGMENTS

This work was financially supported by the Natural Science Foundation of China project (51879284,51609266).

REFERENCES

Bidding documents reference of civil engineering and installation of metal structures of the left bank water diversion and power plant in Baihetan hydropower station[R]. Hangzhou: Power China Huadong Engineering Corporation Limited,2013. (In Chinese)

Chen DH, Hua XZ. Impact of Insitu Stress on Layout Direction of Deep Typical Gateways[J]. Chinese Journal of Underground Space and Engineering [J].2018,14(04):1122–1129. (In Chinese)

Du XL. Research On Evolution and Its Application of Rock Pressure Arch in Coal Mining. (In Chinese)

HUANG RQ, HUANG Da, DUAN SH et al. Geomechanics mechanism and characteristics of surrounding rock mass deformation failure in the construction phase for the underground powerhouse of Jinping I hydropower station[J]. Chinese Journal of Rock Mechanics and Engineering, 2011, 30 (1):23–35. (In Chinese)

JIANG Q, FAN YL, FENG XT et al. Unloading break of hard rock under the high geo-stress condition: inner cracking observation for the basalt in the Baihetan underground powerhouse[J]. Chinese Journal of Rock Mechanics and Engineering, 2017, 36(5):1076–1087. (In Chinese)

JIANG Q, FENG XT, LI SJ et al. Cracking-restraint design method for large underground caverns with hard rock under high geostress condition and its practical application[J]. Chinese Journal of Rock Mechanics and Engineering,2019,38(6):1081–1101. (In Chinese)

LI ZP. Study on EDZ Dynamic Evolution Mechanism Hard Rock Mass in Large Underground Caverns under High Geostress (in Chinese)

LIU J, ZHU ZH, CAI H. Study on deformation and failure characteristics of surrounding rock at the top arch of super large underground caverns [J]. Chinese Journal of Geotechnical Engineering, 2018, 40(7):1257–1267. (In Chinese)

LU Bo, WANG JM, DING XL et al. Study of deformation and cracking mechanism of surrounding rock of Jinping I underground powerhouse[J]. Chinese Journal of Rock Mechanics and Engineering, 2010, 29 (12):2 429–2 441. (In Chinese)

MENG GT, He SH, Chen JL et al. Mechanism of deep deformation of the underground powerhouse roof arch at the right bank of Baihetan Hydropower Station [J]. Chinese Journal of Rock Mechanics and Engineering, 2020,42(03),576–583. (In Chinese)

Ministry of Water Resources of the People's Republic of China. GB 50487-2008 Code for engineering geology investigation of water resources and hydropower[S]. Beijing: China Planning Press, 2009.

PENG Q, WANG DK, DENG JH et al. Analysis of surrounding rock deformation characteristics in underground powerhouse[J]. Chinese Journal of Rock Mechanics and Engineering, 2007, 26 (12):2 583–2 587. (In Chinese)

Preliminary feasibility study-engineering geological investigation report on dam line selection of underground cavern in Baihetan hydropower station located on Jinsha River[R]. Hangzhou: Power China Huadong Engineering Corporation Limited, 2007. (In Chinese)

Safety appraisal Volume II-design self-inspection report on Excavation and support projects in Baihetan hydropower station located on Jinsha River[R]. Hangzhou: Power China Huadong Engineering Corporation Limited, 2019. (In Chinese)

Safety appraisal Volume III-Construction self-inspection report on Excavation and support projects in Baihetan hydropower station located on Jinsha river[R]. Beijing IWHR Technology Co., Ltd. 2019. (In Chinese)

WANG P, HONG WB, SONG G. Size effect and in-situ stress influence of columnar jointed basalt relaxation zone[J] Chinese Journal of Geotechnical Engineering, 2018, 40(1):139–146. (In Chinese)

WEI JB, DENG JH, WANG DKI et al. Characterization of deformation and fracture for rock mass in the underground powerhouse of Jingping I hydropower station[J]. Chinese Journal of Rock Mechanics and Engineering, 2010, 29 (6) :1 198–1 205. (In Chinese)

YU J, Lin ZB, LI YH. Analysis on Deformation of Surrounding Rock of Deep Tunnel Based on Transparent Rock. Chinese Journal of Underground Space and Engineering [J].2017,13(04):943–949. (In Chinese)

ZHANG YH, LU YR, ZHOU HM et al. Study of the relationship between failure properties of surrounding rock and direction of geostress[J]. Chinese Journal of Rock Mechanics and Engineering, 2010, 29(2):3526–3535. (In Chinese)

ZHOU YJ, HE SM, YANG XL. Study on prestress loss of anchor cables under long-term loading [J]. Rock and Soil Mechanics, 2006, 27 (8):1353–1356. (In Chinese)

ZHU HY, SUN HY, WANG HB et al. Analysis of prestressing state of cable applied to rock slope reinforcement[J]. Chinese Journal of Rock Mechanics and Engineering, 2004, 23(16): 2 756–2 760. (In Chinese)

Advances in Urban Engineering and Management Science – Khalil & Yang (Eds)

Leakage analysis and treatment scheme design of dam foundation of Mengjiagou reservoir in Gaomi city

Zhaozhen Wei*
Research Services Office, Shandong Jiaotong University, Ji'nan, China

Lili Liu
Exploration and Design Department, Water Resources Research Institute of Shandong Province, Ji'nan, China

ABSTRACT: Mengjiagou reservoir is located in Mengjiagou village, Baicheng town, Gaomi city, upstream of the main stream of Jiaohe river. The loose layer of reservoir dam foundation is mainly sandy soil with large permeability coefficient, which belongs to strong permeable or extremely strong permeable layer. If anti-seepage measures are not taken, the reservoir water will leak seriously through the loose layer of dam foundation. This paper analyzes the leakage of the dam foundation without seepage interception measures, and through scheme comparison, selects the anti-seepage design scheme of adopting plastic concrete seepage interception wall in the dam section downstream of the sand consolidation dam and concrete seepage interception groove in the dam section upstream of the sand consolidation dam, so as to ensure the normal exertion of engineering benefits.

1 INTRODUCTION

Mengjiagou reservoir is located in the upstream of the main stream of Jiaohe river in Mengjiagou village, Baicheng town, Gaomi city. The dam site is located in the west of Mengjiagou village, with a control basin area of 459.1 km^2. It is a comprehensive reservoir with multiple functions such as irrigation, water supply, flood control, ecology and tourism.

Mengjiagou reservoir is a river type reservoir. The dam is built on both existing banks of Jiaohe river. The maximum dam height is 12.33 m, the profit water level is 33.0 m, the design water level is 35.03 m, the dead water level is 24.5 m, the storage capacity is 1.73×10^7 m^3, and the total storage capacity is 2.406×10^7 m^3.

Mengjiagou reservoir is located at two natural bends in the reach of Wangfengtun-Lijiaying highway bridge, the main stream of Jiaohe river. The hydro-junction project is mainly composed of enclosing dam, flood discharge gate, outlet culvert gate, irrigation diversion pipe, etc.

The enclosing dam, arranged at the stake number 45+730-52+730 of the central line of Jiaohe river main stream, is a geomembrane impervious sand gravel dam with a total length of 10315 m, a dam crest elevation of 36.5–38.2 m and a crest width of 8.0 m.

Lijiaying railway bridge is located at Jiaohe river stake number 51+300. In order to prevent the scouring and undercutting of the downstream channel of Lijiaying railway bridge from affecting the bridge safety, a sand consolidation dam is set at the river channel about 100 m away from the bridge downstream. In downstream of the sand consolidation, the enclosing dam of Mengjiagou reservoir (stake number W0+000-W3+350, E0+000-E3+746) is built on newly excavated bedrock, the enclosing dam (stake number W3+350-W4+933, E3+746-E5+382) is built along the current embankment of Jiaohe river in upstream of the sand consolidation dam.

*Corresponding Author: weizhaozhen@sdjtu.edu.cn

DOI 10.1201/9781003305026-25

After the impoundment of reservoir, leakage will inevitably occur, but excessive leakage will raise the saturation line of the dam, increase the uplift pressure of the foundation, and then endanger the safety of the dam. Due to the large permeability coefficient of dam foundation loose layer of Mengjiagou reservoir, which belongs to strong permeable or extremely strong permeable layer, the reservoir water will seriously leak through the dam foundation loose layer after impoundment, so it is very necessary to take anti-seepage measures.

In this paper, we intend to explain the necessity of anti-seepage measures for Mengjiagou reservoir through the leakage analysis of the dam foundation without anti-seepage measures, select the anti-seepage design scheme through scheme comparison, and analyze the leakage of the dam after anti-seepage measures taken to verify its effectiveness.

2 GEOLOGICAL ANALYSIS OF DAM FOUNDATION

No large structural fracture zone is found in the reservoir area, only Chaigou-Jiaozhou break is developed in the backwater area, and there is no reservoir leakage caused by fracture zone in the reservoir area.

The reservoir dam base layer is divided into Quaternary loose layer and Mesozoic clastic sedimentary rock. Most of the loose layer of the dam foundation is sandy soil. Only a small part of the left bank dam (stake number W1+399-W1+880) is single-layer loam foundation with weak or moderate permeability. Other dam sections on the left bank and the dam foundation on the right bank are double-layer or single-layer sandy soil dam foundations with large permeability coefficient, belonging to strong permeable or extremely strong permeable layer, after impoundment, the reservoir water will leak seriously through the loose layer of the dam foundation.

The underlying bedrock of the dam foundation is sandstone and sandy conglomerate, mostly is sandstone, and the sandy conglomerate coexists with sandstone only in the form of lens or interbedding. Sandstone is intensively weathered and can be divided into full weathering, strong weathering and medium weathering, in which the medium weathering is not exposed. Sandy conglomerate is slightly cemented by clay, and the weathering zone is not obvious. Completely weathered sandstone has weak ~ medium water permeability. The water permeability of strongly weathered sandstone and sandy conglomerate is less than 10lu, belonging to weak permeable layer. The leakage of reservoir water through dam foundation rock mass is small, and strongly weathered sandstone and conglomerate can be regarded as impermeable layer.

Therefore, the water in the reservoir mainly leaks outward through the loose layer of the dam foundation and becomes the recharge source of the regional aquifer (Zhang 2021).

3 SEEPAGE ANALYSIS OF THE DAM WITHOUT ANTI-SEEPAGE MEASURES

According to the topographic characteristics and enclosing dam foundation geological conditions, dam sections W2+724, W4+870, E2+481 and E4+736 are selected for seepage calculation and seepage stability analysis without anti-seepage measures. The seepage calculation adopts the finite element method, and the calculation program is the soil slope seepage stability analysis program (Autobank6.1) developed by Hehai University. According to the geological survey report, the physical mechanics indexes of soil materials at the calculation section are shown in Table 1. The calculation results without seepage interception measures are shown in Table 2, and the flow network distribution of each calculation section without seepage interception measures is shown in Figures 1–4.

Through calculation, without anti-seepage measures the dam foundation maximum leakage is 209.95 $m^3/m/d$ at profit water level, and the reservoir profit capacity of 1.73×10^7 m^3 will leak out in 12 days. Obviously, if anti-seepage measures are not taken after the construction of the reservoir, the reservoir will not be able to store water at all.

Table 1. Soil mechanical indexes of dam foundation and dam body at each section.

section	soil layer	Index			
		Dry density (g/cm^3)	Wet density (g/cm^3)	Saturation density (g/cm^3)	Permeability coefficient (cm/s)
W2+724	Dam sand gravel	1.81	1.96	2.13	2.74×10^{-1}
	Dam sandy loam	1.69	1.94	2.06	1.25×10^{-3}
	Dam loam	1.63	1.94	2.07	5.68×10^{-4}
	Dam foundation sand gravel	1.57	1.69	1.98	2.74×10^{-1}
	Dam foundation sand pebble	1.80	1.91	2.10	1.87×10^{-1}
	Dam foundation completely weathered sandstone	1.75	1.93	2.10	2.42×10^{-4}
W4+870	Dam sand gravel	1.81	1.96	2.13	1.50×10^{-1}
	Dam foundation sand gravel	1.57	1.69	1.98	1.50×10^{-1}
	Dam foundation sand pebble	1.80	1.91	2.10	2.25×10^{-1}
	Dam foundation completely weathered sandstone	1.75	1.93	2.10	2.42×10^{-4}
E2+481	Dam sand gravel	1.81	1.96	2.13	1.95×10^{-1}
	Dam foundation sand gravel	1.57	1.69	1.98	1.95×10^{-1}
	Dam foundation sand pebble	1.80	1.91	2.10	2.74×10^{-1}
	Dam foundation completely weathered sandstone	1.75	1.93	2.10	2.42×10^{-4}
E4+736	Dam sand gravel	1.81	1.96	2.13	1.84×10^{-1}
	Dam foundation sand gravel	1.57	1.69	1.98	1.84×10^{-1}
	Dam foundation sand pebble	1.80	1.91	2.10	3.05×10^{-1}
	Dam foundation completely weathered sandstone	1.75	1.93	2.10	2.42×10^{-4}

Table 2. Seepage calculation results without seepage interception measures.

Calculation section	Calculation condition	Seepage discharge ($m^3/m/d$)	Escape hydraulic gradient
W2+726	profit level	209.95	0.26
	design flood level	404.35	0.45
W4+870	profit level	52.44	0.47
	design flood level	88.13	0.62
E2+481	profit level	134.78	0.29
	design flood level	232.42	0.35
E4+736	profit level	77.50	0.52
	design flood level	146.02	0.44

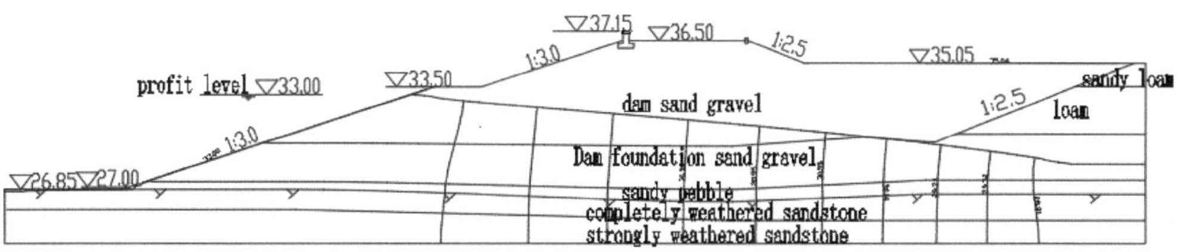

Figure 1. Seepage network diagram of W2+724 section without anti-seepage measure at the profit water level.

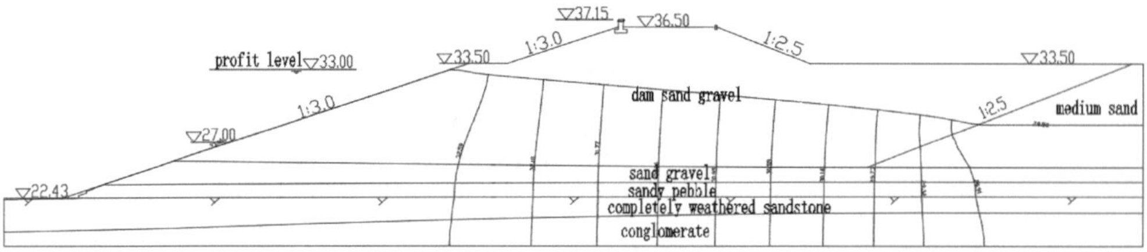

Figure 2. Seepage network diagram of E2+481 section without anti-seepage measure at the profit water level.

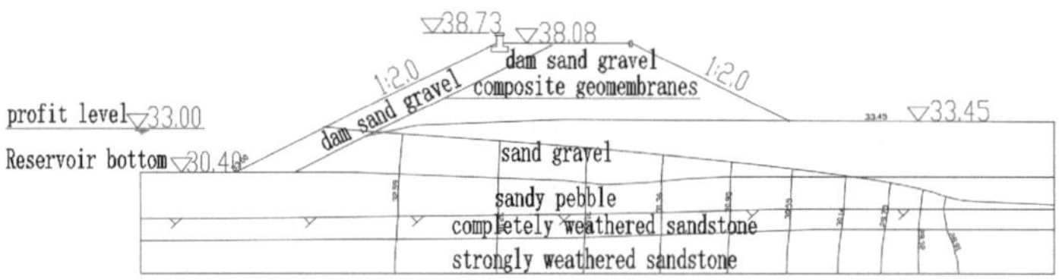

Figure 3. Seepage network diagram of W4+870 section without anti-seepage measure at the profit water level.

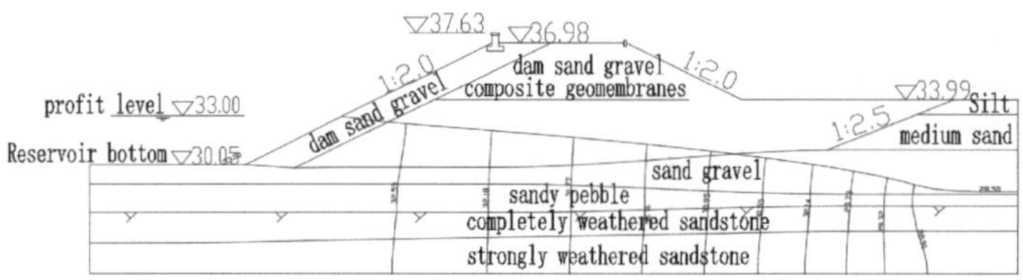

Figure 4. Seepage network diagram of E4+736 section without anti-seepage measure at the profit water level.

According to the geological survey report, the seepage deformation type of dam foundation loam is flow soil, and the allowable hydraulic gradient is 0.43; the seepage deformation type of sand in dam foundation is flowing soil, and the allowable hydraulic gradient is 0.35; the seepage deformation type of gravel layer is transitional, and the allowable hydraulic gradient is 0.16. Under various working conditions, the escape gradient of the sand gravel layer of the dam foundation is greater than the allowable gradient, and the seepage stability of the dam foundation does not meet the design requirements.

In conclusion, without anti-seepage measures, the reservoir leakage is serious and the seepage stability of the dam foundation does not meet the requirements, so the dam foundation seepage interception must be carried out.

4 SEEPAGE INTERCEPTION DESIGN OF DAM FOUNDATION

4.1 *Dam foundation seepage interception scheme*

1) Dam section in downstream of the sand consolidation: According to the stratum structure of the dam foundation and combined with the experience of similar projects in China, three dam foundation seepage interception schemes, including composite geomembranes seepage interception layer, vibration grooved seepage prevention plate wall and plastic concrete seepage interception wall, are selected for economic and technical comparison.

Although the cost of plastic concrete seepage cut-off wall is slightly higher, it has guaranteed wall quality, good seepage cut-off effect, high reliability and strong durability, and it can better solve the connection problem with bedrock. In recent years, with the rapid development of plastic concrete cut-off wall technology, great improvements have been made in hole making machines and tools, plastic concrete cut-off wall materials and reducing project cost. It is widely used in the anti-seepage treatment of dangerous dam body and dam foundation, and there are a large number of successful engineering examples (Wen et al. 2021; Yu et al. 2019; Zheeng et al. 2020; Zhou et al. 2021).

In order to ensure the reliability of the dam foundation seepage interception project and ensure the safe operation of the reservoir, combined with the characteristics of the geological conditions of the project, the dam section in downstream of the sand consolidation dam adopts the scheme of plastic concrete seepage interception wall [], The wall thickness is 0.3 m, the wall top elevation is 32.5 m, and the seepage cutoff wall is located inside the berm of the upstream slope.

2) Dam section in upstream of the sand consolidation: Due to the low height of the dam section and the small buried depth of bedrock below the river bottom (about 3.5 m on average), the schemes of plastic concrete seepage cut-off wall and concrete seepage cut-off groove are mainly compared for the seepage interception of the dam foundation. Because of small investment, the concrete cut-off groove scheme is adopted for the dam section in upstream of the sand consolidation dam, with a groove width of 0.8 m, and the cut-off groove is located at the dam toe of the upstream slope; The bottom of the cut-off wall (groove) is 0.5 m deep into strongly weathered sandstone and 1.0 m deep into conglomerate, connecting with the anti-seepage of dam and dam foundation in downstream of the sand consolidation through concrete tooth-groove and composite geomembrane.

4.2 *Dam seepage calculation after seepage interception*

Select sections W2+724, W4+870, E2+481 and E4+736 for seepage calculation under seepage interception condition, the seepage calculation method is the same as that under natural state, and the permeability coefficient of plastic concrete cut-off wall is 1.0×10^{-6} cm/s. See Table 3 for the calculation results of seepage flow and Table 4 for the escape gradient of dam foundation. And the flow network distribution of each calculation section with seepage interception measures is shown in Figures 5–8.

Table 3. Calculation results of seepage flow after taking seepage interception measures.

Calculation section	Reservoir water level(m)	Seepage flow(m^3/m/d)
W2+724	33.00	0.793
	32.00	0.455
	31.00	0.321
	30.00	0.180
	29.00	0.049
W4+870	33.00	0.597
	32.00	0.463
	31.00	0.330
E2+481	33.00	0.606
	32.00	0.402
	31.00	0.252
	30.00	0.132
	29.00	0.037
E4+736	33.00	0.732
	32.00	0.568
	31.00	0.405
	30.00	0.159

Table 4. Calculation results of escape gradient of dam foundation after taking seepage interception measures.

Calculation section	Calculation condition	Escape hydraulic gradient
W2+726	profit level	0.0019
	design flood level	0.0020
W4+870	profit level	0.0056
	design flood level	0.0255
E2+481	profit level	0.0051
	design flood level	0.0444
E4+736	profit level	0.0210
	design flood level	0.0252

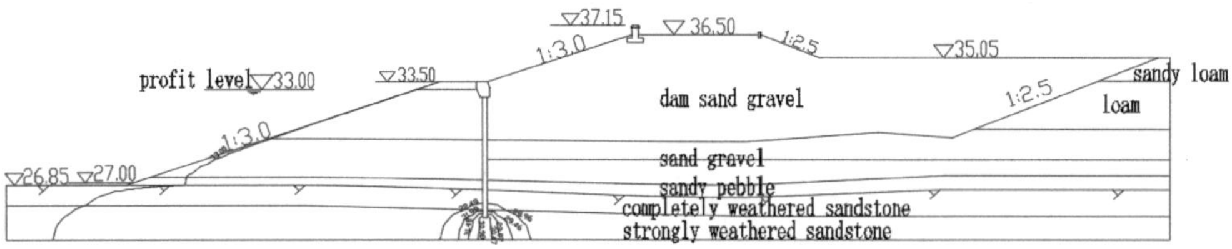

Figure 5. Seepage network diagram of W2+724 section with anti-seepage measure at the profit water level.

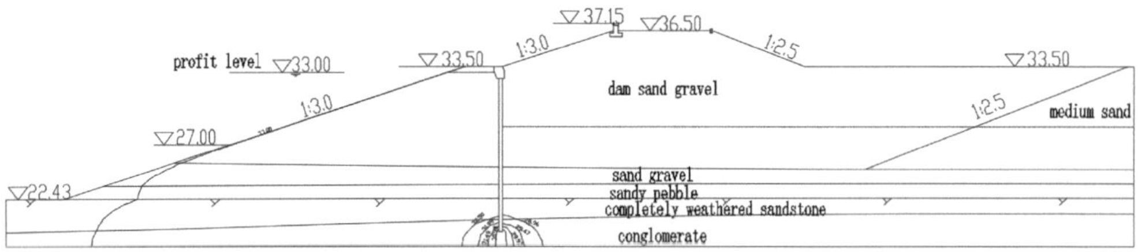

Figure 6. Seepage network diagram of E2+481 section with anti-seepage measure at the profit water level.

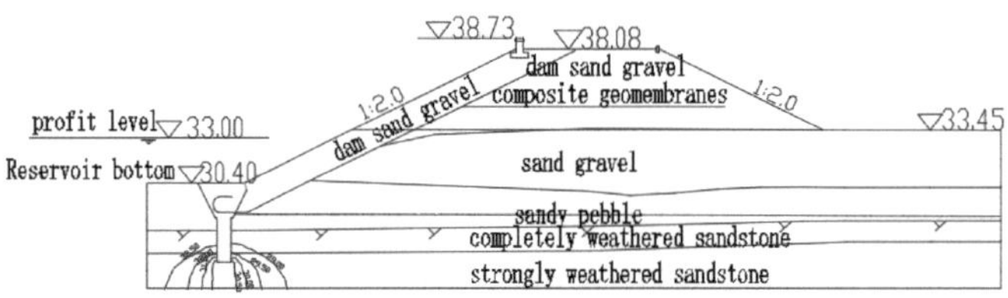

Figure 7. Seepage network diagram of W4+870 section with anti-seepage measure at the profit water level.

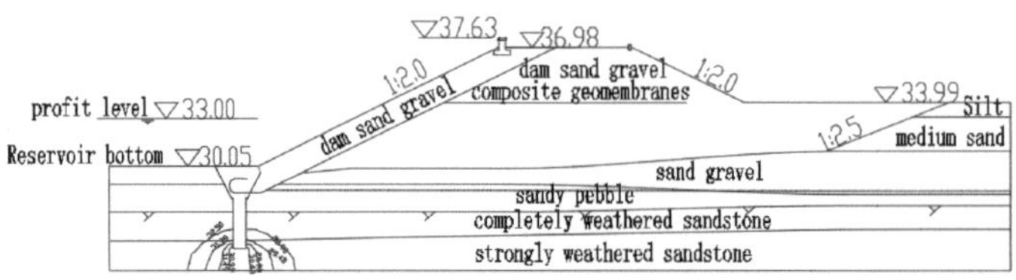

Figure 8. Seepage network diagram of E4+736 section with anti-seepage measure at the profit water level.

According to the calculation results of single width seepage flow, combined with the duration of different reservoir water levels in the process of optimization regulation, after the seepage interception measures are taken for the dam, the annual leakage is about 1.203×10^6 m^3, accounting for 6.95% of reservoir utilizable capacity, and the seepage interception effect is remarkable. In the calculation of optimization regulation, the annual leakage loss of the reservoir is about 1.203×10^6 m^3, which is consistent with the seepage calculation results.

It can be seen from the data in Table 3 that the escape gradient of the sand gravel layer of the dam foundation under various working conditions is less than the allowable gradient of 0.16, meeting the design requirements.

5 CONCLUSION

Through scheme comparison, the anti-seepage design schemes of plastic concrete seepage cut-off wall in the downstream dam section of sand consolidation dam and concrete seepage cut-off groove in the upstream dam section of sand consolidation dam are selected in this paper. Four sections are selected to calculate the dam seepage flow before and after the anti-seepage measures are taken. After the anti-seepage measures are taken, the seepage flow per unit width of dam and the escape hydraulic gradient are both significantly reduced, indicating that the effectiveness of reservoir water storage and the dam safety can be guaranteed.

Mengjiagou reservoir is one of the projects listed in the rainwater and flood resources utilization plan of Shandong Province. It is a medium-sized reservoir in hilly area. The anti-seepage of dam foundation is very important, which directly affects the profit water storage. Combined with Mengjiagou reservoir dam project, this paper briefly introduces the dam leakage analysis method, and explains the effectiveness of seepage interception measures, which has guiding significance for the construction of similar projects.

ACKNOWLEDGMENT

Shandong Jiaotong University Doctoral Research Fund (NO: BS201901059), Provincial water conservancy research projects in Shandong Province (Grant No. SDSLKY201804).

REFERENCES

WB Zhang, ZZ Shen, GY Chen, et al, Optimization design and assessment of the effect of seepage control at reservoir sites under karst conditions: a case study in Anhui Province, China, Hydrogeology Journal, 2021. PP 1–25.

Wen Lifeng, LI Yanlong, CHAI Junrui. Statistical analysis of mechanical properties of dam foundation concrete cutoff wall, Journal of Hydraulic Engineering, 2021,52(02): 241–254.

YL Zhou, XW Gao, HL Ran. Design of anti-seepage treatment scheme for dam foundation of Malu reservoir, Technical Supervision in Water Resources, 2021, (05): 178–182.

Yun Yu, Dongxia Zhang, Caiyun Hou. Analysis on seepage interception technology scheme of Shuangwangcheng reservoir in South-to-North Water Diversion Project, Shandong Water Resources, 2019, (07): 11–13.

Zheng Peixi. Preliminary analysis on seepage control effect of dam body, foundation and connection of Qiaoqi dam, Water Conservancy Construction and Management, 2020,40(01): 35–41+45

Advances in Urban Engineering and Management Science – Khalil & Yang (Eds)
© 2023 The Author(s), ISBN 978-1-032-30426-7

Modal analysis of typical masonry structure of Anhui rural residence in Tan Lu fault zone

Tao Wu*
Laboratory of Building Structure of Anhui Higher Education Institutes, Anhui Xinhua University, Hefei, China

Sijie Xu
School of Urban Construction, Anhui Xinhua University, Hefei, China

ABSTRACT: Taking the rural buildings in Anhui section of Tan Lu fault zone as the research object, based on the field investigation and detection of seismic performance, the typical masonry structure scheme of rural buildings is designed. The numerical model of the basic structure and the variable parameter comparative analysis model considering structural layout, material strength and numerical model grid control are established by using the finite element software ABAQUS. The modal analysis of each model is carried out, and the vibration shape, vibration mode component and vibration mode period are selected as the comparative analysis indexes. The analysis shows that the vibration characteristics of masonry structures arranged according to the conventional layout in village and town buildings are not conducive to structural earthquake resistance. The performance degradation of masonry materials not only reduces the bearing capacity of the structure, but also amplifies the seismic action of high-order out of plane vibration components, which intensifies the probability of external flash collapse of the bearing wall in the earthquake. If measures are taken to realize that the basic vibration mode of the structure is translational along the transverse direction, the vibration characteristics show that the load-bearing and lateral resistance of the transverse wall can be effectively brought into play, and the high-order vibration mode of out of plane vibration is suppressed, which improves the seismic capacity of the structure. In addition, the grid control requirements of ABAQUS for shell elements along the thickness direction are also proposed in this paper.

1 INTRODUCTION

Since the Wenchuan earthquake in 2008, several destructive earthquakes have broken out in inland China, such as Yushu earthquake, Lushan earthquake, Ludian earthquake. The epicenter of these earthquakes is located in villages and towns far away from cities. Although the building density of villages and towns is far less than that of cities, it still caused serious casualties and economic losses. The earthquake damage investigation after the above earthquakes shows that (Civil and Structural Groups of Tsinghua University 2008; Wang 2013), urban buildings built in accordance with relevant specifications have basically withstood the test and played an important role in earthquake relief. Unfortunately, most of the buildings in villages and towns are rural houses. Because they are not included in the standardized construction management, their typical seismic damage performance is regional and complete collapse under the impact of strong earthquakes. It can be seen that the weak link of earthquake prevention and disaster reduction in China mainly lies in village and town buildings, especially rural houses. Village and town buildings have obvious regional distribution characteristics due to their geographical location, functional requirements

*Corresponding Author: taoxia201@163.com

 DOI 10.1201/9781003305026-26

and economic conditions, and their structural characteristics also show strong local characteristics. Therefore, the field investigation on the seismic safety characteristics of village and town buildings, especially rural houses (Zhang 2013) is one of the key basic works of its seismic performance research in recent years.

In order to improve the seismic performance of rural buildings, it is necessary to understand the dynamic characteristics of the structure and its motion response law under earthquake. With the development of numerical technology, with the help of computer simulation method (Ma 2013), the dynamic response process of the structure under earthquake can be obtained objectively and conveniently. In addition, through model variable parameter analysis, the seismic effect of the structure under strong earthquake can be effectively improved.

According to the field investigation and field measured data of Anhui rural residence in Tanlu fault zone, this paper first designs the basic structure scheme of typical rural residence, and then expands and establishes several finite element models in ABAQUS software to analyze the influence of structural parameters and finite element control method on structural dynamic characteristics, and puts forward the reasonable layout and parameter scheme of rural residence structure, It provides a reference basis for subsequent scientific research or reinforcement and reconstruction schemes and the formulation of seismic fortification standards for relevant village and town buildings.

2 BASIC NUMERICAL MODEL OF TYPICAL STRUCTURE

2.1 *Structural scheme and structural parameters*

The seismic research of rural houses in Anhui Province has its unique geological structure background. The famous Tanlu fault zone runs obliquely through the whole province and has always been the key monitoring area of seismic risk (Ye 2012). In addition, Yangzhou-Tongling seismic zone, Macheng-Changde seismic zone and Xuchang-Huainan seismic zone are also distributed in Anhui Province. In recent years, small earthquakes occur frequently and cause disasters for many times, so the seismic fortification work is severe. In order to specifically understand the current situation of seismic performance of rural houses in Anhui, the research group takes the Anhui section of Tanlu fault zone and the surrounding areas with frequent recent seismic activities as the target range, including high intensity areas with fortification intensity of 7 degrees (0.1g) and above and areas significantly affected by the earthquake, including 7 counties such as Si County and Wuhe County, and Lingbi County, Guzhen County and other 11 counties (cities) are taken as general survey areas. The specific investigation protocol and results are presented in the literature (Jiang 2015). The field investigation and field measurement of the village and town buildings were carried out by stages. The investigation of seismic behavior shows that the structural form of Anhui village and town buildings is mainly masonry structure (the proportion of brick concrete structure and brick wood structure in the survey is 89%), the plane layout of rural houses is mainly 2-bay and 3-bay, the longitudinal wall is mainly external wall, and there are many door and window holes, door holes are set on the inner transverse wall, and window holes are occasionally set on the outer transverse wall.

In this paper, the typical 2-storey and 2-bay rural residence (Figure 1) commonly used in the 7 degree (0.1g) seismic fortification area is selected as the object. After appropriate simplification, the model scheme of typical low-rise masonry structure is established. The floor height is taken as the statistical average value of 3.6m according to the survey results. The second floor is provided with an overhanging corridor with a width of 1200mm.

The roof of rural houses in the investigated area generally adopts the superposition form of precast hollow slab and post cast concrete. The thickness of precast slab mainly includes 120mm and 150mm, and the thickness of concrete post cast layer is about 30–50mm. Considering the influence of floor hollow, the converted 120mm thick solid slab is adopted for the roof in this paper. The wall in the model is solid brick wall, the material is ordinary sintered brick, and the thickness of internal and external walls is 240mm. The material performance parameters are

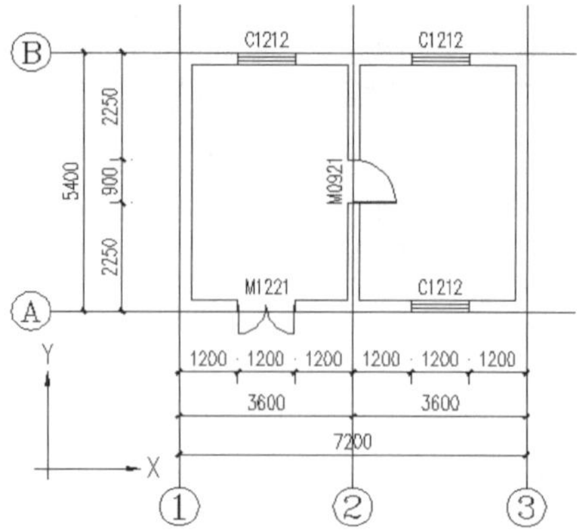

Figure 1. Typical rural residential layout.

determined according to the sample data obtained from the previous field test, in which M5.0 is taken for mortar, MU10 for sintered brick and C20 for concrete.

2.2 *Finite element model of basic structure*

Masonry members are two-phase composite members composed of blocks and mortar. According to the purpose of analysis, there are generally two modeling methods (Zheng 2010), namely separated modeling and overall continuous modeling. Because this paper mainly studies the elastic-plastic dynamic response of masonry structure under strong earthquake, the overall continuous mode is adopted for modeling.

There are generally two modeling ideas for masonry structure in the finite element software ABAQUS. The first is to use the beam element to simulate the ring beam and structural column, the wall and floor adopt the shell element, and the second is that the above components are simulated by the three-dimensional solid element. When the first method is adopted, the contact relationship between components is complex, so the effectiveness of the model needs to be evaluated with the help of measured or experimental data, and the plastic damage criterion of beam element needs to be simulated by subroutine. In view of the above reasons, this paper adopts the second modeling idea, that is, the three-dimensional reduced integral element (C3D8R) provided by ABAQUS software is selected as the solid element. According to the embedded method in ABAQUS software, T3D2 element in the truss element is selected to establish the reinforcement skeleton and be embedded into the concrete as a whole. It is assumed that the wall, floor slab, ring beam and structural column can form an ideal deformation coordination state, so TIE constraints in the software are selected to express this connection effect. According to the above ideas and methods, the basic structural finite element model is established.

The constitutive relationship of concrete is determined according to the concrete stress-strain curve in Appendix C of code for design of concrete structures, and the concrete damage plastic model is adopted. Considering that the strength of reinforcement is relatively higher than that of concrete and masonry, the reinforcement in masonry structure generally will not enter the yield state during earthquake, so the bilinear follow-up hardening model is selected for reinforcement.

The typical failure mode of masonry unit in earthquake is the bond failure of masonry mortar due to shear action, but ABAQUS program cannot reasonably describe this failure mode. Referring to the published research results of masonry constitutive relationship, this paper selects Yang Weizhong's masonry compression constitutive model (Yang 2008), The stress-strain relationship of masonry element in tension is relatively complex, and its tensile failure is mainly controlled by the failure strength of masonry joint, but there are few relevant research results. Referring to the tensile failure mode of concrete proposed by Ma Hongwang (Ma 2013), the tensile constitutive model of masonry is established in this paper.

3 STRUCTURAL DYNAMIC CHARACTERISTIC ANALYSIS

3.1 *Model comparison scheme and modal analysis results*

Combined with the characteristics of field investigation, material performance detection and finite element software, three groups of comparative analysis numerical models are established in the stage of structural dynamic characteristic analysis. The specific model scheme and parameters are shown in Table 1. The bay of all models is 3.6m, and the elastic modulus of masonry is determined according to the method proposed by Yin Yuanyuan (Yin 2011).

Table 1. The parameters of models.

Model scheme	Plane aspect ratio	Wall material and strength grade	Wall unit thickness
M0	1.5	MU10, M3.0	120mm
M1-1	2	MU10, M3.0	120mm
M1-2	1	MU10, M3.0	120mm
M2-1	1.5	MU15, M5.0	120mm
M2-2	1.5	MU7.5, M1.5	120mm
M3-1	1.5	MU10, M3.0	60mm
M3-2	1.5	MU10, M3.0	240mm

The vibration mode, period and participation factor of each model scheme are extracted by subspace iterative algorithm. The basic period of masonry structure is 0.141s according to empirical formula. Except M3-2, the basic period of each model is close to the empirical period value, and the results are at a quantitative level with the measured frequency of similar structures in literature (Wei 2010), indicating that all models except M3-2 are reliable.

3.2 *Vibration form*

Table 2 shows the first three order vibration forms of each model scheme. Limited to space, this paper only gives the first three order vibration mode diagrams of the basic structure model (Figure 2) and M3-2 model (Figure 3), as well as the high-order vibration mode diagrams controlled in each model scheme (Figure 4).

Table 2. The first three vibration modes of models.

Model	First order	Second order	Third order
M0	X-direction translation	Y-direction translation	Torsion
M1-1	Y-direction translation	X-direction translation	Torsion
M1-2	X-direction translation	Y-direction translation	Torsion
M2-1	X-direction translation	Y-direction translation	Torsion
M2-2	X-direction translation	Y-direction translation	Torsion
M3-1	X-direction translation	Y-direction translation	Torsion
M3-2	Local vibration	Local vibration	Local vibration

Figure 2. The first three vibration modes of M0.

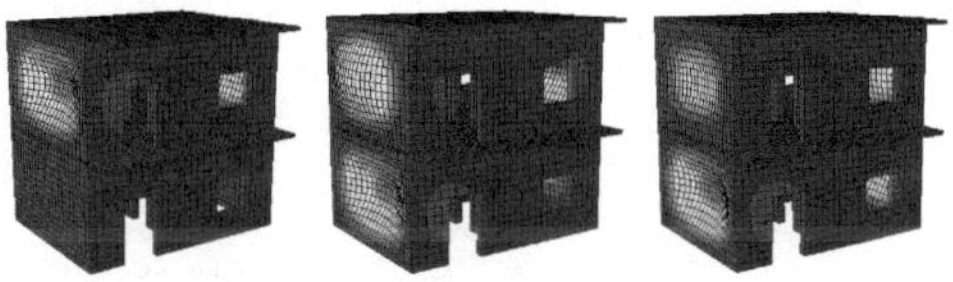

Figure 3. The first three vibration modes of M3-2.

Figure 4. The controlling higher vibration mode of models (Note: M3-2 model is not included).

In Table 2, the vibration patterns of two models are different from other models, which is worthy of attention: (1) the first vibration mode (basic vibration mode) of M1-1 model is Y-direction translation; (2) The first three vibration modes of M3-2 model are not overall vibration.

The low-rise masonry structure in village and town buildings usually adopts the cross wall bearing scheme, and the main door and window openings are generally arranged on the longitudinal wall, forming a typical imbalance of longitudinal and transverse lateral stiffness, and the necessary seismic conceptual design is not carried out in the structural layout to improve this imbalance. Moreover, when the number of cross walls increases with the number of bays, this imbalance will be more significant. Therefore, their basic vibration modes are translational along the longitudinal (X direction) in most cases, as shown in Table 2 and Figure 2. This vibration form is obviously not conducive to the seismic of the structure. In the earthquake, the longitudinal wall is first damaged due to stress concentration at the entrance, and the transverse wall is controlled by out of plane displacement, its bearing capacity and lateral resistance are obviously weakened, which intensifies the damage degree of lateral resistance components. If the prefabricated slabs assembly scheme is adopted for the floor, it is easy to fall and collapse due to the out of plane displacement of the transverse wall. Therefore, the following seismic enlightenment can be obtained from the vibration form of M1-1: as shown in Figure 4, the coupled seismic wall technology (Figure 5) is adopted to adjust the lateral stiffness distribution of the structure in different directions(Wu 2015), so that the stiffness of the structure in the two main axes is close, which will effectively solve the unreasonable vibration problem existing in the conventional layout of the above structures and effectively play the in-plane anti lateral displacement role of the transverse wall. That is to improve the seismic performance of the structure by changing the vibration form and failure mechanism of the structure.

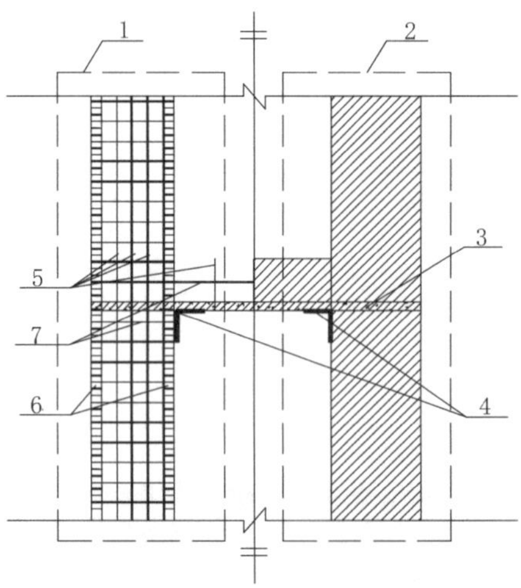

Figure 5. Composite coupling beam restrained masonry coupling seismic wall
1-reinforcement view, 2-block view, 3-floor beam, 4-mild steel damper, 5-structural core column, 6-restrained core column, 7-tie bar.

As for M3-2 model, ABAQUS software does not divide the wall element into elements in the thickness direction. The calculation results show that the low-order vibration mode oscillates along the wall thickness direction, and its vibration mode nephogram is shown in Figure 3. The vibration form is inconsistent with the actual structure, and its vibration mode period and vibration mode participation factor distribution also show significant abnormalities, so it is determined as an invalid model. It is also found that when the floor is not divided into elements in the thickness direction, the oscillation in the thickness direction also occurs. The above anomalies indicate that ABAQUS software should divide at least two elements along the thickness direction for plate and shell components, otherwise abnormal vibration modes and unreasonable calculation results will appear.

3.3 *Mode component*

The participation factors of the first 10 modes of each model in the direction of basic modes are shown in Figure 6.

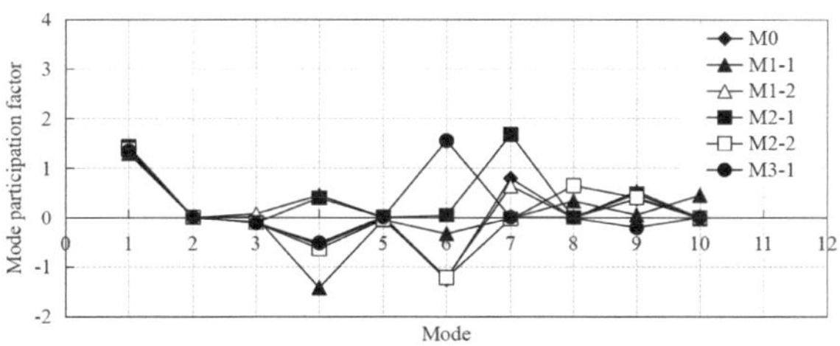

Figure 6. The distribution of mode participation factor (Note: M3-2 model is not included).

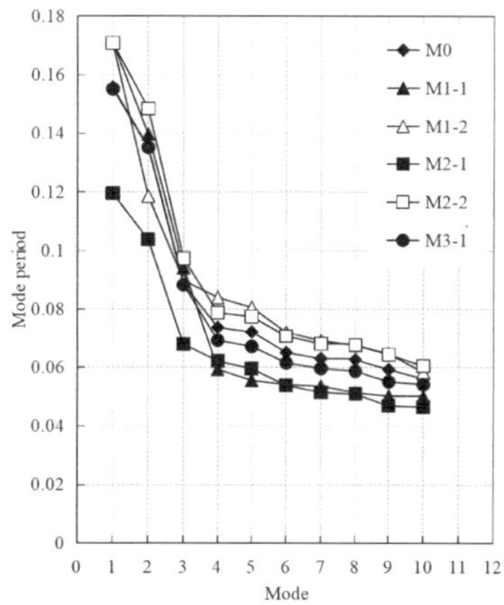

Figure 7. Comparison of the first 10 orders of vibration period (Note: M3-2 model is not included).

It can be seen from Figure 6 that the dynamic characteristics of each effective model are controlled by the basic vibration mode and a high-order vibration mode. According to Figure 4, except M1-1, the high-order vibration modes of other models are out of plane vibration of the wall, which explains the vibration mechanism that the bearing wall is prone to external collapse in previous earthquakes. The high-order vibration form of M1-1 shows that the main structure makes a second-order translational motion along the transverse (Y-direction) and superimposes the vertical vibration of the corridor cantilever plate. It shows that when the basic vibration mode of the structure is translational along the transverse direction, its high-order control vibration mode is still dominated by the overall vibration of the structure, and its dynamic characteristics are significantly better than the local vibration of the out of plane vibration of the wall. It can also be found from Figure 4 that the out of plane vibration peak of the perforated wall has been suppressed to a certain extent. Therefore, in order to improve the dynamic characteristics of village and town buildings and improve their seismic performance, the load-bearing scheme of transverse wall can be transformed in sections on the premise of controlling the basic vibration mode as transverse translation. That is, a strengthening area is set at the end, and the middle part is treated according to the infilled wall construction technology to form a coupled seismic transverse wall (Su 2013), which can significantly reduce the high-order vibration mode component of the out of plane vibration of the local wall in the structure, so as to effectively solve the seismic damage caused by the out of plane vibration of the wall and the resulting overall collapse.

3.4 *Mode period*

Table 3 shows the first three modal periods of each model, and the comparison of the first ten modal periods of each model is shown in Figure 7.

Table 3. The first three vibration periods of models (s).

Mode	M0	M1-1	M1-2	M2-1	M2-2	M3-1	M3-2
1	0.156	0.171	0.171	0.119	0.171	0.155	0.434
2	0.135	0.140	0.119	0.104	0.149	0.135	0.432
3	0.089	0.094	0.090	0.068	0.097	0.088	0.432

From the parameter comparison of wall element thickness in Figure 7, it can be found that when the wall is divided into 2 units (M0) and 4 units (M3-1) along the thickness direction, the calculation results of vibration mode period of the model are very close. Therefore, considering the calculation accuracy and efficiency, the plate and shell components in the structural model are divided into two elements along the thickness direction, which can meet the needs of the research task.

From the parameter comparison of material strength in Figure 7, it can be found that with the decrease of material strength, the period of each order of vibration mode decreases in the order of M2-2, M0 and M2-1, and the regularity is obvious. Combined with the field inspection and investigation of the research group, it shows that the measured strength of materials shows a strong correlation with the construction age of rural houses. Therefore, the component of higher-order modes of out of plane vibration with control effect is amplified with the increase of house age. Therefore, it can be further found that for building materials without standardized management or walls without surface protection, the deterioration of material performance is a factor that cannot be ignored in the earthquake resistance of rural buildings.

From the parameter comparison of the plane aspect ratio in Figure 7, it can be found that the period of the corresponding model has no obvious regularity at the low order, but after the vibration mode number exceeds the fourth order, the period of M1-1 is less than M0 and M1-2, indicating that when the basic vibration mode of the structure vibrates along the transverse direction, the seismic action of the high-order control vibration mode will be further reduced.

4 CONCLUSION

The seismic damage survey and the seismic property survey of village buildings show that masonry structure is the commonly used structural form of village buildings in China at present and in the long term, and the lack of seismic protection or the unreasonable setting of seismic measures is the prominent problem of seismic resistance. In order to improve the seismic performance of village buildings, it is necessary to study the dynamic response of low-rise masonry structures first, and then research and develop simple and effective structural forms and structural systems that are compatible with regional characteristics, taking into account local building forms and economic conditions. Based on the modal analysis method of finite element software ABAQUS, this paper explores the dynamic characteristics of low-rise masonry structures in village buildings by taking typical farm houses in Anhui section of Tanlu Fracture Zone as the object, and conducts a comparative analysis of the effects of factors such as structural plane aspect ratio, material strength and mesh accuracy on the vibration pattern, vibration component and vibration period of the structure, and the results show that:

(1) If the masonry structure of the village building in the survey area adopts the conventional arrangement scheme, its basic vibration pattern is generally shown as a whole flat movement along the longitudinal direction, meanwhile, the higher-order vibration component of the wall out-of-plane vibration is not negligible. Therefore, once encountered destructive earthquake, the hole position of the longitudinal wall will first produce typical earthquake damage due to

stress concentration effect, and then the horizontal wall will rapidly lose bearing capacity and lateral displacement resistance due to excessive out-of-plane displacement, and out-of-plane collapse will inevitably occur in part of the wall.

(2) If reasonable and effective measures are taken to adjust the lateral stiffness distribution of the structure, so that the stiffness of the structure along the two main axes is relatively balanced, the basic vibration pattern of the structure can be realized as a more reasonable transverse plain mode, and at the same time, it can also significantly reduce the external vibration component of the wall surface, which can effectively improve its seismic performance.

(3) According to the investigations and measurements conducted on the village buildings in the Anhui section of the Tanlu Fault Zone, there is a significant correlation between the strength of masonry materials of in-service buildings and their construction age. The deterioration of materials leads to the decrease of structural load-bearing capacity and also increases the higher-order vibration period, the horizontal seismic action will be amplified and the seismic damage will be relatively increased.

ACKNOWLEDGMENT

This work was financially supported by Excellent Young Talents Fund of Colleges and Universities in Anhui Province (gxyq2019130), Research Team Project of Anhui Xinhua University(kytd201901), Innovative Training Program for College Students of the Ministry of Education (202012216052), Innovative Training Program for College Students of Anhui Province (AH201912216183).

REFERENCES

Civil and Structural Groups of Tsinghua University, Xinan Jiaotong University and Beijing Jiaotong University. (2008). Analysis on Seismic Damage of Buildings in the Wenchuan Earthquake[J]. Journal of Building Structures, 29(4): 1–9. (In Chinese)

Jiang Min, Cheng Hua, Wu Tao, et al. (2015). Investigation on Seismic Performance of Rural Buildings in Anhui Section of Tan-Lu Fault Zone[J]. Building Structure, 45(8): 35–39. (In Chinese)

Ma Hong-wang, Wang Lan-min, Chen Long-zhu, et al. (2013). Numerical Simulation of Earthquake Damage in Rural Masonry Buildings[J]. China Earthquake Engineering Journal, 35(2): 232–239. (In Chinese)

Su Qi-wang, Xu Hu, Wu Hao, et al. (2013). Research on Inter-story Displacement Angle of Brick Masonry Structures [J]. China Civil Engineering Journal, 46(S1): 111–116. (In Chinese)

Wang Lan-min, Wu Zhi-jian. (2013). Earthquake Damage Characteristics of the Minxian-Zhangxian MS6.6 Earthquake and Its Lessons[J]. China Earthquake Engineering Journal, 35(3): 401–412. (In Chinese)

Wei Wen-hui, Song Wen-jing, Chen Li. (2010). The Test of Dynamic Characteristics and Analysis of Seismic Behavior on the Rural Residences[J]. Journal of Wuhan University of Technology, 32(3): 42–25,49. (in Chinese)

Wu Tao, Wang Hao, He Xiao-yu, et al. (2015). The invention relates to a coupled seismic wall composed of reinforced pouring block and energy dissipation composite coupling beam and a manufacturing method thereof[P]. China: ZL 201510096823.3, 2017-09-15.

Yang Wei-zhong. (2008). Constitutive Relationship Model for Masonry Materials in Compression[J]. Building Structure, 38(10): 80–82. (In Chinese)

Ye Xian-guo, Wang Ke, Cao Jun-feng, et al. (2012). Seismic Fortification Special Research and Earthquake Damage Prediction of Rural Dwellings in the Northeast of Anhui Province[J]. Earthquake Resistant Engineering and Retrofitting, 34(5): 126–131. (In Chinese)

Yin Yuan-yuan. (2011). Evaluation of Seismic Behavior on Masonry Structure Confined by Core-tie-column[D]. Chongqing: Chongqing University. (In Chinese)

Zhang Shou-jie, Zhu Ze, Ha Hui, et al. (2013). Analysis of Typical Cases of Earthquake Damage in Rural Areas of China and Earthquake Safety Projects for Rural Houses in Recent Years[J]. China Earthquake Engineering Journal,35(4): 936–944. (In Chinese)

Zheng Ni-na. (2010). Research on Seismic Behavior of Masonry Structures with Fabricated Tie-columns[D]. Chongqing: Chongqing University. (In Chinese)

Advances in Urban Engineering and Management Science – Khalil & Yang (Eds)
© 2023 The Author(s), ISBN 978-1-032-30426-7

Study on height selection and operation mode of surge tank in long-distance water supply project

T.C. Zhou*, G.H. Li* & W.J. Danzhen*
Huadong Engineering Corporation Limited, Powerchina, Zhejiang, China

ABSTRACT: Surge tanks are common measures for preventing water hammer damage in long-distance water supply project, and its height selection is very important and complex. The working principle of surge tank and running rules of pumps and pressure regulating valves was introduced. Based on a real engineering example, the water level in the tank was calculated and analyzed. According to the results, the tank height was selected and the operating mode in different water level and discharge combinations were given. Two principles for the elevation selection of surge tank were summarized: first, in high water level and small flow condition, the pump is closed and the regulating valve is working, the tank height should meet gravity flow requirement; second, in low water level and large flow condition, the pump is working and the regulating valve is not working, the tank height should meet pressurized water supply requirement. The conclusion provides reference for similar water supply projects.

1 INTRODUCTION

The geographical distribution of water resources in China is seriously uneven. In order to alleviate this uneven distribution, a large number of large-scale water supply projects have been constructed or are under construction. Most of the long-distance water supply projects is carried out by means of pumps, such as Dahuofang water supply project (Liu 2011), Qiandaohu water distribution project, and Oujiang water supply project. For long-distance pressurized water supply projects, protective measures are required to prevent water hammer damage and ensure the safety and stable operation of the system (Ma 2009).

Surge tank is a commonly used water hammer protection measure in pressurized water supply projects. Its free water surface can effectively reflect the pressure wave generated by the pump stopping (Chaudhry 2013; Yu 2011). Besides, the large amount of water stored in the tank can flow to the pipeline in time, which relieves the negative pressure in the pipeline. In addition. There are many waters supply projects that use surge tank as water hammer protection measures and many studies on this topic (Jin 2004; Wang 2015; Yang 2009; Zhang 2018). Most of the current researches are focused on the effectiveness of surge tank as a water hammer protection measure, as well as the comparison with other protective measures, while the specific principles for the setting of surge tank height and the operation mode of surge tank and pump are relatively scarce.

In this paper, a real long-distance water supply project is calculated. Based on the operating principle of "no pumping while the pressure regulating valve working, and no pressure regulating while pumping", the appropriate surge tank height is selected, the operation mode of the pump and the regulating valve in front of the surge tank was briefly formulated to achieve the goal of economy. Finally, two principles for setting surge tank height were summarized, which provides references for similar water supply projects.

*Corresponding Authors: 981416334@qq.com, 498938310@qq.com and 3538098386@qq.com

 DOI 10.1201/9781003305026-27

2 PROJECT EXAMPLE

A water supply project with a total length of 164 km includes three water plants, and two pump stations. The 1# water plants and 2# water plants are supplied with pressurized water by the 1# pump station, and the 3# water plant is supplied with pressurized water by the 2# pump station. In order to reduce the water hammer pressure, a surge tank with the diameter of 8 m is installed at the node between the two pump stations. The layout of the water delivery system is shown in Figure 1.

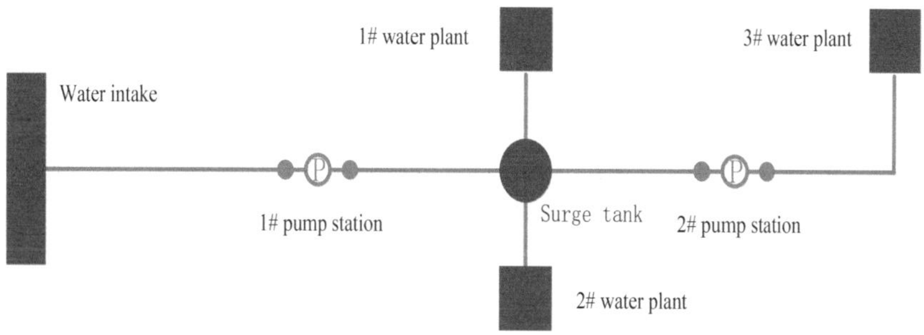

Figure 1. The layout of the water delivery system.

The parameters and design flow of each section of the pipeline are shown in Table 1.

Table 1. Pipeline parameters and design flow.

Pipe	Length Km	Diameter M	Roughness /	Flow m^3/s
water intake -1# pump	68.32	2.4	0.012	6.6 (Qr)
1# pump-surge tank	23.73	2.4	0.012	6.60
surge tank -1# water plant	18.26	1.4	0.012	1.70
surge tank -2# water plant	15.72	1.6	0.012	2.40
surge tank -2# water plant	15.72	1.6	0.012	2.50
2# pump -3# water plant	22.34	1.6	0.012	2.50
water intake -1# pump	68.32	2.4	0.012	6.6 (Qr)
1# pump-surge tank	23.73	2.4	0.012	6.60

The water level of the water intake and the water plant, and the changes in the flow rate of the water supply are as follows:

The water level of the upstream water intake: 50~66 m.
The water head in front of the pump when the pump is running: >2 m.
The water level of the water plant: >4 m.
The water flow rate of the water plant: between 0.68 to 1.0 times the design flow (Qr) rate.

3 HEIGHT SELECTION OF SURGE TANK

3.1 *Operation principle of surge tank and pump*

The upstream water level and flow vary greatly, and there are two operation conditions: gravity flow and pump pressurization. Under the working condition of 0.68 times design flow (0.68Qr), the whole line can supply water by gravity. As the flow increases, the 2# pump station shall opine first, the water plant behind the 2# pump station will be provided pressurized water, and the water plant before the 2# pump station will be provided gravity water. Under the working condition of

1.0Qr, both pump stations shall open and pressurized water supply is provided for the whole line. Considering that the water level at the surge tank may be high when gravity flows, a regulating valve is set on the pipeline in front of the surge tank to reduce the water level in the surge chamber.

In order to reduce energy waste, when the water pump pressurization is needed to supply water, the regulating valve is required to be fully opened, that is, all the energy provided by the pump must be consumed along the pipeline. This operating principle can be referred to simply as "no pumping while the pressure regulating valve working, and no pressure regulating while pumping". So, there are three types of conditions in this project:

a) Under the maximum water level and minimum flow conditions: the whole line can flow by gravity with no need to pumping, but in order to prevent overflow in the surge tank, the regulating valves need to be working to lower the water level in the tank.
b) Under higher water level and minimum flow, or maximum water level and smaller flow conditions: the whole line can flow by gravity with no need to pumping, and the water level in the tank is lower than the overflow elevation with no need to open the regulating valves.
c) Under the minimum water level and maximum flow conditions: the system cannot supply water by gravity, so all pump stations need to opening, and the regulating valve is not working, that is, the water level in the surge tank is lower than the overflow elevation.

There is a minimum elevation of the surge tank. When the surge tank elevation is lower than this elevation, for condition 3, it may happen that both the pumps and regulating valve need to working. For working condition 1, surge tank overflows may happen and the water pressure is not met in the downstream. It is particularly important to choose a suitable surge tank height.

3.2 *Calculation of pump head*

The head in front of 1# pump is calculated according to the pipeline parameters and flow between water intake and 1# pump station. The calculation formula is as follows.

$$H_2 = H_1 - h_f \tag{1}$$

$$h_f = \frac{n^2 L Q^2}{A^2 R^{4/3}} \tag{2}$$

where: H_2 is the front head of pump 1#; H_1 is the intake water level; h_f is the water head loss; n is the pipeline roughness; L is the pipeline length; Q is the pipeline flow; R is the wetted perimeter of the pipeline. The calculation results are shown in Table 2.

Table 2. Head before pump in 1# pumping station under different conditions.

	Qr								
H1 (m)	0.68	0.72	0.76	0.80	0.84	0.88	0.92	0.96	1.00
70	50.86	48.55	46.10	43.52	40.80	37.95	34.97	31.86	28.62
68	48.86	46.55	44.10	41.52	38.80	35.95	32.97	29.86	26.62
66	46.86	44.55	42.10	39.52	36.80	33.95	30.97	27.86	24.62
64	44.86	42.55	40.10	37.52	34.80	31.95	28.97	25.86	22.62
62	42.86	40.55	38.10	35.52	32.80	29.95	26.97	23.86	20.62
60	40.86	38.55	36.10	33.52	30.80	27.95	24.97	21.86	18.62
58	38.86	36.55	34.10	31.52	28.80	25.95	22.97	19.86	16.62
56	36.86	34.55	32.10	29.52	26.80	23.95	20.97	17.86	14.62

The head after the pump meeting the flow supply requirements is calculated. The calculation formula is as follows:

$$H_1 = H_2 + h_f \tag{3}$$

Table 3. Head needed after pump in 1# pumping station under different conditions.

	Qr								
H1 (m)	0.68	0.72	0.76	0.80	0.84	0.88	0.92	0.96	1.00
56–70	16.66	18.19	19.81	21.52	23.32	25.20	27.17	29.23	31.38

The calculation results are shown in Table 3.

Then the pump head of the 1# pump station can be calculated, as shown in Table 4. (The "/" means that the water pump does not need to be opened.)

Table 4. Pump head of 1# pumping station under different conditions.

	Qr								
H1 (m)	0.68	0.72	0.76	0.80	0.84	0.88	0.92	0.96	1.00
70	/	/	/	/	/	/	/	/	2.76
68	/	/	/	/	/	/	/	/	4.76
66	/	/	/	/	/	/	/	1.37	6.76
64	/	/	/	/	/	/	/	3.37	8.76
62	/	/	/	/	/	/	0.20	5.37	10.76
60	/	/	/	/	/	/	2.20	7.37	12.76
58	/	/	/	/	/	/	4.20	9.37	14.76
56	/	/	/	/	/	1.25	6.20	11.37	16.76

The pump head of 2# pump station can be calculated by the same method, and the results are shown in Table 5.

Table 5. Pump head of 2# pumping station under different conditions.

	Qr								
H1 (m)	0.68	0.72	0.76	0.80	0.84	0.88	0.92	0.96	1.00
70	/	/	/	/	/	/	5.53	11.88	15.75
68	/	/	/	/	/	1.44	7.53	13.88	15.75
66	/	/	/	/	/	3.44	9.53	14.51	15.75
64	/	/	/	/	/	5.44	11.53	14.51	15.75
62	/	/	/	/	1.63	7.44	13.33	14.51	15.75
60	/	/	/	/	3.63	9.44	13.33	14.51	15.75
58	/	/	/	0.08	5.63	11.44	13.33	14.51	15.75
56	/	/	/	2.08	7.63	12.19	13.33	14.51	15.75

3.3 *Height selection of surge tank*

The water head behind the 1# pump station can be obtained by 1# pump station head, which is equal to the larger value of Tables 2 and 3, as shown in Table 6.

The head in front of the surge tank can be calculated by the head behind the pump in Table 6 and the water loss between 1# pump station and the surge tank, as shown in Table 7.

According to the pump head, all conditions can be divided into two parts: gravity flow condition on the left and pumping pressurization condition on the right.

For the pump pressurization conditions, the condition of the highest water head in front of the tank can be find out. That is, the condition of 68 m water level and 0.88Qr flow. The water head in

Table 6. Head after pump in 1# pumping station under different conditions.

	Qr								
H1 (m)	0.68	0.72	0.76	0.80	0.84	0.88	0.92	0.96	1.00
70	50.86	48.55	46.10	43.52	40.80	37.95	34.97	31.86	31.38
68	48.86	46.55	44.10	41.52	38.80	35.95	32.97	29.86	31.38
66	46.86	44.55	42.10	39.52	36.80	33.95	30.97	29.23	31.38
64	44.86	42.55	40.10	37.52	34.80	31.95	28.97	29.23	31.38
62	42.86	40.55	38.10	35.52	32.80	29.95	27.17	29.23	31.38
60	40.86	38.55	36.10	33.52	30.80	27.95	27.17	29.23	31.38
58	38.86	36.55	34.10	31.52	28.80	25.95	27.17	29.23	31.38
56	36.86	34.55	32.10	29.52	26.80	25.20	27.17	29.23	31.38

Table 7. Water head before surge tank under different conditions.

	Qr								
H1 (m)	0.68	0.72	0.76	0.80	0.84	0.88	0.92	0.96	1.00
70	44.22	41.10	37.79	34.32	30.66	26.82	22.81	18.61	17.01
68	42.22	39.10	35.79	32.32	28.66	24.82	20.81	16.61	17.01
66	40.22	37.10	33.79	30.32	26.66	22.82	18.81	15.99	17.01
64	38.22	35.10	31.79	28.32	24.66	20.82	16.81	15.99	17.01
62	36.22	33.10	29.79	26.32	22.66	18.82	15.01	15.99	17.01
60	34.22	31.10	27.79	24.32	20.66	16.82	15.01	15.99	17.01
58	32.22	29.10	25.79	22.32	18.66	14.82	15.01	15.99	17.01
56	30.22	27.10	23.79	20.32	16.66	14.07	15.01	15.99	17.01

front of the tank is 24.82 m in this condition. Therefore, as long as the height of the surge tank is greater than 24.82 m, the surge tank will not overflow.

For the gravity flow conditions, the regulating valve is working under some gravity flow conditions. For small flow conditions, such as 70 m water level and 0.68Qr flow, the water level in the surge chamber is 44.22 m. Obviously, the water level of the surge chamber does not need to be so high, and the system can also flow by gravity. In the table, the maximum flow of gravity flow part is 0.88Qr. At this condition, the head in front of the surge tank is 26.82 m, while the actual required outlet head of the surge tank is $25.20 - 11.13 = 14.07$ m. Therefore, as long as the surge tank height is greater than 14.07 m, the surge chamber under all gravity flow conditions will not overflow.

In summary, the overflow elevation of the surge tank only needs to be higher than 24.82 m. The surge tank height can be selected as 26 m. The water level is controlled below 25 m under all conditions (a safety margin of 1 m height).

3.4 *Operation mode of the project*

Due to the surge tank water level is controlled below 25 m under all conditions, the water pump head and the regulation loss of regulating valve under each water level and flow condition can be calculated, as shown in Table 8.

In the table, the left part is the working conditions of pump closing and regulating valve working, and the number is the head loss of the regulating valve. The corresponding opening of the valve can be calculated by the valve overflow curve.

Table 8. Operation mode under different conditions.

	Qr								
H1 (m)	0.68	0.72	0.76	0.80	0.84	0.88	0.92	0.96	1.00
70	19.22	16.10	12.79	−9.32	−5.66	−1.82	5.53	11.88	15.75
68	17.22	14.10	10.79	−7.32	−3.66	1.44	7.53	13.88	15.75
66	15.22	12.10	−8.79	−5.32	−1.66	3.44	9.53	14.51	15.75
64	13.22	10.10	−6.79	−3.32	/	5.44	11.53	14.51	15.75
62	11.22	−8.10	−4.79	−1.32	1.63	7.44	13.33	14.51	15.75
60	−9.22	−6.10	−2.79	/	3.63	9.44	13.33	14.51	15.75
58	−7.22	−4.10	−0.79	0.08	5.63	11.44	13.33	14.51	15.75
56	−5.22	−2.10	/	2.08	7.63	12.19	13.33	14.51	15.75

The middle is the working conditions of pump closing and regulating valve not working. In this condition, the whole line before the 2# pump station can flow by gravity, and the head in front of the surge tank is lower than the overflow elevation, so the regulating valve is no need to work.

The middle is the working conditions of pump is opening and the valve is not working. In this condition, the water plant can no longer rely on gravity for water supply, and the pump station needs to be opened. The number is the minimum head of the 2# pump station, which can be found out according to Table 4. Since the head in front of the surge tank is lower than the overflow elevation, the regulating valve is not working.

4 CONCLUSIONS

The working principle of surge tank as a prevention for water hammer in long-distance water supply projects was summarized. The surge tank height and operation mode of surge tank and pump station in a real project was analyzed. The principles for surge tank selection are summarized as follows:

1. Under the condition of high-water level and low flow rate, the pump is closed, and the regulating valve is working, the surge tank should be higher than the minimum water level at the surge tank that meets the water supply requirements;
2. Under the condition of low water and large flow, the water pump is opened, and the regulating valve is not working, the surge tank should be higher than the minimum water level at the surge tank that meets the water supply requirements.

REFERENCES

Chaudhry, M. H. (2013). Applied hydraulic transients (Third Edition). New York: Springer, 21–23.

Jin, Z. & N. C. Jiang (2004). Pump stop water hammer and protection (Third Edition). China Architecture & Building Press, 56–59.

Liu, G. C. & G. H. Chang (2011). Process design features of supporting project for water supply from Shenyang Dahuofang reservoir. J. ZHONGGUO JISHUI PAISHUI, 27(22), 48–53.

Ma, S. B. & J. Zhang (2009). Study on protective measures of pump stop water hammer in long-distance water conveyance project. J. Yangtze River, 40(1), 85–86.

Wang, J. P. & Y. Z. Du (2015). Study on water hammer protection scheme of long-distance pressurized water conveyance project. J. Water Technology, 9(6), 43–48.

Yu, X. D. & J. Z. Zhang (2011). Critical superposition instant of surge waves in surge tank with long. J. Canadian Journal of Civil Engineering, 38(3), 331–337.

Yang, Y. S. & Y. Y. Xu (2009). Research on water hammer prevention in high-lift hilly and long-distance water transmission pipeline. J. Water Supply and Drainage, 45(4), 108–111.

Zhang, X. Y. & W. X. Pan (2018). Study on water hammer protection of large flow and long-distance water supply project. J. SMALL HYDRO POWER, 2, 20–23.

Advances in Urban Engineering and Management Science – Khalil & Yang (Eds)
© 2023 The Author(s), ISBN 978-1-032-30426-7

Research on research methods and protective measures of rockfall disasters on slope

Chunhui Ma, Tiaohao Zhao*, Jie Yang & Lin Cheng
School of Water Resources and Hydropower, Xi'an University of Technology (State Key Laboratory of Ecological Water Conservancy in Northwest Arid Areas, Jointly Established by the Provincial and Ministerial Departments), Xi'an, Shaanxi, China

Zhiquan Zeng
China Power Construction Group East China Survey, Design and Research Institute Co., ltd., Hangzhou, Zhejiang, China

Yong Huang
East China Survey and Design Institute of China Power Construction Group Co., Ltd., Hangzhou, Zhejiang, China

ABSTRACT: Rock-falling slope disaster is a common geological disaster in mountainous areas of China, which is characterized by wide distribution, concealment and suddenness, etc. It is difficult to implement all-round treatment, and once the disaster occurs, it will cause serious loss of life and property. At present, the three main problems of slope rolling stone disaster are this: 1. Study on the characteristics of rolling stone movement, including the trajectory, impact force and protective measures of rolling stone. 2. Monitoring and early warning of slope rolling stone disaster, summarizing the commonly used monitoring methods and instruments, and pointing out the development direction of slope rolling stone monitoring in the future; 3. Combing common protection technologies of slope rolling stone disaster. On the basis of summarizing the above research problems, the existing problems and future development trends in the research of rolling stone disaster are pointed out: 1. When using numerical simulation to study the motion characteristics of rolling stones, the factors such as the crushing of rolling stones, the shape of rolling stones and the shape of the earth's surface should be considered, so as to better fit the actual motion situation of rolling stones and make the simulation results more accurate; 2. In the monitoring and early warning of slope rolling stone disaster. We should pay attention to the research on the separation stage of the rock-rolling disaster on the slope, and consider the index system based on the combination of dynamic index and static index to judge the risk degree of the rock-rolling disaster on the slope.

1 INTRODUCTION

Rolling stones refer to the dynamic evolution process in which individual rocks lose their stability from the surface of geological bodies for some reason, and then move down the slope rapidly through one or a combination of several movement modes, such as falling, spring back, jumping, rolling or sliding, etc., and finally stop at a relatively gentle zone or near obstacles (Zhang et al. 2004). Rock slope is one of the common disasters in mountainous areas. Although the scale is generally small, the damage caused by a rock slope is equivalent to that caused by large-scale landslide to a certain extent due to its sudden and random characteristics. In the United States, Italy, Canada and other countries with more mountainous areas, the slope rolling stone disaster has become one of the major disasters causing death (Evans & Huanger 1993).

*Corresponding Author: zhaotianhao@stu.xaut.edu.cn

 DOI 10.1201/9781003305026-28

Rolling stone disaster is very common in highway, water conservancy, bridge and tunnel engineering, and it is one of the important problems that need to be faced in engineering life cycle. Once a disaster occurs, it will not only affect the opening of roads and navigation channels, but also cause serious loss of life and property, which will seriously affect the development of society and economy. According to table 1, the slope rolling stone disaster seriously threatens people's life and property safety and the normal transportation safety of the country. Although the theory of prevention and control of geological disasters and treatment measures in China is maturing day by day, the proportion of slope rolling stone disasters is still increasing year by year. Therefore, research, monitoring and protection of slope rolling stone disaster is the key problem that must be solved in current engineering construction, and the research of slope rolling stone disaster protection will become the research focus of future geological disasters in China.

Table 1. Typical cases of slope rolling stone disaster in China.

time	Site	Disaster details	lose
May 2005.	Baozhu Road, Mianzhu County, Sichuan Province.	A boulder of more than ten tons hit a bus.	Eighteen people were killed and 14 seriously injured.
November 2007.	Hubei Yiwan Railway.	The rock mass at the tunnel entrance collapsed and a passing bus was buried.	More than 30 people died on the spot.
July 2009.	Min Jiang you an di Guan bridge	Broken boulder completely closes the No.8 Bridge soil of the bridge.	Twelve people were injured and six died.
July 2014.	Within Maoxian County of National Highway 213.	Rolling stones fell from the top of the mountain at 1000 meters.	19 people were injured and 11 died.
March 2015.	Guangxi Diecai Mountain Scenic Spot.	The boulder rolled down and hit many tourists.	Seven people were killed and 25 injured.
October, 2018	Paiba Village, Chenggan Township, Lushui City, Nujiang Prefecture, Yunnan Province.	Rolling stones hit villagers' houses.	Two people were killed and three were injured.
July 2019.	Minjiang Township, Songpan County, Aba Prefecture, Sichuan Province.	Rolling stone hits tourist bus.	Eight people were killed and 16 injured.

2 RESEARCH PROGRESS OF MOTION CHARACTERISTICS OF ROLLING STONES

The trajectory and impact force of rolling stone is important movement characteristics of rolling stone, and scientific and reasonable arrangement of rolling stone protection measures is the fundamental purpose of carrying out research on slope rolling stone disasters. In order to determine the moving distance of rolling stones and the influence range of rolling stones, a study of rolling stone trajectory can be performed. Computer numerical simulation and other methods can be developed to help determine if the infrastructure or buildings are within the influence range of rolling stones. On this basis, the impact force of rolling stone is further analyzed, which provides guidance for the selection of protective materials, the size formulation of protective structures and structural design.

2.1 *Research on the trajectory of rolling stone*

The research on the trajectory of rolling stone is an important basis for determining the influence range of rolling stone and further developing the research on the impact force of rolling stone.

At present, the research methods of rolling stone trajectory are roughly divided into four types: field investigation, model test, theoretical calculation and numerical simulation. Field investigation and field test are often affected by many factors such as complex slope topography. It is difficult to carry out it comprehensively and systematically, and when the observation range is large. It will consume a lot of manpower and material resources. Compared with the field experiment, the model experimental method is more convenient, but it cannot accurately restore the geological environment and the actual shape of rolling stone, resulting in certain errors. Before numerical simulation technology was mature, theoretical calculation was the most used research method. It is another important theoretical basis for the development of numerical simulation methods.

In the aspect of speculative calculation, most scholars simplify the rolling stone into a particle model, so as to deduce the equation of the moving speed and trajectory of the rolling stone, such as Yang Haiqing (Yang & Zhou 2009). The rolling stone is regarded as an ellipse approximately, and the influence of the shape of a rolling stone on the trajectory of the rolling stone is considered. Tang Hongmei et al. (Tang & Yi 2003) The motion process of precarious rock is clearly divided into four stages: initial motion process, collision process, sliding process and rolling process, and the motion trajectory equations of each stage are analyzed. Lv Qing et al. (Lv et al. 2003) In this paper, the main forms of rolling stone movement on slope are analyzed, and the main influencing factors for controlling rolling stone movement are considered, and the calculation formula for calculating the trajectory of rolling stone movement on slope is derived from the engineering point of view. Feng Jinfu et al. (2011) According to the three stages of the whole process of rolling stone movement, namely, rolling (sliding) movement stage, parabolic movement stage and collision stage with slope, the calculation formulas of the movement speed of each movement stage are obtained by using the subsection circulation algorithm. Muhammad et al. (Muhammad. 2017) The motion is separated into three parts: jumping stage, rolling (sliding) stage and collision stage, and the formulas for calculating the speed of each stage of motion are put forward respectively. Richie (Ritchie 1963) This paper puts forward a simple idea for calculating the trajectory of rolling stones, deduces the mathematical formula for calculating the velocity and distance of the collapsed rolling stones, and puts forward that the trajectory of rolling stones is a combination of motion modes such as free fall.

In numerical simulation, Huang Junguang et al. (Huang & Zhang 2021) using the rolling stone particle model and ANSYS/LS-DYNA finite element software. The trajectory characteristics of rolling stone are studied, and the deflection angle of the two methods at the variable slope point is compared and analyzed. (Liu et al. 2020). The kinetic energy and trajectory of rolling stone are analyzed by using three-dimensional discontinuous deformation analysis method, and the influence of the platform on the trajectory of rolling stone is studied. Xiao Feng et al. (Feng et al. 2013) The rolling stone analysis software is used to simulate the rolling stone movement trajectory, and the potential rolling stone movement trajectory in the dangerous rock section is analyzed, and the application of numerical simulation in the treatment design of rolling stone is discussed. Wang Yu et al. (Wang & Cao 2018) Through Rocfall simulation software, the disaster of different quality rolling stone was simulated, and the influence range of slope rolling stone disaster was roughly predicted. Xu Jingjing et al. (2019) The energy tracking method is used for the first time to study the influence of rock fragmentation on the trajectory of rolling stone on the slopes. Wang Xiang et al. (2020) Rocfall software is used to analyze the motion characteristics of rock mass collapse and rolling stone with different slope angles on the same structural plan, such as translation speed and jumping height, and the damage to the bridge structure. Dong Wang et al. (2021) Rockfall Analyst software is used to numerically simulate the three-dimensional motion trajectory of five kinds of high-position dangerous rock masses with different sizes in the upper part of the tunnel. Bi Ran et al. (2016) based on the traditional impulse response method, energy tracking method, the self-made discrete element program is utilized to simulate the rolling stone movement of a slope around Wenchuan earthquake. Franck et al. (et al. 2009) the random impact model related to the data collection method of the target field is developed, and the springback calculation program is integrated into the 3D rolling stone simulation model, and good simulation results are obtained.

At present, there are more than 10 kinds of software for simulating the trajectory of rolling stone. Guzzetti et al. (2002) common ones are: Rocfall, CRSP, STONE, PFC, etc. Table 2 analyzes and compares them in various aspects. Most of them are only applicable to the simulation of rolling STONE trajectory in a small range, and Stone's simulation of rolling stone in a larger range is outstanding.

Table 2. Comparison of Stone, CRSP, RocFall and PFC (Huang et al. 2010).

software		STONE (Guzzetti, Crosta, Detti, et al. 2002)	CRSP (Jones, Higgins, Andrew 2000)	Rocfall (Alejano, Pons, Bastante, et al. 2007)	PFC
types of models		Centralized mass type	lumped mass Mixed with rigid body.	Concentrated mass and Rigid body mixed type	
Analytical dimension		three-dimensional	two dimensions	two dimensions	2D/3D
Terrain import type		Digital terrain model	Slope profile	Slope profile	Digital terrain model
Analytical method	freely falling body	Centralized mass method	Centralized mass method "Centralized mass method	Centralized mass method	
	Collision/ rebound	Centralized mass method	Rigid body method	Centralized mass method	Newton's second law
	roll	Centralized mass method	Rolling calculation is not considered.	Rolling calculation is not considered.	Centralized mass method
	slide	Sliding calculation is not considered. "Sliding calculation is not considered.	Centralized mass method	Considering sliding calculation.	
Rolling stone shape		particle	Spherical, cylindrical, disc-shaped.	Particle or sphere.	Models with different shapes can be established as required.
other		The initial angle can be changed.	The roughness of the terrain at the collision point is determined by the size of rolling stones.	The initial angle, energy recovery coefficient and dynamic friction coefficient can be changed.	The initial angle and various coefficients can be set by self-contained programming language.

At present, the following problems still exist in the research on the trajectory of rolling stones: ① Most of the current theoretical calculations divide the motion state of rolling stones into four states: sliding, free fall, bouncing and rolling, and the analysis assumes that the rolling stones are in a single motion state, but in practice the motion of rolling stones is a random combination of multiple motion states. How to determine its actual motion state is part of the difficult problems in the analysis of slope rolling stone disaster characteristics; ② at present, the commonly used rolling stone trajectory simulation software assumes that the rolling stone is a block with a regular shape, but in fact, the shape of the rolling stone block is extremely irregular, which will also affect the accuracy of simulation results.

2.2 *Study on impact forces of rolling stone*

The research on the impact force of rolling stone is an important part of the slope protection against rolling stone, and the impact force of rolling stones is directly related to the selection of protective structural materials. Size formulation and structural design. The factors affecting impact force include a series of complex factors such as shape, mass, mechanical parameters and impact velocity of rolling stones, Even the moment of collision, the shape and position of the contact point of rolling stones will significantly affect the impact force.

Rolling stone impact is actually a conflict behavior, and the classical calculation methods of collision problems are Hertz theory based on elastic theory and Thornton theory based on ideal elastic-plastic materials (Thornton & Ning 1998). Japan Road Association (Kawahara & Muro 2006) the rolling stone is regarded as a rigid sphere, while the protective structure is a completely plastic body. Based on Hertz elastic contact theory and experimental data of rolling stone impact force, the empirical formula of rolling stone influence force is obtained. Delhomme et al. (2005) empirical formulae is obtained from Meyer's law. Pichler et al. (2005) The final expression of the impact force was obtained by combining experiment with dimensional analysis. Labiouse et al. (Vincent et al.1996) through the field test of rolling stone impact, the empirical formula for calculating the impact force of rolling stone is established. Yang Qixin et al. (Yang & Guan 1996) According to the laboratory test results, the empirical formula based on impulse theorem is established, and the influence of different buffer layer thickness is considered (Shen et al. 2009). It is considered that the soil reaction when rolling stones fall is a function of penetration depth, and the formula of the impact force in the process of rolling stone impact is derived. Wang Xing et al. (2019) by using LS-DYNA software, the variation of impact force and penetration depth with impact time of rolling stones was studied. Some foreign scholars (Jones & Alves 2004; Kishi et al. 2002; Mougin et al. 2005) taking the protective structure as a beam-slab structure, a formula for calculating the impact force of the cantilever beam and simply supported beam based on material mechanics is proposed.

At present, the application of numerical simulation software to study the trajectory and impact force of rolling stone has become the mainstream means to study the slope rolling stone disaster. However, how to accurately consider the influence of air resistance, geological conditions, rolling stone shape and other factors on the trajectory and impact force of rolling stone needs further research.

2.3 *Study on protective measures for rolling stone*

The impact force of rolling stone is the core issue in the research of protection measures for rolling stone, and numerical simulation is the mainstream means to study the impact force of rolling stone. At present, experts and scholars at home and abroad mainly use the finite element method and the discrete element method to simulate the protection measures against rolling stones.

In terms of finite element method, Ronco et al. (2009) the dissipation process of impact energy of rolling stone and the deformation characteristics of reinforced earth retaining walls are analyzed, and the design principles of protective retaining walls are put forward. Toe et al. (2018) using the results of finite element simulation, a meta-model is established to predict the success probability of intercepting the rolling stone event by the rolling stone protection structure and the dissipation degree of impact kinetic energy after the rolling stone hits the retaining wall. Hu Xiewen et al. (2019) The finite element and infinite element coupling method are used to analyze the dynamic response of protective retaining wall structure under impact load. Because the finite element method uses interconnected elements to simulate complex geometry, it is sometimes difficult to reflect mechanical phenomena and mechanisms such as rotation of soil particles and large deformation of soil.

At present, the discrete element method is used extensively to study the mechanical behavior of granular materials (Xu et al. 2012). Plassiard et al. (2009) The impact process of Gebin reinforced earth retaining wall was studied by SDEC software, and the influence of parameters such as diameter, mass and velocity of rolling stones, and geometric and mechanical characteristics of retaining wall on the structural response of the retaining wall was evaluated. Dugelas et al. (2019) The discrete element model of flexible the fence is established, and the response of fence under the impact of rolling stone is analyzed.

3 MONITORING AND EARLY WARNING OF SLOPE ROLLING STONE DISASTER

Generally, the occurrence of slope rolling stone disasters will go through four stages: gestation stage, separation stage, accelerated failure stage and stopping the accumulation stage. The specific development process is illustrated in Figure 1.

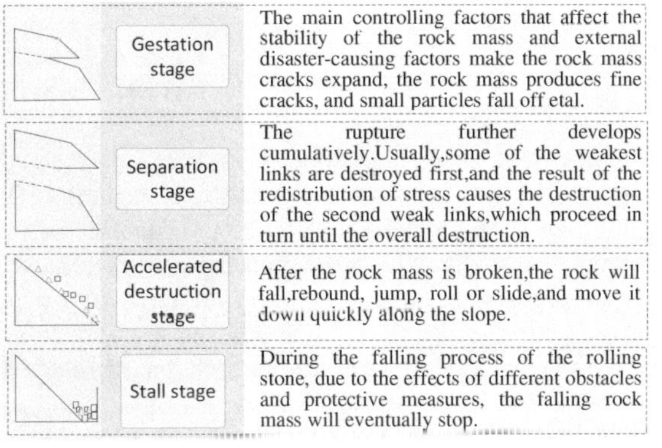

Figure 1. Four stages of slope rolling stone disaster.

3.1 *Slope rolling stone disaster early warning research and monitoring mean*

The traditional early warning method is aimed at the accelerated failure stage in the later stage of the disaster, but the duration of the accelerated failure stage of the slope rolling stone disaster is short, and the early warning time is too close to the disaster occurrence time, which makes it difficult for people to have enough time to take preventive measures or take emergency refuge before the slope rolling stone disaster, which can easily cause serious loss of life and property. Therefore, in order to achieve accurate and timely warning of the slope rolling stone disaster, it is necessary to conduct in-depth research on the separation stage of the rolling stone to overcome the characteristics of high concealment and the sudden occurrence of the slope rolling stone disaster. Du Yan et al. (2019) The experiment shows that the early warning thought based on accelerated failure can only judge the disaster itself, but cannot really predict and prevent the slope rolling stone disaster, while the early warning of the slope rolling stone disaster based on the stage of rock mass separation and failure makes it possible.

To sum up, aiming at the brittle failure of slope such as rolling stone. It is one of the effective means to realize the early warning of brittle failure. (Du et al. 2019)

3.2 *Monitoring and early warning index system of slope rolling stone disaster*

At present, most of the monitoring theories and methods are based on the traditional static indicators (stress, deformation, etc.) and environmental scalars (rainfall, temperature, etc.) to judge the hidden degree of rock mass, so as to judge the threat degree of disasters. Although these indicators can effectively identify risks to a certain extent, there is still a certain time lag and uncertainty. In fact, brittle failure of rock mass is mostly dynamic failure caused by system instability, as well as damage and fracture propagation of main control structural plane (Chen et al.2015). Therefore, the introduction of dynamic monitoring index can help to realize the stability evaluation and early warning of rock blocks (Qiu et al. 2019; Vlentin et al. 2017) In addition, although a single monitoring index can play a certain early warning role in the monitoring process, in practical engineering, it is impossible to make an accurate early warning for disasters only by a single index, and it is necessary to rely on a comprehensive system of multiple indexes to make a more accurate and scientific judgment for disasters.

From the above, it can be seen that the multi-index monitoring and early warning system can bring better early warning effect, and developing a set of multi-index monitoring index system suitable for early warning of collapse disaster is one of the mainstream directions of slope rolling stone disaster monitoring and development at present (Du et al. 2019). Uehan et al. (2012) the monitoring indexes such as the natural vibration frequency, damping ratio and cumulative amplitude are introduced, and a multi-index rock slope stability evaluation system is developed. The safety evaluation method based on dynamic monitoring indexes is also developed, which combines the actual engineering measurement parameters with safety evaluation indexes and enriches the index system in engineering safety monitoring.

4 PROTECTION TECHNOLOGY TO SLOPE ROLLING STONE DISASTER

Common protective measures of slope rolling stone disaster can be divided into two categories, namely, active protection technology and passive protection technology. The protection concept of active protection technology is to enhance the stability of dangerous rock mass and prevent it from caving without causing disaster. The protection concept of passive protection system is to effectively intercept or protect the threatened object from harm after the occurrence of rolling stone (Ye 2017). The classification is shown in Figure 2, and the applicable conditions of active protection measures and passive protection measures are detailed in Tables 3 and 4.

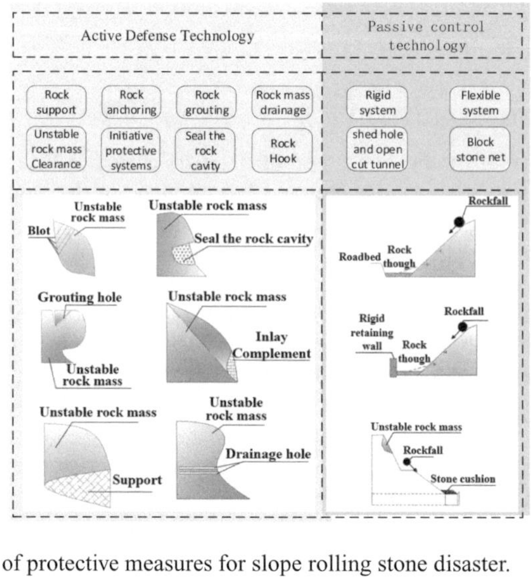

Figure 2. Classification of protective measures for slope rolling stone disaster.

Table 3. Selection of active control technology (Ye 2017).

web technology	Dangerous rock scale	Dangerous rock position	Applicable conditions or requirements
clear away	All kinds of	All kinds of	When there are construction conditions on the spot, it is preferred to use them, but when unloading cracks are relatively developed, they are easy to produce chain instability after being clear, they should not be used. Manual and small mechanical removal, suitable for medium and low dangerous rocks.
support	Medium and small.	low post	It is suitable for suspending the bottom of toppling and falling collapse bodies, and using the supporting structure to transfer the force on the stable foundation.
Embedding and filling	Medium and small.	low post	It is suitable for small cavities at the bottom of toppling and falling collapse bodies, and the embedded foundation should be set on the bedrock to reinforce and prevent the cavities from weathering and expanding.
anchor	All kinds of	Medium and high position	Mainly aiming at toppling and sliding collapse bodies, the deformation and displacement of the collapsed bodies are restrained by connecting anchors with stable rock and soil bodies.
Active protective net	All kinds of	Medium and high position	The fastening force of the net and anchor system on the steep cliff surface is suitable for strengthening the dangerous rock mass on the shallow surface. Or limit the movement range and path of dangerous rock mass after caving.
Grouting engineering	All kinds of	All kinds of	Flexible selection when grouting construction is feasible and economical; Avoid adverse effects of drilling construction and high-pressure grouting on rock mass stability.
drainage works	All kinds of	All kinds of	Surface and underground drainage measures can be flexibly selected according to actual conditions, and are suitable for dangerous rock mass affected by rainfall, water infiltration, water flow and other factors.

Table 4. Characteristics and applicable conditions of common passive protection systems (Ye 2017).

Technology type	Column energy level (kJ)	Applicable conditions or requirements.
Rock-blocking shed	10~200	Applicable to interception of scattered block caving, or emergency rescue, danger removal, and temporary protective measures during construction. When used as a permanent measure, it is necessary to evaluate the impact force of falling rocks and carry out structural calculation.
Semi-rigid retaining wall	100~500	Pile-slab structure system is adopted, which occupies a little less land than rigid stone retaining wall, has flexible layout and enhanced interception ability, and can be designed according to the impact force.
Falling stone wall	30~6000	In theory, even large-scale collapses can be intercepted by rockfalls large enough. However, rockfalls of a certain size are often set on the restricted surface of the site and used in combination with other passive measures.
Passive protective net	20~5000	Flexible layout, short construction period, economy, good site adaptability and convenient maintenance; Durability of metal components.
Shed or open hole.	20~300	It is used for linear projects such as roads, railways and oil pipelines, and the impact force of falling rocks is the main design load.

5 SUMMARY AND PROSPECT

In this paper, the research contents and methods, monitoring and early warning and disaster prevention technology of slope rolling stone disaster are summarized, and the development direction of monitoring and early warning of slope rolling stone disaster is emphatically discussed, and the following conclusions are drawn:

1. The research on the trajectory, impact force and protective measures of rolling stones still faces the following problems: ①Most numerical simulations of rolling stones assume that they only have a single motion state, but in fact, the motion of rolling stones is the synthesis of multiple motion states; ②Most scholars will regard stones as spheres or cubes, and the ground as completely elastic or completely plastic. The results are very different from the actual situation. Most numerical simulation of rolling stone does not consider the crushing of rolling stone during operation.
2. With regard to the monitoring and early warning of rolling stone, the traditional theoretical methods have very limited functions, and there is still a certain time lag in the actual monitoring and early warning. In the follow-up study, based on the traditional theory, we should strengthen research on the separation stage of the slope rolling stone disaster and establish a complete and reliable dynamic monitoring index.

REFERENCES

Alejano, L, R. Pons, B. Bastante, F, G.et al. (2007). Slope geometry design as a means for controlling rockfalls in quarries. J. International Journal of Rock Mechanics & Mining Sciences, 2007, 44: 903–921.

BI, R. TANG, X.LIU, B, J. (2016). An implused-based energy tracking method for the analysis of slope stability of falling rocks. J. The Chinese Journal of Geological Hazard and Control,2016,27(02):14–19.)

C.Thornton, Z. Ning. (1998). A theoretical model for the stick/bounce behavior of adhesive, elastic-plastic spheres. J. Powder Technology, (99):154–162.

CHEN, W. XU Z, M., LIU W, L. (2015). Mechanical model and failure mechanism of unstable cantilevered rock blocks due to differential weathering. J. Rock and Soil Mechanics,36(01):195–204.

Delhomme, F. Mommessin, M. Mougin, J.et al. (2005). Behavior of a structurally dissipating rock-shed: experimental analysis and study of punching effects. J. International Journal of Solidsand Struetures.42:4204 4219.

DU, Y.XIE M, W., JIANG Y, J.et al. (2019). Research progress on dynamic monitoring index for early warning of rock collapse. J. Chinese Journal of Engineering,41(04):427–435.

DU, Y.XIE, M, W. JIANG Y, J.et al. (2015). Methods for determining early warning indices Based on natural frequency monitoring. J. Rock and Soil Mechanics,36(8):2284.

Dugelas, L. Coulibaly, JB. Bourrier, F.et al. (2019). Assessment of the predictive capabilities of discrete element models for flexible rockfall barriers. J. International journal of impact engineering,133, DOI: 10.1016/j.ijimpeng.2019.103365.

Evans, S. Huanger O. (1993). The assessment of rockfall hazard at the base of talus slopes. J. Canadian Geotechnical Journal,30(3):620–646.

FENG, J, F. ZHANG J, H.ZHU B.et al. (2011). Segmented loop algorithm of theoretical calculation of trajectory of rockfall. J. The Chinese Journal of Geological Hazard and Control,22(04):96–101.

FENG, X.GUO, J.LIU, J. (2013). Numerical Simulation on Rockfall Trajectory Based on Probability and Statistics Theory. J. Subgrade Engineering, (06):89–93.

Franck, B. Luuk, D. François, N.et al. (2009). Toward objective rockfall trajectory simulation using a stochastic impact model. J. Geomorphology, Volume 110, Issue 3. PP 68–79.

Guzzetti, F. Crosta, G. Detti, R.et al. (2002). STONE: a computer program for the three-dimensional simulation of rock-falls. J. Computers&Geosciences,2002,28:1079–1093.

HU X, W., MEI X, F. YANG Y, et al. (2019). DYNAMIC RESPONSE OF PILE-PLATE ROCK RETAINING WALL UNDER IMPACT OF ROCKFALL. J. Journal of Engineering Geology,27(01):123–133.

HUANG Y.SUN Q D. XU, Q. (2010). New development of rockfall kinematics study. J. Journal of Vibration and Shock,29(10):31–35+248–249.

HUANG, J, G. ZHANG S. (2021). Research on trajectory of rockfall based on finite element method. J. Journal of Yangtze River Scientific Research stitute:1-8[2021-03-23]. http://kns.cnki.net/kcms/detail/42.1171.TV.20210107.1111.002.html.)In

Jones, C, L. Higgins, J, D. Andrew, R, D. (2000). Rockfall Simulation Program Version 4.0. Colorado Department of Transportation. Colorado Geological Survey.2000,127.
Jones, N, Alves, M. (2004). Postoseverance analysis of impulsively loaded beams. J. International Journal of Solids and Structures.41:6441–6463.
Kishi, N. KonnoH.Ikeda, K.et al. (2002). Prototype impact tests on ultimate impact resistance of PC rock-sheds. J. International Journal of Impact Engineering.27:969–985.
LIU, G, Y.LI, J, J., KANG F. (2020). Study on the Protective Effect of Rockfall Platform Based on 3D DDA Method. J. Journal of Disaster Prevention and Mitigation Engineering,40(05):679–689.
LV, Q.SUN, H, Y.HAI, S, K.et al. (2003). Evaluation models of rockfall trajectory. J. Journal of Natural Disasters, (02):79–84.
Mougin, J, P. Perrotin, P. Mommessin, M.et al. (2005). Rock fall impact on reinforced concrete slab: an experimental approach. J. International Journal of Impact Engineering.3:169–183.
Muhammad, I. Chen, Y, L. (2017). Segmented Loop Algorithm of Theoretical Calculation of Trajectory of Rockfall. J.Geotechnical and Geological Engineering,Volume 35, Issue 1. PP 377–384.
Pichler, B. Hellmich, C. Mang, H, A. (2005). Impact of rocks ontogravel Design and evaluation of experiments. J. International Journal of Impact Engineering,31:559–578.
Plassiard, J P. Donze, FV. (2009). Rockfall impact parameters on embankments: a discrete element method analysis. J. Structural Engineering International,19(3):333–341.
Qiu, L, M. Song, D Z, Li Z H, et al. (2019). Research on AE and EMR response law of the driving face passing through the fault[J]. Safety Science,17:184–193.
Ritchie A. Evaluation of rockfall and its control[J]. Highway Research Record.1963,17:13–28.
Ronco, C. Oggeri, C. Peila, D. (2009). Design of reinforced ground embankments used for rockfall protection. J. Natural Hazards and Earth System Sciences,9(4):1189–1199.
S. Kawahara, T. Muro. (2006). Effects of dry density and thickness of sandy soil on impact response due to rockfall. J. Joual of Terramechanics.43:329–340.
SHEN, J. He S, M. WU Y. (2009). Study on the impact Properties of Rock-fall onto Cushion Material. J. Journal of Anhui Agricultural Sciences,37(17):8286–8288.
TANG H M, YI P Y. (2003). Research on Dangerous Rock Movement Route. J. Journal of Civil and Environmental Engineering, (01):17–23.
Toe, D. Mentani, A. Govoni, L.et al. (2018). Introducing meta-models for a more efficient hazard mitigation strategy with rockfall protection barriers. J. Rock Mechanics and Rock Engineering,51(4):1097–1109.
Uehan, F. Saito, H. Ma, G. (2012). Fundamental study on the remote vibration measuring system for evaluating rock slope stability//15th World Conference on Earthquake Engineering, Lisbon.
Vincent, L. Francois, D. Sara, M. (1996). Experimental study of rock sheds impacted rock blocks. J. Structural Engineering Intermational,3(1):171–175
Vlentin, J. Capron, A, Jongmans, D.et al. (2017). The dynamic response of prone-to-fall columns to ambient vibrations: comparison between measurements and numerical modelling. J. Geophysical Journal International,208(2):1058–1076.
WANG X. ZHOU T, Y.SHI J, T.et al. (2019). Theoretical and LS-DYNA simulation study of based on the theory of free-fall rockfall's impact on soil layer. J. Journal of Beijing Jiaotong University,43(04):9–17.)
WANG, D. WANG, J, F.LI, T, F.et al. (2021) Analysis of three-dimensional movement characteristics of rockfall: A case study at a railway tunnel entrance in the southwestern mountainous area, China. J. Journal of Geomechanics,27(01):96–104.)
WANG, X. P, L, F. WU, G, G. (2020). Simulation analysis and experimental study on the damage of bridge structure caused by tilt collapse and rockfall on the slope of Lalin railway. J. Chinese Journal of Rock Mechanics and Engineering,39(08):1622–1633.
WANG, Y.CAO, Y, J. (2018). Rock Fall Simulation Study and Mitigation Measures. J. Soil Engineering and Foundation,32(05):545–548.
XU, J, J. TANG, X, H.LIU H, S.et al. (2019). Investigation on trajectory of rolling rock affected by rock fragmentation based on energy tracking method. J. Rock and Soil Mechanics,40(S1):541–548.
XU, P, H. HUANG, R, Q. DENG, H. (2012). ADVANCES IN FRACTURES OF PARTICLES WITH DISTINCT ELEMENT METHOD. J. Journal of Engineering Geology,2012,20(03):410–418.
YANG H Q, ZHOU X P. (2009). A new approach to calculate trajectory of rockfall. J. Rock and Soil Mechanics,30(11):3411–3416.
YANG, Q, X. GUAN, B, S. (1996). Test and research on calculating method of falling stone impulsive force. J. Journal of the China Railway Society, (01):101–106
YE, S, Q. (2017). Study and Mitigation of Rockfall Hazards. China Communications Press Co., Ltd, Beijing.
ZHANG, L, Q. YANG, Z, F. XU, B. (2004). ROCK FALLS AND ROCK FALL HAZARDS. J. Journal of Engineering Geology, (03): 225–231.

Advances in Urban Engineering and Management Science – Khalil & Yang (Eds)

Research on the flood control safety design strategies of the riverside community based on the BIM analysis

Mingkun Qiu
The General Hospital of the PLA, Beijing, China

Yonghui Lei*, Yuzheng Lv & Yuan Cui
Institute of Defense Engineering, AMS, PLA, Beijing, China

ABSTRACT: Based on the application of the BIM related software in flood control safety design, this paper discusses the flood control safety design in combination with the planning and design of a riverside community. This paper puts forward the application of flood risk analysis based on BIM, including terrain analysis, mountain flood risk analysis and flooding analysis. According to the analysis results, the flood control safety design strategy of riverside community is proposed visioning the aspects of planning and layout, road and evacuation, site drainage, flood control buffer space and so on.

1 INTRODUCTION

This example is a newly built community, locating in a river in mountainous area, with a planned land area of about 20 hectares (Figure 1) and a total construction area of more than 47,000 square meters. The main functional rooms of the community include office rooms, reception rooms, sanitary rooms, apartment rooms, complex buildings, storeroom and other ancillary rooms. The community is close to the river, and there are mountain torrent gullies around, the terrain is relatively complex, the community planning and design should focus on flood control safety.

1.1 *Flood control design objectives*

The purpose of flood control safety design of riverside community is to reduce the loss caused by the flood. Through reasonable planning and design, the community can resist the large flood, so as to ensure the normal order of the community and reduce the economics damages by the flood. Specific to this project, the objectives are as follows:

- To ensure the safety of community buildings and properties in the event of a once-in-20-year flood;
- To ensure the personal safety of community personnel in the event of a once-in-50-year flood;
- To ensure the smooth evacuation of people or launch flood rescue operations in the event of a once-in-a-century flood. (Qingzhou W 1995)

*Corresponding Author: 523870509@qq.com

 DOI 10.1201/9781003305026-29

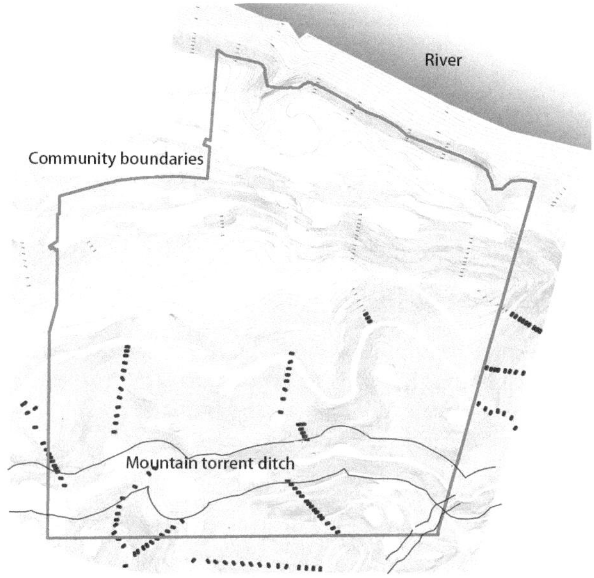

Figure 1. Terrestrial and surrounding environment of the camp.

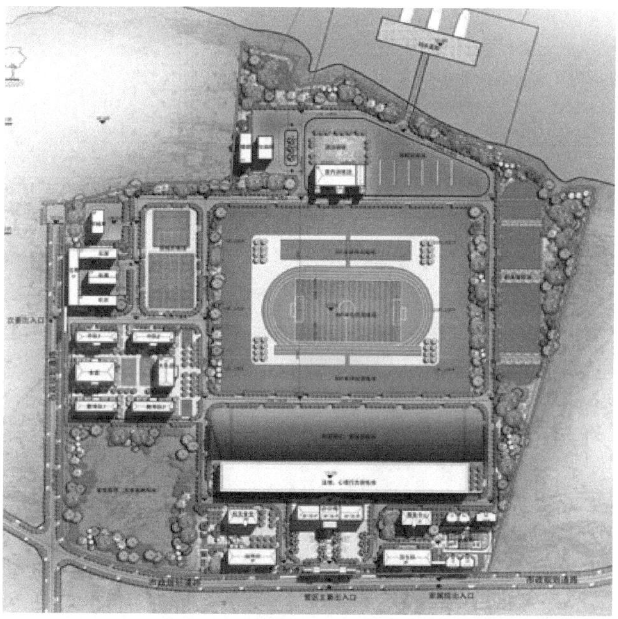

Figure 2. The original planning scheme of a community.

1.2 *Flood control design basis*

The ideas and strategies of flood control safety design have been discussed in the foregoing. However, in the specific design, it is necessary to follow the relevant norms, mainly in the following aspects:

- The planning and design of the community should be in strict accordance with the contents of the local flood control planning.

- According to the actual needs of community flood control works, some flood control works can be designed according to the Code for Design of Urban Flood Control Works (CJJ50-90).
- The drainage project for the community should be carried out in accordance with the relevant drainage design codes (Tao & Yongqin 2011)

2 FLOOD RISK ANALYSIS AND RESPONSE STRATEGIES

2.1 *Introduction to the original scheme of the community*

There is a planning scheme (Figure 2), which was designed using traditional 2D software without BIM and GIS for relevant analysis. Therefore, the scheme is not well combined with the terrain environment, and the flood control problem is also lack of consideration. There are some problems in the scheme, such as great damage to the original terrain and landform, large amount of earthwork, occupation of flood discharge channels, no flood control buffer space etc.

Correlation analysis of flood control safety design

2.2 *The risk analysis of mountain flood mainly includes terrain analysis, flood risk analysis and flood inundation analysis, which provides reference for flood control safety planning and design.*

(1) Terrain analysis of the camp

The BIM software Civil 3D is used to analyze the terrain of the community (Figure 3). It can be seen that the elevation distribution of the community site is high in the middle and low in the north and south. The planning should be based on the principle of reducing the amount of earth and damage to the environment, and make full use of the original topography. The original plan leveled the low-lying area to the south of the community and arranged the main building of the community on this basis. This not only did not make full use of the terrain, resulting in a large amount of earth, but also blocked the floodway, not conducive to flood control. (Shenzhi D, Kai C. 2012)

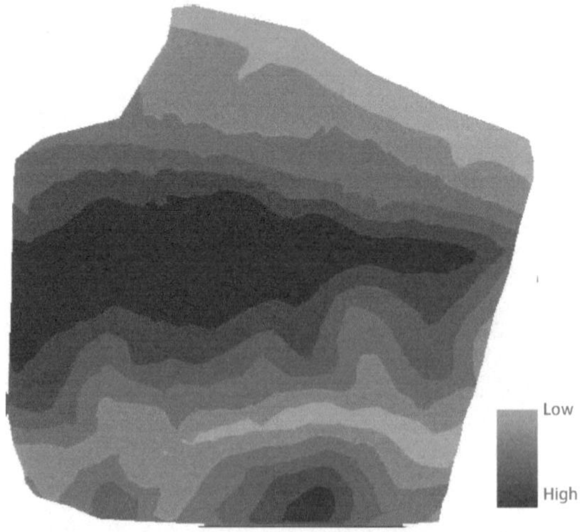

Figure 3. Analysis of barracks terrain elevation.

(2) Analysis of mountain flood risk

In general, the risk of mountain torrent is mainly affected by four factors, namely precipitation, topography, vegetation coverage and river system distribution. For this project, due to the scope of the community have been confirmed, and the area is small, the influence of the rainfall and vegetation coverage for flood risk can be ignored, and simply to the topography of the region and water distribution respectively to analyze the effect of flash flood risk, then overlay can be concluded that the area of flash flood risk map (Figure 4).

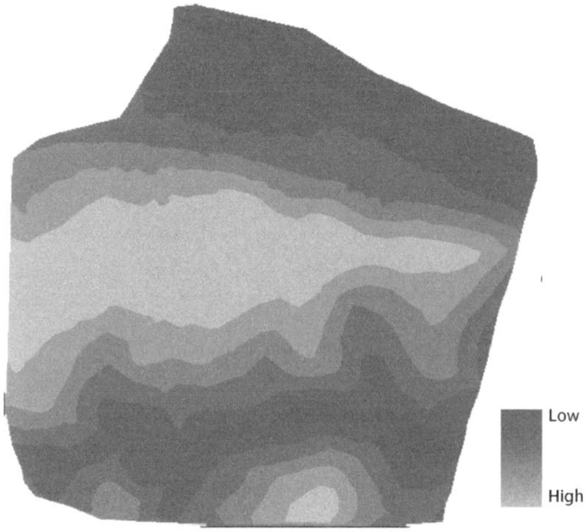

Figure 4. Analysis of mountain flood risk.

As can be seen from the mountain flood risk analysis diagram, the area near the river and mountain flood gully has the greatest risk. The central elevation of the site is high, so the flood risk is low. Moreover, there is a relatively complete land parcel, which is suitable for the main building construction land of the community (Li 2011).

In the original scheme, a sports ground was built in the middle of the most suitable place for construction, and the building was built above the flood spillway, which was not conducive to flood control, and there was a lot of work to fill and excavate.

(3) Analysis of community flooding

Because the community is close to the river, the analysis of the situation when the river water rises and the surrounding mountain flood ditch flooding is of great significance for the safety design of flood control of the community. According to the investigation of the surrounding hydrological conditions and historical floods, the 5-year flood level is 184.3 m, the 10-year flood level is 186.7 m, the 20-year flood level is 188.6 m, the 50-year flood level is 190.9 m, and the 100-year flood level is 192.7 m. According to different water levels, GIS software can be used to conduct flooding analysis for this community (Figures 5–9).

The original plan levelled the higher area in the middle, which was almost the same elevation as the area along the river. Through the flooding analysis, it can be seen that the flood that happened once in 20 years has completely submerged the area along the river of the community, so leveling the middle area is not conducive to flood control.

(4) Analysis of the area suitable for community construction

By using GIS related software to analyze the mountain flood risk and flood inundation of the community, a map of the area suitable for community construction can be obtained comprehensively. Flood control standards of important buildings usually need to be based on the flood standard of once in 100 years, while general facilities can be based on the flood standard of once in 50 years (Figure 10).

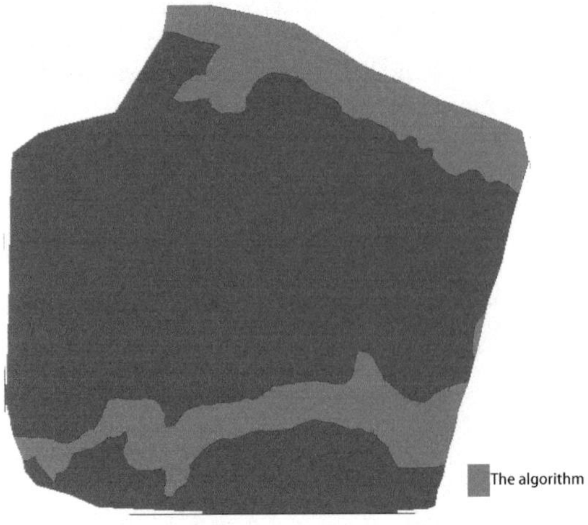

Figure 5. 5-year inundation analysis.

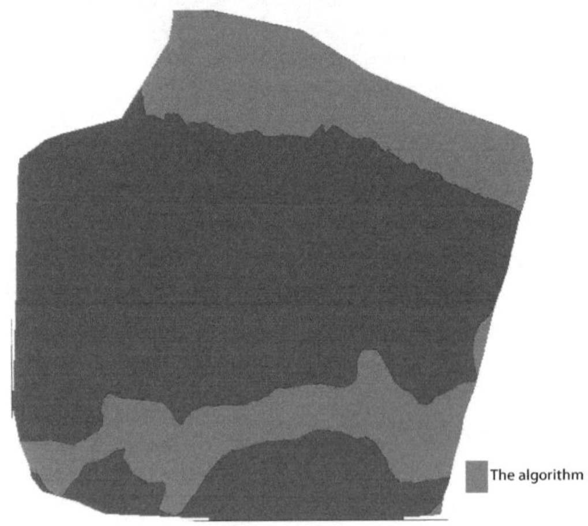

Figure 6. 10-year inundation analysis.

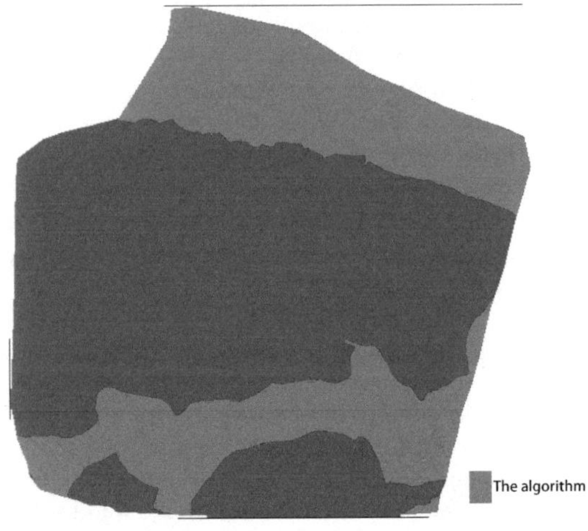

Figure 7. 20-year inundation analysis.

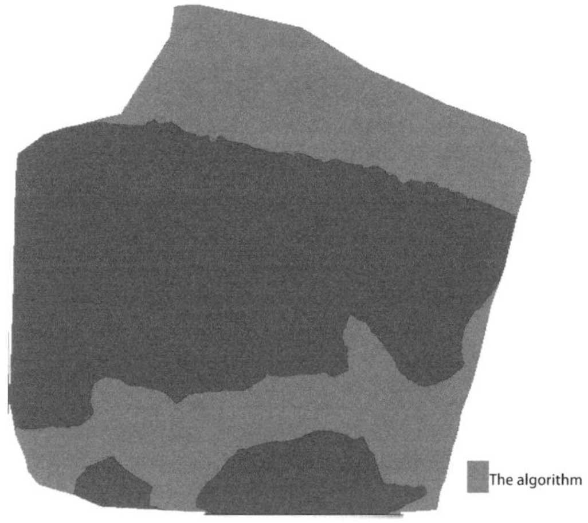

Figure 8. 50-year inundation analysis.

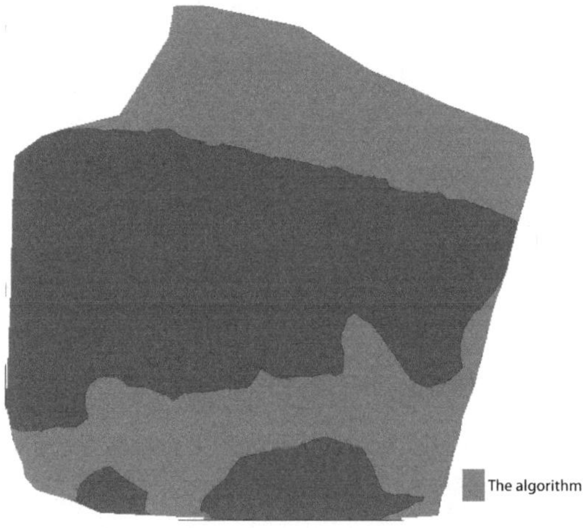

Figure 9. 100-year flood inundation analysis.

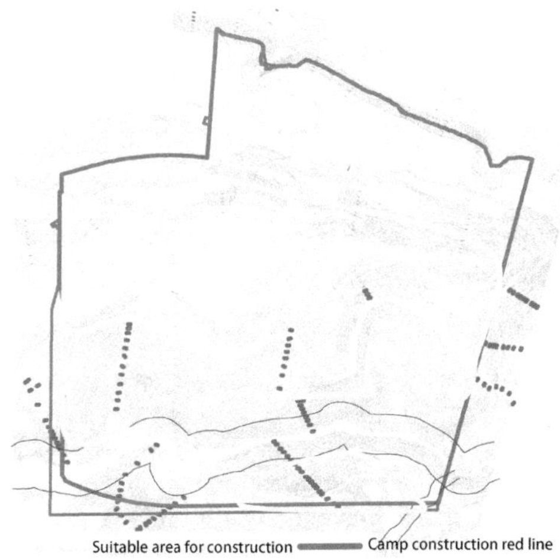

Figure 10. Suitability for community construction.

3 FLOOD CONTROL SAFETY DESIGN STRATEGY

Through the above analysis, it is concluded that the original scheme is very unfavourable to flood control, and the amount of earth is large, and the damage to the environment is greater. According to the flood control design strategy proposed in this paper, the community is re-planned to meet the requirements of function, environment, flood control and other aspects (Chun 2011).

3.1 *Planning zoning and building layout*

In order to realize the flood control safety of the community, it is necessary to start from the planning level and plan the overall layout and zoning of the community according to the mountain flood risk analysis and flooding analysis of the community. In this project, can consider the main building of the community area layout to the high elevation areas, the central for sports venues and other affiliated facilities layout to a flat area, which face the region height is low, the flood risk is higher, the sports venues in the area even flooding don't cause much damage and can increase the hydrophilicity of community environment, People can enjoy the beautiful river view while doing exercise (Figure 11).

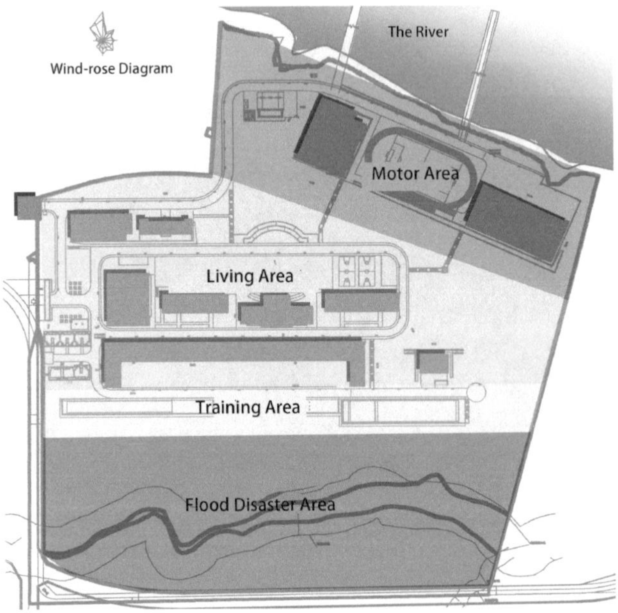

Figure 11. Planning zoning of the project.

As for the layout of the building, since the central terrain is long from east to west and short from north to south, in order to arrange the building as high as possible in the central elevation, the building is arranged in a north-south direction along the east-west direction (Figure 12).

3.2 *Road flood control and evacuation*

Special design is made for the flood control and drainage of the community roads. Due to the complex terrain in this area and the large slope of some roads, the drainage of the roads is not only designed for side ditch drainage, but also set up water cutting grooves perpendicular to the road surface in places with large slope for energy dissipation and preventing excessive rainwater flow and impact force on the road surface (Chengsong & Jinjun 2000).

For flood control evacuation design, the main part of the community road in this project adopts circular driveway, and has two entrances and exits, in order to meet the evacuation of people and

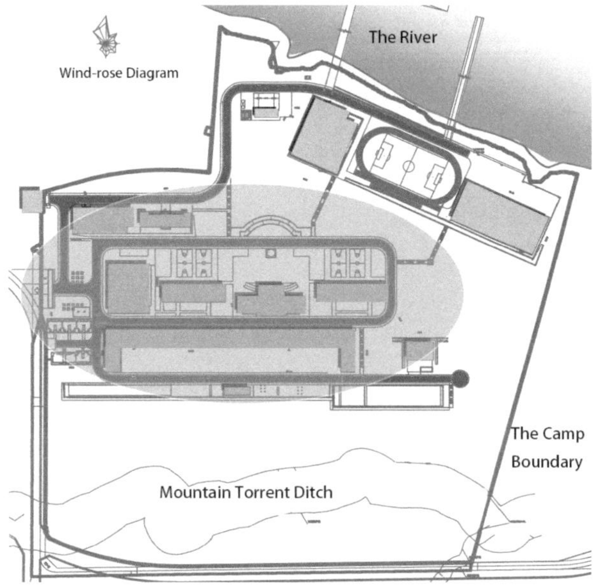

Figure 12. The main building is arranged at a higher elevation.

vehicles in the case of flood. At the same time, for the movement area with low elevation adjacent to the river, in addition to the evacuation channel of the vehicle road, there are also two pedestrian evacuation channels for the rapid evacuation of the flow of people when the flood comes (Figure 13).

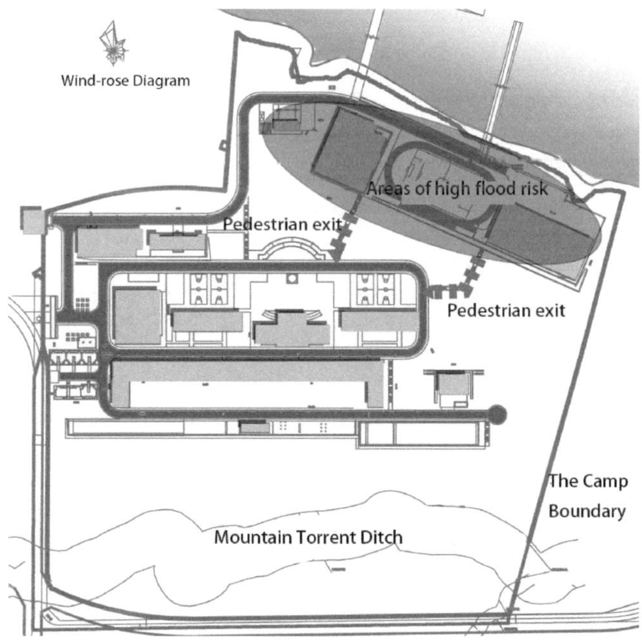

Figure 13. Design of pedestrian flood control evacuation channel in the riverside area of the project.

3.3 *Drainage of the building and site*

In this project, in order to prevent flooding around the buildings under heavy rain, open ditches were also set around each building in addition to organized drainage (Dan 2012).

For the training sites are also designed around the open ditch drainage, in order to be able to drain quickly, to prevent flooding. There is a steep slope between the sports area near the river and the main body of the community. Here, the project specially designed a water cutting ditch to carry out slope drainage and prevent rainwater from washing the sports field (Yuanyuan 2010).

3.4 *Flood control buffer space design*

The project makes full use of the waterfront space according to the topography, using the relatively flat waterfront area for sports venues and ship repair sites. These facilities are less affected by floods and can be built on the riverside, but the design elevation must be higher than the ten-year flood line. When the water level is normal, the riverside playground can be used for People's Daily sports and training, and specially designed with water-loving space. When the flood season comes, the riverside sports field can also act as a buffer zone for the flood without affecting the normal order of the community. For the more important buildings, they were built above the 50-year flood line to ensure their safety (Figure 14) (Qingzhou 2007).

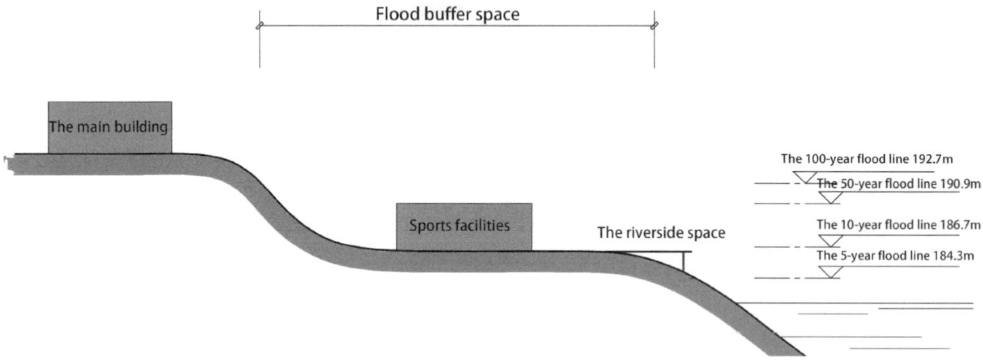

Figure 14. Flood control design of the waterfront space.

3.5 *Flood drainage engineering design*

Because the project is close to the river, the community rainwater pipe network can be arranged according to the terrain, and the rainwater can be directly discharged into the river. For the mountain flood ditch around the community, it is not only reserved, but also dredged and managed, so that it can quickly discharge the flood when the mountain flood occurs (Xiaojing 2016).

4 SUMMARY

This paper discusses how the flood control safety design method proposed in the paper can guide the flood control safety design of the mountain community in the actual project. Because of the complex topographical and hydrological conditions of mountain communities, different flood control measures should be taken according to different conditions in flood control safety design, and the maximum flood control benefits should be achieved with the minimum input.

REFERENCES

Chengsong M, Jinjun M. (2000) Talking about the flood and waterlogging prevention design of buildings. Industrial Construction. 03:24–25.

Chun G. (2011) Research and application of urban ecological flood control system based on GIS. Northwest Sci-Tech University of Agriculture and Forestry, Xi'an.
Dan W. (2012) On the design of flood control and disaster-resistant buildings. Guangdong Science and Technology. 07:243–244.
Li S. (2011) Study on the planning and guarantee ideas of flood discharge system in mountainous cities. Sichuan Building Materials. 04:56–60.
Qingzhou W. (1995) Research on Flood Control in Ancient Chinese Cities. China Building Industry Publishing, Beijing.
Qingzhou W. (2007) Construction Safety. China Construction Industry Press, Beijing.
Shenzhi D, Kai C. (2012) Research on Countermeasures of Urban Flood Control and Drainage in my country. Modern Urban Research. 01:21–28.
Tao Z, Yongqin X. (2011) Analysis and Reference of Flood Discharge and Disaster Prevention in Ancient Chinese Cities. Administrative Management Revolution. 11:67–70.
Xiaojing Y. (2016) Flood control planning and design for a mountain and sea area in Qingdao. Urban Roads, Bridges and Flood Control, 1:84–86.
Yuanyuan L. (2010) Research on some major issues of flood control. China Water Power Press, Beijing.

Advances in Urban Engineering and Management Science – Khalil & Yang (Eds)

Research on the joint upgrading scheme of series shallow tunnels of rainwater system in built-up areas

Chenyu Liu* & Nianchao Luo
Shanghai Water Engineering Design and Research Institute Co., Ltd., Shanghai, Shanghai, China
Flood Prevention and Disaster Mitigation Engineering Technology Research Center of Shanghai Water Affairs Bureau, Shanghai, Shanghai, China

Hongkuan Liu
Shanghai Water Engineering Design and Research Institute Co., Ltd., Shanghai, Shanghai, China
Shanghai Beach Coastal Engineering Technology Research Center, Shanghai, Shanghai, China

ABSTRACT: In order to improve the ability of urban disaster prevention and mitigation and environmental governance and to improve the efficiency of urban renewal, it is necessary to upgrade the rainwater system in the built-up areas. The MikeFlood one-dimensional, two-dimensional, and pipe network "three-in-one" numerical simulation method is used to analyze the heavy rain in the Zhangjiang area, which is based on the Shanghai waterlogging prevention system and water environment governance system. waterlogging scenario is simulated, and a joint upgrading plan for shallow tunnels is proposed in this paper. The new scheme not only improves the feasibility of system upgrading and reconstruction, but also realizes the dual requirements of long-term drainage and waterlogging prevention and initial rain pollution control in the Zhangjiang area, providing a new idea for the renewal of rainwater systems in subsequent built-up areas.

1 INTRODUCTION

The upgrading and transformation of rainwater drainage systems in urban built-up areas is a hard task in urban renewal. The areas with an earlier development time basically have the problems of insufficient drainage capacity and direct flow of initial rain into the river (Wu et al. 2016). The main reason is that the surface development is generally later than the laying of the pipe network, which leads to a higher cost for the pipe network system update. At the same time, the shortage of land space at the end of the system makes the interception and pollution control devices unable to formulate a landing.

In order to improve this status quo, major cities are exploring different forms of upgrading schemes (Zhang 2015). Under the concept of low-impact development and sponge city worldwide, various combined solutions for the difficulty in upgrading rainwater systems in built-up areas have been proposed. The drainage system outside the capital circle completed and put into use in Tokyo, Japan in 2006 went through the severe rainstorm flood through the deep tunnel project, stores water through a large underground reservoir, and then discharges it into Edogawa through the pump station after the rainstorm (Urban Underground River Design Manual 1974). The Chicago Tunnel Drainage System (TARP) uses a combination of deep tunnels, reservoirs, and shafts to store combined sewage from urban pipelines, as well as to regulate and store short-term heavy rainfall floods (Changnon 2010; Song et al. 1988); the domestic Guangzhou Haoyong deep tunnel system regulates, stores and controls the overflow combined sewage (Wang et al. 2016).

In 2020, the Shanghai Municipal Water Affairs Bureau and the Shanghai Municipal Planning and Natural Resources Bureau jointly issued the "Shanghai Urban Rainwater Drainage Plan (2020–2035)." The plan proposed the new idea of "green, gray, blue, and pipe" for the realization of

*Corresponding Author: chenyu_liu@arcplus.com.cn

 DOI 10.1201/9781003305026-30

standard enhancement and pollution control underpinned by the transformation from gray single into multiple measures, through cutting the peak in green sources, gray process storage and discharge, and absorbing at the blue end, improving the quality and efficiency in management. This concept aimed to achieve two goals: blue facilities that increase the receiving capacity of downstream rivers and lakes; and gray facilities such as rainwater pipe networks, large-scale storage facilities, and rainwater pumping stations; and improve the refined and intelligent management measures of pipe network detection, repairment, and impoundment.

However, in the process of transforming the drainage system in the urban built-up area, there are difficulties in the technical and policy dimensions for the new transformation concepts and measures in the implementation of the local projects: Technically, the land attribute of the built-up area is clear, the surface shape is mature, and the transformation of green facilities such as sunken green space and rainwater gardens is quite difficult. At the same time, Shanghai has built a complete layout of connected water systems and river networks, and ensuring drainage by controlling the water level of the river in local areas violates the overall pattern of Shanghai's overall flood control and waterlogging prevention, which poses greater disadvantages than advantages. On the whole, there is an embarrassing situation in the built-up areas of large cities because the drainage pressure of the reformable areas is not large, while the areas with high drainage pressure cannot be reformed.

Therefore, for the built-up areas, consideration should be given to the principle of upgrading and transformation of "gray as the mainstay and blue and green as the supplements" to ensure that the hard indicators of the drainage system in the built-up areas meet the standards and normal operation during heavy rains so as to prevent urban paralysis and reduce unnecessary loss. This study took the Zhangjiang area as an example. By combining this with the assessment and analysis of the current situation, a gray facility construction plan using shallow tunnels has been proposed. This study, combined with the layout of the sponge city green facilities, completed the task of upgrading and reforming the rainwater system in the Zhangjiang area and improved the prevention of waterlogging while solving the problem of interception of the initial rain.

2 MATERIALS AND METHODS

2.1 *Research area*

The Zhangjiang unit belongs to the Pudong New Area of Shanghai, with its central location at 31°09′N and 121°36′E. It is adjacent to Chuansha Town and Tang Town to the east, Huamu Street and Beicai Town to the west, Shenjiacao and Kangqiao Town to the south, and Jinqiao Town to the north. Zhangjiang is the core area of the "Science and Innovation Center" and the location of the Zhangjiang High-Tech Park, known as the "Silicon Valley of China". The overall development of the region starts early, and the development level is high. It is the gathering place for high-tech enterprises and talents. In 2020, the total output value of industrial enterprises above the designated size will be 3033.5 billion yuan. The main industries include biomedicine, integrated circuits, and software industry chains.

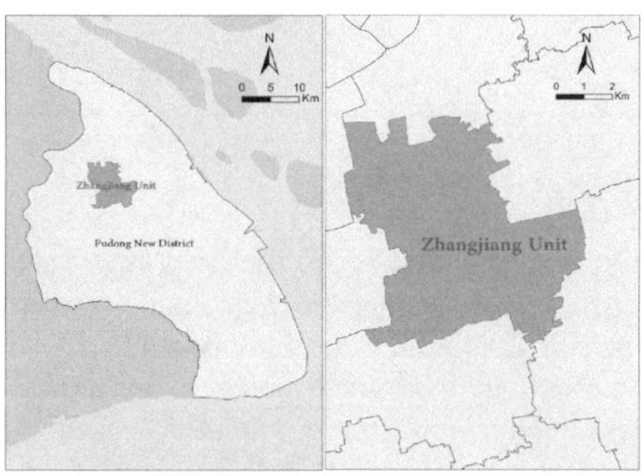

Figure 1. The location of Zhangjiang unit.

2.2 *Current status of drainage system*

The upgrading and reconstruction area mainly involves seven drainage systems on the north side of the Chuanyang River in the Zhangjiang area, namely Jingniu, Dangui, Xiangnan, Zhangdong, Keyuan, Kejiaodong, and Zhangjiangji Town. The total area of the seven systems is 19.79 km^2. The drainage system is a diversion system. The drainage mode is dominated by strong drainage and supplemented by self-drainage. The design drainage standard is once-a-year.

Table 1. Basic information table of seven systems in Zhangjiang.

Number	Drainage System	Pumping Station	Drainage Capacity (m^3/s)	System Area (hm^2)	Drainage Type	Drainage Mode	Current capacity
1	Jingniu	Jingniu	21.45	465	Diversion system	Strong drainage	1
2	Dangui	Dangui	11.20	232		Strong drainage	
3	Xiangnan	Xiangnan	11.84	168		Strong drainage	
4	Keyuan	Keyuan	14.10	254		Strong drainage	
5	Kejiao Dong	Zhangjiangji	29	306		Strong drainage +self drainage	
6	Zhangjiangji			256		Strong drainage +self drainage	
7	Zhangdong	Jidiangang	17.52	298		Strong drainage +self drainage	

In terms of space, the seven systems form a similar closed rectangle (Longdong Avenue, Sanba River, Chuanyang River, and Outer Ring Expressway), and the pumping stations are scattered. There are two large rivers in the large system, namely, Majiabang in the east-west direction and Lujiabang in the north-south direction, among which the pumping stations of Xiangnan, Dangui, Kejiaodong, and Zhangjiangji drainage systems are discharged to Majiabang; the pumping stations of Keyuan and Zhangdong drainage systems are discharged to Lujiabang; and the Jingniu drainage system is discharged to the Sanba River in the west of the drainage system.

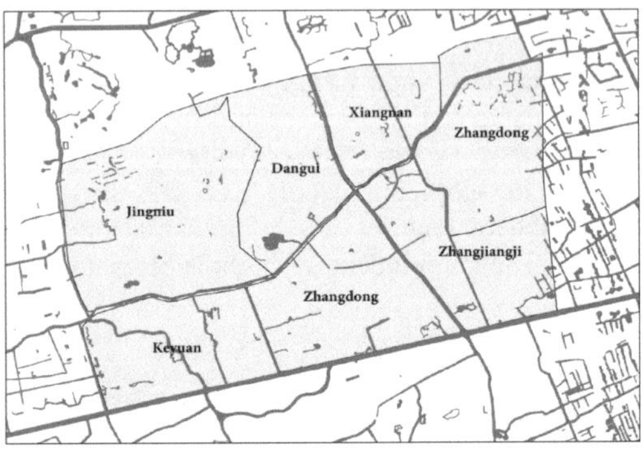

Figure 2. The seven drainage systems in Zhangjiang.

From the perspective of the current situation on the land, the system is basically in the built-up areas, dominated by industrial enterprises, residences, and commercial offices. Among them, Jingniu, Dangui, Kejiaodong, and Zhangdong are mainly centered on high-tech enterprises and are responsible for high-end technology industries such as biopharmaceuticals and semiconductor manufacturing; Keyuan is mainly led by universities and middle schools, such as Fudan University Zhangjiang Campus, Shanghai Jiaotong University School of Information Safety and Engineering, etc.; Xiangnan and Zhangjiangji are mainly used for residential, villa, and commercial offices. The blue advantage of the study area is obvious, criss-crossed with rivers. There is one city-managed river in the surrounding area, the Chuanyang River, and five district-managed rivers. The water surface rate within the range is 5.28%.

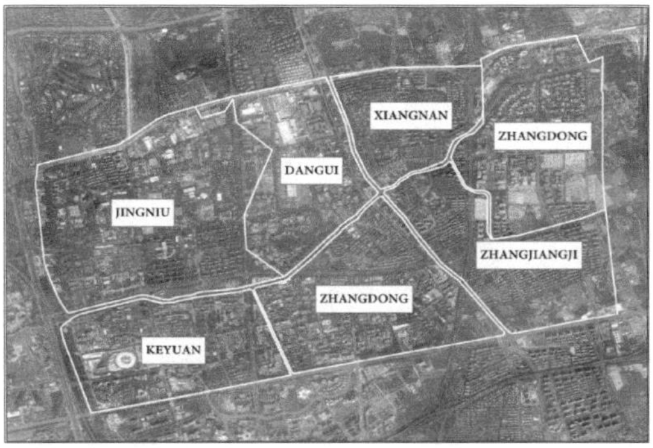

Figure 3. Land properties of research area.

2.3 *Model construction*

Mike Flood's hydrodynamic software is used to establish a waterlogging model for the seven systems of the Zhangjiang River. By integrating Mike Urban's rainwater pipe network model, Mike11 one-dimensional river network model, and Mike21 two-dimensional surface model, they conduct coupling simulations to the whole process of water cycle: rainfall-surface overflow-pipe network outflow-riverway, which reflects the urban waterlogging process in a more actual and comprehensive way. This technology has been successfully applied to waterlogging simulations at various scales (Luan et al. 2021; Zhang et al. 2014)

The MIKE11 one-dimensional, one-layer (vertical homogeneous) hydraulic model uses the Abbott-Ionescu six-point center implicit scheme to calculate the Saint-Venant equation groups of one-dimensional unsteady flow.

$$\frac{\partial Q}{\partial x} = \frac{\partial A}{\partial t} \tag{1}$$

$$\frac{\partial Q}{\partial t} + \frac{\partial(\alpha \frac{Q^2}{A})}{\partial x} + g \cdot A \cdot \frac{\partial h}{\partial x} + \frac{gQ|Q|}{C^2 A \cdot R} = 0 \tag{2}$$

Where A refers to the cross-section area; Q represents the cross-section flow; x denotes coordinates along the way; t means time coordinate; h is the water level of cross-section; C indicates the coefficient of XieCai; R is specified as hydraulic radius.

The Mike21 two-dimensional overflow model uses the ADI method, and the calculation area uses a rectangular grid to calculate the two-dimensional Saint-Venant equation groups.

$$\frac{\partial \zeta}{\partial t} + \frac{\partial p}{\partial q} + \frac{\partial q}{\partial y} = \frac{\partial d}{\partial t} \tag{3}$$

$$\frac{\partial p}{\partial t} + \frac{\partial}{\partial x}(\frac{p^2}{h}) + \frac{\partial}{\partial y}(\frac{pq}{h}) + gh\frac{\partial \zeta}{\partial x} + \frac{gp\sqrt{p^2+q^2}}{C^2 \cdot h^2} - \frac{1}{\rho_w}\left[\frac{\partial}{\partial x}(h\tau_{xx}) + \frac{\partial}{\partial y}(h\tau_{xy})\right] - \Omega_q - fVV_x + \frac{h}{\rho_w}\frac{\partial}{\partial x}(p_a) = 0 \tag{4}$$

$$\frac{\partial p}{\partial t} + \frac{\partial}{\partial y}(\frac{p^2}{h}) + \frac{\partial}{\partial x}(\frac{pq}{h}) + gh\frac{\partial \zeta}{\partial y} + \frac{gp\sqrt{p^2+q^2}}{C^2 \cdot h^2} - \frac{1}{\rho_w}\left[\frac{\partial}{\partial y}(h\tau_{yy}) + \frac{\partial}{\partial x}(h\tau_{xy})\right] - \Omega_p - fVV_y + \frac{h}{\rho_w}\frac{\partial}{\partial y}(p_a) = 0 \tag{5}$$

Where h(x,y,t) denotes water depth; ζ (x,y,t) indicates ater level; p,q(x,y,t) represents single width flow in x,y direction; C(x,y) is Xie Cai friction coefficient; f(v) refers to wind friction coefficient;

V,Vx,Vy(x,y,t) means wind speed and x,y direction components; Ω(x,y) is Coriolis force; Pa(x ,y,t) means atmospheric pressure; $\tau\_xx, \tau\_xy, \tau\_yy$ are specified as shear stress component.

MikeUrban pipe network model calculation still uses Saint-Venant equation groups:

$$\frac{\partial Q}{\partial x} = \frac{\partial A}{\partial t} \tag{6}$$

$$\frac{\partial Q}{\partial t} + \frac{\partial(\alpha \frac{Q^2}{A})}{\partial x} + gA\frac{\partial h}{\partial x} = gAI_f \tag{7}$$

In the process of calculation, the momentum equation of the Saint-Venant equation groups can be transformed as follows:

$$\frac{\partial Q}{\partial x} + \frac{Q}{\rho} \cdot \frac{\partial \rho}{\partial x} + \frac{aA_0}{a^2} \cdot \frac{\partial y}{\partial t} = 0 \tag{8}$$

Where I_f denotes friction slope; α refers to pipe velocity distribution coefficient; Q is flow rate; A means water flow area; y represents water depth.

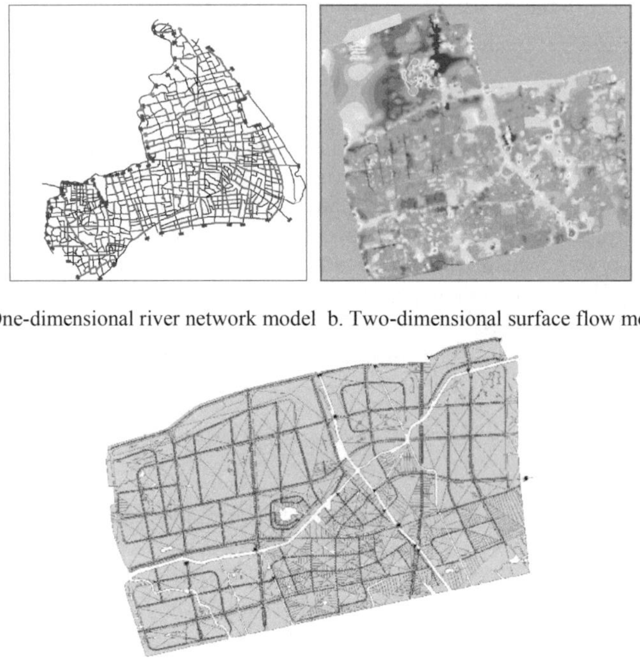

a. One-dimensional river network model b. Two-dimensional surface flow model

c. Pipe network model

Figure 4. The three networks of seven systems in one model.

3 RESULTS & DISCUSSION

3.1 *Current situation assessment*

The current situation assessment design rainfall selects the Chicago rain pattern applicable to Shanghai. The specific formula is:

$$q = \frac{1600(1 + 0.846lgP)}{(t + 7)^{0.656}} \tag{9}$$

Where q is the design rainstorm intensity; P refers to the design rainstorm return period; t denotes the design rainstorm duration.

The waterlogging under the design rainstorm with a return period of one year, three years, and a five-year return period is simulated. The specific rainfall is shown in Figure 5, and the simulated waterlogging is shown in Figure 6.

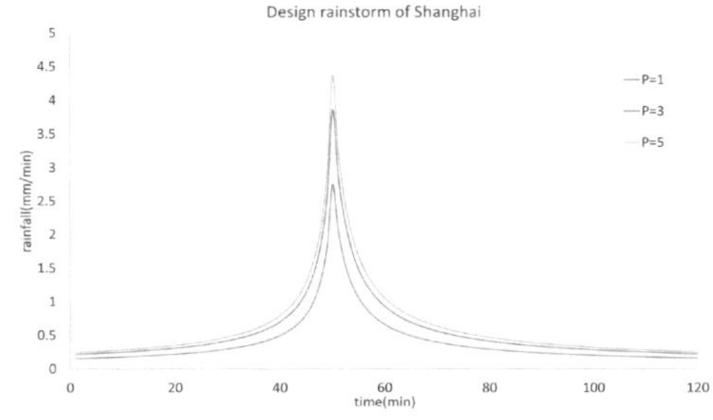

Figure 5. Design rainstorm in Shanghai (once in a year, 3, and 5 years).

a. Once in a year

b. Once in three years

c. Once in five years

Figure 6. The waterlogging situation under different rainstorms.

In the event of a once-a-year rainstorm, there was no accumulation of water in the drainage systems of Jingniu, Keyuan, and Zhangdong, and a small amount of accumulation was found in the drainage systems of Kejiaodong, Dangui, Xiangnan, and Zhangjiangji Town; part of the accumulation of water appeared in the front of the system in the event of a once-in-three-year rainstorm; when the system encountered heavy rain once every five years, there was a large area of accumulation of water. None of the seven systems meet the once-in-three-year drainage standard, and the Kejiadong, Dangui, Xiangnan, and Zhangjiangji towns do not meet the once-a-year drainage standard.

3.2 *Comparison and selection of upgrading and reforming methods*

Taking into account of the importance of the Zhangjiang area, the rainwater drainage capacity of the regional plan is once in five years. First, the two methods of single-system upgrading and multi-system joint upgrading are compared as follows.

3.2.1 *Independent upgrading of single system*

The main idea of a single system is to occupy shallow spaces such as concentrated green space in parks and the community, so that a sunken green space with a water storage capacity is formed or a shallow regulation tank is built to complete the system upgrading. This method can be used for those areas that are difficult to renovate by replacing or expanding the pipe network.

Such projects are unrelated to one another in terms of standardization and pollution control, and the substance of transformation is dispersed. To manage non-point source pollution, early rain regulation and storage facilities must still be built at each system's finishing pumping stations. After careful consideration, the actual plan is constrained by the spatial distribution of centralized green space. Ponding at the spot where regulation and storage facilities may be built is not problematic, and the improving impact of the transformation on the weak regions with hard ground surrounding it is not visible. It is still essential to finish the system's standard upgrading transition by replacing pipes with greater diameters. Simultaneously, there is less accessible land surrounding the present pumping station, and the feasibility of constructing initial rain control and storage facilities is low.

Following an examination of the existing pipelines and land properties in the Zhangjiang area, it has been determined that at least ten new green storage tanks and seven new initial rain storage tanks are required, as well as the reconstruction of a large number of existing rainwater pipelines for each system.

3.2.2 *Joint upgrading of multi-system*

The typical multi-system joint upgrading strategy is to re-divide the catchment area following partial pipe network transformation to boost total rainfall drainage capacity (Zhang 2017). The major body of this article offered a multi-system collaborative upgrade approach centered on shallow tunnel constructions. The basic concept is to create a centralized system for upgrading and pollution management that isolates the original rainwater and subsequent waterlogging from the two dimensions of place and time.

When encountering low-standard rainfall (P 1), the rainwater of each system is still discharged to the current river channel through the original system, preserving the existing drainage pattern; when encountering high-standard rainfall (P 1), the system spatially reconstructs and adjusts the service scope of the existing pump station to meet the drainage standard of 5-year return period. A new upstream rainfall main branch pipe will carry regional rainwater to the main pipe and release it into the river from the new pump station.

In terms of pollution control, centralized linear regulation and storage is adopted for the initial rainwater of each system. The current rainwater system is used to collect the initial rainwater. A new regulation and storage main pipe is connected in series with the rainwater pump station at the end of each system to collect the initial rainwater.

The extensive cooperative upgrading of multi-systems occupies the shallow and middle-level zone along the river's land control line. To enhance additional non-compliant locations, the regulation and storage branch pipes, the regulation and storage main pipe, and the new pumping station are erected on the premise of lowering the service scope of the original drainage system. Simultaneously, new pipes will link each system's original end pumping stations, collecting initial rainfall in a unified way and storing it in the regulation and storage pipeline for timely delivery to the sewage system.

3.2.3 *Comparison of upgrading schemes*

The single-system upgrade is appropriate for locations with superior drainage characteristics in the current drainage system. As an example, in Shanghai, the design standard for the system's pipe network is relatively high (generally designed with drainage capacity at least once every three years). The plot is colored blue and green. A single upgrading approach can be considered for plots with evident blue and green benefits and there is still area for building surrounding the downstream pumping station, and soft measures such as "green and blue" can be employed to raise drainage capacity to the once-in-five-year norm. The drainage system in Shanghai's built-up regions is established earlier, and the majority of them are planned with a once-a-year drainage capacity in mind. "green and blue"

Table 2. Comparison table of upgrading schemes.

Pros and cons	Single-system upgrading	Multi-system joint upgrading
Land use demand	Scattered occupancy of shallow space in concentrated green areas such as parks and communities;	concentrated occupancy of shallow and mid-level space around river land control lines
Content	Relatively scattered	Relatively concentrated
Site selection	Local regulation and storage of each system	Multi-system centralized linear regulation and storage
Pollution control	Use on-site storage facilities to effectively treat non-point source pollution	Use centralized linear storage to effectively treat non-point source pollution
Implementation progress	in fragments, overall moderate and	systematically advanced, overall faster
Operation management	The facilities are scattered, and the management cost is relatively high.	The facilities are centralized, and the management cost is relatively low.

3.3 *Joint upgrading scheme for series shallow tunnels*

The vast system is separated into east-west sub-systems by Majiabang as its border. A 2–2.8 m rectangular shallow tunnel is being built in the western regions of Dangui Road, Jinke Road, and Zhangheng Road. A 2.2–3 m rectangular shallow tunnel is being built in the eastern regions of Dangui Road, Zhangdong Road, and Fangchun Road. At the front end of the original system, the two tunnels are used to collect and store rainfall rainwater.

Two empty pumping stations with capacities of 28 m^3/s and 15 m^3/s are installed along the Lujiabang line of the Jingniu system and the east side of Fangchun Road in the Zhangdong system, respectively, and are mostly used to discharge water within the shallow tunnel.

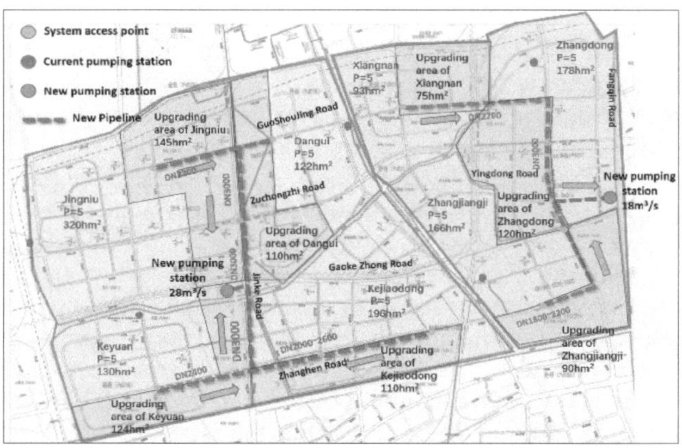

Figure 7. Joint improvement plan for the rainstorm shallow tunnel project.

On this premise, a shallow tunnel with a diameter of 3 m is constructed on the west side of Majiabang along the Lujiabang line (inside the land control line), and on the east side of Majiabang along Zhangjiang Road and Dangui Road. Gaoke Middle Road is linked to the rainfall shallow tunnel on the west side, and all existing drainage pumping stations are linked in series simultaneously. This shallow tunnel can address the issue of not having enough room surrounding the current pumping station to build early rain control and storage facilities. At the same time, it will function as a rainfall shallow tunnel, improving the whole system's drainage capacity during the excessive downpour.

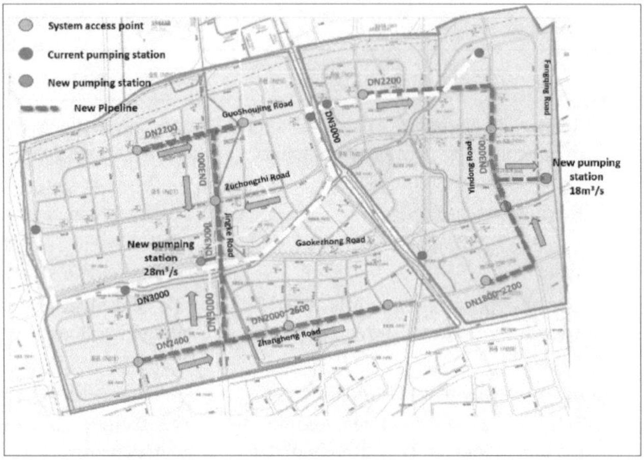

Figure 8. Joint improvement plan for the first rain shallow tunnel project (combined with rainstorm shallow tunnel).

The shallow tunnel uses branch road construction to avoid excavation of the main road and link to the original system pipe network by overflowing with pipes. This technology is simple to connect with the existing pipe network and can immediately lessen the project's difficulties. According to studies, the rainwater pipe network in Shanghai is primarily dominated by half-full or full-pipe flow. The full pipe overflow approach may help prevent silt entering the shallow tunnel project, reducing the complexity of cleaning the shallow tunnel.

3.4 *Post-evaluation of drainage capacity*

According to the Shanghai Criteria for Waterlogging Control, the Shanghai drainage system's anticipated rainfall capacity must exceed the planning standards for 50 to 100 years of waterlogging prevention. Zhangjiang area, as an economically developed region and key development region, plans the waterlogging control standard once every hundred years.

Water buildup shall not exceed 15 cm, according to the Technical Code for the Prevention and Control of Waterlogging in Cities and Towns. Shanghai specifies municipal roadways for no more than 1 hour and no more than 50 m^2 of stagnant water, according to the Code for Design of Outdoor Drainage. As a result, achieving the criteria for waterlogging prevention and control at the same time implies satisfying the requirements for waterlogging prevention and control.

According to model estimates, when the shallow tunnel project is completed, under the conditions of once-in-a-hundred-year rainfall, stagnant water only arises in a tiny region, primarily concentrated in the original system's self-draining area (the northeast side of Science and Education and the northeast side of Zhangzhou). In comparison to the existing state, there has been a major improvement, and the system now fulfills the criteria of the waterlogging prevention and control regulations.

a. The current situation of the accumulation of water under the once in a hundred years rain in 24 h

b. The waterlogging situation of series shallow tunnels multi-system joint upgrading encountering rain and waterlogging once in a hundred years

Figure 9. The water accumulation situation of rainfall once in a hundred years.

4 CONCLUSIONS

Using the Zhangjiang region as an example, this study employs a pipe network, ground surface, and river network connection model. It is quantitatively proved that the series shallow tunnel project is carried out in the built-up region based on the simulation of the entire rainfall process from the surface through the pipe network to the river channel. The following are the findings on the feasibility and efficacy of system upgrading and transformation:

1. For urban built-up areas, if the current drainage system has poor drainage capacity and the surface reconstruction space is limited, on the basis of implementing sponge measures, it is necessary to cooperate with gray facilities to upgrade the standard to ensure that the hard target of the regional rainwater drainage capacity meets the standard, while ensuring the effective control of the initial rain pollution.
2. For rainwater systems with scattered pumping stations, the mode of joint upgrading may be considered, and the service scope of the pumping station may be re-delineated. For areas far away from the pumping station, additional pumping stations or collection of unified storage and drainage methods may be considered.
3. Through overflow, the series shallow tunnel project is linked to the present rainfall pipe network. During the low return period of a rainstorm, it serves as the initial rain storage and collection pipe; during the high return period of a rainstorm, the initial rain is first sent from the shallow tunnel project to the initial rain regulation and storage tank, and the shallow tunnel project continues to serve as the rainwater collection and storage pipe in the area far from the pumping station. The benefits of early rain control and rainwater storage are achieved by separating time and location.

ACKNOWLEDGMENTS

This paper is supported by Scientific research project of Shanghai Municipal Commission of Science and Technology (Grant No. 19DZ1201500); and Scientific research project of Shanghai Municipal Commission of Science and Technology (Grant No. 20DZ1204400).

REFERENCES

Changnon S A. (2010). Stormwater management for a record rainstorm at Chicago[J]. Journal of Contemporary Water Research and Education,146;103–109.

Japan Institute of Land Development Technology. Urban Underground River Design Manual (1974). Institute of Land Development Technology

Luan Zhenyu, Jin Qiu, Zhao Siyuan, Jiang Jiao, Lu Xiang. (2021) Urban waterlogging simulation based on MIKE FLOOD coupling model. Water Resources Protection, 37(02):81–88.

SONG C C S, LIN W, GONG C (1988). Hydraulic Transient Modeling of TARP System [C].

Wang Guanghua, Li Wentao, Chen Yilong, et al (2016). Preliminary study on deep tunnel drainage engineering at Donghao creek in Guangzhou. China Water & Wastewater, 32(22):7–13

Wu Jia, Ding Min, Jin Xin. (2016) Research on the use of hydraulic models in the planning and design of drainage system reconstruction projects. China Water & Wastewater, 32(11):137–140.

Zhang Dongdong, Yan Denghua, Wang Yicheng, et al. (2014) Research progress of urban waterlogging disaster risk assessment and comprehensive response. Disaster Science, 29(1):144–149.

Zhang Linwei (2015) Introduction to sponge city construction. Water & Wastewater Engineering, 41(6):1–7

Zhang Qinzao (2017). Research on the joint upgrading and transformation plan of multiple drainage systems that have been built. China Water & Wastewater, 33(15): 116–121.

Advances in Urban Engineering and Management Science – Khalil & Yang (Eds)
© 2023 The Author(s), ISBN 978-1-032-30426-7

Research on design strategy and application of prefabricated tourism residential building – Taking Chongqing agricultural tourism complex project "Heshan four seasons" as an example

Yihui Du*, Zijin Zhang*, Zixiang Yin* & Yi Li*
Architectural and Urban Planning Department, Chongqing College of Architecture and Technology, ChongQing, China

ABSTRACT: Ecotourism is an important means to maintain the sustainable development of scenic spots, and the residential building of ecotourism is an indispensable part. Prefabricated housing has the characteristics of fast construction, energy saving and emission reduction, flexible and comfortable, which is an important development direction of ecological residential buildings in the future. The paper combs out the problems existing in the prefabricated architecture of ecological sojourn, and applies the design strategy of prefabricated architecture of sojourn to the actual case of "Heshan four seasons", striving to provide the reference for the development of ecological sojourn architecture in Southwest China.

1 INTRODUCTION

1.1 *Characteristics of ecological sojourn buildings*

In recent years, with the awakening of people's environmental awareness, green movement and green consumption are gradually advocated, and ecotourism is more and more recognized by the public.

The connotation of eco-tourism has two aspects: one is "return to nature tourism" and "green tourism"; The second is "protection tourism" and "sustainable development tourism" (Pacificwb 2013). In 1990, the International Ecotourism Association defined it as a kind of tourism behavior to protect the environment and improve the welfare of local residents in a certain natural area. This kind of behavior should be carried out without disturbing the natural region, protecting the ecological environment, reducing the negative impact of tourism and providing beneficial social and economic activities for the local population.

Ecological sojourn building is an extension of the concept of ecotourism. Guided by the connotation of ecotourism, ecotourism has the following characteristics: ecological protection, fast construction, flexible space, beautiful decoration, perfect function, high safety and strong sustainability (Wang 2021; Zhang 2021).

1.2 *Problems in prefabricated tourism residential buildings*

Southwest China is rich in mountains and hills and tourism resources. With the improvement of living standards, people's tourism behavior is increasing year by year, and the demand for the diversification of residential construction products is very strong. In the scenic spots and villages that go deep into the nature, the ecological residential building products are more sought after and favored by people. In addition to the traditional brick and concrete structure buildings, the common prefabricated building products of ecological sojourn in Southwest China include prefabricated wooden house, container house, passive limo, light steel integrated house.

---

*Corresponding Authors: 27059823@qq.com, 397817847@qq.com, 196268049@qq.com and cqfdc2008@163.com

DOI 10.1201/9781003305026-31

Through the research of the design team, it is found that the common prefabricated residential buildings in Southwest China have the following problems:

It is inconvenient to build and causes a certain degree of damage to the ecological environment. The construction of some residential buildings is complicated and the construction period is long, which increases the construction waste and pollution emissions, and has a great impact on the ecological environment.

The structure of the interior space is single, and additional interior decoration is needed. There are great limitations in the combination and scale of interior space of sojourn buildings. At the same time, after the completion of the construction, more interior space decoration is needed, and the construction period is extended.

The building safety and weatherability are uneven, and the residential function is not perfect. Many types of residential buildings have poor weatherability, poor comfort and bad experience for consumers.

The economy is poor and the sustainability of building resource utilization is insufficient. There are many common building types, which have the characteristics of high cost, high maintenance requirements and low reuse rate. In the case of changes or adjustments in the scenic area planning, it is difficult to reorganize and use, and the green ecological sustainability is poor.

2 DESIGN STRATEGY OF PREFABRICATED RESIDENCE IN SOUTHWEST CHINA

2.1 *Respect nature and combine tradition*

Southwest China is characterized by high mountains and steep slopes, rugged roads, subtropical climate, great climate change, cold winter and hot summer. Most of the dwellings in this area originated from the dry fence architecture and traditional courtyard. Plane layout: Diaojiaolou, concave shape, a seal and a word type. The houses are built according to the mountain, and the layout is flexible without seeking the direction and axis symmetry; Paying attention to ecology and nature embodies the concept of "harmony between man and nature" (Cheng 2019; Fu 2016).

2.2 *Ecological protection and recycling*

In the process of construction and use, the damage to the ecological environment should be reduced as much as possible; The construction period should be short and the construction and decoration should be convenient; Building safety should be excellent, and provide comfortable living experience; The sustainability of building resources recycling is strong. In addition, considering the differences of the construction conditions of each scenic spot, it should also have the characteristics of standardized design and mass production (Luo 2021).

2.3 *Abundant space and perfect functions*

The use function of sojourn building should have the living room, bedroom, kitchen, toilet and balcony modules. The modules can be independently spliced with the theme of appearance design, which can be extended and widened horizontally, but also increased vertically. At the same time, the building itself light control system, water supply system and sewage treatment system, and other functions should be complete.

2.4 *High safety and diversified decoration*

The main body of the building structure should have good mechanical performance, replaceable and fatigue resistance; The roof / wall has low thermal conductivity and good thermal insulation performance. In order to achieve comfortable and beautiful effect, at the same time, reduce the cost and construction waste caused by secondary decoration, and avoid environmental pollution. In addition, through the selection of different style combinations, the transformation of appearance design, material and interior decoration can be realized, forming a personalized and independent style effect.

3 ARCHITECTURAL DESIGN AND APPLICATION PRACTICE OF "HESHAN FOUR SEASONS "IN AGRICULTURAL TOURISM COMPLEX PROJECT

Heshan four seasons is an agricultural tourism complex project located in Heshanping, Jiangjin, Chongqing. The main body of the building is a small prefabricated residential building, which embodies the advantages of fast and environmental protection, free and efficient, safe and comfortable, controllable cost, and strong sustainability. Taking the project as an example, this paper expounds the specific application of small prefabricated residential buildings in southwest eco-tourism.

Table 1. Proportion of prefabricated residential construction cost of the project

Section	Compose	Proportion of expenses
Steel structure	Main structure system.	31.28%
Roof part	Polyurethane roof board, PVC external hanging board, PVC thatch, asphalt tile, resin tile, integrated board, roof waterproof, etc.	12.12%
Wall part	Wood grain wallboard, sprayed wallboard, integrated board, etc.	8.13%
Ground part	Base slab, floor tile, squatting toilet, wood plastic, etc.	7.63%
Door and window	Doors, windows, top glass, etc.	12.09%
Water and electricity part	Water and electricity materials, switches, sockets, lamps, etc.	2.81%
Auxiliary materials	Decorative parts, stairs, waterproof coiled materials, angle steel, structural adhesive, etc.	16.72%
Other parts	Installation fee, long-distance transportation fee and material handling fee.	9.22%

3.1 *Constructing is fast and convenient, and protect the natural environment*

From design to production, the small prefabricated residential buildings in Heshan four seasons implement the design concept of modularization and lightweight, maximize the degree of industrialization, realize convenient transportation and rapid assembly, and cope with the terrain characteristics of high mountains and steep slopes in Southwest China.

The architectural structure design imitates the traditional dry column building in Southwest China, and applies the traditional building frame structure system to the prefabricated building. Based on the independent small pile and the steel square tube (heavy steel) as the structural frame, it forms an overall overhead building form, which does not harden the ground and occupy the surface green space, so as to adapt to various complex landforms in Southwest China. In the four seasons project of Heshan, if the overhead height of the building is enough, the overhead floor can be cultivated continuously, or a semi open flexible space can be formed for tea drinking and rest after work, which further excavates the essence of sojourn on the basis of protecting the natural environment.

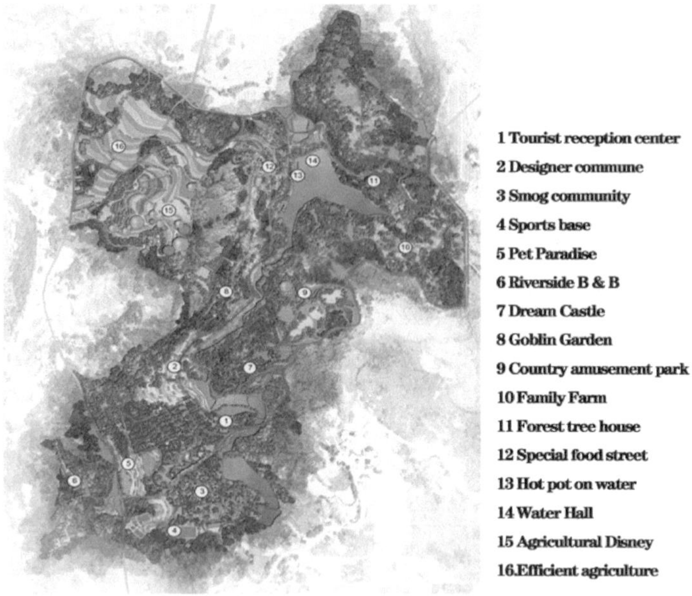

Figure 1. General layout of the project.

3.2 *Free combination of space, easy and diversified decoration*

The use function of the small prefabricated residential building of the project can be divided into five basic modules: living room module, bedroom module, kitchen module, toilet module and balcony module. Combined with the appearance design theme, the horizontal and vertical independent splicing can be carried out. In addition, the internal and external decoration insulation composite board is used as the maintenance system, which is basically formed through one-time installation (both internal and external wall decoration are completed). While ensuring comfort and beauty, avoid the cost pressure and environmental pressure brought by secondary decoration.

At the same time, in the process of design and production, combined with the local cultural characteristics, it shows the charm of local cultural characteristics and becomes another beautiful cultural landscape in Jiangjin area.

Figure 2. Courtyard renderings 1.

Figure 3. Courtyard renderings 2.

3.3 *The house has high security and perfect living function*

The small prefabricated buildings of Heshan Siji are mainly made of steel structure, which has good mechanical performance, replaceable and fatigue resistance; The surface of all steel structures is sprayed with thermosetting powder for anti-corrosion treatment, anti-oxidation and corrosion, which can guarantee the design service life of more than 50 years.

The building itself integrates the intelligent light control system, water supply system and sewage treatment system of the whole house. At the same time, it can integrate the energy (solar energy, wind energy), heating (floor heating, radiator), weak current (remote control system), artificial intelligence and other scientific and technological details according to the demand. The intelligent camera head and intelligent door lock can be integrated into the building's IOT system, which is convenient and safe. After the installation, you can move in, and easily realize comfortable and intelligent living.

Figure 4. Real scene of project courtyard.

Figure 5. Interior decoration.

3.4 *Resource recycling is highly sustainable*

The whole system is composed of frame structure system, wall system, roof system and ground system, each of which is an independent system, and the structure is a large frame structure with bolt connection, few connection points and convenient disassembly. The main body is made of heavy steel, which has larger size, thickness and strength to ensure the integrity rate of disassembly.

According to statistics, the damage rate of floor and roof decoration materials is about 10%, and the damage rate of floor and roof base plate is less than 5%. Therefore, the overall demolition damage rate is about 10%, which reflects the advantages of relocation and reuse.

4 CONCLUSION

Based on the analysis of the common problems of ecological residential building products in Southwest China, this paper puts forward the design strategy of small prefabricated residential building suitable for Southwest China. Taking Chongqing Jiangjin agricultural tourism complex project "four seasons in Heshan" as an example, this paper expounds that the design of small prefabricated residential buildings should meet the construction and use requirements of fast and convenient construction without damaging the environment, flexible combination of interior space and interior decoration, high housing safety, perfect living function, and strong sustainability of resource recycling, to provide ideas for the development of ecological sojourn architecture in Southwest China.

REFERENCES

Cheng Y. H. (2019). Study on integrated planning and design of living and providing for the aged in Southwest Mountain Resort -a case study of Shangtianchi resort in Jiuba Town, Guizhou Province. Chongqing University.

Luo N. (2021). Analysis of architectural design strategies under the concept of low carbon. Management and technology of small and medium sized enterprises. 8, 130–131.

Pacificwb. (2013). The concept and connotation of Ecotourism. https://wenda.so.com/q/1363265921061811.

Wang T. (2021). Research on the core power unit of integrated residential building. City Building. 18, 24–28.

Zhang X. Y. (2021). Research on the design of ecological livable residence based on the concept of green building. Metal structure of Chinese Architecture. 6, 78–79.

Advances in Urban Engineering and Management Science – Khalil & Yang (Eds)

A study on the correlation of vertical greening design on the outside facade of coastal city office building

Zhang Wu* & Kim Chul-Soo*
Pukyong National University, Nam-gu, Busan, Korea

ABSTRACT: It is pointed out that the construction of coastal cities in South Korea will be affected by typhoons, especially in Busan, which will lead to certain losses. This paper concludes that the urban heat island effect also affects buildings. However, facade protection cannot only strengthen the office building but also reduce its energy consumption. The use of vertical greening facades can further reduce the consumption of cooling and heating in high-rise buildings. In this study, PHOENICS was used to measure and analyze the wind environment and thermal radiation environment of buildings. In addition, it was utilized to simplify the simulation environment of municipal buildings and obtain the analysis results. Preservation of old buildings can also improve their appearance. We believe that through this study, the vertical greening design of facades can be paid attention to and applied to green buildings.

1 GENERAL INSTRUCTIONS

According to the data (Table 1) of the National Typhoon Center (NTC), South Korea is a country that often suffers from typhoons, and the old buildings that are near the seaside should have had resistance. In this research, we are recommending using the vertical greening of the facade as a solution to upper problems. According to the report by Korea International News this year, disaster compensation has reached 185.6 billion won, which also attaches great importance to disaster construction (Van der Geer et al. 2010).

Table 1. Statistics on typhoon occurrence status (General year: 1991–2020).

month	1	2	3	4	5	6	7	8	9	10	11	12	Sum
the normal year	0.3	0.3	0.3	0.6	1.0	1.7 (0.3)	3.7 (1.0)	5.6 (1.2)	5.1 (0.8)	3.5 (0.1)	2.1	1.0	25.1 (3.4)
this year	0(0)	1(0)	0(0)	1(0)	1(0)	1(0)	–	–	–	–	–	–	4(0)

※ The number in parentheses indicates the number of typhoons that have affected Korea. (Source: Korea Meteorological Administration Weather Nuri.)

In the modern world, it is very difficult for people to be close to nature, especially now. Therefore, “green design” has become an important tool for solving social problems through design. In the case that coastal cities cannot provide insufficient ecological construction areas while improving coastal natural disasters, urban three-dimensional design planning can be carried out to efficiently carry out the ecological design.

*Corresponding Authors: 852755086@qq.com and kimcsoo@pknu.ac.kr

DOI 10.1201/9781003305026-32

2 GENERAL INSTRUCTIONS

2.1 *Materials and methods*

PHOENICS was used to evaluate and analyze the wind environment and thermal radiation environment of a building, and the simulation environment of urban buildings was simplified. The model can now be used to predict urban heat islands and assess the ambient temperature distribution outside cities.

The whole simulation region was split into a spatial grid of 11010052, and the structural grid distribution was used, with the area where the buildings were located being treated with local grid encryption. Figure 1 depicts the meshing results.

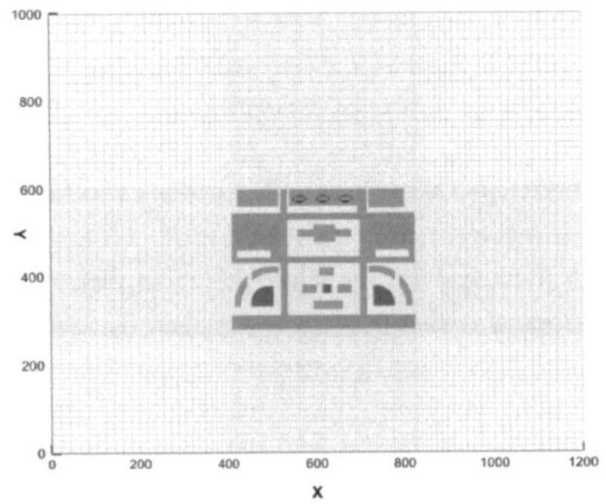

Figure 1. Calculate area grid diagram (top view)

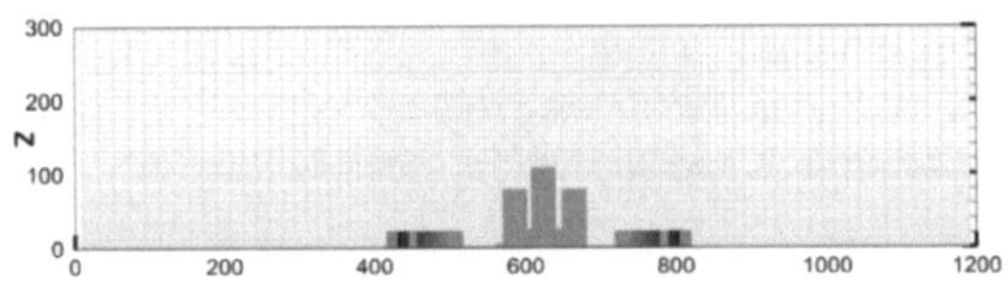

Figure 2. Calculates the region grid side view

As wind flows across the ocean and cities, it creates friction, reducing wind energy and speed, as well as the airflow's structure. Such as turbulence, vortex scale, and so on.

The computer should be used to analyze architectural designs. The wind environment inside and outside the building, the height of the pedestrian spacing surrounding the building at 1.5 m, the wind speed less than 5 m/s, and the windless region and vortex area are all controlled.

Table 2. Simulation analysis 1

Environmental analysis: Wind Pressure Distribution on Windward Surface of Building	summary of problems
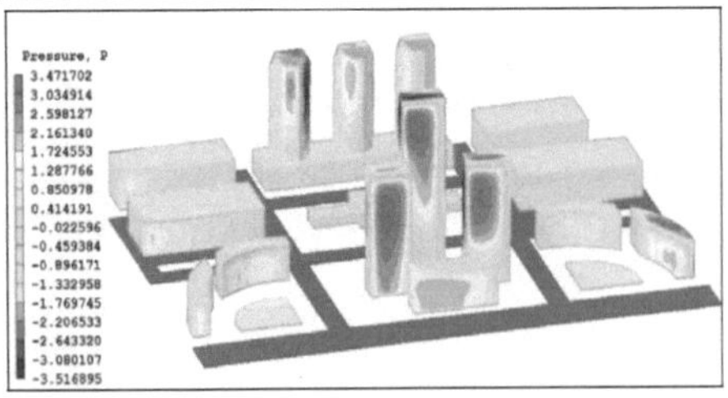	1.Lack of protection of coastal high-rise buildings 2.There is an urgent need to reduce the heat island effect in coastal cities and promote environmentally sound solutions 3.The red part of the coastal building, as illustrated, needs to be addressed 4) Urban coastal buildings are close to the seaside

2.2 *Selection of governing equation formulas*

In general, the power-law can be used for the wind speed profile at the surface layer:

$$\overline{u}(z) = \overline{u}_1 \left(\frac{z}{z_1}\right)^n \tag{1}$$

Where is the average wind speed, z represents the ground and is the mean wind speed at height z1. The exponent n is determined by the hierarchy's stability. The value of neutral layered n is 0.25. The unstable stratification n is less than the stable stratification n.

Assuming that the flow on the outflow surface has been fully developed and that the flow has reverted to normal flow without the formation of obstructions, the outlet boundary is interpreted locally unidirectionally. Given that the airflow surrounding the building is typically incompressible low-speed turbulence, as predicted by Boussinesq's theory, and that the airflow contacts the building to produce a confined flow, the conventional k-model has a favorable influence on the restricted flow (with wall constraints). Furthermore, the typical k-model has a low computation cost and an accurate forecast. The k-model is used to simulate and solve issues in this article. This study focuses on the incompressible flow issue, and the control equations are as follows.

$$\text{Continuity equation}: \frac{\partial u_l}{\partial x_l} = 0$$

Dissipation rate ε equation:

$$\frac{\partial \varepsilon}{\partial t} + u_j \frac{\partial \varepsilon}{\partial x_j} = \frac{\partial}{\partial x_j}\left(\frac{v_t}{\delta_\varepsilon}\frac{\partial \varepsilon}{\partial x_j}\right) + \frac{\varepsilon}{k}\left(C_1 P_k + C_3 C_k + C_2 \varepsilon\right)$$

3 RESULTS & DISCUSSION

Figure 3. Effect diagram of building facade after reconstruction 1.

Figure 4. Effect diagram of building facade after reconstruction 2.

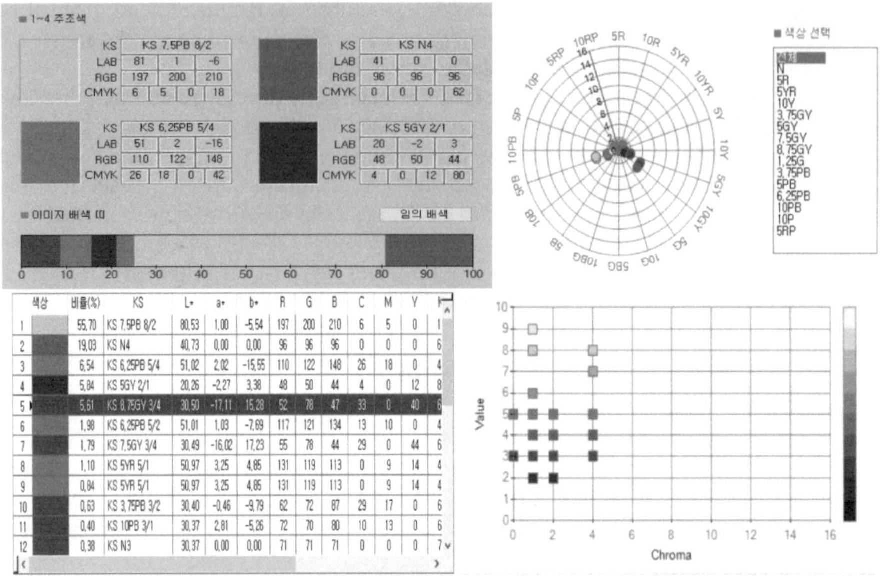

Figure 5. Color analysis of building facade after reconstruction.

Table 3. Evaluation and analysis of building wind environment

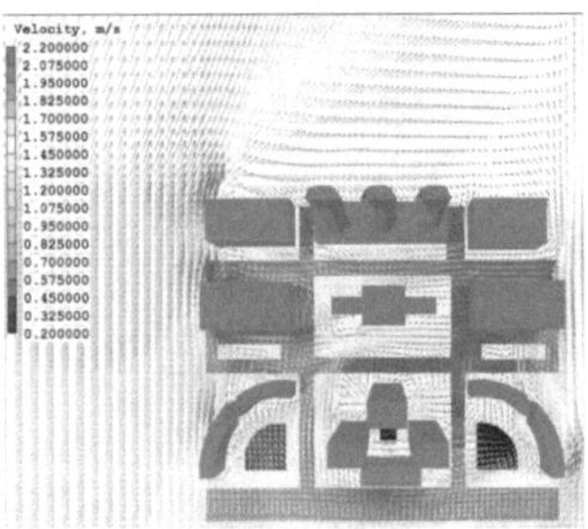	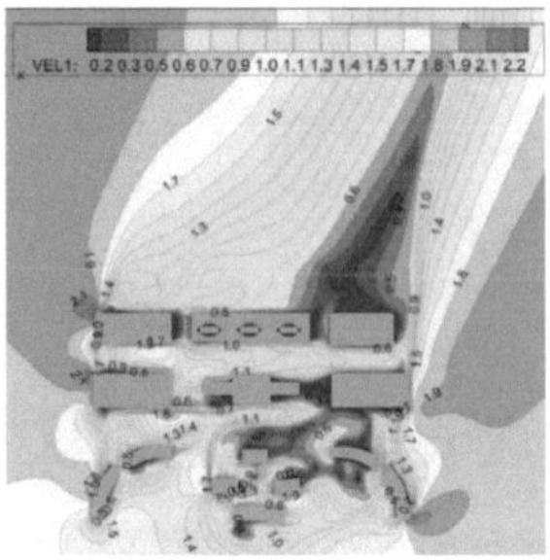	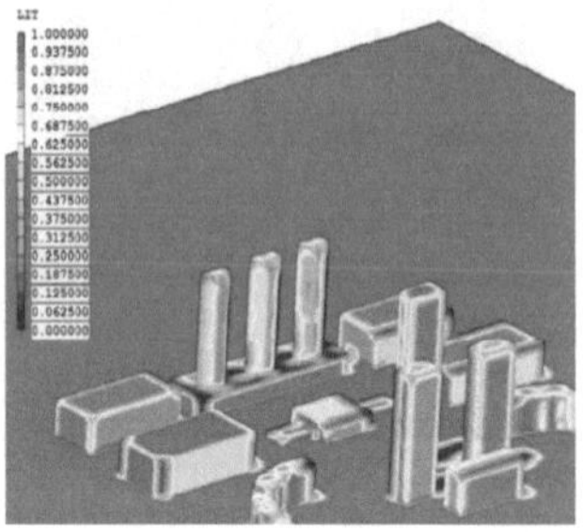
Vector diagram of summer wind speed	Summer wind speed cloud chart	Thermal analysis1

- The overall structure uses corrosion-resistant Monel alloy materials to form a streamlined structure. This streamline refers to the automobile streamline design, which reduces air resistance and wind to the greatest extent. Furthermore, it improves the overall office building during bad weather, such as sea breeze and typhoons (Yang Xu 2016.01) (Yuan Yunkai & Gu Mingyuan, 1991).
- Although it affects visibility, it is a particularly suitable attachment carrier for climbing plants. The vertical greening is used for shading and has a good shading effect, despite winter or summer.
- Reasonable use of color shading panels through color contrast can better foil the characteristics of the building. However, bright shading facilities with dynamic information enrich people's perception of color.
- The texture in shading design of smooth corrosion-resistant metal materials makes people feel light and delicate psychological feelings. Furthermore, grid shading design had a good conducive effect on light (Huang Haijing & Liu Yanfei, Landscape China, 2021.5.17).

Table 4. The earth trajectory.

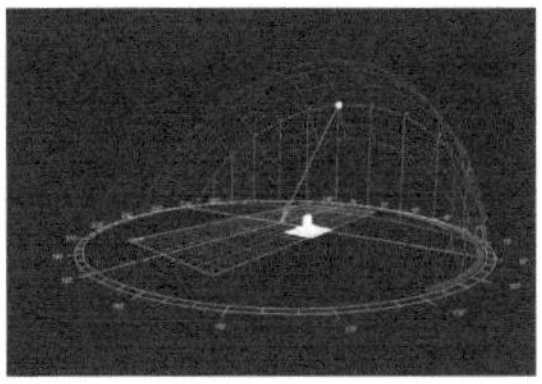

Table 5. Thermal analysis 2.

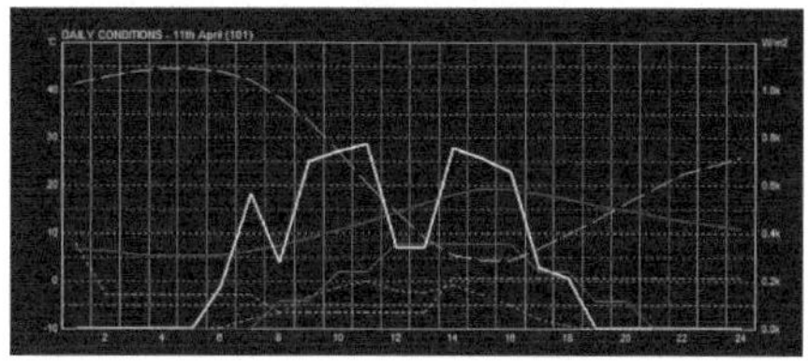

- Figures 3 and 4 demonstrate color analysis of the green. The experimental results showed that the product shape after the golden ratio correction had a better visual aesthetic feeling (Huang Danny, Cui Tianjian, 2014.09).
- In Table 4 we can watch the solar trajectory for the movement of thermal radiation. While in Table 5 we illustrated the corresponding heat change curve. Showing a weakening trend proves that vertical greening is effective for building surface heat.

4 RESULTS

PHOENICS is used to measure and analyze the wind environment and thermal radiation environment of the reconstructed building. From the simulation results, this design reduces air resistance and takes away heat. The wind environment in urban is an important factor in energy-saving design for architecture, which has a significant effect on regional climate and area microclimate (Ashrant Aryal, Burcin Becerik-Gerber 2020).

Vertical greening of the facade is only suitable for coastal cities in the sunshine time as shown in Table 4. Furthermore, it can resolve turbulent conditions of the office building facade in coastal cities and reduce thermal radiation to meet the requirements of greener buildings. The overall structure uses corrosion-resistant Monel alloy materials to form a streamlined structure. This streamline refers to the automobile streamline design, which reduces air resistance. Moreover, it will improve the overall office building in the face of bad weather, such as typhoon weather, and can protect the building to the greatest extent.

5 CONCLUSION

The wind pressure and thermal radiation on the windward surface of urban buildings is the highest.

The protective facade can solve the problem, and the vertical greening facade can be utilized to design. In other words, during severe typhoons, it has certain protection for coastal cities' office buildings, and can also directly improve and beautify the old or unattractive office buildings. In addition, vertical greening can enhance the greening area of the urban environment.

ACKNOWLEDGMENTS

This work was supported by a grant from Brain Korea 21 Program for Leading Universities and Students (BK21 FOUR) MADEC Marine Designeering Education Research Group.

REFERENCES

Ashrant Aryal, Burcin Becerik-Gerber (2020)Thermal comfort modeling when personalized comfort systems are in use: Comparison of sensing and learning methods[J] Building and Environment.

Huang Danny, Cui Tianjian (2014.09), Product Shape Correction Method Research Based on Gold Ratio, School of Art, Southeast University, Mechanical Design. Core of Peking University

Huang Haijing & Liu Yanfei, Landscape China, (2021.5.17) Building facade design based on shading form https://mp.weixin.qq.com/s/wUxCXt79TSXVxWH-pjbEYQ

Van der Geer, J., Hanraads, J.A.J., Lupton, R.A. (2010) The art of writing a scientific article. J. Sci. Commun., 163: 51–59.

Yang Xu. (2016.01) Based on the optimization of ProCAST Monel alloy impeller casting method, China Petrochemical Group Jinling Petrochemical Company Alkylbenzene Factory, Contemporary Chemical Research, Beijing, China.

Yuan Yunkai&Gu Mingyuan. (1991) Social Dictionary of Science and Technology · Physics: Zhejiang Education Press.

Construction technology and civil engineering management

Advances in Urban Engineering and Management Science – Khalil & Yang (Eds)

HFACS-based analysis of accidents related to lifting operations in construction

Xiaowei Lv*, Yan Liu & Zhifei Song
North China University of Technology, Beijing, China

ABSTRACT: Lifting operations have become ever more important with the rapid development of the prefabricated building. In order to effectively reduce the occurrence of accidents during lifting operations, this paper first selected 58 construction-related lifting accidents that were statistically calculated by using the HFACS model to find out the indicators with a higher occurrence frequency and then utilized SPSS to calculate the ratios to obtain the relationship between different levels. Finally, countermeasures and suggestions were discussed from the perspectives of enterprises, individuals, and relevant departments based on the higher frequency indicators.

1 INTRODUCTION

To promote the sustainable development of the construction industry, the most crucial thing is to ensure safety during construction. The prefabricated building (Chen et al. 2019), different from the traditional construction, is characterized by the production of pre-made components in the factory that are then transported to the site for assembly through lifting equipment. Therefore, lifting operations are particularly important not only for traditional construction but also for prefabricated buildings.

Many domestic scholars have analyzed the lifting operations with respect to both prefabricated buildings and traditional construction. For example, Sun Pengwei et al. (Sun 2014) analyzed the causes of different accidents during lifting operations and put forward corresponding countermeasures and suggestions. Chen Xiaobo et al. (2019) adopted DEMATEL to analyze the lifting efficiency of prefabricated buildings from the aspects of the external environment, components, mechanical facilities, methodology, and personnel and put forward corresponding suggestions. Chang Chunguang, Wu Xi, et al. (Chang & Wu 2018) proposed a set of criteria for safety risk assessment suitable for the lifting operations in prefabricated buildings by using the improved LEC method. Lu Yan (Lu 2016) analyzed some common accident types in lifting operations by using the Fault Tree Analysis (FTA) and propounded corresponding preventive measures. Peng Jingqi (Peng 2014) believed that the root cause of lifting accidents was due to the lack of proper management on the site after a detailed analysis based on the Fishbone Diagram and the Safety Pyramid Model developed by ConocoPhillips Marine.

It is therefore can be concluded that lifting accidents are caused by many reasons, and many scholars have studied various factors in lifting accidents by using Fishbone Diagram and FTA to find that the most prevalent reason is improper management. But nowadays, most of the literature on lifting accidents revolves around discourse, without data to support it. Therefore, this paper used HFACS to study the human factors in lifting accidents based on the frequency of different types of accidents and SPSS to analyze the relationship of various indicators, and finally put forward some countermeasures to prevent accidents.

*Corresponding Author: 2532826518@qq.com

DOI 10.1201/9781003305026-33

2 HFACS FRAMEWORK FOR LIFTING ACCIDENTS

2.1 *HFACS introduction*

The Human Factors Analysis and Classification System (HFACS) was proposed by FAA expert Shappel and Wieg-mann of the University of Illinois, and its framework mainly includes four levels, namely the lack of organizational management, unsafe supervision, the premise of risky behaviors, and risk behaviors (Chen et al. 2013). Later, Reinach and Viale (2006) added external factors to it, forming a five-level HFACS framework (Reinach & Viale 2006).

2.2 *HFACS framework for lifting accidents*

This paper uses five different levels that are divided into eleven categories. The HFACS framework of lifting accidents adopted in this paper is obtained in combination with the characteristics of accidents on the basis of HFACS.

2.2.1 *Risky behaviors*

Risky behaviors are divided into two categories, namely operating errors and operating fouls, and the former are further classified into four indicators, namely failure to install proper equipment at the lifting point, failure to lift according to the instruction, failure to calculate the lifting load or lifting in spite of overloading, and use of unsafe equipment, while the latter are also further classified into four indicators, namely continuing to work after an accident, improper use of protective tools, illegal lifting operations, and no tower crane commander or improper command. So, there are a total of 8 indicators.

2.2.2 *Premise of risky behaviors*

The premise of risky behaviors mainly refers to the quality of personnel and the status of operators, which is classified into staff management and includes two indicators, namely personnel training/technical disclosure and unlicensed construction.

2.2.3 *Unsafe supervision*

Unsafe supervision is divided into three categories, namely no supervision, no correction, and improper supervision. No supervise is further divided into two indicators, namely no supervision personnel for safety management, and insufficient supervision by safety personnel; no correction is further divided into two indicators, namely not discovering hidden risks, and discovering hidden risks without not correcting in time; improper supervision is also further divided into two indicators, namely ineffective supervision by related departments, and failure to report to superiors or related departments. So, there are a total of six indicators.

2.2.4 *Lack of organizational management*

The lack of organizational management is divided into three categories, namely insufficient resource management, insufficient organizational climate management, and insufficient organizational process management. Insufficient resource management is further divided into four indicators, namely no safety warning signs, poor equipment maintenance, and poor equipment installation; insufficient organizational climate management is further divided into three indicators, namely organizing dangerous operations without authorization, insufficient safety awareness by the managers, incomplete or poorly implemented production safety system, and poorly implemented safety responsibility system; the organizational process management is further divided into one indicator, namely forging documents. So, there are a total of eight indicators.

2.2.5 *External factors*

External factors are divided into two categories, namely insufficient external supervision and climate. Insufficient external supervision is further divided into three indicators, namely poorly implemented supervision responsibility system, illegal subcontracting, and failure to strictly supervise safety management during subcontracting. Climate is further divided into one indicator, namely dim weather. So, there are a total of four indicators.

3 DATA COLLECTION AND RESULT ANALYSIS

3.1 *Data collection*

The paper includes 76 construction-related lifting accidents from 2014 to 2020 that are collected from the safety management website and provincial emergency management bureaus, among which 58 accidents with complete information are selected for research. There are a total of 124 deaths, 68 injuries, and 58 lifting accidents (including 28 accidents of falling objects, 24 accidents of tower cranes overturning, and 1 squeezing accident) caused by those 58 accidents. Table 1 shows detailed information. It thereby can be seen that the accidents of falling objects and tower crane overturning are the main factors causing lifting accidents.

Table 1. Statistics of casualties.

Type of accident	Number of occurrences		Number of deaths		Number of injuries	
Falling from high places	5	8.62%	13	10.48%	6	8.82%
Tower crane overturning	24	41.38%	79	63.71%	59	86.76%
Hanging objects falling	28	48.28%	31	25.00%	3	4.41%
Squeezing accident	1	1.72%	1	0.81%	0	0

3.2 *Indicators of accident causes*

According to the investigation report based on the 58 accidents, this paper summarizes 28 accident cause indicators by utilizing the five indexes of HFACS, namely risky behaviors, the premise of risky behaviors, unsafe supervision, lack of organizational management, and external factors. The specific indicators are shown in Table 2.

3.3 *Analysis of accident causes*

According to the investigation report based on the analysis of the causes of 58 lifting accidents, the following conclusion is made. Risky behaviors account for 23.88%, among which illegal operations are the most prevalent, accounting for 55.02%. The premise of risky behaviors accounts for 19.03%, among which personnel training and technical disclosure are the most prevalent indicators, accounting for 69.09%, Therefore, more attention should be paid to personnel training and technical disclosure before lifting operations. Unsafe supervision accounts for 28.37%, which is the highest among the five levels. Therefore, more attention should be paid to the supervision of personnel during the lifting operations. No hidden safety hazards discovered in time are the most prevalent indicator, accounting for 29.27% of unsafe supervision; in addition, the number of times of safety management without arranging supervision personnel is 22, which is only two times less than that of no safety hazards discovered in time for prevention, making both of them equally

Table 2. Statistics of human factors indexes for lifting accidents.

Level	Category	Index	Frequency	Percentage
Risky behaviors	Operation errors	Poor installation at the lifting point	1	0.35%
		Not lifting according to the construction method	1	0.35%
		Lifting without calculating the lifting load or overloading	5	1.73%
		Using unqualified equipment	2	0.69%
	Improper operations	Continuing to work after an accident	1	0.35%
		Improper use of protective tools	2	0.69%
		Illegal lifting operations	38	13.15%
		No tower crane commander or improper command	19	6.57%
The premise of risky behaviors	Personnel management	Personnel training/technical disclosure	43	14.88%
		Unlicensed construction	12	4.15%
Unsafe supervision	No supervision	No supervision personnel being arranged for safety management	22	7.61%
		Inadequate supervision by security personnel	13	4.50%
	No rectification	No safety hazards being discovered in time	24	8.30%
		No timely rectification after the discovery	7	2.42%
	Improper supervision	Inadequate supervision by relevant departments	15	5.19%
		Failure to report to superiors or related departments	1	0.35%
Lack of organizational management	Insufficient resource management	No safety warning signs	3	1.04%
		Inadequate maintenance of equipment	11	3.81%
		Inadequate installation of equipment	1	0.35%
	Insufficient organizational climate	Unauthorized organization of dangerous operations	1	0.35%
		Insufficient safety awareness of managers	3	1.04%
		Incomplete or poorly implemented production safety system	13	4.50%
		Poorly implemented safety responsibility system	16	5.54%
	Insufficient organizational process management	Forging documents	2	0.69%
External factors	Insufficient external supervision	The poorly implemented supervision responsibility system	22	7.61%
		Illegal subcontracting	6	2.08%
		Failure to strictly supervise the safety management during subcontracting	4	1.38%
	Climate	Dark weather	1	0.35%

important. Lack of organizational management accounts for 17.3%, among which a poorly implemented safety responsibility system is the most prevalent indicator, accounting for 32%, and the gap between incomplete or poorly implemented production safety system and the highest indicator is 13 times. External factors account for 11.42%, among which the number of times of poorly implemented supervision responsibility system is the most prevalent indicator (22 times).

From the above data, it can be seen that personnel training/technical disclosure, the number of hidden safety hazards not discovered in time, without arranging supervision personnel for safety management, and poorly implemented supervision responsibility system are the four indicators that appear more frequently. Among them, personnel training/technical disclosure is the highest of all indicators, making it the main cause of lifting accidents.

4 LEVEL CORRELATION ANALYSIS

In order to study the causes of lifting accidents in more detail, this paper uses SPSS to calculate the odds ratio in addition to frequency. If the value p is less than 0.05 while the odds ratio is greater than 1 (Zhang & Liu 2019) at a certain level, it is significantly correlated to the previous one. The specific research results are shown in Table 3.

Table 3. Lifting accidents model hierarchy.

Lifting accidents model levels	Odds ratio	95% confidence interval
Unlicensed construction (Level 2) * No tower crane commander or improper command (Level 1)	1.391	0.484-4.003
Personnel training/technical disclosure (Level 2) * Illegal lifting operations (Level 1)	1.618	0.797-1.712
Personnel training/technical disclosure (Level 2) * Use of unsafe equipment (Level 1)	2.867	0.191-43.031
Personnel training/technical disclosure (Level 2) * Lifting without calculating the lifting load or overloading (Level 1)	1.911	0.353-10.356
Inadequate supervision by relevant departments (Level 3) * Unlicensed construction (Level 2)	2.571	0.67-9.866
Poorly implemented safety responsibility system (Level 4) *No timely rectification after discovery (Level 3)	9.091	1.548 53.392
Inadequate supervision by security personnel (Level 3) * Falsification (Level 4)	1.182	0.937-1.49
Inadequate supervision by relevant departments (Level 3) * Falsification (Level 4)	1.154	0.946-1.407
Incomplete or poorly implemented production safety system (Level 4) *Illegal subcontracting (Level 5)	4.2	0.735-23.991
Illegal subcontracting (Level 4) *Poorly implemented safety responsibility system (Level 5)	6.667	1.085-40.975
Poorly implemented supervision responsibility system (Level 5) *Forging Documents (Level 4)	1.1	0.964-1.255

The correlation between level 1 and level 2: The indicator with the highest value between level 1 and level 2 is personnel training, and the correlation factor between technical disclosure and use of unsafe equipment is 2.867. Therefore, personnel who have not been trained and technically explained are more likely to use unsafe equipment.

The correlation between level 2 and level 3: The most obvious correlation between level 2 and level 3 is ineffective supervision by related departments and unlicensed construction, with a ratio of 2.571. Therefore, the probability of unlicensed construction is 2.571 times more likely if there is not adequate supervision by related departments.

The correlation between level 3 and level 4: The most obvious correlation between level 3 and level 4 is a poorly implemented safety responsibility system and no timely rectification after discovery, with a ratio of 9.091. Therefore, if the safety responsibility system is poorly implemented, there will be a high probability that hidden hazards occur without timely rectification.

The correlation between level 4 and level 5: The most obvious correlation between level 4 and level 5 is illegal subcontracting and a poorly implemented safety responsibility system, with a ratio of 6.667. If a project is illegally subcontracted, the possibility that the safety responsibility system will not be implemented is 6.667 times more likely. Moreover, the ratio of the poorly implemented production safety system and illegal subcontracting is 4.2, which also plays a relatively leading role.

5 COUNTERMEASURES AND SUGGESTIONS

Based on the analysis of different accident causes, it can be seen that inadequate management of enterprises, personnel, and related departments are the main causes of accidents. Therefore, as for a company, the safety management of enterprises should be strengthened, professional safety personnel should be provided, managers should be regularly trained, a safety system should be improved, and a safety responsibility system should be properly implemented. As for personnel, they should not work in violation of relevant regulations, keep in mind the operating rules, and undergo regular examinations. The equipment should be maintained on a regular basis. As for relevant departments, it is necessary to strengthen daily supervision, intensify regulation enforcement, strictly investigate and punish illegal activities such as qualification frauds, minimize accidents, and consolidate the relationship between the Ministry of Housing and Urban-Rural Development and other departments.

6 CONCLUSION

A HFACS mode was established based on 58 lifting accidents. The causes of frequent accidents were confirmed according to the accident investigation report. It, therefore, can be concluded that personnel training/technical disclosure, improper lifting operations, no timely discovery and prevention of hidden safety hazards, and poorly implemented supervision responsibility system are the main causes of most accidents, while insufficient personnel training and poor supervision are the main reasons for the occurrence of lifting accidents.

According to the odds ratio of the five levels calculated by using SPSS for later analysis, it can be concluded that a poorly implemented safety responsibility system is the main reason for unsafe supervision, while poorly implemented supervision responsibility system and illegal contracting are the main reasons for the lack of organizational management.

Based on the high-frequency indicators and discussion in terms of enterprises, personnel, and related departments, the countermeasures related to strengthening personnel training and management, supervision of managers, equipment maintenance, strict investigation and punishment of qualification frauds, affiliation, and other illegal activities, as well as supervision of safety management of each project are proposed.

REFERENCES

Chang Chunguang, and Wu Xi. Research on Safety Risk Assessment of Lifting Operations in Prefabricated Building Based on Improved LEC[J]. Liaoning Economics, 2018(05): 72–73.

Chen Xiaobo, Gan Kaixin, and Cui Ping. Analysis of Factors Affecting Lifting Efficiency of Prefabricated Concrete Components Based on DEMATEL [J]. Journal of Engineering Management, 2019, 33(06): 111–115.

Chen Zhaobo, Zeng Jianchao, Dong Zhui, et al. Analysis of Human Factors in Coal Mine Accidents Based on HFACS [J]. Chinese Safety Science Journal, 2013, 23(7): 116–121.

Lu Yan. Risk Analysis of Lifting Operations in Construction Based on Fault Tree Analysis [J]. Modern Industrial Economy and Information Technology, 2016, 6(13): 37–39.

Peng Jingqi. Analysis of Lifting accidents Causes and Systematic Preventive Measures [J]. Hoisting and Conveying Machinery, 2014 (02): 91–96.

Reinach S, Viale A. Application of Human Error Framework to Conduct Train Accident/incident Investigations [J]. Accident Analysis and Prevention, 2006, 38:396–406.

Sun Pengwei. A Brief Discussion on Causes and Countermeasures of Lifting Accidents [J]. Chemical Management, 2014(03): 48.

Zhang Yingyu, and Liu Lin. Research on China's Serious Coal Mine Accidents Based on HFACS[J]. Safety, 2019, 40(9):56–63.

Advances in Urban Engineering and Management Science – Khalil & Yang (Eds)
© 2023 The Author(s), ISBN 978-1-032-30426-7

Sensing strategy of autonomous emergency braking based on V2X in non-line-of-sight scenarios

Xiaoyi Guo* & Changhao Piao
Chongqing University of Posts and Telecommunications, Chongqing, China

ABSTRACT: Aiming at the problem of ineffective detection of the autonomous emergency braking system based on radar and camera in non-line-of-sight bend scenes, a sensing strategy based on the scene of hazardous location warning in vehicle-to-everything is proposed. This strategy separates vehicle perception from control and remotely obtains target information in non-line-of-sight scenarios through cellular-vehicle-to-everything communication technology. On board unit calculates the actual distance from the target to the vehicle, and completes the perception of non-line-of-sight targets, thereby further improving the reliability of the autonomous emergency braking system.

1 INTRODUCTION

At present, with the increasing incidence of traffic year by year, Advanced Driving Assistance System emerged (Peden et al. 2004). Among them, the AEBs (Automatic Emergency Braking system) based on single sensor or multi-sensor fusion (Cicchino 2017) has greatly improved the active safety of the vehicle. In sensor-based perception, the camera has the advantages of high resolution, a large amount of information transmission, low cost, etc., and deep learning technology is often used to effectively improve the ability of target classification and recognition (Dong 2019); the millimeter-wave radar sensor can adapt to various working environments, and based on the Doppler effect, they can obtain the target's movement status such as distance, speed, and azimuth by sending out chirp signals (Hasch et al. 2012); lidar has the characteristics of strong three-dimensional perception ability and can build accurate three-dimensional models for the target through methods such as clustering (Zhang 2020). The sensor-based AEBs are effective in LOS (line-of-sight) scenes, but in NLOS (non-line-of-sight) scenes, target information cannot be obtained smoothly, and there is a risk of accidents.

V2X (vehicle-to-everything) is a high-reliability and low-latency communication method between vehicles and other objects on the Internet of Vehicles. Everything respectively represents the vehicle, infrastructure, pedestrian, cloud, and other traffic participants (Wedel & Radusch 2010). The current mainstream V2X communication standards include DSRC (Dedicated Short-Range Communication) (Kenney 2011), and C-V2X (Cellular-vehicle-to-everything) communication standards based on cellular networks (Gonzalez-Martin et al. 2019). Among them, C-V2X includes LTE-V2X (Long Term Evolution-Vehicle to Everything) and 5G-V2X (5th Generation Wireless Systems-Vehicle to Everything) (Wei & Wang 2018; Wu & Xiao 2018).

The emergence and development of V2X technology provide a new solution for target perception and recognition in NLOS scenarios. Aiming at the problem of ineffective detection of AEBs based on radar and camera in NLOS bend scenes, a sensing strategy based on the scene of HLW (hazardous location warning) in V2X is proposed. This strategy proposes a new perception method to realize AEBs in NLOS scenarios, which further improves the reliability of AEBs and avoids the risk of collision.

*Corresponding Author: 382254036@qq.com

 DOI 10.1201/9781003305026-34

2 FUNCTIONAL DESIGN

2.1 *Software design*

In order to implement the strategy proposed in this paper, the following software process is specially designed.

In the beginning, a RSU (Road Side Unit) with DSRC capability periodically broadcasts map information and road hazard warning information. Then, after receiving the information broadcast by the RSU, the vehicle equipped with an OBU (On Board Unit) and with the V2X communication function turned on, and ttc, the distance to the dangerous target based on its own location information, map information, and road hazard warning information were calculated. And the information was output to the vehicle control terminal to complete the perception of the AEBs for subsequent decision-making control of AEBs.

2.2 *Data flow*

The strategy mentioned in this article is used in the perception layer of the AEBs. The specific perception process is shown in Figure 1.

Figure 1. Data flow.

The OBU, as the main body of data receiving and data processing, gets map information and target information from the RSU, and then accesses speed and vehicle position information from the car terminal; secondly, it analyzes and processes the received information, and calculates the distance between the vehicle and the non-line-of-sight target and the time to collision; finally, calculation result will be sent to the vehicle terminal.

3 INSPECTION AND ANALYSIS

3.1 *Test scenario*

The proof experiment in this article will be carried out on a self-built networked unmanned vehicle platform. The test site is the intersection of the southern campus of Chongqing University of Posts and Telecommunications. The test road has a width of 4 meters and a turning radius of 10 meters. Microsoft Visio Professional 2019 is used to draw the coordinate system established with the center of the curve as the coordinate origin, as shown in Figure 2.

Where point A is the head of the vehicle carrying OBU, point B is the intersection center, point C is the target center, point D is the location of RSU, and point O is the axis of the turning radius. Before the test, record points were set at the same distance on the road, and data were recorded when vehicles passed these points. The test records are shown in Table 1.

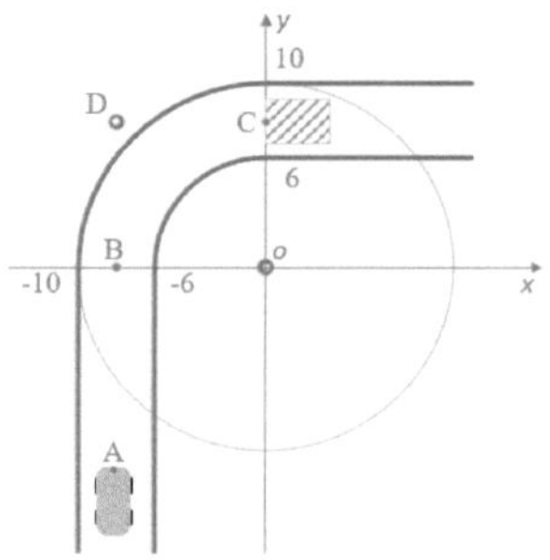

Figure 2. Test scenario.

Table 1. Test records.

Speed (kph)	Point C	Point A	Distance(m)			Average (m)	Actual-value (m)	Relative error(%)
20	(0,8)	(−8,−100)	112.8	115.3	114.2	114.1	112.5	1.4
		(−8, −80)	92.6	93.3	93.7	93.2	92.5	0.8
		(−8, −60)	72.9	74.8	71.9	73.2	72.5	1.0
		(−8, −40)	53.1	54.1	53.8	53.7	52.5	2.2
		(−8, −20)	32.7	34.0	33.8	33.5	32.5	3.1
		(−8, 0)	12.9	11.9	12.6	12.5	12.5	−0.3
	(4,8)	(−8, −100)	115.3	118.7	117.6	117.2	116.5	0.6
		(−8, −80)	96.3	95.3	94.4	95.3	96.5	−1.2
		(−8, −60)	75.4	76.1	76.9	76.1	76.5	−0.5
		(−8, −40)	54.9	54.5	55.6	55.0	56.5	−2.7
		(−8, −20)	36.5	35.3	39.8	37.2	36.5	1.9
		(−8, 0)	17.4	14.2	15.5	15.7	16.5	−4.8
40	(0,8)	(−8, −100)	112.1	119.5	117.3	116.3	112.5	3.4
		(−8, −80)	92.9	94.2	94.0	93.7	92.5	1.3
		(−8, −60)	69.0	71.1	74.1	71.4	72.5	−1.5
		(−8, −40)	57.7	53.3	53.7	54.9	52.5	4.6
		(−8, −20)	37.9	34.0	34.2	35.4	32.5	8.8
		(−8, 0)	11.5	10.5	13.3	11.8	12.5	−5.9
	(4,8)	(−8, −100)	117.4	118.4	119.3	118.4	116.5	1.6
		(−8, −80)	91.7	96.2	98.9	95.6	96.5	−0.9
		(−8, −60)	79.0	72.1	76.5	75.9	76.5	−0.8
		(−8, −40)	55.9	54.3	56.4	55.5	56.5	−1.7
		(−8, −20)	38.1	34.5	37.5	36.7	36.5	0.5
		(−8, 0)	17.6	15.5	18.5	17.2	16.5	4.2

3.2 *Results analysis*

The test data in Table 1 when the target is (0,8) is taken as the main analysis object. It is needed to calculate the error between the perceived distance recorded by vehicles at different recording points and the actual distance, sort out and draw the line chart of the error and average error of each recording point, as shown in Figure 3.

In order to better describe the effect of the strategy proposed in this paper, according to the data in Table 1, the relative error of perceived distance passing through each recording point at each speed was calculated, as shown in Figure 4.

It can be seen from Figures 3 and 4 that:

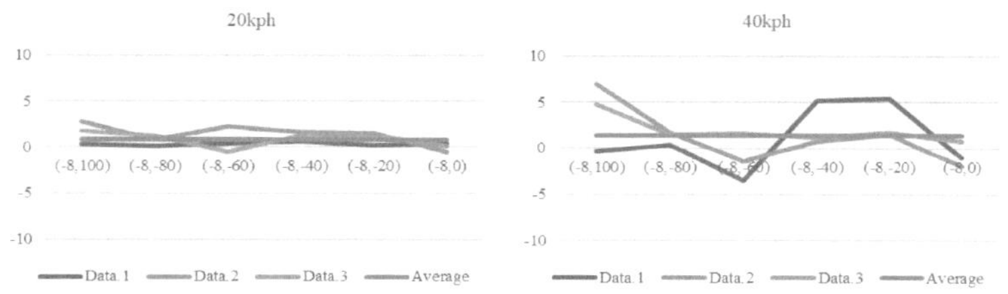

Figure 3. Line chart of error values at different speeds.

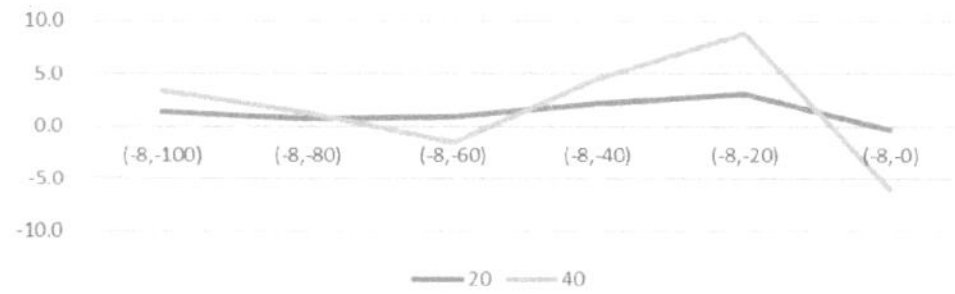

Figure 4. Relative errors.

When the target obstacle is in the non-line-of-sight scenario of the vehicle, the absolute value of the perceived relative error of the target is less than 10% when the vehicle is within 100m or less before the turning intersection and the speed is less than 40 kph.

With the increase of the actual vehicle speed, the calculation error of obstacle distance increases gradually. When calculating the distance between the vehicle and the obstacle target perceived by the policy at the decision layer, a dynamic threshold that changes according to time should be added.

4 CONCLUSION

This paper proposes a V2X-based automatic emergency braking perception strategy in non-line-of-sight scenarios. It uses the hazardous location warning function of V2X to replace the sensor perception scheme in the traditional automatic emergency braking function. The target detection work is stripped from the vehicle and handed over to the roadside unit for detection. The identification of automatic emergency braking targets can be accomplished effectively in non-line-of-sight scenarios. In this paper, the feasibility of the above strategy is proved by real vehicle experiments. The results show that the adoption of the perception strategy can solve the problem of AEBs perception failure in non-line-of-sight scenarios.

Since the road conditions used in the test conducted in this paper are relatively simple compared with the actual road conditions, more different test scenarios will be added in the subsequent study to cover different conditions and requirements. In addition, the distance calculation algorithm is improved to reduce the calculation error and reduce the risk of perception error.

REFERENCES

B Schünemann, Wedel, J. W., & Radusch, I. (2010). V2x-based traffic congestion recognition and avoidance. Journal of Applied Science and Engineering, 13(1), 63–70.

Cicchino, J. B. (2017). Effectiveness of forward collision warning and autonomous emergency braking systems in reducing front-to-rear crash rates. Accident Analysis & Prevention, 99(PT. A), 142–152.

Gonzalez-Martin, M., Sepulcre, M., Molina-Masegosa, R., & Gozalvez, J. (2019). Analytical models of the performance of c-v2x mode 4 vehicular communications. IEEE Transactions on Vehicular Technology, 68(2), 1155–1166.

Hai Wu, & Ziyu Xiao. (2018). Research on technical architecture and key technologies of cellular vehicle networking. Telecom Engineering Technics and Standardization | Telecom Eng Techn Stand, 31(6), 6.

Hasch, J., Topak, E., Schnabel, R., Zwick, T., Weigel, R., & Waldschmidt, C. (2012). Millimeter-wave technology for automotive radar sensors in the 77 GHz frequency band. IEEE Transactions on Microwave Theory & Techniques, 60(3), 845–860.

Kaihang Zhang. (2020). Environment sensing and path planning based on camera and Lidar [D]. Southwest University.

Kenney, & J., B. (2011). Dedicated short-range communications (dsrc) standards in the united states. Proceedings of the IEEE, 99(7), 1162–1182.

Peden, M., Scurfiled, R., Sleet, D., Mohan, D., Hyder, A. A., & Jarawan, E. (2004). World report on road traffic injury prevention. World Health Organization.

Xuyang Dong. (2019). Automotive AEB System Simulation Based on ADAS Experimental Platform. Liaoning University of Technology.

Yao Wei, & Qingyang Wang. (2018). Analysis and development status of C-V2X Cellular Vehicle Networking standard. Mobile Communications, 42(10), 4.

Advances in Urban Engineering and Management Science – Khalil & Yang (Eds)
© 2023 The Author(s), ISBN 978-1-032-30426-7

Study on progress management based on building information modeling and plan-do-check-action cycle

Muhammad Faisal Khan* & Chunlong Nie*
School of Civil Engineering, University of South China, Hengyang, China

ABSTRACT: Building Information Modeling (BIM) technology enables engineering and technical personnel to correctly understand and efficiently respond to various Building Information through the integration of data and information models of buildings, and provides a basis for collaborative work for design teams and construction subjects including construction and operation units. It plays an important role in improving production efficiency, saving costs, and shortening the time limit. This paper takes the Hengyang High-tech future City project as the backing project, establishes a BIM model based on Glodon software, and adds time and capital factors to form the BIM5D model. Plan-Do-Check-Action Cycle (PDCA) schedule control based on the BIM5D model is studied. The research results show that the BIM model is conducive to the visualization and overall management of schedule control, and improving the efficiency of management. This study can be used as a reference for similar projects.

1 INTRODUCTION

1.1 *Put up the problem*

My Research Plan is to solve the problem with the help of BIM and PCDA to monitor the progress of the construction project, and this passage allows the least possible project improvement and management with the prospective of BIM technology in a Project. In this BIM technology, all the tasks and goals are updated with time, new drawings, tasks, goals, and achievements.

1.2 *General methods for progress management*

Project management methodology is the collection of different ideas and logic to get the best plan to control and develop the construction project successfully. The Project management makes a platform to achieve every construction goal according to the desired plan. The methodology is the way to achieve and perform the activities systematically (Ge & Zhu 2020; Qin 2020).

Here are some general methods (Chen et al. 2011; Guan et al. 2015; Heigermoser et al. 2019):

(1) CPM, Critical Path Method. In this method, all tasks are divided into different sequences and then the least time sequence is selected.
(2) Lean, in this method of management, we increase the value to the customers and decrease the waste of resources.
(3) Six Sigma, this method was created by Motorola, in the method, the production process is improved and defects are reduced.
(4) CCPM, critical chain project management.
(5) SCRUM, in this method, the team uses the software and makes the plan for 30 days, and the project is divided into 30 days, the bud.

*Corresponding Authors: faisalgul93@gmail.com and chlnie@usc.edu.cn

DOI 10.1201/9781003305026-35

1.3 *BIM technology*

For the application of BIM technology, GuanglianDA was used. GuanglianDA is very friendly to use, which has better visualization of construction activities, communications, uninterrupted flow of information, gives better understanding, automatically detects failures and faults, and saves time and costs. No other software is required, in this software, a 3d model is created, then the time and cost are calculated. and the progress can be managed according to the time (Lou et al. 2017).

1.4 *PDCA cycle*

PDCA cycle consists of 4 parts 1. Plan 2. Do 3. Check 4. Act (Rokooei 2015; Dmitriev et al. 2019).

(1) Plan. in this case, an activity is planned in BIM according to my expected time and output for the improvement of the project.
(2) Do. In this case, it is available to start work on that activity and complete that according to plan time.
(3) Check. In this case, it can be observed whether the activity is done in the planned time or not and I get my output according to the plan.
(4) Act. If my activity is done according to the planned time and I get my output also according to the plan, my planning in the BIM is correct.

This PDCA cycle will be repeated for all the activities.

2 PROJECT CASES

This Project is HI-Tech Future City, which is situated in Hengyang City, Hunan Province, China. Cultural innovation valley Project covers an area of 102.91 m^2, with a total construction area of about 262,000 m^2, and a total planned number of 1877 households, including 696 residential households, 1181 apartment households, and about 150,000 m^2 of the international commercial complex. The first innovative entrepreneurship service demonstration community in Hengyang city is created. This project is oriented to a shopping center and residential building. The project cost is 100 million yuan. The life of the building is 100 years. The building is stored and the time for construction is 36 months.

3 CREATION OF 3D MODEL

To start the BIM, the base step is to make the 3D model of the Project by GuanglianDA. The model should be very simple and easy to use, the file can be easily edited and updated. This model can be used by all types of operating systems, which can support 3D. The model should also meet international standards which are simple and can be improved. All the parts of the building, foundation, structure work, and walls are drawn. The model should not be separated from the building model. Furthermore, to stay off from the mixing and conflict of every floor plan, the axis of each layer is generated, while the GuanglianDA is allowed to draw by materials and their details sizes according to the actual design of the project and can be modified according to the actual project function (Doumbouya at al. 2016; Liu et al. 2016).

(1) Creation of the parts of the building. These parts include beam, column, slab, shades, roofs, stairs, elevators, etc. To draw these parts, designs on the wall are relatively similar, you can change them any time according to a given parameter but neglecting the project reference is not convenient without communication with the director.
(2) Integration of the model of the basement pipeline. Heavy construction of the underground pipeline integration is a very difficult and complex task. However, it's not only important but professionals are needed to complete this task. It is not completed by pipelines professionals only but civil engineers, electric engineers, water supply and drainage, and all other

professionals are required. It is not only for task completion but for modeling of 3D, and professionals are also required, so it will be easy to communicate, integrate, and create a platform.

There are some points for attention.

(1) The construction model is the base step for all models. The other parts, such as wall column floor stairs doors and windows are also should be made accurately. The size, shape, height, and thickness should be drawn according to the design and needs.
(2) The height structural beam is the base for all the work of consequent pipes, the Beam height should be measured accurately and not in an inclined direction. The other parts of the beam like bar bending, placing, and fixing should be accurate otherwise it will be useless to use BIM, the GuanglianDA is mostly used to create a 3D model. Attention should be given to time and cost control, and project management.

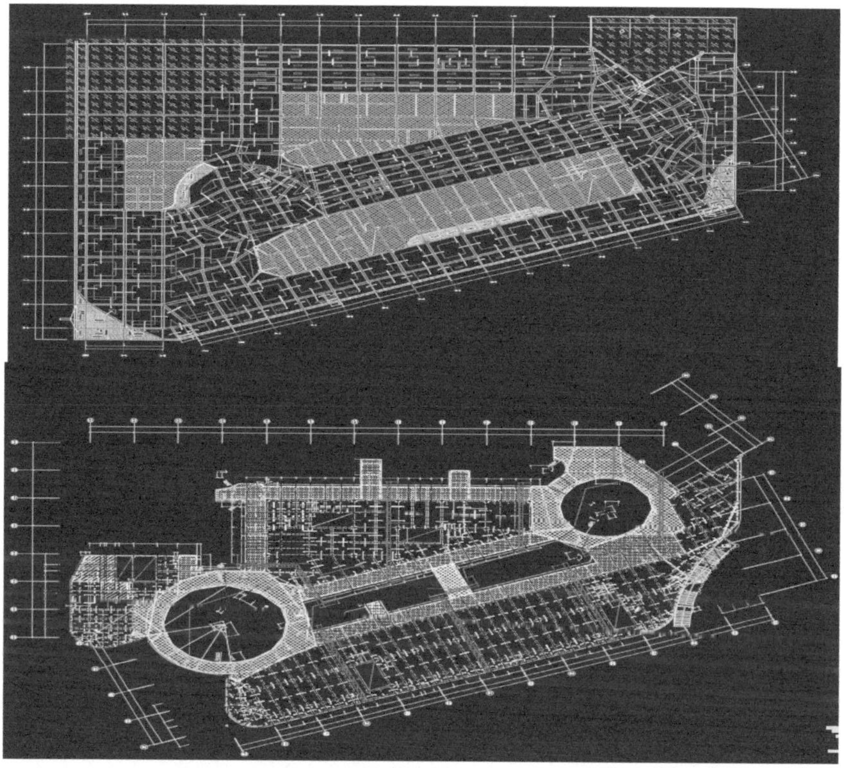

Figure 1. The map with the area of the project covered.

When the PC end of the BIM 5D task starts, for example, the concrete is poured, and all the departments are linked together for a large amount of concrete.

(1) Safety. Two days before the beginning of concrete, pouring the safety and security task will be notified, and received on mobile phones. The data regarding security and safety will be fed into the BIM 5D after completion.
(2) Technical Department. The technical department solves all the information and changes in the drawing to proceed with the change in the BIM 5D library
(3) Quality Department. This department creates the video practices of key nodes on-site to create different nodes and makes the 2-dimensional code which is created by BIM 5D at PC and delivered to all the cellphones. Thus, the site staff can easily get the new data just by scanning the QR Code of the BIM 5D APP on the Mobile Phone.

(4) Engineering Department. This department finalizes all the data and response forms from every department and manages all the tasks with details on BIM 5D mobile App.

4 PROGRESS MANAGEMENT BASED ON PDCA CYCLE

4.1 *Step of "Plan"*

The schedule is made as shown in Figure 3. We can't put the whole schedule in this paper. In this schedule, all the tasks are listed and configured according to time and tasks. Different types of tasks are denoted with different colors of lines. Each month and each building schedule are differentiated for easy understanding. This model is designed by integrating the data from the project team, owner, architects/engineers, equipment, resources, and other venders are associated with the building construction. In Figure 2, the BOQ of steel accords with the requirements of the time period. If there is any change detected during the installation, it is placed in this step then forward to the next step, which is the same as the collection of pipeline detection and reschedules.

4.2 *Step of "do"*

The schedule and drawing are directed to all the staff according to their responsibilities through the BIM5D program. The site supervisors supervise all activities according to schedule and drawing, all the changes in the redesign or reschedule or change detected during the installation, all of which are also directed by BIM5D. Similarly, we have the problem of the collection of pipelines, which also needs to be rescheduled.

A	B	C	D	E	F	G	H	I	J	K	L	M	N
Component type	Total (t)	Level	6	8	10	12	14	16	18	20	22	25	30
Column	1.39	Φ		1.39									
	178.853	Φ̶		71.105	2.355	0.796		6.257	13.278	60.205	3.524	21.333	
Embedded column	5.501	Φ̶		1.973				0.547	2.981				
Constructional column	1.179	Φ	1.179										
	8.646	Φ̶	1.952		6.694								
Masonry wall	14.677	Φ̶	14.677										
lintel	0.457	Φ	0.457										
	0.741	Φ̶		0.167	0.003		0.558		0.013				
Beam	11.505	Φ	9.558	1.947									
	504.149	Φ̶	8.036	56.3	23.554	60.954	5.529	19.831	18.693	42.619	78.142	189.979	0.512
Ring beam	0.13	Φ	0.13										
	24.472	Φ̶	1.663	6.094	9.494	7.221							
Cast-in-place plate	164.151	Φ̶	6.766	145.671	11.083	0.327	0.304						
Staircase	7.807	Φ̶		1.353		1.354	5.1						
Post-cast strip	0.784	Φ̶	0.028	0.631	0.011	0.114							
Custom lines	0.549	Φ		0.549									
Others	6.976	Φ̶		1.864	1.034	0.613	3.465						
Total (t)	15.21	Φ	11.324	3.886									
	916.759	Φ̶	33.122	285.158	54.228	71.378	14.957	26.636	34.965	102.824	81.666	211.313	0.512

Figure 2. Bill of quantity (example of steelwork).

Task Name	% Comple	Duration	Start	Finish
Prep & Release Pile & Foundation Detail Drawings of CWP Area	0%	15 days	06-02-18	20-02-18
Prep & Release of Pile & Foundation Detail Drawings for HRSG Feed Pumps	0%	45 days	14-04-18	28-05-18
Prep & Release of Pile & Foundation Detail Drawings for Fire Water Pump	0%	45 days	30-03-18	13-05-18
Prep & Release of Pile & Foundation Detail Drawings for GIS Area	0%	45 days	28-12-17	10-02-18
Prep & Release of Pile & Foundation Detail Drawings for GT-GSU Transformer	0%	45 days	26-03-18	09-05-18
Prep & Release of Pile & Foundation Detail Drawings for ST-GSU Transformer	0%	45 days	21-01-18	06-03-18
Prep & Release of Pile & Foundation Detail Drawings for Raw Water Intake	0%	45 days	24-02-18	09-04-18
Prep & Release of Pile & Foundation Detail Drawings for Oil Unloading Jettys	0%	60 days	11-12-17	08-02-18
Prep & Release of Pile & Foundation Detail Drawings for Main Gate Complex	0%	60 days	25-01-18	25-03-18
Prep & Release of Pile & Foundation Detail Drawings for Guard House &	0%	60 days	09-02-18	09-04-18
Mechanical - Detail Engineering	0%	75 days	03-12-17	15-02-18
Electrical - Detail Engineering	0%	75 days	28-11-17	10-02-18
C&I - Detail Engineering	0%	90 days	24-12-17	23-03-18
Preparation & Release of As Built Drawings	0%	90 days	08-05-19	05-08-19
Site Preparation & Test Piling	**0%**	**136 days**	**21-09-17**	**03-02-18**
Area Grading & Land Filling	0%	90 days	21-09-17	19-12-17
Test Pile Construction	0%	21 days	20-12-17	09-01-18
Load Test of Test Pile	0%	15 days	20-01-18	03-02-18
Electrical & Power Evacuation	**0%**	**636 days**	**21-09-17**	**18-06-19**
EHV Power Cable and Termination Kit	**0%**	**447 days**	**31-12-17**	**22-03-19**
EHV Power Cable and Termination Kit Ordering	**0%**	**96 days**	**31-12-17**	**06-04-18**
EHV Power Cable and Termination Kit - Release of Tech Spec, Enquiry	0%	26 days	31-12-17	25-01-18

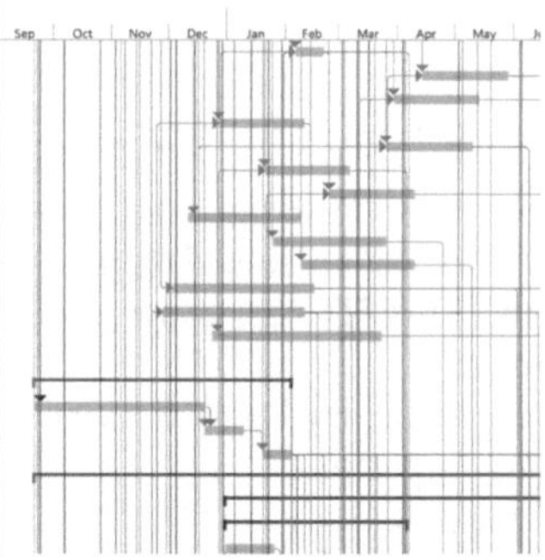

Figure 3. Part of the worksheet

4.3 *Step of "check"*

The site supervisors will inspect all the installation process which is going on or done every day and update the progress and will also send the pictures through the BIM5D app. The person will update all the completed tasks and count them. Meanwhile, the problem discussed in the paper of pipes lines collision will also be inspected according to the new schedule.

The collision of all pipes lines in the basement and the water supply line were at the main loop of the project time detected during the installation on-site which is redesigned by the GuanglianDA on the pointing of collision. Thus, in the new design, the collision disappeared and is installed according to the new design.

4.4 *Step of "action"*

After the completion of pipelines installation, the site supervisors will turn up the project progress conference and manage the time for the non-critical loop of the project. Other activities are scheduled according to the design schedule after lack the time. Thc quick detection of the collision of pipelines and quick response can avoid delaying reducing by 5 days and the total time difference of critical path reducing by 5 days. Thus, the cycle of the PDCA will complete its circle.

5 CONCLUSIONS

Project management is a crucial and tedious task in the construction industry. BIM is the systematic way to make sure the progress of a construction project. The knowledge about every task including time requirements is defended in BIM. The BIM should be active on-site without any mishap, the information should be well done by the technical team, and the complete task should be checked and confirmed by the inspection team as guided by PCDA. Also, uploading on BIM and saving for the onward used/ the saved data should be analyzed further by BIM, so the BIM will be able to make further decisions in this process called PCDA.

In this paper, the research of PCDA and project management by using BIM is conducted. The study of this paper is enabled to understand the workflow and project management by using BIM base project model. From the previous study, it can be concluded that BIM is a powerful tool for construction project management, better communication among the staff, and the people involved in the construction Project.

REFERENCES

Chen Lisong, Wang Xiaming, Qin Kang, etc. Research on Construction Progress Management Based on BIM and PDCA. Construction Technology, 2019.

Daniel Heigermoser, Borja García de Soto, Ernest Leslie Sidney Abbott, David Kim Huat Chua. BIM-based Last Planner System tool for improving construction project management, Automation in Construction, Volume 104, 2019:125–130.

Dmitriev AN, Vladimirova IL, Kallaur GY, Tsygankova AA. Approaches to Classifying Building Innovations while Implementing Information Modeling and Project Management. Journal of Engineering Science & Technology Review. 2019 Mar 1:12–19.

Doumbouya L, Gao G, Guan C. Adoption of the Building Information Modeling (BIM) for construction project effectiveness: The review of BIM benefits. American Journal of Civil Engineering and Architecture. 2016 Apr,4(3):74–79.

Ge Yi, Zhu Yongxiang. PDCA Law Is Studied in The Application of BIM Project Management in School-Enterprise Cooperation. Anhui Architecture, 2020:7–10.

Guan Wenxue, Lujiao. Based on PDCA Quality Control Research in The Process of BIM Architectural Design. China's Real Estate Industry, 2015(9):68–69.

Junying Lou, Jiang Xu, Kun Wang. Study on Construction Quality Control of Urban Complex Project Based on BIM, Procedia Engineering, Volume 174, 2017:35–42.
Liu J, CAI J. Construction Cost Control Based on PDCA Cycle. International Journal of Simulation–Systems, Science & Technology, 2016:17–19.
Qin Kang. Construction Progress Management Research based on BIM and PDCA (Master's thesis, Guangxi University).
Rokooei S. Building Information Modeling in Project Management: Necessities, Challenges, and Outcomes. Procedia-Social and Behavioral Sciences. 2015 Dec 2(210):87–95.

Advances in Urban Engineering and Management Science – Khalil & Yang (Eds)
© 2023 The Author(s), ISBN 978-1-032-30426-7

The control and improvement of anchorage construction quality under complex and deep soil filling conditions

Z. Li*, N. Jia, J.X. Tong, J.F. Li & B.B. Ma
State Key Laboratory of Building Safety and Built Environment, Beijing, China
Institute of Foundation Engineering China Academy of Building Research, Beijing, China

ABSTRACT: With the development of the national economy, more and more construction projects involve complex filling areas, the soil instability and physical and mechanical properties brought by complex filling conditions are not easy to predicate, the reliability of foundation pit and slope support and the stability of support in the excavation process is increasingly prominent. A slope staining project in Beijing has a thick soil filling and a large difference in engineering properties. This project aims to control and improve the construction quality of anchorage under the condition of complex soil filling. The project adopts two ways of double pipe split grouting and intermittent grouting. The results show that the double pipe split grouting can effectively improve the construction quality of anchorage, while the intermittent grouting has no obvious improvement on the construction quality of anchorage.

1 INTRODUCTION

Complex soil filling is particularly disadvantageous in foundation excavation and slope support because of its loose soil quality, low soil property index, and large porosity. In recent years, there are still many problems caused by complex soil filling in foundation excavation and slope pile-anchor support systems. Its main properties are as follows: the filling property is not uniform and the thickness is different; the filling property is not uniform and the thickness is different; complex soil filling is generally in the state of under consolidated, porosity is large, and water permeability is strong. At the same time, the material composition is complex and relatively soft, resulting in large deformation of soil; complex soil filling generally has higher compressibility and lower strength. At present, in view of complex soil filling, anchorage construction mostly adopts grouting, and once split grouting, compared with other soil layers, there is no special treatment, anchorage construction quality is not easy to guarantee.

In this paper, the construction quality of anchorage under the condition of complex soil filling is studied in combination with engineering examples, in order to improve the construction quality of this project and to provide a reference for similar projects to form a set of construction technology of anchorage construction under the condition of complex soil filling. This project adopts two methods of double-pipe split grouting and intermittent grouting to carry out multi-stage super-tensile tests respectively and the quality improvement effect of anchorage is analyzed comprehensively through test data.

2 PROJECT PROFILE

A slope retaining project in Beijing is located in Mentougou District, Beijing, which is a residential project with 4 floors above ground and 1~3 floors underground. The project is a permanent slope

*Corresponding Author: lz417130724@126.com

DOI 10.1201/9781003305026-36

retaining project, including supporting piles, permanent anchorages, many kinds of retaining walls, and other forms.

The proposed site was originally a gully with complex terrain and a large elevation difference. The site has been filled with soil, the thickness of the site is about 20m, and the backfilling time is about one year from the project construction. The site has been treated by the DDC method, and a permanent slope of 9-12m is formed after backfilling. On the southern center of the site, there are ancient trees and Bridges, while a cemetery and the current mountain are on the northern side.

The slope staining of this project adopts the pile-anchor support system, and 3~5 anchors are set according to the height of the slope, the anchor aperture is 200mm, anchor length is 20~38m, materials are 4 or 5 steel strands, construction angle is 35~40°. The standard value of tensile force is 290~570kN and the lock-off load is 390~770kN. The anchorage uses P.O 42.5 cement and the water-cement ratio is 0.5~0.55, the construction technology is casing follow up anchor drill hole, secondary split grouting.

3 ENGINEERING GEOLOGICAL CONDITION

The special rock and soil of this project are artificial fill soil and weathered bedrock.

Artificial soil filling mainly consists of two layers of miscellaneous fill layer and a gravel fill layer. The thickness of this layer is 0.5~20.0m. According to the investigation, it was formed when the site was flat and the backfilling time was about one year.

Miscellaneous fill layer: variegated, slightly dense, slightly wet, containing a lot of gravel, house slag, ash slag, plant roots, local stone, etc.

Gravel fill layer, variegated, slightly dense, slightly wet, mainly gravel, clay soil filling, locally filled with fine sand, containing a small number of brick chips, locally containing block stone.

Fully weathered sandstone layer: brownish-red and medium to fine grain structure. The core has been weathered into the soil, dry drilling can be drilled, hand pinch was broken, and local clip weathered gravel.

Strongly weathered sandstone layer: brownish-red ~ cyan gray, medium ~ fine grain structure, medium thick-bedded structure. The main components are feldspar, quartz, calculous cementation is the main, part is argillaceous cementation, drilling core is fragmented, local is a short column, and section length is 4 ~ 10 cm, with joint crack development.

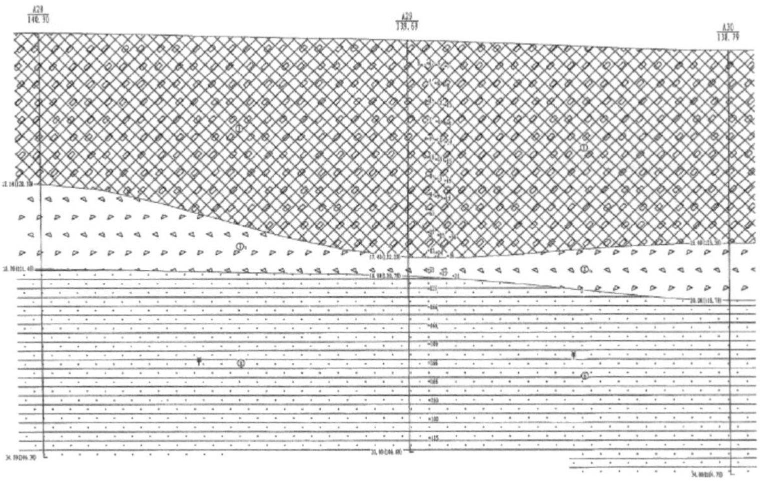

Figure 1. Section diagram of the typical geological.

According to the hydrogeological data collected, the highest buried depth of groundwater level in the plot in recent 3-5 years is about 20.0m, and the highest buried depth of groundwater level in history is about 10.0m. Under the influence of the atmosphere, upper stagnant water and bedrock fissure water may be formed locally in this area. The upper stagnant water is mainly supplied by atmospheric precipitation, which mainly occurs in the artificial soil layer and is mainly discharged by evaporation. The groundwater level changes irregularly and is greatly affected by atmospheric rainfall and human activities.

4 PLAN FORMULATION AND IMPLEMENTATION

4.1 *Plan formulation*

In recent years, with the development of urbanization in China, complex soil filling problems are often encountered in the process of old city reconstruction. The pile-anchor support structure has the advantages of good supporting effect, strong adaptability, simple construction, and good economic effect, which makes it widely used in deep foundation excavation and slope staining. Experts and scholars domestic and overseas have also conducted very in-depth and comprehensive research on this.

By referring to a large number of domestic and foreign literature, this paper understands the reasons for pile and anchor construction quality accidents caused by complex soil filling and explores better measures to improve the quality of anchor construction. According to the condition that the backfill soil is thick and the soil is loose and easy to collapse, two ways of double-pipe split grouting and intermittent grouting are adopted to carry out the experimental study.

(1) Double pipe split grouting technology is to add a second split slurry pipe on the basis of the original primary grouting and secondary split slurry, through the effect of two split slurries to compensate for the loss of cement slurry in loose soil, so as to enhance the holding force of anchor solid and the overall improvement of anchorage strength. The process flow is as follows:

Drilling emplacement → Correct position and Angle → Drill to design depth → Insert the anchor rod body → Filling cement slurry (with filling) →The first split pipe splits the cement slurry→The second split pipe splits the cement slurry→Waist beam production→Anchorage installation→The anchor cable tensioning→Anchor cable lock→The anchor head sealing

(2) Intermittent grouting is also to make up for the loose soil slurry erosion problem, considering loose soil by means of piecewise grouting shape on the fragile hole to form a better filling effect.

The intermittent grouting process is as follows:

Drilling emplacement→Correct position and Angle→Drill to design depth→Insert the anchor rod body→Filling cement slurry (stage-filling, multiple intermittent filling, multiple filling) →Split and fill cement slurry→Waist beam production→Anchorage installation→The anchor cable tensioning

In this test, ordinary Portland cement P.O 42.5 was used for mixing, and the water-cement ratio was 0.5~0.55. The first grouting pressure was 0.5 MPa, and the second high-pressure grouting was carried out after an interval of about 3~4 h, with the second grouting pressure being 2.0~2.5 MPa. When grouting, the grouting pipe should be inserted 250~500 mm away from the bottom of the hole, and the mixed cement slurry should be injected into the bottom of the hole and poured outward from the bottom of the hole. Due to the shrinkage of cement slurry, grouting should be done in time to make the hole section full grout.

Double pipe split grouting, the interval between two split pulp is also controlled in 3~4 hours. For intermittent grouting, each grouting interval shall not be less than 2 hours. The test blocks under the same condition were retained in the two test sites, and the test control was carried out according to the actual situation.

4.2 *Plan implementation*

In view of the thick backfill soil and loose soil which easy to collapse holes, two methods of double-pipe split grouting and intermittent grouting were adopted to carry out experimental research.

The fourth anchorage in the double-row pile area was selected for the double pipe split grouting and intermittent grouting tests. The anchorage is composed of four 17.8 (1860MPa) steel strands. The anchor fixed length is 15m, the anchor free length is 7m, the standard value of tensile force is 320kN, and the lock-off load is 430kN.

The maximum load in the acceptance test required by the drawing is 1.5Nak, and the maximum additional load in the acceptance test of this anchorage is 480kN. In the process of on-site tensioning lock acceptance, multi-stage overstretching test results will be verified according to the actual tensioning condition.

According to the anchor displacement calculation formula of CECS 22 ‘Technical specification for ground anchors’, the formula is as follows:

$$\Delta s = \frac{P*(Lf+\frac{La}{2})}{E*A} = \frac{480*(0.8*7)}{1.95*4*193} \sim \frac{480*(7+15/2)}{1.95*4*193} = 1.79 \sim 4.62\ \text{cm}$$

According to the provisions of this article 9.4.6, “The total measured displacement of tensile anchor under the maximum test load should exceed 80% of the theoretical elastic elongation value of the free segment length of the anchor under the load, and be less than the theoretical elastic elongation value of the sum of the free segment length of the anchor and 1/2 of the length of the anchor segment”. The displacement of the anchorage in this project is between 1.79~4.62 cm when the maximum added load value is 480kN.

(1) Test results of double pipe split grouting
In order to verify the test results, the maximum added load value is increased to 1.95Nak in the tensile process. The load-displacement curves and data are shown in the figure below.

Table 1. General anchorage Q-S data.

Q(kN)	48.0	96.0	192.0	288.0	336.0	384.0	432.0	480.0	528.0	576.0	624.0
s(mm)	0.00	6.26	10.80	15.23	18.51	21.87	25.32	28.77	32.28	35.93	39.70
Unloading rebound	14.00	/	/	/	/	/	/	/	/	/	39.70

Table 2. Double pipe split grouting Q-S data.

Q(kN)	48.0	96.0	192.0	288.0	336.0	384.0	432.0	480.0	528.0	576.0	624.0
s(mm)	0.00	3.53	6.61	9.79	12.10	14.48	16.97	19.59	22.30	25.10	28.18
Unloading rebound	12.05	/	/	/	/	/	/	/	/	/	28.18

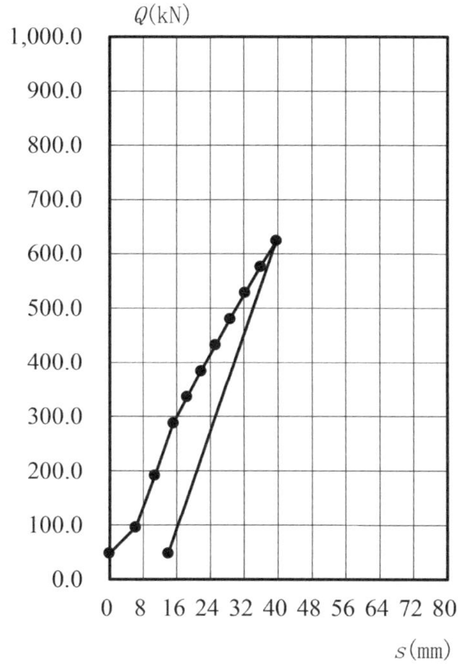

Figure 2. Q-s curve of ordinary anchor.

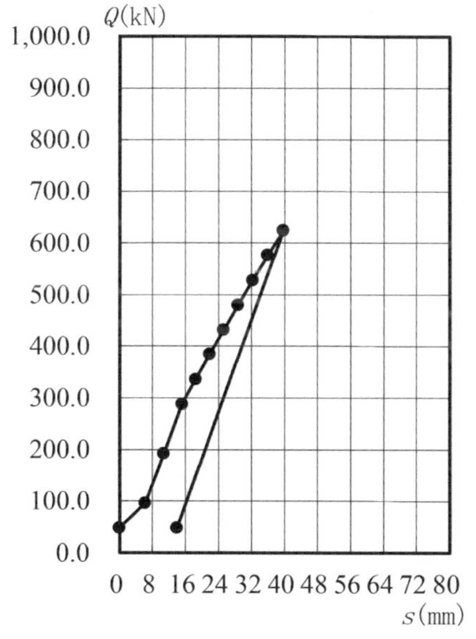

Figure 3. Q-S curve of double-pipe split grouting.

In order to verify the improvement effect of double pipe split grouting on anchorage, the Q-S curve of ordinary anchorage and double-pipe split grouting is compared and analyzed, the data is as follows.

Table 3. Anchor Q-S data difference table.

Q(kN)	48.0	96.0	192.0	288.0	336.0	384.0	432.0	480.0	528.0	576.0	624.0
Ordinary (mm)	0.0	6.26	10.80	15.23	18.51	21.87	25.32	28.77	32.28	35.93	39.70
Doubles (mm)	0.0	3.53	6.61	9.79	12.10	14.48	16.97	19.59	22.30	25.10	28.18
Differences (mm)	0.0	2.73	4.19	5.44	6.41	7.39	8.35	9.18	9.98	10.83	11.52
Ratio (%)	0.0	43.61	38.80	35.72	34.63	33.79	32.98	31.91	30.92	30.14	29.02

According to the above comparison data, it can be seen that the displacement and deformation of double pipe split grouting are about 30% less than that of ordinary anchorage. For loose soil and other cave-prone strata, the construction quality of anchorage can be improved by a double pipe split.

(2) Results of intermittent grouting test

The load-displacement curve and data of intermittent grouting bolt are shown in the figure below.

Table 4. Q-s datasheet of intermittent grouting.

Q(kN)	48.0	96.0	192.0	288.0	336.0	384.0	432.0	480.0
s(mm)	0.00	7.10	13.25	20.52	26.60	32.78	38.86	44.99
Unloading rebound	1.93	/	/	/	/	/	/	44.99

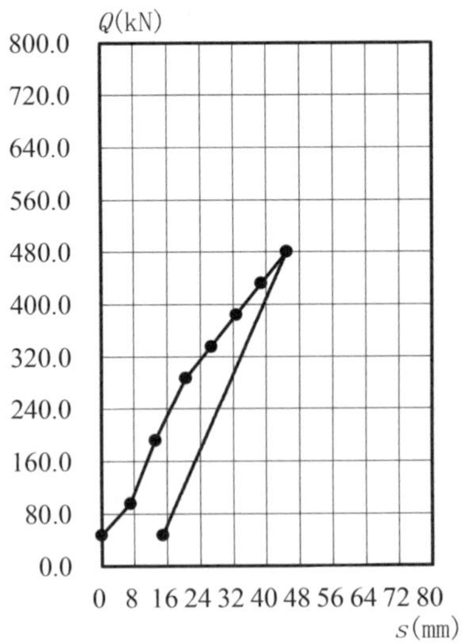

Figure 4. Q-s curve of intermittent grouting.

It can be seen from the Q-S curve that when the load is increased to 1.5Nak, the displacement is very close to the elastic deformation limit, indicating that the intermittent grouting does not achieve the desired effect. In order to avoid further damage to the anchorage, no further over tension test is carried out.

Through the review and analysis of the whole construction process of intermittent grouting, intermittent grouting is difficult to control in the actual construction process, which has high requirements on workers' operation level, time interval, grouting amount, and other technological parameters, and time cost is also high. At the same time, due to the existence of water and loose soil, it is easy to form a partition between sub-grouting bodies, which affects the overall anchorage effect.

5 CONCLUSION

In this paper, double pipe split grouting and intermittent grouting method are adopted to study the construction quality of anchorage under the condition of complex soil filling. The main conclusions can be summarized as follows:

(1) Complex fill has many problems in foundation excavation and slope staining because of its loose soil quality, low soil property index, and large porosity.
(2) For the pile-anchor support system in the complex soil filling area, the displacement and deformation of the double pipe grouting are about 30% less than that of the ordinary anchorage. For loose soil and other porous strata, the construction quality of the anchorage can be improved by double pipe grouting.
(3) For the pile-anchor support system in the complex soil filling area, intermittent grouting is difficult to control in the actual construction process, which has high requirements on workers' operation level, time interval, grouting amount, and other technological parameters, and time cost is also high. At the same time, due to the existence of water and loose soil, it is easy to form a partition between sub-grouting bodies, which affects the overall anchorage effect. The reinforcement effect of double pipe split grouting is better than intermittent grouting for complex and deep filling soil.
(4) In future work, it is necessary to strengthen the in-depth study of double-pipe split grouting under complex filling conditions.

ACKNOWLEDGMENT

This work was financially supported by the Youth Fund project of China Academy of Building Research Co., LTD (20201602331030027).

REFERENCES

CECS-22. (2005). Technical specification for ground anchors. S. Beijing: China planning press.
GB-50086. (2015). Technical code for the engineering of ground anchorages and shotcrete support. S. Beijing: China planning press.
GB-50330. (2013). Technical code for building slope engineering. S. Beijing: China building industry press.
Liu, L. (2011). Numerical analysis study on internal force and deformation of deep miscellaneous fill foundation with pile anchor supporting. D. Shandong Jianzhu University.
Wang, P.J. (2017). Study of supporting scheme for deep excavation engineering in miscellaneous soil-filling area. D. Taiyuan university of technology.
Wu, M.Z. (1993). Application of miscellaneous fill foundation. J. Chinese Journal of Geotechnical Engineering. 06, 103–109.

Advances in Urban Engineering and Management Science – Khalil & Yang (Eds)
© 2023 The Author(s), ISBN 978-1-032-30426-7

Civil engineering construction in the concrete structure construction technology research

Liu Wei*
Tianjin University of Technology, Tianjin China

ABSTRACT: In the field of civil engineering application of concrete structure, not only certain durability but also plasticity improvement are the focus, which has a vital significance. Therefore, in the process of civil engineering construction, attention should be paid to the construction of the concrete structure, following the construction technical standards and requirements, ensuring the construction quality of concrete structure, effect, strength, and stability. It is needed to give full play to the advantage and value of advanced construction technology, for the construction of civil engineering buildings and the development of a solid foundation.

1 INTRODUCTION

The relevant concrete structure in civil engineering construction is very important, and its quality directly affects the construction effect of the building. Therefore, in the specific field of civil engineering construction, we should pay attention to the application of concrete structure construction technology, reasonable use of preparation technology, mixing technology, pouring technology, and so on. In the application of various technical measures in the process of ensuring the structure, strength, and stability, attempts should be made to fundamentally avoid prevention-related engineering quality problems and play the role of advanced technology advantages (Chen 2021).

2 CIVIL ENGINEERING CONSTRUCTION OF THE CONCRETE STRUCTURE

2.1 *Concrete structure plasticity*

Concrete structure plasticity is strong, mainly reflected in the composition of concrete materials including cement, sand, stone, and water and admixture mixing fusion image of gelling materials (Chen 2005). The construction process accords with the control of different strength grades and the dosage of the cementing materials, as shown in Table 1. When the concrete material under the condition of water sufficient flow to be arbitrary, and according to actual design and construction need to shape, buildings in the actual construction work in the process of casting and perfusion of operation process can be very smooth, to avoid the related problems in the process of casting, ensure the quality of molding speed and concrete structure, so as to achieve the objectives of the project construction (Fu 2021).

2.2 *The durability of concrete structure*

In concrete structure, inside the casting of concrete around the steel bar, steel bar with good tensile and pull-out characteristics, in the combination of steel and concrete in the structure of the entity,

*Corresponding author: lw18802296956@163.com

 DOI 10.1201/9781003305026-37

Table 1. The total amount of cementitious material used for concrete of different strength.

The strength grade of concrete	<C40	C40~C50	$\geq$ C60
Use total gelled material	400	440	440–500 (The pumping)

contains the advantages of steel and concrete. When after its construction meets the requirements of quality standards, the process of the actual application to maintain the integrity of the time is very long. Compared to other material structures, it is characterized by obvious advantages of high quality, long use time, good concrete structure durability, and low possibility of quality problems.

2.3 *Fire resistance of the concrete structure*

The construction of concrete structures has good fire resistance, the composition of concrete raw materials is mainly water, cement, sand, and gravel material are nonflammable. At the time of the fire, concrete can stop the fire from spreading, so as to improve the structure of the fire performance and safety performance. In addition, the concrete raw material sources with low price, low production, transport, and storage requirement are widely used in the construction industry of building materials.

3 THE INFLUENCING FACTORS OF CIVIL ENGINEERING CONSTRUCTION TECHNOLOGY OF CONCRETE STRUCTURE

3.1 *The configuration of concrete*

The strength of concrete is mainly affected by the proportion of various raw materials and the maintenance of the latter two aspects. Different types of civil engineering for the proportion of concrete raw materials have different provisions, so before making concrete, we must fully understand the actual situation of civil engineering, determine the requirements of concrete hardness, and then make concrete. And after the completion of production, the focus must be on the strength of the concrete testing, once it was found that its strength does not meet the specific requirements of construction, this concrete cannot be put into normal use. In fact, in the process of concrete production, some enterprises did not realize the importance of concrete hardness, concrete production process requirements are not strict and standard, and after the completion of concrete production, it disaccords with scientific and reasonable maintenance methods for conservation, resulting in poor conservation effect.

3.2 *Concrete crack*

The structural crack of concrete is one of the most common problems in civil engineering. The crack of concrete will adversely affect the safety of the whole project. There are many reasons for concrete cracks, which are mainly caused by two aspects. On the one hand, in the process of concrete pouring, its specific operation is not completely in strict accordance with the provisions, resulting in the quality of concrete does not meet the requirements of construction; on the other hand, there are problems in the process of curing concrete in the later period, and the way of curing is not scientific, resulting in cracks. At the same time, in the process of concrete mixing materials of hydration, cement can evaporate a large quantity of heat, but in the process of heating, cement evaporates the upper concrete barriers, leading to concrete internal large amounts of heat to send out, so the internal heat will be gathered in the concrete. To this end, the concrete inside and outside temperature difference is bigger, due to the influence of the temperature of concrete cracks.

In addition, in the concrete structure, we often use a large number of coal cinder or additives, which affect the quality of concrete structure crack.

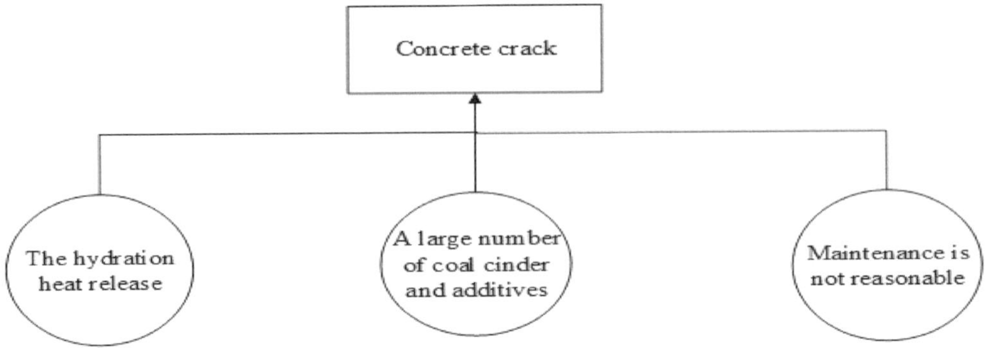

Figure 1. Factors of concrete cracks in civil engineering.

3.3 *Shrinkage of concrete structures*

Concrete belongs to compound material, which is composed of a variety of raw materials, but because of different kinds of raw materials, their physical characteristics are larger different. The cement contained in concrete, admixture, and so on can cause a certain influence on the performance and quality of concrete. For example, in the process of concrete pouring, the temperature outside the concrete will change substantially, and the moisture inside the concrete will evaporate in great quantities. Once the increase of moisture is too large, it is likely to lead to rapid contraction of the concrete structure, and then it is likely to cause the situation of cracks in the concrete. The shrinkage of the concrete structure can be controlled by adding mineral admixtures, but the proportion of mineral admixtures must meet the actual demand.

4 CIVIL ENGINEERING CONSTRUCTION IN THE CONCRETE CONSTRUCTION TECHNOLOGY MANAGEMENT

4.1 *The active adoption of materials processing technology*

In general, there are certain differences in the characteristics of different cement materials. At present, cement materials such as silicate and slag silicate are mainly used in civil engineering. The parameters of the hydration heat of silicate slurry with the strength of 43MPA can be controlled at 250 kilojoules per 1000 grams, and the hydration heat of silicate slurry with slag is 180 kilojoules per 1000gramst kilojoules. So, in the process of selecting materials, you can use slag Portland cement to prevent cracks because of high hydration heat. In the process of selecting aggregates, we should form a correct concept of consciousness, reasonable selection, mainly because of aggregate and transpiration structure quality, shrinkage, and forming effect. There is a direct relationship that if the bone cannot be guaranteed, the quality and strength of the material will affect the quality of the overall structure. Therefore, during the selection of aggregate, according to the actual situation of civil engineering, it is required to choose 0.38mm aggregate, reduce the heat of hydration, and ensure the construction quality of the structure (Wang 2021).

4.2 *Reasonable use of preparation and mixing technology*

Specific preparation work should be in accordance with the specific characteristics of the concrete strength, quality, and cost characteristics, such as application materials necessary to clear the mixture ratio. At the same time, in accordance with the clean sand material situation and situation of aggregate and cement, structure, quality factor, and so on, in the actual preparation process, the rational use of the corresponding technical measures can ensure that the material preparation proportion plan is at its best, and concrete indicators shall be observed in the process of configuration, as shown in Figure 2. In the strict control of the material under the condition of mixture ratio, the quality of preparation is guaranteed and the standard is met (Li 2021).

Second, we should focus on application-related stirring technology measures, combined with the slump of concrete structure, the mixing time and equipment capacity, and other characteristics, it is needed to choose high capacity during construction of mixer equipment, guarantee its length and stirring time, slump values match each other, to ensure the quality of the structure at the same time, the prevention of civil engineering construction, stability, and the quality of life.

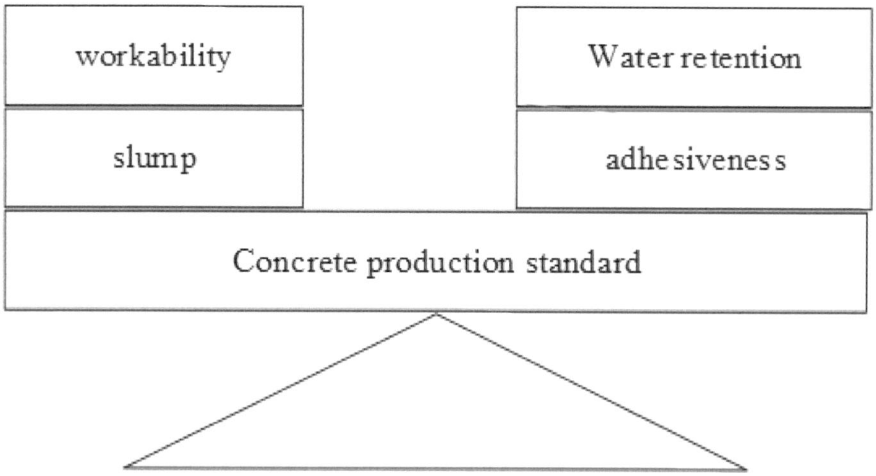

Figure 2. The basic requirement of concrete production.

4.3 *The reasonable application of casting technical measures*

Advanced technical measures should be actively applied in the concrete casting process to strictly manage pouring thickness and effectively control slump (Li 2021). First of all, in the process of setting thickness, research should be carried out in accordance with the structure type and parameter information processing. If it is to carry on the mass of the structure of the casting, jump warehouse can be used, which first will be divided into many small parts, large structure by using the method of interval casting, casting of stratification, each a small percentage of completion of pouring. Combining them into a large part can not only prevent the quality problems of the post-pouring strip but also ensure that the quality of the structure meets the standard. Secondly, in the pouring process of the shear wall structure, the thickness of the initial stage should be controlled at 50 mm, and then it is combined with the standard depth of 60 cm for layered treatment. During the concrete construction of slab structure and beam structure, the initial pouring thickness shall be controlled at 100mm, and continuous and systematic construction activities shall be carried out according to the standard of 200mm. Finally, the reasonable slump control process should also be in accordance with the characteristics and requirements of the structure, with a focus on setting the parameter information. For example, in the process of carrying out the pouring of a high-density reinforced concrete structure, there is a need to strictly control the slump, so it is controlled at 60 mm. In the process of casting the structure without reinforcement, the slump should also be controlled at 20 mm, so as to ensure the quality of the engineering structure. In addition, in the civil engineering beam structure and slab structure pouring treatment, it is necessary to strictly control the slump, so that it is controlled at 40 mm, so as to prevent and avoid the quality defects and stability of the structure.

4.4 *The active adoption of quality detection technology*

In order to ensure that the structure of the related quality is in conformity with standards between engineering, quality problems, or defects, a timely grasp structure should actively use quality testing technology measures to ensure the quality standard at the same time and the prevention of quality defects from the root. First of all, the focus should be put on the application of advanced technology for structural testing internal defects, quality problems, clear judgment about the material, the structure with and without quality problems, testing whether the structure is a crack phenomenon, cavitation, damage phenomenon and so on. Through the ultrasonic testing technology measures, related technical personnel can use low-frequency ultrasound equipment, access relevant spring back data value, grasp the quality problem of the structure, and defects, and make technical personnel can better develop related repair work.

Secondly, in the process of concrete test, measures should be made to perfect the relevant inspection technology scheme and model, unify quality test standards and content, use scientific means for quality management and control, fully display the various aspects of check function and advantage technology, the testing work of the relevant effects, guarantee the quality of concrete structure conforms to the standard requirements, so as to meet the current basic standards of quality management and control (Wang 2020).

5 CONCLUSION

From what has been discussed above, we can see in the civil engineering construction process that the quality of concrete will exert a certain influence on the quality of civil engineering. At the same time, the quality of concrete for the whole building construction plays a crucial role. So, in the relevant field of civil engineering construction, technical measures should be adopted, and for the concrete structure, great attention should be paid to the rational use of mixing technology, preparation technology, casting technology, quality testing technology, etc., to ensure the engineering quality and maintain the stability of the structure, building structure and construction requirements.

REFERENCES

Chen Chuangqun. Configuration and Application of High Strength Concrete [J]. Jiangxi Building Materials,2005(01):21–23.

Chen Yuan. The construction technology of concrete structures in civil engineering construction analysis [J]. China's residential facilities, 2021 (02) 108–109

Fu. Chunmei Civil building materials of high-performance concrete materials and green materials [J], 2021 (16)31–32.

Li Qiang. Research on Construction Technology of Concrete Structure in Civil Engineering Building [J]. China Building Metal Structure,2021(08):20–21.

Wang Meng. Construction Technology analysis of Concrete Structure in Civil Engineering Building [J]. Residential,2021(01):81–82.

Wang Wei. Civil engineering construction in the construction technology of concrete structure analysis [J]. Journal of Sichuan cement, 2020 (10) 46–47.

Advances in Urban Engineering and Management Science – Khalil & Yang (Eds)
© 2023 The Author(s), ISBN 978-1-032-30426-7

Analysis of the structure safety of block 0 with tower crane

Xiaolei Li*, Junlong Zhou*, Huijie Ma & Penglin Xie
China Construction Sixth Engineering Bureau Co., Ltd., Tianjin, China

ABSTRACT: In order to study whether the stress state of block 0 with the tower crane meets the requirements during construction. This paper analyzes the safety of block 0 structure of cable-stayed bridge with tower crane and establishes a 3D solid model of block 0 for numerical simulation by finite element method. The stress states under various working conditions and combined loads are calculated, and the tensile stress of block 0 structure under the action of the tower crane is analyzed. The results show that the maximum tensile stress at the connection between tower crane column foot and base foundation is 6.9MPa, exceeding the limit value by 119%. The tensile stress in the base foundation decreases rapidly along the horizontal direction and depth direction. The tensile stress in block 0 structure meets the limit value requirements, and the safety of block 0 structure meets the requirements under tower crane working load.

1 INTRODUCTION

The Cable-stayed bridge has excellent span performance and good stiffness, and the cable-stayed bridge has been widely used in highway engineering and railway engineering. The tower crane must be set up in cable tower construction. When the cable-stayed bridge crosses a deep mountain canyon and navigable water area, the bridge deck is as high as the ground (or river bed) or the foundation treatment is difficult, so it is not convenient to set up the traditional floor tower crane. In fact, it is a convenient, fast, scientific, and economic scheme to fix the tower crane on block 0 of the cable-stayed bridge. In this paper, the influence of the tower crane on the safety of the block 0 structure is analyzed.

2 ENGINEERING SITUATIONS

2.1 *Description of bridge structure*

The main span of the extradosed cable-stayed bridge is (100+165+100) m and a rigid frame system, with a total length of 365m. The main girder is a cantilevered prestressed concrete continuous box girder. The main tower is a herringbone tower. The lower main pier is a double limb thin wall pier and the foundation of a bored pile.

The main girder section adopts the single box and triple cells with an inclined belly board, a three-way prestressed system, and the concrete grade of C55. The width of the bridge deck is 28.75m, the cantilever length is 6 m, the beam height at the pedestal is 6.5m, and the middle beam height across is 3.5m. The thickness of the main beam roof is 28cm, and the thickness of the bottom plate at the pedestal is 100cm. The thickness of the inclined belly board varies by 50cm, 60cm, and 70cm from the mid-span pedestal. The middle belly board is straight with a thickness of 50cm.

The cross beam of block 0 corresponds to the double thin-wall pier, with a thickness of 1.6m.

Limited by the construction site, the tower crane is arranged on the roof of block 0. In order to ensure the construction safety and the structural quality of block 0, the finite element method is used to analyze and check the calculation.

---

*Corresponding Authors: 2514663099@qq.com and junlong1215@163.com

DOI 10.1201/9781003305026-38

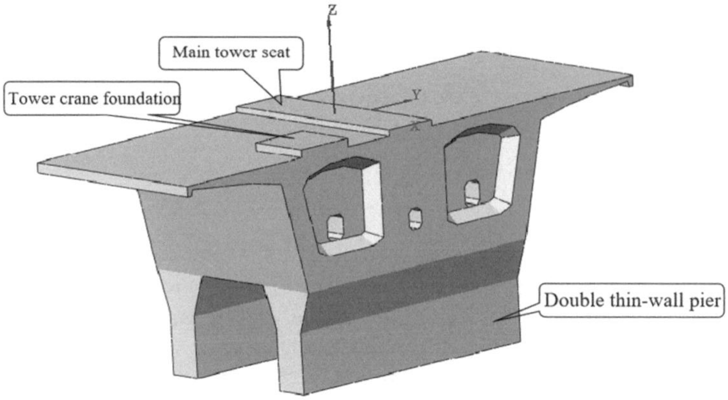

Figure 1. Model of block 0.

2.2 *Layout of tower crane*

The tower crane model is QTZP160, the rated lifting moment is 160t·m and the maximum lifting weight is 10t. The fixed tower crane's maximum lifting height is 45.5m and the height of the main tower is 38m. The tower crane for construction is installed on the roof of block 0, which is located on the north side of the main tower. The center of the tower crane is offset 3.5m from the longitudinal bridge of block 0, and 5.5m from the transverse bridge. The tower crane foundation adopts the fixed foundation with embedded support legs, located above the belly board and diaphragm plate. The dimensions of concrete foundation: 3m (length) $\times$3m (width) $\times$0.4m (thickness).

2.3 *Calculation model*

Midas FEA3.7.0 was used for structural calculation, and the solid model was established according to the construction drawings. The model includes blocks 0, a tower crane foundation, a partial double thin-walled pier, and the main tower seat. For block 0 and tower crane foundation, a whole complete model is established, consisting of roof, bottom slab, belly board, flange plate, and the joint hunched. The tower crane foundation is a rectangle concrete, for 3 m (length) * 3 m (width) * 0.4 m (thickness). For double thin-wall piers and the main tower, only the stress affect area is established, the double thin-wall pier extends 5m from the bottom of block 0, and the main tower seat is built from the top of block 0. The tower seat is set along the longitudinal bridge with a width of 2.75m and a height of 0.3m. The main pier, main beam, and tower crane foundation material are C55 grade concrete. The model mesh is tetrahedral. The control indexes include size, shape coefficient, and optimization coefficient, and the mesh size converges empirically. The boundary condition is a consolidation constraint on the bottom of the double thin-wall pier.

2.4 *Design load*

Load conditions include structure weight, construction loading, and tower crane loading. The structure weight includes the weight of the model structure and the additional load of the main tower. The weight of the model structure is automatically loaded by defining the density of concrete materials, and the weight of the main tower is transformed into surface load applied on the main tower seat. The construction loading is 4 kN/m^2 according to the *Load code for the design of building structures* and applied to the top surface area except for the tower and tower crane foundation. The tower crane loading is selected according to the technical parameters provided by the manufacturer. The force exerted by the tower crane on the foundation in the working state and non-working state is shown in Table 1.

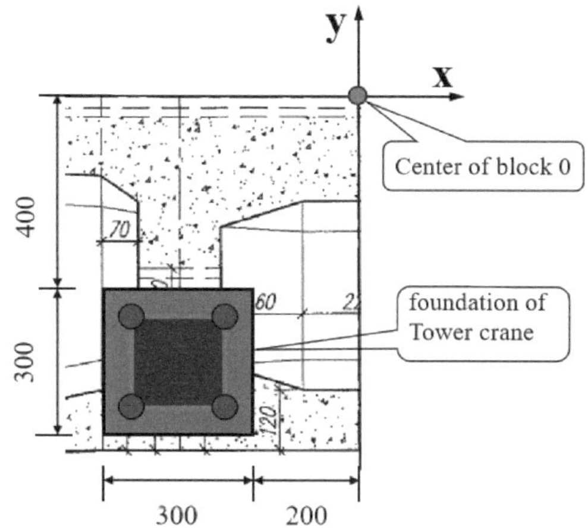

X - longitudinal direction, Y - transverse direction.

Figure 2. Plane position of tower crane (unit: cm).

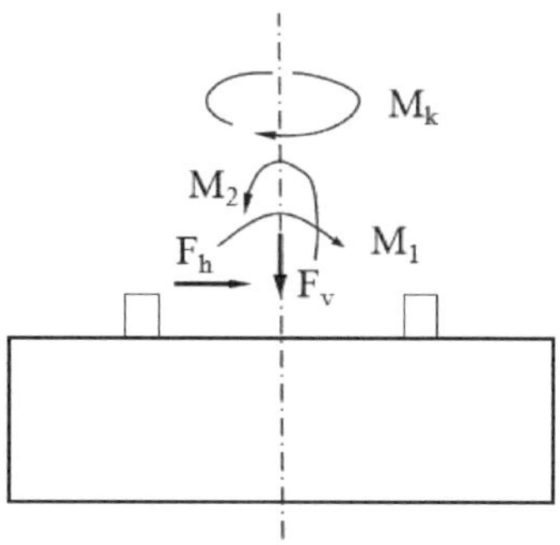

F_v -- vertical load on the foundation, F_h -- horizontal load on the foundation
M_1 -- transverse bridge bending moment on the foundation
M_2 -- longitudinal bridge bending moment on the foundation
M_k -- torque on the foundation.

Figure 3. The loading on the tower crane foundation.

Table 1. The loading on the tower crane foundation.

Condition Loading	F_v (kN)	F_h F_h (kN)	M_1 (kN·m)	M_2 (kN·m)	M_k (kN·m)
Working state	878	50	1948	1800	605
Non-working state	718	106	2790	0	0

2.5 *Load combination*

According to the *Load code for the design of building structures*, load combination is carried out according to the serviceability limit state and the ultimate limit state. The partial coefficients for the loads are shown in Table 2:

Table 2. The partial coefficients for the loads.

State (Combination) Loading	Structure weight	Construction loading	Tower crane loading
Serviceability limit state (Nominal combination)	1.0	1.0	1.0
Ultimate limit state (Fundamental combination)	1.3	1.5	1.5

3 RESULTS AND DISCUSSIONS

3.1 *Results and discussions of the whole model*

The calculation results include the different working states and the corresponding loading combination. The longitudinal direction normal stress, transverse direction normal stress, and vertical direction normal stress of block 0 are shown in Figures 4–7. The tensile stress is positive and the compressive stress is negative.

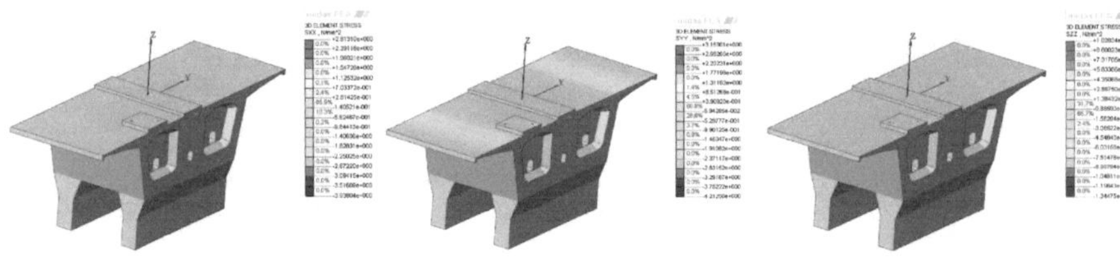

(a) Longitudinal direction stress. (b) Transverse direction stress. (c) Vertical direction stress.

Figure 4. Normal stress under a fundamental combination of tower crane working state.

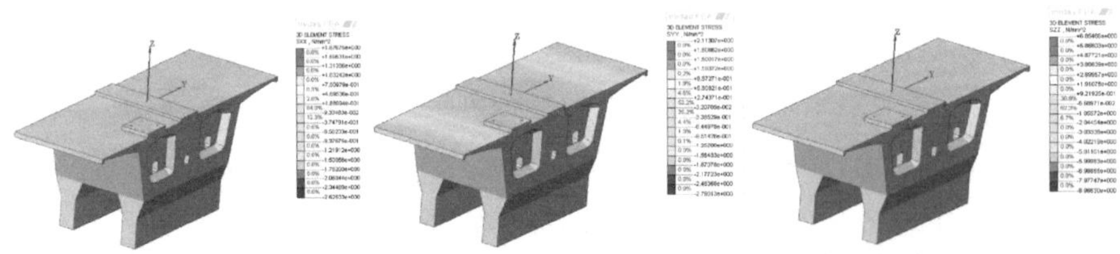

(a) Longitudinal direction stress. (b) Transverse direction stress. (c) Vertical direction stress.

Figure 5. Normal stress under a nominal combination of tower crane working state.

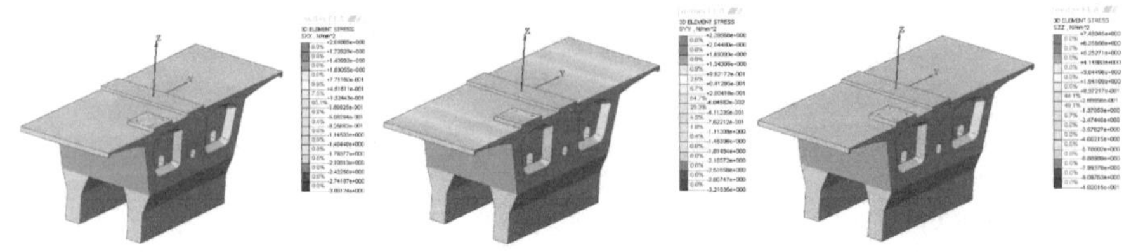

(a) Longitudinal direction stress. (b) Transverse direction stress. (c) Vertical direction stress.

Figure 6. Normal stress under a fundamental combination of tower crane non-working state.

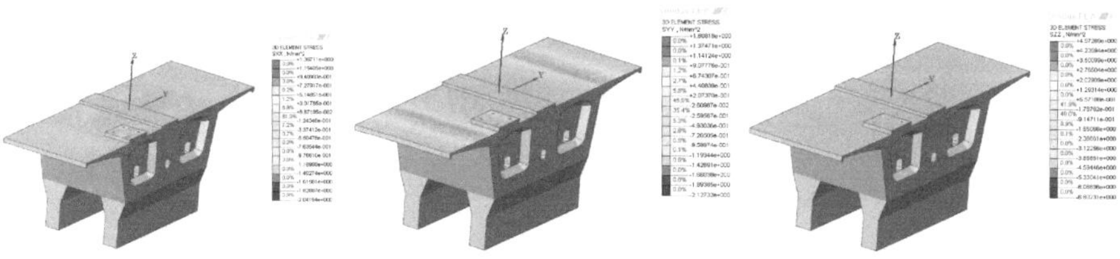

(a) Longitudinal direction stress. (b) Transverse direction stress. (c) Vertical direction stress.

Figure 7. Normal stress under a nominal combination of tower crane non-working state.

Table 3. Three-direction normal stress under various conditions.

NO.	Conditions	Longitudinal direction (MPa)	Transverse direction (MPa)	Vertical direction (MPa)
1	Fundamental combination of working state	−3.9~2.8	−4.2~3.1	−13.4~10.3
2	Nominal combination of working state	−2.6~1.9	−2.8~2.1	−8.9~6.9
3	Fundamental combination of non-working state	−3.06~2.04	−3.2~2.4	−10.2~7.5
4	Nominal combination of non-working state	−2.04~1.37	−2.1~1.6	−6.8~4.9

According to the normal stress cloud diagram of each working condition, the normal stress of block 0 in the working condition of the tower crane in all directions is greater than that in the non-working condition of the tower crane. Therefore, the working condition of the tower crane should be taken as the control condition to study its influence on the structural safety of block 0. The maximum compressive stress in the vertical bridge direction of block 0 under the fundamental combination of working conditions is 13.4MPa. The standard value of axial compressive strength of C55 concrete is 35.5MPa. According to article 7.2.8 of the *Code for Design of Highway Reinforced Concrete and Prestressed Concrete Bridges and Culverts*, the compressive stress shall not be greater than 0.7 times the standard value of the axial compressive strength of concrete: -13.4≤24.85=0.735.5. The results show that the compressive strength of block 0 under the fundamental combined load effect meets the requirements.

The vertical bridge tensile stress is the largest in the triaxial normal stress of block 0 under the nominal combination of working conditions. The maximum tensile stress occurs at the connection between the tower crane foundation and the underground joint of the tower crane. The maximum tensile stress of the vertical bridge is 6.9MPa, the maximum tensile stress of the transverse bridge is 2.1MPa, and the maximum tensile stress of the longitudinal bridge is 1.9MPa. The standard value of the axial tensile strength of C55 concrete is 2.74MPa. According to article 7.2.8 of the *Code for Design of Highway Reinforced Concrete and Prestressed Concrete Bridges and Culverts*, the tensile stress shall not be greater than 1.15 times the standard value of the axial compressive strength of concrete, that is, 3.15MPa. The results show that the tensile stress in transverse and longitudinal directions meets the requirements of the code, while the maximum tensile stress in the vertical direction does not meet the requirements of the code, exceeding the limit value by 119%.

According to the stress cloud diagram of the model, the maximum vertical tensile stress appears at the connection between the column foot of the bottom section of the tower crane and the foundation and spreads around and deep from the top of the tower crane foundation. The tensile stress diffusion is analyzed in the next section.

3.2 *Results and discussions of the tower crane foundation*

The results in the previous section show that the tensile stress of block 0 exceeds the design allowable value. Now, the local analysis of the tower crane foundation and the influence plane area range and influence depth of column foot is carried out to verify whether block 0 is affected.

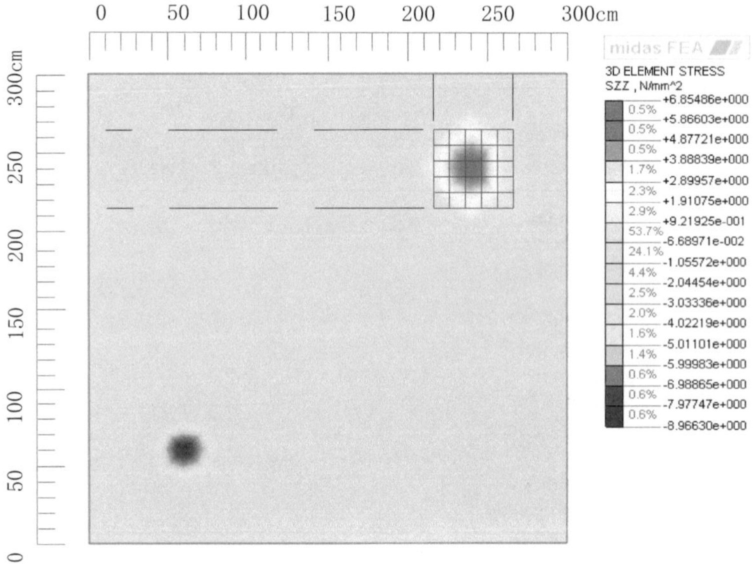

Figure 8. Plan of vertical direction stress under a nominal combination of tower crane working state.

The plan of vertical direction tensile stress under a nominal combination of tower cranes' working state shows that the horizontal diffusion range of tensile stress at the column foot is about

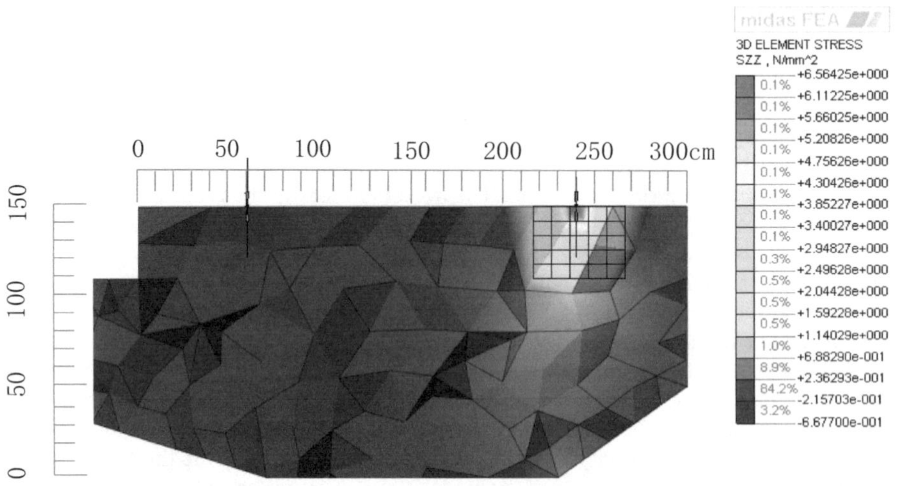

Figure 9. Profile of vertical direction stress under a nominal combination of tower crane working state.

50cm × 50cm, and it decreases uniformly in a circular shape around the center of the column foot. The maximum tensile stress at the edge of the range is 1.9MPa, and the stress state outside the influence range is compressive stress.

The profile of the vertical direction stress under a nominal combination of tower cranes' working state shows that the vertical diffusion range of the tensile stress at the column foot is about 50cm × 40cm, and uniformly decreases in a semicircular manner around the center of the column foot. The maximum tensile stress at the edge of the range is 1.6MPa, and the stress state outside the influence range is compressive stress.

The analysis of the local tensile stress at the foot of the tower crane shows that the distribution area of tensile stress is 50cm (transverse) × 50cm (longitudinal) × 40cm (vertical). It can be seen that the distribution area of tensile stress is only in the tower crane foundation and has not diffused to block 0, so there is no tensile stress in the main structure, and the tower crane has no influence on the safety of block 0.

4 CONCLUSION

Based on the results and discussions presented above, the conclusions are obtained as below:

(1) The results show that the compressive strength of block 0 under the fundamental combined load effect meets the requirements.
(2) The results show that the tensile stress in transverse and longitudinal directions meets the requirements of the code, while the maximum tensile stress in vertical direction does not meet the requirements of the code, exceeding the limit value by 119%.
(3) The distribution area of tensile stress is 50cm (transverse) × 50cm (longitudinal) × 40cm (vertical) in the tower crane foundation, and the tower crane has no influence on the safety of the block 0.

In this study, the beneficial effect that is reinforced and prestressed in block 0 resistance to tensile stress is not considered, and the calculation result is rather conservative. In the next research work, the reinforced concrete and prestressed effect should be considered. To verify the feasibility of a larger tower crane based on block 0, the lifting demand of bridge construction should be met.

ACKNOWLEDGMENTS

This work was financially supported by CSCEC "Study on construction technology of superstructure of long-span cable bearing bridge" (Grant No. CSCEC-2021-Z-30).

REFERENCES

Du Pingan. Basic Principle of Finite Element Mesh Division [J]. Machinery Design & Manufacture, 2000(1):3.
Industrial standards of the People's Republic of China. JTG3362-2018 Code for Design of Highway Reinforced Concrete and Prestressed Concrete Bridges and Culverts[S].China Communications Press Co., Ltd.2018
National Standards of the People's Republic of China. GB50009-2012 Load code for the design of building structures[S]. Beijing: China Architecture Publishing & Media Co., Ltd. 2012.
Sun Lingfu, SONG Guizhen, HOU Yuanping. (2015)Mechanical performance analysis of new crane assembly foundation based on ANSYS[J].Journal of Taiyuan University of Technology.46(05):583-586+591.
Wang Mao, Gao Fei, Tan Shibing, et al. Construction technology for combining temporary tower crane foundation with permanent basement floor [J]. China Harbour Engineering, 2019, 39(12):5.
Yin Xinsheng, Huang Chunhua. (2017) Tall building construction tower crane based on ANSYS calculation[J]. Building Structure.47(S1):1185-1187.
Zeng Pan. Finite Element Analysis and Application [M]. Tsinghua University Press, 2004.

Advances in Urban Engineering and Management Science – Khalil & Yang (Eds)
© 2023 The Author(s), ISBN 978-1-032-30426-7

Quality control of emery wear-resisting curing construction for underground garage

Ming Li*
Gansu Chengruju Real Estate Development Co., Ltd., Lanzhou, Gansu, China

ABSTRACT: Emery wear-resisting concrete surface is made up of a silicon carbide mixture of aggregate, special cement, and additives. Quality problems such as dust, delamination, hollow drums, and cracks may occur during construction. Therefore, a method to control the construction quality of emery wear-resistant maintenance for underground garages is designed. The first technical control chart of the construction group is designed, and then the underground garage construction quality management system is built. Thus, the quality control of emery wear-resisting curing construction is realized. The results show that the method of this paper can effectively control the construction quality and has effectiveness and application value.

1 INTRODUCTION

With the rapid development of the modern construction industry, the scale of industrial engineering is bigger and bigger, the concrete wear-resisting ground in large areas is more and more common, but the quality of the concrete is generally having a detrimental effect on the durability and aesthetics of the concrete floors. There are many quality points that need to be controlled (Wu et al. 2020). When laying concrete according to the elevation control line, the laser leveler is used to check the flatness of the ground, and a concrete vibration ruler is used to scrape and vibrate along with laying. If the elevation part (Bendikiene et al. 2019; Malikongwa et al. 2021; Zhang et al. 2019) is not reached, it is needed to timely use concrete material to fill, vibrate dense, and then rub flat with a wooden trowel, use mechanical pulping, mechanical inconvenient construction column, corner and other parts rub flat pat slurry with a wooden trowel. After the initial setting of the concrete, the initial paving work can be started (Peng et al. 2019). To ensure appearance quality, the emery should be applied evenly. It is not possible to shovel emery directly with a shovel, and a special screening tool should be used (Feng et al. 2021; Kumar et al 2020; Masoumi et al. 2019).

According to the construction requirements, in the division of warehouses, it is necessary to calculate the use of emery wear-resistant materials and the amount of spraying in each warehouse in advance (Park & Oh 2019). After spreading the material evenly on the surface of the concrete with a sieve, it should be immediately planned with a stick or float. Then it is processed by an optical disc finishing machine (Cigolini et al. 2020). After many times of processing, the color becomes darker and thicker, and the emery is ground and dispersed in combination with the foundation concrete slurry. To ensure emery ground quality, relevant personnel must ensure the approach of corundum and concrete quality. The construction of concrete must be conducted before the test, the optimum mixture ratio according to site construction requirements, the determination of the initial setting, and the final setting time of concrete. When arranging on-site construction, a corresponding proportion of concrete needs to be supplied. Relevant personnel shall ensure the consistency of concrete entry, appropriate aggregate gradation, and sand rate, control the amount of fly ash and avoid concrete segregation, bleeding, and other phenomena, so as not to reduce the concrete surface strength and ultimately affect the ground quality. Emery's spreading operation should be performed uniformly. It is needed to strictly control the emery spreading time, accurately

*Corresponding Author: 18893929889@sina.com

DOI 10.1201/9781003305026-39

insert the spreading operation, and use special sieving tools to sieve the diamond evenly to ensure the appearance quality.

2 DESIGN OF QUALITY CONTROL METHOD FOR EMERY WEAR-RESISTING CURING CONSTRUCTION

2.1 *Designing and constructing a group technical control diagram*

With the development of project quality control, a growing number of statistical analysis methods are applied to engineering quality control, thus greatly improving the project quality control ability. There are several kinds of conventional control methods, and the first is a histogram also known as the quality of distribution, rectangle, frequency histogram, or frequency distribution. The stability of the process can be judged according to the distribution shape of the straight square and the tolerance limit. The causality analysis diagram is also known as the Ishikawa diagram, fishbone diagram, or characteristic factor diagram. This method is an effective tool to start from the results and find the causes of quality problems in reverse order. It is characterized by the results and the causes as clues and connects them with arrows to indicate causality. The causality diagram is shown in Figure 1 below.

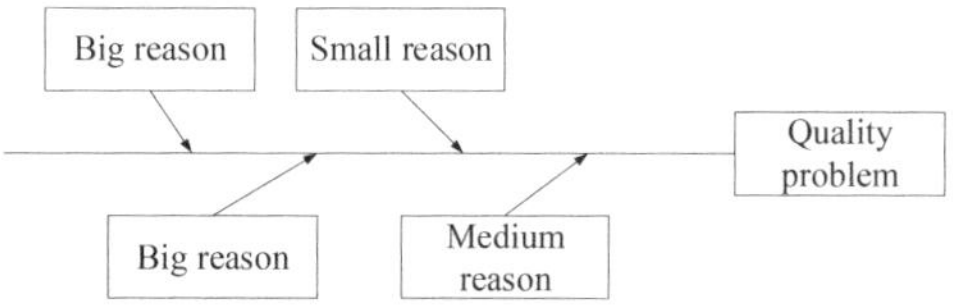

Figure 1. Schematic diagram of causality diagram.

As can be seen from Figure 1, looking for a general quality problem will lead to the identification of the main causes in five areas: personnel, method, machinery, materials, and environment. Then it is decomposed step by step until the concrete reasons are found and some countermeasures are given. The control chart provides production process parameters and time stability, which is important for the design and management of production processes. Only when the process is in a steady-state can the control line of the analysis control chart be extended as the control chart. Control charts are used to keep the process steady and prevent nonconforming products. The process of implementing the above control charts for analysis and control is actually a process of continuous quality improvement. The basic types of control charts are shown in Table 1.

Table 1. Basic types of control charts.

Distribution	Control chart
Normal distribution	Mean-range control chart
	Mean-standard deviation control chart
	Median range control chart
	Single value – moving range control diagram
The binomial distribution	Nonconforming product rate control chart
	Nonconforming drawing of nonconforming products
Poisson distribution	Unit nonconformity control chart
	Nonconforming control chart

As can be seen from Table 1, it is necessary to judge whether the process is in a stable state by applying the anomaly judgment criterion. If one of the following conditions is met, the process is considered to be in an abnormal state. According to the first spread of the thickness of the appropriate adjustment of local material, it is important to achieve a surface layer with uniform material. The operation should be carried out horizontally and horizontally in a uniform and orderly manner to prevent material aggregation. In this case, the original construction statistical data should be collected, and the random variables in this way are shown in Formula (1) below.

$$Y_n = \frac{y - \sum_{i=1}^{n} \mu_i}{\sqrt{\sum_{i=1}^{n} i\sigma^2}} \tag{1}$$

In Formula (1), n represents the number of random variables, μ_i represents the mean value of random variables, σ represents variance, and y represents the initial function. In this case, data need to be standardized, as shown in Formula (2) below.

$$Z_{ij} = \frac{X_{ij} - M_i}{T_i} \tag{2}$$

In Formula (2), Z_{ij} represents standardized data, X_{ij} represents sample measured data, M_i represents tolerance center, and T_i represents tolerance value. There are many methods for data transformation. Generally, the quality characteristic data with similar processes are taken as the transformation object and converted into the same distribution index, so as to expand the sample size. Usually, mean and standard deviation are selected as statistical variables, and the specific algorithm is shown in Formula (3) below.

$$y_{ij} = \frac{X_{ij} - \mu_i}{\sigma} \tag{3}$$

In the transformation of Formula (3), distribution parameters are needed, but they are unknown and cannot be directly applied. In this paper, specification boundaries are used for data conversion, and the conversion formula is shown in Formula (4) below.

$$Z_{ij} = \frac{X_{ij} - \frac{1}{2}(USL_i + LSL_i)}{USL_i - LSL_i} \tag{4}$$

In Formula (4), *USL* represents the upper specification limit and *LSL* represents the lower specification limit. Different processes in the same process family should be similar processes with different construction sites, so they should have the same source of quality variation. The data similarity test is to test whether the variance of normalized data after classification is consistent with the mean value.

2.2 *Constructing underground garage construction quality management system*

The construction of the underground garage project needs to comply with the constraint conditions of quality inspection and evaluation standards for construction and installation engineering issued by the central government of China, to ensure that the construction quality of the sub-project can achieve a series of standards of high-quality construction stipulated in the provisions, and achieve everything to meet the expectations of end-users. According to the actual situation of the construction progress of the underground garage project, the overall quality management objectives of the construction project can be decomposed, as shown in Table 2.

Table 2. Classification of project quality objectives.

Part of project Name	Project goals
Waterproof engineering	High quality
Building project	Qualified
Template engineering	High quality
Reinforcing bar engineering	High quality
Concrete engineering	High quality
Backfill engineering	Qualified

As can be seen from table 2, before building a quality system, the quality policy of the whole project should be clarified first. The construction of the underground garage project construction quality management system needs to take the concept of all parties involved in the construction process as the starting point, not limited to certain quality management involved in the participants. The framework of the designed construction quality management system is shown in Figure 2 below.

Figure 2. Construction quality management system framework.

Figure 2 shows that TQM – ISO quality management system in the project construction framework, where management organization of quality management system is the top priority of the underground garage project construction quality management, also serving as the underground garage project construction quality management system of the design personnel. The construction project quality management system is the key position. When the quality management plan and quality management guarantee are fully realized, the rest is the quality management operation process.

2.3 *To achieve the quality control of emery wear-resistant curing construction*

Emery wear-resisting curing construction quality control also need clear quality control points, in the columns, the pop-up building elevation line on metope, and elevation lines after the popup should carry on the review. If there are multiple teams, construction should adopt unified elevation control points, and apply the same instrument of measure, to prevent the uneven ground joint caused by instrument error. In addition, to ensure the slope and control the elevation error, elevation control is carried out by setting a gray cake with a line of 2~3m to ensure that the requirements are met.

To ensure the overall beauty of the wear-resisting floor, if the column foot, corner, and other parts have size deviation. Then, before construction, measures such as chiseling, painting, and repairing are taken to ensure vertical smoothness. Before construction, it is needed to adopt mechanical grinding, plane washing, cleaning, and other ways of carborundum floor cleaning up at the grass-roots level. It shall be ensured that has been cleaned up at the grassroots level surface dusting, empty drum and no loose, peeling phenomenon, at the grass-roots level surface smoothness, firmness, elevation and profile size comply with the design requirements, such as grassroots clean, especially paying attention to the column and a few a spot as yet untouched parts. After the floor is cleaned, it is needed to ensure that the thickness of the concrete surface layer is no less than 80mm. Before the formal construction, the plain cement slurry is used as a scraper slurry to enhance the bonding force of concrete and reduce the empty drum.

Through the level and ruler with inspection, setting up a good level, through the level to observe the ruler reading to check the surface of the angle steel flatness, if unevenness needs to be adjusted again, the distance between the level and the ruler is less than 20m. Concrete pouring needs to send to a person at any time on the spot to check the concrete slump, slump should be controlled between 110~140mm. Concrete vibration can not be leaked or polarized. In the strengthening process of concrete, seepage, segregation or excessive shrinkage caused by vibration should be avoided.

According to the working environment, climate, and other factors, it is required to confirm the initial setting time of basic-level concrete. Generally, it is appropriate to step on the foot to sink less than 5mm. The mechanical trowel with a disc installed is used to polish the concrete as a whole and gradually draw in from the periphery to play the role of leveling and slurry raising. The specified amount of about 2/3 of wear-resistant material is evenly spread on the surface of the concrete in the initial setting stage until the material absorbs the moisture in the concrete, the surface appears damp, the color becomes gray, the first leveling, using the disc polishing machine grinding so that the emery and the basic concrete slurry together.

The total station is used to set out the elastic line to the tire rack platform according to the position demarcated in the in-depth design drawing, and the positioning centerline is marked with a black line. After the steel truss members are assembled, the control points of each member are measured by the total station instrument again. For the bars that exceed the error, jacks, chain reversals, and electric hoists are used to adjust to ensure that the overall dimensional accuracy error of the steel truss after installation meets the design requirements. After the first unit slips in place and is positioned accurately, it can be positioned. After the positioning, the size of the docking site is checked again, and the construction size of the docking site of the sub-unit is guided to ensure the accurate docking of the two units to the maximum extent. A unified command system should be used during the docking to avoid collisions.

3 THE EXAMPLE ANALYSIS

3.1 *Overview and preparation*

The frame structure is selected as the workshop of construction engineering technology training base in a school. This area belongs to the coastal soft soil area, and the soil quality of the site is shown in Table 3.

Table 3. Foundation soil index.

The serial number	The soil	The thickness (m)	Natural gravity (KN/m^3)
1	Miscellaneous fill	0.456	17.5
2	Cohesive soil	0.964	18.1
3	Silt	9.464	15.2
4	Clay	3.046	19.5

As can be seen from Table 3, the total area of the workshop is about 1500 square meters, and the original ground is fine stone concrete. The cracking and settlement of the ground are obvious, and the relative height difference is as high as 9cm. At the same time, the ash on the ground is serious, which greatly affects the use of the workshop. In the workshop ground transformation, after the decision to maintain the original ground does not damage the case, then it is needed to pour emery wear-resistant concrete.

3.2 *Application effect and discussion*

The emery wear-resisting curing construction quality control method and the traditional construction quality control method designed in this paper are used to control the internal column cracking size after the two methods are controlled. The application effect is shown in Table 4 below.

It can be seen from Table 4 that after the application of the method designed in this paper, the cracking of the inner column is effectively alleviated, which proves that the method designed is effective and has a certain application value.

Table 4. Application effect.

Measurement point	The internal column cracking size after the emery wear-resisting curing construction quality control method designed in this paper (cm)	Internal column cracking size controlled by traditional emery wear-resisting curing construction quality control method (cm)
1	0.236	2.665
2	0.114	3.942
3	0.362	5.164
4	0.095	4.369
5	0.003	3.169
6	0.396	3.495

4 CONCLUSION

Emery's wear-resistant ground construction period is short, with the characteristics of integrity, leveling, wear resistance, corrosion resistance, impact resistance, and so on. Combined with the construction process and other related literature records, we should pay attention to the emery wear-resistant ground construction process control, otherwise, it is prone to empty drum cracking, uneven ground, water, appearance color, and other quality defects. Relevant personnel should consider the site, concrete supply, and other conditions, reasonable warehouse operations, arrangements, and personnel, to ensure the quality of emery ground forming. Emery concrete wear-resistant ground construction technology is being used in various site construction, especially in the application of high-strength site construction. For example, large mechanical equipment storage warehouse, or goods transport warehouse, now basically using emery concrete wear-resistant ground. This type of floor operation ensures that the floor is more resistant to abrasion, compression, and bearing capacity. Most importantly, this floor operation makes it easier to clean the floor, thus saving a lot of time. In the construction of emery concrete wear-resistant ground, it is necessary to carefully choose the materials for ground construction and strictly control the specific construction technology. Only in this way can high-quality construction of sandy concrete foundations be realized.

REFERENCES

Bendikiene R, Ciuplys A, Kavaliauskiene L(2019). Circular economy practice: From industrial metal waste to production of high wear-resistant coatings. Journal of Cleaner Production, 229(AUG.20):1225–1232.

Cigolini R, Gosling J, Iyer A, et al(2020). Supply chain management in construction and engineer-to-order industries. Production Planning and Control, 2020(3):1–8.

Feng S, Liu S, Jing L, et al(2021). Quantification of the Environmental Impacts of Highway Construction Using Remote Sensing Approach. Remote Sensing, 13(7):1340.

Kumar R, Antonov M, Holovenko Y, et al(2020). Erosive Wear Resistance of Nature-inspired Flexible Materials. Tribology Letters, 68(2):1-8.

Malikongwa K, Tlotleng M, Olakanmi E O(2021). Optimization of the wear resistance properties of laser cladding T-800 coatings. The International Journal of Advanced Manufacturing Technology, 114(1):481–496.

Masoumi M, Echeverri E, Tschiptschin A P, et al(2019). Improvement of wear resistance in pearlitic rail steel via quenching and partitioning processing. Scientific Reports, 9(1):7454.

Park H Y, Oh S H(2019). Design range of the damper of a T-stub damage-controlled system. Journal of Constructional Steel Research, 162(Nov.):105719.1–105719.15.

Peng F L, Dong Y H, Wang H L, et al(2019). Remote-control technology performance for excavation with pneumatic caisson in soft ground. Automation in construction, 105(SEP.):102834.1–102834.12.

Wu H, Zhao Y, Zhang D (2020). The automatic control method of construction quality of geological survey in geotechnical engineering. Arabian Journal of Geosciences, 13(17):836.

Zhang Q, Liu T, Zhang Z, et al(2019). Unmanned rolling compaction system for rockfill materials. Automation in Construction, 100(APR.):103–117.

Advances in Urban Engineering and Management Science – Khalil & Yang (Eds)
© 2023 The Author(s), ISBN 978-1-032-30426-7

Multi-factor monitoring method for construction of fabricated continuous rigid-frame bridge based on BIM

Jianwei Zhang*
Road & Bridge International Co., Ltd., Beijing, China

ABSTRACT: Aiming at the problem of the poor structural quality of fabricated continuous rigid-frame bridges, a multi-factor monitoring method of fabricated continuous rigid-frame bridge construction based on BIM is proposed. Based on BIM technology, the influencing factors and degree of rigid frame bridge construction quality are analyzed. The construction quality management evaluation algorithm is designed. The real-time monitoring of the construction of the prefabricated continuous rigid frame bridge is realized. Finally, experiments show that the multi-factor monitoring method of fabricated continuous rigid-frame bridge construction based on BIM has high practicability in the process of practical application and fully meets the research requirements.

1 INTRODUCTION

With the advancement of science and technology and the improvement of the industrial level, the curved continuous rigid frame bridge is widely used because of its small size and large gap under the bridge (Zhang et al. 2020). The structure of a curved continuous rigid frame adopts the consolidation form of a pier beam, which eliminates the constraint of support in support design and system transformation. However, with the wide application of curved continuous rigid frames, the uneven stress distribution of this kind of bridge has become a difficult point in the study. It is particularly necessary to study the design and construction of a curved continuous rigid frame, combined with different boundary conditions and load conditions on-site, and optimize the structure through finite element calculation, so as to make the alignment of the completed bridge meet the design requirements (Sang et al. 2020). The problems involved in this paper include boundary conditions for side piers and boundary conditions for non-pre-bending and pre-bending piers. This article also involves the calculation of the entire construction process and the calculation of the bridge model. The content of this paper also includes the influence of temperature, curvature, and other factors on the bridge. The above factors will produce large internal force and deformation (Dao & Lee 2020). The bending torsion coupling effect caused by the curved structure is reduced by means of pre-eccentricity of the support and pre-bending of the pier body. In this paper, through the analysis and research of various factors, the internal force and deformation of the corresponding bridge structure are obtained, and the results are used to guide the construction process, set and adjust the formwork elevation with the stage, and is gradually adjusted (Lee et al. 2019), which can make the alignment of the completed bridge meet the design requirements.

2 MULTI-FACTOR MONITORING OF RIGID FRAME BRIDGE CONSTRUCTION

2.1 *Factors affecting construction quality of rigid frame bridge*

The main purpose of monitoring bridge construction is to make the actual state of the bridge under construction as close to the ideal state as possible through monitoring. Given that the torque per unit length of the curved beam is m and the uniformly distributed vertical load is Q, three equilibrium

*Corresponding Author: bmw0932@139.com

DOI 10.1201/9781003305026-40

equations are needed to solve the bending moment (M), torque V and vertical reaction (s). A curved beam micro element is taken as d, and the arc length P from point a to point b is the positive direction. The natural coordinate system is selected. The arc length n direction specifies points a to b. its unit vector is consistent with the positive direction, n is the external normal direction, and the positive direction of the secondary normal vector B shall make the three vectors K, $L$and R form α right-handed system. Consideration is given to balancing in direction.

$$V = Q^{-}\left(V + \frac{M}{ds}\right) - Pn(a - b) \tag{1}$$

The moment is taken for the tangent of the point EI_W, and its balance condition is:

$$T = \left(L + \frac{M}{ds}\right)\cos d\alpha - \left(M + \frac{1}{R}\right)\sin ds - \cos\frac{R}{2} \tag{2}$$

For the straight beam bridge, it is not affected by the torsional load, while for the curved bridge, due to the influence of its curvature, the structure produces bending and torsion, and the two factors affect each other to form an affinity effect.

$$\frac{EI_W}{R}T/Pn(a - b) - \phi = E \tag{3}$$

BIM structural reliability is defined as the probability that the structure will complete the predetermined function within the specified time and under the specified conditions. It includes three aspects: safety, applicability, and durability. If the random variable of the structure is set to x, then$\phi$is called a functional function. Z > 0 represents structural reliability, and the reliability probability is represented by P, which is the structural reliability; Z < 0 indicates structural failure, and the failure probability is expressed. If U represents capacity and s represents load, BIM structural reliability is defined as the probability that the structure will complete the predetermined function within the specified time and under the specified conditions. It includes three aspects: safety, applicability, and durability. If the random variable of the structure is set to x, then Z is called a functional function. Z> 0 represents structural reliability, and the reliability probability is represented by P, which is the structural reliability; Z < 0 indicates structural failure, and the failure probability is expressed. S represents capacity and s represents load, then

$$P_r = E[Z/(R - S) - A] \tag{4}$$

$$P_f = E[Z/R - S - U] \tag{5}$$

Obviously, P_r and P_f are complementary:

$$P_r + P_f = 1 \tag{6}$$

According to the knowledge of probability theory, if the joint probability density function of the basic random variable $F_X(x)$ in the structure is $f_X(x)$ and the joint cumulative distribution function is dx, the failure probability of the structure is:

$$\eta = \int_{Z\leq 0} dxF_X(x) - P_r/P_f \int_{Z\leq 0} f_X(x)dx \tag{7}$$

Combined with the relevant BIM technical specifications, the construction loads in sectional construction are divided into three categories: structural dead load, construction live load, and other loads.

For the temperature difference effect of the bridge structure, it is mainly caused by environmental factors. Different regional lighting and climate make different environmental temperature changes, which brings various effects to the bridge structure. For static structures, expansion displacement is generated due to temperature change. This scientific research is coordinated through bridge deck expansion joints, bearing displacement, and flexible piers, while for statically indeterminate structure. When the displacement of its structure is limited, it will produce temperature secondary

Table 1. Classification of construction load.

Number	Classification	Load name	Function description
A	Structural	Block gravity	Static gravity,dynamic
B	dead load	Gravity deviation of block	3%–6% of block gravity
C		Unbalanced load of block	One block difference
D	Construction live load	Distributed live load	600pa
E		Concentrated live load	As the case may be
F		Dynamic live load	5%–15% of block gravity
G	Other loads	Prestress or cable force	Prestressing tendon shape, prestressing force
H		Influence of shrinkage and creep	Structural deformation and structural lining

internal forces, such as arch bridge, continuous rigid frame, and so on. Under the local temperature of sunshine radiation and temperature difference between day and night, the temperature secondary internal force or temperature secondary stress of the structure is the main reason for the cracks of the structure.

2.2 *Bridge structure risk identification based on BIM Technology*

The whole process of bridge risk assessment includes risk communication, risk definition, risk identification, risk estimation, risk evaluation, and risk control. Various risk assessment applications, although the specific methods may be different, this basic process should be obeyed. The basic process of risk assessment is shown in Figure 1 below.

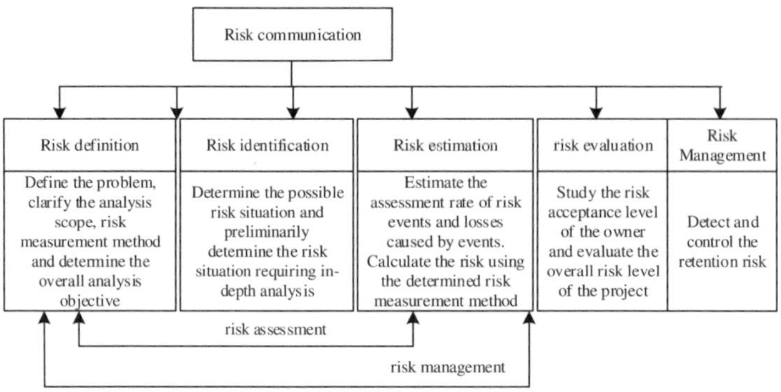

Figure 1. Basic process of risk assessment.

According to the classification of bridge construction risk loss, the purpose of risk loss estimation, and the establishment principle of the estimation model, combined with BIM Technology and various actual situations in the bridge construction stage, this paper establishes the overall model of construction risk loss estimation as follows:

$$I = g(v)(I_Z + I_J + I_R + I_H + I_S) \text{ -}N \tag{8}$$

Where: I_Z is the total risk loss. For a single risk situation, I_J is the direct economic loss, I_R is the indirect economic loss, I_H is the life loss, I_S is the environmental loss, and N is the social loss. In this paper, in order to intuitively display the influence degree of each parameter change on the structural alignment, it is expressed by the deflection change rate.

$$\theta = \frac{\sigma}{f'\omega} - K \tag{9}$$

Where: q is the deflection value after parameter change, D is the original theoretical deflection value; s is the whole construction stage of the bridge, and the stress control of the main bridge is mainly to control the maximum stress of the section. In this paper, it is expressed by the stress change rate.

$$\mu = \frac{Ds - \phi}{\varphi(q - v)} - \theta\Delta\sigma \tag{10}$$

Of which: φ is the stress value after parameter change; ϕ is the original theoretical stress value; $\Delta\sigma$ is the difference between the changed stress value and the theoretical stress. The elevation measurement is carried out on the section of the measuring points arranged in the bridge structure after tensioning and prestress. The difference between the measured and theoretical values is compared to determine whether the prestress tensioning value meets the design requirements.

2.3 *Realization of construction factor monitoring of rigid frame bridge*

This paper briefly analyzes and introduces the factors in the whole bridge state and construction process of the bridge, and finds out the differences, so that we can find better boundary conditions close to the actual bridge and simplify the model in the theoretical modeling. The closed-loop control method is mostly used for the construction control of long-span bridges with the complex structural systems, because the structural stress and linear error will gradually accumulate with the construction in the construction process, so once the construction error occurs, it must be strictly adjusted. Even if the bridge structure can not reach the ideal perfect state according to the design requirements, the structure can be close to the optimal state according to some optimal principles. The main feature of the BIM method is to estimate the calculation parameters required in the finite element model, and then modify the relevant parameters according to the actual construction situation of the project. For segmental construction with strong regularity of repetition, especially cantilever construction, there is a certain deviation between the relevant parameters in the model and the actual parameters.

3 ANALYSIS OF EXPERIMENTAL RESULTS

The specific methods and steps of sensitivity analysis of bridge structural parameters can be summarized as follows: it is attempted to change the design parameters in the finite element model to make them change to a certain extent, and then calculate the resulting structural displacement and internal force change. Finally, according to the influence degree of each parameter on the structural state, the main design parameters and the influence table of main structural parameters are as follows in Table 2.

The cast-in-situ side span section of the bridge is divided into 11# side span cast-in-situ section and 9# pier side span cast-in-situ section. At the same time, since the side span cast-in-situ section

Table 2. Main structural parameters of the bridge.

Main structural parameters	Girder linearity	Bending moment of the main beam	Axial force of the main beam	Main beam stress
Deadweight of the main beam	yes	yes	yes	yes
Elastic modulus of the main beam	yes	yes	yes	yes
Prestress related parameters	yes	yes	yes	yes
Effect of shrinkage and creep	yes	yes	yes	yes
Effect of temperature change	yes	yes	yes	yes

begins to prepare for the corresponding closure section after pouring, the tensioning of the side span closure bundle is not started until the closure. Therefore, when evaluating the side span closure section, its alignment is measured after the section pouring. The measured alignment mainly includes section elevation and beam section axis deviation. The specific evaluation results are shown in Tables 3 and 4 below.

Table 3. Linear measurement results of a cast-in-situ section of the pier-side span.

Construction beam section	Resurvey of pouring concrete elevation				Axis deviation error (mm)
	Measuring point	Theoretical elevation (m)	Retest elevation (m)	Error (mm)	
Left span of 9# pier 16# section (18e)	A	102.658	102.659	1	1
	B	102.782	102.785	3	
	C	102.965	102.968	3	
	D	100.265	100.267	2	
	E	100.325	100.326	1	
Left span of 9# pier 16# section (19e)	A	102.585	102.587	2	3
	B	102.658	102.659	1	
	C	102.865	102.868	3	

Table 4. Alignment measurement results of a cast-in-situ section of the pier-side span.

Construction beam section	Resurvey of pouring concrete elevation				Axis deviation error (mm)
	Measuring point	Theoretical elevation (m)	Retest elevation (m)	Error (mm)	
Left span of 9# pier 16# section (18e)	A	102.583	102.587	4	2
	B	102.925	102.922	−3	
	C	102.982	102.988	6	
	D	100.335	100.339	4	
	E	100.335	100.339	4	
Left span of 9# pier 16# section (19e)	A	102.625	102.629	4	−3
	B	102.745	102.748	3	
	C	102.873	102.879	6	

According to the field measured data, the elevation alignment and axis deviation values of the pier and 11# pier side span cast-in-situ section are mostly within the initial evaluation threshold proposed in this paper, and the elevation alignment score of the pier side span cast-in-situ section is 99.3 points. The evaluation results after the monitoring of other price indicators are 100 points, and the overall monitoring evaluation result is obtained. The measured stresses of 1.022mpaa, 3.245mpa, 2.978mpa, and 4015 mpa are taken as the original stress data, which are accumulated and compared with those before accumulation.

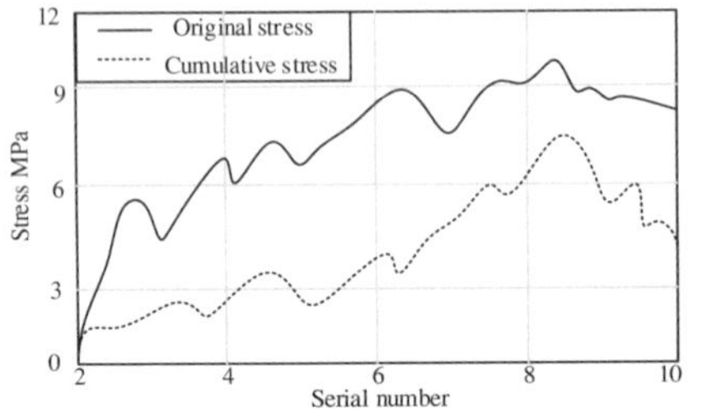

Figure 2. Comparison between accumulated stress and original stress of bridge.

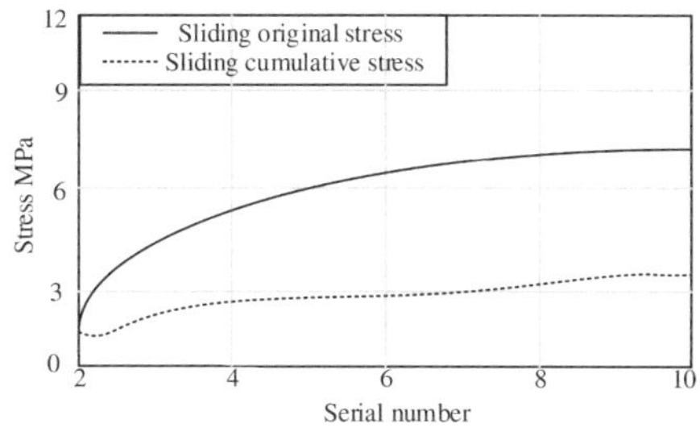

Figure 3. Comparison between cumulative stress and original stress sliding average treatment of bridge.

It can be seen that the original data has the nature of fluctuating up and down. It is difficult to see its internal law from the original data, and the generated data shows an obvious increasing trend. Although there is a turning point, it can be regarded as approximately obeying the law of exponential distribution. The data generated after smoothing is more regular than the data generated before smoothing, and the degree of turning is greatly reduced, which is closer to the exponential distribution curve.

4 CONCLUSION

Through different boundary conditions, such as different boundary conditions with or without side piers, the supports of straight piers without pre-deflection values and pre-deflection piers with pre-deflection values are set eccentrically. The finite element calculation in the construction process is completed at one time, and the structural behavior of the curved continuous rigid frame is analyzed under the influence of factors such as temperature, shrinkage, creep, and radius of curvature. It can be found that boundary conditions and load cases should be fully considered in the finite element calculation of curved continuous rigid frames. During the construction of the gondola, different boundary conditions and environmental factors will cause structural deformation and redistribution of internal forces. From the side span closing to the mid-span closing to the final bridge completion stage, it has undergone many system conversions. Thus, higher requirements are put forward for our on-site construction monitoring. Only through closer to the actual finite element calculation, on-site real-time monitoring of various factors, comprehensively considering the influence of various factors, and giving a reasonable formwork elevation, can the bridge meet the design alignment requirements after completion.

REFERENCES

C Zhang, AX Zhu, L Zhou, M Ch, T Qiu. 2020, Constraints for Improving Information Integrity in Information Conversion from CAD Building Drawings to BIM Model. IEEE Access, PP(99):1–1.
Dao N D, Lee D C. 2020, Modulation of Bidirectional AC/DC Converters Based on Half-Bridge Direct-Matrix Structure. IEEE Transactions on Power Electronics, PP(99):1–1.
IP Sang, SH Lee, A Almasi, JH Song. 2020, Extended IFC-based strong form meshfree collocation analysis of a bridge structure. Automation in Construction, 119(4):103364.
SH Lee, SH Song, HB Jo, HJ Ryoo. 2019, Solid-State Bipolar Pulsed-Power Modulator Based on a Half-Bridge Power Cell Structure. IEEE Transactions on Plasma Science, PP(99):1–7.

Advances in Urban Engineering and Management Science – Khalil & Yang (Eds)
© 2023 The Author(s), ISBN 978-1-032-30426-7

Study on bearing capacity of reinforced sleeve grouting connected concrete shear wall after fire

Jiahao Zheng, Junhua Li* & Jiawei Chen
School of Civil and Environment Engineering, Ningbo University, Ningbo, China

ABSTRACT: In order to study the bearing capacity of reinforced sleeve grouting-connected concrete shear walls after the fire, the finite element software ABAQUS is used to establish a numerical analysis model for the seismic performance of reinforced concrete shear walls connected by grouting reinforced concrete shear walls at room temperature and the seismic performance of reinforced concrete shear walls after the fire. The skeleton curve obtained from the simulation is in good agreement with the experimental skeleton curve. The rationality of the finite element model is verified. On this basis, the numerical simulation of reinforced sleeve grouting connected concrete shear wall after the fire is carried out, and the effects of fire time, axial compression ratio, and longitudinal stressed reinforcement diameter on the bearing capacity of sleeve grouting connected shear wall are analyzed. The results show that the fire does great harm to the bearing capacity of the reinforced sleeve grouting connected concrete shear wall. After 60 min and 120 min of fire, the peak load of the precast shear wall decreases by 10.94% and 40.59% respectively compared with that at room temperature; compared with the specimens with an axial compression ratio of 0.06, the peak loads of the specimens with an axial compression ratio of 0.12, 0.18 and 0.24 increased by 10.79%, 19.63%, and 27.24% respectively; compared with the specimens with the diameter of 14mm, the peak loads of 16mm, 18mm and 20mm specimens increased by 3.19%, 7.73%, and 12.17% respectively.

1 INTRODUCTION

In recent years, the research and application of prefabricated concrete buildings in China have gradually heated up. The prefabricated concrete structure can not only achieve the goal of building energy conservation and environmental protection, but also meet the development needs of housing industrialization. The prefabricated shear wall structure is a new building structure system (Huang, Tian 2010; Wang et al. 2016). The high temperature of the fire will degrade the material of the fabricated shear wall structure, which will have an adverse impact on the bearing capacity of the fabricated shear wall structure. However, at present, a lot of research has been done on the bearing capacity analysis of cast-in-situ reinforced concrete shear walls after a fire in China. Wang Y.Z. established the numerical analysis model of reinforced sleeve grouting connected concrete shear wall and studied the effects of shear span ratio and axial compression ratio on the seismic performance of grouting sleeve connected shear wall (Wang 2020). Zhang X.H. used ABAQUS to establish the numerical analysis model of sleeve grouting connected concrete shear wall and studied the mechanical performance of shear wall when the vertical stressed reinforcement is distributed in quincunx (Zhang 2020). There are few reports on the bearing capacity of prefabricated shear wall structures after the fire. Therefore, the numerical analysis model of reinforced sleeve grouting connecting concrete shear wall and reinforced concrete shear wall after the fire is established by ABAQUS. The simulated skeleton curve is in good agreement with the experimental skeleton curve, which verifies the rationality of the finite element model. On this basis, the numerical

*Corresponding Author: lijunhua@nbu.edu.cn

 DOI 10.1201/9781003305026-41

model of reinforced sleeve grouting connected concrete shear wall after the fire is established, and the effects of fire time, axial compression ratio, and longitudinal stressed reinforcement diameter on the bearing capacity of reinforced sleeve grouting connected concrete shear wall after a fire are analyzed, which provides a reference for the research on the bearing capacity of prefabricated shear wall structure after the fire to a certain extent.

2 MODEL VERIFICATION OF REINFORCED SLEEVE GROUTING CONNECTED CONCRETE SHEAR WALL AT ROOM TEMPERATURE

2.1 *Model overview*

The TW1 specimen tested by Peng Y.Y. (Peng 2010) of Qinghua University is used as the verified numerical analysis model. The size and reinforcement of the specimen are shown in Figure 1. The parameters are set by using the measured mechanical properties of TW1 specimen concrete, grouting material, and reinforcement in the document (Peng 2010). The measured cube compressive strength of concrete is 40.8Mpa, the measured cube compressive strength of grouting material is 76.6Mpa, and the measured mechanical properties of reinforcement are shown in Table 1.

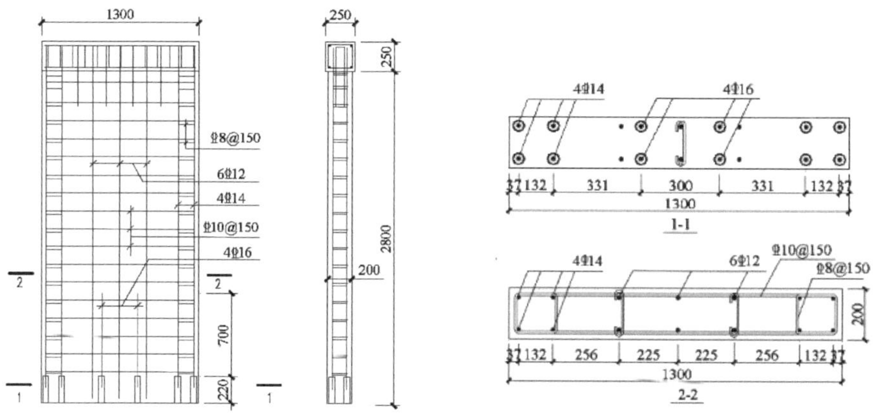

Figure 1. TW1 specimen size and reinforcement diagram.

Table 1. Measured the value of reinforcement strength.

d/mm	f_y/MPa	f_u/MPa	$\varepsilon_y/10^{-6}$
8	427.4	626.7	2137
10	483.2	658.8	2416
12	455.4	608.4	2277
14	458.3	625.8	2292
16	439.9	649.0	2200

2.2 *Unit type and material constitutive selection*

2.2.1 *Concrete*

The three-dimensional solid element (C3D8R) is selected as the concrete element type, the concrete material model is the concrete plastic damage model, and the concrete strength is the measured value of concrete cube compressive strength of 40.8Mpa. The stress-strain relationship and damage factor expression of concrete under uniaxial compression and uniaxial tension are determined based on the

Code for Design of Concrete Structures GB50010-2010 (Beijing: China Construction Industry Press, 2010). The elastic modulus of concrete is 32660N/mm^2, Poisson's ratio is 0.2, the concrete density is 2500kg/m^3, and the expansion angle is ψ take 38°. Eccentricity ε takes 0.1, the ratio of biaxial ultimate compressive strength to uniaxial ultimate compressive strength is 1.16, the ratio of the second stress invariant on the tensile meridional plane and the compressive meridional plane K_c = 0.667, and the viscosity parameter is 0.005.

2.2.2 *Grouting material*

The three-dimensional solid element (C3D8R) is selected as the grouting material unit type, the plastic damage model is adopted as the material model, the measured value of cube compressive strength of grouting material is 76.6Mpa, the stress-strain relationship and damage factor expression of grouting material are the same as that of concrete, the elastic modulus is 37660N/mm^2, the Poisson's ratio is 0.2, and the density is 2500kg/ m^3, and the definition of plastic damage model parameters of grouting materials in ABAQUS is the same as that of concrete.

2.2.3 *Rebar*

The two-node linear three-dimensional truss element (T3D2) is selected as the reinforcement element outside the sleeve, and the three-dimensional solid element (C3D8R) is selected as the reinforcement element inside the sleeve. The stress-strain constitutive relationship of the reinforcement adopts the bilinear model, and the strength of the reinforcement adopts the yield strength and ultimate strength of the reinforcement measured in the literature test (Peng 2010). The tensile and compressive elastic modulus are the same, taking $E_s = 2.06 \times 10^5$Mpa. Poisson's ratio is 0.3, and the tangent modulus of steel after yield is $E_t = 0.01E_s$.

2.2.4 *Sleeve*

The sleeve unit type is a three-dimensional solid unit (C3D8R). The grouting sleeve material adopts the ideal elastic-plastic model. According to JG/ T398-2012 *Grouting Sleeve for Reinforcement Connection* (JG/T 398-2012), the grouting sleeve material adopts 45# high-quality carbon structural steel, with a tensile strength of 600MPa, yield strength of 355MPa, and elastic modulus E_s= 2.1 × 10^{5}Mpa, and Poisson's ratio is 0.3.

2.3 *Contact settings*

The model adopts separate modeling, which does not consider the bond-slip relationship between reinforcement, concrete, and grouting materials, nor the bond-slip relationship between sleeve, concrete, and grouting materials. The reinforcement and sleeve are embedded in the concrete in the form of a built-in area. The inner side of the sleeve and the outer side of the grouting material are restrained by "Tie", and the reinforcement and grouting material in the sleeve are also restrained by "Tie". The end of the reinforcement outside the sleeve (T3D2) is built into the reinforcement inside the sleeve. The upper interface of the grouting material layer and the lower interface of the wall body are restrained by contact, and the tangential direction of the contact unit is Coulomb friction. The friction coefficient is 0.4, and the normal direction adopts hard contact. The lower interface of the grouting material layer and the upper interface of the ground beam are constrained by "Tie", and the lower interface of the loading beam and the upper interface of the wall body are constrained by "Tie".

2.4 *Boundary conditions and load settings*

The ground beam of the shear wall model is fully consolidated, and the top restricts the displacement and rotation outside the plane of the shear wall, and restricts the displacement and rotation in six directions. A reference point is set at the center of the top section of the loading beam. The reference point is connected with the top surface of the loading beam by coupling constraints. At the same time, as the loading point, constraints are imposed on the coupling node to restrict the out-of-plane displacement and rotation angle of the shear wall. The load condition is the combined action of vertical load and horizontal load. The application of load is divided into two steps: the first step is to apply axial force at the loading point on the top of the shear wall. The axial force is taken as 500kN according to the test and remains constant; the second step is to apply horizontal load at the loading point on the top of the shear wall, with the loading mode shown in Figure 2. In order to facilitate calculation and convergence, 50kN, 100kN, and 200kN controlled by forces are converted into 1mm, 2mm, and 4mm to realize the horizontal load loading mode controlled by displacement.

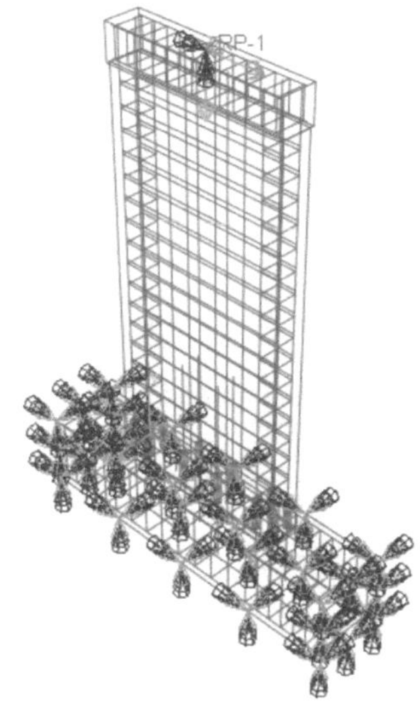

Figure 2. Loading mode of TW1 specimen.

2.5 *Model validation*

The above modeling method and constitutive relationship are used to establish the finite element analysis model, and ABAQUS is used for numerical simulation analysis. As shown in Figure 3, the skeleton curve obtained by finite element analysis is compared with the skeleton curve of the TW1 test in literature

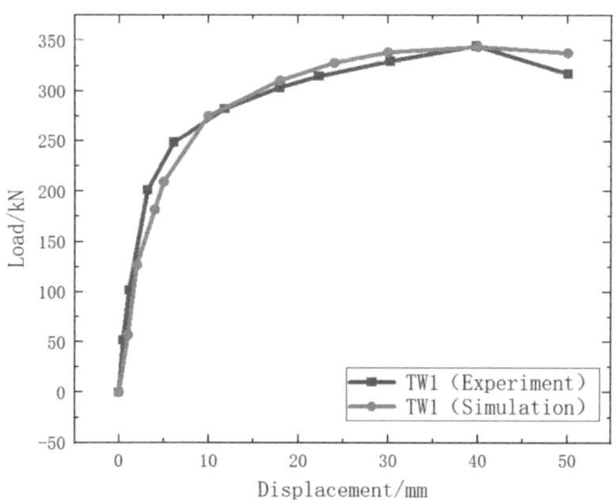

Figure 3. Comparison between TW1 simulated skeleton curve and test skeleton curve.

(Beijing: China Construction Industry Press, 2010). The comparison shows that the simulated skeleton curve is in good agreement with the experimental skeleton curve, which proves that the modeling method, parameters, and constitutive relationship in the finite element analysis model are more accurate, and verifies the rationality of the model.

3 MODEL VERIFICATION OF REINFORCED CONCRETE SHEAR WALL AFTER FIRE

3.1 *Model overview*

The Q5 specimen tested by Hu C.P. (Hu 2014) of Huaqiao University is used as the verified numerical analysis model. The size and reinforcement of the specimen are shown in Figure 4.

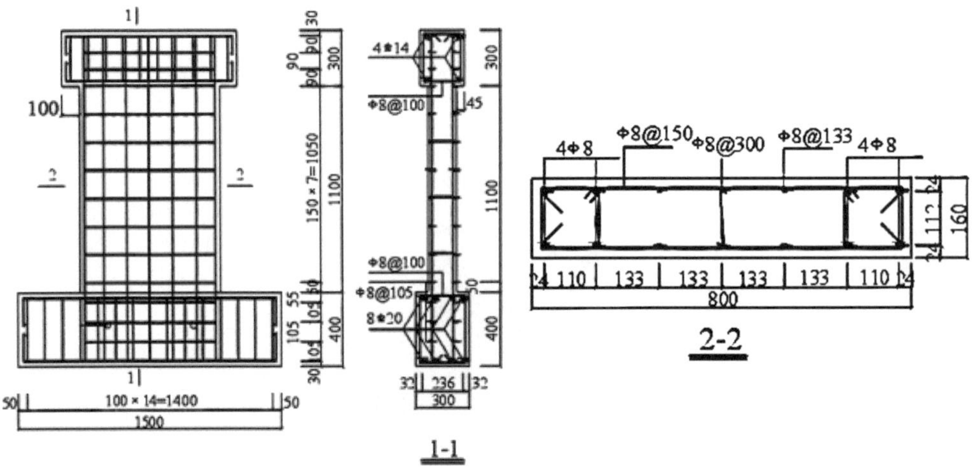

Figure 4. Q5 specimen size and reinforcement drawing.

The temperature rise curve of the Q5 test piece adopts the ISO834 standard temperature rise curve, and the curve expression is as follows:

Temperature rise section:

$$(t \leq t_h) : T_g = T_g\,(0) + 345\,\lg\,(8t + 1) \quad (1)$$

Cooling section:

$$\begin{aligned} &(t > t_h) : \frac{dT_g}{dt} = -10.417 && t_h \leq 30\text{min} \\ &\frac{dT_g}{dt} = -4.167\,(3 - t_h/60) && 30\text{min} < t_h \leq 120\text{min} \\ &\frac{dT_g}{dt} = -4.167 && t_h > 120\text{min} \end{aligned} \quad (2)$$

3.2 *Temperature field verification*

3.2.1 *Thermal parameters*

The thermal parameters of concrete adopt the thermal parameter formula proposed by Lie (Lie & Celikkol 1991).

Thermal conductivity W/(m ·°C):

$$\lambda_c = \begin{cases} -0.00085T + 1.9 & (0°\text{C} < T \leq 800°\text{C}) \\ 1.22 & (T > 800°\text{C}) \end{cases} \quad (3)$$

Product of specific heat and bulk density J/(m^3·°C):

$$\rho_c C_c = \begin{cases} (0.005T + 1.7) \times 10^6 & (0°C < T \leq 200°C) \\ 2.7 \times 10^6 & (200°C < T \leq 400°C) \\ (0.013T - 2.5) \times 10^6 & (400°C < T \leq 500°C) \\ (-0.013T + 10.5) \times 10^6 & (500°C < T \leq 600°C) \\ 2.7 \times 10^6 & (T > 600°C) \end{cases} \quad (4)$$

The unit weight of concrete is 2400kg/m^3.

The thermal parameters of reinforcement adopt the thermal parameter formula proposed in the European code (Li et al. 2006).

Thermal conductivity W/(m·°C):

$$\lambda_s = \begin{cases} 54 - 3.33 \times 10^{-2} T_s & (20°C \leq T_s \leq 800°C) \\ 27.3 & (800°C < T_s \leq 1200°C) \end{cases} \quad (5)$$

Specific heat J/(kg·°C):

$$C_s = \begin{cases} 425 + 7.73 \times 10^{-1} T_s - 1.69 \times 10^{-3} T_s^2 & \\ +2.22 \times 10^{-6} T_s^3 & (20°C \leq T_s \leq 600°C) \\ 666 - \dfrac{13002}{T_s - 738} & (600°C < T_s \leq 735°C) \\ 545 + \dfrac{17820}{T_s - 731} & (735°C < T_s \leq 950°C) \\ 650 & (950°C < T_s \leq 1200°C) \end{cases} \quad (6)$$

The unit weight of reinforcement is 7850kg/m^3

3.2.2 *Load and boundary condition setting*

According to literature (Hu 2014), the room temperature of the Q5 specimen during the fire test is 34°C, so the value of the predefined temperature field is set to 34°C. Q5 specimen is exposed to fire on four sides. For the four exposed surfaces of the wall body, the heat convection heat transfer coefficient is taken as 25W/(m^2·°C), and the comprehensive radiation coefficient is taken as 0.5W/(m^2·°C). The heat convection heat transfer coefficient of the backfire surface is taken as 9W/(m^2·°C), and the comprehensive radiation coefficient is taken as 0W/(M2·°C). In addition, the absolute zero value is set in the model as – 273.15°C, and the Boltzmann constant value is set as 5.67×10^{-8}.

3.2.3 *Verification and post-processing of temperature field results*

The temperature field model is submitted for the calculation to obtain the comparison between the simulated node temperature at the No. 1 thermocouple of the Q5 specimen and the test node temperature, as shown in Figure 5. The comparison shows that the simulated node temperature of Q5-1 is in good agreement with the test node temperature. Because the influence of water migration is not considered when the finite element model is established, there is no platform section in the simulated node temperature curve. The temperature field simulation results are post-processed with Python software, and the maximum temperature of each node is extracted as the initial condition for static field import to realize pushover numerical simulation. The maximum temperature of the Q5-1 simulation node is 301°C, the maximum temperature of the test node is 297°C, and the error is 1.33%. Therefore, the established temperature field model is reasonable.

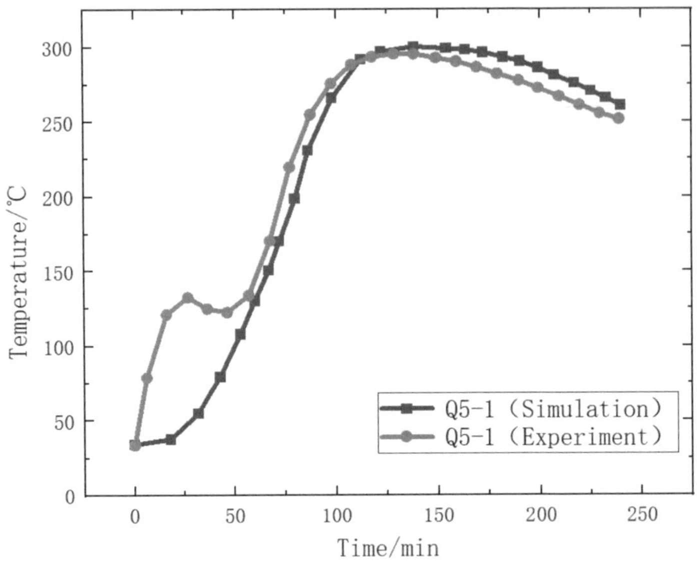

Figure 5. Q5-1 comparison diagram of simulated node temperature and test temperature.

3.3 *Pushover process verification*

3.3.1 *Element type and material constitutive*

C3D8R unit is used for concrete and the T3D2 unit is used for reinforcement. The plastic damage model is adopted for concrete, and Poisson's ratio is taken as 0.2. The elastic modulus of concrete after high temperature adopts the relationship suggested by Xu Y. (Xu et al. 2000):

$$E_{Tm} = \begin{cases} (100 - 0.0744T_m) \times 10^{-2}E & (0°\text{C} < T_m \leq 300°\text{C}) \\ (127.54 - 0.1662T_m) \times 10^{-2}E & (300°\text{C} < T_m \leq 700°\text{C}) \\ 0.11E & (700°\text{C} < T_m \leq 900°\text{C}) \end{cases} \tag{7}$$

The definition of plastic damage model parameters of concrete after the high temperature in ABAQUS is the same as that at room temperature.

The ideal elastic-plastic model is adopted for the reinforcement after high temperature, the Poisson's ratio is taken as 0.3, and the elastic modulus and yield strength are defined according to the recommendations of literature (Li et al. 2006; Wu 2003):

$$\begin{aligned} &E_{s,T_m} = (100.53 - 0.0265T_m) \times 10^{-2}E_s \; 20°\text{C} < T_m \leq 900°\text{C} \\ &\frac{f_{y,T_m}}{f_y} = \begin{cases} (100.19 - 0.0159T_m) \times 10^{-2} \; 20°\text{C} < T_m \leq 600°\text{C} \\ (121.39 - 0.0512T_m) \times 10^{-2} \; 600°\text{C} < T_m \leq 900°\text{C} \end{cases} \end{aligned} \tag{8}$$

3.3.2 *Boundary conditions and load settings*

Except for the introduction of the temperature field, the setting method of boundary conditions is the same as that at room temperature. According to the loading method of the document (Beijing: China Construction Industry Press, 2010), the vertical load is 866kN, and the horizontal force loading system is controlled by displacement angle. The amplitude of displacement angle is 1/800, 1/400, 1/300, and 1/200 respectively. Each level of displacement angle is loaded repeatedly once; when the displacement angles are 1/150, 1/125, 1/100, 1/75, 1/50, 1/35, and 1/25, the reciprocating loading cycle shall be carried out three times at each displacement angle. The displacement angle control is converted into displacement control according to the structure height to facilitate the parameter definition in ABAQUS.

3.3.3 *Result analysis and verification*

The model is submitted for the calculation to obtain the comparison between the simulated skeleton curve of the Q5 specimen and the test skeleton curve, as shown in Figure 6. The comparison shows that the simulated skeleton curve of the Q5 specimen is in good agreement with the test skeleton curve. The peak load obtained from the test is 305kN, and the simulated peak load is 296kN, with an error of 2.95%. It verifies the rationality of the modeling method, parameters, and constitutive model of this model.

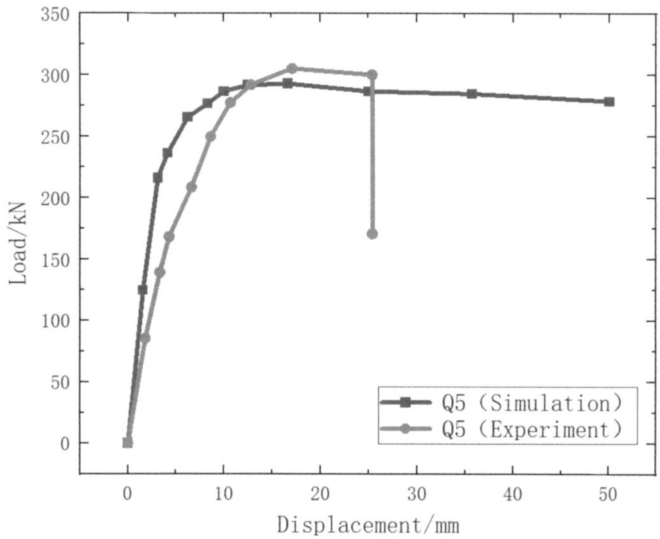

Figure 6. Comparison diagram of Q5 simulated skeleton curve and test skeleton curve.

4 BEARING CAPACITY ANALYSIS OF SLEEVE GROUTING CONNECTED CONCRETE SHEAR WALL AFTER FIRE

4.1 *Model and specimen parameter setting*

Based on the above-verified modeling method and constitutive model, ABAQUS is used to calculate the skeleton curve of reinforced sleeve grouting connected concrete shear wall after the fire, and the effects of fire time, axial compression ratio, and longitudinal stressed reinforcement diameter on the bearing capacity of reinforced sleeve grouting connected concrete shear wall after a fire are studied. The specific design parameters are shown in Table 2.

The model data adopts the TW1 specimen data in Section II, with 2.1 for details. The three-dimensional solid element (C3D8R) is selected for concrete, grout, and sleeve. The two-node linear three-dimensional truss element (T3D2) is selected as the reinforcement element outside the sleeve, and the three-dimensional solid element (C3D8R) is selected as the reinforcement element inside the sleeve. Firstly, the temperature field is simulated, the thermal parameters of concrete and grouting material are selected as Formulas (3) and (4), and the thermal parameters of reinforcement and sleeve are selected as Formulas (5) and (6) in this paper, and then it is needed to post-process the temperature field simulation

Table 2. Specific design parameters of the test piece.

Number	Fire time t/min	Axial compression ratio	Reinforcement diameter d/mm
JLQ 1	0	0.12	16
JLQ 2	60	0.12	16
JLQ 3	120	0.12	16
JLQ 4	60	0.06	16
JLQ 5	60	0.18	16
JLQ 6	60	0.24	16
JLQ 7	60	0.12	14
JLQ 8	60	0.12	18
JLQ 9	60	0.12	20

results with Python software, extract the maximum temperature of each node as the initial condition for static field introduction to realize pushover numerical simulation. The stress-strain relationship and compressive strength of concrete and grouting materials after high temperature are defined according to the recommendations of literature (Wu 2003):

$$y=\begin{cases} 0.628x+1.741x^2-1.371x^3 & x\leq 1 \\ \dfrac{0.674x-0.217x^2}{1-1.326x+0.783x^2} & x>1 \end{cases} \tag{9}$$

$$x=\varepsilon/\varepsilon_{0,\mathrm{T_m}};y=\sigma/\sigma_{0,\mathrm{T_m}}$$

$$\frac{\varepsilon_{0,\mathrm{T_m}}}{\varepsilon_0}=\begin{cases} 1.0 & T_m\leq 200°\mathrm{C} \\ 0.577+2.352\left(\dfrac{T_m-20}{1000}\right) & T_m>200°\mathrm{C} \end{cases} \tag{10}$$

$$\frac{\sigma_{0,\mathrm{T_m}}}{\sigma_0}=\begin{cases} 1.0-0.582\left(\dfrac{T_m-20}{1000}\right) & T_m\leq 200°\mathrm{C} \\ 1.146-1.393\left(\dfrac{T_m-20}{1000}\right) & T_m>200°\mathrm{C} \end{cases} \tag{11}$$

The expression of stress-strain relationship and damage factor of reinforcement and sleeve after the high temperature is the same as that at room temperature, which is determined based on *Code for Design of Concrete Structures GB50010-2010* (Beijing: China Construction Industry Press, 2010). The contact settings, boundary conditions, and load settings are the same as those in 2.3 and 2.4 of this paper. The specific loading mode is shown in Figure 2.

4.2 *Affected by fire time*

The axial compression ratio of specimens JLQ1, JLQ2, and JLQ3 is 0.12, the diameter of longitudinal stressed reinforcement is 16mm, and the fire time is 0, 60, and 120min respectively. According to the test, the peak load of JLQ1 is 345kN, and the peak load of specimen JLQ1 in finite element simulation is 344.136kN, with an error of 0.3%, indicating that the simulation results are in good agreement with the test results. As shown in Figure 7, the peak load of the shear wall after 60 min and 120 min of fire is 305.414kN and 205.539kN respectively, which is 10.94% and 40.59% lower than that at room temperature.

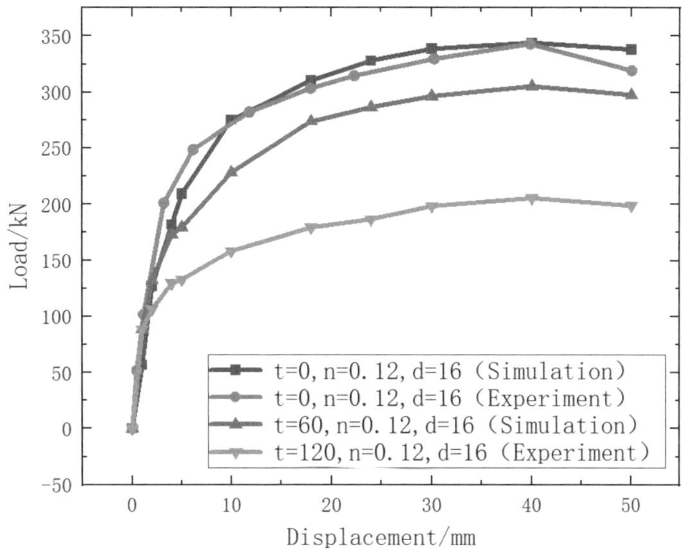

Figure 7. Comparison diagram of skeleton curve of the shear wall after different fire time.

4.3 *Influence of axial compression ratio*

The firing time of specimens JLQ4, JLQ2, JLQ5, and JLQ6 is 60 min, the diameter of longitudinal stressed reinforcement is 16 mm, and the axial compression ratio is 0.06, 0.12, 0.18, and 0.24 respectively. As shown in Figure 8, the peak loads of specimens with axial compression ratios of 0.06, 0.12, 0.18, and 0.24 are 275.667kN, 305.414kN, 329.781kN, and 350.759kN respectively. Compared with the JLQ4 specimen with an axial compression ratio of 0.06, the peak load of specimens with an axial compression ratio of 0.12, 0.18, and 0.24 is increased by 10.79%, 19.63%, and 27.24% respectively.

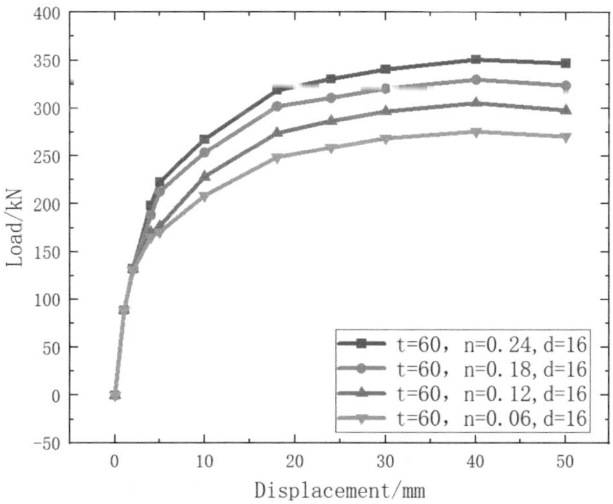

Figure 8. Comparison diagram of skeleton curve of the shear wall under different axial compression ratio.

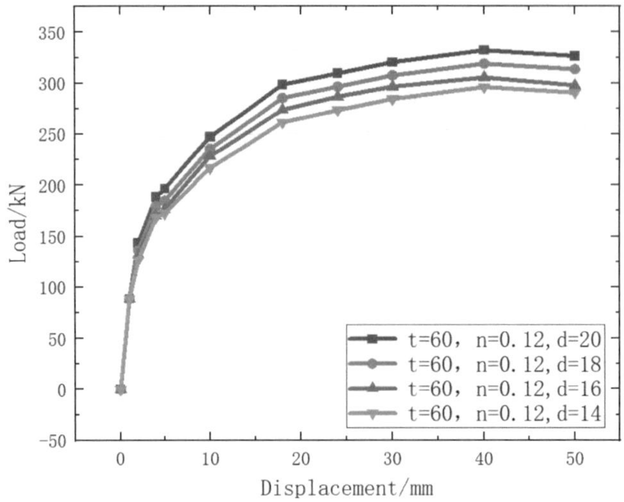

Figure 9. Comparison of skeleton curves of the shear wall under different stress reinforcement diameters.

4.4 *Influence of diameter of longitudinal stressed reinforcement*

The longitudinal reinforcement of the fabricated shear wall is divided into distributed reinforcement and stressed reinforcement. The stressed reinforcement is connected with the ground beam through the sleeve. When calculating the reinforcement ratio at the joint, because the stressed reinforcement passes through the joint and the distributed reinforcement does not pass through the joint, only the reinforcement ratio of the stressed reinforcement is calculated, and the diameter of the longitudinal stressed reinforcement directly affects the reinforcement ratio of the stressed reinforcement. The firing time of test pieces JLQ7, JLQ2, JLQ8, and JLQ9 is 60 min, the axial compression ratio is 0.12, and the diameter of longitudinal stressed reinforcement is 14mm, 16mm, 18mm, and 20mm respectively. As shown in Figure 9, the peak loads of specimens with diameters of 14mm, 16mm, 18mm, and 20mm are 295.947kN, 305.414kN, 318.823kN, and 331.964kN respectively. Compared with the JLQ7 specimen with a diameter of 14mm, the peak load of the specimens with the diameter of 16mm, 18mm, and 20mm increased by 3.19%, 7.73%, and 12.17% respectively.

5 CONCLUSION

In this paper, the numerical analysis model of seismic performance of reinforced sleeve grouting connected concrete shear wall at room temperature and reinforced concrete shear wall after the fire is established by ABAQUS. The simulated skeleton curve is in good agreement with the experimental skeleton curve, which verifies the rationality of the finite element model. On this basis, the numerical simulation of reinforced sleeve grouting connected concrete shear wall after the fire is carried out, and the effects of fire time, axial compression ratio, and longitudinal stressed reinforcement diameter on the bearing capacity of reinforced sleeve grouting connected concrete shear wall after a fire are analyzed.

The following conclusions are obtained:

(1) The skeleton curves of reinforced sleeve grouting connected concrete shear wall and reinforced concrete shear wall after fire obtained by ABAQUS numerical simulation are in good agreement with the experimental results, which verifies the rationality of the modeling method and material constitutive model in this paper.
(2) After 60 min and 120 min of fire, the peak load of reinforced sleeve grouting connected concrete shear wall decreased by 10.94% and 40.59% respectively compared with that at room temperature, which

reflects that fire does great harm to the bearing capacity of reinforced sleeve grouting connected concrete shear wall.

(3) After 60 minutes of fire, compared with the specimens with an axial compression ratio of 0.06, the peak load of the specimens with an axial compression ratio of 0.12, 0.18, and 0.24 increased by 10.79%, 19.63%, and 27.24% respectively, basically showing a linear growth trend.

(4) After 60 min of fire, the peak load of concrete shear wall connected by reinforcement sleeve grouting is increased by 3.19%, 7.73%, and 12.17% respectively compared with the specimens with longitudinal stressed reinforcement diameter of 14 mm, indicating that increasing the diameter of longitudinal stressed reinforcement can improve the bearing capacity of reinforced sleeve grouting connected concrete shear wall to a certain extent.

This paper also has some shortcomings:

(1) When the finite element model was established, the "Tie" contact was set between the grouting material and the sleeve, and the spring element was not used to simulate the slip between the grouting material and the sleeve.

(2) In this paper, only the effects of fire time, axial compression ratio, and vertical stressed reinforcement diameter on the bearing capacity of reinforced sleeve grouting connected concrete shear wall after a fire are considered, but other factors are ignored.

In view of these shortcomings, it is attempted to further study the slip between the grouting material and the sleeve and consider the influence of more parameters on the bearing capacity of reinforced sleeve grouting connected concrete shear wall after a fire in the next step.

ACKNOWLEDGMENT

This work was financially supported by the Natural Science Foundation of Zhejiang Province (LZ22E08002) and the Natural Science Foundation of Fujian Province(2021J01541).

REFERENCES

China Academy of Building Sciences Code for the design of concrete structures: GB50010-2010 [S] Beijing: China Construction Industry Press, 2010:2–3

Hu C.P Study on seismic performance of reinforced concrete short leg shear wall after fire [D] Overseas Chinese University, 2014

Huang X.K., Tian C.Y. Study on prefabricated concrete structure [J] Housing industry, 2010 (09): 28–32

JG/T 398-2012, grouting sleeve for reinforcement connection [S]

Li G.Q, Han L.H, Lou G.B, et al Fire resistance design of steel structure and steel-concrete composite structure [M] Beijing: China Construction Industry Press, 2006.124 ~ 140

Lie T T, Celikkol B. Method to calculate the fire resistance of circular reinforced concrete columns. ACI Materials Journal, 1991, 88(1):84~91.

Peng Y.Y. Experimental study on seismic behavior of precast reinforced concrete shear wall [D] Tsinghua University, 2010

Wang N, Gao L., Su Y.P, Chen H.B., Han Y.T. Finite element simulation of vertical reinforcement connection performance of fabricated shear wall [J] World Earthquake Engineering, 2016,32 (02): 217–222

Wang Y.Z. Experimental study on seismic behavior of prefabricated reinforced concrete shear wall connected by grouting sleeve [D] South China University of technology, 2020 DOI:10.27151/d.cnki. ghnlu. 2020.003703.

Wu B. Mechanical properties of reinforced concrete structures after fire [M] Beijing: Science Press, 2003

Xu Y., Xu Z.S., Zhu M. Experimental study on strength and deformation of concrete after high temperature [J] Changsha Railway Institute News, 2000 (02): 13–16+21.

Zhang X.H. Finite element study on mechanical properties of reinforced grouting sleeve connection of fabricated shear wall [D] Qingdao University of technology, 2020 DOI:10.27263/d.cnki. gqudc. 2020.000146.

Advances in Urban Engineering and Management Science – Khalil & Yang (Eds)

Construction technology of large span double-deck cast-in-situ beam in the deep water area

Weiliang Qiang, Junlong Zhou, Yian Shi*, Tao Bo & Penglin Xie
China Construction Sixth Engineering Bureau Co., Ltd, Tianjin, China

ABSTRACT: Taking the Daqingzhou bridge as the background, the construction technology of a double-layer cast-in-situ beam in a deep water area is studied. A rotatable support system is designed. It not only solves the problem of space collision between steel support and deep-water cofferdam, but also ensures that a set of support is shared during the construction of double-layer cast-in-situ beams. The calculation of the finite element method provides a technical guarantee for construction. The engineering practice shows that the designed support system meets the specification requirements, which provides some practical experience and reference value for similar projects in the future.

1 INTRODUCTION

With the large-scale construction of bridges across rivers, lakes, and seas, bridge construction is facing more complex environmental and technical problems. Generally, water operation conditions are limited, geological conditions are complex and the meteorological environment is changeable. The design and construction of a cast-in-place beam support system are more complex than ordinary bridges. How to ensure the reliability of the support system in water and the quality and safety of cast-in-place beam construction is the key and difficult point of construction. At present, China has rich practical experience in water cast-in-situ bridge construction. Among the construction methods used, the Bailey beam support system is more common. Among the existing technical achievements, the construction method of a single-layer cast-in-situ beam is mostly studied, and the engineering cases and technical achievements of a double-layer cast-in-situ beam in deep water are less summarized. Based on the above practice, this paper systematically analyzes the characteristics of the double-layer cast-in-situ beam of the Daqing Hou bridge and studies the construction technology of the double-layer cast-in-situ beam in deep water.

2 PROJECT OVERVIEW

Daoqingzhou bridge is a fast track connecting Fuzhou and Changle. It is a highway-rail bridge with a total length of about 6.82 km. The upper highway is a 6-lane class I highway with a design speed of 60 km/ h, while the lower layer is a double-track subway line.

The total length of the bridge on the south bank is about 1365.5 m, which is divided into upper and lower layers, with a total of 10 connections. The first link crosses the Minjiang River and extends to the land. The riverbed in this area is hard granite with a maximum water depth of 20.50m. The upper highway beam adopts a prestressed concrete continuous beam with a width of 25 m, a height of 2.3 m, and a span of 39.0 m. The lower track beams are prestressed concrete simply supported box girders, with a height of 2.4 m and a width of 10.0 m. The substructure shares piers with highway bridges, as shown in Figure 1.

*Corresponding Author: 374638297@qq.com

DOI 10.1201/9781003305026-42

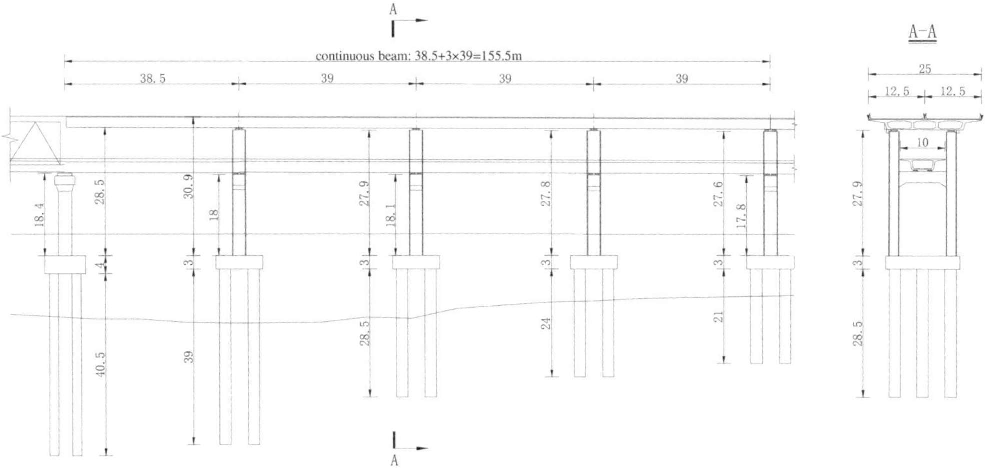

Figure 1. River crossing bridge on the south bank.

3 SCHEME COMPARISON AND SELECTION

The key to the construction of the double-layer cast-in-situ beam of the project lies in the design of the support system. The support system shall meet its own strength, stiffness, stability, and other technical indicators and also consider the construction sequence of upper and lower beams, hydrogeological conditions, construction period requirements, and other factors. The supporting systems available for the project include single-span type, three-span type, multi-span type, etc., as shown in Figure 2.

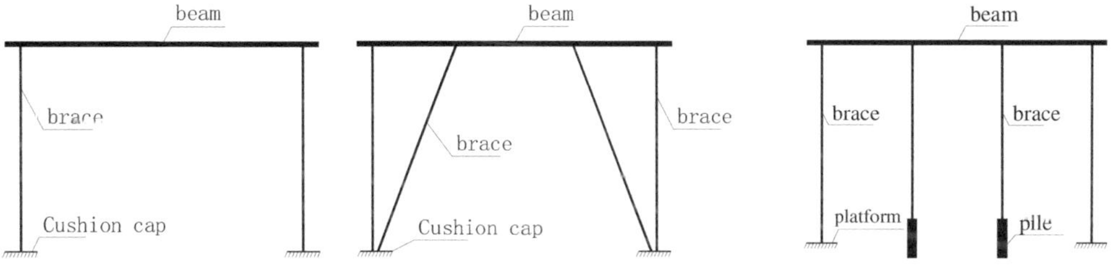

(a) Single-span type. (b) Three-span type. (c) Multi-span type.

Figure 2. Schematic diagram of support system scheme.

(1) Multi-span support system requires temporary support in the water. However, the site water depth is large, and the covering layer is thin, so the construction of temporary support is difficult. The scheme will undoubtedly increase additional costs and affect the construction period.
(2) The single-span support system is based on the bearing platform and does not need to be supported in the water. However, the cast-in-place beam has a large span and high requirements for the steel beam of the support system, so it is difficult to ensure the alignment. At present, the practical experience of a single-span support system with the same long span is less.

(3) The inclined leg three-span support system can arrange the foundation on the bearing platform, but the bearing platform is located in the cofferdam with a height of 15.6m and the inclined leg will collide with the cofferdam in space. Moreover, the double-layer cast-in-situ beam needs to share a set of support systems, so the traditional inclined leg support system can not be implemented.

Through the above comparative analysis, it can be seen that the above conventional support system is difficult to ensure the implementation of the project. Based on the inclined leg three-span support system, a rotatable support system is designed in this paper. It not only solves the space collision problem between the inclined leg and the deep-water cofferdam but also ensures that the double-layer cast-in-place beam shares a set of support during construction, which provides an effective technical guarantee for the smooth implementation of the project.

4 SUPPORT SYSTEM DESIGN

4.1 *Structural system*

The support system adopts a rotatable three-span Bailey beam composite system. The structure is composed of a formwork system, Bailey beam, transverse distribution beam, Bailey truss, main beam, steel pipe column, and rotatable support from top to bottom.

The support system of the upper beam is 28.5 m high and 26.0 m wide. The support system of the lower beam is 18.5 m high and 12.9 m wide. The steel pipe column is fixed with the bearing platform by bolts. The rotatable support is connected with the bearing platform through the hinge system. The support system is shown in Figures 3 (a) and (b).

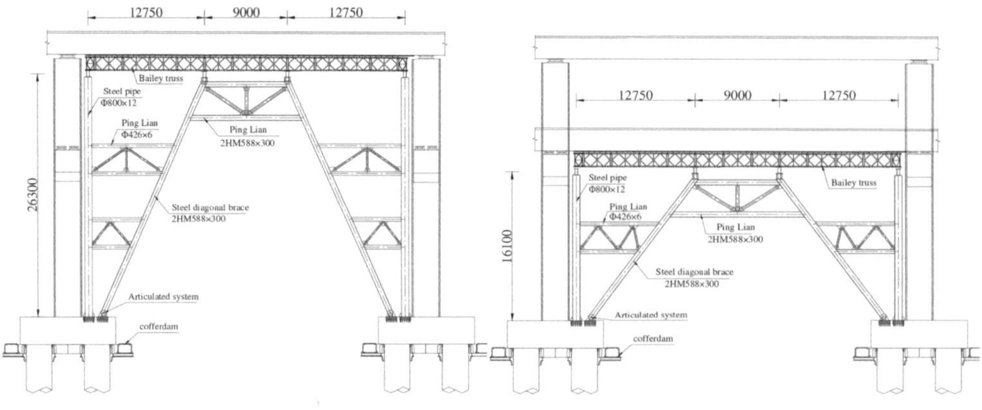

(a) Upper beam construction. (b) Construction of lower beam.

Figure 3. Elevation of rotatable support system.

4.2 *Articulated system*

The articulated system includes four parts, namely, steel diagonal brace, base, pin shaft, and embedded parts (as shown in Figure 4). After the cofferdam is removed, the articulated system is submerged below the water level. In addition to bearing large loads, it will also be eroded by the river water for a long time. Therefore, the reasonable design of an articulated system is particularly important.

The specification of section steel of steel diagonal brace is 2HM500× 300. The base consists of an ear plate, base plate, and stiffener plate. There are 3 ear plates, 50 mm thick and 20 mm thick base plate, which are connected by double-sided fillet welds. The outer diameter of the pin shaft is 100 mm and the inner diameter of the pinhole is 101 mm. All components are made of Q345 steel. The embedded parts are composed of steel plates and anchor bars, and the reinforcement mesh cushion is used to improve the local bearing strength of the bearing platform.

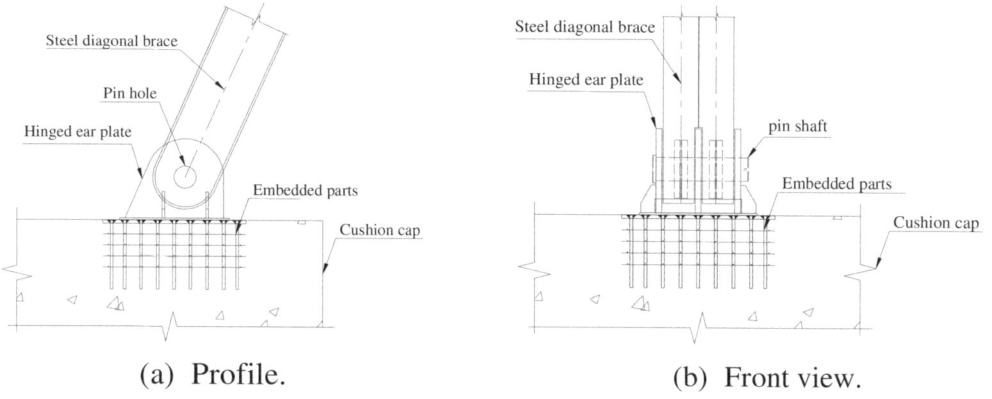

Figure 4. Hinge system.

4.3 *Construction procedure*

The construction sequence of the double-layer cast-in-place beam is from top to bottom. Both Bailey stringers share one set of support. The construction process mainly includes the following five key processes. Each process is shown in Figure 5.

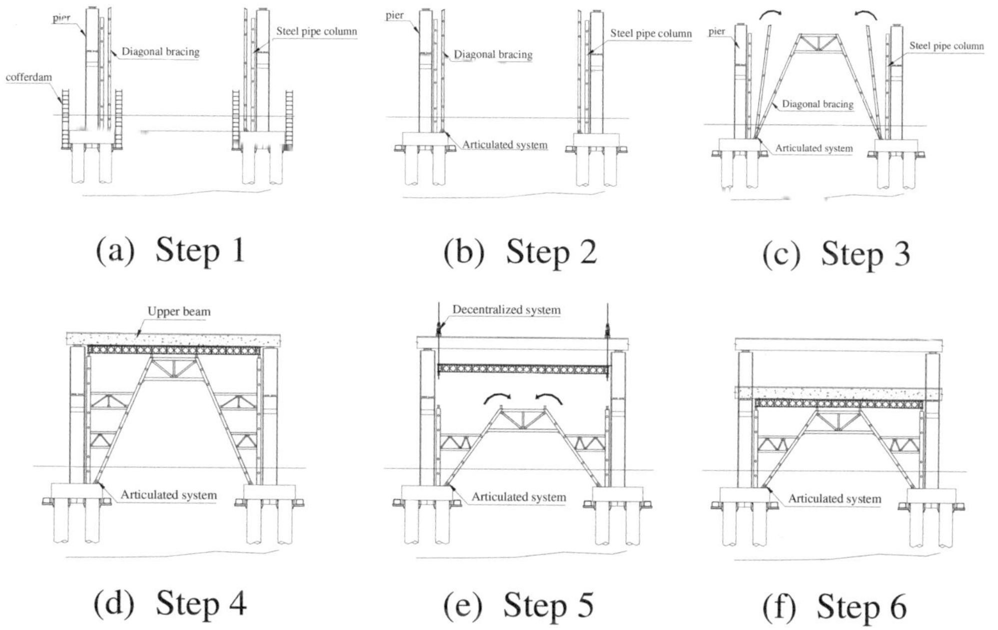

Figure 5. Construction procedure.

(1) After the pier is completed, the steel pipe column shall be installed and fixed with the bearing platform through the column base. The steel diagonal brace is hoisted and connected with the bearing platform through the articulated system. The rotatable steel diagonal brace shall be fixed vertically first.
(2) The steel cofferdam is removed to provide space for the rotation of steel diagonal bracing.
(3) The steel diagonal brace shall be vertically turned to the design position for the first time, and the connection system between them shall be welded.
(4) It is required to weld the connection system between the steel pipe column and steel diagonal brace, install the Bailey truss, and complete the cast-in-situ beam support system of the upper layer. After the preloading of the support system is completed, the upper cast-in-situ beam shall be constructed.
(5) After the construction of the upper cast-in-situ beam is completed, it is needed to install the vertical jack and temporarily fix the Bailey truss through the finish rolled deformed steel bar. The second vertical rotation of the steel diagonal brace is set to the design angle. At the same time, the steel diagonal brace and steel pipe column are cut to the design elevation.
(6) Use the jack to lower the Bailey truss and complete the cast-in-situ beam support system of the lower main beam. After the preloading of the support system is completed, the lower cast-in-situ beam shall be constructed.

5 STRUCTURAL CALCULATION

5.1 *Support structure calculation*

(1) Model and boundary strip

The structure adopts Midas civil finite element software. Each component is simulated by a beam element. The steel pipe column is fixed at the bottom end. The bottom end of the steel diagonal brace is longitudinally hinged. The calculation model is shown in Figure 6.

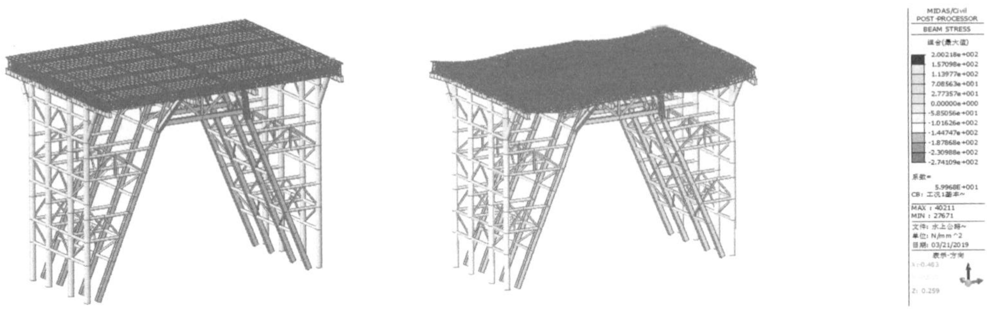

(a) computational model (b) computational model

Figure 6. Hinge system.

Under the combined action of concrete load, construction load, and wind load, the maximum combined stress design value of Q235B material in each member is 153.6mpa, which is less than the strength design value of 190 MPa, so the strength meets the requirements. The maximum axial force of the chord of the Bailey truss is 345.9 kn, which is less than the allowable value of 560 kn; the maximum axial force of the vertical bar is 185.7 kn, which is less than its allowable value of 210kN; the maximum axial force of the inclined rod is 162.3 kn, which is less than its strength of 171.5kn; therefore, the strength of Bailey truss meets the requirements.

The maximum vertical of the Bailey truss is 20.6 mm, which appears in the middle of the span and meets the service requirements.

5.2 *Calculation of hinge system*

The three-dimensional static contact solid finite element model of the hinge system is established by ABAQUS. As shown in Figure 7, the calculation model includes three parts: diagonal brace, base, and pin shaft. The bottom surface of the base is consolidated. The contact surface between the pin shaft and the inclined leg and the base is friction contact, and the friction coefficient is 0.2. The mesh generation of each component adopts an advanced algorithm, and the element type is a hexahedral element (C3D8R). According to Saint Venant's theorem, the length of the inclined leg is 3 times, the section width is 1.5m, and its internal force is applied on the reference point coupled with its end section, with a size of 350t. The stress nephogram of each component under the action of internal force of the inclined leg is shown in Figures 8 ~ 10.

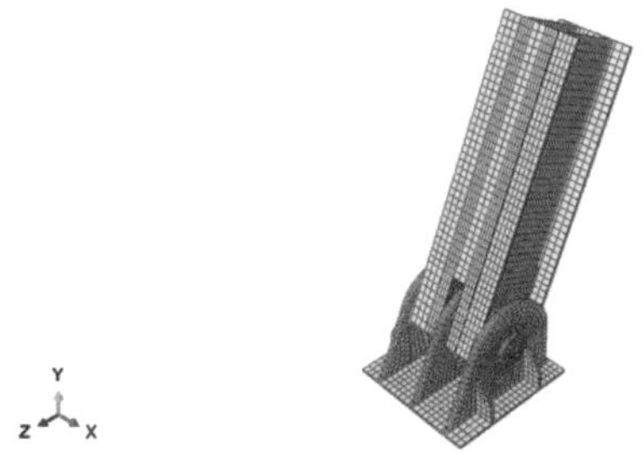

Figure 7. Model of the hinge system.

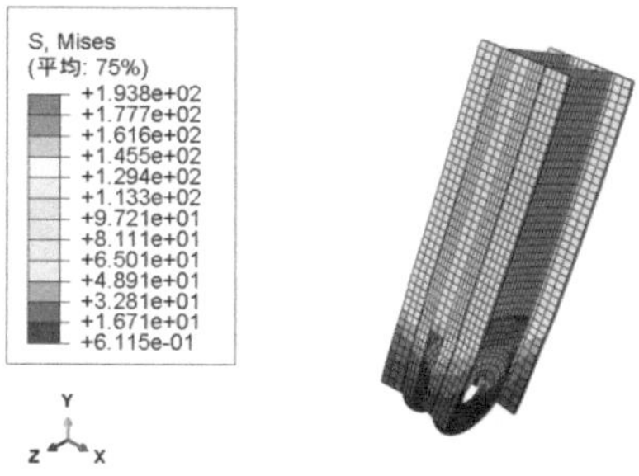

Figure 8. Diagonal brace stress (unit: MPa).

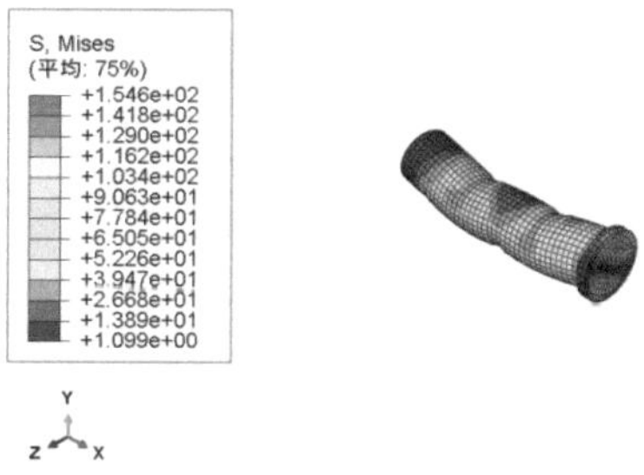

Figure 9. Stress of pin shaft (unit: MPa).

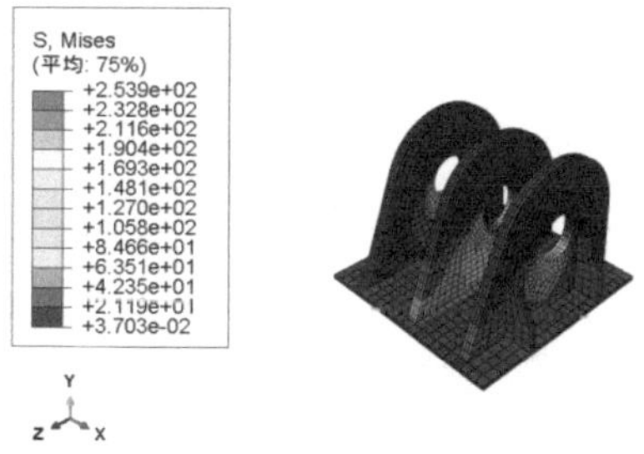

Figure 10. Stress of ear plate (unit: MPa).

Table 1. Stress of hinge system.

Component name	Material quality	Maximum stress (MPa)	Unfavorable position	Strength design value (MPa)
Diagonal bracing	Q235	193.8	Strengthen the position where the steel plate pinhole contacts the shaft	290
Pin shaft	Q345	154.6	Contact between pin shaft and pinhole	290
Base	Q345	253.9	Contact between middle padeye pinhole and shaft	290

6 CONCLUSION

Through the engineering practice of the double-layer cast-in-situ beam of the Daqing Hou bridge, the conclusions are as follows:

(1) The designed support system avoids the erection of temporary support in water and solves the problem of space collision between diagonal bracing and cofferdam.
(2) The overall strength, stiffness, and stability of the support structure meet the specification requirements. At the same time, the articulated system also has sufficient strength and applicability, which provides a technical guarantee for the smooth implementation of the project.
(3) The double-layer cast-in-situ girder adopts the construction process from top to bottom. Through the two vertical rotations of "hinged" steel diagonal bracing, effective support for the construction of a two-layer cast-in-situ girder is provided, the construction process is simplified, good economic benefits are created, the smooth construction of the bridge is ensured, and experiences for the construction of long-span cast-in-situ girder project in deep water area are successfully accumulated, thus providing a useful reference for the follow-up project construction.

ACKNOWLEDGMENTS

This work was financially supported by CSCEC's "Study on construction technology of superstructure of long-span cable bearing bridge"(Grant No. CSCEC-2021-Z-30).

REFERENCES

Chen Fo-ci, Application of a New Truss to Beam-Type Support for Cast-in-Situ Box Girder[J]. World Bridges, 2016, 44(6): 36–40.
Fang Ke, Liu. Construction Technique of Casting Prestressed Concrete Box Girders on Outrigger Scaffolds for Railway Deck of South Approach Bridge of Wuhu Changjiang River Rail-cum-road Bridge, Bridge Construction, 2020, 50(6): 116–121.
GB 50017-2017 Standard for the design of steel structures [S]. Beijing: China Construction Industry Press, 2017.
JTG D60-2015 General code for design of highway bridges and culverts[S], Beijing: China Communications Press, 2015.
Wang Qiang, LIU Ai-lin. Key Techniques for Construction of 40.7 m-Span Simply-Support Girders of South Approach Bridge of Wuhu Changjiang River Rail-cum-Road Bridge, 2019, 49(1): 7–11.
Wang Tong-min, REN Wen-hui, Key Techniques for Construction of East Approach Bridge of Yinchuan Binhe Huanghe River Bridge over Water[J]. World Bridges, 2017, 47(3): 105–110.
Yang Guangwu, Zheng Ya-peng. Overall Design and key Techniques of Daoqingzhou Bridge in Fuzhou[J], Bridge Construction, 2020, 50(S2): 62–68, in Chinese
Zou Min, Zheng Ya-peng. Design of Varying-width Curved Steel Truss GirderAccommodating Two Decks for Daoqingzhou Bridge[J], World Bridges, 2020, 48(3): 7–11.

Advances in Urban Engineering and Management Science – Khalil & Yang (Eds)
© 2023 The Author(s), ISBN 978-1-032-30426-7

One-off tensioning construction control technology for tied-arch bridge

Penglin Xie*, Junlong Zhou, Xiaomin Liu & Xu Zhao
Construction Sixth Engineering Bureau Ltd., Tianjin, China

Jian Sun
China State Construction Bridge Corp., Ltd., Chongqing, China

ABSTRACT: The tensioning operation of a tied-arch bridge is mainly divided into two steps: initial tensioning and secondary cable force adjustment, where the latter is difficult due to the interaction between suspenders and the adjustment amount of cable force. In this study, the thinking method of the one-off tensioning construction control for tied-arch bridges was introduced. With a 30+130+30 m fly-bird-type tied-arch bridge taken as an example, the one-off suspender tensioning construction control was explored. In the theoretical calculation phase, the deviation and precision control criteria for cable force were proposed, and fast convergence was achieved in the iterative computations by taking the initial tensile force approximate to the dead weight of the midspan steel box girder. In the completed bridge phase, the suspenders with the over-ranging deviation of the measured cable force from the target cable force of the completed bridge accounted for about 11%. In the suspender tensioning process, the temporary loads on the bridge deck would change the unstressed length of the suspenders, which was the primary cause for the deviation of cable force. However, the cable forces of all suspenders met the requirements through local fine adjustment. Therefore, the overall one-off tensioning control effect of suspenders was satisfying, thus providing a reference for similar structures.

1 INTRODUCTION

The cable forces of suspenders will change with the external load, the structural system, and the tensioning adjustment of suspenders (Qin 2007), and the practical construction process can hardly be consistent with the theory. Therefore, the suspenders of fly-bird-type tied-arch bridges, which are featured by the girder construction first, followed by the arch construction, generally experience two steps: initial tensioning and secondary cable force adjustment. In practical engineering, a major difficulty is usually considered in the cable force adjustment. In general, tensioning should be conducted multiple times so that the final cable force of completed bridges is as reasonable as possible (Zhong et al. 2019). To reduce the number of tensioning times of suspenders/stay cables, the one-off tensioning of cable-stayed bridges with large-tonnage cable forces has been investigated by some scholars. For instance, by theoretically analyzing, actually measuring, and comparatively analyzing the completed bridge tower, girder, and cable force through the one-off and twice cable tensioning methods, Zhong W concluded (Zhong et al. 2019) that both tensioning methods could reach the design conditions very well, but the cable force adjustment should be still done once in case of the one-off tensioning method used. Shi H P (Shi et al. 2013) performed a finite element simulation analysis of one-off tensioning for an over thousand-tonnage stay cable and pointed out that the one-off tensioning construction of stay cable was feasible. In the completed bridge phase,

*Corresponding Author: 846756502@qq.com

DOI 10.1201/9781003305026-43

the actual cable force remains unknown. However, the one-off tensioning of tied-arch bridges has been less explored. Tian Z C (Tian et al. 2004) and Han Y (Han et al. 2018) explored the cable force tensioning problem of cable-stayed buckles in the main arch ring of an arch bridge and realized one-off tensioning of cable-stayed buckles, *i.e.*, the perfect closure of the main arch ring. Yu J C (Yu & Ye 2000) conducted the one-off tensioning model test on rigid suspenders and stated that the one-off tensioning process was feasible for suspenders, but after the tensioning, the measured suspender cable force differed a lot from the theoretical value. Most of the existing studies have concentrated on the influence of the one-off tensioning process of suspenders/cables on the structural force-bearing and stability, while the difference between the cable force of a completed bridge after one-off tensioning from the expected value has been less probed.

In this study, the one-off tensioning construction control of a fly-bird-type tied-arch bridge with the girder-first and arch-second construction process will be explored, expecting to meet the design state of the completed bridge just by the initial tensioning of all suspenders without secondary cable force adjustment.

2 DETERMINATION OF INITIAL TENSION

According to research findings, the initial tension of suspenders can be rapidly solved using the difference value method (Xu et al. 2019), which followed such a basic idea: It is assumed that the suspender is initially tensioned according to a group of adjusted cable force $\{T_1\}$, followed by the forward calculation based on the established construction process until the cable force $\{F_1\}$ of the completed bridge is obtained in the completed bridge phase. $\{F_1\}$ deviates from the target cable force under the reasonable state of the completed bridge by $\{\Delta F_1\} = \{F_m\} - \{F_1\}$. To gradually approach the target cable force, a group of adjusted cable force $\{T_2\} = \{T_1\} + \{\Delta F_1\}$ is updated for the next round of forwarding calculation to obtain the cable force $\{F_2\}$ of the completed bridge. By such a circulation, the difference value$\{\Delta F_i\} = \{F_m\} - \{F_i\}$ between the cable force of the completed bridge and the target cable force is finally obtained. The calculation is stopped when the difference value is converged until meeting specific requirements (Xu, Cai, Li 2019). In this study, the cable force was even not adjusted in order to realize the one-off tensioning of the suspender and reduce the number of adjustment times for the cable force, and the accuracy control was conducted according to the following criteria:

$$\begin{cases} \omega = \dfrac{1}{n}\sum_{i=1}^{n} \dfrac{|\Delta F_i|}{F_i} \leq 0.6\% \\ \max\limits_{1 \leq i \leq n} \dfrac{|\Delta F_i|}{F_i} \leq 2\% \end{cases} \tag{1}$$

3 ANALYSIS OF CALCULATION EXAMPLE

3.1 *Brief introduction of a calculation example*

A three-span half-through tied-arch bridge was taken as an example, and the bridge span was set as (30+130+30) m. The main arch ribs were parallelepiped steel box-arched ribs, with a calculated span of 130 m, the rise of 35 m, the rise to span ratio of 1/3.714, and the arch axis coefficients of m=1.756 and k=1.163. The suspender cables were 13 pairs of vertical suspenders spaced by 7.5 m. 1#, 13# suspenders were made of 121ϕ7high-strength steel wire, and 2#-12# suspenders were made of 85ϕ7high-strength steel wire, with the standard strength of f_{pk}=1,670 MPa. The lengths of suspenders ranged from 7.5 to 22.8 m, while their two ends were anchored using chilled cast anchors. A total of 6 19ϕ^s15.2 steel strands (standard strength: f_{pk}=1,860 MPa) were set on two sides as the tie bars. The bridge layout is shown in Figure 1.

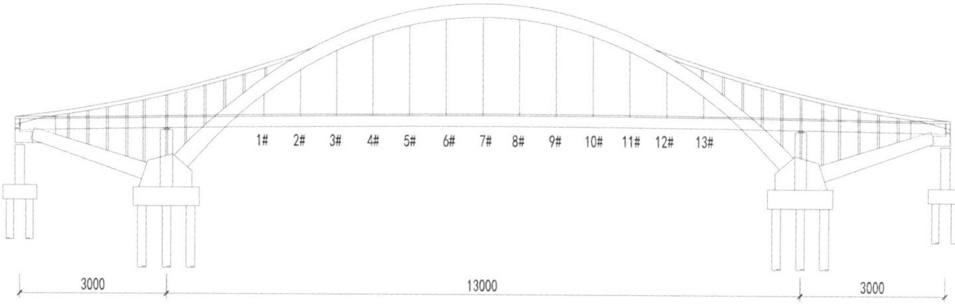

Figure 1. General layout of tied-arch bridge (unit: cm).

Table 1. Main construction procedures.

Phase	Construction item	Phase	Construction item
1	Arch foot pouring and main girder installation	9	Installation and tensioning of 1# and 3# suspenders
2	Initial tensioning of tie bars	10	Installation and tensioning of 6# and 8# suspenders
3	Installation of steel arch ribs	11	Installation and tensioning of 4# and 10# suspenders
4	Removal of arch rib stands	12	Installation and tensioning of 2# and 12# suspenders
5	Installation and tensioning of 7# suspender	13	Girder installation
6	Installation and tensioning of 5# and 9# suspenders	14	Tensioning of tie bars for the third time
7	Tensioning of tie bars for the second time	15	Pavement of main girder and attachments on the bridge deck
8	Installation and tensioning of 3# and 11# suspenders		

The girder-first and arch-second construction method was adopted for this bridge. The construction procedures are depicted in Table 1.

3.2 *Initial tension calculation of suspenders*

A finite element model (Figure 2) of this arch bridge was constructed via Midas Civil finite element program to assist in the initial tension calculation. The structure was discretized totally into 1,150 nodes and 1,503 elements. The suspenders and tie bars were simulated using truss elements and the arch ribs and steel girders were simulated by beam elements. The consolidation constraints were used for the main pile cap, the vertical constraints for abutment pier bearings, and the master-slave constraints for the junctions of suspenders with arch ribs and main girder. Except that the suspenders are made of steel strand, the rest are steel of Q345.

First, the forward analysis of 1#–13# suspenders was performed at the initial tension of N=500 kN (the sum of initial tension was slightly smaller than the dead weight of the midspan steel box girder) to obtain the cable force $\{F_1\}$ in the corresponding completed bridge phase. The target cable force of the completed bridge was the designed cable force $\{F_m\}$ of the completed bridge, and the mean and maximum deviation values of the suspender cable force were 7.24% and 13.13%, respectively, as seen in Table 2.

The iterative computations were conducted according to the difference value method in Section 1. The construction control criteria were met just upon the completion of the first iteration. The mean and maximum deviations of the suspender cable force were 0.6% and 1.4%, respectively,

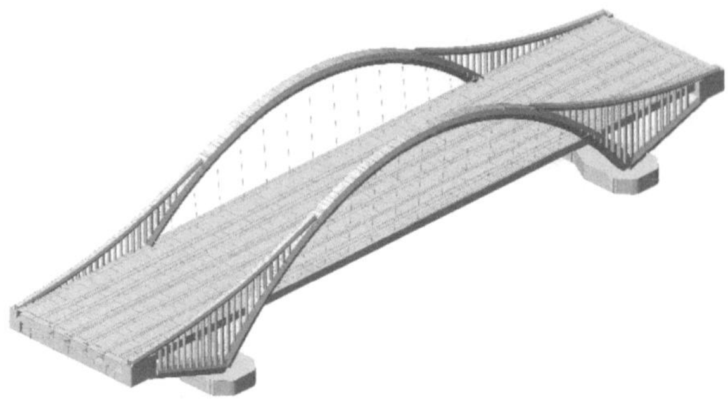

Figure 2. Finite element model of tied-arch bridge.

Table 2. Initial tension calculation of suspenders.

Suspender No.	Target cable force F_m/kN	Cable force 1 of the completed bridge F_1/kN	Deviation of cable force $\Delta F/F_m$	Iterative cable force 1 T_2/kN	Cable force 2 of the completed bridge F_2/kN	Deviation of cable force $\Delta F/F_m$
1	1275	1297	1.74%	478	1271	0.34%
2	982	1071	9.06%	411	976	0.61%
3	990	860	13.13%	630	988	0.18%
4	1014	1064	4.94%	450	1005	0.86%
5	1011	958	5.23%	553	1006	0.45%
6	1027	1093	6.44%	434	1013	1.35%
7	1006	875	13.00%	631	998	0.82%
8	1027	1093	6.44%	434	1013	1.34%
9	1011	958	5.23%	553	1006	0.45%
10	1014	1064	4.94%	450	1005	0.85%
11	990	860	13.13%	630	988	0.18%
12	982	1071	9.06%	411	976	0.61%
13	1275	1297	1.74%	478	1271	0.34%

thus meeting the relevant requirements. The iterative computations showed a high convergence rate when the initial tension was approximate to the dead weight of the midspan steel box girder.

3.3 *Comparison of cable forces of suspenders in the completed bridge*

1#-13# suspenders were initially tensioned according to the adjusted cable force obtained through iterative computations, and the construction was done step by step until the completed bridge phase. Then, the fundamental frequency of suspenders was tested via KINGMACH JMM-268 dynamic cable force tester. Subsequently, different cable force formulas (Sheng & Wang 2013) were chosen based on the length of the suspenders. The measured suspender cable force of the completed bridge and the errors are listed in Table 3.

As seen in Table 3, there were three upstream and downstream suspenders with the measured cable force error exceeding 15%, accounting for about 11%, and the cable force errors of other suspenders basically satisfied ≤10%, thus generally meeting the relevant requirement and achieving an overall satisfying one-off tensioning control effect. The errors between the measured cable force and the target cable force of the completed bridge were generated mainly for the following reasons: First, the vehicles, temporary loads, and vibration on the bridge deck could hardly be radically cleared away in the suspender tensioning phase, which directly impacted the unstressed cable

Table 3. Measured data of suspender cable force.

Suspender No.	Target cable force kN	Upstream measured value kN	Upstream error %	Downstream measured value kN	Downstream error %
1	1275	1064	−16.6	1336	4.8
2	982	885	−9.8	939	−4.4
3	990	926	−6.4	942	−4.9
4	1014	893	−11.9	894	−11.8
5	1011	1100	8.8	944	−6.6
6	1027	1025	−0.2	962	−6.3
7	1006	1009	0.3	927	−7.9
8	1027	1007	−2.0	953	−7.2
9	1011	982	−2.9	1063	5.2
10	1014	822	−18.9	906	−10.6
11	990	934	−5.6	972	−1.8
12	982	877	−10.7	1005	2.3
13	1275	1397	9.5	1558	22.2

length of tensioning suspenders, and this was the primary cause for the deviation. Second, the four suspenders were synchronously tensioned using four sets of jacks as commanded and coordinated by a specially-assigned person with a high-frequency intercom, so the tensioning synchronicity was difficult to guarantee. Third, the stiffness of structures like steel arch ribs and suspenders was different, to some extent, from the members leaving steel structure processing plants, so they could not be accurately simulated. Fourth, there are various theoretical testing methods for cable force, and no unified testing theory has been formed yet.

During the construction control process, the importance should be attached to the following two points: First, sufficient communication with the construction unit should be done before important process steps (like suspender tensioning), ensuring that the state of the bridge deck is consistent with the theoretical calculation. Second, the trial tensioning should be performed before the initial suspender tensioning, and the cable force parameters should be identified by combining the pressure gauge calibration method, pressure sensor method, frequency method, etc.

4 FINE ADJUSTMENT OF LOCAL CABLE FORCE

According to the principle of influence matrix, the calculation method for adjusted cable force was given in the reference (Xie 2016):

$$\{T\}=[C]^{-1}(\{F_m\}-\{F\}) \tag{2}$$

$$C=\begin{bmatrix} C_{11} & C_{12} & \cdots & C_{1n} \\ C_{21} & C_{22} & \cdots & C_{2n} \\ \vdots & \vdots & \vdots & \vdots \\ C_{n1} & C_{n2} & \cdots & C_{nn} \end{bmatrix} \tag{3}$$

Where F_m is the target suspender cable force; F represents the measured suspender cable force; C stands for the cable force influence matrix of suspenders; C_{ij} denotes the actual cable force of the suspender i when the unit nominal tensile force is applied to the suspender j; T refers to the adjusted cable force of each suspender.

Table 4. Adjusting scheme of suspender cable force.

Step I		Step II		Step III	
Suspender	Adjusted value	Suspender	Adjusted value	Suspender	Adjusted value
S1	1275	S4	980	S2	1030
S13	1330	S10	950	S3	950
X1	1275	X4	960	S12	900
X13	1275	X10	960	X12	1020

Note: S represents the upstream side and X denotes the downstream side, the same below.

Table 5. Measured values after the adjustment of suspender cable force.

	Step I		Step II		Step III			
Suspender	Matrix calculated value	Measured value after cable force adjustment	Matrix calculated value	Measured value after cable force adjustment	Matrix calculated value	Measured value after cable force adjustment	Target value	Error (%)
S1	1272	1292	1281	1324	1201	1206	1275	−5.4
S2	805	812	794	837	1007	1022	982	4.1
S3	897	892	861	879	902	945	990	−4.5
S4	885	873	981	986	946	928	1014	−8.5
S5	1099	1098	1070	1062	1044	1033	1011	2.2
S6	1026	1031	1015	992	987	978	1027	−4.8
S7	1010	1013	999	976	975	963	1006	−4.3
S8	1008	1011	993	971	972	954	1027	−7.1
S9	984	995	964	929	929	918	1011	−9.2
S10	825	835	951	949	944	933	1014	−8.0
S11	943	946	911	907	896	893	990	−9.8
S12	899	884	864	858	899	904	982	−7.9
S13	1314	1341	1329	1341	1316	1320	1275	3.6

According to the above equation, the cable force adjustment scheme was obtained as seen in Table 4. The test results after the adjustment as per the adjustment scheme are listed in Table 5.

After the suspender cable force was locally adjusted three times, the difference value between the measured cable force value and the target value of the completed bridge was kept within ±10%, indicating a high control accuracy of suspender cable force in the arch bridge and the conformity with the relevant code requirement.

5 CONCLUSION

The methodology of one-off suspender tensioning construction control for tied-arch bridges was introduced. A 30+130+30 m fly-bird-type tied-arch bridge was taken as an example to study and apply the one-off suspender tensioning construction control, and some beneficial conclusions were drawn:

(1) The convergence is fast in the iterative computations if the initial tension is approximate to the dead weight of the midspan steel box girder.

(2) The number of suspenders with the over-ranging deviation of the measured cable force from the target cable force of the completed bridge accounts for about 11%. After the cable force is locally adjusted three times, the cable forces of all suspenders meet the requirements, manifesting the satisfactory one-off tensioning construction control effect. Hence, this study can provide a reference for similar projects.
(3) The causes for the deviations existing in the measured cable force values are analyzed. It appears that the unstressed length of suspenders is directly impacted by the vehicles and temporary loads on the bridge deck in the suspender tensioning process, being the primary cause for the deviation of cable force. Given this, sufficient communication with the project department in clearing away the aforementioned objects should be done before the suspender tensioning during the construction control. In this study, the one-off suspender tensioning construction control fails to reach the target cable force without the cable force adjustment of all the suspenders, so this remains to be further perfected.

ACKNOWLEDGMENT

This work is supported by CSCEC's "Study on construction technology of superstructure of long-span cable bearing bridge" (Grant No. CSCEC-2021-Z-30).

REFERENCES

Han Y, Qin D Y, Zheng J. Optimal computation method for cable-stayed buckle construction of CFST arch bridges [J]. Highway, 2018 (01): 100–104.
Qin S Q. Bridge Construction Control—Theory and Practice of Unstressed State Method [M]. Beijing: China Communications Press, 2007.
Sheng H Y, Wang G H. Some problems in tensile force detection and error analysis of measurement for short suspender [A]. The 22nd National Academic Conference on Structural Engineering, 2013: 288–291.
Shi H P, Hu S Z, You X P, Peng C M. Application of one-off tensioning technology for super-tonnage stay cables in highway-railway co-constructed cable-stayed bridges with steel truss girders [J]. Journal of China & Foreign Highway, 2013, 33 (05): 187–189.
Tian Z C, Chen D L, Yan D H, Chen Z Q. Study of key technology with construction control of steel box basket-handling arch bridge [J]. China Journal of Highway and Transport, 2004, 17 (03): 46–50.
Xie P L. Computational analysis and research on tied-arch bridge with concrete-filled steel tubes [D]. Hefei University of Technology, 2016.
Xu H Z, Cai C W, Li H S. Suspender force calculation methods of the tied-arch bridge [J]. Journal of Chongqing Jiaotong University (Natural Science), 2019, 38 (04): 23–28.
Yu J C, Ye Z Q. Study on one-time full tension model test of tied arch prestressed concrete rigid suspension cable of Taizhou Bridge [J]. Journal of Highway and Transportation Research and Development, 2000, 17 (06): 41–44.
Zhong W, Li B, Yu G. Applied research of "one-off tensioning method" of stay cables in the construction of single pylon cable-stayed bridge with large cable force [J]. Highway, 2019 (07): 160–165.

Advances in Urban Engineering and Management Science – Khalil & Yang (Eds)
© 2023 The Author(s), ISBN 978-1-032-30426-7

Research on influencing factors of urban viaduct collapse accident based on DEMATEL

Zhu Tian*
School of Management, Shenyang Jianzhu University, Shenyang, China

ABSTRACT: In recent years, urban viaduct collapse accidents have occurred from time to time, which not only caused serious economic losses and casualties but also had a huge social impact. Based on the analysis of literature and cases, this paper firstly constructs an index system of influencing factors of urban viaduct collapse accidents, and then uses the DEMATEL method to analyze the influencing factors, and calculates the degree of influence, degree of influence, centrality, and cause of each influencing factor. The key influencing factors were found, and corresponding countermeasures and suggestions were put forward in a targeted manner.

1 INTRODUCTION

With the continuous acceleration of urban modernization, the construction speed and scale of urban viaducts in China have achieved leapfrog progress. The pursuit of construction speed has become a trend, which has led to the frequent occurrence of urban viaduct collapse accidents and huge losses. In recent years, many scholars have done relevant research on urban viaduct collapse accidents and the formulation of management measures, but they rarely analyze the influencing factors of urban viaduct collapse accidents. Therefore, it is necessary to analyze the influencing factors of urban bridge collapse accidents, find out the key influencing factors of the urban viaduct collapse accident, and take corresponding countermeasures to the risk source, so as to prevent the recurrence of the accident and reduce the loss caused.

2 DETERMINATION OF INFLUENCING FACTORS

2.1 *Analysis of influencing factors*

In recent years, many scholars at home and abroad have made relevant research on urban bridge collapse accidents. For example, Min Dan (2015) analyzed the safety risk of large-scale transportation accidents on urban highways and bridges to find out the cause of the accident; Zheng Jian (2017) et al. conducted statistics and analysis of urban overpass construction accidents to find out the cause of the accident; Wang Feng (2020) et al. analyzed the causes of bridge collapse accidents at home and abroad in the past three years, found out the causes of the accidents, and learned lessons. Based on this, statistics and analysis of relevant documents and cases of urban bridge collapse accidents

*Corresponding Author: 1085524142@qq.com

 DOI 10.1201/9781003305026-44

that occurred from 2000 to 2020 are carried out, and final,y the stage and the cause of the accident are shown in Figures 1 and 2.

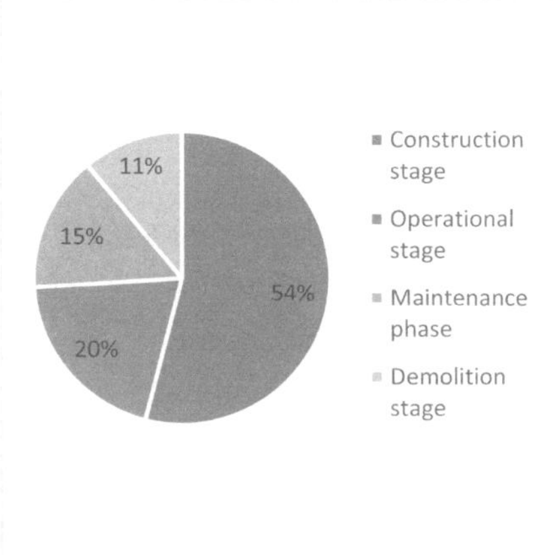

Figure 1. The stage of the collapse accident.

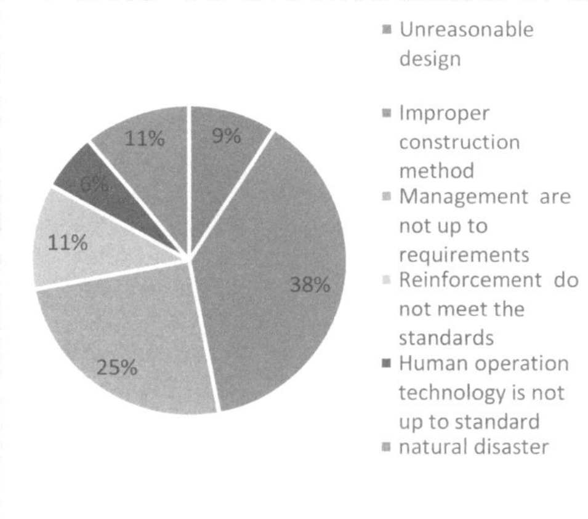

Figure 2. The cause of the collapse accident.

2.2 *Determination of the index system of influencing factors*

Through literature and case analysis, the influencing factors of urban viaduct collapse accidents are designed as a questionnaire survey, the opinions of relevant experts are widely solicited, and the influencing factor indicators are summarized and summarized. Finally, the determined influencing factor system of the urban viaduct collapse accident is shown in Table 1.

Table 1. Influencing factor system.

First level indicator	Secondary indicators	Factor index	Factor description
External factors A1	Policy aspect B1	Mature Disaster Prevention and Mitigation Deepening Design Specification C1	Countermeasures to respond to disaster attacks
		Safety Management System C2	Norms and early warning mechanisms to prevent safety accidents
	Personnel aspect B2	Practitioner technical ability C3	Construction operation ability of construction personnel
		Managerial competence C4	The degree of implementation of safety management by managers
	Natural environment aspect B3	Natural disaster C5	Floods, earthquakes, mudslides, etc.

(*continued*)

Table 1. Continued.

First level indicator	Secondary indicators	Factor index	Factor description
Internal factor A2	Design stage B4	Design structure rationality C6	Reasonable bridge design
		Construction drawing accuracy C7	The construction drawing design and calculation are correct
	Construction stage B5	Material quality C8	Construction materials passed inspection
		Construction method C9	Proper construction method
		Construction Management Safety Education C10	Safety awareness of construction management
		Management of construction equipment C11	Repair and maintenance of construction machinery and equipment
	Operation and maintenance stage B6	Whether the vehicle is overloaded C12	Vehicle overload frequency
		Curing degree C13	Bridge maintenance
		Detection system C14	Inspection of bridge safety and health

3 ANALYSIS OF INFLUENCING FACTORS

The DEMATEL method was proposed by the scholars of the American National Laboratory in 1971. It mainly calculates the "four degrees" value of each factor by establishing a direct relationship matrix, namely, the degree of influence, the degree of influence, the degree of centrality, and the degree of cause, for which the matrix used in combination with the graph theory to quantitatively analyze the logical relationship between various factors. Analyzing the influencing factors of urban viaduct collapse accidents through the DEMATEL method can help managers discover hidden dangers in advance and take preventive measures (Li et al. 2013).

3.1 *Establish a direct influence relationship matrix*

The most important step in the DEMATEL method is to directly affect the determination of the relationship. This paper invited 10 experts to use the 0-4 scale method to score the degree of influence of a certain factor on other factors through the form of expert interviews, and a direct relationship matrix is obtained.

$$X=\begin{bmatrix} & C1 & C2 & C3 & C4 & C5 & C6 & C7 & C8 & C9 & C10 & C11 & C12 & C13 & C14 \\ C1 & 0 & 0 & 0 & 1 & 2 & 2 & 1 & 4 & 2 & 2 & 2 & 1 & 0 & 3 \\ C2 & 0 & 0 & 0 & 2 & 2 & 3 & 2 & 3 & 2 & 0 & 4 & 3 & 2 & 4 \\ C3 & 0 & 0 & 0 & 3 & 3 & 2 & 1 & 4 & 0 & 2 & 2 & 3 & 2 & 2 \\ C4 & 0 & 0 & 0 & 2 & 2 & 1 & 1 & 1 & 1 & 2 & 4 & 1 & 2 & 0 \\ C5 & 0 & 1 & 0 & 0 & 0 & 0 & 0 & 0 & 0 & 0 & 0 & 0 & 0 & 0 \\ C6 & 0 & 2 & 3 & 1 & 1 & 4 & 1 & 2 & 1 & 1 & 1 & 2 & 2 & 2 \\ C7 & 0 & 0 & 0 & 2 & 3 & 2 & 0 & 4 & 1 & 1 & 3 & 1 & 1 & 2 \\ C8 & 0 & 0 & 0 & 0 & 3 & 3 & 2 & 3 & 2 & 1 & 4 & 1 & 1 & 4 \\ C9 & 2 & 2 & 0 & 4 & 3 & 3 & 1 & 2 & 2 & 4 & 4 & 1 & 1 & 3 \\ C10 & 0 & 0 & 0 & 2 & 0 & 2 & 2 & 2 & 2 & 1 & 3 & 1 & 1 & 2 \\ C11 & 0 & 0 & 0 & 1 & 2 & 2 & 1 & 3 & 2 & 2 & 3 & 0 & 2 & 2 \\ C12 & 0 & 0 & 0 & 2 & 3 & 2 & 3 & 0 & 3 & 4 & 4 & 2 & 3 & 4 \\ C13 & 0 & 0 & 0 & 4 & 4 & 0 & 4 & 4 & 4 & 2 & 2 & 3 & 4 & 3 \\ C14 & 0 & 0 & 0 & 1 & 1 & 2 & 1 & 3 & 3 & 3 & 0 & 2 & 2 & 2 \end{bmatrix}$$

Table 2. "Four degrees" summary table.

Influencing factors	Influence F_i	Sort	Affected E_i	Sort	Centrality M M=Fi+Ei	Sort	Causedegree *N* N = Fi - Ei	Sort
C1	1.5591	10	0.1825	14	1.7416	14	1.3766	3
C2	2.2061	4	0.4706	12	2.6767	11	1.7355	1
C3	1.8617	7	0.2882	13	2.1499	13	1.5735	2
C4	1.3177	13	1.9101	8	3.2278	9	−0.5924	10
C5	0.0943	14	2.2643	5	2.3586	12	−2.1700	14
C6	1.9603	5	2.2657	4	4.2260	6	−0.3054	8
C7	1.5160	11	1.6131	10	3.1291	10	−0.0971	7
C8	1.8675	6	2.6917	2	4.5592	3	−0.8242	11
C9	2.4621	2	2.1024	6	4.5644	2	0.3597	6
C10	1.5006	12	2.0653	7	3.5659	8	−0.5647	9
C11	1.5906	9	2.8672	1	4.4578	4	−1.2766	13
C12	2.3781	3	1.5380	11	3.9161	7	0.8401	4
C13	2.7153	1	1.8950	9	4.6103	1	0.8203	5
C14	1.7039	8	2.5794	3	4.2833	5	−0.8754	12

3.2 *Calculate the "four degrees" of influencing factors*

According to the calculation steps of the DEMATEL method, the normalized direct influence matrix is calculated, and then Excel is used to calculate the degree of influence, degree of influence, centrality, and cause a degree of each influencing factor (Li 2016), as shown in Table 2.

3.3 *Determine the causality diagram of influencing factors*

Based on the centrality and cause a degree of the influencing factors of urban viaduct collapse accidents, with centrality as the horizontal axis and cause degree as the vertical axis, the causality diagram of the influencing factors is drawn (Chen 2019), and the coordinate positions of the 14 influencing factors are marked, as shown in Figure 3.

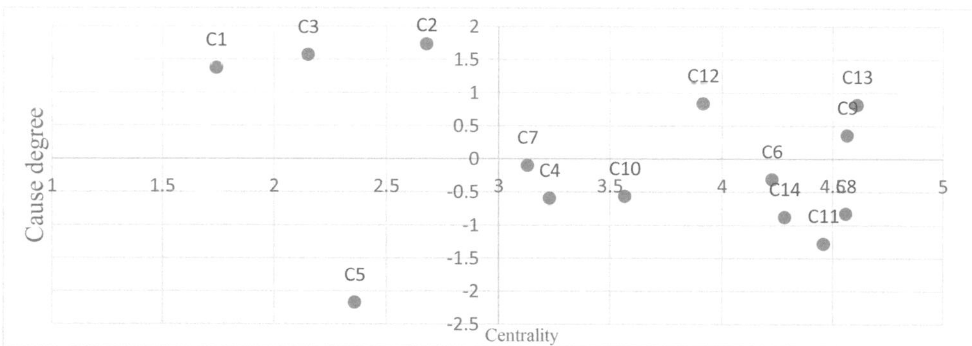

Figure 3. The causality diagram of the key influencing factors of the urban viaduct collapse accident.

3.4 *Result analysis of influencing factors*

3.4.1 *Centrality analysis*

According to the centrality ranking table of the influencing factors of urban viaduct collapse accidents, the top six importance of each factor are ranked from large to small: maintenance degree

C13> construction method C9> material quality C8> construction equipment management C11> inspection system C14 >. The design structure is reasonable C6. The greater the value of the centrality, the greater the sum of the degree of influence and the degree of influence (Li 2016), which has an important influence on determining the key influencing factors of the viaduct collapse accident. The degree of maintenance is an important factor affecting the collapse of an urban viaduct. This factor has the greatest connection with other factors. The management of the viaduct operation and maintenance phase should be strengthened, and real-time control should be carried out through the corresponding detection system.

3.4.2 *Cause analysis*

According to the ranking table of the influencing factors of urban viaduct collapse accidents, the top six factors in order of importance are as follows: safety management system C2> technical competence of employees C3> mature disaster prevention and mitigation deepening design specifications C1> vehicles. Whether it is overloaded C12>cure degree C13>construction method C9. The greater the positive value of the cause degree, the stronger the influence on the overall factor, the larger the negative value, the greater the influence of this factor by other factors on the whole. The safety management system has the largest impact on the overall factors, followed by standardized design requirements. The process level of the construction stage is greatly affected by other factors.

3.4.3 *Causality analysis*

According to the causality diagram, combined with the centrality and cause a degree of the influencing factors of the urban viaduct collapse accident, whether the vehicle is overloaded C12, the maintenance degree C13 and the construction method C9 are located in the first quadrant, and the correlation and cause are high, which are the cause of the urban viaduct collapse accident. The key influencing factors should be listed as priority items. The management ability of managers C4, the rationality of design structure C6, the accuracy of construction drawings C7, material quality C8, construction management safety education C10, construction equipment management C11, inspection system C14 are located in the fourth quadrant, which is an influencing factor of urban viaduct collapse accidents. The core points of the essay are most susceptible to the influence of other factors on the whole.

4 DETERMINE THE KEY INFLUENCING FACTORS

According to the causality diagram shown in Figure 3, combined with the centrality and cause a degree of the factors influencing the urban viaduct collapse accident shown in Table 3, it can be known whether the vehicle is overloaded C12, the maintenance degree C13, the construction method C9, the management ability of the manager C4, the rationality of the design structure C6, the accuracy of the construction drawing C7, material quality C8, construction management safety education C10, construction equipment management C11, inspection system C14, safety management system C2, etc., are the key factors affecting urban viaduct collapse accidents.

5 COUNTERMEASURES

Through the analysis of the influencing factors of urban viaduct collapse accidents, the key influencing factors of urban viaduct collapse accidents are obtained. Therefore, in order to prevent the probability of urban viaduct collapse accidents, these key influencing factors can be controlled.

5.1 *Establish a safety early warning management system*

The safety management system should be improved, the built urban viaducts should be maintained and modified, and disaster prevention management should be carried out. A scientific and quantitative database should be established for the design durability and service life of the bridges (Cai & Jia 2018), a scientific evaluation system should be established, and the disaster warning ability and defensive capabilities should be continuously improved. At the same time, we need to establish a bridge health detection system (Zhang & Li 2016) to continuously detect the built bridges, find problems, and take timely response management and control plans for the bridges to avoid serious safety accidents.

5.2 *Strengthen safety management education for staff*

The construction unit should strengthen the safety management and training of the staff (Sun 2015) so that the construction staff can master the technical specifications of safe production, and hold special seminars to concentrate on learning and solving the processes and construction links that have hidden safety hazards. Training should be strengthened for the use of new technologies and new processes, in this regard, construction personnel can master various operating skills and better guide actual work. At the same time, attention is given to improving the sense of responsibility of on-site management personnel, strengthening supervision and control, and resolving existing safety hazards in a timely manner to ensure construction safety.

5.3 *Pay attention to the application of modern information technology*

An attempt is made to vigorously promote the application of the latest technologies such as big data, and attach importance to the information management of safety work. It is necessary to establish an engineering safety information platform and a database, grasp the situation in the process of bridge construction and operation in a timely manner, and take timely improvement measures when construction safety problems are found, so as to lay a theoretical foundation for improving the level of safety management.

6 CONCLUSION

The influencing factors of highway bridge collapse accidents are multifaceted. The paper qualitatively analyzes the logical relationship among the related factors of urban bridge collapse accidents through expert interviews, finds out common key influencing factors, and puts forward countermeasures and suggestions in a certain way. The theoretical and practical significance of later studies can be constructed based on the influencing factors of urban viaduct collapse accidents, and the maturity evaluation of the prevention of urban viaduct collapse accidents can provide a certain reference for ensuring the safe operation of urban bridges.

REFERENCES

Cai Tangtao, Jia Yanwu. Analysis of Bridge Collapse Accidents[J]. Road Machinery and Construction Mechanization, 2018, 35(05): 180–184.

Chen Sunlight. Analysis of key factors of green supply chain based on dual-theme DEMATEL method [J]. Logistics Technology, 2019, 38(09): 119–124.

Li Kebai, Qi Baoku, Wang Huan. Analysis of the restrictive factors in the development of prefabricated buildings based on DEMATEL[J]. Housing Industry, 2013(08): 49–51.

Li Xiaolong. Analysis and Strategy Research on Influencing Factors of Economic Operation of Wind Power Projects [D]. North China Electric Power University, 2016.

Sun Tiejun. Causes of highway bridge collapse and quality control countermeasures[J]. World of Transportation (Transport. Vehicle), 2015(07): 114–115. DOI: 10.16248/j.cnki.cn11-3723/u.2015.07 .051.

Zhang Zhijie, Li Qilong. Research on the causes of bridge collapse accidents[J]. Jiangxi Building Materials, 2016(06):184+188.

Advances in Urban Engineering and Management Science – Khalil & Yang (Eds)
© 2023 The Author(s), ISBN 978-1-032-30426-7

Study on horizontal spacing of H-shaped steel piles for HU combined sheet pile support system

Junwei Chen*
School of Civil Engineering and Architecture, University of Jinan, Jinan, Shandong, China

Junyan Liu*
The Engineering Technology Research Center for Urban Underground Engineering, Supporting and Risk Monitoring of Shandong Province, China

Quanming Zhen
Shandong Bureau of China Metallurgical Geology Bureau, Shandong, China

ABSTRACT: HU combined sheet pile support system is a new type of foundation support system with traditional Larsen sheet pile and H-beam pile, there is no mature theoretical basis for the horizontal spacing of H-beam piles set behind the Larsen sheet pile wall. The formula for calculating the spacing of H-beam piles satisfies the static equilibrium condition and strength condition and is corroborated by MADIS GTS/NX 3D numerical simulation. Finally, an effective calculation method for the maximum value of horizontal spacing of H-beam piles in the HU combined sheet pile support system is obtained to guide the actual project.

1 INTRODUCTION

In the context of rapid economic development, the development of urban underground space has become an inevitable trend of modern urban development. Larsen sheet piles are being widely adopted due to many advantages such as convenient construction and relatively low cost and can be pulled out and reused (Li 2016). However, in deep foundation pit projects, Larsen sheet piles as a flexible support system in many cases cannot meet the stiffness requirements of the support structure. The HU combined sheet pile support system, in which a row of H-beams is constructed on the inner side of the sheet pile as the main load-bearing member to meet the stiffness and stability of the whole support structure, can be a good solution to this problem (Sun 2014). Deguo Kong combined engineering examples to prove that the application of MADIS GTS/NX software for the simulation of SP-IV Lassen sheet piles has good results (Kong 2019). Xiangming Xiao proposed a HSW method pile foundation support structure, whose structure mainly includes Larsen sheet piles and H-shaped steel piles. The openings of two adjacent Larsen sheet piles are arranged in reverse, and H-shaped steel piles are arranged on the inner side of the Larsen sheet piles. Such structure achieves large structural stiffness, and has the function of retaining soil and stopping water, which is suitable for the purpose of excavating deep foundation pits (Xiao 2010). At present, the practical application of HU combined sheet pile support structure in deep foundation pits is relatively small, and the study on the calculation method of H-beam setting spacing after Larsen sheet pile can provide a reference for similar projects in the future, which has certain theoretical and practical significance.

*Corresponding Authors: 2474415905@qq.com and liujunyan1220@163.com

 DOI 10.1201/9781003305026-45

2 THEORETICAL DERIVATION OF MAXIMUM PILE SPACING OF H-SHAPED STEEL PILES

When a part of the supporting soil body yields, the yielding soil body will move out from its original position, and the relative displacement of the yielding soil body and the adjacent stationary soil body will be hindered by the shear stress between the two parts of the soil body. As the shear stress hinders the tendency to retain the yielded soil body to its original position, the soil pressure in the yielded region decreases while the soil pressure in the adjacent stationary soil body increases. This phenomenon of soil pressure transferring from the yielding region to the neighboring stationary region is usually called the soil arch effect (Yang 2014).

In geotechnics and engineering, compressive stresses are generally specified as positive and tensile stresses as negative, and the Uniform Strength Theory can be expressed in the following form:

$$F = \alpha\sigma_1 - \frac{b\sigma_2 + \sigma_3}{1+b} = \sigma; \sigma_2 \le \frac{\sigma_3 + \alpha\sigma_1}{1+\alpha} \tag{1}$$

$$F' = \frac{\alpha}{1+b}(b\sigma_2 + \sigma_1) - \sigma_3 = \sigma_1; \sigma_2 \ge \frac{\sigma_3 + \alpha\sigma_1}{1+\alpha} \tag{2}$$

where b is the unified strength parameter, which varies from 0 to 1. The unified strength theory can be translated into various criteria, and a series of new damage criteria are also generated. Among all the externally convex limit surfaces, the limit surface of the single shear strength theory at b=0 is the smallest; the limit surface of the double shear strength theory at b=1 is the largest (Yu 2020). If the cohesive force c and the angle of internal friction Φ, which are commonly used in geotechnical engineering, are used as the basic parameters, the unified strength theory can be expressed as follows:

$$\sigma_1 = \frac{2C\cos\varphi}{1+\sin\varphi} + \frac{1-\sin\varphi b\sigma_2 + \sigma_3}{1+b)(1+\sin\varphi}; \sigma_2 \le \frac{\sigma_1(1-\sin\varphi) + \sigma_3(1+\sin\varphi)}{2} \tag{3}$$

$$\sigma_1 = \frac{(1+b)(2C\cos\varphi + \sigma_3 - \sigma_3\sin\varphi)}{1+\sin\varphi} - b\sigma_2; \sigma_2 \ge \frac{\sigma_1(1-\sin\varphi) + \sigma_3(1+\sin\varphi)}{2} \tag{4}$$

Since the earth arch is powerful before the arch, the earth arch must maximize its efficiency and form a reasonable arch axis. According to the aforementioned assumptions, the earth arch is a symmetrical structure, and its semi-structured can be taken for study, and the simplified calculation model is shown in Figures 1 and 2.

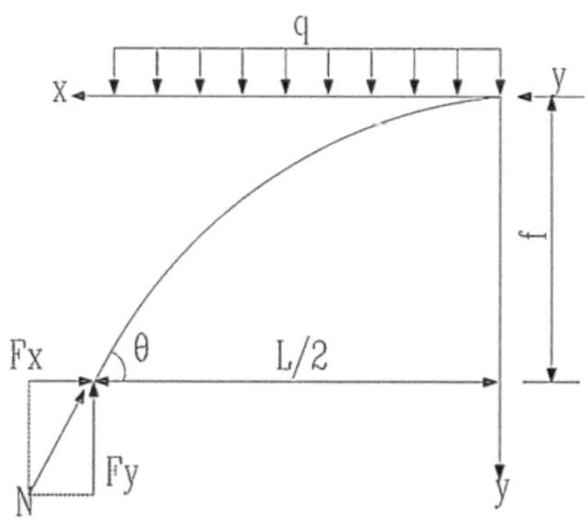

Figure 1. Sketch of soil arch calculation.

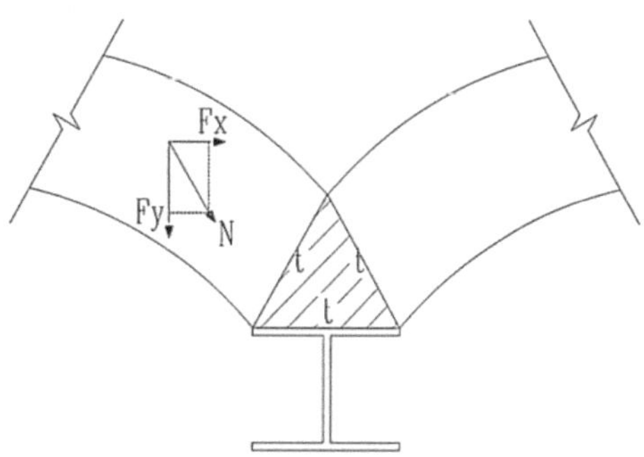

Figure 2. Soil arch foot force diagram.

To ensure the normal function of the soil arch between two adjacent piles, the soil arch among the piles should meet the static equilibrium conditions. Where the soil arch span is L, the arch ring thickness is t, the arch vector height is $f = 1/4 \tan\phi$ (Jia 2004). The distributed pressure acting on the soil arch at depth z is q, and the reaction forces at the foot of the arch are F_x and F_y. Since the loads on the adjacent soil arches are roughly equivalent, the x-direction will not produce a form of damage due to insufficient friction at the back of the pile, and according to the force and moment equilibrium conditions of the soil arch, it is obtained that:

$$F_x = \frac{ql^2}{8f}dz; F_y = \frac{ql}{2}dz \tag{5}$$

Then the axial force in the tangential direction to the foot of the arch is:

$$N = \sqrt{F_x^2 + F_y^2} = \frac{ql}{8f}\sqrt{l^2 + 16f^2}dz \tag{6}$$

The force in the normal direction at this location is zero. In addition, the location of the arch foot of the earth arch is also affected by the self-weight stress. After analysis, the positive stress along the tangent direction of the arch foot is the major principal stress σ_1, the self-gravity stress is the intermediate principal stress σ_2, and the minor principal stress σ_3 is the force in the normal direction and has zero sizes. Then there are:

$$\sigma_1 = \frac{N}{tdz} = \frac{ql}{8ft}\sqrt{l^2 + 16f^2}; \quad \sigma_2 = \gamma_z; \quad \sigma_3 = 0 \tag{7}$$

Bringing Equation (7) into Equation (3), Equation (4) is organized to yield:

$$\frac{ql}{8ft}\sqrt{l^2 + 16f^2} = \frac{\gamma_z(1 - \sin\phi)b + 2c(1 + b)\cos\phi}{(1 + b)(1 + \sin\phi)}; \gamma_z \geq \frac{ql(1 - \sin\phi)\sqrt{l^2 + 16f^2}}{16ft} \tag{8}$$

$$\frac{ql}{8ft}\sqrt{l^2 + 16f^2} = \frac{2c(1 + b)\cos\phi}{1 + \sin\phi} - b\gamma_z; \gamma_z \leq \frac{ql(1 - \sin\phi)\sqrt{l^2 + 16f^2}}{16ft} \tag{9}$$

When calculating the load q on the soil arch, it can be found from the soil pressure part behind the pile according to the Rankine soil pressure formula. When the soil body c, φ, and the soil arch uniform force are known, the span l of the soil arch between the H-shaped steel piles can be obtained from Equations (8) and (9), and thus the maximum pile spacing of H-shaped steel piles can be obtained when Larsen sheet piles are not used. Assuming a single powder clay soil layer, soil capacity of 19.6kN/m^3, cohesion of 27.8kpa, internal friction angle of 16.8°, excavation depth of 7.5 m, and H-pile support, the maximum net distance between H-piles is 1.79 m according to the above derivation formula.

3 THREE-DIMENSIONAL FINITE ELEMENT SIMULATION AND ANALYSIS OF RESULTS

3.1 *Computational modeling*

MADIS GTS/NX is used to establish the three-dimensional calculation model, the in-plane calculation model size is taken 3-5 times the excavation depth of the pit, the calculation range taken in this case is 50 m×40 m×25 m (X×Y×Z), the excavation depth of the pit is 7.5 m, the length of H-beam and Larsen sheet pile are 15m, H-beam is chosen as HN700×300 narrow flange H-beam, and Larsen sheet pile is chosen as SP-IV. The steel sheet pile is equivalent to 116.2 mm thick homogeneous steel plate by the "equivalent flexural stiffness method (Yang 2016)". The load at the top of the pit is set at 30kpa, 2 m from the edge of the pit, and in this area, the soil units, H-beams, and Larsen sheet piles in the excavation area are activated and passivated to simulate the support structure and its surrounding soil displacement after excavation. A single powder clay stratum is selected as the stratum in the 3D calculation model. The soil body adopts Moore-Coulomb elastoplastic principal model.

3.2 *Model boundary constraints*

The numerical simulation load is the self-weight load and foundation pit perimeter load. The boundary conditions of the 3D computational model are: Z-directional constraint for the bottom surface of the model, Y-directional constraint for the front and back of the model, and X-directional constraint for the left and right of the model.

3.3 *Simulation of construction working condition setting*

The foundation support form is simulated in four cases. Each support form only simulates the final working condition of the pit excavation to the end, and each supported form is zeroed out after the calculation is completed, and the other working conditions are continued to be simulated. The corresponding working conditions of each type of support form are shown in Table 1.

Table 1. Simulated working conditions.

Work conditions	Construction instructions
Work conditions1	Larsen sheet pile after H-beam inserted one jump one arrangement (spacing 800mm)
Workconditions2	Larsen sheet pile after H-beam insert a jump three arrangement (spacing 1600mm)
Work conditions3	Larsen sheet pile after H-beam inserted a jump five arrangement (spacing 2400mm)
Work conditions4	Larsen sheet pile after H-beam inserted a jump seven arrangement (spacing 3200mm)

3.4 *Three-dimensional finite element simulation results and analysis*

By changing the spacing of H-beam pile after Larsen sheet pile, the calculation model of different spacing of H-beam after Larsen sheet pile wall is established, and the horizontal displacement deformation of a pile of Larsen sheet pile when the spacing of H-beam pile increases is analyzed by comparing the model calculation results (positive horizontal displacement of pile moves towards the pit).

Collating the model calculation results, we can get that when H-beam is inserted into one jump one arrangement, the maximum horizontal displacement of the Larsen sheet pile between the H-beam piles is 11.83 mm. When H-beam inserted one jump three arrangement the maximum horizontal displacement of Larsen sheet pile between H-beam piles is 16.52 mm. When H-beam inserted one jump five arrangement, the maximum horizontal displacement of the Larsen sheet pile between H-beam piles is 20.43 mm. When the H-beam inserted a jump seven arrangement, the maximum horizontal displacement of the H-beam pile between the Larsen sheet pile is 33.56 mm.

By comparing the horizontal displacement of the H-beam Larsen sheet pile under four kinds of H-beam spacing conditions, it can be seen that the spacing of the H-beam is smaller in working condition one and working condition two, and the H-beam and Larsen sheet pile share the post-pile soil pressure under the effect of inter-pile soil arch, and the deformation of the whole support structure is more uniform. The spacing of the H-beam in working conditions three and four is larger than the maximum pile spacing of the H-beam pile calculated by the derivation formula, which is the critical width of the soil arch between H-beam piles. However, because the Larsen

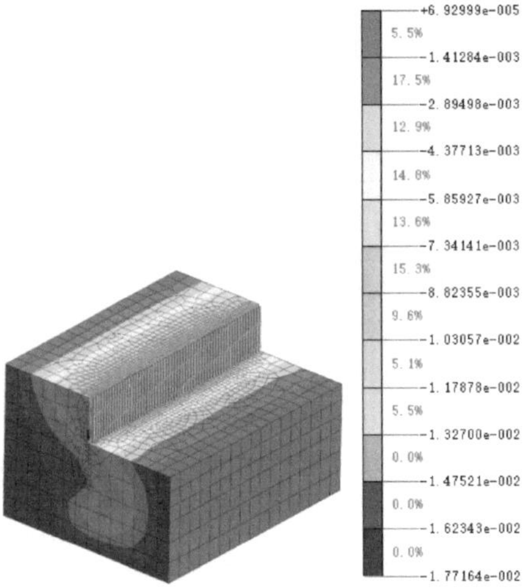

(a) H-beam insert one jump one arrangement.

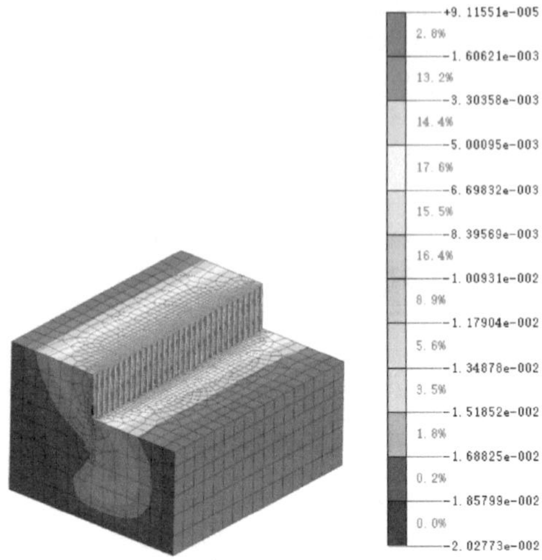

(b) H-beam insert one jump three arrangement

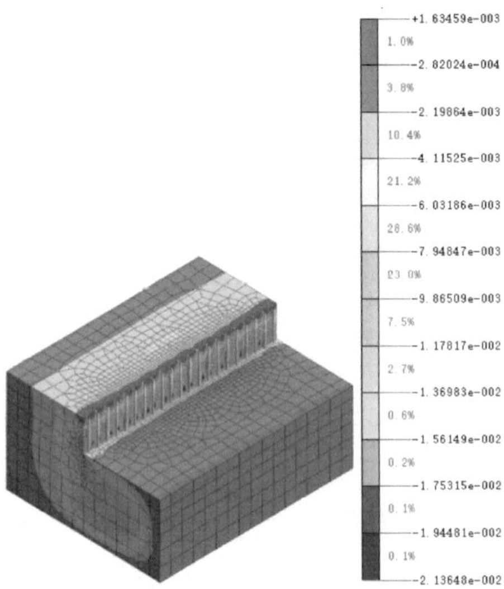

(c) H-beam insert one jump five arrangement.

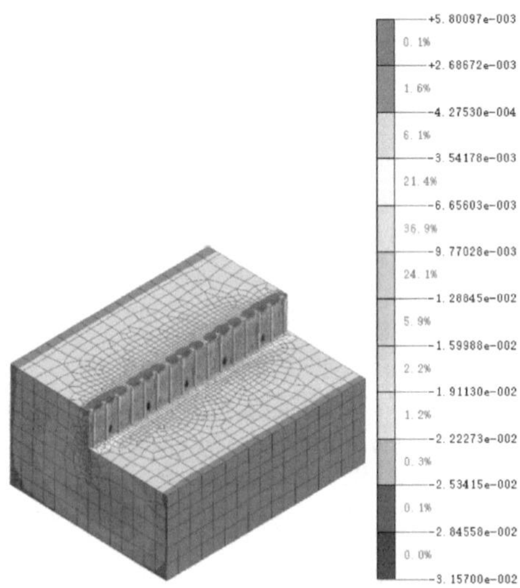

(d) H-beam insert one jump seven arrangement.

Figure 3. Pile horizontal displacement cloud map (m).

sheet pile itself has certain stiffness, the support structure deformation in working condition three is still within the control range. In case four, the horizontal spacing of the H-beam reached 3200 mm and the soil arch among the piles was destroyed. The Larsen sheet pile was a flexible support structure with insufficient stiffness to bear the soil pressure between the H-beam piles, and due to the restraint of H-beam piles with greater stiffness on both sides, the Larsen sheet pile between the H-beam piles produced horizontal and vertical bending in two directions. The maximum horizontal displacement was generated near the bottom of the pit, reaching 33.56 mm which exceeded the deformation control value of the Larsen sheet pile.

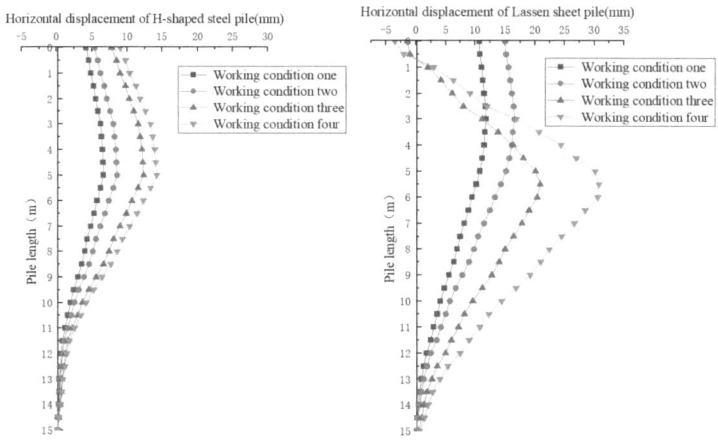

Figure 4. Horizontal displacement diagram of the pile for each working condition.

4 CONCLUSION

Due to the addition of the H-beam, the overall stiffness of the Larsen steel sheet pile support system is greatly improved so that it can be more safely applied to deep foundation support projects. The calculation results of the MADIS GTS/NX three-dimensional model can be obtained from the unified strength theory derived from Formulas (8) and (9) to calculate the maximum pile spacing of H-beam piles of the HU combined sheet pile system, which meets the specification requirements and has a certain safety redundancy. For the current HU combined steel sheet pile system in the actual engineering application of H-type steel piles in the dense construction, the method proposed in this paper can be calculated in the actual project to reduce the number of H-type steel piles after the insertion of Larsen sheet piles, effectively reducing the cost of foundation pit enclosure structure construction. In summary, the formula has a certain guiding significance for the actual project.

ACKNOWLEDGMENT

Key R&D Projects of Shandong Province (project number 2019GSF111064).

REFERENCES

Jia, H L. (2004) Pile spacing analysis of anti-slip piles and retaining wall piles based on soil arch effect. J. Journal of Engineering Geology, 12(01):98–103.

Kong, D G. (2019) Numerical simulation of double-row Larsen sheet piles based on Midas GTS/NX and its engineering application. J. Engineering Construction, 51(07): 39–42+60.

Li, Q Y. (2016) Application of Lassen sheet pile support construction technology. J. Anhui Construction, 23(06): 105–106.

Sun, H T. (2014) Research on the application of green combined steel sheet piles in deep foundation pits. D. University.

Xiao, X M. (2010) Application of steel sheet pile in the design of a large punching pit support structure. J. Shanxi Construction, 36(4): 117,164.

Yang, L. (2016) Design of combined sheet pile (HSW method) support structure and its engineering application. D. University.

Yang, X Q. (2014) Analysis of soil arch effect between anti-slip piles and its soil arch mode. J. Chinese Journal of Highways, (01): 30–37.

Yu, M H. (2020) A new method for determining soil damage criterion. J. Journal of Xi'an Jiaotong University, 54(08):1–10.

Advances in Urban Engineering and Management Science – Khalil & Yang (Eds)
© 2023 The Author(s), ISBN 978-1-032-30426-7

Simulation study on vibration frequency of monolayer cable net glass curtain wall

Guangyang Xu & Gang Chen*
Zhejiang Zhongtianfangyuan Curtain Wall Co., Ltd., HangZhou, Zhejiang, China

ABSTRACT: In the vibration frequency simulation of glass curtain walls, the extraction of natural vibration characteristics is inaccurate, which affects the simulation accuracy. A vibration frequency simulation method of monolayer cable net glass curtain wall is proposed. Assuming that all materials are in an elastic working state, the finite element model of the monolayer cable net structure of the glass curtain wall is established, and the natural vibration characteristics of the single-layer cable net are extracted according to the static balance equation of the curtain wall. When the fundamental frequency of the main structure is close to the cable network, the shape of the maximum displacement response curve of the cable network node is close to the calculated result of the monolayer cable network. According to the linear correlation between the bearing capacity of the curtain wall panel and natural frequency, the critical frequency of wind load is calculated. The concept of standard working conditions is introduced to simulate the vibration frequency of the glass curtain wall, and the structural vibration mode and frequency are obtained. Under the condition of an intact cable net, the relative error measured by this method is 0.162%, which is 0.172% and 0.244% lower than that of the vibration frequency simulation method of monolayer cable net glass curtain wall-based on target mode and cable interaction. This method can improve the simulation accuracy and is conducive to the monitoring of vibration frequency in practical engineering.

1 INTRODUCTION

As a kind of external enclosure structure of modern buildings, the glass curtain wall is composed of supporting skeletons and glasses. In large-scale public buildings, the single-layer cable net structure of the glass curtain wall is widely used because it is favored by most designers because of its permeability, livability, beauty, and a strong sense of technology (Guo et al. 2019). Although the application of the structure of the glass curtain wall is increased, the problems brought by the structure of the glass curtain wall are also gradually aggravated. The single-layer cable net structure of the glass curtain wall system is a concave flexible cable mesh support system composed of a large suspension cable and a monolayer orthogonal cable net system. The monolayer plane cable net structure cross braided by horizontal and vertical cables resists external load and belongs to a prestressed geometric nonlinear flexible system (Zhang 2019). When the glass curtain wall is put into use, all kinds of failure forms will be produced. The wind resistance and seismic performance of glass curtain wall components cannot meet the requirements, and they will be damaged and changed too much under the complex effects of component weight, external wind load, and temperature change, so as to hinder their use. The whole structural system supporting it can be easily seen through glass, so it can not only be used for support but also play a dual role in expressing structural beauty (Guo & Liu 2019). In the structural form of cable net glass

*Corresponding Author: chengang@zjztfymqyxgs.wecom.work

 DOI 10.1201/9781003305026-46

curtain wall, the external envelope and the main structure form a hole through connection, and they interact and influence each other (Ashrith et al. 2019; Zhang et al. 2021). This difference in displacement will cause the deformation of components of the outer envelope curtain wall structure (Pan et al. 2020). At the same time, the curtain wall components will also be constrained by the lower floors, resulting in an additional horizontal force, which eventually leads to damage to the cable net curtain wall structure system (Wang & Zhou 2021). The vibration frequency simulation of monolayer cable net glass curtain wall can give full play to the mechanical properties of steel cable and contribute to the analysis of the overall stress of curtain wall engineering. Therefore, this paper studies the vibration frequency simulation of monolayer cable net glass curtain wall so that the frequency method can be better applied to the test of cable force of a single cable curtain wall, to provide a reference for later engineering practice.

2 SIMULATION METHOD OF VIBRATION FREQUENCY OF MONOLAYER CABLE NET GLASS CURTAIN WALL

2.1 *The finite element model of monolayer cable net glass curtain wall*

The prestress of the cable is the life of the cable. If the cable wants to have the necessary stiffness and stability, it must be prestressed. To simulate the vibration frequency, it is necessary to analyze the prestress of the cable (Gong 2019; Wang & Deng 2021). The stability of steel members can be achieved by relying on geometric characteristics such as the section shape of the members and the wall thickness of the members (Yvonne et al. 2019). In this paper, the cooling method is used to apply the pretension of the cable to reduce the temperature of the steel cable, to achieve the effect of applying the pretension. The cross-section of the cable is twisted by multiple stainless-steel wires. The relationship between elastic modulus and strain ratio can be expressed as:

$$W = \alpha PS \tag{1}$$

In Formula (1), W represents the pretension applied by the cable; α is the initial strain; P represents elastic modulus; S represents the cross-sectional area of the cable. Formula (1) is deformed and simplified to obtain the required cooling value. The calculation formula is:

$$\Delta\tau = \frac{W}{\alpha PS} \tag{2}$$

In Formula (2), $\Delta\tau$ represents the temperature before and after the change. After determining the pretension value applied to the cable, the temperature change value is calculated, and then the temperature load is determined through numerical simulation to realize the application of pretension. After the initial size defect is applied to the glass panel, the gradually increasing horizontal shear force is applied to the glass panel after the size is updated to check the mechanical properties of the glass panel in the process of gradually increasing load. The deformation of the gasket between the supporting steel claw joint and the glasses was ignored. All connection points between the steel claw joint and the glass panel are assumed to be ideal friction-free hinged joints. The buckling mode shape presents an antisymmetric buckling shape with the midpoint of the tensile diagonal as the axis and the tensile diagonal remain in the plane.

2.2 *Extraction of natural vibration characteristics of monolayer cable network*

The vibration of the structure under wind load is not only related to the characteristics of wind load but also closely related to its own natural vibration characteristics. In order to simulate the vibration frequency of the monolayer cable net glass curtain wall, it is necessary to understand its natural frequency and vibration mode characteristics. In the finite element analysis software, different numerical calculation models are established by changing the parameters, so as to study the change law of the natural vibration characteristics structure under the change of parameters. Based on the finite element model, the static equilibrium equation of the curtain wall can be obtained on the premise of considering the gravity of the curtain wall itself. The natural vibration of the cable is a slight free vibration near the static wind balance position. The balanced equation of the glass curtain wall in the horizontal direction can be expressed as:

$$F_a + bgdc = \frac{da}{dc} + \frac{\partial}{\partial c}\left(\frac{da}{dc}\right) dca \tag{3}$$

In Formula (3), a is the displacement in the horizontal direction when the vibration starts from the equilibrium position; F_a represents the tensile force along the height direction; b is the mass per unit length of cable; g represents gravitational acceleration; c represents the cable force increment caused when the cable vibrates from the equilibrium position. Similarly, the balanced equation of glass curtain wall in the vertical direction can be expressed as:

$$H_y = Gdy + \frac{dy}{dc} + \frac{\partial}{\partial c}\left(\frac{dy}{dc}\right) dc \tag{4}$$

In Formula (4), H_y represents the tensile force along the horizontal direction; y is the displacement in the vertical direction when the vibration starts from the equilibrium position; G represents the average wind load acting on the cable along the cable length within the unit width. Due to the small horizontal and vertical spans of the cable net, the free vibration dynamic increment of the transverse and longitudinal undamped cable structure system can be obtained by ignoring the second-order micro term. This also leads to more and more errors if we continue to assume that the change of vertical cable tension caused by glass self-weight is not considered to analyze the natural vibration frequency and vibration mode of the structure.

2.3 *Critical frequency for calculating wind load*

Strong wind is an important external factor affecting the safety state of the curtain wall panels. The wind pressure resistance capacity of the glass curtain wall is an important technical parameter, and its size depends on the quality of the panel itself and the strength of constraints. With the increase of the applied prestress, the out-of-plane displacement of the cable network decreases, indicating that it can increase the out-of-plane stiffness. In order to calculate the critical frequency of wind load, it is necessary to define the standard working condition. The condition of curtain wall glass with maximum deflection under specific wind pressure is defined as a standard working condition, and its calculation formula is:

$$k = \frac{\gamma \beta \varphi^4}{u} \tag{5}$$

In Formula (5), k is the maximum deflection under wind load; γ is the deflection coefficient; β represents the reduction factor; φ is the length of the short side; u represents the bending stiffness of the glass panel. When the deflection of curtain wall glass under wind load is greater than the maximum deflection, it indicates that the constraint condition of actual working condition is worse than that of standard working condition. At this time, the natural frequency of the glass curtain wall is less than the critical frequency.

2.4 *Simulating the vibration frequency of glass curtain wall*

When deriving the vibration equation, considering that the effect of damping on its natural vibration characteristics is very small, the influence of damping on the structure system is ignored in the analysis. In the tensioning stage of a single cable curtain wall, the cable tension is tested, and the free vibration equation without considering the influence can be obtained as follows:

$$[Z]\{v_1\} + [D]\{v_2\} = \{0\} \tag{6}$$

In Formula (6), $[Z]$ represents the mass matrix of cable net curtain wall structure; $\{v_1\}$ represents the displacement vector of the structure; $[D]$ represents the stiffness matrix in the tangent direction when the cable net curtain wall is in static equilibrium; $\{v_2\}$ represents the acceleration vector of the structural system. To obtain free vibration with finite amplitudes, it must make the determinant of the coefficient in Formula (6) zero, that is:

$$\left|[Z] - \vartheta^2[D]\right| = 0 \tag{7}$$

Formula (8) calculates the natural frequency of curtain wall structure, and ϑ is the natural vibration circle frequency. The vibration of the structure of the glass curtain wall is mainly in the form perpendicular to the plane, the first vibration mode of the structural system is perpendicular to the front and rear vibration of the curtain wall plane, and the vibration direction is relatively single.

3 TEST ANALYSIS

3.1 *Test model and preparation*

In this study, a vibration frequency simulation method of monolayer cable net glass curtain wall is proposed. In order to verify the effectiveness of the method, the test model is designed and tested. In the design of the test model, in addition to drawing lessons from previous experience, the model size is verified by finite element software. The design model is mainly composed of the glass panel, outer frame, diagonal brace, and cross brace. The external dimension of the selected model is 2.2 m×1.5 m. The connecting claw adopts 100 series, the vertical and horizontal cables are fixed on the outer frame beam through anchors, and the cable nodes are connected with the glass panel through the connecting claw. The size of the glass plate is 500 mm×500 mm, the thickness is 10 mm, and the diameter of the steel cable is 10.5 mm. The model material table is shown in Table 1.

Table 1. Test model materials.

Category	Dimension (mm)	Category	Design number
Ordinary tempered glass	500×500×10	Single-hole connecting claw	1001
Diagonal brace	50×50×5	Double holes connecting claw	1003
Cross brace	100×50×6	Four holes connecting the claw	1003
Side beam	100×100×10	Anchorage	DM pier head anchorage

In the experimental model, the joint connection mainly includes the joint connection between transverse cable and longitudinal cable and the fixation of the tempered glass panel, as well as the connection between cable and outer frame beam. In this experiment, to simulate the vibration response of the monolayer cable net in the structure of the glass curtain wall system, the force hammer is used as the excitation device. In order to verify the effect of vibration frequency simulation of glass curtain walls under different working conditions, two different working conditions are set for the test, namely, intact cable network and cable prestress loss of 30% and 60%.

3.2 *Results and analysis*

The method designed in this paper is compared with the vibration frequency simulation method of monolayer cable net glass curtain wall based on target mode and cable interaction. The evaluation index is the relative error between the simulated frequency and the measured frequency, and the calculation formula is:

$$\varepsilon = \frac{f_1 - f_0}{f_0} \times 100\% \tag{8}$$

In Formula (8), ε is the relative error; f_1 and f_0 correspond to the analog frequency and measured frequency respectively. The relative error-index is used to compare the difference between the simulated frequency and the measured frequency of cables with different methods. The comparison results of the three simulation methods under the working conditions of intact cable net and cable prestress loss of 30% and 60% are shown in Figure 1.

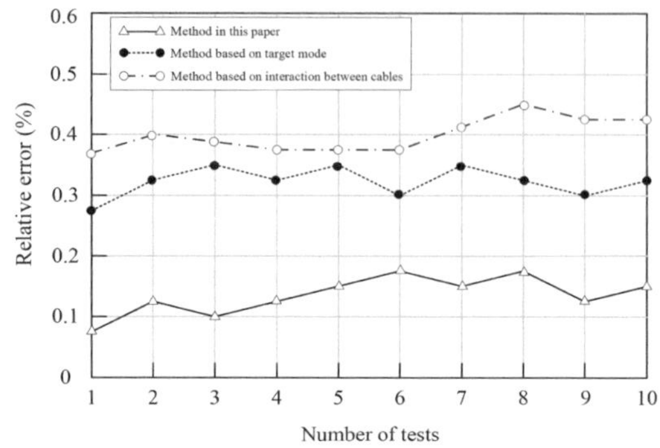

(a) Intact cable net.

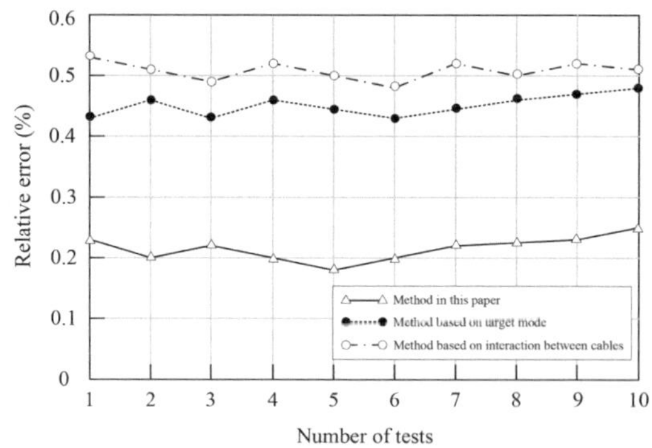

(b) Cable prestress loss 30%.

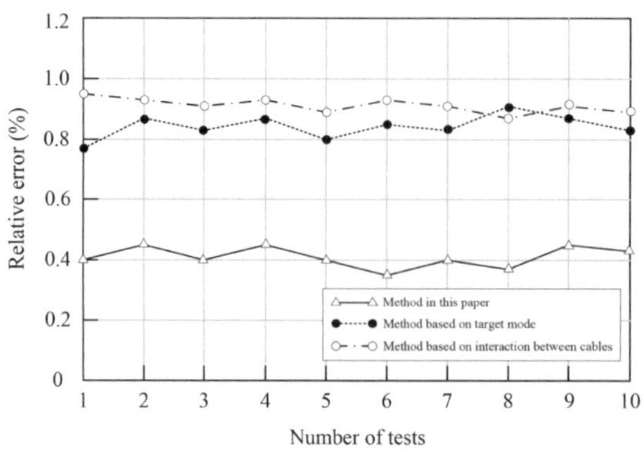

(c) Cable prestress loss 60%.

Figure 1. Comparison of results of different working conditions.

It can be seen from Figure 1 that under the working conditions of 30% and 60% prestress loss of intact cable network and cable, there is an obvious difference between the relative error between the simulated frequency and the measured frequency of each method. With the increase of cable prestress loss, the relative error of each simulation method increases accordingly. Taking the working condition of an intact cable net as an example, the relative error of this method is 0.162%, which is 0.172% and 0.244% lower than that of the vibration frequency simulation method of monolayer cable net glass curtain wall based on the target mode and cable interaction. Therefore, the error between the simulated frequency data and the measured data is within the allowable range and can improve the simulation accuracy, which is conducive to the smooth implementation of the project.

4 CONCLUSION

The cable net structure of the glass curtain wall is obtained by combining the structure of the glass curtain wall and the cable net structure, which represents the development trend of the large-scale curtain walls. In the single cable glass curtain wall system, stainless steel prestressed cable is an important component. It can form the stress system of the curtain wall and bear external load only after appropriate pretension is applied. Due to the characteristics of cable net structure such as lightweight and small damping, the effect of wind load on it can not be ignored. The relative error measured by this method under the working condition of an intact cable net is 0.162%, which is 0.172% and 0.244% lower than the vibration frequency simulation method of monolayer cable net glass curtain wall based on target mode and cable interaction. This method has a good application effect. The research in this paper still has some limitations. In the follow-up, further analysis can be made on the factors affecting the actual vibration frequency of curtain wall cable, including the length diameter ratio of cable, end connection structure, support constraints, and additional mass on cable such as connecting claw, transverse cable, etc.

REFERENCES

Ashrith H S, Doddamani M, V Gaitonde. (2019) Effect of wall thickness and cutting parameters on drilling of glass micro balloon/epoxy syntactic foam composites[J]. Composite Structures, 211(MAR.):318–336.

Gong Jie. (2019) Simulation of Wind Induced Stress Detection Algorithm for Point-Supported Building Glass Curtain Wall[J]. Computer Simulation, 36(11):356–359,395.

Guo Weihua, Liao Jie, Wang Bin. (2019) Design of corner plan for a large-span monolayer cable net glass curtain wall[J]. Building Structure, (10):18–21.

Guo Yurun, Liu Junhong. (2019) Construction and simulation of condensate water cooling glass facade[J]. New Building Materials, 46(11):82–86.

Pan Fengqun, Jiang Xiangjun, Iang XiangjunFAN Yesen, et al. (2020) Optimization Design of Structural Surface Precision of Shape Memory Cable Mesh[J]. Journal of Mechanical Engineering, 56(9):1–8.

Wang Xintao, Deng Hua. (2021) Target modes testing strategy for monitoring the key stiffness of cable net structures[J]. Journal of Vibration and Shock, 40(2):204–212.

Wang Xiu-li, Zhou Lei. (2021) Dynamic response analysis of CFRP strengthened debris flow flexible cable net system[J]. Journal of the Lanzhou University of Technology, 47(1):136–143.

Yvonne R, Morari M, Smith R S . (2019) Control of an Architectural Cable Net Geometry[J]. IEEE Transactions on Control Systems Technology, PP(99):1–15.

Zhang Bai-zhen. (2019) Research on Design and Construction of Single-layer Cable Net Curtain Wall Structure in High-rise Buildings[J]. Building Technique Development, 46(2):12–13.

Zhang Yuanzhi, WEI Mingyu, HUANG Youbang, et al. (2021) Test research and numerical simulation on vibration test method of glass curtain wall[J]. Sichuan Building Science, 47(1):16–23.

Advances in Urban Engineering and Management Science – Khalil & Yang (Eds)

Research on influencing factors of prefabricated construction cost based on DEMATEL

Yanjing Zhao* & Daiyan Mei
Department of Urban Construction, Anhui Xinhua University, Hefei, China

Min Cheng
Anhui sanjian Engineering CO., LTD, Hefei, China

ABSTRACT: The cost of prefabricated building is the key to restricting the development of the prefabricated building. This paper decomposes the construction process of prefabricated buildings, uses the decision laboratory method (DEMATEL) to build a model, finds out the key factors affecting prefabricated buildings, points out the relationship between various factors, and then achieves the purpose of cost control by controlling the key cost influencing factors, so as to provide a reference for the development of prefabricated buildings in the future.

1 INTRODUCTION

Building industrialization is the core content of the housing industrialization construction mode. The basic characteristics of traditional buildings are high energy consumption, high pollution, and high labor power. In recent years, China has been encouraging and promoting prefabricated buildings. It has the advantages of energy conservation, environmental protection, shortening construction period, and noise reduction, and effectively solves the disadvantages of traditional reinforced concrete buildings. However, the high cost of prefabricated buildings has become a key obstacle to the development of prefabricated structures in China. The research on the influencing factors of prefabricated construction cost has important theoretical and practical significance to promote the further development of prefabricated components, and also conforms to the general direction of the development of the construction industry in the new era (Qi, Zhu, Ma 2016). Many domestic scholars have studied the influencing factors of prefabricated construction costs. Zheqi LV (Lv, Zhang 2020) used a fuzzy analytic hierarchy process to study the influencing factors of prefabricated construction costs from three aspects of production, transportation, and construction. Kai Liu et al. (Liu, Ding 2019) used DEMATEL to analyze the formation of influencing factors of the life cycle cost of prefabricated buildings. Yuanyue Chen (Chen, Li 2020) used the influence coefficient analysis method to analyze the influencing factors of the production cost of prefabricated buildings. Wei Chen et al. (Chen, Wu 2019) ua sed structural equation to establish a model to analyze the factors affecting the cost of prefabricated construction from three aspects: production, transportation, and on-site installation of prefabricated components. Weishu Zhao (Zhao, Peng 2019) made a quantitative analysis of the cost of prefabricated construction by the AHP method. Xiao Yin (Yin 2020) analyzed the composition of PC component cost from the perspective of system dynamics. Junchao Bao (Bao 2020) used the 5M1E theory to collect the influencing factors of prefabricated construction cost from six aspects. Based on the research of the above scholars, this paper summarizes the factors affecting the cost of prefabricated buildings, uses the DEMATEL method for quantitative analysis, finds out the relationship between the factors, and determines

*Corresponding Author: 460955524@qq.com

 DOI 10.1201/9781003305026-47

the key influencing factors, so as to provide some references for the development of prefabricated buildings in China in the future.

2 RESEARCH METHOD

2.1 *DEMATEL*

Decision laboratory method (DEMATEL) is a research method to analyze and judge the mutual restriction relationship when dealing with complex elements (Chen, Gan, Cui 2019). When using this method, we need to conduct relevant consultation and investigation for expert groups and collect and analyze a series of data. The specific steps are as follows:

(1) Construct the direct influence matrix S among each factor. Each factor is compared in pairs and measured on a five-level scale, i.e., levels 0, 1, 2, 3, and 4, respectively representing no impact, small impact, general impact, large impact, and great impact. In the formula, S_{ij}: direct influence degree of factor i on factor j (i is an integer, $1 \leq i \leq n$; j is an integer, $1 \leq j \leq n$). When the influence of the factor itself is not considered, that is, when $i = j$, $S_{ij} = 0$

$$S = \begin{bmatrix} 0 & S_{12} & \cdots & S_{1n} \\ S_{21} & 0 & \cdots & S_{2n} \\ \vdots & \vdots & & \vdots \\ S_{n1} & S_{n2} & \cdots & 0 \end{bmatrix} = \left(S_{ij}\right)_{n\times n} \tag{1}$$

(2) Normalize the direct matrix S according to the following formula to obtain the normalized matrix X.

$$X = S/\max_{1\leq i\leq n}\sum_{j=1}^{n} s_{ij} = (x_{ij})_{n*n} \tag{2}$$

(3) Establish a comprehensive matrix Z. The normalized matrix X is calculated by Formula (3), wherein i is a unit matrix.

$$Z = X^1 + X^2 + \cdots + X^{\mathrm{n}} = X(I - X)^{-1} \tag{3}$$

(4) The influence degree (E_i), affected degree (F_j), center degree (M), and cause degree (U) of each element are calculated from the comprehensive influence matrix. The center degree (M) represents the total degree of influence and is affected by a factor, so $M = E + F$. The cause degree (U) indicates the degree to which a factor belongs to cause and effect, so $U=E-F$. A positive value indicates that the factor is the cause type, and a negative value indicates that the factor is the result type.

$$\mathrm{E}_i = \sum_{j=1}^{n} Z_{ij}(i = 1, 2, \ldots, n); \quad F_j = \sum_{i=1}^{n} Z_{ij}(j = 1, 2, \cdots, n) \tag{4}$$

2.2 *Establishment and description of influencing factors*

There are many factors affecting the cost of prefabricated buildings. By summarizing many domestic kinds of literature (Lv, Zhang 2020, Liu, Ding 2019, Chen, Li 2020, Chen, Wu 2019, Zhao, Peng 2019, Yin 2020, Bao 2020, Peng, Zhou 2021) and further interviewing experts in the prefabricated industry, the factors affecting the cost of prefabricated construction are finally divided into 11 factors from the four stages of design, production, transportation, construction, and installation (see Table 1 for details).

Table 1. Influencing factors of prefabricated construction cost

Primary indicators	Secondary indicators	explanation or description
design stage	designed standardization (S_1)	The components are standardized according to certain standards and modules so that they have the advantages of universality, convenient processing, saving materials, and ensuring quality.
	prefabrication rate(S_2)	The volume ratio of the concrete consumption of the prefabricated part in the main structure and envelope above the outdoor floor of the building to the total concrete consumption of the corresponding components.
	detailed design of components (S_3)	According to the structure and characteristics of the project, it is needed to split the prefabricated components, deepen the design, assembly node design, detail node design, etc.
production stage	industrial-scale(S_4)	Industrial-scale refers to the production capacity, large-scale production, and corresponding industrial chain of assembly component manufacturers.
	technical system(S_5)	The technical system for manufacturing PC components includes the technological process, the advanced nature of the machinery, the technical level of the operator, and the related technical configuration.
	material and mold cost(S_6)	The cost of materials and molds refers to the cost of materials, components, and molds used in the production of PC components.
transportation stage	logistics cost(S_7)	In the stage of component transportation, the distance and the choice of transportation mode play a decisive role in the transportation cost.
	transport efficiency (S_8)	Transportation efficiency refers to scientifically and reasonably arranging component distribution and optimizing resource allocation according to the construction progress.
construction and installation stage	technical level of professionals(S_9)	The technical level of professionals refers to the technical professionalism of construction personnel on the construction site and the coordination and cooperation among workers.
	machinery cost(S_{10})	Machinery cost refers to the mobilization and demobilization, installation, and demolition costs of various machinery used in the construction and installation stage.
	project management (S_{11})	Project management refers to the expenses of management personnel on the construction site and the expenses incurred in quality management, schedule management, and cost management.

3 EMPIRICAL ANALYSIS

3.1 *Construction of direct impact matrix*

According to the survey, there are 12 teachers and industry experts in this specialty, including 5 teachers in prefabricated architecture, 4 front-line constructors, and 3 experts in the prefabricated industry. Through pairwise comparison among all factors and expert scoring methods, the direct influence degree of each factor is divided into five grades from 0 to 4. Then the scores of several experts and round of the data are averaged to obtain the direct impact matrix S, as shown in Table 2.

Table 2. Direct relationship matrix of influencing factors

Factors	S_1	S_2	S_3	S_4	S_5	S_6	S_7	S_8	S_9	S_{10}	S_{11}
S_1	0	4	3	3	2	3	2	2	1	1	2
S_2	4	0	4	2	3	3	3	2	1	2	2
S_3	1	3	0	3	4	3	0	0	2	3	2
S_4	3	3	3	0	3	4	3	3	2	2	3
S_5	2	3	3	1	0	2	2	2	1	1	2
S_6	2	2	3	3	3	0	2	2	1	1	1
S_7	1	2	1	2	2	2	0	4	0	1	3
S_8	1	2	1	2	2	3	4	0	0	1	3
S_9	2	2	1	0	0	0	1	1	0	3	4
S_{10}	2	2	1	0	0	0	1	1	3	0	3
S_{11}	2	1	2	0	0	0	2	2	4	3	0

3.2 *Calculation of synthetic matrix and synthetic relation parameters*

According to the DEMATEL method and MATLAB programming software, the comprehensive matrix and comprehensive relationship parameters of influencing factors are calculated. Tables 3 and 4 show the details.

Table 3. Comprehensive matrix of prefabricated building cost influencing factors

Factors	S_1	S_2	S_3	S_4	S_5	S_6	S_7	S_8	S_9	S_{10}	S_{11}
S_1	0.174	0.326	0.291	0.243	0.241	0.275	0.237	0.227	0.162	0.188	0.265
S_2	0.308	0.223	0.334	0.226	0.283	0.287	0.277	0.240	0.174	0.230	0.283
S_3	0.191	0.272	0.174	0.215	0.271	0.243	0.150	0.141	0.187	0.234	0.242
S_4	0.297	0.333	0.322	0.172	0.295	0.331	0.298	0.288	0.218	0.246	0.335
S_5	0.203	0.258	0.251	0.153	0.140	0.207	0.202	0.194	0.138	0.159	0.227
S_6	0.213	0.243	0.264	0.223	0.248	0.159	0.214	0.206	0.144	0.166	0.210
S_7	0.163	0.215	0.179	0.174	0.193	0.200	0.139	0.254	0.101	0.149	0.250
S_8	0.170	0.222	0.187	0.180	0.200	0.235	0.267	0.139	0.105	0.154	0.256
S_9	0.159	0.170	0.133	0.070	0.079	0.084	0.124	0.119	0.084	0.187	0.241
S_{10}	0.154	0.165	0.127	0.067	0.077	0.081	0.119	0.114	0.171	0.087	0.206
S_{11}	0.166	0.154	0.169	0.080	0.091	0.095	0.163	0.158	0.211	0.196	0.134

Table 4. Comprehensive relationship parameters of prefabricated building cost influencing factors

Factors	Influence degree (E_i)	Affected degree (F_j)	center degree (M)	cause degree (U)
S_1	2.629	2.198	4.827	0.431
S_2	2.865	2.581	5.446	0.284
S_3	2.32	2.431	4.751	−0.111
S_4	3.135	1.803	4.938	1.332
S_5	2.132	2.118	4.25	0.014
S_6	2.29	2.197	4.487	0.093
S_7	2.017	2.19	4.207	−0.173
S_8	2.115	2.08	4.195	0.035
S_9	1.45	1.695	3.145	−0.245
S_{10}	1.368	1.996	3.364	−0.628
S_{11}	1.617	2.649	4.266	−1.032

3.3 *Draw cause and result analysis diagram*

Taking the center degree as the abscissa and the cause degree as the ordinate, the factor values at the corresponding positions in the coordinate system are marked to obtain the cause result distribution diagram, as shown in Figure 1.

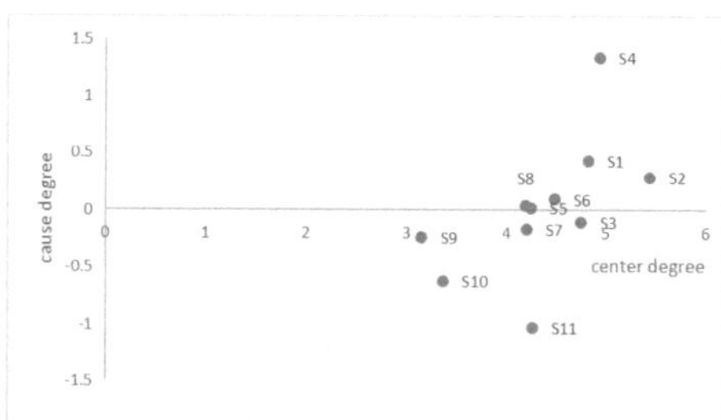

Figure 1. Cause and result analysis of influencing factors of prefabricated construction cost.

3.4 *Results analysis*

3.5 *Center degree analysis*

In the DEMATEL method, the centrality index focuses on the evaluation factors. Among them, prefabrication rate, industrial scale, designed standardization, and detailed design of components are the main factors affecting the cost of prefabricated construction. The technical level of professionals and machinery cost is in the last two, indicating that these two factors do not have a great impact on the cost of prefabricated construction.

3.6 *Cause degree analysis*

Through the calculated cause degree data, these 11 values are divided into two categories, positive and negative, in which positive values represent cause factors and negative values represent result factors.

(1) Causal factors
The factors at the top of the abscissa of the cause result analysis chart are cause factors. The order of cause factors from large to small is $S_4 > S_1 > S_2 > S_6 > S_8 > S_5$. These factors are easy to affect other factors. On an industrial scale, design standardization and prefabrication rate are the main factors.

(2) Outcome factors
The factors located at the bottom of the abscissa of the cause result analysis diagram are the result factors, mainly including project management, machinery cost, the technical level of professionals, logistics cost, and detailed design of components. These factors are easily affected by other factors, especially when other factors change, and the project management and machinery costs are affected.

4 CONCLUSIONS AND SUGGESTIONS

4.1 *Conclusions*

Prefabricated building is the inevitable trend in the development of China's construction industry. According to the model established by DEMATEL, there are many factors affecting the cost of prefabricated buildings, but the action direction and degree of various factors are different. The influence degree of the industrial scale (S_4) ranks first, the center degree ranks second, and the cause factors rank first in the cause degree. Therefore, both the influence degree and the importance degree are factors that can not be ignored, and also affect other factors. The influence degree of prefabrication rate (S_2) is ranked second, the center degree is ranked first, and the causal factors in the cause degree are ranked third, indicating that the prefabrication rate has the greatest impact on the cost of prefabricated buildings, and also affects other factors to a large extent, which can be considered as the key influencing factor. The influence degree of designed standardization (S_1) ranks the third, the center degree ranks the third, and the cause factor ranks the second in the cause degree, indicating that this factor has a great influence on other factors and can also be regarded as a key influencing factor. The influence degree of project management (S_{11}) is ranked first, and the resulting factor is ranked first, indicating that this factor is easily affected by other factors. Once other factors change, this factor will change greatly. Therefore, it can also be considered a key factor affecting the cost of prefabricated construction.

4.2 *Suggestions*

Based on the above analysis conclusions, the following suggestions are put forward:

(1) For the production of prefabricated buildings, due to the unbalanced development of prefabricated buildings and inconsistent starting times, it is difficult to form large-scale production.

Expanding industrial scale is an inevitable trend in the development of prefabricated buildings. The government and production enterprises should go hand in hand to build a harmonious and good market and form a mature industrial chain and produce scale effect, so as to reduce the cost of prefabricated buildings.

(2) In the design stage, we should improve the prefabrication rate of prefabricated buildings and reasonably split components. At the same time, we should optimize the design technical system, take standardization and generalization as the goal of prefabricated components, and reduce the specifications and types of components, so as to reduce the cost of prefabricated buildings.

(3) Different design schemes will lead to changes in project management. In the construction and installation stage, project management personnel is not only required to have a high professional level, strong communication, and coordination ability but also to dynamically manage the cost objectives and schedule objectives of the project to ensure the quality of construction.

ACKNOWLEDGMENTS

This work was financially supported by the 2020 National College Students' Innovation and Entrepreneurship Training Program (202012216061), Anhui College Students' Innovation and Entrepreneurship Training Program (AH201912216188), Building Information Model Research Institute (yjs202102), Engineering Management Teaching Team (2016jxtdx02).

REFERENCES

Baoku Qi, Ya Zhu, Bo Ma. (2016) Study on comprehensive benefits analysis method of prefabricated building. Construction Technique 45: 39–43.

Junchao Bao. (2020) Analysis and research on influencing factors of cost in the construction stage of prefabricated building. Lanzhou University of Technology, Lan Zhou.

Junlong Peng, Jing Zhou. (2021) Hidden cost analysis of prefabricated buildings based on ISM-BN. Journal of Civil Engineering and Management, 38: 37–43.

Kai Liu, Xiaoxin Ding. (2019) Study on influencing factors of assembled building cost based on the life cycle. Journal of Neijiang Normal University, 34: 51–56.

Wei Chen, Yashuai Wu. (2019) Analysis of influencing factors of the construction cost of prefabricated buildings based on SEM. Journal of Civil Engineering and Management, 36:50–55.

Weishu Zhao, Hao Peng. (2019) Analysis of influencing factors of prefabricated building cost based on AHP. Journal of Tangshan University, 6:88–94.

Xiao Yin. (2020) Cost control and analysis of prefabricated buildings based on system dynamics. Jiangsu university, Zhen Jiang.

Xiaobo Chen, Kaixin Gan, Ping Cui. (2019) Analysis of influencing factors of hoisting efficiency of prefabricated concrete members based on DEMATEL method. Journal of Engineering Management, 33: 111–115.

Yuanyue Chen, Weiqing Li. (2020) Analysis on influencing factors of production cost of prefabricated buildings. Journal of Southwest Normal University, 45:68–72.

Zheqi Lv, Dan Zhang. (2020) Study on influencing factors of prefabricated construction cost in Qinghai Province. Journal of Qinghai University, 38: 56–63.

Advances in Urban Engineering and Management Science – Khalil & Yang (Eds)
© 2023 The Author(s), ISBN 978-1-032-30426-7

Study on the optimization method of multi-component construction scheme based on the empirical mode decomposition method

Da Xie
State Grid Hebei Electric Power Co., Ltd., Hebei, Shijiazhuang, China

Zhengping Wang
State Grid Hengshui Electric Power Supply Company, Hebei, Hengshui, China

Kai Zhang
State Grid Hebei Electric Power Co., Ltd., Hebei, Shijiazhuang, China

Liya Chen & Duo Zhang*
State Grid Hengshui Electric Power Supply Company, Hebei, Hengshui, China

ABSTRACT: Under the new situation, improving the installation and construction efficiency of multiple components such as tower assembly hoisting and steel structure building beam-column hoisting is of great significance for the implementation of the development strategy of "improving quality and efficiency" of power grid enterprises. Therefore, this paper takes the transmission line tower assembly hoisting construction as the research object. Based on the analysis of the influencing factors of construction operation efficiency, combined with the relevant theoretical principles of the empirical mode decomposition method, the optimization design is carried out for the typical multi-component construction and installation scheme of tower assembly hoisting, which improves the construction operation efficiency.

1 INTRODUCTION

The iron tower is an important part of the transmission line and plays the role of supporting the wires. Under the new situation, building an "international leading energy Internet enterprise with Chinese characteristics" has become an important development strategy goal of my country's power grid companies. Therefore, improving the efficiency of the installation and construction of tower components has become one of the important issues facing power grid companies.

The steel structure has the advantages of high strength, high reliability, and a high degree of factory production, and it occupies an absolute proportion of the type of transmission line tower structure. However, due to the uneven capabilities of tower steel structure design units, the mismatch between construction procedures and on-site resources, and the simplicity of on-site organization and management of construction methods, the construction efficiency of steel tower steel structures in power grid enterprises needs to be further improved.

Through combing the relevant literature, relevant scholars pay more attention to the management dimensions of power transmission and transformation engineering construction safety, quality, schedule, and cost, while their attention to the optimization of construction technology schemes is obviously weak. For example, Li (2018) analyzed the safety management issues of key links in the process of power engineering construction and the construction of related safety management

*Corresponding Author: 1182306191@ncepu.edu.cn

 DOI 10.1201/9781003305026-48

and control systems. Chen et al. (2020) mainly analyzed the schedule management of power transmission and transformation projects and their importance and put forward a scientific and reasonable schedule management method based on the actual situation. Zhang (2018) combined the author's practical work experience and relevant references to study the current status of power transmission and transformation projects and existing specific problems and proposed effective solutions. Song et al. (2021) analyzed the internal connection and mutual influence of the cost management of each stage and link of the project construction and studied the establishment of a standardized model for the cost management of power transmission and transformation projects.

In summary, considering the fact that related scholars have paid insufficient attention to the optimization of the construction plan of power transmission and transformation projects, this paper selects the installation of the tower components of the power transmission and transformation project as the research object and analyzes the premise of the main factors affecting the efficiency of construction operations. Next, combined with the relevant theories of the empirical mode decomposition method, optimized design is carried out for the construction operation plan.

2 ANALYSIS OF THE STATUS QUO OF THE INSTALLATION PROCESS OF TRANSMISSION LINE TOWER COMPONENTS AND THE INFLUENCING FACTORS OF CONSTRUCTION EFFICIENCY

2.1 *Status quo of installation process of iron tower components*

At present, the installation process of iron tower steel structure in transmission line engineering mainly includes the following contents: Measurement (elevation, axis), preparation of components in place → hoisting of steel columns → correction, and temporary fixing → hoisting of primary and secondary beams → initial tightening of high-strength bolts → final fixing of columns → final tightening of high-strength bolts → roof panel installation.

(1) On-site inspection of steel components. After the steel components enter the construction site, the specifications, models, and quantities of the components should be checked, and the deformation during transportation should be checked and corrected to ensure the quality of the components, and at the same time, it should be reported to the supervisory unit for inspection.
(2) Foundation re-inspection. The steel structure installer should promptly handle the intermediate handover information with the civil engineer and use the level and theodolite to re-measure the foundation axis and elevation.
(3) Inventory of components, unloading and stacking components to the site, the delivery list will be carried with the vehicle. The on-site installer unloads the truck according to the factory delivery list, counts the quantity, and signs on the material list to confirm the quantity; when unloading, the plant floor Lay and compact prepare a certain number of skids in advance and places the components on the sleepers before they are in place to avoid damage to the paint of the steel components. The components on the floor plan are marked according to the shipping details, and the component armor is placed on the armor. Material B is placed in place B, and it is attempted to avoid the second re-shipment of materials. According to the delivery list, the on-site installers mark the incoming materials on the floor plan with a color pen. This method can keep abreast of the site's incoming material status.
(4) Installation and construction method of steel components. The steel column feet are connected to the foundation with anchor bolts. To smoothly insert the anchor bolts when the steel column is in place, the vertical lifting method is usually used for hoisting. When hoisting the steel column, special attention should be given to protecting the anchor bolts and slings. To protect the slings from being damaged by the sharp edges and corners of the steel column, a corner rubber is placed at the binding place.
(5) Steel column correction. The plane position is corrected, and the displacement is strictly controlled when the column is in place. The correction method adopts the screw jack, and the verticality of the steel column is corrected. The verticality of the steel column is tested by

erecting two theodolites on the vertical and horizontal axes. If there is any deviation, a screw jack is used to correct it.

(6) Installation of primary and secondary beams. According to the position of the corresponding components, the bolt holes of the beam are hoisted and aligned with the bolt holes of the column flange plate, and high-strength bolts are introduced for connection and fixation.

(7) Bolt tightening. After the steel structure columns and beams are installed, the anchor bolts and high-strength bolts are uniformly reinforced and tightened.

2.2 *Analysis of influencing factors on construction efficiency of iron tower components*

The tower is an important part of the transmission line and plays the role of supporting wire, high voltage line commonly used independent angle steel tower, independent steel tower, with each engineering base, single base component more (and number only), and bolt connection node. In the construction process, it has more operation surface, scattered, topography and traffic conditions, and high operational characteristics.

After a systematic analysis of the whole process of an existing design, processing, transportation, and construction, the factors restricting construction efficiency have the following aspects:

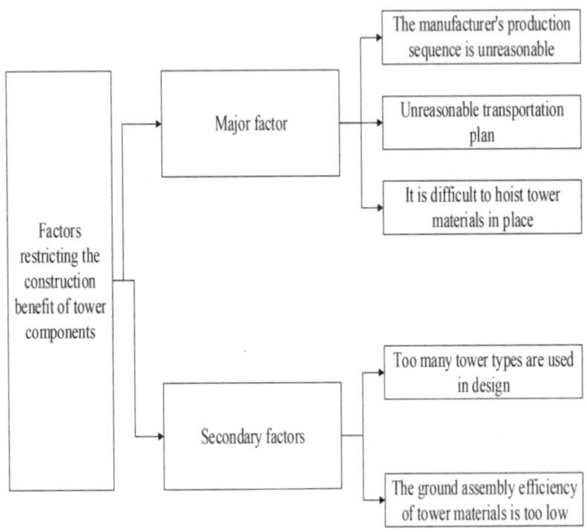

Figure 1. Factors restricting the construction benefit of tower components.

2.2.1 *Material cause*

The manufacturer has an unreasonable production order. In order to speed up the production progress, some manufacturers adopt the unified processing of the same tower and fail to produce tower parts according to the construction needs. To this end, the tower materials for first processing and delivery cannot be organized, and the tower positions can be organized without tower materials.

The transport scheme is unreasonable. In order to save transportation costs, some manufacturers mixed tower materials, leading to a large number of centralized stacking on the site tower materials, which greatly increased the workload of site picking, but also lead to the tower materials and bolts in place.

The tower material is difficult to lift in place. The processing quality of tower materials is poor, resulting in problems such as misplaced connection holes and improper setting of hanging points, leading to difficulties of tower materials in place and even construction stagnation.

2.2.2 *Secondary cause*

There are too many tower types for design applications. According to the climate conditions, cost reasons, and other factors, to control the cost technical indicators, we need to carry out the fine design of the tower type, leading to the selection of tower types and increasing the drawing release time of the tower manufacturer, which also has a certain impact on the manufacturer's product in mass.

Tower ground assembly is too inefficient. The construction level of the team is uneven, some teams lack operation experience, the ground assembly is mainly convenient, and the assembly order is not fully considered, resulting in improper tower plate stacking pressure and position hoisting.

3 OPTIMIZATION OF TOWER COMPONENT INSTALLATION OPERATION PLAN BASED ON EMPIRICAL MODE DECOMPOSITION METHOD

Empirical mode decomposition is an adaptive data decomposition method based on the characteristics of the signal itself, which can decompose any type of complex data into a limited number of components, which can accurately represent the local characteristics of the original data at different time scales (Wang et al. 2021).

3.1 *EMD decompose*

In the formula, n represents the number of IMF_s decomposed. All sequences IMF_s must meet the following two conditions: (1) the number of extreme points and zero crossings of the whole sequence must be equal or at most one different; (2) at any time point, the mean value of the upper envelope determined by the local maximum point and the lower envelope determined by the local minimum point must be zero.

The empirical mode decomposition algorithm can adaptively decompose the original sequence $x(t)$ into multiple essential mode functions IMF_s with different frequencies and a residual sequence $r(t)$ through the "screening process". The formula is as follows:

$$x(t) = \sum_{i=1}^{n} IMF_i(t) + r(t), \quad t = 1, 2, \ldots, T \tag{1}$$

In the formula, it represents the number of decompositions. All sequences must meet the following two conditions: (1) the number of extreme points of the whole sequence must be equal or one different from the number of excessive zero points; (2) at any time point, the mean of the upper envelope and the local maximum point determined by the local minimum point must be zero.

The EMD decomposition process mainly includes the following five steps: 1) identify all local extreme points of the signal sequence; 2) Connect all maximum points or minimum points with cubic spline function, fit the upper envelope $e_{\max}(t)$ and lower envelope $e_{min}(t)$, and calculate the average of them to form the average envelope sequence $m(t)$, that is:

$$m(t) = (e_{\max}(t) + e_{\min}(t))/2 \tag{2}$$

3) Subtract the average envelope $m(t)$ from the original signal sequence $x(t)$ to obtain a new data sequence $x'(t)$ without low frequency, that is:

$$x'(t) = x(t) - m(t) \tag{3}$$

4) Detect the characteristics of $x'(t)$: if $x'(t)$ is not a IMF (does not meet the two conditions of sequence IMF_s), $x'(t)$ is taken as the signal sequence to be processed and repeat the above operation; if $x'(t)$ is a IMF (meets the two conditions of sequence IMF_s), $x'(t)$ is regarded as the first IMF, that is, $r(t) = x(t) - x'(t)$ at the same time, record the residual $x(t)$ as new;

5) Repeat the above steps. When $IMF_1(t)$ is less than a preset value or the residual term is a monotone function and IMF cannot be filtered out, the decomposition process ends.

3.2 *IMF_s synthesis*

Through EMD decomposition, two signal sequences of index variables $IMF_i(t)$ and $IMF_i(t)$ can be obtained. The synthesis formula of IMF_s is as follows:

$$IMFH_j = \sum_{i=1}^{i=k-1} IMF_{i,j}(t), \quad t = 1, 2, \ldots T \tag{4}$$

$$IMFL_j = \sum_{i=k}^{i=n} IMF_{i,j}(t), \quad t = 1, 2, \ldots T \tag{5}$$

4 EMPIRICAL ANALYSIS

This paper selects the installation project of the tower group in a mountain area and optimizes the construction operation as follows:

4.1 *Construction method planning stage*

The overall construction deployment should fully consider the constraints such as site, resources, technology, and overall progress, and in accordance with the principle of "zoned construction, streamlined operations, and sequential advancement", and reasonably divide areas to carry out parallel construction.

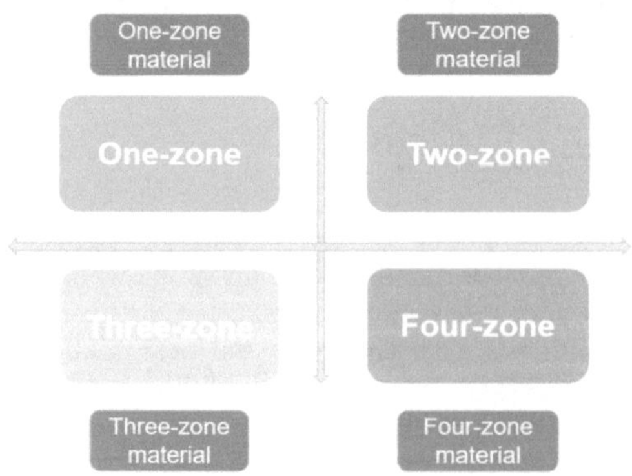

Figure 2. Schematic diagram of material stacking and construction in sub-zones.

The construction of steel structures in each area should be divided into construction sections reasonably according to the construction sequence of "pillars first, beams first, mains first, secondary buildings first, structures first, enclosures first, exterior installations and then interiors", and flow-through construction should be organized.

4.2 *Tower material transportation stage*

4.2.1 *Pre-transport inspection*

The tower material inventory shall be completed at the material station or ropeway entrance. If the key tower material is missing or another iron tower may not be smoothly organized, it shall not be transported temporarily, and if the components have deformation, they shall be corrected or replaced within the material station.

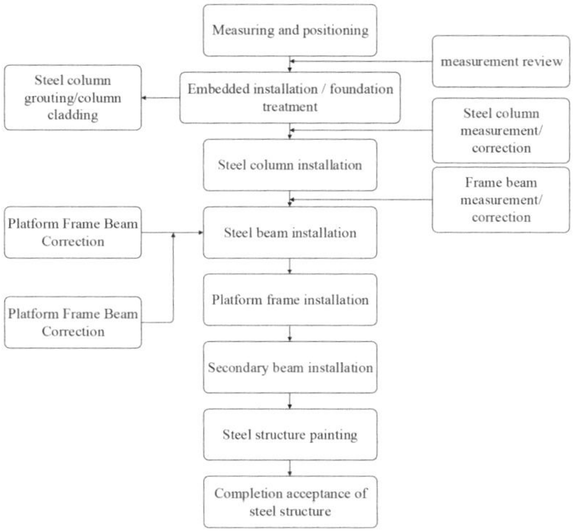

Figure 3. Schematic diagram of the installation process of the main body of the steel structure.

4.2.2 *Medium transport stage*

Once transported from the material station near the pile position, a suitable unloading site should be selected in advance. After transportation to the unloading point, a special person should be provided to prevent loss. In principle, single base transportation should be adopted. When the single base weight is light, the components of different pile numbers should be stacked by edges or in layers to avoid confusion.

4.2.3 *Small stage*

When the tower material is transported from the unloading point near the pile position, the appropriate transportation method is selected according to the transportation quantity, the distance of the tower, and mechanical transportation after road repair; when the road construction conditions are not met, the following transportation methods can be adopted:

Mule and horse transportation: it is suitable for tower positions with small 110 kV and 35 kV lines and short transportation distance. The transportation channel should be cleared in advance.

Ropeway transportation: it is suitable for the transportation of various groups of towers in mountainous areas. The erection and acceptance of the ropeway shall be completed before transportation to ensure safety and reliability.

When the operation surface of the tower is sufficient, all shall be transported to the site; the operation surface is small. The transportation team shall communicate with the tower team, and the transportation demand plan shall be transported in sections according to the demand order of the tower team.

When transportation in mountainous areas, it is also necessary to consider the transport capacity and the ability of a team group to match each other, to ensure that the needs of the group are met and the transportation team does not work.

The above construction shall be prepared for the tower transportation group plan according to the requirements of "one foundation, one plan".

4.3 *Iron tower hoisting stage*

4.3.1 *Main construction tools*

UHV tower shall be arranged; low voltage tower shall adopt suspension holding rod. With the conditions of the external pull line, an internal suspension external pull line shall be adopted.

When selecting the suspension lever, it is required to select the appropriate lever section and length according to the tower type and lifting height to improve the lifting capacity and reduce the number of lifting levers.

Ground anchors, pile anchors, etc., are set up in advance according to the construction plan;

When using the ropeway transportation, the location of the ropeway end unloading point should communicate with the ropeway installation team in advance;

When the operation surface is small and needs to be transported in sections, the ropeway team should be informed of the group order and transportation plan in advance;

After transportation to the site, the tower materials shall be counted and checked again. The missing parts and wrong holes shall be reported in time.

4.3.2 *Ground assembly*

According to the possible lifting height and lifting capacity of the holding rod, the sections of the lifting components shall be reasonably determined. In principle, the lifting components should be assembled in sections. When the lifting capacity is met, the two sections can be lifted into one piece to reduce the lifting times.

The layout position of the ground assembly tower sheet is determined according to the site terrain, and in situ location, so as to avoid the intersection of ground assembly and hoisting.

When the terrain-limited tower pieces need to be stacked, attention should be paid to the stacking order, hoisted and then assembled, as for the upper layer.

If the tower pieces need to be strengthened, the reinforcing wood shall be installed in the ground assembly stage.

4.3.3 *Member binding*

The binding point of the hanger and control rope is warped in advance to protect the galvanizing coating.

Special lifting point tools should be used, or special lifting ears should be set up in the design stage to reduce the workload of lifting sleeve binding and removal.

4.3.4 *Hoisting in place*

A tower plate-in-place guide device should be adopted.

5 CONCLUSION

Based on the construction efficiency improvement demand of iron tower and steel structure buildings of power grid enterprises, the construction scheme optimization method based on the empirical mode decomposition method is proposed. The method can further improve the construction efficiency of power transmission and transformation projects and enhance the operating efficiency of enterprises.

REFERENCES

Chen Jianhua, Xi Zhaocai, Ding Guowei.The Application of Project Progress Management in Power Transmission and Transformation Project Management [J].China Management Information Technology, 2020,23 (14): 154–155.

Li Qiang.Analysis of Safety Control System Construction of Power Transmission and Transformation Project [J].Communication World, 2018 (10): 167–168.

Song Xingbin, Chen Qiannan, Chen Yunlin, Wang Yue.Analysis of cost management of power transmission and transformation Project [J].Management of China Electric Power Enterprises, 2021 (03): 74–75.

Wang Junhao, Wang Chong, Jackie Chan, Li Guoqing, Gao Jiaxing.Photovoltaic power combinations predictions based on ensemble empirical modal decomposition and deep learning [J/OL].High Voltage Technology: 1–10 [2021-10-27]. https://doi.org/10.13336/j.1003-6520.hve.20210971.

Zhang Penggang. Some measures to improve the construction quality of the Power Transmission and Transformation Project [J].Scientific and Technological Innovation Guide, 2018,15 (01): 122 + 126.

Advances in Urban Engineering and Management Science – Khalil & Yang (Eds)

The study of the relationship between natural forests and modern wooden dwellings

Ruixi Chen*†
Vanke Meisha Academy at Shenzhen, Shenzhen, China

Chenghao Dong†
Beijing New Oriental Foreign Language School at Yangzhou, Yangzhou, China

ABSTRACT: In modern society, the population tends to move to cities and the center of the country. Fewer and fewer people choose undeveloped areas. Nowadays, under social pressure, seeking for peaceful dwelling place becomes attractive. The purpose of this study is to create a building that is able to provide different experiences of modern and stressful life. The use of materials and the site choice can bring the effect we expect to have. We choose to establish our design near the rivers at the foot of the mountains. The main materials that form most of the visually viewable parts of the building are wood and glass. Research shows that most of the interviewees believe that woodenware and furniture can make people feel joyful. Most people use wooden furniture in their individual spaces. Wooden furniture is believed to contain the meaning of private and mental relaxation. The site choice also follows the rule of some basic, usual psychological hint, since most people believed that forest and river are representations of clam and non-emotional choice. Therefore, we deiced on our site and materials, since they can create the building that is closest to our concept of being able to make a connection with the natural world and enable people to feel relaxed.

1 INTRODUCTION

Since living areas are strongly connected to mental health, a large amount of anxiety, depression, and mood disorders were caused by being isolated from nature. Those are all health-related factors. Studies show that interaction with nature can promote psychological recovery, improve mood, and concentration, and reduce stress and anxiety.

In the 21st century, the population tends to move to urban and rural centers. However, urban life might be harmful to the people, and possibly cause health problems, since the population that lives in urban areas usually live under high social pressure with a heavy workload. Those are all problems that convenient city life brings to people. Therefore, finding a peaceful place to live becomes attractive to many people. Treatment using environmental factors might be a way out. A study shows that the more quality green land a community owns, the lower rate of depression the community possibly will have. That shows the importance of the natural environment.

And architectural design may affect microorganisms in the building environment and human health and wellbeing. The impact of human beings on microbial community structure through architectural design, more directly through human living and the use patterns of microorganisms in different spaces and space types, has an indirect guiding impact on microbial diversity in buildings. The impact of design decisions on the structure of the indoor microbiome provides the possibility

*Corresponding Author: chenruixi@stu.vma.edu.cn
†These authors contributed equally.

DOI 10.1201/9781003305026-49

to use ecological knowledge to shape our buildings and to select indoor microbiomes to promote our life, health, and wellbeing.

The study aims to create a building that could provide a different experience from modern, stressful, and healthy life, and help people get away from stressful lifestyles and workplaces. The design is based on the concept of creating a natural environment and nature-related living area. This is helpful to both people's mental health and the development in suburban areas.

2 MECHANISM

2.1 *Location*

With the development of the economy and society, the land price in the urban areas has increased a lot to be not affordable to most people who get normal wages anymore. Compared with land in the cities, the land in the mountain areas is relatively cheaper. It means that choosing mountain areas as a building location would guarantee people have places to live in. Also, because of their wide space, the mountain areas can easily adapt to all kinds of houses. And as transportation develops like the establishment of more expressways, living in the mountain areas becomes quite convenient. The most important thing is that the sights in the mountains are quite beautiful, including trees, flowers, and birds, to make people feel very relaxed. With much sunlight, for example, people may get health benefits for their bones (Steven W. Kembel, James F. Meadow, Timothy K. O'Connor, Gwynne Mhuireach, Dale Northcutt2,5, Jeff Kline, Maxwell Moriyama, G. Z. Brown, Brendan J. M. Bohannan, Jessica L. Green). As a result, away from the noise made by cities' cars and factories, people can get a good experience while living in mountain areas.

The experience of nature provides many mental health benefits, especially for people who have a hard-living experience in urban areas. The natural characteristics of urban residential communities are likely to be the key determinants of how happy people will be and how satisfied with their work or living experience every day; Reference (Daniel et al. 2017) demonstrated that vegetation coverage and the number of birds in the afternoon were positively connected to the people's lower prevalence of depression, anxiety and stress in the natural characteristics of the five communities tested in Reference (Daniel T. C. Cox et al. 2017). Secondly, the dose-response model in Reference (Cox et al. 2017) told a threshold response, in which the prevalence of people with mental health problems was significantly lower than that of people who live with the minimum of neighborhood vegetation coverage with not too many flowers or trees (depression more than 20%, anxiety more than 30%, stress more than 20%). The results show that there is a quantifiable correlation between mental health and nearby natural characteristics that people actually experience. So, if people are able to live in the mountain areas, it is beneficial to their mental health.

According to Reference (Frederick J. Swanson et al. 1998), the roots of trees have specific properties, such as organization, dynamics, and control factors. These characteristics of root channels lead the roots to have a great influence on water or river flow. Root structure analysis is helpful to upgrade the process from root scale to hillside scale. The living and dead components of roots can promote or dissipate soil water pressure, which is dependent on structure, direction, and interconnection between each other. As a result, these roots may improve or even reduce the possibility of shallow landslides during storms. Although people have begun to pay much attention to discovering the benefits of mechanically strengthening roots, the potentially harmful impact of roots on groundwater flow should be given more consideration if buildings are built in the mountain areas where there is much rain and land as well as kinds of roots.

In Reference (Michael F Holick), a wave of the rivers in the mountain areas has characteristics, like waveform, amplitude, wavelength, and frequency, which are according to recognized hydrological quantitative descriptions, such as flow shape, size, return period, and frequency. In addition, the position on the crest, trough, and rising or falling branch is equivalent to the position on the process line: flood peak, base or zero flow, rising or falling flow. Overall, the concept of wave explains the remarkable characteristics of river hydrology. Learning river hydrology would make living in the mountain areas with rivers more easily.

Figure 1. Mountain areas.

2.2 *Material*

Wood is employed to build the most of the design for several benefits. First of all, based on References (Murielle Ghestem et al. 2011), wood is renewable and degradable to be economic-efficient. Secondly, it is easy for people to get wood materials from nearby forests. This accessibility reduces the need for material transportation to save costs. Compared with materials like concrete, wood enables the construction workers to construct more easily the house while spending less time. Choosing wood as the main material even saves the need for air conditioning since wood is warm in winter and cool in summer to provide a comfortable indoor temperature. This advantage further decreases the cost to reduce people's financial burden. Finally, the wood can better fit into local natural landscapes compared with concrete or steer.

Figure 2. Expressway.

There are always strong floods in the mountain areas, like what Reference (Paul Humphries et al. 2014) says. These serious floods are the result of forest-cutting and strong rain around the mountains. These floods are strong enough to cause the design to break down. In order to avoid this terrible thing from taking place, two key points should keep in mind. Firstly, when people cut the mature trees in the forests nearby, they need to plant new seeds right away to keep the forest intact. As we know, trees play an important role in preventing floods and land from sliding. This one must be taken to people who live in this design. Also, if people choose to live in these houses in the mountain areas, they are required to check the weather report all the time. Otherwise, while they do not pay attention to what the weather reports say, a bad accident would happen like a flood destroying the house.

Glass is also considered in construction. By using glass, transparent windows can be created. Such windows can guarantee enough strong sunlight to illuminate the whole house. This would not only provide enough light to eliminate the need for many artificial bulbs, but also heat the house a bit. And, with glass windows, harmful insects can be prevented from entering the house. Since there is always strong wind and rain in mountain areas, the glass window plays an important role in protecting the house's interior from these weather conditions. And people are able to enjoy the beautiful natural views through the transparent windows, which can be very pleasant. Therefore, glass has a strong function in construction.

And the glass windows would allow air to enter the house. In reference (Franz Schmithüsen & Stephan Wild-Eck 2000) and reference (Roger S. UlrichScience 1984), architecture is a complex ecosystem, interacting with trillions of microorganisms. Understanding the determinants of the architectural environment, the composition diversity of the microbial community, and the ecological and evolutionary process of the indoor microbial community is very important to understanding the relationship between architectural design, biodiversity, and human health. Bacterial communities in indoor environments contain many taxa that do not exist or are rare outdoors, including taxa closely related to potential human pathogens. Architectural properties, especially ventilation air source, airflow rate, relative humidity, and temperature, are related to the diversity and composition of the indoor bacterial community. The relative abundance of bacteria closely related to human pathogens is higher indoors than outdoors, and the bacterial diversity in the air is higher in rooms with a low airflow rate and relative humidity. The observed relationship between architectural design and bacterial diversity in the air shows that the indoor environment can be managed by changing the microbial species community that may colonize the indoor human microbiome through architectural design and operation.

2.3 *Structure*

The first step in the design is to use small blocks like LEGO. With the small blocks, different shapes are arranged. This way enables the outline of the design to become clear. Then, total small blocks form a sloping ground floor with an attic.

3 RESULTS

3.1 *First floor*

When people enter the house from the front door, there would be a door mattress to help house owners clean the soil on the shoes to avoid smearing the floor. An umbrella stand is necessary to put umbrellas because heavy rain always occurs in wet mountain areas. From the door to the entrance of the dining room, a display cabinet and a clothes stand are provided for owners to put their personal staff. And windows are put along the wall to allow the path to depend on the sunlight to be illuminated instead of relying on artificial bulbs. At the end of the path, a staircase from the first floor to the attic is established.

The entrance of the dining room is designed in Japanese-style *Shôji*. Such entrance can more beautifully isolate space between the path and dining room to protect the owner's privacy than common doors. Also, some of the sunlight can go through the entrance, which emphasizes the designs' key point of approaching nature.

Through the entrance, a piano producing the relaxing sounds is set to function as the amusement equipment. From Figure 3, above the piano is a bathroom with a shower. And the main bedroom is

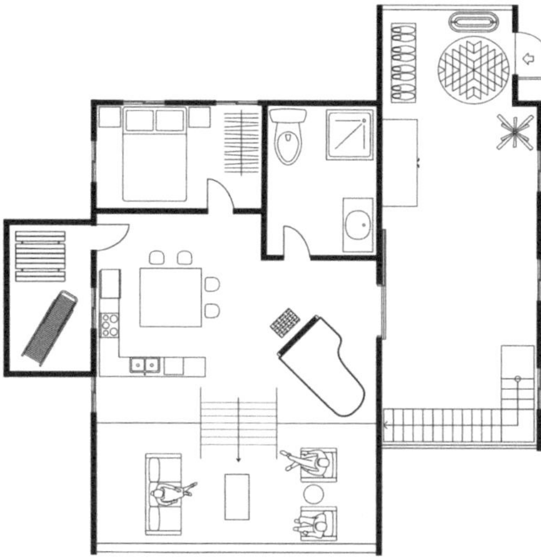

Figure 3. First floor.

next to the restroom. So, in this way, if people want to go to the used toilet while sleeping, they just need to spend seconds which is quite convenient. Three windows are set in the bedroom to allow sufficient sunlight and people can directly enjoy the outside natural views through these windows.

Compared with a traditional kitchen set in a separate room, an open room is employed. An open room has an obvious advantage of using space more efficiently. The smell of delicious dishes would be full of the house under this situation, which can be quite relaxing to people who give favor for food. A balcony is also provided for people to go outside to breathe fresh air after having a nice meal.

And then if people want to reach the living room, they first need to pass a staircase since the living room is on a slope. The most attractive is that the wall of the living room is made entirely of glass. Through a glass, while chatting with friends, the lake and trees would relax people and the sunshine would warm people in the cold winter.

3.2 *Attic*

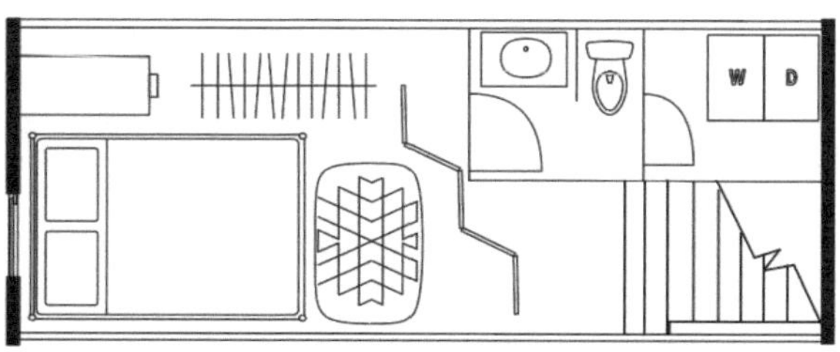

Figure 4. Attic.

Through the staircase from the path, the owners of the house are able to get to the attic. The attic is made up of a secondary bedroom, a bathroom as well as a lounge room. A screen is put between the bedroom and the staircase to protect privacy. Establishing a bathroom in the attic is to eliminate the need for going to the first floor to use the restroom while falling asleep. About the walls on both sides of the attic, the glass material is used instead of using wood walls. This would allow sunlight to be filled with the attic too.

4 CONCLUSION

It is quite enjoyable for people to live in the mountain areas. With green trees and beautiful flowers around, people are able to live a quiet and relaxed life. So, living in the mountain areas should be a good choice for people who are busy with their jobs.

And while designing buildings, suitable materials should be considered. For instance, in the mountain areas, wood should be chosen as the main material for the design. By thinking carefully about construction materials, the design can fit better into the local circumstance.

The wonderful design depends on the effective design and reasonable construction materials.

REFERENCES

Architectural Design Drives the Biogeography of Indoor Bacterial Communities Steven W. Kembel 1,2,39, James F. Meadow2,34s, Timothy K. O'Connor2,3,4, Gwynne Mhuireach2,5, Dale Northcutt2,5, Jeff Kline2,5, Maxwell Moriyama2,5, G. Z. Brown2,5,6, Brendan J. M. Bohannan2,3Jessica L. Green2,3,71 Département des sciences biologiques, Université du Québec à Montréal, Montréal, Québec, Canada, 2 Biology and the Built Environment Center, University of Oregon, Eugene, Oregon, United States of America, 3 Institute of Ecology and Evolution, University of Oregon, Eugene, Oregon, United States of America, 4 Department of Ecology and Evolutionary Biology, University of Arizona, Tucson, Arizona, United States of America, 5 Energy Studies in Buildings Laboratory, University of Oregon, Eugene, Oregon, United States of America, 6 Department of Architecture, University of Oregon, Eugene, Oregon, United States of America, 7Santa Fe Institute, Santa Fe, New Mexico, United States of America

Daniel T. C. Cox, Danielle F. Shanahan, Hannah L. Hudson, Kate E. Plummer, Gavin M. Siriwardena, Richard A. Fuller, Karen Anderson, Steven Hancock, Kevin J. Gaston, Doses of Neighborhood Nature: The Benefits for Mental Health of Living with Nature, BioScience, Volume 67, Issue 2, February 2017, Pages 147–155, https://doi.org/10.1093/biosci/biw173

Flood Disturbance in a Forested Mountain Landscape: Interactions of land use and floods Frederick J. Swanson, Sherri L. Johnson, Stanley V. Gregory, Steven A. Acker Author Notes BioScience, Volume 48, Issue 9, September 1998, Pages 681–689, https://doi. org/10.2307/1313331

Sunlight and vitamin D for bone health and prevention of autoimmune diseases, cancers, and cardiovascular disease Michael F Holick Boston University

The Influence of Plant Root Systems on Subsurface Flow: Implications for Slope Stability Murielle Ghestem, Roy C. Sidle, Alexia Stokes Author Notes BioScience, Volume 61, Issue 11, November 2011, Pages 869–879, https://doi.org/10.1525/bio.2011.61.11.6

The River Wave Concept: Integrating River Ecosystem ModelsPaul Humphries, Hubert Keckeis, Brian Finlayson BioScience, Volume 64, Issue 10, October 2014, Pages 870–882, https://doi.org/10.1093/biosci/biu130

Uses and Perceptions of Forests byPeople Living in Urban Areas:Findings from Selected EmpiricalStudiesFranz Schmithüsen and Stephan Wild-EckPublished in Forstwissenschaftliches Centralblatt:Cbl. 119 (2000), 395–408. Zurich 2001 Working Papers International Series" Forest Policy and Forest Economics Department of Forest Sciences

View through a Window May Influence Recovery from Surgery Roger S. UlrichScience, New Series, Volume 224, Issue 4647 (Apr. 27, 1984), 420—421.

Advances in Urban Engineering and Management Science – Khalil & Yang (Eds)
© 2023 The Author(s), ISBN 978-1-032-30426-7

Research on the application of the parametric design method in special-shaped buildings

Boyuan Shi
Innovative Entrepreneurship Institute, Lanzhou Jiaotong University, Anning, Lanzhou, Gansu, China

Kaili Hao*
Department of Environmental Design, Lanzhou Jiaotong University, Anning, Lanzhou, Gansu, China

ABSTRACT: The digital technologies in the era of big data have brought other high-quality technologies. The application of the parametric design method in architecture has been promoted, becoming the most widely used design tool among digital technologies and penetrating all areas of design. At present, the parametric design method is widely used in irregular and special-shaped buildings. As a kind of digital programming software, parameterization is another medium for designers to transform their thinking. Through logical deduction and rigorous calculations, designers are required to input numbers like programmers to convert the model. Rational thinking is adopted at the beginning of the design, while emotional works are created at the end of the design. Nowadays, cross-border integration and development are heatedly discussed such as intelligent building and smart cities. In fact, unlike other modeling software and methods, the parametric design, regardless of its excellent modeling ability or fast and efficient drawing speed, falls into the category of intelligent technologies in the field of design, for its core operation mode is different from other modeling software from the source. It is important to embrace possibilities brought by technological changes to architecture and welcome the innovation and development of architecture. In this context, special-shaped buildings have sprung up in all the corners of cities thanks to technologies, and the intervention of the parametric design method helps the special-shaped buildings to be perfectly presented to the public.

1 INTRODUCTION

In the wave of post-modern design, the parametric design method has long been known in architecture. As a design method, its core idea is to turn all the elements of architectural design into a variable of a certain function, and by changing the function or the algorithm, different architectural design schemes can be obtained[1]. This means that designers are at full liberty to design imaginary buildings through changes in digital calculation. The term parameterization was originally a concept in geometry, the branch of mathematics in the study of public space relations. This idea, first appearing in the 20th century, has been discovered and used in the intelligent development of computers, and later in automobiles, ships, etc.

Parametric architectural design, emerging in the mid to late 1990s, was taught in the American School of Architecture of Columbia University, the Massachusetts Institute of Technology School of Architecture, and the British Architectural Association School of Architecture at that time, and has been applied by pioneer architects to projects (Zhou et al. 2020). In China, as economic growth varies from city to city, special-shaped buildings have first emerged in Shanghai, Beijing, Shenzhen, and other first-tier cities, becoming landmarks with regional characteristics as soon as

*Corresponding Author: 1695708419@qq.com

DOI 10.1201/9781003305026-50

they are constructed. People have been attracted by their beautiful curves and unique appearance, as they act as attractive sights in cities. However, it is difficult to build these strange-looking buildings in the process of modeling because of their irregular layout and the complexity of the internal space. To accurately model, it is time-consuming and complicated, so a whole team needs to collaborate. At present, with the advent of the era of big data, complex engineering designs have been realized through the establishment of effective order with the help of science and technology, thus being applied to architectural design and construction (Shao WP. 2019). The purpose of design is to meet material and aesthetic requirements, while technology is the means to achieve the end because only by meeting the functional requirements and complying with economic principles can technology yield twice the result with half the effort (Jiang F. 2019). The new architectural model is no longer dominated by the designers' ideas but integrates their thinking and ideas into the technology, which is more scientific, effective, and rigorous, and promotes mutual development whether the initial ideas or the final models.

In recent years, with the material abundance and economic development, people call for an improved artistic atmosphere in cities. Many distinctive buildings appear to the public, but their internal construction needs to be refined and improved due to the lack of basic research. Similar designs or operation processes should be avoided. If designers only pay attention to the early stage of the design but ignore the accomplished building, this will lead to an inconsistency between ideas and the final products, which is the most common problem in special-shaped buildings. For architects, to construct a good building, they must authentically apply their drawings to the construction in addition to finding the best option among countless possibilities, which is a task that needs to be done through collaboration. The Guangzhou Opera House designed by Zaha Hadid Architects can be taken as an example. In the early design stage, only preliminary discussions were conducted, but the feasibility of building materials was not evaluated on the spot. In the later stage of the project, as the construction team did not communicate well with the design team, materials were directly cut at will, which eventually led to the discrepancy between the finished building and its proposed architectural design. This project is very innovative in terms of design concepts and design methods, but there exist many problems, such as details and material selection. The local construction team worked without guidance, which resulted in uneven wall facades, inadequate stitching between the curtain wall and window frames, and gaps of varying widths. In addition, some window frames with large curvatures are handled simply and carelessly, and the turning points of the triangular slabs were not connected, just like patches. For the volume construction of irregularly curved buildings, a slate curtain wall was not a good choice compared with the materials used in other good architectural practices. Besides Guangzhou Opera House, there are other buildings that can only be viewed from afar in China, which also reflects problems in the modeling and construction of domestic special-shaped buildings. As the architectural design in the public space mirrors the urban culture, their design methods visually influence their later construction. Therefore, designers should consider how to perfectly present the architectural design of special-shaped buildings in the early stage of their design.

2 DEVELOPMENT OF SPECIAL-SHAPED BUILDINGS

2.1 *The intervention of parameterization*

The Intervention of Parameterization is not new in the entire design community, since parametric design methods are used in products or environmental design. In terms of architecture, many special-shaped buildings are designed through parameterization. The most familiar buildings are those designed by Zaha Hadid Architects and the newly established MAD Architects in China, reflecting the magic of parametric. In laymen's minds, nonlinear design is mostly regarded as parametric design. However, they are quite different as the latter is far more powerful and compatible than the former, and the model generated by the parametric design is not limited to the nonlinear surface structure. The non-linear design is a scientific one that is not linear, irregular, and unorganized, which is included in the field of parametric design. Thanks to parameterization, the shape of models can be changed at will, whether a curve or a straight line; it also pays attention to the scientific and logical nature of the design and the continuity of the design. Designers' ideas are transformed into parameters and variables controlled by programming, thereby changing the shapes. The currently used intelligent tool for the parametric design is the Grashhopper plug-in. For example, in the process of modeling, to change the shape of the building is to change the digital variable in Grashhopper, and the result will change as the variable changes (see Figure 1).

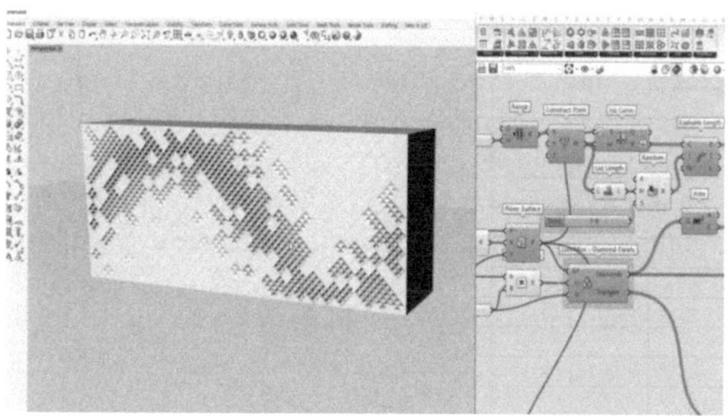

Figure 1. The interface of Grasshopper.

2.2 *Seamless connection from design to construction*

For a long time, attention has been paid to the breakthroughs in techniques and ideas in the process of drawing models. With the intervention of parameterization, various spatial relationships and morphological evolution can be calculated. Both design and the early-stage construction can be simulated and checked in the programming software, and finally, a reasonable density scale and material performance can be obtained (see Figure 2).

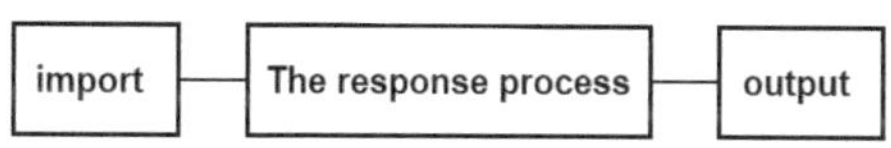

Figure 2. Operation process of Grasshopper.

The core concept of the parametric design is logic which includes three factors in the modeling: parameters/variablcs-operational relationships-generating models. Parameters and relationships can be changed at any time, and changes in variables affect the size, relationship, and position of the model accordingly (see Figure 3). Parameterization can not only help to generate complex shapes, but also improve the speed and efficiency of modeling. Besides the field of architecture, accurate analysis can be conducted and sketches can be produced for landscape topography, etc.

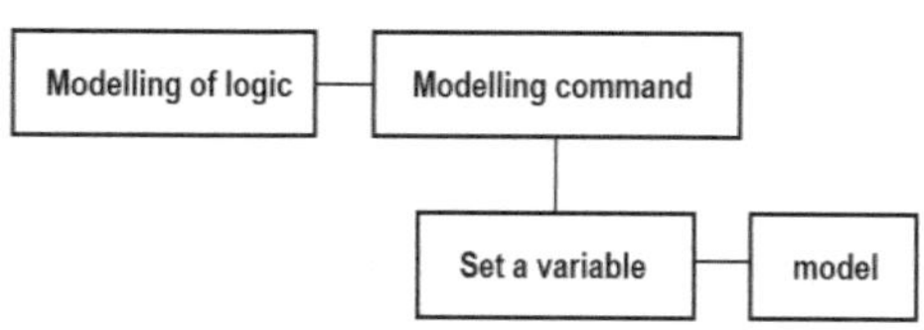

Figure 3. Operation process of Grasshopper.

3 APPLICATION OF PARAMETRIC DESIGN IN SPECIAL-SHAPED BUILDINGS

3.1 *RHINO+Grashhopper operation generation analysis*

To put it simply, Grashhopper is a plug-in that uses procedural algorithms to generate building models in the Rhino software. Unlike other plug-ins of Rhino, it can directly produce the desired model through the logic process without much programming language. Therefore, its greatest value lies in that it is different from other programming software, for it records the whole process of modeling and model generation with its unique symbolic language, and to change the shapes of models is to change the parameters.

After the Rhino software is opened, the plug-in name is entered in the command bar and the button Enter is pressed or the Grashhopper icon in the toolbar is directly clicked to open the view window. Grashhopper is a viewport independent of Rhino, for plug-ins of Grasshopper can also be installed in addition to the built-in basic modules. After Grashhopper is opened, various battery packs can be created. Then the battery icon is dragged to the panel for use, and the selected battery will turn green. After the battery pack is connected and the logic is correct, the battery pack will work. The preliminary model corresponding to the battery pack can be displayed in the Rhino view (see Figure 4).

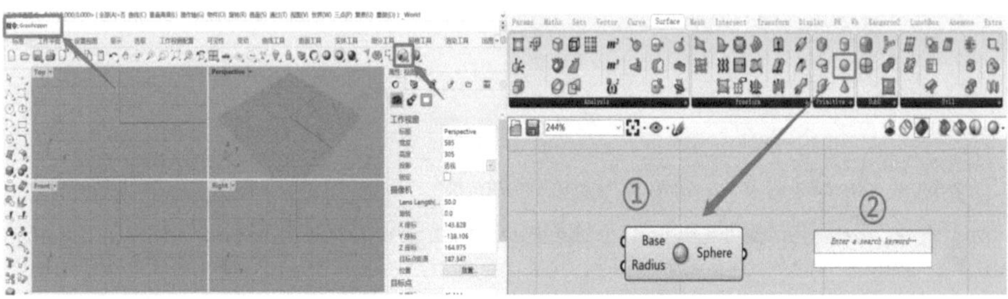

Figure 4. The interface of Grashhopper.

Next, the operation is verified by using the parametric design of Grashhopper and other plug-ins such as auxiliary data extraction and output. First of all, the convex intersections of all fixtures are sorted according to the internal sorting method of a single fixture section, and the number and unit of the feature point of coordinate data are set (Yang et al. 2020) as follows: first, the shape is determined through the sketch design, and the overall shape skeleton lines are drawn on the original Rhino interface. These splines are important, as they determine the shape of the surface, and timely adjustments must be made to make them accurate. During the process, great attention must be paid to the sequence order, otherwise, the final interface will be chaotic and not formed; the determined splines will generate a surface through the stakeout option; based on this, the next step is to combine the number of structural lines in the UV direction, to weave the physical surface, and to control the weaving order through the number of warps and wefts. Secondly, through rigorous quantitative calculations, the algorithm is used to simulate the rods with the aid of programming control points. With the help of the lofting function, core nodes are generated, and the overall modeling is adjusted through specific parameters. Small rectangular frames perpendicular to the surface are made from points to lines to represent the cross-section of bamboo weaving. The innovation in the shape of the bamboo structure device can be made to the greatest extent; afterward, the grashhopper programming algorithm is used to simulate the weaving of bamboo strips, and the entire shape is weaved from the outside to the inside, from the big to the small. The rectangular points are connected to get the same shape as the bamboo strips through lofting; their width, length, and height are adjusted by digital sliding to determine the structure line in the UV direction, and finally, the model is generated. Attention must be paid to the direction, otherwise, the simulation can go wrong (see Figure 5).

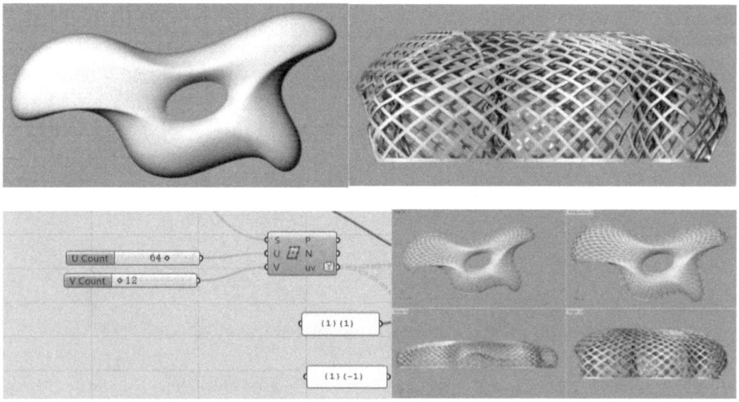

Figure 5. The modeling process of Grashhopper.

3.2 *Advantages of parametric design methods in special-shaped buildings*

In the traditional architectural design, as computer drawings are separated from on-site construction, teams need to work together, which will lead to the above-mentioned inconsistency between design and construction. Such discrepancy results in low efficiency, and when any part of the design is flawed, the entire team needs to make corresponding changes, which takes a lot of time to complete a building. At the same time, this brings about a vicious circle in which the delay of time leads to insufficient consideration of all aspects and the final design can only be modified based on the designer's experience, so, this design model can hardly guarantee the desired design.

Nowadays, in the modeling of special-shaped buildings, the parametric design method improves the work efficiency and ensures the perfect restoration of the completed model. Various shapes can be quickly and accurately generated by modifying the variables, and finally, the best option can be chosen from them. As a result, more time is left for designers to communicate well with engineers. The accurate model can be built and demonstrated in Grashhopper, and to modify the part in the model is to directly modify the variables. In addition, a table can be set up for building construction to facilitate subsequent departments to collect data.

Except for these advantages, parametric software can also be used in collaboration with other BIM software, and even make the Rhino function like BIM through the Grashhopper plug-ins. It can not only give full play to the unique modeling language of the Rhino, but also realize the basic information management of the model, which undoubtedly saves a lot of troubles for the architectural design, and finally optimizes the building.

4 CONCLUSIONS

The parametric design is the product of the combination of modern new technology and new design art, and in turn, it introduces interdisciplinary technology into the field of architecture, thus forming a new architectural design technology. (Xu YY. (2014). With the continuous advancement of science and technology, parametric design is widely used in various fields and can be closely integrated with each part of the design. Also, it can conduct optimal, specific analysis ranging from illumination, terrain to temperature, elevation, etc., to present an entire intelligent building process aided by technology. This design process has offered designers a new experience that the shapes generated by rational digital calculation are different since they have the beauty of lines and rhythm that are different from traditional architecture. This also testifies that the development of special-shaped buildings is the product of the perfect integration of technology and art, and indicates the main direction of the development of post-modern architecture.

ACKNOWLEDGMENT

Gansu Province Department of Science and Technology soft discipline special—Research on the resource protection and sustainable evolution path of Gansu cultural heritage of historical building (21CX6ZA071); Science Fund Project for Distinguished Young Scholars, Lanzhou Jiaotong University – Research on the protection and development strategy of the Long March (Gansu Section) based on the concept of linear cultural heritage (2021042).

REFERENCES

Jiang F. (2019). The applied practice of parametric design in public buildings. J. ZHUANGSHI. 2019 (08), 110–113.

Shao WP. (2019). Freedom and Order Digital Construction of OS-10B Building. J. Architectural Journal. Issue 4, 2019.

Xu YY.(2014). Parameterization·architectural space——the transformation of architectural space forms under the influence of parametric design technology. J. Beauty and Times (Mid). 2014 (09), 100–101.

Yang FX, Wang LW, Duan M, Wang CL, Wang XZ.(2020). Research and application of Rhino+Grashhopper parameterization in special-shaped buildings. J. Proceedings of the 2020 Academic Annual Conference of the Chinese Society of Civil Engineering: 237–245.

Zhou WQ, Deng FD, Wang J.(2020). Discussion on Technique Path of Parametric Architecture Design. J. Architecture Technique. 2020(05), 119–12.

Advances in Urban Engineering and Management Science – Khalil & Yang (Eds)
© 2023 The Author(s), ISBN 978-1-032-30426-7

Risk assessment of building damage caused by undercrossing shield metro construction using Bayesian network

Xinyue Lu, Chengshun Xu* & Xiuli Du
Key Laboratory of Urban Security and Disaster Engineering, Beijing Univ. of Technology, Beijing, China

Jianxun Zhu
Beijing Agiletech Engineering Consultant Co., Ltd., Beijing, China

Yichen Wang
Key Laboratory of Urban Security and Disaster Engineering, Beijing Univ. of Technology, Beijing, China

ABSTRACT: Undercrossing shield tunnels often cause damage to aboveground structures, which are high-risk projects. Reasonable risk assessment is crucial for construction decision-makers and safety managers. In this paper, the risk of building damage caused by an undercrossing Beijing shield metro is assessed using the Bayesian network. The noisy-or model and the Noisy-Max model are used to establish the Bayesian network model. The model is applied to predict the risk probability of building damage induced by the undercrossing shield metro by forwarding reasoning. The results are prepared with the statistical risk probability derived by building settlement monitoring value, which verifies the effectiveness of the model. The key risk factors affecting building safety were diagnosed and analyzed, which can provide a reference for project risk management.

1 INTRODUCTION

The urban rail transit system has increasingly become an essential component of transportation infrastructure due to its advantages such as large traffic capacity, low energy consumption, and low ground space occupation. Since metro construction is mostly located in densely constructed urban areas, it is inevitable for tunnels to approach or underpass existing buildings, which may cause uneven settlement of the ground and affect the safety of the adjacent buildings (Zhu 2021). The shield construction method is widely used in tunnel construction, however, the frequent risk events of adjacent structure damage caused by shield tunnel construction have attracted people's attention. In 2009, the shield construction of Guangzhou Metro Line 2 and line 8 caused the inclination of adjacent buildings, and the residents of 5 adjacent buildings were evacuated urgently; In 2008, the shield construction of Foshan Metro Line 2 caused the collapse of a tunnel and pavement, which brought great potential safety hazards to the surrounding buildings. To reduce the probability of shield-induced risk events, it is of great significance to evaluate the damage risk of adjacent buildings induced by shield tunneling construction.

The existing methods for evaluating the construction risk of underground structures can be divided into qualitative analysis methods and quantitative analysis methods. The qualitative analysis methods include the analytic hierarchy process (Li 2013), risk matrix, and fuzzy comprehensive evaluation (Chu 2017), of which the evaluation results are greatly affected by people's subjective factors. The quantitative analysis methods include fault tree analysis (Hyun 2015), event tree analysis (Špaèková 2013), Monte Carlo simulations, etc., which require a large number of statistical data on the construction site. Due to the complex interaction mechanism between shield tunnels and

*Corresponding Author: xcs_2017@163.com

DOI 10.1201/9781003305026-51

adjacent buildings, the strong uncertainty of risk factors and the limitation of statistical data lead to limitations in the application of traditional risk assessment methods. With the development of artificial intelligence, the Bayesian network (BN) has been extended to construction risk analysis. The Bayesian network can integrate the prior knowledge and historical data of experts in many fields, which provides a theoretical basis for the risk assessment of damage to adjacent buildings induced by shield tunnel construction.

This paper assesses the risk of building damage caused by an undercrossing Beijing shield metro using the Bayesian network. A shield tunnel under crossing residential buildings in Beijing was selected to conduct a case study. The risk probability of damage to residential buildings was predicted, and the main risk factors causing building foundation settlement were diagnosed and analyzed. The evaluation results were compared with the measured data to verify the correctness of the proposed BN-based risk assessment method.

2 METHODOLOGY

2.1 *Bayesian network*

The Bayesian network consists of a direct acyclic graph and several conditional probability distributions. In the direct acyclic graph, the variables are represented by nodes, and the relationships between the variables are represented by the direct edges. The nodes starting from an arrow are cause variables, named parent nodes; the nodes pointed by arrows are result variables, named child nodes. The conditional probability table (CPT) is used to quantitatively express the interaction among the random variables.

A possible value of a discrete variable is called a state of a node. The joint probability can be simplified according to the conditional independence assumption, and the edge probability of a node can be determined through marginalization, as shown in Eq. (1). It can be used to predict the probability of risk events. According to the Bayesian theorem, when the probability of risk event $Y=y$ is known, the probability of risk factor X can be diagnosed and analyzed using Eq. (2).

$$\mathrm{P}(Y)=\sum_{X_1,X_2,\ldots,X_n}\mathrm{P}(X_1,\ldots,X_n,Y)=\sum_{X_1,X_2,\ldots,X_n}\prod_{i=1}^{n}\mathrm{P}(Y|X_i) \tag{1}$$

$$\mathrm{P}(X_n=x_n|Y=y)=\frac{\mathrm{P}(X_n=x_n)\mathrm{P}(Y=y|X_n=x_n)}{\mathrm{P}(Y=y)} \tag{2}$$

2.2 *Noisy-or model*

In the process of building a Bayesian network for complex engineering, the number of items in CPT increases exponentially with the number of parent nodes. Noisy-or model can make the number of items in CPT increase linearly with the number of parent nodes, which greatly reduces the workload of experts and has a low error rate. The Noisy-or model assumes that: (1) all nodes in BN are in two states; (2) each risk factor can cause the risk events independently. When all risk factors do not occur, the risk events do not occur (Feng 2020).

The conditional probability of risk event is calculated by Eq. (3), where p_i represents the probability of Y caused by the occurrence of X_i only, i.e., $p_i=\mathrm{P}(Y|\overline{X_1},\overline{X_2},\ldots,X_i,\ldots,\overline{X_n})$ X_T represents the occurrence of X,

$$\mathrm{P}(Y|X_1,\ldots,X_n)=1-\prod_{i:X_i\in X_T}(1-p_i) \tag{3}$$

Because the Noisy-or model can only calculate variables with two states, researchers have extended the model to the Noisy-Max model, which can calculate variables with multiple states (Zagorecki 2013). The new model assumes that variable Y must be sequential variables and that the risk factors are independent of each other. The conditional probability of risk events can be calculated by Eq. (4) and Eq. (5), where y and X_i represent the values of nodes Y and X_i.

$$\mathrm{P}(Y \leq y|X) = \prod_i \left[\sum_{Y \leq y} \mathrm{P}(Y = y|X = x_i) \right] \tag{4}$$

$$\mathrm{P}(Y|X) = \begin{cases} \mathrm{P}(Y \leq y|X) - \mathrm{P}(Y \leq y-1|X) & y \neq y_{\min} \\ \mathrm{P}(Y \leq y|X) & y = y_{\min} \end{cases} \tag{5}$$

3 CASE STUDY AND RESULT ANALYSES

3.1 *Engineering background*

A shield section of the Beijing metro was excavated by the earth pressure balance shield method, which passes through a residential building group. 43 settlement monitoring points were installed in the main load-bearing walls at the interval of 10 meters between each survey point, as marked by green triangles in Figure 1. The crown level of the tunnel is 14m to 15m, and the net distance to the nearest part of the building is about 12.8m. According to the geological survey report, the tunnel was excavated primarily within fine silty sand and silt. Underground water is abundant in the area of tunnels passed below the buildings. The Phreatic water layer is over the crown level of each tunnel. The Interlayer water to the Confined water layer is over the foundation level of each tunnel. Figure 2 shows a summary log of Borehole QHSC01.

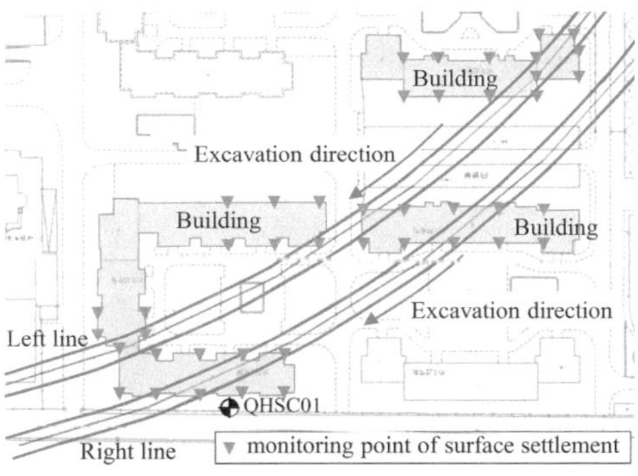

Figure 1. Schematic diagram of the metro tunnel.

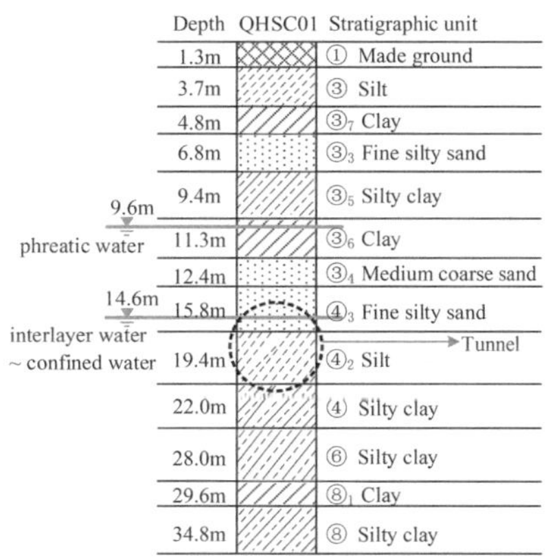

Figure 2. Geological profile from Borehole QHSC01.

3.2 *BN model for risk assessment*

For the shield tunnel under crossing existing buildings, the safety performance of the buildings is affected by the construction environment, construction technology, construction machinery, construction management, and so on. Based on literature research and expert experience, this work selects 12 risk factors from the four mentioned aspects to establish the BN-based risk assessment model, as shown in Figure 3. The state of each risk factor is shown in Table 1. Note that, the five states of R correspond to the five-level division of tunnel engineering risk in the codes, and level I represents the greatest risk (Eskesen 2004).

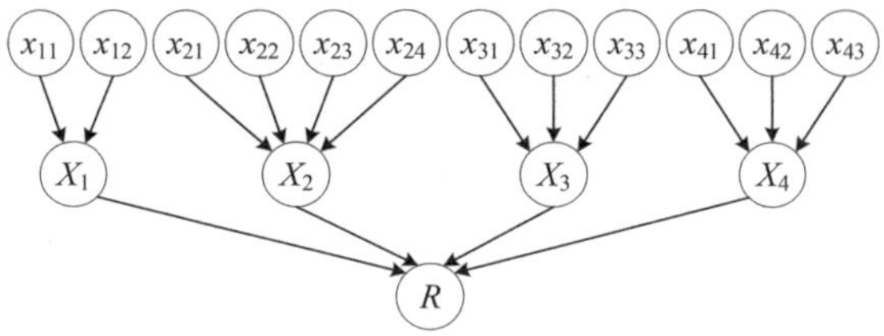

Figure 3. BN model of building damage induced by tunnel.

Table 1. Risk factors state of building damage induced by undercrossing shield tunnel construction.

Variables	State 1	State 2	Variables	State 1	State 2
Geological conditions x_{11}	Cohesive soil stratum	Sandy cobble stratum	Abnormal wear of the cutter x_{31}	Yes	No
Groundwater stability x_{12}	Unstable	Stable	Failure of the main shaft of the cutter head x_{32}	Yes	No
Not timely grouting x_{21}	Yes	No	Failure of the driving system of the shield machine x_{33}	Yes	No
Not timely rectification of shield machine x_{22}	Yes	No	Insufficient experience of constructors x_{41}	Poor	Good
Improper earth pressure x_{23}	Yes	No	Lack of safety awareness x_{42}	Poor	Good
Improper assembly of the tube x_{24}	Yes	No	Poor management organization x_{43}	Poor	Good
Variables	**State 1**	**State 2**	**State 3**	**State 4**	**State 5**
Geological and hydrological conditions X_1	Cohesive soil and stable groundwater	Cohesive soil and unstable groundwater	Sandy cobble and stable groundwater	Sandy cobble and unstable groundwater	\
Construction technique X_2	Poor	Good	\	\	\
Construction equipment X_3	Fault	Well	\	\	\
Construction management X_4	Poor	Good	\	\	\
Risk of building damage induced by the undercrossing tunnel construction R	Level I	Level II	Level III	Level IV	Level V

The CPTs of two-state nodes in the BN model are determined by the Noisy-or model, and the CPTs of multiple-state nodes are determined by the Noisy-Max model. Through investigation and interviews with engineering experts, the parameters of these two models are calculated by the expert experience method. The weight coefficient description of experts' experience probability is shown in Table 2. Using Eq. (6), the probability distribution of risk factors can be calculated from the survey results, herein, P_j^i represents the probability that the state of risk factor j is i, j is the number of risk factors, i represents the i-th state of a risk factor, n is the number of states of a risk factor, ω_{ijk} represents the weight coefficient of the k-th expert who gives P_j^i.

Taking the conditional probability calculation of the "construction management $X4$" node as an example, experts determine the probability (p_{43}) of "Poor management organization x_{43}" leading to poor construction management. In this study, 16 experts were investigated. Five experts believe that the occurrence of only x_{43} will lead to a poor construction management level, including 1 I-expert, 1 II-expert, 1 III-expert, and 2 IV-experts according to the classification rules shown in Table 2. From Eq. (6), p_{43} can be calculated as: $p_{43} = (1 * 1 + 0.9 * 1 + 0.8 * 1 + 0.7 * 2)/(1 * 3 + 0.9 * 4 + 0.8 * 5 + 0.7 * 4) = 0.3$. Similarly, $p_{41} = 0.35$ and $p_{42} = 0.35$. The conditional probability is shown in Table 3.

$$\mathrm{P}_j^i = \sum_{k=1}^{16} \omega_{ijk} \Bigg/ \sum_{i=1}^{n} \sum_{k=1}^{16} \omega_{ijk} \tag{6}$$

Table 2. Weight of experts.

Category	Introduction	Weight coefficient	Number of people
I	Well-known experts in tunnel construction Construction Project Manager Construction or supervision technicians with senior title	1.0	3
II	Designers or researchers with senior professional titles Construction or supervision technicians with intermediate title	0.9	4
III	Designers or researchers with intermediate title Construction or supervision technicians with the primary title	0.8	5
IV	Designers or researchers with the primary title	0.7	4

Table 3. Conditional probability of X_4.

x_{41}	x_{42}	x_{43}	$X_4 =$ Poor	$X_4 =$ Good
Poor	Poor	Poor	0.704	0.296
Good	Poor	Poor	0..545	0.455
Poor	Good	Poor	0.545	0.455
Good	Good	Poor	0.3	0.7
Poor	Poor	Good	0.577	0.423
Good	Poor	Good	0.35	0.65
Poor	Good	Good	0.35	0.65
Good	Good	Good	0	1

3.3 *Result analyses*

NETICA software is used for modeling to predict the risk probability of structural damage induced by shield tunnel construction. The calculation results are shown in Figure 4. The damage risk of

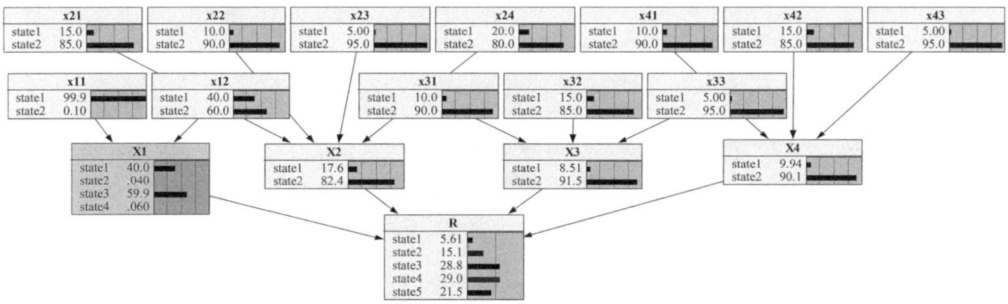

Figure 4. Probability of the building damage risk.

residential buildings caused by undercrossing shield tunneling is most likely to be at Level 3 and Level 4, and the corresponding risk probabilities are 28.8% and 29%, respectively.

The monitored data of 43 settlement monitoring points around the affected buildings were sorted out. Considering the soil conditions and construction risk control requirements in Beijing, the cumulative ground settlement value (S) of buildings is set to five states, corresponding to the three-level warning values of yellow, orange, and red. For S exceeds the control value ($S_{max} = 15$mm), it is set as risk Level 1, moreover, it is Level 2 for $0.85S_{max} < S < S_{max}$, Level 3 for $0.70\ S_{max} < S < 0.85S_{max}$, Level 4 for $0.33S_{max} < S < 0.70S_{max}$, and Level 5 for $S < 0.33S_{max}$. The statistical risk probabilities obtained from monitoring data corresponding to different risk levels are shown in Table 4. The adjacent buildings have the highest probability of being at risk in Levels 3 and 4, reaching 27.9% and 37.5% respectively. It demonstrated that the prediction results of the BN model are consistent with the monitoring results, which verifies the effectiveness of the proposed BN-based risk assessment method.

Table 4. Risk level classification and statistical risk probability from monitoring data.

Risk level	Settlement monitoring value S	Probability
1	$S \geq 15$ mm	0
2	12.75 mm $\leq S <$ 15 mm	9.3% (=4/43)
3	10.5 mm $\leq S <$ 12.75 mm	27.9% (=12/43)
4	5 mm $\leq S <$ 10.5 mm	39.5% (=17/43)
5	0 mm $\leq S <$ 5 mm	23.3% (=10/43)

The posterior probability of risk factors can be calculated using Eq. (2), and the risk factors that are most likely to cause serious damage to the building can be diagnosed and analyzed. Figure 5 presents the posterior probability of $x_{21} \sim x_{43}$ corresponding to State 1 (serious damage to the building). It is shown that "Improper assembly of the tube x_{24}" is a high-risk factor leading to building damage, thus this factor should be strictly controlled during the construction process.

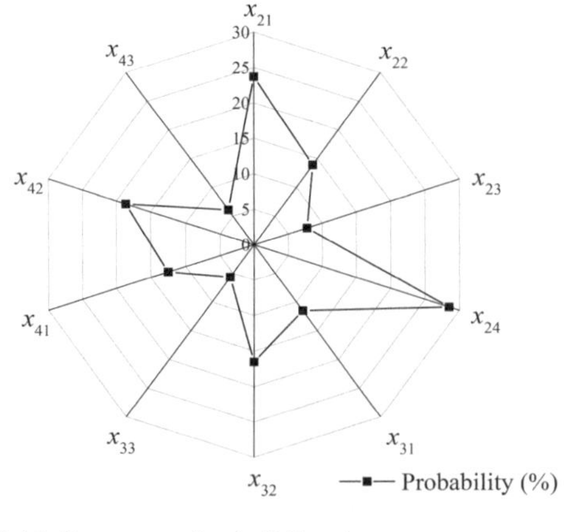

Figure 5. Fault diagnosis of risk factors causing building damage.

4 CONCLUSION

The shield tunnel constructions under crossing ground structures are high-risk projects. This paper assesses the risk of residential buildings damage caused by an undercrossing Beijing shield metro using the Bayesian network. The predicted building damage risk level matches well with the early warning level derived from the measured settlement, indicating the effectiveness and applicability of the proposed method. The diagnostic analysis identified the risk factor most likely to cause severe damage to the building, i.e., the "Improper assembly of the tube x_{24}". The proposed risk assessment method can provide engineers with reliable risk-based decision-making support.

REFERENCES

Chu H. D., Xu G. L., Yasufuku N., Zhang Y., Liu P. & J. F. Wang (2017). Risk Assessment of Water Inrush in Karst Tunnels Based on Two-Class Fuzzy Comprehensive Evaluation Method. Arab. J. Geosci. 10, 179.

Eskesen S. D., Tengborg P., Kampmann J. & T. Holst Veicherts (2004). Guidelines for Tunnelling Risk Management: International Tunnelling Association, Working Group No. 2. Tunn. Undergr. Sp Technol. 19(3), 217–237.

Feng X., Jiang J. C. & W. F. Wang (2020). Gas Pipeline Failure Evaluation Method Based On a Noisy-Or Gate Bayesian Network. J. Loss Prevent. Proc. 66, 104175.

Hyun K., Min S., Choi H., Park J. & I. Lee (2015). Risk Analysis Using Fault-Tree Analysis (FTA) and Analytic Hierarchy Process (AHP) Applicable to Shield TBM Tunnels. Tunn. Undergr. Sp Technol. 49, 121–129.

Li F. F., Phoon K. K., Du X. L. & M. J. Zhang (2013). Improved AHP Method and its Application in Risk Identification. J. Constr. Eng. Manage. 139(3), 312–320.

Špaèková O., Novotná E., Šejnoha M. & J. Šejnoha (2013). Probabilistic Models for Tunnel Construction Risk Assessment. Adv. Eng. Softw. 62–63, 72–84.

Zagorecki A. & M. J. Druzdzel (2013). Knowledge Engineering for Bayesian Networks: How Common are Noisy-Max Distributions in Practice. IEEE Trans. Syst., Man and Cybern. 43(1), 186–195.

Zhu. C. (2021). Surface Settlement Analysis Induced by Shield Tunneling Construction in the Loess Region. Adv. Mater. Sci. Eng. 5573372.

Advances in Urban Engineering and Management Science – Khalil & Yang (Eds)
© 2023 The Author(s), ISBN 978-1-032-30426-7

Research on the engineering technical optimization of Village planning at Zhejiang J village based on the collaborative concept

Yuhua Zhong, Lei Tong*, Fei Su & Xiaoqian Zhu
School of Tourism and Urban-Rural Planning, Zhejiang Gongshang University, Hangzhou Zhejiang China

ABSTRACT: At present, the outmoded engineering technology in village planning has led to long-term problems such as high time and space costs, inaccurate information transmission, and insufficient planning toughness. This paper puts forward a set of engineering technology to optimize village planning, optimizes the technology of investigation, engineering planning, and engineering implementation in detail, and takes J village in Zhejiang Province as an example. Finally, the collected evaluation data are statistically analyzed. The results show that this set of engineering technology has a better application effect than mainstream technology, and has a certain optimization effect on the above problems.

1 INTRODUCTION

In recent years, China has vigorously carried out digital projects. From digital city construction to digital village construction, the planning field has formed a strong trend of digital transformation (Ding & Sun 2000; Yang 2012). In this trend, how to apply digital technology in the field of planning has become a hot topic in the planning field. Digital technology has a series of advantages such as high efficiency, low time and space cost, visualization, and so on, and has great potential to solve the multi-agent coordination problem in the planning process. In the mainstream planning mode, the time and space cost of coordination between the main bodies of village planning is too high, which leads to the lack of resilience in village planning and other problems. These problems affect and restrict the development of the village for a long time. Therefore, an efficient coordination planning method is urgently needed to solve these constraints and drive the development of the village. Digital technology has been an emerging technology in recent years, since the introduction of the planning field, got rapid development, and planning management, urban design has played a huge role in areas. Its fast, convenient advantages can just make up for the current village planning time high-cost disadvantage, plan for the future development direction for village provides a good way of thinking.

At present, many scholars have carried out research on the application of digital technology in the field of planning, showing a trend of rapid growth, and presenting diversified characteristics in the research direction and content. Existing researches mainly focus on planning technology and planning management. For example, Zheng Nan discussed the significance, implementation approaches, and key technologies of three-dimensional simulation technology in the whole process of planning management (Zheng et al. 2013). Other scholars have studied the construction and application of the three-dimensional auxiliary systems of urban and rural planning management based on the GIS system, so as to explore the method of public participation in planning. The above research mainly discusses the content of digital technology-assisted management and does not consider the application to rural environments. In terms of collaborative planning, there are still few studies on the application of digital technology, which mainly focus on improving public

*Corresponding Author: zjutonglei@126.com

 DOI 10.1201/9781003305026-52

participation with the help of digital technology. For example, Wang Peng has developed a complete set of public participation modes for the urban planning community based on new media by integrating the advantages of WebGIS and WeChat (Wang & Zhao 2015). At the theoretical level, Yao Jingyi proposed the use of AR technology to improve public participation in planning (Yao et al. 2018). Most of the above studies focus on the construction of theoretical models without putting their theories into practice for testing. In general, the application of digital technology has brought huge benefits to planning technology, management, and other stages, and also played a good role in promoting the quality of planning. But in the field of collaborative planning, research on the application of digital technology is more focused on the theoretical level, how the digital technology in collaborative planning method and application of technology in the specific process of exploration is still insufficient, at the same time, the digital technology application in the collaborative planning related theory mostly lack practice test, in village planning and construction of actually played a leading role in promoting. Based on this, taking J Village in Zhejiang as an experimental case, this paper intends to conduct experimental discussion and verification on the technical methods and specific processes applied by digital technology in village cooperative planning, in order to provide some beneficial ideas for the efficient integration of digital technology into the whole process of village planning and the promotion of real-time multi-agent collaboration in village planning.

2 GENERAL SITUATION OF J VILLAGE

J Village is located in Tiantai County, Taizhou City, Zhejiang Province, consisting of a central village and two small scattered settlements. The village was built in the Ming Dynasty, and there are many historical buildings in the village. In addition, J village is rich in biological resources, which is a typical ancient village in the Jiangnan area. But in recent years, due to traffic, terrain, and other conditions, the village has been unable to find a breakthrough for development, which is a representative mountain village in the current development stage. Thanks to the construction of the "digital countryside", Tiantai County, where J village is located, has become the pilot and demonstration county of digital countryside in Zhejiang Province, pointing out a new direction for the development of J Village.

In 2020, the research group selected J village for the collaborative planning experiment, and the research area was the central village area of J Village. Through the field investigation of the current situation of J Village, it decided to introduce the collaborative planning technology method based on virtual reality, big data, and other digital technologies and the "lianxiang" platform developed by the research group for the compilation and implementation of village planning. In order to provide some reference for the digital technology-based collaborative planning to promote rural revitalization in mountainous areas.

3 APPLICATION PRACTICE OF DIGITAL TECHNOLOGY IN J VILLAGE COOPERATIV-E PLANNING

3.1 *Application scheme of digital technology in J Village Cooperative planning*

The digital technology in this paper mainly refers to virtual reality technology and big data technology. Virtual reality technology (VR) rose in the 1980s, is a fusion of computer image processing technology, network and multimedia technology, simulation technology, and other technologies, but also different from the above technology immersion, interaction, and other features of the emerging comprehensive information technology (Hu 2000). With the emergence of big data technology in the early 21st century, the academic community generally believes that big data is characterized by large data scale, variety, speed, authenticity, and sparse data value density (Han et al. 2018; Ye et al. 2014). Virtual reality technology is mainly applied in the construction of village scenes, which builds the basis for villagers' cognition and participation in planning. Big data analysis technology is mainly applied in the collection and analysis of survey data such as opinions of villagers, government, and developers, and builds a bridge for communication with villagers.

In the mainstream village planning model, the village planning process can be roughly divided into research, planning, and implementation stages, although the entire process is relatively rigorous, but did not pay attention to the villagers in the process of the underlying interest groups, all kinds of value orientation have no coordination between subject, led to a lot of contradictions appeared in the process of planning, Finally, it is difficult to achieve main body coordination (Meng et al. 2015). Therefore, this article decided on this basis, through the research, plan, implementation of the three links in the digital technology, to solve the problem of the part is mainly to achieve in the planning of villages in the form of online interactive multi-agent efficient coordination, planning and to solve the current village in collaborative space and time cost is high, the information transmission problems such as low precision, inadequate planning, and toughness.

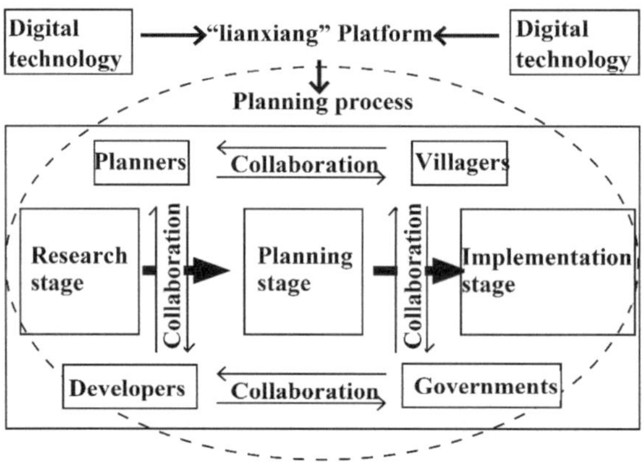

Figure 1. Schematic diagram of improved planning technical scheme.

3.2 *The specific process of digital technology application in J village cooperative planning*

3.2.1 *Investigation stage*

In mainstream programming mode, the village planning research mainly to the scene reconnaissance, the villagers' representatives meeting, focus-group discussion, in-home interviews, investigation questionnaire form is given priority, but because of time and space costs higher a variety of factors, the research method becomes a mere formality, leading to widespread distortion of villagers to participate in the breadth and depth and survey data, the problem such as low quality of information feedback, The quality of subsequent village planning was seriously affected. To solve the above problems, the new method builds a multi-online real-time interactive collaborative planning research system based on the "lianxiang" platform. The system connects villagers, planners, governments, developers, and other stakeholders, the villagers, decision-makers, developers, and other major subjects to express an individual opinion with the help of the platform to the planner's needs, the platform through the analysis of big data technology powerful computing capacity automatic collecting and classifying relevant opinions demand, and then transferred to the account of the planner, The planners can make the next plan according to the collected demand and opinion information, saving them time and space cost of research and greatly enhancing the breadth and depth of public participation. In addition, this method also forms a good "top-down" multi-directional interactive data and information transmission mechanism (as shown in Figure 2), avoiding the formal research phenomenon caused by the single "top-down" elite planning in the mainstream mode.

3.2.2 *Engineering planning stage*

In the planning stage, there are some problems in the current planning model, such as the lack of participation path of villagers, the high threshold of professional participation of villagers, and

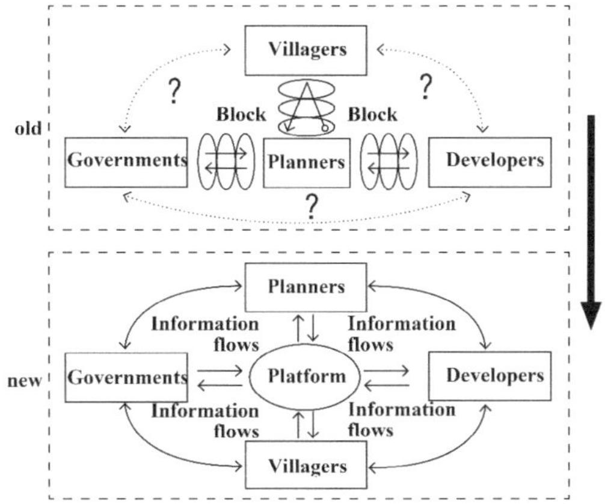

Figure 2. Schematic diagram of research technology mode improvement.

the form of evaluation. In view of the above problems, the technical methods of the planning scheme preparation stage are improved by introducing live broadcasting of the evaluation process and enriching the form of planning results. By using the powerful scene modeling ability of virtual reality technology, as well as the characteristics of immersion and interaction, the village planning scene model is constructed so that villagers can intuitively understand the expected effect of planning from the first-person perspective. At the same time, in the past, the determination of key projects in village planning mostly depended on the discussion of planners, the government, the village committee, developers, and other "upper" elite groups, especially the government and other institutions have almost absolute decision-making power, and villagers are usually unable to participate in this process. The reason is usually caused by the lack of participation mechanism and path. Therefore, the collaborative planning scheme adopts the form of online real-time discussion and communication to form a multi-body discussion mechanism with real villagers' participation.

In planning results form, using the "big data analytics + various kinds classification accurate delivery", designed for different groups of a different form of planning achievements project, for example for the villager's groups use easy-to-read graphic and focus on the needs to solve, government agencies are focused on the economic and technical index, village development prospect, Professional planning texts are uploaded to the platform for users to download freely, which lowers the professional threshold to participate in planning and improves the enthusiasm of villagers to participate in planning. In addition, the platform can also provide villagers with other planning information viewing functions, so that villagers can view the planning information according to their own needs, which not only cultivates villagers' awareness of participating in the planning, but also increases the coverage of planning information, greatly improving the accuracy of information transmission of planning results. At the level of planning review, the mainstream planning review process is relatively closed, and villagers are unable to know the specific review content, form, and other information, which has a certain negative impact on planning. Collaborative planning scheme based on digital technology improves the link, through the chain township platform realizes the geared to the needs of the villagers live public review link, the villagers on the platform will be able to participate in the planning and review, let the villagers know more about the village planning, at the same time the review can absorb more widely and advice, make planning more targeted and practical.

3.2.3 *Engineering implementation stage*

In the mainstream village planning mode, village planning pays more attention to the compilation of planning schemes than to the implementation of planning schemes, which is inconsistent with the current positioning of planning focusing on the implementation and effectiveness. The lack of implementation supervision also impacts the resilience of village planning to a certain extent. Based on this, it is planned to optimize and improve the implementation stage of village planning with the help of the characteristics of fast and convenient interaction of the Internet. At this stage, digital technology was mainly applied in digital supervision, and the digital supervision system of village planning was constructed by using the "lianxiang" platform. We compared with the mainstream planning implementation tracking method, the tracking method based on digital technology is more advantageous in time and space cost and has higher authenticity. In terms of the specific process of implementation and supervision, villagers and other practical stakeholders can check the implementation of planning schemes in real-time and upload and share the implementation of planning schemes using mobile phones and other terminals.

In the implementation stage of the process, the villager's practical interests subjects such as being able to live pictures, text, and other content via mobile phone, tablet, and another terminal to upload to share with other people, other people don't need to the site will be able to check the situation of the implementation of the planning scheme, planner according to the implementation of planning scheme applicability, timely and reasonably planning scheme for revision, Promote the planning scheme to serve the village development and construction for a long time, so as to enhance the resilience of the village planning scheme.

4 EVALUATION OF THE J VILLAGE SCHEME

4.1 *Construction of evaluation index system*

In view of the main problems that need to be solved in this paper, two specific indicators are selected from three categories of data, namely space-time cost, planning coverage, and planning satisfaction, which can fully express their excellent status to constitute the evaluation index system of this scheme. The spatial and temporal cost index includes two specific indexes: villagers' communication frequency and spatial cost. The planning coverage index includes villager participation and planning information transmission efficiency. The planning satisfaction index includes the satisfaction of decision-makers and the satisfaction of villagers. The score range is divided into five levels: low, medium-low, medium, medium-high, and high. The data source is the "lianxiang" platform, which is more objective and authentic than offline collection.

4.2 *Comparison and evaluation of J village cooperative planning scheme and mainstream scheme*

As shown in Table 1, through the implantation of virtual reality, big data, and other digital technologies, the village planning scheme has been improved to a certain extent in terms of spatial and temporal cost, planning coverage, and planning satisfaction. In space and time cost, the adoption of online interaction, in the form of various characters to realize the online real-time interactive

Table 1. Statistical table of comparison and evaluation results between collaborative planning scheme and mainstream scheme.

Category indicators	Specific indicators	Collaborative planning results	Mainstream planning results
spatial and temporal cost	Communication frequency	Medium-high	Low
	Spatial and temporal cost	Medium-low	Medium-high
planning coverage	Efficiency of planning information transmission	High	Middle
	Villagers' Participation	Medium-high	Low
planning satisfaction	Decision-maker satisfaction	Medium-high	Medium-high
	Villagers' satisfaction	Medium-high	Medium-low
Overall effect	—	Medium-high	Middle

communication, compared to mainstream village planning pattern distance cost to save a lot of time and space, a higher degree of freedom in the form of online communication at the same time, to promote the efficiency of communication between the main body has also played a role in promoting, It provides a good foundation for the improvement of planning effect. In terms of planning coverage, due to the adoption of online communication and interaction planning, villagers and other figures can participate in planning anytime and anywhere through mobile phones, tablets, and other terminals, which greatly improves the breadth and depth of villagers' participation. At the same time, the information transmission mechanism combining "bottom-up" and "top-down" and diversified precise transmission forms has greatly promoted the improvement of planning information transmission efficiency, cultivated villagers' active participation in planning, and effectively expanded the coverage of planning. In terms of planning and satisfaction, the villagers in major planning mode of vulnerable groups can be free to participate in the planning process, make the planning content accords with the practice of village development, government policymakers can more conveniently view the various indexes such as completion and village development prospect, which laid a foundation for planning and satisfaction.

5 SUMMARY AND PROSPECT

Digital construction is an important way to make up for the shortcomings of rural development, improve the level of rural governance and promote rural revitalization. Under the background of the collaborative concept, this paper puts forward optimization methods for the engineering technology of village planning according to the existing problems and puts forward specific strategies for the three links of investigation, engineering planning, and engineering implementation. At the same time, J village in Zhejiang Province is taken as a case to test the technology, which plays an important role in promoting the optimization of village planning engineering technology. However, the promotion cost of this planning method is still high, and it is difficult to adapt to the weak economic foundation in rural areas. In addition, due to the impact of urbanization, the left-behind villagers are mostly elderly groups, and their difficulty in accepting digital products is also a major obstacle to the implementation of the program. In addition, whether the "lianxiang" platform can truly integrate planners and other "foreign groups" into the environment of rural autonomy to serve the vast rural construction and development needs further demonstration in subsequent practice.

ACKNOWLEDGMENTS

Project supported by National Natural Science Foundation of China (Grant No. 51908498)

REFERENCES

Ding, l. Y. & Sun, J. (2000). Digital City: changes in urban planning. Planner, (06): 21–23.

Han, W. H. Jia, Y. & Zhou, B. (2018). Key technologies and challenges of big data analysis. Information technology and network security, 37 (04): 7–10.

Hu, M. X. (2000). Virtual reality technology and its application in urban planning. Planner, (06): 19–20 + 18.

Meng, Y. Dai, S. Z. & Wen, X. F. (2015). Problems and Countermeasures in current rural planning practice in China. Planner, 31 (02): 143–147.

Wang, P. & Zhao, L. H. (2015). The transformation of urban planning is promoted by big data and new media technology. In: The 17th annual meeting of China Association for science and technology: Proceedings of the Symposium on 16 big data and Urban-Rural Governance. Guangzhou. pp. 49–55

Yang, T. (2012). Digital city and Space Syntax: a way of digital planning and design. Planner, 28 (04): 24–29.

Yao, J. Y. He, Z. Y. & Xu, F. J. (2018). Application of augmented reality technology in public participation in urban planning. Urban architecture, (34): 114–118.

Ye, Y. Wei, Z. C. & Wang, H. J. (2014). Urban planning response in the era of big data. Planner, 30 (08): 5–11.

Zheng, N. Chen, F. X. & Zhang, J. L. (2013). Application of 3D simulation in the whole process management of urban and rural planning. Information technology of civil construction engineering, 5 (04): 93–98.

Advances in Urban Engineering and Management Science – Khalil & Yang (Eds)
© 2023 The Author(s), ISBN 978-1-032-30426-7

Research on influencing factors of prefabricated construction supply chain management based on ISM-AHP

Zhu Tian*
School of Management, Shenyang Jianzhu University, Shenyang, Liaoning, China

ABSTRACT: At present, the supply chain management of prefabricated buildings is in the development stage, and many risks are yet to be resolved. This article takes the influencing factors of prefabricated building supply chain management as the research object, uses a questionnaire survey method, expert interview method, and other qualitative methods to determine 18 influencing factors, and then adopts the method of ISM-AHP combination to clarify the internal factors among the influencing factors. Following the logical relationship and degree of importance, corresponding countermeasures and suggestions are proposed for key influencing factors.

1 INTRODUCTION

In recent years, as the concept of supply chain management has matured, its application in prefabricated buildings has also increased. In 2020, the rapid accomplishment of construction of Huoshenshan hospitals and Leishenshan hospitals highlighted the superiority of prefabricated building supply chain management and made the entire construction industry realize that prefabricated building supply chain management is significant in promoting the development of prefabricated buildings.

2 DETERMINATION OF INFLUENCING FACTORS

First, 20 influencing factors of prefabricated building supply chain management were selected and confirmed from the literature, and then the opinions of relevant experts were extensively solicited through the form of a questionnaire survey to classify and confirm the influencing factor indicators (Fan Yuncui & Shan Tong 2021). Finally, an indicator system of factors that affected the supply chain management of prefabricated buildings was established from three aspects: internal environmental factors, participating parties, and external environmental factors as shown in Table 1.

3 ANALYSIS OF INFLUENCING FACTORS

3.1 *ISM model analysis of factors affecting the supply chain management of prefabricated buildings*

3.1.1 *Build adjacency matrix*

According to determined index system of influencing factors in the supply chain management of prefabricated buildings, an adjacency matrix A is established. Element aij indicates whether Si has an effect on Sj. If Si has an effect on Sj, element aij=1; if Si has no effect on Sj, element aij=0.

*Corresponding Author: 1085524142@qq.com

 DOI 10.1201/9781003305026-53

Table 1. Index system of factors affecting supply chain management of prefabricated buildings.

Influencing factors	Criteria for influencing factors	Influencing factor index	Factor description
Internal environmental factors	Manufacturing stage	The operability of the construction process S_1	Feasibility of construction process
		Technical ability of construction personnel S_2	Professional competence of construction personnel
	Assembly stage	Project schedule control ability S_3	Duration control ability
		Assembly quality and safety S_4	Strict control of assembly quality
	Transportation and delivery stage	Component transportation protection S_5	Component transportation protection
		Timeliness S_6	Components arrive on time
		Delivery product quality S_7 (Wang Yingchao, 2020)	Delivery product quality standards
	Design and procurement stage	Design feasibility S_8	The design scheme meets the requirements
		Intensified design of prefabricated components S_9	Deepen the design to meet the requirements
		Procurement plan S_{10}	Correct Procurement plan
		Quality of purchased products S_{11}	The quality of purchased products meets the standards
Influencing factors of all parties involved	Influencing factors of all parties involved	Information sharing mechanism S_{12}	Information sharing platform
		Benefit distribution S_{13}	Profit distribution among enterprises
		Trust mechanism S_{14} (Chen Chao 2020)	Trust among supply chain members
External environmental factors	External environmental factors	Natural environment S_{15}	Earthquakes and other natural disasters
		Market demand S_{16}	Supply and demand inequality
		Policy standard S_{17} (Zhang Chuang 2020)	National standards and specifications
		Industry chain S_{18}	Integrity of the industrial chain

According to this principle, the adjacency matrix of the following factors can be obtained (Tang et al. 2020).

$$
A = \begin{bmatrix}
0 & 0 & 0 & 0 & 1 & 0 & 0 & 0 & 0 & 0 & 0 & 0 & 0 & 0 & 0 & 0 & 0 & 0 \\
0 & 0 & 0 & 0 & 0 & 1 & 0 & 1 & 0 & 0 & 0 & 0 & 0 & 0 & 0 & 0 & 0 & 0 \\
0 & 0 & 0 & 0 & 0 & 0 & 0 & 0 & 0 & 1 & 0 & 0 & 0 & 0 & 0 & 0 & 0 & 0 \\
0 & 0 & 0 & 0 & 0 & 0 & 0 & 0 & 1 & 0 & 1 & 0 & 0 & 0 & 1 & 0 & 0 & 0 \\
0 & 0 & 0 & 0 & 0 & 1 & 0 & 0 & 0 & 0 & 0 & 0 & 0 & 0 & 0 & 1 & 0 & 0 \\
0 & 0 & 0 & 0 & 0 & 0 & 1 & 1 & 0 & 0 & 0 & 0 & 0 & 0 & 0 & 0 & 0 & 0 \\
0 & 1 & 0 & 0 & 0 & 0 & 0 & 0 & 0 & 0 & 0 & 0 & 0 & 0 & 0 & 0 & 0 & 0 \\
0 & 0 & 0 & 0 & 0 & 0 & 0 & 0 & 0 & 1 & 0 & 0 & 0 & 0 & 0 & 0 & 0 & 0 \\
0 & 0 & 0 & 0 & 0 & 1 & 0 & 1 & 0 & 0 & 0 & 0 & 0 & 0 & 0 & 0 & 0 & 0 \\
0 & 0 & 0 & 0 & 0 & 0 & 0 & 0 & 0 & 0 & 0 & 0 & 0 & 0 & 1 & 0 & 0 & 0 \\
0 & 0 & 0 & 0 & 0 & 0 & 0 & 1 & 0 & 0 & 0 & 1 & 0 & 0 & 0 & 0 & 0 & 0 \\
0 & 0 & 0 & 1 & 0 & 1 & 0 & 0 & 0 & 0 & 0 & 0 & 0 & 0 & 0 & 0 & 0 & 0 \\
0 & 0 & 1 & 0 & 0 & 0 & 1 & 0 & 0 & 0 & 0 & 0 & 0 & 0 & 0 & 0 & 0 & 0 \\
1 & 0 & 0 & 0 & 0 & 0 & 0 & 0 & 0 & 0 & 0 & 0 & 0 & 0 & 0 & 0 & 0 & 0 \\
0 & 0 & 0 & 0 & 0 & 0 & 0 & 1 & 1 & 1 & 0 & 0 & 0 & 0 & 0 & 0 & 0 & 0 \\
0 & 0 & 0 & 0 & 1 & 0 & 0 & 0 & 0 & 0 & 0 & 1 & 0 & 1 & 0 & 0 & 0 & 0 \\
1 & 0 & 0 & 0 & 0 & 0 & 0 & 1 & 0 & 0 & 0 & 1 & 0 & 0 & 0 & 0 & 0 & 1 \\
0 & 1 & 0 & 0 & 0 & 0 & 0 & 0 & 0 & 0 & 0 & 0 & 0 & 0 & 0 & 0 & 1 & 0
\end{bmatrix}
$$

3.1.2 *Find the reachable matrix*

The matrix adopts Boolean algebra operations. When the adjacency matrix A satisfies $(A+I)^{n-1} \neq (A+I)^{n} = (A+I)^{n+1} = M$, where A represents the adjacency matrix and I represents the identity matrix, MATLAB is used to calculate the reachable matrix M (Guo et al. 2020).

$$
M = \begin{bmatrix}
1 & 1 & 0 & 1 & 1 & 1 & 1 & 1 & 1 & 1 & 1 & 1 & 0 & 1 & 1 & 1 & 0 & 0 \\
0 & 1 & 0 & 0 & 0 & 1 & 1 & 1 & 1 & 1 & 0 & 0 & 0 & 0 & 1 & 0 & 0 & 0 \\
0 & 0 & 1 & 0 & 0 & 1 & 1 & 1 & 1 & 1 & 0 & 0 & 0 & 0 & 1 & 0 & 0 & 0 \\
0 & 1 & 0 & 1 & 0 & 1 & 1 & 1 & 1 & 1 & 1 & 1 & 0 & 0 & 1 & 0 & 0 & 0 \\
1 & 1 & 0 & 1 & 1 & 1 & 1 & 1 & 1 & 1 & 1 & 1 & 0 & 1 & 1 & 1 & 0 & 0 \\
0 & 1 & 0 & 0 & 0 & 1 & 1 & 1 & 1 & 1 & 0 & 0 & 0 & 0 & 1 & 0 & 0 & 0 \\
0 & 1 & 0 & 0 & 0 & 1 & 1 & 1 & 1 & 1 & 0 & 0 & 0 & 0 & 1 & 0 & 0 & 0 \\
0 & 0 & 0 & 0 & 0 & 1 & 1 & 1 & 1 & 1 & 0 & 0 & 0 & 0 & 1 & 0 & 0 & 0 \\
0 & 1 & 0 & 0 & 0 & 1 & 1 & 1 & 1 & 1 & 0 & 0 & 0 & 0 & 1 & 0 & 0 & 0 \\
0 & 1 & 0 & 0 & 0 & 1 & 1 & 1 & 1 & 1 & 0 & 0 & 0 & 0 & 1 & 0 & 0 & 0 \\
0 & 1 & 0 & 1 & 0 & 1 & 1 & 1 & 1 & 1 & 1 & 1 & 0 & 0 & 1 & 0 & 0 & 0 \\
0 & 1 & 0 & 1 & 0 & 1 & 1 & 1 & 1 & 1 & 1 & 1 & 0 & 0 & 1 & 0 & 0 & 0 \\
0 & 1 & 1 & 0 & 0 & 1 & 1 & 1 & 1 & 1 & 0 & 0 & 1 & 0 & 1 & 0 & 0 & 0 \\
1 & 1 & 0 & 1 & 1 & 1 & 1 & 1 & 0 & 1 & 0 & 1 & 0 & 1 & 0 & 1 & 0 & 0 \\
0 & 1 & 0 & 0 & 0 & 1 & 1 & 1 & 1 & 1 & 0 & 0 & 0 & 0 & 1 & 0 & 0 & 0 \\
1 & 1 & 0 & 1 & 1 & 1 & 1 & 1 & 1 & 1 & 1 & 1 & 0 & 1 & 1 & 1 & 0 & 0 \\
1 & 1 & 0 & 1 & 1 & 1 & 1 & 1 & 1 & 1 & 1 & 1 & 0 & 1 & 1 & 1 & 1 & 1 \\
1 & 1 & 0 & 1 & 1 & 1 & 1 & 1 & 1 & 1 & 1 & 1 & 0 & 1 & 1 & 1 & 1 & 1
\end{bmatrix}
$$

3.1.3 *Decompose the relationship between various factors*

R(Si) represents the reachable set of the reachable matrix, Q(Si) represents the antecedent set of the reachable matrix, and A(Si) represents the intersection of the reachable set and the antecedent set. The results are shown in Table 2.

Table 2. Analysis of the relationship between influencing factors.

Influencing factors S_i	Reachable set $R(S_i)$	Antecedent set $Q(S_i)$	$A(S_i)=R(S_i)\cap Q(S_i)$
S_1	$S_1,S_2,S_4,S_5,S_6,S_7,S_8,S_9,S_{10},S_{11},S_{12},S_{14},S_{15},S_{16}$	$S_1,S_5,S_{14},S_{16},S_{17},S_{18}$	S_1,S_5,S_{14},S_{16}
S_2	$S_2,S_6,S_7,S_8,S_9,S_{10},S_{15}$	$S_1,S_2,S_3,S_4,S_5,S_6,S_7,S_8,S_9,S_{10},S_{11},S_{12},S_{13},S_{14},S_{15},S_{16},S_{17},S_{18}$	$S_2,S_6,S_7,S_8,S_9,S_{10},S_{15}$
S_3	$S_2,S_3,S_6,S_7,S_8,S_9,S_{10},S_{15}$	S_3,S_{13}	S_3
S_4	$S_2,S_4,S_6,S_7,S_8,S_9,S_{10},S_{11},S_{12},S_{15}$	$S_1,S_4,S_5,S_{11},S_{12},S_{14},S_{16},S_{17},S_{18}$	S_4,S_{11},S_{12}
S_5	$S_1,S_2,S_4,S_5,S_6,S_7,S_8,S_9,S_{10},S_{11},S_{12},S_{14},S_{15},S_{16}$	$S_1,S_5,S_{14},S_{16},S_{17},S_{18}$	S_1,S_5,S_{14},S_{16}
S_6	$S_2,S_6,S_7,S_8,S_9,S_{10},S_{15}$	$S_1,S_2,S_3,S_4,S_5,S_6,S_7,S_8,S_9,S_{10},S_{11},S_{12},S_{13},S_{14},S_{15},S_{16},S_{17},S_{18}$	$S_2,S_6,S_7,S_8,S_9,S_{10},S_{15}$
S_7	$S_2,S_6,S_7,S_8,S_9,S_{10},S_{15}$	$S_1,S_2,S_3,S_4,S_5,S_6,S_7,S_8,S_9,S_{10},S_{11},S_{12},S_{13},S_{14},S_{15},S_{16},S_{17},S_{18}$	$S_2,S_6,S_7,S_8,S_9,S_{10},S_{15}$
S_8	$S_2,S_6,S_7,S_8,S_9,S_{10},S_{15}$	$S_1,S_2,S_3,S_4,S_5,S_6,S_7,S_8,S_9,S_{10},S_{11},S_{12},S_{13},S_{14},S_{15},S_{16},S_{17},S_{18}$	$S_2,S_6,S_7,S_8,S_9,S_{10},S_{15}$
S_9	$S_2,S_6,S_7,S_8,S_9,S_{10},S_{15}$	$S_1,S_2,S_3,S_4,S_5,S_6,S_7,S_8,S_9,S_{10},S_{11},S_{12},S_{13},S_{14},S_{15},S_{16},S_{17},S_{18}$	$S_2,S_6,S_7,S_8,S_9,S_{10},S_{15}$
S_{10}	$S_2,S_6,S_7,S_8,S_9,S_{10},S_{15}$	$S_1,S_2,S_3,S_4,S_5,S_6,S_7,S_8,S_9,S_{10},S_{11},S_{12},S_{13},S_{14},S_{15},S_{16},S_{17},S_{18}$	$S_2,S_6,S_7,S_8,S_9,S_{10},S_{15}$
S_{11}	$S_2,S_4,S_6,S_7,S_8,S_9,S_{10},S_{11},S_{12},S_{15}$	$S_1,S_4,S_5,S_{11},S_{12},S_{14},S_{16},S_{17},S_{18}$	S_4,S_{11},S_{12}
S_{12}	$S_2,S_4,S_6,S_7,S_8,S_9,S_{10},S_{11},S_{12},S_{15}$	$S_1,S_4,S_5,S_{11},S_{12},S_{14},S_{16},S_{17},S_{18}$	S_4,S_{11},S_{12}
S_{13}	$S_2,S_3,S_6,S_7,S_8,S_9,S_{10},S_{13},S_{15}$	S_{13}	S_{13}
S_{14}	$S_1,S_2,S_4,S_5,S_6,S_7,S_8,S_9,S_{10},S_{11},S_{12},S_{14},S_{15},S_{16}$	$S_1,S_5,S_{14},S_{16},S_{17},S_{18}$	S_1,S_5,S_{14},S_{16}
S_{15}	$S_2,S_6,S_7,S_8,S_9,S_{10},S_{15}$	$S_1,S_2,S_3,S_4,S_5,S_6,S_7,S_8,S_9,S_{10},S_{11},S_{12},S_{13},S_{14},S_{15},S_{16},S_{17},S_{18}$	$S_2,S_6,S_7,S_8,S_9,S_{10},S_{15}$
S_{16}	$S_1,S_2,S_4,S_5,S_6,S_7,S_8,S_9,S_{10},S_{11},S_{12},S_{14},S_{15},S_{16}$	$S_1,S_5,S_{14},S_{16},S_{17},S_{18}$	S_1,S_5,S_{14},S_{16}
S_{17}	$S_1,S_2,S_3,S_4,S_5,S_6,S_7,S_8,S_9,S_{10},S_{11},S_{12},S_{13},S_{14},S_{15},S_{16},S_{17},S_{18}$	S_{17},S_{18}	S_{17},S_{18}
S_{18}	$S_1,S_2,S_3,S_4,S_5,S_6,S_7,S_8,S_9,S_{10},S_{11},S_{12},S_{13},S_{14},S_{15},S_{16},S_{17},S_{18}$	S_{17},S_{18}	S_{17},S_{18}

3.1.4 *Divide the levels and establish an ISM analysis model for the influencing factors of prefabricated building supply chain management*

From the results in Table 4, we can obtain the top-level element set L1 = [S_2, S_6, S_7, S_8, S_9, S_{10}, S_{15}], the second level element set L2 = [S_3, S_4, S_{11}, S_{12}], the third level element set L3=[S_1, S_5, S_{13}, S_{14}, S_{16}], the fourth layer element set L4=[S_{17}, S_{18}]. After the hierarchy is divided, the ISM analysis model of prefabricated building supply chain management is established after reduction and hierarchical processing. As shown in Figure 1.

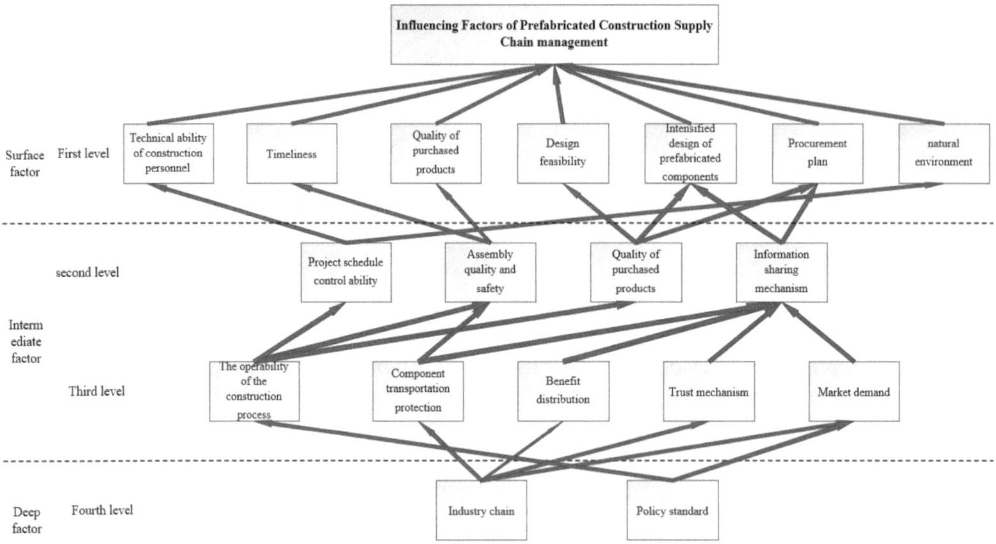

Figure 1. ISM model of prefabricated building supply chain management.

Through analysis of the ISM model, the factors affecting the supply chain management of prefabricated buildings are divided into three levels: surface layer, middle layer, and deep layer. The first-level surface factors are timeliness, quality of the delivered products, the feasibility of the design plan, procurement plan, natural environment, etc. These factors are the direct factors that affect the supply chain management level of the prefabricated building. The intermediate factors are operability of the construction process, trust mechanism, benefit distribution, market demand, information sharing mechanism, etc. These factors are affected by deep-level factors and also act on surface factors. The third-level deep-seated factors are influencing factors, such as policy standards and industrial chain, which are the main influencing factors affecting the supply chain management of prefabricated buildings.

3.2 *Analysis of AHP method of influencing factors of prefabricated building supply chain management*

Based on the result of analysis of the influencing factors of prefabricated building supply chain management based on the abovementioned ISM interpretation structure model, each influencing factor in the model is layered as the index system of AHP, and the judgment matrix is constructed according to the scaling method and the expert scoring method. The importance of the factors is compared and judged in pairs, and then MATLAB is used to calculate the maximum eigenvalue λmax and eigenvector of the judgment matrix, and then normalize them to calculate the weight coefficient Wi of each element, and then perform a consistency test on it to ensure the accuracy of the weight coefficient effectiveness. The results are shown in Table 3.

According to the weights of various factors calculated by the AHP method, it can be seen that prefabricated building supply chain management is mainly affected by policy standards, industrial chain, trust mechanism, benefit distribution, operability of construction technology, market demand, and technical capabilities of construction personnel. And the weight of policy standards, industrial chain, trust mechanism, and technical capabilities of construction personnel is far greater than other factors.

Table 3. Statistical table of weights of influencing factors.

	Consistency test index value											Feature vector	Comprehensive weight	Sort
Surface	λ max=7.7245		S_2	S_6	S_7	S_8	S_9	S_{10}	S_{15}					
factor	CI=0.1208	S_2	1	2	5	1	2	2	5			0.2692	0.089747	1
judgment	CR=0.0888<0.1	S_6	1/2	1	3	1/3	1/3	1/2	1			0.0908	0.030266	12
matrix		S_7	1/5	1/3	1	1/3	1/4	1	1			0.0631	0.021017	13
		S_8	1	3	3	1	1/3	1/2	1			0.1449	0.048293	9
		S_9	1/2	1	4	3	1	2	1			0.1949	0.06498	5
		S_{10}	1/2	2	1	2	1/2	1	2			0.1437	0.047904	17
		S_{15}	1/5	1	1	1	1	1/2	1			0.0934	0.031125	18
			S_3	S_4	S_{11}	S_{12}	S_1	S_5	S_{13}	S_{14}	S_{16}			
Intermediate	λ max=9.8505	S_3	1	1/3	1	1/2	1/3	1/2	1/3	1	1/2	0.0514	0.017126	8
factor	CI=0.1063	S_4	3	1	3	3	1	2	1	4	3	0.1984	0.066139	2
judgment	CR=0.0728<0.1	S_{11}	1	1/3	1	1/3	1/4	1	1/2	1	1	0.0619	0.020646	16
matrix		S_{12}	2	1/3	3	1	1/3	1/3	2	1	1/2	0.09	0.030015	3
		S_1	3	1	4	3	1	2	1	2	3	0.1915	0.063849	11
		S_5	2	1/2	1	3	1/2	1	2	2	1/3	0.1147	0.038224	4
		S_{13}	3	1	2	1/2	1	1/2	1	1	1	0.1078	0.035949	15
		S_{14}	1	1/4	1	1	1/2	1/2	1	1	1/2	0.0648	0.021603	10
		S_{16}	2	1/3	1	2	1/3	3	1	2	1	0.1193	0.039782	7
Deep factor			S_{17}	S_{18}										
judgment	λ max=2	S_{17}	1	1/3								0.250	0.083333	14
matrix	CI=0	S_{18}	3	1								0.750	0.25	6
	CR=0<0.1													

4 COUNTERMEASURES AND SUGGESTIONS

4.1 *Establish a system of relevant standards and specifications*

Relevant national departments should actively formulate relevant standards, specifications, and requirements, further improve relevant policies, guide and standardize the construction of relevant application platforms, and provide basic platform support for prefabricated building supply chain management. To improve the supply chain management capabilities of prefabricated buildings, it plays a fundamental role in guaranteeing (Wang Wei 2020).

4.2 *Improve the construction of the prefabricated construction industry chain*

Currently, the development of the prefabricated construction industry has always been driven by policies with a low level of social participation, and a specialized division of labor has not yet been formed. To better promote the benign development of prefabricated buildings, it is necessary to improve the construction of the entire industry chain of prefabricated buildings, ensure the coupling between main bodies of various industrial chains, and improve the level of cooperative management of the main bodies (Wang et al. 2021). At the same time, it is important to strengthen investment in information construction, so that the information between the participating entities can be shared, and better utilize the advantages of prefabricated building supply chain management.

4.3 *Intensify the training of talents related to the supply chain management of prefabricated buildings*

Promote the implementation plan for talent training, and improve the training system for talents related to the supply chain management of prefabricated buildings. Also, increase training of employees, maximize the potential of employees, improve the level of professionalization of employees, and cultivate compound talents that are more in line with the needs of enterprises and projects, thereby improving the efficiency of prefabricated construction supply chain management.

5 CONCLUDING REMARKS

This article uses a combination of qualitative and quantitative methods to analyze the influencing factors of prefabricated building supply chain management, which improves the accuracy of analysis and provides a reference for strengthening prefabricated building supply chain management. At this stage, as the country vigorously promotes the development of prefabricated buildings, the prefabricated building supply chain management model will surely play an important role in upgrading and transforming the construction industry.

REFERENCES

Chen Chao. Research on Partnership Management of Prefabricated Building Supply Chain[D]. Beijing Jiaotong University, 2020.DOI:10.26944/d.cnki.gbfju.2020.000996.

Fan Yuncui, Shan Tong. Research on the application status and countermeasures of prefabricated building supply chain management model[J]. Intelligent Buildings and Smart Cities, 2021(04):33–34.DOI:10.13655/j.cnki.ibci.2021.04 .009.

Guo Ting, Lu Shuran, Ding Xiaoqiang. Research on risk factors of high-rise building construction based on ISM and AHP analysis[J]. Building Safety, 2020, 35(05): 51–54.

Tang Xiaoyu, Li Wenying, Cao Guiyin, Xu Mengjie. Research on Influencing Factors of Service Competitiveness of Logistics Enterprises Based on ISM and AHP[J]. Journal of Xiangnan University, 2020, 41(05): 56–62.

Wang Xiang, Zhou Shiyu, Guo Yibin, Duan Yonghui. Research on the Supply Chain Risk of Prefabricated Buildings Based on EFA and CFA[J]. Mathematics in Practice and Knowledge, 2021, 51(12): 286–299.

Wang Wei. Research on the sustainable development of prefabricated buildings from the perspective of supply chain management[J]. Sichuan Cement, 2020(12): 55–56.

Wang Yingchao. Research on risk management of prefabricated building supply chain based on immune principle[D]. North China University of Technology, 2020.DOI:10.26926/d.cnki.gbfgu.2020.000430.

Zhang Chuang. Research on Risk Management of Green Supply Chain of Prefabricated Buildings [D]. Northeast Forestry University, 2020.DOI:10.27009/d.cnki.gdblu.2020.000801.

Advances in Urban Engineering and Management Science – Khalil & Yang (Eds)
© 2023 The Author(s), ISBN 978-1-032-30426-7

Analysis of the optimization model of the renovation project management in colleges and universities based on time constraint theory

Qianshan Wen*, Jiangtao Liu & Jie Ren
Beijing Jiaotong University New Campus Construction Office, Beijing, China

ABSTRACT: Via analyzing the current management model and characteristics of renovation projects in colleges and universities, for the multi-project management of renovation projects, this paper introduces the time constraint theory of immediate pre-construction constraint, immediate post-construction constraint, intermediate constraint, and intermittent constraint of renovation projects in colleges and universities. With a renovation project in a Beijing's university as a case, an in-depth analysis is conducted from different aspects such as project function, scope, initiation, design, schedule, and process. The optimization model of the management for renovation projects in colleges and universities based on time constraint theory is innovatively proposed, which provides a basis for the project management of renovation projects in colleges and universities and has a high reference value.

1 INTRODUCTION

In line with the development of higher education, universities have gradually shifted from external growth to internal development. The quality of higher education cannot be improved without the support of physical resources such as space and environment, which are required to meet the educational concepts such as collaborative innovation, sharing and openness, human-centeredness, and school-city integration.

With the expansion of the school scale, the problems of backward school buildings and supporting facilities have become prominent, which has led to an increase in campus infrastructure construction projects year by year, and the scale and number of renovation projects have increased significantly (Peng 2013). Under this situation, new construction projects for colleges and universities in Beijing are limited, and the focus of capital construction has mostly shifted to renovation projects. Existing buildings and facilities in colleges and universities need to meet the requirements of modern education development, and the renovation and transformation of old buildings and facilities should be strengthened.

At present, the state and local governments have increased their investment in education, especially in the renovation and transformation of buildings and infrastructure in colleges and universities. A large amount of money is invested every year. The Ministry of Education has set up a special fund project for "Improving Basic Educational Conditions" to increase support for universities directly under the central government, with a total amount of about 10 billion yuan per year. It is mainly used for the renovation of houses, infrastructure transformation, construction projects, and facilities acquisition, of which the first three categories are dominated by renovation projects with a high proportion of funds. However, the current situation of the management level of the renovation projects in China's colleges and universities is worrying, and the progress and quality cannot be guaranteed, resulting in a lot of waste of time and energy (Lou 2020). Therefore, there is an urgent need to deeply study the management strategy of the renovation project in colleges and

*Corresponding Author: qswen@bjtu.edu.cn

DOI 10.1201/9781003305026-54

universities, establish an optimized management model, scientifically allocate school resources, and improve the management efficiency of the renovation projects.

2 OVERVIEW OF RENOVATION PROJECTS IN COLLEGES AND UNIVERSITIES

The renovation projects in colleges and universities are cumbersome and complicated system projects, involving civil engineering, facilities, and equipment, decoration, landscape greening, etc., with the aim to ensure the school's daily personnel training, scientific research, life support, and logistical services (Sun 2021). They are renovation and transformation projects of auxiliary basic school buildings, teaching and research instruments and equipment, and public infrastructure, which have the characteristics of regularity, repetition, concentration, contingency, and specialization (Li & Zhang 2020). The specific characteristics are as follows:

1. Large quantity of projects and multiple types. Based on the average volume of 500,000–1,500,000 square meters of basic school buildings in colleges and universities, the annual workload of maintenance works due to daily depreciation, technical equipment renewal, and scientific research needs to reach tens of thousands of square meters, which is a large number of works and involves a wide variety of works.
2. Strong engineering expertise. In addition to a series of professional technical work such as construction, structure, water, heating, electricity, gas, etc., the project also involves professional knowledge in load-bearing loads, electromagnetic compatibility, and hazardous materials disposal related to scientific research equipment, which puts forward higher professional requirements for construction management.
3. Lower project cost. A university in Beijing is taken as an example. In 2017, there were 23 renovation projects with an average cost of 1.87 million yuan and 11 projects with a single project below 1-million-yuan accounting for about half.
4. Complicated operation interface. The renovation works are mostly carried out in existing buildings. Thus, other non-construction parts need to be maintained and the impact on the surrounding environment needs to be reduced to ensure the normal operation of other non-renovation parts; the site for stockpiling and construction channels is relatively small and higher safety construction requirements are also encountered, which entails higher requirements for construction organization and site management.
5. The suddenness and urgency of the project. The renovation projects in colleges and universities have sporadic characteristics. At the same time, due to the frequent large-scale activities in schools, many temporary projects will be increased. In order to ensure the normal operation of the school, possible problems must be repaired in a timely manner at any time, hidden dangers must be eliminated, and the construction organization must respond with higher efficiency.
6. Higher time requirement. In order not to affect the daily teaching work and life order of the school, the implementation time of renovation projects is often concentrated in the winter and summer vacations when the students go home from vacation. The construction period is short and there are many constraints before and after.

3 RESEARCH ON TIME CONSTRAINT OF RENOVATION PROJECTS IN COLLEGES AND UNIVERSITIES

1. Immediate pre-construction constraint: it refers to the constraint of project beginning time when the project can begin formal construction on project management, generally representing the last time when the project's preliminary work must be completed. As a renovation project for colleges and universities, in order not to affect the normal teaching order, the renovation work can only be carried out after the housing is vacated, the equipment is out of service, and the infrastructure is suspended. Before the start of construction, it is necessary to coordinate with the user unit and related management units to complete the filing or approval procedures.

2. Immediate post-construction constraint: it refers to the constraint of project completion time on project management, which determines the time node for the completion of the project and affects all procedures of inverted processes. The completion time requirements of renovation projects in colleges and universities are often stricter than those of other types of infrastructure projects, and they need to be specific to the day. For projects implemented during the holidays, the start of the school year is the time when the renovation work must be completed; the start of a major equipment research program is the time when the renovation of equipment must be completed. Hence, the immediate post-construction constraint of the renovation projects in colleges and universities affects the organization and arrangement of all work procedures.
3. Intermediate constraint: it refers to the constraint of the overall construction time on project management. Summer and winter vacations are the time when the teaching and research tasks are centrally finished and the living and logistic facilities can be intensively rested, usually lasting one to two months. The duration of the holiday time affects the overall construction time of the renovation project, and other activities during the holiday will also have an impact on the construction time, thereby affecting the time that the renovation project can be implemented. Some important rooms, facilities, and equipment of the school, such as water and electricity systems, student dormitories, canteens, bathrooms, sports fields, etc., are also in operation during winter and summer vacations. In addition, the shutdown of network servers, major scientific research equipment, etc., will cause a greater impact or cause greater restart costs. Therefore, the renovation works of this type of project are faced with the need for a special time window to ensure the construction time.
4. Intermittent constraint: it refers to the constraint caused by the influence of intermediate constraints, and it is generated when a single time period of the project is not enough to complete the entire renovation work. Due to the scale of the project or the complexity of the construction, the renovation project cannot be finished within a complete time period, resulting in the same renovation object being divided into different projects. In addition, some renovation projects need to be completed in different time periods because a long period of time is required for renovation and the centralized development has a greater impact on campus life. Intermittent constraint requires that the project management process be more refined, and the connection between various time periods should be done well.

4 OPTIMIZATION MODEL OF THE RENOVATION PROJECT MANAGEMENT IN COLLEGES AND UNIVERSITIES UNDER TIME CONSTRAINTS

The renovation projects in colleges and universities embody the characteristics of time schedule constraints, which require improved management efficiency, preparation for daily project response, and control of key time nodes in the process of project implementation.

4.1 *Optimization of project functions*

1. Fund management party: According to different sources of funds for engineering projects, the school's financial department generally manages the funds as a whole. However, the financial department has less experience in the implementation of engineering projects. It is advisable to assign most of the functions of project management to the infrastructure department and other engineering management departments, and fully respect the characteristics of the life cycle of the engineering project in terms of fund control, avoid adopting the same management mode as that of equipment procurement projects.
2. User side: Most renovation projects in colleges and universities have a clear user side. They are the proponents of the project and the final user of the project. Therefore, the renovation project should involve the user side in the whole process.
3. Project management party: The renovation projects in colleges and universities are generally in the charge of infrastructure, logistics, and other departments. They have certain management

experience and technical teams. They generally assume the role of Party A's representative in the renovation project and undertake the primary responsibility for the progress, quality, investment, and safety of the project.

4. Project implementer: In addition to the construction unit, the project implementer also includes the designer, the supervisor, and the consulting party, all of which are required to obey the overall management of the project management party in accordance with the requirements of the contract.

In view of the current situation of multi-project cross-promotion of renovation projects in colleges and universalities, all participants should optimize the division of labor plans, refine responsibilities, and coordinate the promotion of management work. It is recommended that the management model in Figure 1 be adopted.

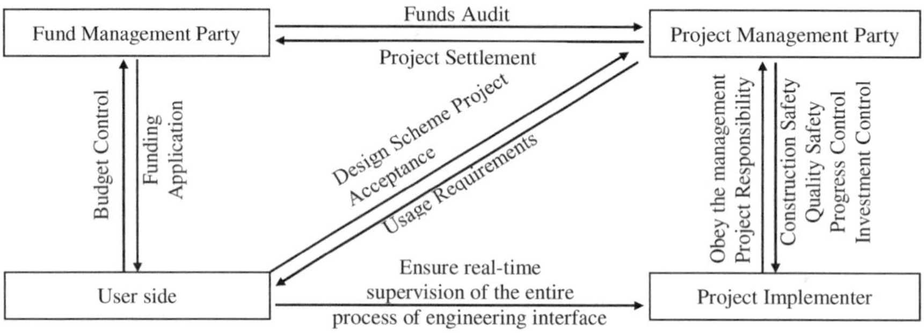

Figure 1. Optimization of participants in renovation projects in colleges and universities.

4.2 *Optimization of project scope*

Renovation projects in colleges and universities are subject to time constraints and space constraints. The scope of the project is chaotic, and the constraints of objective conditions make the project inconsistent with the law of project implementation, resulting in artificial disconnection of the project, causing patches to be added and dropped, and inefficient use of funds. To solve this problem, the following strategies are proposed:

1. For the building with few time constraints, long periods, and easy-to-satisfy construction interfaces, priority is given to the overall renovation of the building.
2. Projects with less vertical system transformation, more difficult turnover of construction interface and certain guarantees of construction time (such as summer vacation) are mostly based on the transformation of student dormitories, office buildings, and laboratories, which can be implemented in stages.
3. The renovation of water, heating, electricity, gas, and firefighting systems, and buildings that are easy to split vertically, can be implemented in stages in the vertical direction.
4. Partial sub-projects or infrastructure renovation projects may take partial renovation measures.

4.3 *Optimization of project establishment*

Renovation projects in colleges and universities are highly professional, and users are limited in their professional backgrounds and often do not have the ability to independently organize the initiation of projects. To this end, the initiation of renovation projects in colleges and universities generally needs to go through the process that users propose the requirements, and the project management unit organizes the preliminary demonstration. Through repeated research, necessity and feasibility are combined to determine the scope and investment of the project.

The project initiation strategy can be divided into three steps: necessity auditing (Figure 2), preliminary feasibility auditing (Figure 3), and feasibility review (Figure 4).

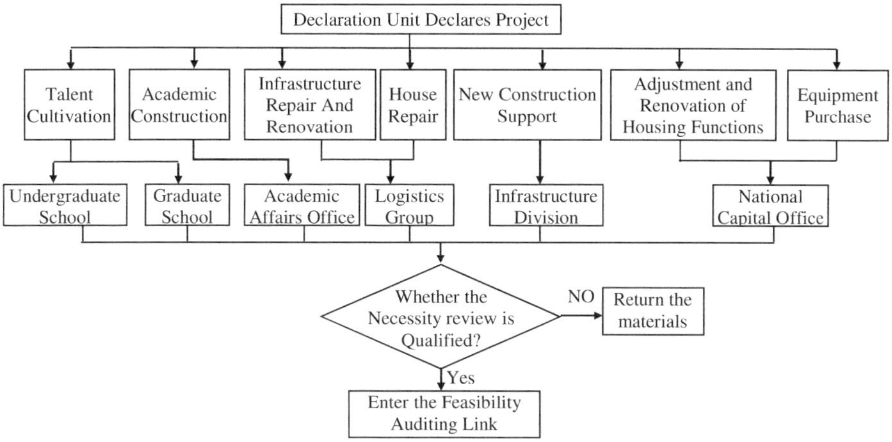

Figure 2. Necessity auditing link (led by the finance department).

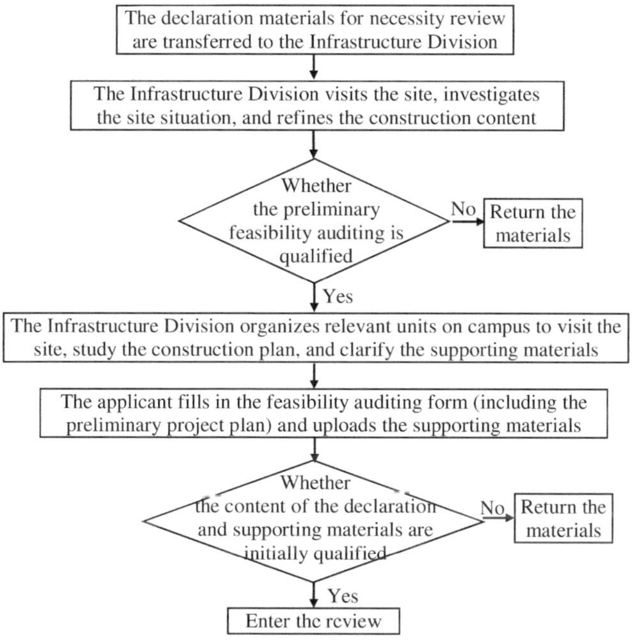

Figure 3. Preliminary feasibility auditing link (led by the infrastructure department).

4.4 *Optimization of project design*

After completing the preliminary review of the necessity and feasibility of the project, it is available to enter the preliminary design stage. Generally, the project management unit organizes professional designers to participate in the implementation. It is mainly to form the preliminary plan of the project and complete the rationality review. The follow-up can be combined with the preliminary design budget and be revised according to the review opinions of the superior department, and then the formal project decision can be submitted. The specific process strategy is shown in Figures 5 and 6.

For renovation projects with a large investment amount, it is necessary to improve the design plan in order to meet the drawing approval requirements and go through the start-up procedures.

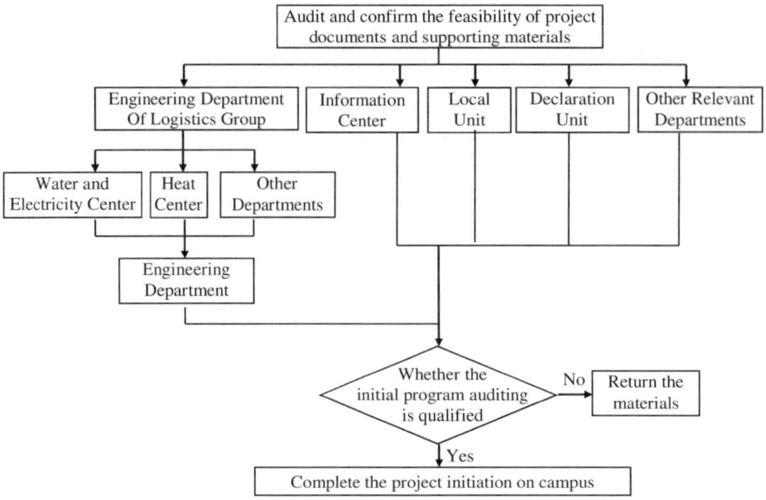

Figure 4. Feasibility review link (led by the infrastructure department).

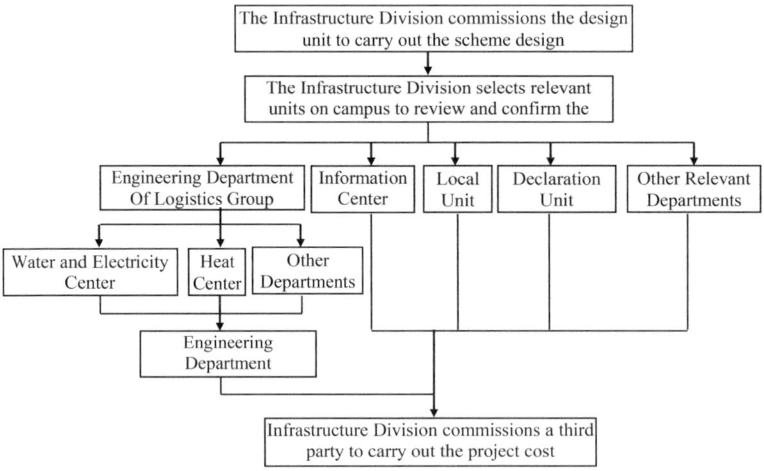

Figure 5. Preliminary rationality auditing link (led by the infrastructure department).

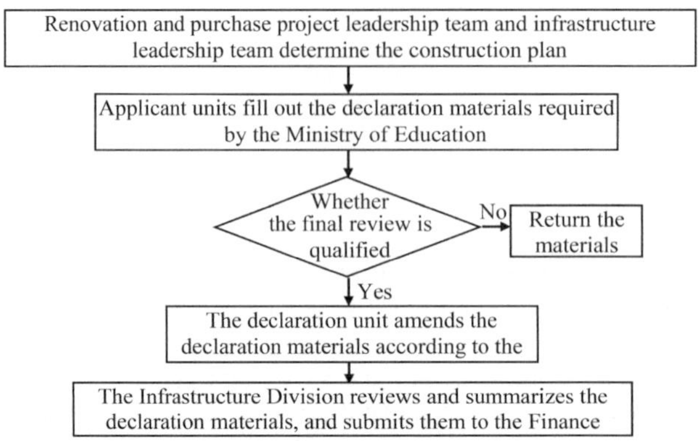

Figure 6. Rationality review link (led by the finance department).

4.5 *Optimization of project schedule*

To optimize the project schedule, firstly, it is required to clarify the classification of renovation projects in colleges and universities under time constraints.

4.5.1 *Project classification*

According to the time constraint requirements of renovation projects in colleges and universities, it can be divided into six categories: teaching period limitation, working day limitation, interval, intermittent category, specific limitation, and non-limitation. Taking the renovation project of a university in Beijing as an example, the distinction is as follows:

Table 1. Classification of the renovation project.

No.	Project name	Duration (days)	Time constraint	Constraint category
1	Teaching building Fire Renovation	35	Non-teaching period and non-working day	Teaching period and working day limitation
2	Power capacity expansion	30	Power outage constraint 0 o'clock–5 o'clock	Special limitations
3	Renovations of underground pipelines and building heating	40	Non-teaching period, non-working days, and non-heating season	Teaching period limitations, working day limitation, and special limitations
4	Main office building Elevator replacement	25+25	Four elevators are replaced on non-working weekends and in the evening	Working day limitation, intermittent category, and special limitations
5	Renovation of Experiment Center	30	Non-teaching period	Teaching period limitation
6	Replacement of roof waterproofing	25	Non-limitation	Non-limitation
7	Renovation Project of No.4 Canteen for Students	35	During summer vacation, from July 20th to August 28th	Interval category and space constraints
8	Renovation and transformation of students' apartment	35	During summer vacation, July 20th–August 28th	Interval category
9	Renovation and transformation of the stadium	35	Non-teaching period	Teaching period limitation
10	Renovation and transformation of No. 9 Office building	100	From July 17–October 7, Suspended during the day and three important meetings,	Interval category and special limitations
11	Renovation of the School of Science Office Building	45	Non-working day	Working day limitation
12	East Campus Bathroom renovation	35	Early summer vacation, July 22–August 14 Military training use	Interval category and special limitations

4.5.2 *Establishment of a Gantt chart of project schedule*

From the perspective of resource optimal allocation, the renovation projects of colleges and universities are optimized according to the principle of reducing project peaks and maintaining a certain degree of concentration, and it is advisable to arrange the project schedule according to the following steps:

1. The "super-rigid" projects are determined, that is, the schedule requirements cannot be met under time constraints. Items 10-12 are all "super-rigid" projects, but Item 10 is relatively large and can be implemented by shortening the construction period, and the project is retained. Item 11 is to change the process, with a focus on the renovation of the indoor part and the ground part of the public space in the summer, and the finishing work is implemented on the non-summer weekends, and the project is retained. Item 12 was canceled due to the inability to meet the time constraints through coordination.
2. The rigid items are determined, that is, the requirements of the construction period cannot be met under time constraints, and it is executed according to time constraints, such as items 1, 3, 5, 7, and 8.
3. Adjustable projects: There is ample time for implementation. By optimizing the construction period, the peak value of the project is reduced and a certain degree of concentration is maintained. According to the priority of the nature of the project, that is 2, 4, 9, 6 four items. Item 2 has a small intersection with other projects, and the examination period and the beginning of school should be avoided; Item 4 should avoid the examination period, National Day, and other projects; Item 9 can start early because the physical education class ends in July; Item 6 should be completed before the onset of the rainy season.

The project schedule Gantt chart obtained through the above optimization strategy is shown in Table 2.

Table 2. Gantt chart of project schedule.

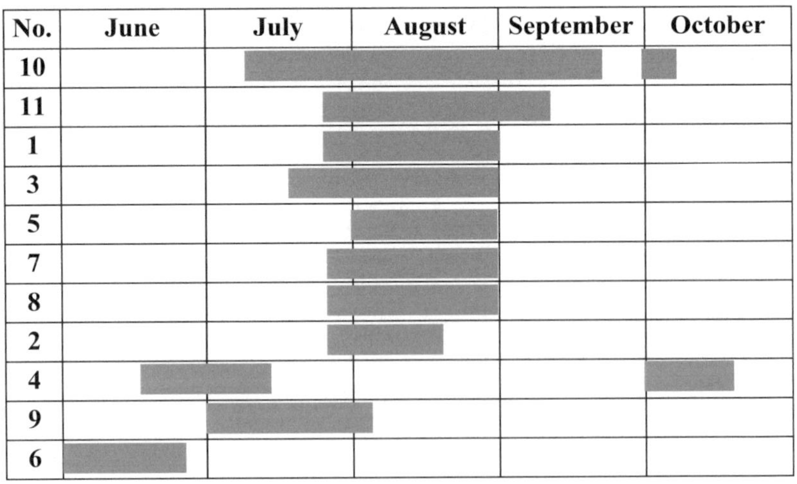

4.6 *Optimization of the project process*

Buildings in colleges and universities are mostly distributed in units of schools and departments. According to the age and renovation conditions of different buildings, the overall renovation and transformation of buildings as a unit is the priority of the renovation project, which is conducive to the overall planning of resources and centralized construction, thus realizing maximum benefits. In the face of time constraints, such projects should adhere to the principles of "overall design, phased implementation, systematic transformation, and resource pooling", and it is needed to conduct demonstration and design six months to one year in advance, to study and formulate a work plan for implementation by layer, parallel old and new systems and barrier-free operation of interfaces. It is advisable to complete the coordination plan of multiple systems such as civil engineering, heating power, fire protection, monitoring, strong electricity, communication, and wired and wireless network. Commitment is made to coordinate national allocation, school self-raised, college self-raised funds, coordinate college, information, and security, Infrastructure, state-owned assets, buildings, hydropower, thermal power, and other users and management departments, with an effort to strengthen the comprehensive scheduling of design, construction, and supervision, boost the coordination of external approval, supervision, and acceptance departments, and ensure the overall smooth progress of the project.

5 CONCLUSION

The renovation projects in colleges and universities have the characteristics of a large number of projects, various types, strong engineering expertise, low cost of a single project, complex operation interface of the renovation project, engineering suddenness, urgency, and high time requirements. It places high demands on project management level, project responsiveness, and project management efficiency, as well as good quality control to ensure project safety.

Using the optimization model of renovation project management in colleges and universities based on the time constraint theory proposed in this paper, the efficiency of renovation project management can be effectively enhanced and the overall duration can be shortened, thus saving the cost of school renovation and reducing the impact on the overall operation of the school.

REFERENCES

Li X X, Zhang L Y (2020). Study on the Development of Infrastructure and Renovation Projects in Colleges and Universities under the New Situation. J. Logistics Research in Colleges and Universities. 8, 35–36.

Lou W (2020). Analysis of the Current Status and Improvement Measures of Engineering Renovation Management in Colleges and Universities. J. Value Engineering. 39(20), 53–54.DOI:10.14018/j.cnki.cn13-1085/n.2020.20.026.

Peng L (2013). The Model Innovation and Practice of the Management of Large and Medium-sized Renovation Projects in Colleges and Universities: Taking the Introduction of Social Project Management Institutions into the Renovation Project of the Renmin University of China as an example. J. University Logistics Research. 4, 70–71+41.

Sun P Q (2021). Research on the Management Problems and Countermeasures of University Renovation Project from the Perspective of Internal Control. J. Engineering Economics. 31(10), 31–34.DOI: 10.19298/j.cnki.1672-2442.202110031.

Advances in Urban Engineering and Management Science – Khalil & Yang (Eds)
© 2023 The Author(s), ISBN 978-1-032-30426-7

Research on the control of the settlement of buildings by automatic groundwater recharge device

J.J. Yan*
PowerChina Huadong Engineering Corporation Limited, Hangzhou, China

Y.M. Yao*
Ningbo Rail Transit Group Corporation Limited. Ningbo, China

Y.W. Zang* & C.B. Liu*
PowerChina Huadong Engineering Corporation Limited, Hangzhou, China

H.W. Ying*
Hohai University, Nanjing, China

Z.Y. Luo*
Ningbo University, Ningbo, China

ABSTRACT: A method of settlement control by recharging underground water with an automatic device is proposed in this paper, The on-site monitoring data shows that by recharging groundwater into the foundation, the groundwater level can gradually rise, and the settlement of the building induced by water pumping reduces when the underground water level rise. And the test result demonstrates that the automatic device for pumping and recharging designed in this paper can work as designed.

1 INTRODUCTION

With the continuous and rapid development of urbanization in China, the contradiction between the rapidly increasing urban population and limited urban land resources has become increasingly prominent. The development and utilization of the urban underground space have become an important means and trend to solve the urban traffic congestion and the land resources tension. As the development of underground space develops deeper, the risk of being affected by the confined water during the construction of underground space is also increasing, especially in the southeast coastal areas of China. At present, measures of decompression and precipitation are usually adopted in actual projects to reduce or eliminate the construction risk of the impact of confined water on the development of the deep underground space. A large number of existing engineering cases show that the sudden construction accidents in the development of the underground space are often related to the improper handling of the groundwater (especially confined water) problems. Excessive extraction of the confined water will waste and pollute the precious groundwater resources, and at the same time will cause significant consolidation and settlement of the soil layer foundation, which will cause varying degrees of damage to the roads, underground pipelines, and other structures within the area affected by precipitation. When the precipitation does not down to the design water level, damage accidents such as foundation pit surges and water penetration will often occur.

A large number of existing engineering measured data indicate that the precipitation will cause the foundation settlement in the area near the precipitation and adversely affect neighboring buildings (Zhang & Zhao 2008). This impact is related to the depth of the precipitation, the mechanical

*Corresponding Authors: yan_jj@hdec.com, yaoyanming77@126.com, zang_ywj@hdec.com, liu_cb@hdec.com, ice898@zju.edu.cn and luozhanyou@163.com

 DOI 10.1201/9781003305026-55

behavior of the foundation soil, and the rate of precipitation. According to the effective stress principle proposed by Terzaghi (Terzaghi et al. 1943), the deformation mechanism of the foundation caused by the precipitation is the increase in the effective stress of the foundation soil caused by the precipitation, which causes the soil to produce compaction deformation as the consolidation developed. Gibso et al. proposed a nonlinear compaction theory based on the influence of permeability and compressibility changes in the compaction process (Gibson et al. 1967, 1981). Bear combined the traditional Terzaghi consolidation theory with the groundwater flow equation under the assumption that the liquid has compressibility, derived the two-dimensional and three-dimensional groundwater flow equations of Terzaghi-Jocob unidirectional consolidation, and gave the physical meaning of water storage coefficient (Bear 1983). In 1941, Biot considered the influence of the coordinated deformation conditions on the consolidation process of the soil, taking the total stress and the pore pressure of the saturated soil as the state variables, and deduced a three-dimensional consolidation that accurately reflects the relationship between the dissipation of the pore pressure and the deformation of the soil skeleton. The equations laid the foundation for the fluid-solid coupling of porous media (Biot 1941).

Xie Kanghe (Xie et al. 2002) studied the changes in the soil stress caused by the precipitation in the excavation of foundation pits in layered soils and the solution to the settlement of the surrounding ground surface. Under the assumption that the seepage caused by precipitation is one-dimensional and vertical, the calculation formula for effective stress and surface settlement in the soil around the foundation pit. The research results show that the effective stress change in the soil caused by the dewatering of the foundation pit and the resulting seepage is the fundamental reason for the settlement of the ground around the foundation pit.

At present, there have been many types of research on the settlement of foundations caused by precipitation, but there are few related types of research on how to control and reduce the settlement of foundations caused by precipitation. Based on the dewatering of a certain subway station foundation pit in the city of Ningbo, this paper studies the method of using groundwater recharge to control the ground settlement caused by dewatering during subway foundation pit excavation, field monitoring was also carried out which indicated that this method can control the settlement of the buildings effectively, and which can provide a reference for similar projects.

2 PROJECT OVERVIEW

In this paper, the Children's Park Station foundation pit which is influenced by the underground water is studied. The Children's Park Station is a "T" type transfer station for the Lines 3 and 4 of the Ningbo metro, with Line 3 under Line 4, the foundation pit is buried 20 m below the ground surface. The two lines will be implemented at the same time. The station part of line 3 is a three-story underground station, located on the northwest side of the intersection of Xingning Road and Zhongxing Road. Most of the metro station is located in the Children's Park, and the west side of the station is the current Zhongxing Road. The settlement of the road should be strictly controlled. The station part of Line 4 is arranged along the north side of Xingning Road and is a two-story underground station. There are lots of old buildings near the station. The station foundation pit adopts the open-cut construction method. The structure of the station top is buried 3.2 m below the ground, and the total excavation depth of the station foundation pit is 18–21 m. The layout of the foundation pit and the surrounding environment are shown in Figure 1 and Figure 2, and it can be seen that there is a building very near to the foundation pit, the distance is about 15 m.

The stratum of the pit site in the depth range of 85 m below the ground is mainly composed of cohesive soil, clayey soil, silty soil, and deep sand, and the stratum distribution is very complex and the underground water level is about 1 m below the ground. According to the characteristics, genesis, and physical and mechanical properties of the foundation soil, the soil layers within the exploration depth can be divided into 9 large layers and several sublayers. The confined water of this site is located in the sandy silt and fine silt layer. The depth of the confined water level is about 2 m below the ground, the permeability coefficient of the confined water layer is about

Figure 1. Surrounding environment of the foundation pit.

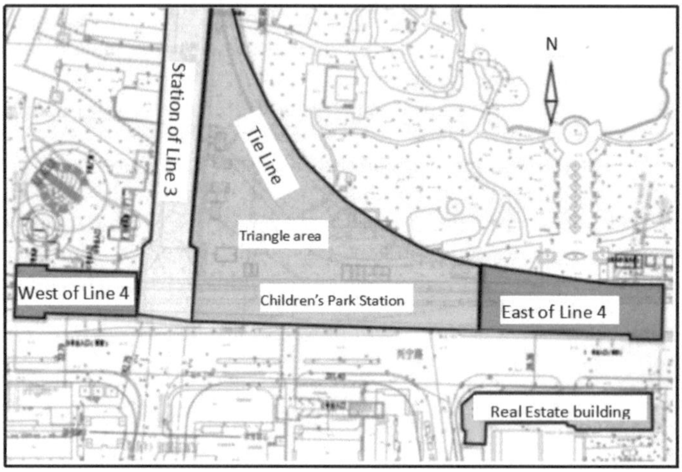

Figure 2. The layout of the foundation pit.

10^{-2} cm/s, the water permeability is moderate, the water volume is abundant, and the single well water production volume is 1500~1800 m^3/d.

After the excavation of the foundation pit, as the thickness of the overlying soil layer over the confined aquifer becomes thinner, the pressure of the overlying soil layer decreases. When the pressure of the overlying soil is less than or equal to the top supporting force of the confined aquifer, the confined water may uplift the bottom surface of the foundation pit and the foundation pit damage will occur as shown in Figure 3.

Formula (1) is usually used to determine whether the foundation pit is in a stable (safe) state against the inrush of the confined aquifer at the bottom (hereinafter referred to as "anti-surge") after excavation.

$$\frac{P_s}{P_w} = \frac{\sum h_i \times \gamma_{si}}{H \times \gamma_w} \geq FS \tag{1}$$

Where, P_w is the water pressure, H is the height of the water, γ_w is the water severity, γ_s is the soil severity, h_i is the thickness of the ith layer of soil FS is the safety factor.

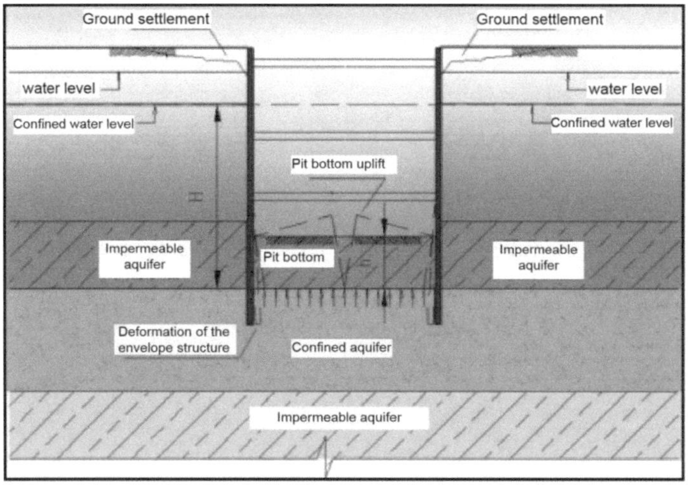

Figure 3. The section of the foundation pit.

Because the existence of confined water affects the overall stability of the foundation pit, it is necessary to lower the confined water level in the foundation pit, or the depth of the foundation pit retaining structure should be lengthened below the confined water layer to isolate the influence of the confined water on the excavation of the pit. The factor of safety will be significantly increased when the depth of retaining structure extent or lower the confined water level. While the cost of the retaining structure is large and when the confined water layer is very deep it is hard to isolate the confined water. So, lowering the confined water level by dewatering is always an economic and safe method, the main problem is the control of the ground and building settlement during dewatering. And a method of settlement control by recharging underground water is proposed in this paper.

3 INTRODUCTION OF GROUNDWATER RECHARGE METHOD

During the dewatering process, it will cause settlement of the foundation within the radius of the dewatering well, which will affect the safety of surrounding buildings. Since there are buildings that are sensitive to the settlement deformation around the foundation pit of this project, in order to reduce the impact of foundation pit dewatering on the surrounding buildings, it is necessary to take measures to reduce the impact of dewatering on the settlement. The mechanism of the ground settlement induced by dewatering is the increase of the effective stress of the soil skeleton, which will cause the consolidation of the soil layer.

This paper proposes to control the settlement caused by the foundation pit dewatering by recharging groundwater near the building. The principle of this method is to inject water into the recharge well, and the groundwater level around the recharge well will continue to rise. The rising water level is called the recharge water level. Because there is a gap between the recharge water level in the recharge well and the groundwater level, the water injected into the recharge well may seep into the aquifer. When the seepage flow and the injection volume are in balance, the recharge water level will no longer continue to rise and will stabilize. At this time, a rising cone of water level is formed around the recharge well. Its shape is very similar to the descending funnel of pumping, but the direction is just the opposite, as shown in Figure 4.

The process of dewatering and recharging should be dynamically controlled and adjusted, so an automatic device for pumping and recharging is designed for the settlement control as shown in figure 5. This device contains two wells, one is for water recharging and the other is for monitoring

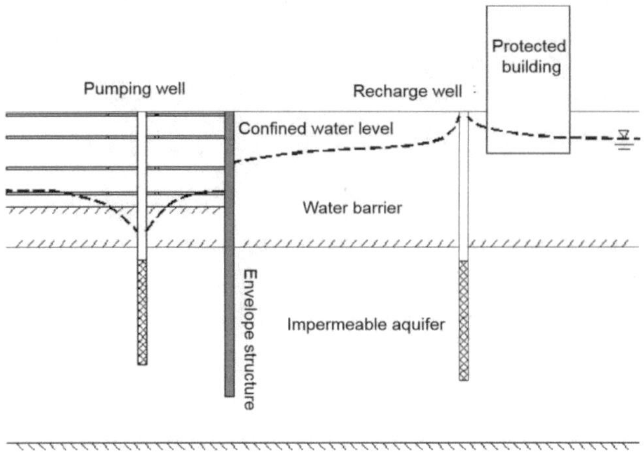

Figure 4. The principle of water recharge.

the water level, which is called the observation well. When the underground water is lower than the minimum level, the water recharging pump will be started recharging water into the ground, and the underground water level will rise, when it reaches the target level, the signal will be delivered to the pump to stop working. The recharging water wells are evenly arranged around the protected building as shown in Figure 5.

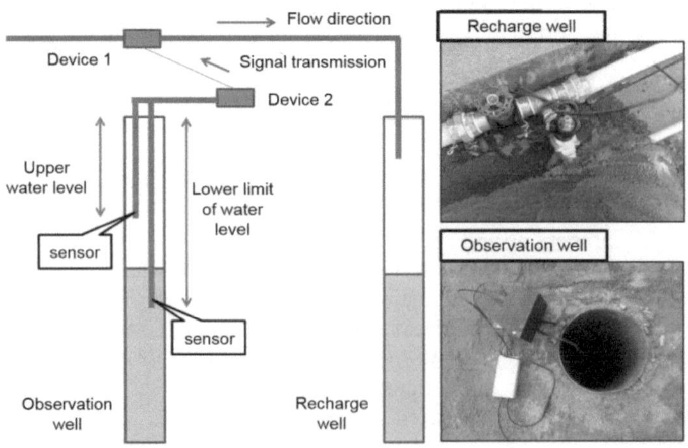

Figure 5. Stress state of hollow cylinder specimen and element.

The concept of the integrated technology of foundation pit pumping and irrigation is to set up recharge wells outside the foundation pit while dewatering and depressurizing the foundation pit project and to ensure the depth of the pressured water level in the foundation pit through the reasonable arrangement of pumping wells and recharge wells meet the requirements of anti-surge stability while reducing or even eliminating the water level changes in the surrounding protection area caused by the foundation pit precipitation, and using the extracted groundwater as the recharge water source, not only can guarantee the recharge amount, but also effectively protect the groundwater resources free from pollution.

4 APPLICATION OF THIS METHOD

The Children's Park Station of Ningbo Rail Transit Line 3 is the transfer station of Lines 3 and 4. The station of Line 4 was excavated in October 2017. There are 20 decompression wells in the foundation pit with a depth of 36~38 meters and recharge wells are also arranged outside the pit. The 15 recharge wells are 43 meters deep, and all the recharge wells are set on the side of buildings such as the Ningyan Building and Qiubi Resettlement House on the south side of Xingning Road. The internal pumping and external irrigation are used to achieve the equilibrium state of the confined water layers to ensure the stability of the building's safety.

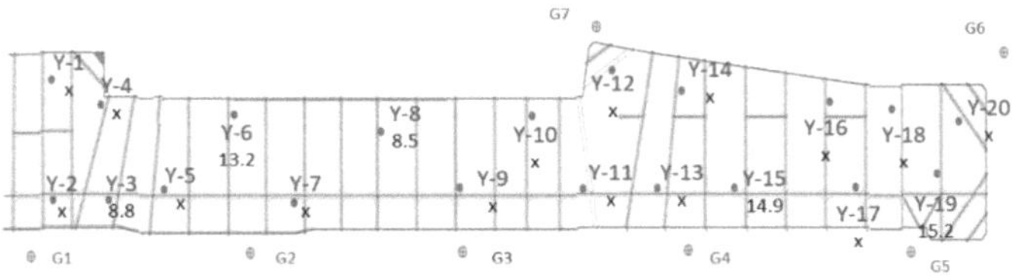

Figure 6. The layout of the wells.

During the above construction period, the deformation data of 4 measuring points (JC111, JC113, JC115, and JC118) near the building and the variation of the underground water level were selected for analysis. The settlement of the building is shown in Figure 7, and Figure 8 shows the water level recovered during recharging with time.

It can be seen that before the water recharging, the largest settlement of the building reaches 6 mm. When the water recharging started, the underground water level rises significantly from 8 m to 4 m during 120 hours, after that, the underground water level remained basically stable as the automatic control device worked as shown in Figure 8. At the same time, the settlement of the building also reduced significantly and can be controlled stably.

The on-site monitoring data shows that by recharging groundwater into the foundation, the groundwater level can gradually rise. This demonstrates that the automatic device for pumping and recharging designed in this paper can work as designed. In order to check the settlement control the monitor of the deformation of the nearby building is shown in Figure 7.

Field measurements show that the integration of pumping and irrigation technology can not only meet the decompression and precipitation requirements of foundation pits in this area, but also that the recharge on the side of the building can better control the settlement of the building, also ensure the safe excavation of the foundation pit, and reduce the impact of depressurized water on surrounding buildings.

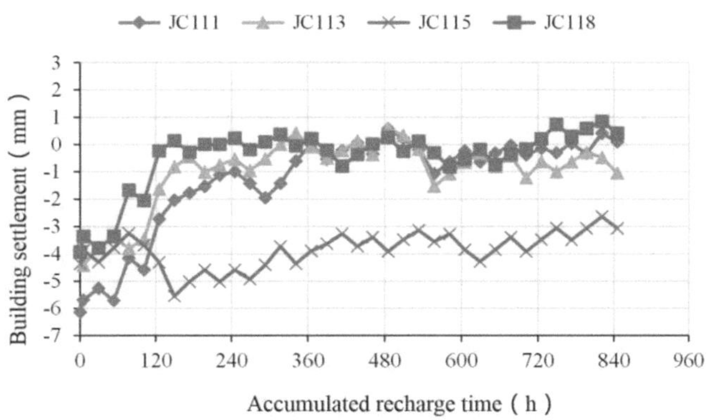

Figure 7. The layout of the wells.

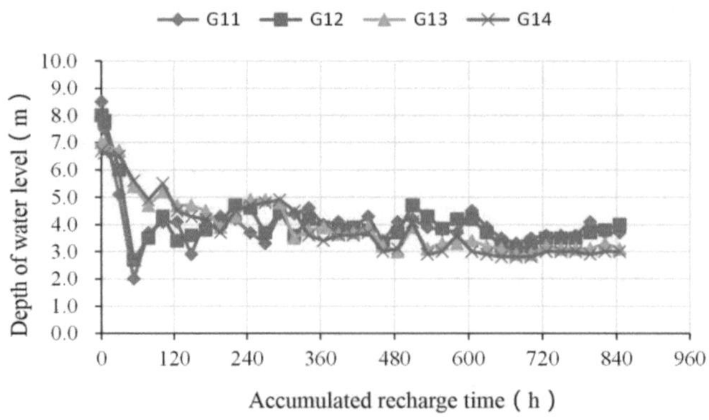

Figure 8. The layout of the wells.

5 CONCLUSION

The effect of underground water on the construction of underground space structures cannot be avoided, lower underground water level is often used to keep the safety of the construction from the influence of underground water. While the underground water pumping always induces the settlement of buildings nearby the well. A method of settlement control by recharging underground water is proposed in this paper, and an automatic device for pumping and recharging is designed for settlement control. The on-site monitoring data shows that by recharging groundwater into the foundation, the groundwater level can gradually rise, and the settlement of the building induced by water pumping reduces when the underground water level rises.

REFERENCES

Bear J., Yavuz M. Mechanics of fluids in porous media. 1983: 109–110.

Biot M A. Consolidation Settlement Under a Rectangular Load Distribution[J]. Journal of Applied Physics, 1941, 12(5):426–430.

Gibson, R.E., England, G.L., Hussey, M.J.L., The theory of one-dimensional and consolidation of saturated clays, I. Finite nonlinear consolidation of thin homogeneous layers. Geotechnique, 1967. 73(3):261–273.

Gibson RE, Schiffman RL, Cargill KW. The theory of one-dimensional soil consolidation of saturated clays: II Finite non-linear consolidation of thick homogeneous layers. Canadian Geotechnical Journal 1981,18:280–93.

Terzaghi K., Theoretical Soil Mechanics, John Wiley & Sons, New York,1943.

Xie, K.H., Xie, X.Y., Jiang, W., A study on one-dimensional nonlinear consolidation of double-layered soil. Computers and Geotechnics, 2002. 29(2):151–168.

Zhang Y., Zhao Y. Y. Real-time prediction of land subsidence caused by foundation pit dewatering. Rock and Soil Mechanics, 2008, 29(6):1593–1596.

Advances in Urban Engineering and Management Science – Khalil & Yang (Eds)
© 2023 The Author(s), ISBN 978-1-032-30426-7

Urban distributions of producer services in building industries in China based on information entropy models: A case of top 100 construction supervision enterprises

Qi Zhang* & Yu-jie Zhang*
School of Civil Engineering, Xi'an University of Architecture and Technology, Xi'an, China

ABSTRACT: The producer services in the building industry in China is a high-order industrial sector with an important impact on the national economy. Therefore, their urban distributions are carefully discussed in this paper through a case study employing the Construction Supervision Enterprises (CSEs) as the representative research object. Firstly, the information entropy models are established for time series detection of the M-distribution system which shows the spatial characteristics of CSEs on national metropolitan agglomerations, as well as for the detection of the R-distribution system which shows the hierarchical characteristics of CSEs in national city rankings of administration. Moreover, the empirical tests of these models are analyzed through the data of the top 100 CSEs from 2006 to 2014. The results of this study indicate that: (1) The top 100 CSEs were apt to cluster in cities with higher administrative rank but this hierarchical characteristic was also fluctuant during the studying period. (2) The spatial characteristics of the top 100 CSEs in each category of metropolitan agglomerations were relatively balanced and stable. (3)The relationship between spatial characteristics of urban distributions of top 100 CSEs and hierarchical ones was changed from antagonism to coordination in 2011.

1 INTRODUCTION

Since China's reform and opening-up, massive fixed asset investments and real estate developments have been done throughout the country and the building industry has also become one of the fundamental impetus to the rapid level rising of China's industrialization and urbanization. Correspondingly, the producer services in the building industry (PSBI) which provides various knowledge and technological supports (designing, management, consulting, etc.) to the whole constructing activities have been thriving in the past two decades. According to the data from "China Statistical Yearbook in 2015", there are 3 main business organizations of PSBI, including survey and design enterprises, construction supervision enterprises (CSEs), and project bidding agencies. It is suggested that the PSBI is regarded as a high-order producer service sector with an important impact on China's national economy in recent years by both the increased amounts of output values and diversified business organizations.

The urban distributions of China's PSBI will be further explained in this paper through a case study that employs the top 100 CSEs as the representative research object. Both the spatial and hierarchical characteristics of the urban distribution of the top 100 CSEs during recent several years are carefully discussed and compared to find the geographical acting regularities for this high-order producer service sector, as well as to obtain useful guidance for improving urban and regional developments of China's building industry.

*Corresponding Authors: zhangqi-xauat@163.com and 976579108@qq.com

DOI 10.1201/9781003305026-56

2 METHODOLOGIES

2.1 *Principles of information entropy model used in urban distributions*

Information entropy is a variable for the measurement of the complexity and balance of a system in Physics originally, which means that the higher the information entropy value is, both the lower order degree and the higher chaos of the corresponding system will be. For this study, both the urban distribution system based on the category of national metropolitan agglomerations (M-distribution system) and the system based on the category of rankings of urban administration (R-distribution system) are established in accordance with the given rules at first. And then the annual structural characteristics of these 2 systems can be shown up through proportional relation of CSEs' scale (or amount) distributed on each urban category under a specific system. Moreover, the evolution and transformation degrees of the distribution systems during a certain period can be put forward and dynamically detected by information entropy of CSEs' structural characteristics so that the spatial distributions patterns of CSEs and the hierarchical ones can be gained together, while their comparisons can be done at the same time.

2.2 *M-distribution system and its information entropy model*

Metropolitan agglomeration is a specific geographic area consisting of megalopolis and large cities, in which the economical activities and spatial organizations are closely connected by well-developed infrastructure networks to reach highly urbanized and integrated urban groups. As being one of the most important approaches to fulfill the tasks of new industrialization and new urbanization in China, the Metropolitan agglomeration provides fundamental guidance to the urban distributions of the high-order producer service sectors, including CSEs and so on. Therefore, 5 metropolitan agglomerations with nationwide influences and the most developed economic & social levels in the eastern, central, and western China are selected to establish the M-distribution system in this paper, which is used to measure the spatial characteristics of urban distributions of CSEs. As is shown in Table 1.

Table 1. Categories of M-distribution systems of CSEs.

Category code	Metropolitan agglomeration	Containing cities
M1	Yangtze River Delta agglomeration	16 cities containing Shanghai, Nanjing, Hangzhou, etc.
M2	Pearl River Delta agglomeration	9 cities containing Guangzhou, Shenzhen, Zhuhai, etc.
M3	Beijing-Tianjin-Hebei agglomeration	10 cities containing Beijing, Tianjin, Shijiazhuang, etc.
M4	The middle reaches of the Yangtze River agglomeration	28 cities containing Wuhan, Nanchang, Changsha, etc.
M5	Chengdu-Chongqing agglomeration	13 cities containing Chongqing, Chengdu, Mianyang, etc.
M6	Cities besides the above 5 agglomerations	Other cities with CSEs are distributed.

Note: Only cities in the Chinese Mainland are covered except Hong Kong, Macao, and Taiwan regions.

For the M-distribution system, when the total size of CSEs' scale is set as m and each size of 6 categories in the system is set as m1$\sim$m6 respectively, the size distribution probability of CSEs on certain agglomeration categories can be described as follows.

$$p(m_i) = m_i/m, i = 1, 2, \ldots, 6, \tag{1}$$

Since the normalization is satisfied on condition that $\sum p(mi) = 1$, the information entropy calculation formula of the M-distribution system is:

$$H(M) = -\sum p(m_i) * \ln p(m_i), i = 1, 2, \ldots, 6 \tag{2}$$

In formula (2), the information entropy H(M) is used to characterize the complexity and balance of the size distribution of CSEs in various agglomeration categories. If equal size distribution of CSEs appears in each category, the system achieves the ideal conditions that p(mi) = 1/6 (i = 1,2,. . .,6) and the entropy reaches its maximum value that H(M)max = ln(6). On the contrary, if there is an extreme situation in which no size distribution of CSEs can be found on any agglomeration category in the system, the information entropy reaches its minimum value that H(M)min = 0. As a result, the closer the size scale appears on various agglomeration categories, the larger the calculated value of information entropy will be.

By the way, nothing may distribute on a specific agglomeration category at some time nodes for the dynamic urban distribution studies of continuous time series. This will lead to the variances of system volume at different time nodes so that the calculating results of information entropy on different time series can not be strictly compared. In order to deal with the incompatibility of system volume, both the concept of equilibrium and dominance, which makes calculations of the M-distribution system being comparable, are introduced in this paper. The formula for calculating the equilibrium of the system at a certain time node during the studying period is:

$$J_M = H(M)/H(M)_{max} = (-\sum p(m_i) * \ln p(m_i))/\ln(6), i = 1, 2, \ldots, 6 \tag{3}$$

In formula (3), H(M)≤H(M)max, so 0≤JM≤1, and the M-distribution system reaches the most ideal equilibrium state when JM = 1. On basis of the equilibrium calculated, the formula for calculating the dominance of the system can be further given as:

$$I_M = 1 - J_M \tag{4}$$

In formula (4), the value range of IM is also in [0, 1], and the increase in its value reflects that the size scale of CSEs is more concentrated on one or several agglomeration categories. An obvious dominant position in the M-distribution system has been achieved for them at this moment. Conversely, the decrease in IM value reflects that the size scale of CSEs on each agglomeration category becomes more equalizing in the system.

2.3 *R-distribution system and its information entropy model*

The hierarchy of a city in China's urban system is determined by the administrative ranks of its governmental heads and other officials. In the light of relevant studies [8], it is reported that the operations of urban political and economic systems markedly interfere with the hierarchy of cities and there are interactive logical relations between the hierarchy of cities and their development. So administrative powers deriving from the hierarchy of cities have a certain far-reaching impact on the urban distribution of China's servicing industrial sectors, including CSEs. Thereby, 4 categories of administrative ranks based on China's statutory ranking system of urban administration and 2 categories based on political & budgetary peculiarities of the national capital and deputy provincial cities, are selected and combined to establish the R-distribution system in this paper, which is used to measure the hierarchical characteristics of urban distributions of CSEs as shown in Table 2.

For the R-distribution system, when the total size of CSEs' scale is set as r and each size of 6 categories in the system is set as r1~r6 respectively, the size distribution probability of CSEs on certain ranking categories can be described as follows.

$$p(r_i) = r_i/r, i = 1, 2, \ldots, 6 \tag{5}$$

The formulas for calculating the information entropy H(R), equilibrium JR, and dominance IR of the R-distribution system at a certain time node during the studying period are raised in the same way as the M-distribution system designed in 2.2.

$$H(R) = -\sum p(r_i) * \ln p(r_i), i = 1, 2, \ldots, 6 \tag{6}$$

Table 2. Categories of R-distribution system of CSEs.

Category code	Ranking of urban administration	Containing cities
R1	National capital	Beijing city.
R2	Municipalities except for the national capital	3 cities containing Shanghai, Tianjin, and Chongqing.
R3	Both Deputy provincial cities and provincial capitals	10 cities containing Jinan, Shenyang, Xi'an, etc.
R4	Deputy provincial cities only	5 cities containing Qingdao, Xiamen, Ningbo, etc.
R5	Provincial capitals only	17 cities containing Zhengzhou, Hefei, Urumqi, etc.
R6	Cities besides the above 5 rankings	Other cities with CSEs are distributed.

Note: Only cities in the Chinese Mainland are covered except Hong Kong, Macao, and Taiwan regions.

$$J_R = H(R)/H(R)_{max} = (-\sum p(r_i) * \ln p(r_i))/\ln(6), i = 1, 2, \ldots, 6 \tag{7}$$

$$I_R = 1 - J_R \tag{8}$$

Similarly, the value range of IR is also in [0, 1], and the increase in its value reflects that the size scale of CSEs is more concentrated on one or several ranking categories. An obvious dominant position in the R-distribution system has been achieved for them at this moment. Conversely, the decrease in IR value reflects that the size scale of CSEs on each ranking category becomes more equalizing in the system.

In addition, since the variances of system volume are avoided by the calculating methodologies of equilibrium and dominance established above, the time series comparison of IM and IR can be done not only vertically in their systems, but also horizontally between the M-distribution system and R-distribution system.

3 RESULTS AND FINDINGS OF URBAN DISTRIBUTIONS

3.1 *Data preparation of top 100 CSEs*

Data about CSEs from statistical yearbooks in China's building industry is always reported through the calibre of provincial units and hardly adopted in urban distribution studies. So the top 100 CSEs screened out by their annual main business incomes are selected as the representative research object in this paper. Then the fundamental database used in the urban distribution of the top 100 CSEs is established by following steps.

Firstly, access authority to closed online statements of the Construction Supervision Market prepared by the Ministry of Housing and Urban-Rural Development of China is gained with the permission of the China Construction Supervision Association. Secondly, both the amounts and total operating incomes of the top 100 CSEs in every city with their distribution are summarized and checked at the end of each year during the period from 2006 to 2014 with the help of careful inquiries in the closed online statements. Finally, according to the entropy models of the M-distribution system and R-distribution system discussed above, the value of p(mi) and p(ri) for each year of the studying period can be respectively calculated.

3.2 *Calculating results of information entropy models*

With the information entropy models obtained in 2.2 & 2.3, all of the detecting parameters of the M-distribution system can be calculated by inducing the value of p(mi) for each year into formulas (1) to (4), as well as parameters of R-distribution system by inducing the value of p(ri) for each year into formulas (5) to (8). Moreover, the time-varying features of dominance between the M-distribution system and the R-distribution system can be detected through annual series comparison, which are shown in Figure 1.

3.3 *Analysis of urban distributions of top 100 CSEs*

The evolution regularities of urban distribution of top 100 CSEs in the past are analyzed then.

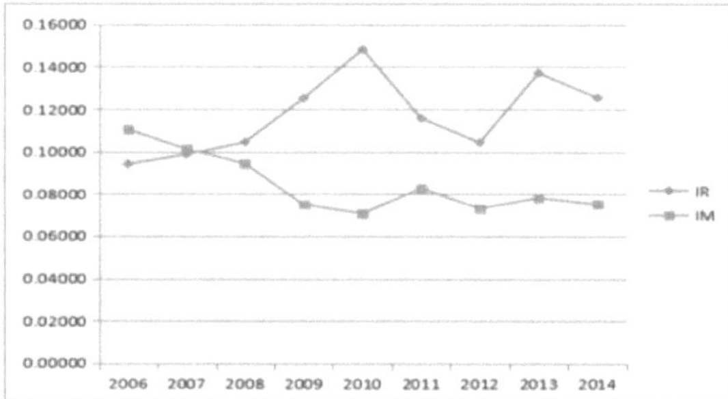

Figure 1. Time-varying curves of IM and IR and their comparisons (2006–2014).

(1) Since the values of I_R had always been greater than the values of I_M from 2007 till the end of the studying period, it is said that a relatively concentrated distributio.n pattern of the top 100 CSEs had occurred in China's rankings of urban administration in the past decade, as well as a relatively balanced distribution pattern of them on metropolitan agglomerations in China. There may be 2 explanations for the patterns forming. On the one hand, most of the products and services businesses of CSEs referring to the construction management and coordination ought to be done only under the allowance of national laws, governmental regulations, or professional and local standards. Therefore, it is obvious that the benefits of understanding and grasping the latest changes in relevant administrative regulations and standards for CSEs can be got and corresponding adjustments to their business content can be made rapidly on the condition that they were distributed in higher administrative urban rankings, such as national capital, municipalities, or deputy provincial cities. On the other hand, local operating strategies were preferred by most CSEs due to the lower transnational and trans-provincial opening degree of China's construction supervision market in the past. Thus a more balanced spatial pattern of top 100 CSEs had been formed on metropolitan agglomerations till 2014 Instead of a gradual decline of concentration from eastern metropolitan agglomerations in China to central and western ones, which was often detected in urban distributions of some other typical producer services and mainly decided by the economical gradient of regional development in China.

(2) It is suggested that the relationship between spatial characteristics of the urban distribution of top 100 CSEs and hierarchical ones were improving from antagonism to coordination in the late studying period because the varying trend of values of I_M and the ones of I_R are always the same after 2011, rather than the inverse trends detected from 2006 to 2010. The reason for this improvement lies in the formation and development of China's metropolitan agglomeration strategies in recent years. Previously, the industrial drivers of China's high-grade central cities were almost dependent on the siphoning of natural resources and other necessary elements from their economic hinterland. Therefore, an antagonistic pattern which means the growth of advanced industrial sectors in central cities but the decline of them in the surrounding hinterland inevitably occurred in past China, such as CSEs. Since 2010, a leading role of radiation and diffusion of China's high-grade central cities was determined by the metropolitan agglomeration strategies, through which more and more resources and elements had been fed back from high-grade central cities to other lower-grade cities within the agglomeration in hierarchical contacting networks to cultivate diversified growth poles. In this context, agencies in producer services would be relocated through careful selection and coordination between original central cities and other new growth poles within the agglomeration according to their operating characteristics and developing needs at the same time, including CSEs.

(3) During the whole studying period, the fluctuation of the value sequence of I_R was always more significant than it of I_M. The reason is that the fluctuation caused by competition was inevitable

but apt to be found in the perspective of China's rankings of urban administration. Firstly, as a typical asset-light servicing firm, there must be fierce competition among individual CSEs so that their business performance may also increase or decrease in a short running time. Therefore, member changes of top 100 CSEs was lasting all the time from 2006 to 2014, as well as the changes in city lists with top 100 CSEs distributed. So the fluctuating calculation results of the value sequence of I_M and I_R were not surprising. Secondly, because the volumes of containing cities in categories M1 to M5 under the metropolitan agglomeration system are often larger than the ones in categories R1 to R5 under the rankings of urban administration system by comparisons between Tables 1 and 2, the frequent time series transformations of cities with top 100 CSEs distributed and their corresponding scale probabilities changes were easier to be recorded and identified by information entropy model of rankings of urban administration.

4 CONCLUSIONS

The main conclusions are as follows.

Because of the administrative requirements in servicing businesses of CSEs and the lower transnational and trans-provincial opening degree of their market in the past decade, a relative concentrated distribution pattern of top 100 CSEs had occurred on China's rankings of urban administration during the studying period, as well as a relatively balanced distribution pattern of them on metropolitan agglomerations in China.

The relationship between spatial characteristics of the urban distribution of top 100 CSEs and hierarchical ones were improving from antagonism to coordination after 2011 with the help of the formation and development of China's metropolitan agglomeration strategies in recent years.

There must be dominance fluctuations of the urban distributions whether in terms of spatial characteristics or hierarchical ones due to inevitable competition among individual CSEs. However, the changes in scale probabilities were easier to be identified by the information entropy model of rankings of urban administration.

ACKNOWLEDGMENTS

This work was financially supported by the Social Science Foundation of Shaanxi Province, China, grant number 2018S39.

REFERENCES

Fang Chuanglin, MAO Qizhi, NI Pengfei. Discussion On the Scientific Selection And Development Of China's Urban Agglomerations[J]. Acta Geographica Sinica, 2015, 70(4): 515–527.

Keeble D, Nacham L. Why do business service firms cluster? small consultancies, clustering, and decentralization in London and Southern England. Transaction of the Institute of British Geographers, 2002, 27 (1):67–90.

P. Guerrieri, V. Meliciani. Technology and International Competitiveness: The Interdependence between Manufacturing and Producer Services. Structural Change and Economic Dynamics, 2005(16): 489–502.

Wang Chengyun, QIN Jian,YANG Sui. Analysis of the cluster mode of R&D industry in the innovative city districts: Taking Beijing, Tianjin, Shanghai, and Chongqing as examples[J]. Acta Geographical Sinica, 2013, 68(8): 1097–1109.

Yang Baojun, CHEN Peng, Lü Xiaobei. Urban and Rural Planning in Transition: Starting From "National New-Type Urbanization Planning[J]. City Planning, 2014, 38(S2): 67–76.

Zhang Qi, LI Tongsheng, HAN Xia, WANG Hui. Theoretical research on the location selection of construction supervision enterprises within the city and its empirical study[J]. Geographical Research,2013,32(11): 2121–2132.

Zhou Jingxiang. The Logic of Sub-Provincial Cities Development: Official configuration and Growth Driving [J]. China Industrial Economics, 2014, (10): 20–32.

Zhou Ziying, Duan Jiannan, LIANG Chunfeng. Temporal-spatial Changes Analysis of Land Use Structure in Changsha City Based on Information Entropy[J]. Economic Geography, 2012, 32(4): 124–129.

Advances in Urban Engineering and Management Science – Khalil & Yang (Eds)
 ISBN 978-1-032-30426-7

Assessment of prefabricated concrete buildings construction safety based on objective and subjective weighting combined with fuzzy comprehensive evaluation

Guozong Zhang*, Xuqiao Fan* & Hai Zhu
School of Urban Economics and Management, Beijing University of Civil Engineering and Architecture, Beijing, China

ABSTRACT: The high risk and complexity of the construction industry have caused frequent safety accidents. China's prefabricated concrete buildings are in the promotion and initial stage. There is a lack of systematic research on the construction safety management of prefabricated concrete buildings. It is urgent to use scientific and reasonable safety evaluation methods to further explore the construction safety management of prefabricated concrete buildings. This paper identifies hazardous and harmful factors in the construction process of prefabricated concrete buildings and constructs six aspects of prefabricated concrete construction safety including human factors, material factors, equipment factors, technical factors, management factors, and environmental factors. Evaluation indicators, combined with the subjective weight of the G1 method and the objective weight of the information entropy weight method, determine the comprehensive weight of the index factors and establish a fuzzy comprehensive evaluation model for the construction safety of prefabricated concrete buildings. The actual case was verified, and the critical path of construction safety of the project was found, which is helpful to improve the construction safety management level of prefabricated concrete buildings.

1 INTRODUCTION

In recent years, China's economy has maintained a high-speed growth, and the construction industry, as a pillar industry of the national economy, has become an important force driving the rapid growth of the national economy. According to statistics, as of 2018, the total output value of China's construction industry has reached 23.5 trillion yuan, an increase of 9.9% year on year. At present, the construction method of the Chinese construction industry is still traditional on-site construction. With the continuous advancement of urbanization, this method has exposed a lot of drawbacks. In terms of energy, the backwardness of traditional building technology has caused a great deal of energy waste, which has created a huge contradiction with the current worldwide resource crisis. In terms of labor costs, with the disappearance of China's demographic dividend, labor shortages on construction sites, especially the loss of young workers, have gradually increased labor costs. Environmentally, the layout of the construction site is messy, especially the serious dust and noise pollution, which leads to the destruction of the ecological environment. Short-board effects in all aspects have emerged, the sustainable development of the construction industry has been restricted, and the upgrading and transformation of the traditional construction industry are imminent.

The prefabricated building has changed the construction and production mode. The production of prefabricated components is mainly carried out in the factory, which greatly avoids the waste of raw materials, saves energy, and produces in a closed environment, reducing environmental pollution and noise pollution, and at the same time, standardization in the factory and large-scale production reduces quality problems, and the streamlined manufacturing process also reduces labor costs.

*Corresponding Authors: zhangguozong@bucea.edu.cn and fanxuqiao@foxmail.com

DOI 10.1201/9781003305026-57

During the construction process, components are mainly assembled on site. According to statistics, compared with traditional buildings, prefabricated buildings can save 31% of the construction period, 30% of energy, 36% of water, 70% of materials, and 40% of labor (Li 2014; Song & Zhu 2014). Prefabricated buildings reflect the tremendous progress of the construction industry in terms of resources, environment, and manpower. These sustainable development advantages reflect the role of building industrialization on social development. At the same time, the national and local governments have also actively promoted and guided them, and have made preferential subsidies and rigid requirements for some specific projects to continuously increase the promotion of prefabricated buildings.

The construction industry is a high-risk industry in China. The high difficulty and high risk of engineering production and construction have caused the construction industry to be extremely prone to safety accidents. Statistics show that an average of 3 people die from construction safety accidents every day (Zhou 2013). The problem is second only to mining, ranking second in China (Xu 2015) At present, Chinese prefabricated concrete buildings use external wall panels and cast-in-place joints. There will be certain wet operations on site (He 2018), so there are construction safety problems in traditional buildings, and prefabricated concrete buildings have obvious advantages over traditional construction processes. The difference is that there are more professional and technical links such as transportation, storage and stacking, hoisting, and installation of prefabricated components. At the same time, there are high requirements for workers' refined work, so there may be other dangerous accidents besides high-altitude work safety and falling of personnel.it requires high precision operations for workers. (Li 1989). At present, experts and scholars in the industry have conducted a lot of research in the field of traditional construction safety evaluation and management, and have achieved many results in practical application and theoretical research. However, prefabricated concrete buildings are in the promotion and initial stage, and the technology and specifications are not perfect enough, and there is a lack of systematic research on the safety issues of the transportation, stacking, hoisting, and installation of unique prefabricated components. Therefore, it is of great theoretical and practical significance to identify the safety factors that affect the construction process of prefabricated concrete buildings, establish safety evaluation indexes, and use scientific and reasonable evaluation methods to deeply explore the construction safety problems of prefabricated concrete buildings.

2 MATERIALS AND METHODS

2.1 *Identification of factors affecting construction safety of prefabricated concrete buildings*

By combing the current research results, the factors affecting the construction safety of prefabricated concrete buildings are summarized. According to the characteristics of prefabricated buildings in the construction process of production, transportation, stacking, lifting, assembly, and acceptance, the rail crossing theory is used to eliminate the mutual influence of various potential safety hazards. At the same time, the correlation test is carried out on the influencing factors of prefabricated concrete construction safety, and the factors with a strong correlation with prefabricated concrete construction safety are retained, and the factors with poor correlation with prefabricated concrete construction safety are deleted. Three levels of prefabricated concrete construction safety evaluation indicators are constructed, with a total of 27 factors, which are divided into six aspects, such as people, technology, and safety. The index factors are shown in Table 1.

2.2 *Determine the comprehensive weight based on G1 method and information entropy*

2.2.1 *G1 method to determine the weight*

G1 method is a subjective weighting method, which was first proposed by Guo Yajun scholars after improving AHP (Guo 2002). The G1 method doesn't need to judge the importance of all indexes like AHP analysis to construct a judgment matrix. Its main point is to compare each layer of indexes in the index system to determine the importance, which is simple and intuitive, has less

Table 1. Construction safety evaluation index system of prefabricated concrete buildings.

Target Layer	Primary Index	Secondary Index
Construction Safety of Prefabricated Concrete Buildings	Human factors S1	Safety awareness of field workers S11 Safety awareness of on-site safety management personnel S12 Number of special operations personnel with certificates S13
	Material factors S2	Assembly stability of prefabricated components S21 Setting and enclosure of operation platform S22 Temporary support bearing strength S23 Difficulty in hoisting prefabricated components S24
	Equipment factorsS3	Mechanical equipment regular safety inspection and maintenance S31 Construction safety protection of hoisting machinery and equipment S32 Selection of special transport vehicles for prefabricated components S33 Overload lifting of hoisting machinery S34 Selection of mechanical equipment and spreader S35
	Technical factors S4	Technical level of prefabrication components hoisting S41 Positioning technology and monitoring technology S42 Composite technology of precast concrete S43 Rationality of lifting point setting S44 Implementation of special construction safety plan S45
	Management factors S5	Safety management organization and responsibility system S51 Investment of safety measures fee S52 Safety education, culture, and training S53 Implementation of safety measures for dangerous goods S54 Implementation of safety inspection system S55 Rationality of accident emergency management scheme S56
	Environmental factors S6	On-site safe and civilized construction working environment S61 Road transport conditions of prefabricated components S62 Layout status of prefabricated component yard S63 Safety standard policy environment S64

computation, and avoids the disadvantage of failing the consistency test. G1 method empowerment steps are as follows:

(1) Determine the importance of indicators and sort them. Invite experts from relevant industries to judge the importance of the set of indicators (S1, S2, …, Sj), select the most important indicators, and then continue to select the most important indicators among the remaining indicators, and arrange the importance order of all indicators according to this method.
(2) Determine the importance of adjacent indicators. Experts make a scientific comparison of the importance of the adjacent indexes Sj-1 and Sj after sorting, and get Rj = Wj − 1/Wj, where j = 1,2,. . .,z). The values of Rj are shown in Table 2.
(3) Calculate the weight of the primary index to the target layer. According to the G1 weighting method, calculate the weight of the z-th first-level index to the target layer, and the formula is as follows:

$$\mathrm{Wz} = \left(1 + \sum_{z=2}^{z} \prod_{i=z}^{z} Ri\right)^{-1} \tag{1}$$

$$Wz - 1 = R_z W_z \; (\mathrm{z} = 2, 3 \ldots, j) \tag{2}$$

Table 2. Rj Assignment reference.

Rj Value	Meaning
1.0	Index Sj-1 is as important as index Sj
1.1	Index Sj-1 is between equal and slightly more important than index Sj
1.2	Index Sj-1 is slightly more important than index Sj
1.3	Index Sj-1 is between slightly important and obviously important than index Sj
1.4	Index Sj-1 is obviously more important than index Sj
1.5	Index Sj-1 is between obviously important and very important than index Sj
1.6	Index Sj-1 is more important than index Sj
1.7	Index Sj-1 is between very important and extremely important than index Sj
1.8	Index Sj-1 is more important than index Sj

(4) Calculate the weight of the secondary index to the target layer. According to the following calculation formula, the weight value of the secondary index to the target layer is obtained:

$$W_j^{'} = W_z \times W_j \tag{3}$$

In the formula, Wj is the weight of the second-level index of item j to the first-level index, and the solution process is similar to the solution of the first-level index weight.

2.2.2 *Information entropy determines the weight*

The concept of "entropy" first appeared in thermodynamics and has been widely used in engineering, economics, and sociology. The principle is that the smaller the information entropy value of index factors is, the more unstable it is, and the greater it plays in the process of comprehensive evaluation. The calculation process is as follows:

(1) Construct the original matrix. M experts are invited to score N indicators according to their own knowledge and experience. The indicators are divided into five evaluation grades: very important, important, general, unimportant, and very unimportant, and the corresponding scores are 5, 4, 3, 2, and 1. The evaluation matrix P = (Pij)m × n of prefabricated concrete building construction safety is obtained.
(2) Blindness analysis. Because there are always deviations in the cognition of different experts, in order to reduce the influence of their uncertainty in the scoring process, it is necessary to conduct a blind analysis of the scoring process of experts, and the calculation formula is as follows.

$$D = D_{\mathrm{ij}} = \ln(\lambda - P_{ij}) / \ln(\lambda - 1) \tag{4}$$

$$\lambda = \max(j) + 2 \tag{5}$$

$$D_{\mathrm{j}} = (D_{1j} + D_{2j} + \cdots + D_{mj})/\mathrm{M} \tag{6}$$

$$T_j = \{[Max(D_{1j}, D_{2j}, \ldots, D_{mj}) - D_j] + [Min(D_{1j}, D_{2j}, \ldots, D_{mj}) - D_j]/2 \tag{7}$$

In which D_{ij} is the information entropy of the ith expert to the jth index. λ is the conversion parameter quantity. D_j is the average recognition degree. D_{mj} is the quantitative transformation of the m-th expert to the j-th index. Tj is cognitive blindness.

(3) Modify the information entropy and determine the weight. The m experts' overall understanding of index j is defined as Pj, and the revised information entropy is obtained by modifying it, the calculation formula is as follows:

$$Q_j = |\ln P_j| = |\ln((1 - T_j) \times D_j)| \tag{8}$$

According to the calculated Q j, an expert evaluation vector set Q={Q1, Q2, …, Qj}. Then, the normalized equation is used to process the data of the evaluation vector, and the calculation formula is as follows:

$$W_j^n = Q_j / \sum_{j=1}^{n} Q_j (\mathrm{j} = 1, 2, \ldots, \mathrm{n}) \tag{9}$$

In the formula, Wj is the weight of each index.

2.2.3 *Combination weight of G1 method and information entropy*

When the G1 method judges the importance of indicators, it is quite subjective for experts to judge and sort them directly. The revised information entropy can be used to judge the importance of indicators. In this way, the uncertainty of expert judgment can be eliminated, so as to objectively rank the importance of indicators. Combining the weights of the two indexes, the final comprehensive weight is calculated, which not only ensures the professional opinions of experts, but also eliminates subjective errors by objective weighting.

The primary problem of combination weighting is how to allocate the proportion of subjective and objective weights. In this paper, the Lagrange extremum method is used to determine the undetermined coefficients of the two weights, and the calculation formula is as follows:

$$\begin{cases} W_j = \alpha W_j' + \beta W_j''(j = 1, 2, \ldots, n) \\ \alpha^2 + \beta^2 = 1(\alpha > 0, \beta > 0) \end{cases} \tag{10}$$

In the formula, α and β are the undetermined coefficients of the subjective and objective weights. The solution of formula (10) is further optimized, and the following formula is obtained:

$$\begin{cases} \max F(\alpha, \beta) = \sum_{i=1}^{m} \sum_{j=1}^{n} (\alpha W_j' + \beta W_j'') \\ s.t. \quad \alpha^2 + \beta^2 = 1(\alpha > 0, \beta > 0) \end{cases} \tag{11}$$

According to the Lagrange extremum method, the following formula is obtained:

$$\begin{cases} \alpha' = \dfrac{\sum_{i=1}^{m} \sum_{j=1}^{n} W_j' P_{ij}}{\sqrt{(\sum_{i=1}^{m} \sum_{j=1}^{n} W_j' P_{ij})^2 + (\sum_{i=1}^{m} \sum_{j=1}^{n} W_j'' P_{ij})^2}} \\ \beta' = \dfrac{\sum_{i=1}^{m} \sum_{j=1}^{n} W_j'' P_{ij}}{\sqrt{(\sum_{i=1}^{m} \sum_{j=1}^{n} W_j' P_{ij})^2 + (\sum_{i=1}^{m} \sum_{j=1}^{n} W_j'' P_{ij})^2}} \end{cases} \tag{12}$$

$$\begin{cases} \alpha = \frac{\alpha'}{\alpha' + \beta'} \\ \beta = \frac{\alpha'}{\alpha' + \beta'} \end{cases} \tag{13}$$

2.3 *Fuzzy comprehensive evaluation*

Various safety evaluation indexes in the construction process of prefabricated concrete buildings are fuzzy, which is not conducive to quantitative evaluation, while fuzzy comprehensive evaluation has the advantage of dealing with uncertainty and imperfect information. Therefore, this paper uses a fuzzy comprehensive evaluation method to evaluate the safety of prefabricated concrete buildings in the construction process. The specific steps are as follows:

(1) Determine The Index Set and Evaluation Set

Firstly, a factor set A = (A1, A2, . . ., Am) is established, where m is the number of index factor sets. Then, build the evaluation set of prefabricated concrete construction B = (B1, B2, . . ., Bn), and set the evaluation set of prefabricated concrete construction safety as B = (B1, B2, B3, B4, B5) = (very safe, very safe, safe, unsafe and extremely unsafe).

(2) Construct Membership Degree and Evaluation Matrix

According to the above weight calculation, the weight set W = (W1, W2, ..., Wm) is summarized. M experts are invited to carry out a fuzzy evaluation on the indicators according to the evaluation set B, and the following evaluation matrix C is constructed according to the evaluation results of the indicators.

$$C = (C_{ij})_{m\times n} = \begin{pmatrix} C_{11} & C_{12} & \cdots & C_{1n} \\ C_{21} & C_{22} & \cdots & C_{21} \\ \vdots & \vdots & \vdots & \vdots \\ C_{m1} & C_{m2} & \cdots & C_{mn} \end{pmatrix} \tag{14}$$

In the formula: M is the number of experts; N is the number of indicators.

According to the evaluation results, the frequency of each index is calculated, so as to obtain the membership degree of each index. Finally, the following fuzzy evaluation matrix R is constructed.

$$R = R_{i5} = \begin{pmatrix} R_{11} & R_{12} & \cdots & R_{15} \\ R_{21} & R_{22} & \cdots & R_{25} \\ \vdots & \vdots & \vdots & \vdots \\ R_{i1} & R_{i2} & \cdots & R_{i5} \end{pmatrix} \tag{15}$$

In the formula: I is the number of indicators.

(3) Fuzzy Comprehensive Evaluation and Analysis

According to the fuzzy comprehensive evaluation algorithm, the index weight set W and the fuzzy evaluation matrix R are combined to obtain the fuzzy comprehensive evaluation result I of the construction safety of prefabricated concrete buildings, and the formula is as follows:

$$I = W \cdot R = (B_1, B_2, \ldots, B_5) \tag{16}$$

The evaluation score of the object is calculated, and the final evaluation result is obtained according to the principle of maximum membership degree, that is, according to the five evaluation grades of safety evaluation of fabricated concrete construction-very safe, very safe, safe, unsafe and extremely unsafe.

3 RESULTS & DISCUSSION

3.1 *Case analysis*

3.1.1 *Project overview*

This project is the first phase of affordable housing in Anyue Jiayuan Community, Baoding City, Hebei Province, China. The project type is a fabricated concrete shear wall structure, with a total of 7 residential buildings with a total construction area of 73750.03m2 and a construction period of 1,092 days. No.1-No.7 residential buildings in this project all have one basement, among which No.1 and No.2 buildings are the highest, with 27 floors. The first floor to the fourth floor above the ground are all cast-in-place concrete structures, and the fifth to twenty-seventh floors are prefabricated concrete component assembly floors. The prefabricated components used in this project are divided into the following six types: the precast concrete outer wall, precast concrete inner wall, precast concrete composite board, precast concrete balcony, precast concrete stairs, and precast concrete air conditioning board. The construction period of single-storey installation is 5–7 days, and the assembly process is selected for kitchen and bathroom, decoration, etc. The assembly rate is as high as 61%, and the installation height and assembly rate are among the highest in China, which has important demonstration significance for promoting the development of assembly buildings in the Beijing-Tianjin-Hebei region.

3.1.2 *Determine the index weight*

Invite the safety management personnel of the construction unit, construction unit, design unit, and supervision unit of this project and 10 experts and scholars in related safety management to score the 27 indicators of this project according to the actual situation of the project and their own knowledge theory. The weight of each index is obtained by formulas (1), (2), (3), (4), (5), (6), (7), (8), (9), (10), (11), (12) and (13), as shown in Table 3.

Table 3. Index weight table.

Target Layer	Primary Index	Combination Weight	Secondary Index	Objective Weight	Subjective Weight	Combination Weight
Construction Safety of Prefabricated Concrete Buildings	Human factors S1	0.2358	S11	0.0625	0.0869	0.0748
			S12	0.0890	0.1078	0.0984
			S13	0.0584	0.0668	0.0627
	Material factors S2	0.1328	S21	0.0445	0.0385	0.0415
			S22	0.0361	0.0267	0.0314
			S23	0.0362	0.0326	0.0344
			S24	0.0284	0.0226	0.0255
	Equipment factorsS3	0.1448	S31	0.0302	0.0230	0.0266
			S32	0.0314	0.0315	0.0315
			S33	0.0244	0.0180	0.0212
			S34	0.0314	0.0281	0.0298
			S35	0.0348	0.0365	0.0357
	Technical factors S4	0.1660	S41	0.0272	0.0255	0.0263
			S42	0.0364	0.0443	0.0404
			S43	0.0281	0.0290	0.0286
			S44	0.0319	0.0331	0.0325
			S45	0.0354	0.0410	0.0382
	Management factors S5	0.2025	S51	0.0311	0.0339	0.0325
			S52	0.0390	0.0573	0.0481
			S53	0.0332	0.0427	0.0380
			S54	0.0308	0.0252	0.0280
			S55	0.0310	0.0209	0.0301
			S56	0.0292	0.0225	0.0258
	Environmental factors S6	0.1181	S61	0.0527	0.0315	0.0421
			S62	0.0315	0.0246	0.0281
			S63	0.0247	0.0193	0.0220
			S64	0.0302	0.0216	0.0259

3.1.3 *Fuzzy comprehensive evaluation and analysis*

Continue to invite the above-mentioned experts to judge the safety influence degree of each index on the project, and then calculate by formulas (14) and (15) to obtain the evaluation matrix R of the assembled concrete construction safety evaluation index. Finally, the fuzzy comprehensive evaluation result I of fabricated concrete building construction safety is calculated by the formula (16).

(1) Safety Evaluation of Project Level Index Factors
Personal Factor

$$I_1 = W_1 \cdot R_1 = \begin{pmatrix} 0.0748 & 0.0984 & 0.0627 \end{pmatrix} \cdot \begin{pmatrix} 0.1667 & 0.3333 & 0.5000 & 0 & 0 \\ 0.1667 & 0.5000 & 0.3333 & 0 & 0 \\ 0 & 0.5000 & 0.5000 & 0 & 0 \end{pmatrix}$$

$$= \begin{pmatrix} 0.0289 & 0.1055 & 0.1015 & 0 & 0 \end{pmatrix}$$

Material Factors

$$I_2 = W_2 \cdot R_2 = \left(0.0415\ 0.0314\ 0.0344\ 0.0255\right) \cdot \begin{pmatrix} 0.3333 & 0.5000 & 0.1667 & 0 & 0 \\ 0 & 0.5000 & 0.5000 & 0 & 0 \\ 0 & 0.5000 & 0.5000 & 0 & 0 \\ 0 & 0.5000 & 0.5000 & 0 & 0 \end{pmatrix}$$

$$= \left(0.0138\ 0.0664\ 0.0526\ 0\ 0\right)$$

Factors of Equipment

$$I_3 = W_3 \cdot R_3$$

$$= \left(0.0266\ 0.0315\ 0.0212\ 0.0298\ 0.0357\right) \cdot \begin{pmatrix} 0.1667 & 0.6667 & 0.1667 & 0 & 0 \\ 0 & 0.1667 & 0.8333 & 0 & 0 \\ 0 & 0.1667 & 0.6667 & 0.1667 & 0 \\ 0.1667 & 0.6667 & 0 & 0 & 0.1667 \\ 0.1667 & 0.5000 & 0.3333 & 0 & 0 \end{pmatrix}$$

$$= \left(0.0153\ 0.0642\ 0.0567\ 0.0035\ 0.0050\right)$$

Technical Factors

$$I_4 = W_4 \cdot R_4 = \left(0.0263\ 0.0404\ 0.0286\ 0.0325\ 0.0382\right) \cdot \begin{pmatrix} 0 & 0.5000 & 0.5000 & 0 & 0 \\ 0 & 0.1667 & 0.8333 & 0 & 0 \\ 0 & 0.1667 & 0.8333 & 0 & 0 \\ 0.1667 & 0.3333 & 0.5000 & 0 & 0 \\ 0.3333 & 0.3333 & 0.3333 & 0 & 0 \end{pmatrix}$$

$$= \left(0.0182\ 0.0482\ 0.0996\ 0\ 0\right)$$

Factors of Management

$$I_5 = W_5 \cdot R_5$$

$$= \left(0.0325\ 0.0481\ 0.0380\ 0.0280\ 0.0301\ 0.0258\right) \cdot \begin{pmatrix} 0 & 0.6667 & 0.3333 & 0 & 0 \\ 0.3333 & 0.3333 & 0.1667 & 0.1667 & 0 \\ 0.3333 & 0.3333 & 0.1667 & 0.1667 & 0 \\ 0.1667 & 0.3333 & 0.5000 & 0 & 0 \\ 0 & 0.6667 & 0.3333 & 0 & 0 \\ 0 & 0.8333 & 0.1667 & 0 & 0 \end{pmatrix}$$

$$= \left(0.0334\ 0.1013\ 0.0535\ 0.0144\ 0\right)$$

Environmental Factors

$$I_6 = W_6 \cdot R_6 = \left(0.0421\ 0.0281\ 0.0220\ 0.0259\right) \cdot \begin{pmatrix} 0 & 0.5000 & 0.3333 & 0.1667 & 0 \\ 0 & 0.3333 & 0.6667 & 0 & 0 \\ 0 & 0.3333 & 0.5000 & 0.1667 & 0 \\ 0 & 0.5000 & 0.3333 & 0.1667 & 0 \end{pmatrix}$$

$$= \left(0\ 0.0507\ 0.0524\ 0.0150\ 0\right)$$

(2) Comprehensive Evaluation of The Project

$$\mathrm{I} = \mathrm{W} \cdot \mathrm{R} = (\mathrm{W}_1\ \mathrm{W}_2\ \mathrm{W}_3\ \mathrm{W}_4\ \mathrm{W}_5\ \mathrm{W}_6) \cdot \begin{pmatrix} \mathrm{R}_1 \\ \mathrm{R}_2 \\ \mathrm{R}_3 \\ \mathrm{R}_4 \\ \mathrm{R}_5 \\ \mathrm{R}_6 \end{pmatrix}$$

$$= (0.1096\ \ 0.4363\ \ 0.4163\ \ 0.0329\ \ 0.0050)$$

3.2 *Research results*

3.2.1 *Analysis of safety influencing factors*

By analyzing the index weights in Table 3, it can be seen that among the first-level index factors, the weight of human factors and management factors is greater than the average weight, which has a great influence on the construction safety of prefabricated concrete buildings. There is little difference between the factor weight and the average weight of technology, which has a moderate influence on the construction safety of prefabricated concrete buildings. The weight of other factors is less than the average weight, which has little influence on the safety of fabricated concrete construction. Through analysis, it can be known that human factors and management factors are important supervision objects in construction safety, and it is necessary to focus on control during project construction.

By analyzing the weights of secondary influencing factors, we can know that the weights of on-site workers' safety awareness, on-site safety management personnel's safety awareness, the number of special operators with certificates, the assembly stability of prefabricated components, positioning technology, and monitoring technology, the implementation of special construction safety schemes, the investment of safety measures fees, safety education and culture training, and on-site safe and civilized construction environment are all larger than the average weight, accounting for about 50% of the total weight, which shows that these secondary influencing factors have a great influence on the construction safety of prefabricated concrete buildings and should be safe in the project.

Summarizing the above analysis, we can get the influence path of safety and toughness of construction engineering (in which the solid line is the key influence path), as shown in Figure 1.

Figure 1. The Impact Path of Construction Safety Indicators.

3.2.2 *Analysis of fuzzy evaluation results*

The maximum membership value of comprehensive evaluation for safety construction of the prefabricated concrete project is 0.4363. According to the maximum membership principle of fuzzy evaluation, the evaluation conclusion of the safety grade is very safe. Among them, the safety level of human factors is very safe, the safety level of material factors is very safe, the safety level of equipment factors is very safe, the safety level of technical factors is safe, the safety level of management factors is very safe, and the safety level of environmental factors is safe. Through field investigation and personnel interviews, it can be known that the project managers of this project are very familiar with fabricated construction and have many years of management experience. Workers in front-line construction also have rich experience in assembly-type construction, and they have been trained to obtain engineering construction certificates before taking up their posts. In particular, special types of workers are certified to work, and there is no separation of witnesses; All the building materials entering the site meet the construction requirements, the overall construction difficulty in the field component assembly process is moderate, and the support system of the assembled components is relatively perfect; Carry out daily maintenance of assembled equipment, and make regular safety inspection during construction; The project management personnel of the project have divided the organizational structure of the project, and put all the responsibilities of the construction into practice. The general contractor and subcontractors have a good cooperation foundation, which can be implemented in time in safety cost input, safety management, training, safety guarantee, and emergency response. However, there are some shortcomings. The construction site is relatively narrow, and there are some difficulties in stacking and transporting materials. At the same time, the technical scheme is the first application in the project, and the positioning monitoring technology is not particularly ideal in practical applications, so there is still room for improvement in technology.

4 CONCLUSION

In this paper, according to the types and characteristics of construction safety risks of prefabricated concrete buildings, the laws, regulations, standards, and specifications related to construction safety and the literature of experts and scholars are analyzed and summarized. The accident causal chain theory and trajectory crossing theory are used to identify the safety influencing factors in the construction of prefabricated concrete buildings, and the correlation analysis method is used to retain the factors with high correlation to establish the final safety evaluation index system. On the basis of the safety index system, the weight of index factors is determined by combining the G1 method and information entropy weight method, and then fuzzy evaluation is used to deal with the fuzzy boundary and uncertainty of the system, and the qualitative index is transformed into quantitative data for project construction safety evaluation, which makes the evaluation result more scientific and reasonable. Finally, through case analysis, the whole model method is effectively tested, and the evaluation result of construction safety management on site is very safe, which is consistent with the actual situation of the whole project safety management. At the same time, the main safety impact paths that affect the safety construction of engineering projects are found. Human factors and management factors are the key influencing factors, which need to be paid attention to.

REFERENCES

Guo, Y. (2002). Comprehensive evaluation theory and method. M. Beijing: Science Press.

He, X. (2018). Research on safety management of PC building construction based on BIM. D. Zhengzhou university.

Li, B. (2014). Present situation and development of prefabricated concrete buildings in China. J. China Science and Technology Information, 07, 114–115.

Li, Y., Li, F, Zou, Y., Ma, X., Lu, Z., Wu, X. (2016). Construction safety and quality assessment of prefabricated concrete buildings. J. Building Technology, 04, 305–309.
Song, D., Zhu, D. (2014). Application of industrialized construction in sustainable housing. J. Housing Science and Technology, 34(08), 31–34.
Xu, Z. (2015). Discussion on safety evaluation and management methods of building construction hazards. J. Construction Science and Technology, 07, 121–122.
Zhou, W. (2013). Study on the Influencing Factors of Housing Industrialization in the Promotion and Application of Public Rental Housing. D. Chongqing Jiaotong University.

Advances in Urban Engineering and Management Science – Khalil & Yang (Eds)
© 2023 The Author(s), ISBN 978-1-032-30426-7

The implication of using modular construction projects on the building sustainability: A critical literature review

Zhenquan Zhou*
School of Architecture and Engineering, Tianfu College of SWUFE, Mianyang, China
Department of Civil Engineering, Faculty of Engineering, Technology and Built Environment, UCSI University, Kuala Lumpur, Malaysia

Deprizon Syamsunur*
Department of Civil Engineering, Faculty of Engineering, Technology and Built Environment, UCSI University, Kuala Lumpur, Malaysia

Xinyi Wang*
School of Architecture and Engineering, Tianfu College of SWUFE, Mianyang, China

ABSTRACT: Construction technologies are classified into cast-in-situ, prefabrication, and 3D printing. For the most part, cast-in-situ construction is an extensively used traditional on-site construction in many developing nations, such as Lebanon and India. In addition, 3D printing is a relatively nascent construction technology projected to capture a sizable market share. However, several civil projects in wealthy nations and some emerging countries have favored prefabrication and modular construction in recent decades. A rising number of scholars and professionals have been acknowledging the benefits of using this advanced technique to address critical challenges in various areas, including energy consumption, project budget and schedule, and construction durability and safety. This article discusses many essential concerns for prefabrication and modular building technologies, including their impact on civil health, project safety, building sustainability, management effectiveness, and stakeholder engagement. The analytical process relies on a comprehensive literature review and secondary data collection. The final results verify the advantages of prefabrication on building sustainability, civil health and safety, construction project management, and stakeholder engagement. The inferred conclusions can support the favorable implication of prefabrication in construction projects, effectively facilitating the sustainable development of the local construction industry in China and other urbanized countries.

1 INTRODUCTION

In recent years, a large number of people have been appealing to the importance of environmental sustainability and eco-friendly thinking for the global construction industry. Some scholars agree to improve building sustainability by adopting advanced construction technologies instead of simply insisting on the traditional way. It is critical to grasp the fundamental ideas behind standard building techniques like cast-in-place, prefabrication, and 3D printing. Regarding cast-in-situ construction, Knaack et al. (2012) note that it is a practical traditional construction approach that involves directly transporting raw materials from suppliers to the construction site and then using these resources to create the structures on-site (Knaack, Chung-Klatte & Hasselbach 2012). In terms of 3D printing, this new developing construction approach utilizes a specialized 3D printer to rapidly create the primary structural components of the structure in the required locations. This technique is only applicable to low-rise structures since the endurance of their constructions should still be evaluated

*Corresponding Authors: 1002164129@ucsiuniversity.edu.my, Deprizon@ucsiuniversity.edu.my and wangxinyi@tfswufe.edu.cn

 DOI 10.1201/9781003305026-58

over a lengthy period (Tech Insider 2018) (Tech Insider 2018). In addition to these two processes, prefabrication is a growing and significant segment of the global construction industry, and is expected to progressively displace cast-in-situ buildings in recent years. Prefabrication, in theory, builds modules and panels in factories and then transfers them to the job site. These finished components will be artificially connected to form the building structure during the subsequent construction phase (Cao et al. 2015) (Cao, Li, Zhu, Zhang 2015). As shown in Figure 1, prefabrication is a form of off-site construction. Specifically, building components are manufactured on a regulated assembly line and then transported to the construction site for assembly. Prefabricated structural systems are classified according to their construction materials and structural configurations. They may be classified as frame systems, cell systems, or panel systems (Alireza, Meltem & Omid 2015) (Alireza, Meltem & Omid 2015). The modular construction, which constructs structures from the inside, is the most comprehensive and appropriate type of prefab structure building. The phrases prefabrication and modular construction are often used interchangeably, including in this article (Knaack et al. 2012). In many practical cases, a rising number of civil projects in industrialized nations are willing to use prefabricated building technologies. For example, many commercial buildings in New York City today are constructed using these technologies. Moreover, the Collins House, a well-known 60-story skyscraper in Melbourne, has been constructed with modular components (The B1M 2018) (The B1M 2018).

Figure 1. Construction phases of a modular house (Knaack, Chung-Klatte & Hasselbach 2012, p. 39).

2 METHODS AND MATERIALS

This paper explores whether prefabrication is more advantageous than traditional cast-in-situ building methods under sustainable development strategies. The proposed research uses both quantitative and qualitative methods to develop the research. In terms of the quantitative considerations, it mainly relies on secondary data collection and analysis. The selected sources include the journal articles focusing on construction engineering and prefabrications, the conference papers related to the civil engineering industry, and a few reliable non-academic sources, such as online-accredited blogs and videos published by construction professionals. On the other hand, the qualitative methods used in this paper refer to the professional comments collected from the university lectures and other reputable scholars in the current global construction industry. These investigations aim to study the implications of prefabrications on civil health, project safety, management effectiveness, building sustainability, stakeholder engagement, and other potential challenges.

3 THE IMPLICATION OF THE CIVIL HEALTH

It is clear that prefabrication has a positive influence on health and safety. As for its importance in health aspects, the primary health diseases related to the construction industry can be decreased by the prefabricated construction method, as illustrated in Table 1. This situation is because prefabrication can play a positive role in preventing and reducing construction pollutants, which will negatively impact the health condition of people. For example, Cao, Li, Zhu, and Zhang (2015), who are reputable environment engineers in China, utilize two accredited assessment models to strictly check the environmental performance of prefabricated residential building (PRB) construction and traditional residential building (TRB) construction. Based on the 10-month research, the investigation figures illustrate that compared with TRB, the cases of health damage in PRB, including respiratory diseases and warming temperature-related illness, are decreased by 6.99% and 7%, respectively. This improvement can be attributed to the prefabrication method because PRB is assembled by building components to produce less dust and greenhouse gases, such as CO_2 and SO_2 (Cao et al. 2015). Second, the off-site construction and prefabrication methods can minimize the noise pollution from the construction site. Another research also states that compared with the traditional construction method, the prefabrication approach could reduce the construction noise by 30%–50%, effectively protecting the hearing of on-site construction workers and surrounding residents (Navaratnam et al. 2018) (Navaratnam, Ngo, Gunawardena, Henderson 2019). It is common knowledge that noise can also impair people's mental health to some extent. In other words, the decrease in noise pollution on the construction site can relieve the negative impacts on the mental health of surrounding stakeholders.

Table 1. The reduction of health damage by modular constructions (Cao et al. 2015, p. 140).

	TRB (*A*)	PRB (*B*)	Difference (*A* − *B*)
Health damage	1.69E+00	1.58E+00	1.12E−01
Climate warming related diseases	3.05E−04	2.83E−04	2.25E−05
Carcinogenesis	0.00E+00	0.00E+00	0.00E+00
Respiratory effect	1.51E+00	1.40E+00	1.05E−01
Circulatory effect	1.83E−01	1.77E−01	6.29E−03

4 THE IMPROVEMENTS IN THE PROJECT SAFETY AND MANAGEMENT

Regarding prefabrication's importance in safety aspects, it is a fact that this method can improve building safety during the construction period and building quality for the end users. Jackson (2021) agrees that the controlled industrial environment can ensure high quality of building components through more accurate trials before building on-site (Jackson 2021). Therefore, the qualified subassemblies can improve the durability of the assembling building for most end users. Additional secondary studies demonstrate that modular structures retain adequate fire resistance, earthquake resistance, and windstorm resistance performance. Notably, the laboratory discoveries that "the reserve strength of Modular Steel Building braced systems was larger than that of conventional braced systems" is also considered in the relevant investigation (Navaratnam et al., 2018, pp. 6–8). Prefabrication can also increase construction safety under the use of automated activities. Dr. Huanyu Wu presented during his lecture on Subject 7048, "Construction Management and Technology II", at the University of Adelaide on May 7, 2019 that most construction laborers can operate more securely on-site when cranes are used rather than cranes operating aloft. It is capable of preventing fatalities during the construction of high-rise structures. Notably, the qualified construction time associated with the prefabricated approach has been reduced by 40% since 80%–90% of on-site construction activities are substituted by the factory assembly line (Navaratnam et al. 2018). Due to the shorter building duration associated with prefabricated tasks, project costs are expected to decrease. According to Jackson (2021), around 60% of 800 prefabricated structures can efficiently lower project timelines and budgets. These findings imply that individuals may administer using the prefabrication process more efficiently. This advanced method can potentially reduce the project budget, aiding the financial sustainability of construction projects.

5 THE ENHANCEMENT OF THE BUILDING SUSTAINABILITY

It is worth emphasizing that prefabrication can improve the sustainability of building projects. Zuo (2019) introduces that project sustainability in construction is generally classified into three major categories: environmental, economic, and social sustainability. Regarding environmental sustainability, it is clear that prefabricated building is more environmentally friendly than traditional construction, particularly in terms of energy consumption and construction waste reduction. According to the environmental impact analysis issued by Cao et al. (2015), PRB construction can significantly minimize energy consumption throughout the building phase. Their report points out that water, coal, and diesel usage can be reduced by 22.48%, 41.02%, and 51.84%, respectively, during PRB construction. It implies that renewable and non-renewable energy could be effectively conserved via prefabricated buildings, benefiting environmental sustainability over time. Prefabrication has an apparent benefit of decreasing construction waste. Construction wastes, by definition, are useless building materials destroyed during the on-site construction process, hence negatively affecting environmental sustainability. In this study, reducing wastes in PRB varies between 25% and 81%, demonstrating the significant improvement in decreasing construction wastes via the prefabrication approach (Cao et al. 2015). In addition, prefab technology may help lower and regulate project expenses, critical for financial sustainability. Prefabrication is the process of manufacturing construction components in factories. It does not rely on the cast-in-situ method, which will considerably reduce the number of construction laborers required throughout the on-site building process (Jackson 2021). Reducing the number of construction employees can drastically decrease the budget since construction labor costs account for a sizable portion of the overall project budget. For example, according to a study completed by Navaratnam et al. (2018), labor expenses are decreased by 25% when prefabricated buildings are used instead of conventional constructions. Braimah (2013) asserts that construction time has been seen as money in civil projects in most situations (Braimah 2013).

6 THE MODIFICATION OF THE STAKEHOLDER ENGAGEMENT

It is significant to comprehend the critical nature of stakeholder management in prefabricated structures. Li et al. (2016) highlighted a typical issue of stakeholder administration in prefabricated home production (PHP), including stakeholder communication and engagement (Li, Hong, Xue, Shen, Xu, Mok 2016). PHP is not required to consider all conventional stakeholders throughout the project lifecycle. It is likely to reduce the interoperability between various stakeholders, making it more difficult to effectively exchange project information and resources to complete the targeted constructions. Based on this situation, it is essential to ensure effective and efficient communication among many parties involved in prefabrication projects, such as construction workers, designers, suppliers, project managers, legal approvers, end users, and nearby neighbors. Indeed, several stakeholders play indispensable roles in prefabrication constructions. For instance, suppliers of materials and legal approvers, such as municipal councils, may substantially impact modular building projects in various aspects. Expressly, prefabricated construction necessitates transporting required building components from manufacturers to job sites, significantly disrupting regular traffic flow. For example, managers in Australia must get permission hours from the City Council for potential transportation. If the City Council authorizes inadequate transportation time, the project must pay the additional funds since delayed transportation would extend the schedules for prefabricated buildings (Zuo 2019). On the other hand, this phenomenon also implies that suppliers are responsible for delivering components to the building site in prefabricated structures in a timely and safe manner (Cao et al. 2015).

7 DISCUSSIONS ON THE PREFABRICATION CHALLENGES

Although there are many positive implications caused by prefabrication on construction projects, it is necessary to recognize the fundamental hindrances. Table 2 lists the significant barriers to prefabrication in a regular construction project (Wuni & Shen 2020) (Wuni, Shen & Mahmud 2019). While assembling components in factories reduces the total burden of laborers, the reality is that the required construction precision will probably be significantly greater than conventional structures

Table 2. Five critical risks in PRB (Wuni, Shen & Mahmud 2019) (Wuni & Shen 2020).

No.	Potential issues while adopting the PRB in practical constructions
1	The complexity of administrating various stakeholder groups
2	Higher incipient financial expenditures compared with TBA
3	Deficient supply chain planning and management
4	The legal regulations on PRB can be improved more accurately
5	Uncertainties embedded during the transportation process

due to the strict and effective safety regulations applicable to assembled projects. This situation is because the jointing and assembly of the individual components significantly impact the safety of the prefabrication process (Navaratnam et al. 2018). Meanwhile, construction workers should assemble the structures using heavy gear, such as cranes. It requires the operational personnel to have recognized training sessions and sufficient experience. Otherwise, the incidents above will likely result in significant casualties and financial damages (Jackson 2021). In prefabricated building projects, civil engineers and designers are often tasked with ensuring the viability of the construction plan, particularly the safety component. For example, many civil projects in China would prefer to adopt the cast-in-situ method over the prefabricated approach, owing to the reluctance of most construction groups to embrace new technologies in the building industry (Wuni & Shen 2020). In other words, numerous engineers in China have insufficient expertise with prefabricated structures. Thus, civil engineers and designers involved in prefabrication projects will be obliged to adhere to stringent structural design standards, ensure model safety, and monitor construction quality. The specific situations of each proposed critical risk will be explored and verified in the further research project.

8 ANALYTICAL RESULTS AND DISCUSSIONS

Based on the secondary data collection and critical analysis of the previous literature, Table 3 compares the general conditions between TRB and PRB. The analytical results indicate that the prefabricated buildings are more advantageous than the cast-in-situ constructions under the sustainable development considerations, including but not limited to environmental sustainability, project management effectiveness, civil health, and building safety. Except for the stakeholder management aspect, which involves the complex and system-thinking tasks, other targeted aspects evidence the positive implications of prefabrication in a modern construction project. However, there are still few flaws involved in this research. First, the collated source is mainly based on secondary data collection, containing more first-hand surveys. As an up-to-date investigation, this research can present pursuable findings under the comprehensive literature review, but first-hand quantitative investigations should be undertaken in subsequent research projects. Moreover, there is still a research gap in whether PRB can better the stakeholder engagement and relevant management performance compared with TRB. This research question could be solved by further desk research and case studies.

Table 3. The critical comparison between TBA and TRB.

Construction type	TRB	PRB
Environmental performance	Not so well	More environmental friendly
Project management effectiveness	More complexities	More efficient
Potential damage to civil health	More	Less
Construction safety	More dangerous	Safer
Stakeholder engagement	TBA	TBA

9 CONCLUSION

Prefabrication is an effective technology that facilitates an eco-friendly urbanization process. For one reason, prefabrication can cause beneficial influences on construction project safety and stakeholder health. Moreover, prefabricated construction projects can significantly improve financial and environmental sustainability than traditional building techniques. Meanwhile, it is essential to consider the importance of stakeholder engagement while implementing this advanced approach. In most situations, the prefabrication construction project involves various stakeholder groups, which can genuinely impact the project designing stage, material transportation, and on-site construction during the whole lifecycle. Most civil projects would prefer traditional cast-in-situ construction techniques in the short term. Nevertheless, prefabrication is a more ecologically responsible building approach. The analytical findings in this paper can support the benefits of using prefabrication in construction projects, facilitating the long-term sustainable development of the construction industry in many industrialized countries. Thus, if individuals can deal with relevant barriers and use prefabricated buildings properly, it is likely to better the sustainable growth of the construction sector. Moreover, a few potential investigations should be developed in the further research, including but not limited to the first-hand data collection and sufficient sample size of desk research on the prefabricated constructions. The proposed further research can more accurately and convincingly validate and verify the current analytical findings.

ACKNOWLEDGMENTS

In terms of this paper's development and completion, it is necessary to show appreciation for the Department of Civil Engineering, Faculty of Engineering, Technology and Built Environment at UCSI University, Malaysia. It is also essential to thank the School of Architecture and Engineering, Tianfu College of SWUFE, China.

REFERENCES

Alireza, B, Meltem, VK & Omid, RB. (2015) Using Prefabrication Systems in Building Construction. The Leadership Quarterly. 24: pp. 44258–44262.

Braimah N. (2013) Construction Delay Analysis Techniques—A Review of Application Issues and Improvement Needs. Buildings (Basel). 3:506–531.

Cao X, Li X, Zhu Y, Zhang Z. (2015) A comparative study of environmental performance between prefabricated and traditional residential buildings in China. Journal of cleaner production. 109: pp. 131–43.

Jackson, S. (2021) Off-Site Modular Construction Improves Quality and Safety of Projects. https://www.constructormagazine.com/off-site-modular-construction-improves-quality-and-safety-of-projects/

Knaack, U, Chung-Klatte, S, & Hasselbach, R. (2012) Prefabricated Systems: Principles of Construction. Walter de Gruyter GmbH, Basel/Berlin/Boston.

Li CZ, Hong J, Xue F, Shen GQ, Xu X, Mok MK. (2016) Schedule risks in prefabrication housing production in Hong Kong: a social network analysis. Journal of cleaner production. 134:482–494.

Navaratnam S, Ngo T, Gunawardena T, Henderson D. (2019) Performance Review of Prefabricated Building Systems and Future Research in Australia. Buildings (Basel). 9(2):38-.

Tech Insider. (2018) 3D-Printed Home Can Be Constructed for under $4,000. https://www.youtube.com/watch?v=wCzS2FZoB-I.

The B1M. (2018) Collins House: Building Melbourne's Prefabricated Skyscraper. https://www.youtube.com/watch?v=mKSMT6sEmrs.

Wuni, IY & Shen, GQ. (2020) Barriers to adopting modular integrated construction: Systematic review and meta-analysis, integrated conceptual framework, and strategies. Journal of Cleaner Production. Vol. 249.

Wuni, IY, Shen, GQP & Mahmud, AT. (2019) Critical risk factors in the application of modular integrated construction: a systematic review. International Journal of Construction Management. pp. 1–15.

Zuo, J. (2019) lecture notes on the topic 7048 Construction Management and Technology II. https://myuni.adelaide.edu.au/courses/40763/files/4378729?module_item_id=1477612.

Advances in Urban Engineering and Management Science – Khalil & Yang (Eds)
© 2023 The Author(s), ISBN 978-1-032-30426-7

Research on the influencing factors of prefabricated building quality based on AHP

Zhan Song* & Sun Chuan*
Department of Management, Shenyang Jianzhu University, Shenyang, China

ABSTRACT: The current urbanization process continues to advance, and the demand for buildings is increasing day by day. In order to improve the efficiency of building construction, prefabricated buildings have been widely popularized. Therefore, this article studies the factors affecting the quality of prefabricated buildings and evaluates the factors affecting the quality of prefabricated buildings through AHP. It is found that the component production quality, design experience, and finished product protection in the construction stage are the main factors affecting the quality of prefabricated buildings. The above-mentioned main influencing factors put forward countermeasures to improve the quality management of prefabricated buildings from the aspects of the management system and technical theory.

1 INTRODUCTION

Prefabricated construction is the key to the construction industry transformation and upgrading, since 2016, the state has vigorously developed the prefabricated buildings, while the research achievements of domestic and foreign scholars about the prefabricated building are rich, but because prefabricated construction is different from the traditional construction industry, its quality management also has the certain difficulty, combined with our country has not yet formed a complete prefabricated construction quality management system. As a result, the domestic prefabricated building is still in its infancy and seriously restricts the development of the domestic prefabricated building. Therefore, this paper mainly takes prefabricated buildings as the research object, using the Analytic Hierarchy Process (AHP) for the prefabricated building in each stage of stratification, weight calculation, and prefabricated building quality factors for evaluation.

2 IDENTIFICATION OF INFLUENCING FACTORS OF PREFABRICATED BUILDING QUALITY

2.1 *Selection principle of influencing factors of prefabricated building quality*

2.1.1 *Architectural design stage*

From the preliminary planning and design stage, there will be unreasonable design, incomplete laws and regulations, incomplete feasibility studies, unreasonable contracts, and other factors to affect the quality of prefabricated buildings. In this paper, the assembly design experience is insufficient, the feasibility study in the early stage of the project is not comprehensive, and the signing of the project contract is not reasonable from three aspects, with a view to analyzing its impact on the assembly building.

2.1.2 *Construction phase*

Material quality is the key factor affecting the quality of the prefabricated construction finishing factor, especially involving the key parts of the prefabricated construction materials. In prefabricated building key materials and the selection of products, each unit of product requirements,

*Corresponding Authors: 1152972258@qq.com and 1364778765@qq.com

DOI 10.1201/9781003305026-59

implementing the subject index in the regulatory process is missing, and then prefabricated construction quality is considered; when each raw material and components enter the factory, whether the site acceptance and inspection are conducted according to the standard standards; concerning prefabricated components in the transport and storage links, there is a probability of damage. If certain protective measures are not taken, the overall quality of the building will be greatly affected.

2.1.3 *Construction completion acceptance stage*

There is no systematic management system in the current domestic engineering management for assembly engineering, and the management measures and awareness of managers are relatively weak, which makes the work efficiency and engineering quality of assembly construction engineering difficult to reach the expected level.

3 ESTABLISHMENT OF AN ASSESSMENT MODEL OF PREFABRICATED BUILDING QUALITY INFLUENCING FACTORS

3.1 *Establishment of Analytic Hierarchy Process model*

The decision of a problem is usually related to multiple levels, among which the most important step is to establish the influencing factors of each level and the influence relationship of these influencing factors to the final decision, which is the basis of correlation analysis and calculation. Therefore, AHP is to decompose a decision into multiple levels, and each element in the same level has certain dominance over its lower elements, while they are also dominated by the upper elements. Therefore, the top layer is the goal layer, also known as the decision goal; The middle layer is the control layer, which is generally the criteria, basis, and factors that need to be referred to or considered for decision-making. It is dominated by decision-making objectives. The lowest layer is the index layer, also known as the scheme layer, which is generally the alternative scheme, factors, indicators, etc. The scheme layer establishes the association with the decision-making objectives through the control layer.

The target layer in this paper is a-prefabricated building quality evaluation factors; The control layer is the B1-building design stage, B2-building construction stage, B3-building completion acceptance stage; indicators are C1-the feasibility study in the early stage of the project is not comprehensive, C2-the assembly design experience is insufficient, C3-the signed project contract is not reasonable, C4-the production quality of the assembly components does not meet the requirements, C5-the transportation of components and parts, Pile up in the process of finished product protection, C6-prefabricated building construction equipment backward, C7-lack of prefabricated building construction quality inspection system, C8-lack of professional quality acceptance personnel, C9-prefabricated building construction quality acceptance standard is not perfect, C10-post quality management mechanism is not sound.

3.2 *Determine the weight of evaluation indicators*

3.2.1 *Establish the judgment matrix*

The key to Analytic Hierarchy Process is to construct a judgment matrix. Sati's 9 importance levels and their assigned values are used to compare the relative importance of the subordinate influencing factors in the same index. The scale of proportional value is shown in Table 1.

Table 1. Importance scale.

Importance scale	Comparison of importance
1	If you compare them in pairs, they are equally important
3	Pi factor is slightly more important than Pj factor
5	Pi factor is more important than Pj factor
7	Pi factor is more important than Pj factor
9	Pi factor is more important than Pj factor
2,4,6,8	Is the median of the two adjacent judgments above
Reciprocal if the Pi factor is compared with the Pj factor, the judgment value is Pji=1 /Pij, Pii=1	

The experts who constructed the index system were invited to construct the judgment matrix of the criterion layer relative to the overall objective layer and the judgment matrix of the index layer to the criterion layer, and the judgment matrix was constructed by the expert consultation suggestion.

3.2.2 *Consistency test*

Sadie defines the consistency index CI to check the consistency of thinking in the construction of a matrix, which is determined by Formula 1.

$$CI = \frac{\lambda_{\max} - n}{n - 1} \tag{1}$$

The CI value is inversely proportional to the consistency of the judgment matrix. The larger the CI value is, the worse the consistency is. Generally, the smaller the CI is, the better the judgment matrix passes the consistency test. RI is the average of enough consistency indicators calculated according to the randomly occurring judgment matrix. In order to more reasonably measure the consistency of the judgment matrix, Satie passed the random consistency indicator RI (Table 2) and constructed the consistency ratio CR for detection.

Table 2. Average random consistency index.

The matrix order	1	2	3	4	5	6	7	8	9
RI	0.00	0.00	0.58	0.90	1.12	1.24	1.32	1.41	1.45

The ratio between the consistency index CI of the judgment matrix and the average random consistency index RI of the same order is called the random consistency ratio, denoted as CR, as shown in Equation 2.

$$CR = \frac{CI}{RI} \tag{2}$$

Similarly, the condition for the judgment matrix to pass the consistency test is CR≤0.1. In this case, its normalized eigenvector can be used as the weight vector. The CR values of each matrix are as follows: CR (A) = 0.0032; CR (B1) = 0.0384; CR (B2) = 0.0349; CR (B3) = 0.0334. The single-layer consistency detection of the four matrices all passed, and the weight of the target layer relative to the criterion layer and the weight of the criterion layer relative to the target layer were calculated successively. Finally, the weight of the indicator layer relative to the target layer was obtained.

3.3 *Total hierarchy sorting*

According to the calculation of hierarchy analysis, the total weight of the target layer can be calculated, and the weight is sorted. (Table 3) The top eight weights are C4prefabricated components whose production quality does not meet the requirements, C2 prefabricated design experience is insufficient, C5 parts transportation, Piled up in the process of product protection, such as backward, C6 prefabricated construction equipment, C10 afterward quality management mechanism is not sound, the C3 project contract is unreasonable, lack of C7 prefabricated construction quality inspection system, imperfect C9 prefabricated construction quality acceptance standard, C1 project prophase feasibility study is not enough comprehensive, and C8 lack of professional quality inspection personnel.

Table 3. Influence factor weight.

The target layer	Control layer	Control the weight	Index layer	Index layer weight	Total target weight
Evaluation of influencing factors of prefabricated building quality	B1- Architectural design stage	0.2299	The preliminary feasibility study of the C1 project is not comprehensive enough	0.0768	0.0176
			Lack of experience in C2 assembly design	0.7370	0.1694
Evaluation of influencing factors of prefabricated building quality			The project contract signed by C3 is unreasonable	0.1863	0.0428
	B2- Construction phase	0.6479	The production quality of C4 assembly components does not meet the requirements	0.5678	0.3679
			C5 parts transportation, stacking, and other processes of finished product protection	0.2593	0.1680
			C6 prefabricated building construction equipment is backward	0.1194	0.0774
			C7 lacks a prefabricated building construction quality inspection system	0.0535	0.0347
	B3- Construction completion acceptance stage	0.1222	C8 lacks professional quality acceptance personnel	0.1062	0.0130
			C9 assembly building construction quality acceptance standard is not perfect	0.2605	0.0318
			C10 afterward quality management mechanism is not sound	0.6333	0.0774

4 ANALYSIS AND COUNTERMEASURE RESEARCH ON THE INFLUENCING FACTORS OF PREFABRICATED BUILDING QUALITY

4.1 *Analysis based on the theoretical results of AHP*

Through the analysis of the factors affecting the quality of prefabricated buildings, we can find that the component production quality, design experience, and finished product protection are the main factors affecting the quality of prefabricated buildings. In addition, it is necessary to improve the rationality of contract signing at the initial stage of the project and the quality management mechanism afterward and improve the quality of prefabricated buildings by formulating complete quality management norms, strengthening communication and cooperation, and promoting technical improvement.

4.2 *Countermeasures and suggestions for quality management of prefabricated buildings*

4.2.1 *Improve the relevant quality management system*

Since the construction of assembly building projects is very systematic, we will gradually establish the quality control system, improve the relevant quality control system, clarify the direction of quality control in each link, and achieve quality control. It is attempted to establish quality control objectives for prefabricated construction projects at all stages, perfect design quality control system, production management system, construction quality control system, and construction quality acceptance system, to ensure measures.

4.2.2 *Establish a complete and unified technical theoretical system*

Prefabricated building is only used for pilot work in some areas, and because of the diversity of prefabricated building design, the production lines of various prefabricated components are often different, so it is very important to build the scientific and technological system of prefabricated building quality control; on the other hand, the connection of building components is the center of the whole building. It can connect components well, so unified connection methods and acceptance standards are adopted. This is also the main factor affecting the quality control of prefabricated buildings.

4.2.3 *Strengthen the application of BIM technology in practice*

BIM technology is adopted to build prefabricated building information model, three-dimensional visualization and three-dimensional design, the project acceptance of all documents, the information feedback back to the information model, the progress, cost, and other control; BIM can also be fully used to control the quality of components with high quality and unity; it is also adopted in the construction preparation or construction, simulated rehearsal, to strengthen management in the actual construction.

5 CONCLUSION

This paper evaluates the factors affecting the quality of prefabricated buildings through AHP and finds that the component production quality, design experience, and finished product protection in the construction stage are the main factors affecting the quality of prefabricated buildings. Some improvement measures are taken for assembly-type quality management. In construction, it is necessary to correctly grasp the factors affecting the quality of prefabricated buildings and discuss their effective control measures to ensure the overall quality of prefabricated building projects and ultimately promote the development of building industrialization.

REFERENCES

Liu Guangchen, Wen Zhendi, He Xueli, Shen Jing. Influence factors of prefabricated building quality based on ISM MICMAC [J]. Journal of civil engineering and management,2019,36(05):33–39.

Rural economy and science and technology,2017,28(18):103–104. Li jiabao.

Wei Qing. Factors affecting the construction quality of prefabricated buildings and control measures[J].Green Building Materials, 2020(03): 208–209. DOI: 10.16767/ j. cnki. 10-1213/tu.2020.03.146.

Zhang Ke, CAI Jinsong, Huang Qingyun. Building Economy, 201,42(10):95–98.

Zhuang Li, LIU Hongwei, Wang Xiaomei. Quality management of prefabricated buildings based on hall 3d model [J]. Engineering economics,2019,29(01):18–22.

Advances in Urban Engineering and Management Science – Khalil & Yang (Eds)
© 2023 The Author(s), ISBN 978-1-032-30426-7

Research on influencing factors of construction economy development based on principal component analysis

Zhan Song* & Sun Chuan*
Department of Management, Shenyang Jianzhu University, Shenyang, China

ABSTRACT: At present, economic and social development is in an important period of opportunity and strategic transition. The key tasks of the development of the 14th Five-year Plan are optimizing the industrial structure, transforming the economic development mode, comprehensively enhancing the overall strength and comprehensive economic competitiveness of the construction industry, and striving to build a modern economic system. This paper expounds on the current situation of the development of the construction economy, pays attention to the existing deficiencies at the practical level, and analyzes the various factors affecting the development of the construction economy by using principal component analysis. The results show that GDP and the total output value of the construction industry have the greatest impact on the development of the construction economy, followed by fixed-asset investment and the proportion of the added value of the construction industry with the least impact on GDP. Based on this, the development strategy is put forward to play the role of the construction economy in the national economy, stimulate domestic demand and expand employment.

1 INTRODUCTION

Economic development promotes the construction of national modernization and the continuous improvement of the comprehensive strength of a big country, while the construction economy has always played an important role in the national economy. With the gradual expansion of the development scale of China's construction industry and people's demand for life and production, the construction economy has also played a significant role. However, there are also some problems in the process of development. Therefore, it is of practical and theoretical significance to explore the development countermeasures of China's construction economic transformation.

2 RESEARCH ON THE DEVELOPMENT OF CHINA'S CONSTRUCTION INDUSTRY

2.1 *Development status of the construction industry*

The biggest characteristic of China's construction industry system is a large degree of complexity, so there is no clear standard for the division of the relevant industry sectors of the construction industry. There are many differences between different standards, but overall, it can be divided into two types of the construction industry, large and small. Both differences reflect whether to include survey units, design units, and other units wait in the building industry. Based on the above division, the broad scope of the construction industry is limited, and then there are two kinds of the construction industry, big and small, which are also the basic definitions of the two kinds of the construction industry.

2.2 *Thinking about the development of the construction industry*

Since the founding of new China, the status of the construction industry has been changing and gradually become an important pillar industry of China's national economy, which plays an important

*Corresponding Authors: 1152972258@qq.com and 1364778765@qq.com

DOI 10.1201/9781003305026-60

role in economic development. People's living standards and quality are closely related. The development of the construction industry has determined the development of China's urban and rural construction and changed the appearance of both cities and rural areas. The process of the development of the construction industry is broad, the characteristics of the construction industry are effectively used, the comprehensive development of the construction industry is promoted, and resource development technology is applied to the construction industry. A commitment is made to clear some important factors affecting the development of the construction industry and determine the effective factors for the development of the construction industry and the usefulness of the progress of other industries, which can drive the country, serving as the focus of this article.

3 PCA MODEL CONSTRUCTION AND ANALYSIS

3.1 *Establish a principal component analysis model*

Since the indicators selected in this study are comprehensive indicators reflecting the development of the construction economy and cover many variable indicators, it is necessary to reduce the dimension of high-dimensional variable indicators before establishing the model and achieve the optimal comprehensive simplification of multi-variable data tables, which is conducive to further analysis of the study.

Table 1. Index correlation coefficient matrix.

Correlation	X1	X2	X3	X4
X1	1.000	0.996	0.772	0.920
X2	0.996	1.000	0.763	0.919
X3	0.772	0.763	1.000	0.612
X4	0.920	0.919	0.612	1.000

3.2 *KMO and Bartlett sphericity tests*

Before using principal component analysis, the applicability of principal component analysis should be determined through the KMO and Bartlett sphericity test on the basis of determining the correlation between variables.

Through KMO and Bartlett sphericity statistical tests in Table 2, the collected variable data was used to obtain the KMO value of 0.775, which was larger than 0.7, indicating that it was very suitable. The Chi-square statistic of the Bartlett sphericity test was 53.672, and the significance was 0.001. The spherical hypothesis was rejected and passed the significance test. It shows that the original variables are not independent and the selected variables are suitable for principal component analysis.

Table 2. KMO and Bartlett tests.

KMO sampling suitability quantity		0.775
Bartlett's sphericity test	Last time I read card squared	53.672
	Degrees of freedom	6
	Significant	0.001

3.3 *Interpretation of total variance*

After determining the applicability of the principal component analysis method for each variable, SPSS software was used to analyze and calculate the characteristic values of each index and the cumulative contribution rate of the index, and determine the number of principal components of the building economic index. The specific test results are shown in Table 3.

Table 3. Explains the total variance.

	Initial eigenvalue			Extract the sum of squares of loads		
Composition	A total of	Percentage of variance	Cumulative %	A total of	Percentage of variance	Cumulative %
1	3.506	87.652	87.652	3.506	87.652	87.652
2	0.417	10.431	98.083			
3	0.073	1.826	99.909			
4	0.004	0.091	100.000			

According to the interpretation results of the total variance in Table 3, it can be found that the cumulative variance contribution rate of the first component only reaches 87.652%, but its characteristic root value is greater than 1. In general, it can cover the information of most of the original variables with less missing information, and it meets the principal component extraction standard, indicating that the principal component analysis has an ideal effect. Therefore, the first component is selected as the principal component of each index of building the economy and named Z_1.

3.4 *Component load matrix*

In order to highlight the representative variables of each component, SPSS software was used to extract the first component and obtain the component matrix, as shown in Table 4. Each load quantity listed in the component matrix represents the correlation coefficient between each principal component and corresponding variables, so the extraction of principal components can be better explained according to the coefficient in the table. The composition of the data in the table shows that 1 in the first, second, and third variable load coefficient for a larger proportion, and suggests that 1 covers the GDP composition, construction industry output, fixed assets investment information, comprehensive to reflect the impact of building economic development situation, including the variable load coefficient of the first largest in composition 1, It shows that the principal component is the main influencing factor.

Table 4. Load matrix and coefficient matrix of principal components and normalized independent variables.

Correlation	Component matrix	Component coefficient matrix
X1	0.911	0.528
X2	0.691	0.527
X3	0.704	0.496
X4	0.668	0.444

According to the matrix coefficients and standardized data, the principal component score function expression can be obtained (1):

$$Z_1 = 0.528X_1 + 0.527X_2 + 0.496X_3 + 0.444X_4 \tag{1}$$

Taking the variance contribution rate corresponding to each principal component as the weight 87, the comprehensive evaluation function can be obtained (2):

$$F = 0.877Z_1$$

$$\text{That is: } F = 0.463X_1 + 0.462X_2 + 0.435X_3 + 0.389X_4 \tag{2}$$

According to the expression of the principal component score above, the importance of influencing factors of each index on building economic development is ranked by the size of the standardized coefficient. The analysis results show that GDP and the total output value of the construction industry have the greatest impact on the development of the construction economy, and the least impact is the proportion of fixed asset investment and construction industry added value in GDP.

4 SUGGESTIONS FOR THE ANALYSIS OF INFLUENCING FACTORS OF CONSTRUCTION ECONOMIC DEVELOPMENT

4.1 *Improve the adaptability of the construction industry supply structure*

Economic activities cannot be separated from the relationship between supply and demand. Low supply capacity and excess low-end capacity in the construction industry conflict with people's high demand, resulting in the quality of products cannot be guaranteed, people's needs cannot be met, and the capacity structure of enterprises is unbalanced. This affects the quality of products and services as well as the improvement of the overall service level of the tertiary industry, making the agglomeration of the construction industry unable to develop further. Based on the unbalanced development of current supply and demand, the transformation of the construction economy has a long way to go.

4.2 *Improve the enterprise management system security system*

The development of a sound management system is the guide to the transformation of the construction economy. The management system is not only production management, but also includes human resources management, environmental management, and other levels. On the one hand, the transformation and development of the construction economy require the concept to precede, and the modernization of the management concept cannot be realized without the boost of the powerful joint force of professional management organizations and the establishment of the scientific and reasonable advanced system by construction enterprises. However, in reality, construction enterprises focus on production and schedule, only to complete the project task. The attention to the management of talents, the introduction of ideas, and the utilization of resources are lacking. On the other hand, the transformation and development of the construction economy mean that the components in the previous system that do not adapt to the modernization construction are constantly updated and the construction management system is constantly improved. We should pay attention to the management system of construction enterprises, find the weak point of enterprise management, and provide an important basis for the perfection of the system.

4.3 *Establish a scientific evaluation system*

It is attempted to establish a scientific evaluation system for the green building economy and establish evaluation index standards to effectively evaluate the impact of green building on the economy. At the same time, green building construction technology is improved. At present, China has made some achievements in green construction technology, such as foundation pit construction precipitation recycling technology, external wall self-insulation technology, paste external wall thermal insulation technology, and so on. But we still need to move on. We should reduce construction waste pollution and building energy consumption emissions and adopt new technologies to promote the development of green buildings and a green building economy.

4.4 *Implement the concept of sustainable development*

Importance should be attached to implementing the concept of sustainable development so that the sustainable development of the construction economy is deeply rooted in the people. The first is to promote a green building economy. Green building economy is also an important part of the sustainable development of the building economy, which is mainly to make the building economy and green economy develop together and implement the unity of benefits in all aspects of building economic benefits. It is necessary to implement the sustainable development of the construction economy and realize the importance of sustainable development of the construction economy.

5 CONCLUSION

This paper expounds on the development status of the construction economy, analyzes the existing shortcomings in practice, and uses principal component analysis to analyze the various factors that affect the development of the construction economy. The conclusion shows that GDP and the total output value of the construction industry have the greatest impact on the development of the construction economy. The proportion of fixed asset investment and added value of the construction industry in GDP are the least, and the strategy for the development of the construction economy is proposed from the aspects of the supply side, system guarantee system, evaluation system, and sustainable development concept, so as to promote the healthy and stable development of construction industry.

REFERENCES

Hossain U, S. Thomas N, Antwi-Afari P, et al. Circular economy and the construction industry: Existing trends, challenges and prospective framework for sustainable construction[J]. Renewable and Sustainable Energy Reviews, 2020130:109948.

Hu Xiaona. Application and Practice of green Economy concept in sustainable Development of Building Economy – Comment on Technical Economics of Building Engineering [J]. Industrial Construction, 201,51(02):214.

Liu Ye. Analysis of the Influencing Factors and Countermeasures of prefabricated concrete building cost based on PCA and FCE [D]. North China University of Water Resources and Electric Power,2019.

Zhang Longyu. Study on the Influencing Factors and Promotion Strategy of Prefabricated building Development in Anhui Province based on System Dynamics [D]. Hefei University of Technology,2018.

Zhou Tong. Research on Economic Evaluation of Green Building Based on Life Cycle [D]. Dalian University of Technology,2012.

Advances in Urban Engineering and Management Science – Khalil & Yang (Eds)
© 2023 The Author(s), ISBN 978-1-032-30426-7

Application of muddy lime soil flushing and backfill method in water conservancy and civil engineering

Yaping Zhu*
Construction and Administration Bureau of South-to-North Water Diversion Middle Route Project, Department of Soil and Water Conservation, Northwest A&F University Beijing, China

Lingyu Xiong
Construction and Administration Bureau of South-to-North Water Diversion Middle Route Project, Department of Water Conservancy, North China University of Water Resources and Electric Power Beijing, China

Xiaoxiao Ji
Construction and Administration Bureau of South-to-North Water Diversion Middle Route Project, Department of Computer Science, North China University of Water Resources and Electric Power Beijing, China

ABSTRACT: At present, there are many types of water conservancy and civil engineering projects. In addition to the frequent occurrence of heavy rains during the flood season, there are many rain and flood disasters. Various holes and pits often appear. China's water conservancy project is taken as an example of the Middle Route of the South-to-North Water Diversion Project, which is known as the "Human Tianhe". The main route of the South-to-North Water Transfer Project is 1432 km long. There are various buildings along the route, including 3,623 bridges, inverted siphons, gate stations, and bridges across the canal. Taking the Yellow Crossing Project as an example, the upper part of the stratum lithology of the excavated canal section on the south bank of the Crossing Yellow River is loess or silty. Loam, the middle and lower parts are loess-like silty loam, and its properties will change greatly after the accumulation of water, especially after the platform surface water, along the vertical joints or vertical micro-fissures, the rapid infiltration makes the loess saturated and disintegrated. The collapsible self-skeletal structure is destroyed, and due to gravity, it is easy to form horizontal and vertical penetrating sinkholes, sinkhole cavities, or partial collapse of the slope.

In order to solve this kind of infiltration or erosion channel defect treatment, and at the same time to be economical and practical, the Yellow Crossing Management Office has developed a muddy lime soil flushing and backfill method after long-term research and successfully applied it.

1 GENERAL SITUATION AND SOIL PROPERTIES OF THE MIDDLE ROUTE OF THE SOUTH-TO-NORTH WATER TRANSFER PROJECT CROSSING THE YELLOW RIVER

The middle route of the South-to-North Water Diversion Project is 1432 km long, with the Yellow River crossing project as an example. As an important water delivery structure and key project across the Yellow River, the Yellow River Crossing Project is a landmark and control project for the entire South-to-North Water Transfer Project. It is located in Xingyang City on the south bank of the Yellow River and Wen County on the north bank of the Yellow River, North Bank Open Channel, Yellow Tunnel, Import and Export Buildings, South Bank Backwater Buildings, North Bank Protective Dike, South Bank Licun Main Canal Aqueduct, North Bank New and Old Mang

*Corresponding Author: 1906595180@qq.com

 DOI 10.1201/9781003305026-61

River Inverted Siphon Project, North and South Zhangqiang Left Bank Inverted Siphon Project, Gubaizui Control and guidance engineering and other components. The channel project has a total length of 13.95 km and consists of open channels on the north and south banks. The open channel on the south bank is 4.63 km long, the bottom longitudinal slope is 1/8000, the bottom width of the trapezoidal section is 12.5 m, the top width is 53.00 m, the slope ratio is 1:2.25, the channel design water depth is 7 m, and all forms are excavated channels. The maximum digging depth is 40.77 m, and the side slope with an open channel elevation above 120.0 m is designed according to the shape of a wide horse track and a multi-level slope. The first level horse track is set for every 10.0 m height difference above the 120.0 m elevation, and the highest is the fourth level horse track, and the side slope is 1:0.7 and 1:1 two kinds.

Due to the special geographical location of the Yellow Crossing Project, especially the deep excavation section of the south bank, the runoff and seepage formed by the heavy rain scouring the slope during the flood season will aggravate the erosion and erosion of the slope, resulting in the gradual expansion of the cavity of the sinkhole. Under the continuous hydrodynamic action, the cavity of the soil will develop and expand downwards along the slope, which will aggravate the damage. For this reason, it is necessary to study the treatment methods of the excavated canal sections and the holes and collapses on the platforms of the South-to-North Water Transfer Project, so as to be fast, durable, and with the lowest cost. After a long-term research, the Yellow Crossing Administration Office has developed a muddy water lime soil flushing backfill method and related technical standards (Cheng & Su 2018).

At present, the research results of the midline are mainly concentrated in the deep excavation section of the platform at all levels and the high filling section of the entrance and exit wing wall section of the water conveyance building. The muddy water lime soil backfill method is a scientific and technological innovation method for repairing holes and collapsed pits obtained through long-term research. This article will discuss the application of the muddy water lime soil flushing backfill method in the actual operation and maintenance of various projects.

2 MUDDY LIME SOIL FLUSHING AND BACKFILL METHOD

The construction of muddy water lime soil flushing and backfill method is to mix water, cement, clay, fly ash, and other diluted mixtures uniformly formed in a certain proportion through a pipe to flow into the hole or collapse pit until it is full. After filling, the settlement is observed for 24 hours. If the sinking is obvious, grouting again, and the sinking is not obvious, you can use cement mortar or fine-stone concrete cover to keep the shape and elevation of the original design consistent.

2.1 *Terms and definitions:*

Muddy lime soil: It is formed by mixing a certain amount of cement and clay with water, that is, a diluted mixture formed by mixing water, cement, clay, and fly ash in a certain proportion. The function is to play the role of adhesion and deep backfill. After solidification, it not only plays the role of mud waterproofing but also is equivalent to low-grade concrete with a certain degree of hardness and strength (Bai, Xu, Cai).

Flushing and backfill: The diluted mixture is formed by mixing water, cement, clay, fly ash, etc. according to a certain proportion, flowing into the hole or collapsed pit through the pipe, which can backfill the deeper holes and the filling is relatively dense (Dong & Deng 2010).

2.2 *Technical standards and requirements*

The muddy lime soil flushing and backfill method use sieved soil and cement to be mixed evenly at 10:1, called cement soil. In order to increase the workability, in the beginning, you can also use fly ash and cement to mix well at 5:1, collectively called cement earth. Then the water and cement soil are mixed according to 2.15:1 in the mud processor and the flushing method is used to introduce them into the holes or collapse pits. Attention is paid to two details during construction. Loose rocks and other debris are removed before backfilling. There is no need to excavate the undisturbed soil. If there is a large amount of grouting at the bottom, grouting must be stopped, and cement mortar should be used to seal the grouting location. Blocking, after solidification, re-filling and backfill. Until it is filled, after filling, it is needed to settle for 24 hours and observed. If the sinking is obvious, it should be grouted again; if the sinking is not obvious, you can use the cement mortar cover to keep the shape and elevation of the original design consistent (Wu 2016). In the selection of materials, the plasticity index of the cohesive fine-grained soil is 12–18, the organic matter content is less than 5%, and the sulfate content is less than 0.8%. The cement is bagged ordinary Portland cement P.O 42.5, and the upper cover is made of natural or machine-made medium sand, with a fine solid modulus of 3.0 to 2.3 and a mud content of less than 5%.

2.3 *Construction requirements*

2.3.1 *Measurement and estimation:*

Before construction, it is necessary to measure and estimate how much earthworks, how much fly ash, how much cement, how much sand, and how many stones will be needed.

2.3.2 *Clean up the base surface:*

Before backfilling, it is needed to remove loose rocks, branches, plastic bags, and other debris. Do not excavate the original soil twice.

2.3.3 *Lime-soil mixing:*

A certain amount of cement, sieving soil, fly ash, etc., is prepared.

The sieved soil and cement are mixed according to the ratio of 10:1, manually mixed, and can be mixed on-site. If fly ash is used, the mining and cement should be mixed in a ratio of 5:1 and mixed manually.

2.3.4 *Mechanical mixing:*

It is required to prepare the mixer in advance (a horizontal mixer is recommended, which can be directly introduced into the pit or hole; if a drum mixer is used, it must be equipped with a special diversion tank for use), a generator (if there is no power supply nearby, it must be self-provided for power generation) Machine) and so on.

Water and cement soil are mixed according to 2.15:1 in a mud processor.

2.3.5 *Flushing and backfill:*

After mixing well, the flushing irrigation method is used to introduce into the hole or collapse pit.

2.3.6 *Plugging cover:*

After the initial muddy water and lime soil have solidified, they are refilled and backfilled until they are full. After filling, it is needed to settle for 24 hours and observed; if the sinking is obvious, it is grouted again; if the sinking is not obvious, you can use cement mortar or fine-stone concrete cover to keep the shape and elevation consistent with the original design (Gu, Zhao 2013).

2.4 *Scope of application*

This standard applies to the treatment of large pits or deeper holes in a small area in various water conservancy and civil engineering areas, such as the backfill of the wing walls on both sides of the building at the entrance and exit of the tunnel through the Yellow River; the top, middle, and bottom pits of the concrete drainage ditch linin or holes; concrete pavement and asphalt pavement collapse; mortar stone drainage ditch sidewalls and bottom collapse pits or holes; mortar stone slope protection collapses or holes; protective concrete lining slabs settlement collapse pits, etc.

2.5 *Safe production and civilized construction*

2.5.1 *Safety in production*

2.5.1.1. *Carry out the entrance and pre-work safety education; purchase accidental casualty group insurance for personnel; purchase qualified safety helmets, seat belts, reflective vests, masks, disinfectants, and other safety protection and epidemic prevention products.*

2.5.1.2. *Designate full-time security personnel to wear red armbands to work for safety monitoring; special operations personnel to work with certificates and report to the record.*

2.5.1.3. *Formulate safety measures for construction use of electricity (water) and report them to the record.*

2.5.1.4. *Do a good job in epidemic prevention, submit a roster of personnel, carry out body temperature measurements, and report to the daily health report of employees.*

2.5.1.5. *Ensure that safety production costs are invested in full and on time.*

2.5.2 *Civilized Construction*

2.5.2.1. *Develop project safety and civilized signs and close the construction area.*

2.5.2.2. *The layout of the site is reasonable, and the materials and tools are neatly stacked.*

2.5.2.3. *The production and domestic garbage are stacked in an orderly manner and cleared and transported in time.*

2.5.2.4. *Do a good job in the disposal of construction waste materials. It is strictly forbidden to pour into the river channel or privately landfill, and achieve "complete construction, site cleaning, and material cleaning".*

2.5.2.5. *Ensure that civilized production costs are invested in full and on time.*

2.6 *Implementation management flow chart*

Implementation management flow chart.

3 CASE ANALYSIS OF MUDDY WATER LIME SOIL FLUSHING AND BACKFILL METHOD

From 21:00 on July 18 to 17:00 on July 19, 2016, the rainfall in Wangcun Town of Xingyang City reached 132 mm, breaking the historical extreme value during the same period. The channels in the Yellow Crossing Administration Office also suffered different degrees of rain erosion. Especially in the deep excavation section of the south bank, there have been many holes and pits at all levels of the platform. The Yellow Crossing Management Office attached great importance to it and reported it to the Henan Branch. The Henan Branch commissioned the Yellow River Survey, Planning and Design Co., Ltd. to carry out a special design and add a drainage ditch vertically. It is required to add a drainage ditch every 30 m in the horizontal direction, which increases the drainage volume, enhances the drainage capacity, and reduces the formation of surface runoff. On January 24, 2017, the Henan Branch organized the bidding work according to the special treatment plan after the review by the Central Bureau. From March 3 to 17, 2017, after a bidding review, the Yellow River

Construction Engineering Group Co., Ltd. was determined to be the winning bidder for the water-damaged defect treatment project on the south bank slope of the Yellow River Crossing Project. After the completion of the construction, the collapse of the horse road at all levels was greatly reduced.

However, during heavy rainfall or long-term rainfall, a large amount of surface water infiltrates along the loess joints and cracks to cause the palaeosoil layer, which will still cause the soil layer on the top of the palaeosoil layer to be saturated and softened. Then a horizontally and longitudinally penetrating sinkhole or partial slope collapse is formed. Finally, the weight of the slope body is affected by heavy rainfall or long-term rainfall, and the weight increases sharply, resulting in large-scale damage to the slope. In addition, there are certain permeable channels inside the slope, which will see through the outer boundary of the washed fine fissures. Part of the initial rainfall seeps and part of it produces runoff on the slope. With the continuous erosion and seepage of the runoff, pit-like water gradually forms on the surface. When the depth of the water reaches a certain level, the infiltration volume will gradually increase, which will also Increase the scour and erosion of the soil in the pit, and then form a connected infiltration or erosion channel, and gradually expand the cavity of the sinkhole. Under the long-term continuous hydrodynamic action, the cavity will develop and expand down the slope, which will aggravate the damage. For example, there were two collapses near the upstream and downstream of the Aspen Production Bridge (stake number 2+400) before the flood season in July 2019, and one near the downstream of Provincial Highway 314 Bridge (stake number 3+700) during the flood season in August 2020. The collapsed pit, as well as a collapsed pit near the exit platform of the control gate in the September 2020 flood season, is about 0.5 m in length, 0.3 m in width, and 4 m in depth for larger holes that are filled with muddy water and lime soil. According to statistics, if the traditional stepped excavation and re-masonry method are used for the defect treatment of the former poplar, the cost per location is about 100,000 yuan, but the muddy water lime soil flushing and backfill method are used for disposal, and each location requires only 5,000 yuan, which is equivalent to each location. 95,000 yuan was saved, and a total of 190,000 yuan was saved in the two places; if the defect treatment of the collapsed pit the downstream of the 314 Provincial Highway Cross Canal Bridge is to adopt the traditional stepped excavation and re-masonry method, the cost of this place will be at least about 60,000 yuan, but muddy water is used. Lime-soil flushing and backfill method requires 4,000 yuan in this place, which saves 56,000 yuan; the defect treatment of the pit near the asphalt pavement of the platform in front of the exit control gate on the north bank, if the traditional stepped excavation and re-masonry method is used, the cost of this place shall be at least About 50,000 yuan, but the muddy water lime soil flushing and backfill method is used for disposal, the site needs 4,000 yuan, which saves 46,000 yuan. The three projects have saved a total of 292,000 yuan. Details are shown in the real figure (combination) 1–4 and the economic comparison analysis table 1 for the treatment of defects in the larger collapsed pit (deeper hole) of the Yellow Crossing Project.

Figure (combination) 1. A close-up view of the collapsed area of the 130 m platform and the location of the hole at the exit platform of the control gate.

Figure (combination) 2. 130 m platform position and control gate exit platform position machinery and material preparation.

Figure (combination) 3. Manual preliminary mixing, mechanical mixing, flushing, and backfill according to the proportion.

Figure (combination) 4. The pit collapse at the 130 m platform of the 314 bridge and the hole defects at the exit platform of the control gate have been processed.

Figure (combination) 4. Continued.

Table 1. Defect treatment of larger collapsed pit (deeper hole) in the Yellow River crossing project. Economic comparison analysis table for traditional excavation and backfill methods and flushing and backfill methods.

					Economic comparison		
Serial number	project	Technical comparison of defect handling methods	unit	quantity	Traditional method to estimate the total price (Yuan)	Estimated total price of irrigation method (Yuan)	Remark
1	On August 13, 2020, it was discovered that the large collapse (deeper hole) defect treatment of the 130 m platform on the left bank of the downstream of the 314 bridge	1. Remove part of the mortar and masonry 2. Mechanical excavation, dig to a certain depth, find the original soil, and backfill in layers; 3. Or manual excavation, dig into steps, backfill in layers; 4. Mortar masonry repair.	Place	1	60000		Excavation and backfill
2	On August 13, 2020, it was discovered that the large collapse (deeper hole) defect treatment of the 130 m platform on the left bank of the downstream of the 314 bridge	1. Clean up surface rubbish; 2. Muddy lime soil flushing and backfill; 3. After reaching the design elevation, use cement mortar or fine stone concrete cover.	Place	1		4000	Slurry lime soil flushing and backfill method
3	On September 22, 2020, it was discovered that the large collapsed pit (deeper hole) defect treatment on the right bank platform of the north bank exit was found	1. Remove part of the asphalt pavement 2. Mechanical excavation, dig to a certain depth, find the original soil, and backfill in layers; 3. Or manual excavation, dig into steps, backfill in layers; 4. Asphalt surface repair.	Place	1	50000		Excavation and backfill
4	On September 22, 2020, it was discovered that the large collapsed pit (deeper hole) defect treatment on the right bank platform of the north bank exit was found	1. Clean up surface rubbish; 2. Muddy lime soil flushing and backfill; 3. After reaching the design elevation, use bitumen cold to patch the cover to restore the original shape.	Place	1		4000	Slurry lime soil flushing and backfill method

(*Continued*)

Table 1. (Continued).

Serial number	project	Technical comparison of defect handling methods	unit	quantity	Economic comparison Traditional method to estimate the total price (Yuan)	 Estimated total price of irrigation method (Yuan)	Remark
5	In the flood season of July 2019, two collapse pits were discovered near the upstream and downstream of the former Baiyang Production Bridge (stake number 2+400), which have formed an infiltration channel with the outer slope, and the defects were dealt with	1. Remove part of the mortar and masonry 2. Mechanical excavation, dig to a certain depth, find the original soil, and backfill in layers; 3. Or manual excavation, dig into steps, backfill in layers; 4. Mortar masonry repair.	Place	1	200000		Excavation and backfill
6	In the flood season of July 2019, two collapse pits were discovered near the upstream and downstream of the former Baiyang Production Bridge (stake number 2+400), which have formed an infiltration channel with the outer slope, and the defects were dealt with	1. Clean up surface rubbish; 2. Muddy lime soil flushing and backfill; 3. After reaching the design elevation, use cement mortar or fine stone concrete cover.	Place	1		10000	Slurry lime soil flushing and backfill method
Total		Description of economic benefits: In July 2019, the defect treatment of the former Baiyang Department saved 190,000 yuan; in August 2020, the defect treatment of the 314 bridge saved 56,000 yuan; in September 2020, the defect treatment of the export control gate platform saved 46,000 yuan. The total savings of the above three projects was 292,000 yuan.			310000	18000	

4 CONCLUSION

The construction of muddy water lime soil flushing and backfill method is to mix a certain proportion of water, cement, clay, fly ash, etc., to form a diluted mixture that is uniformly stirred and flows through a pipe into the hole or collapsed pit until it is filled. After filling, it should be settled for 24 hours and observedd; if the sinking is obvious, it is grouted again; if the sinking is not obvious, you can use cement mortar or fine-stone concrete cover to keep the shape and elevation of the original design consistent. The mud-water lime-soil irrigation method is obviously superior to the traditional stepped excavation backfill masonry method in terms of defect treatment technology and economic benefits. The traditional method requires large-area excavation and requires clay backfill or modified soil layered compaction and backfill. The work is complicated, but the innovative treatment method only needs to clean up surface garbage and loose blocks, and only muddy lime soil is required on the material for mixing (WANG 2000). The innovation is that the traditional stepped excavation and backfill masonry method is no longer used, that is, all excavated and rebuilt. Instead, the muddy water lime soil flushing and backfill method is used for treatment, which saves manpower, material, and financial resources, is efficient and practical, and greatly improves the construction safety. Advantages: Less is affected by the rainy season weather, reducing procedures, saving money, and non-recurring.

At the same time, the location where the muddy water lime soil flushing and filling method is used has the following characteristics: (1) The main structure has not been damaged; that is, the protective shell (masonry, asphalt pavement) is not completely damaged; (2) The area is large (more than 2 m^2) or the aperture Smaller, but the depth is deeper, more than 1 m; (3) The appearance damage is heavier, if it is not handled in time, new and greater damage will occur.

There are the following differences from traditional excavation treatment methods:

- The excavation methods are different. The traditional method requires a large area of excavation and a larger working area. This work method does not require a large amount of excavation, only the surface garbage and loose blocks need to be cleaned up.
- The backfill materials are different. The traditional method requires clay backfill or modified soil layered compaction and backfill; this working method only needs to be mixed in proportion to muddy water and lime soil (GAO 2002).
- The degree of difficulty is different, and the traditional way of work is complicated; This working method is simple to construct and easy to operate.
- Different economic benefits, the traditional way of investment is large, which is a waste of funds; the investment is small and saves money.

In short, the muddy water lime soil flushing and backfill method is used to deal with the defects of larger collapse pits or deeper holes. This method and technical standard have been optimized and applied year by year to achieve the goal of improving work efficiency, ensuring a low rate of anti-repair, and saving costs. At the same time, it has been adopted and applied by the Construction and Management Bureau of the Middle Route of the South-to-North Water Transfer Project and is included in the civil engineering maintenance standardization list. It is promoted and applied across the board. It also allows the maintenance management specialists to have more energy to engage in other on-site work, as well as for various water conservancy and civil engineering projects such as the South-to-North Water Transfer. The safe operation and technical decision-making of the Yellow Project provide strong support.

REFERENCES

Bai Hanyang, Xu Lei, Cai Peng. Application of Lime Soil in Fertilizer Trench Backfill[J]. Chinese Science and Technology Journal Database (Abstract Edition) Engineering Technology: 00051–00052.

Cheng Dehu, Su Xia. Technical progress and demand for the middle route of the South-to-North Water Transfer Project[J]. China Water Resources,2018(10):24–27.

Dong Bo, Deng Changqi. Discussion on the technology and regulations of backfill grouting and consolidation grouting[J]. Sichuan Hydropower, 2010, 29 (s1): 106.

GAO Zhigang, Surface Grouting water Plugging Aggregate Technology [J]. Coal Mining, 2002.

Gu Zhiqiang, Zhao Chong. Analysis of backfill grouting construction plan for a water conservancy project[J]. Heilongjiang Water Conservancy Science and Technology, 2013 (1): 211~212.

WANG Guoji, Theory and Practice of Grouting Technology [M]. Xuzhou: China University of Mining and Technology Press, 2000.

Wu Xinglun.Analysis of grouting construction technology for backfill of hydraulic engineering tunnels[J]. Sichuan Cement, 2016(10): 163.

Advances in Urban Engineering and Management Science – Khalil & Yang (Eds)
© 2023 The Author(s), ISBN 978-1-032-30426-7

The key points of technical management in the construction of high-rise shear wall buildings

Haoxin Jia*
Civil Engineering, Changan University, Xi'an, China

ABSTRACT: With the accelerated pace of urban construction, shear wall structure is widely used in high-rise residential buildings. This paper combines theoretical knowledge and the author's many years of construction experience, for the study and application of shear walls in super tall buildings, a more comprehensive review of the development of shear walls at home and abroad, from three aspects of reinforcement engineering, formwork engineering, and concrete engineering introduced the main points of construction technology management of high-rise shear wall structures. Research shows that high-rise shear wall structure is more and more widely used in the current building system, and its corresponding construction technology is also maturing. The construction process, construction technology, and management technology summarized in this paper have important theoretical and practical significance for the construction technology management of high-rise shear wall buildings.

1 INTRODUCTION

With the continuous development of the construction industry, shear wall structure has increasingly become the focus of attention. Based on the characteristics of the shear wall structure itself, it is highly favored by construction workers in construction projects. The so-called shear wall structure mainly uses reinforced concrete as a beam and column structure, and this form plays a certain breakthrough to the original frame. The shear wall structure has excellent structural characteristics of high stability and large load-bearing capacity, and the shear wall itself has a unique resistance to various external forces, both longitudinal and transverse. Chity L (Chity 1974) proposed the continuous grid method for the elastic analysis of double-limb shear walls as early as the 1940s, Roseman R(Roseman,1964) further investigated this theory in 1964, and Coull A (Coull 1967) performed the elastic analysis of double limb shear walls in 1967, and these results provided the necessary basis for the elastic analysis of early shear wall structures.

Paulay T (Paulay 1997) carried out tests on 7-story double-limb shear walls, Sozen M A (Sozen 1975) completed shaking table tests on open-hole shear wall models, and Salonikios T N (Salonikios 2002) Salonikios carried out experimental studies on low shear walls with diagonal reinforcement and medium-high shear walls. Fang (Fang 1985), a scholar at Tsinghua University, did tests on shear wall specimens under repeated loading. The results showed that: the shear walls with an aspect ratio greater than 2 were bending damage with good ductility; the shear walls with an aspect ratio less than 2 were shear damage with poor ductility, and the bearing capacity decreased significantly after the peak load; the vertical load had a great influence on the strength and deformation of shear walls, and the bearing capacity increased with the increase of axial force, the stiffness increased accordingly, and the ductility decreased. As the axial force increases, the bearing capacity increases, the stiffness increases, and the ductility decreases. The research results played a great role in the preparation of relevant structural design codes in China. Unlike other major cities on the seismically active west coast of North America, structural steel systems and reinforced concrete moment-resisting frame systems are rarely used in Vancouver. Almost all of the high-rise buildings in Vancouver are

*Corresponding Author: 872983233@qq.com

DOI 10.1201/9781003305026-62

concrete shear wall buildings (Adebar 2017). It is evident that shear walls are also superior in the field of seismic resistance.

An in-depth analysis of the shear wall structure shows that the shear wall structure is laminated. It is not only more rigid but also if the structure itself is severely damaged, it will not have a serious impact on the whole building structure. Therefore, the use of shear wall structure in super high-rise buildings is the future development trend of architecture. The shear wall structure is also receiving more and more attention from experts and scholars. Therefore, the construction technology of high-rise shear walls deserves to be studied and explored in depth. Huo et al (Huo 2021) used external shear wall scaffolding as a new technology new form of shear wall structure in the assembly building. Compared with the traditional building perimeter scaffolding protection network, the external hanging shear wall scaffolding has the advantages of construction safety, easy operation, and aesthetics. It has become an important part of the assembled building system, and its construction technology has an important impact on the safety of the building and structure. Zhao et al. (Zhao 2021) reinforced the core shear wall and connecting beam with increased reinforcement and improved the concrete strength grade through the finite element analysis software ANSYS simulation analysis. At the same time, the steel beam of the boom tower crane was strengthened to reduce the load carried by the shear wall, and then the steel beam was pushed back from the bottom with seamless steel pipes and jacks to solve the problem of insufficient bearing capacity of the ultra-thin shear wall of the high-speed tower crane. The traditional ultra-thin shear wall construction of supertall proposed a new technical solution.

In this paper, the advantages and shortcomings of traditional construction technology are studied by introducing the construction process and construction technology of high-rise shear walls to provide a reference direction for the development direction of new technology. By studying the key points of steel structure engineering and concrete construction technology management, the construction and management are controlled globally from both technical and management levels. This paper discusses the key technical issues that need to be paid attention to in the design of super high-rise buildings and gives corresponding examples with engineering applications, in order to provide useful guidance for the construction of modern high-rise shear wall buildings.

2 TECHNICAL ANALYSIS OF THE CONSTRUCTION PROCESS OF HIGH-RISE SHEAR WALL

2.1 *Carefully review the architectural design drawings of the design institute*

In order to better complete the construction project, the construction technicians should carefully check the architectural design drawings and have an understanding of the detailed parameters of the overall layout of the whole project, so that the main architectural information can be implemented correctly and unnecessary mistakes can be avoided in the construction of the building. The technicians should compare the design institute drawings with the on-site construction drawings, because the designers of the design institute design the drawings according to theoretical knowledge, and it is difficult to avoid the appearance of inappropriate design parts during the construction of the project. Therefore, in the process of construction drawing design, designers and constructors need to communicate with each other efficiently, discuss possible problems in construction drawings together, and adopt scientific and effective ways to solve them.

2.2 *Do a good job of technical delivery*

The construction technology handover work is also an important work content that the engineering construction staff needs to pay high attention to. It can be seen from the shear wall construction project that the construction handover work is more difficult, and in the process of construction, the staff should pay more attention to the relevant issues, do it carefully, and cautiously to ensure the quality of construction and promote the efficient implementation of shear wall projects.

3 STEEL ENGINEERING CONSTRUCTION TECHNOLOGY MANAGEMENT POINTS

The amount of reinforcement in the concrete shear wall structure is large, usually for the diameter of less than X12 reinforcement joint using tied connections, while for the diameter of more than X14 reinforcement using welded joints or mechanical connections.

3.1 *Key points of reinforcement tying connection*

The wall uses the steel bar is slim, the upper and lower steel bar is generally used to tie the connection. When pouring concrete, the steel mesh is easy to occur to the internal or external displacement. In order to prevent the occurrence of this phenomenon, the full-length horizontal bar can be spot-welded in the overlap range of the vertical bar to prevent the displacement of the bar due to casting and vibrating.

When the floor reinforcement is bundled, enough pads should be placed. It is better not to use mortar pads, but to use special rotating plastic protective layer pads to ensure that the reinforcement has enough thickness and spacing of the protective layer. When laying wire pipes on the floor, it is needed to design its direction, to prevent 2 or 3 layers of wire pipes from stacked on top of each other, so that the floor in the reinforcing steel online under the spacing is too large, which leads to the floor concrete pouring in the wire pipe overlap at the local super-thickness, which in turn leads to the entire floor leveling and design elevation does not match, to the ground decoration and other next process brings a series of problems.

3.2 *Key points of steel welding joints*

Steel connection using welding can be applied to the more advanced domestic DGQ electric rebar pneumatic welding machine, the technology, and arc lap welding, compared with the advantages of high efficiency, saving steel.

Pneumatic welding process: (1) top pressure will be pneumatic welding clamp on the two bars to be welded, butt weld head, to ensure that the two sections of the butt weld in a straight line, leaving a gap of 2 mm in the middle; (2) it is heated with a pneumatic welding gun (oxygen, acetylene flame) to heat the rebar butt weld head 2 times the diameter range of the rebar, until a thin layer of melting layer on the surface of the rebar (temperature at 1200°C a 1300°C) and the joint gap disappears, swinging the welding gun from the joint is greater than 114 times the diameter of the rebar and lasts 3 a 5S; (3) crimping open the pressure feed valve, according to different rebar diameter pressurized to different pressure gauge readings; (4) it falls into decompression when the bulge size of the pressure joint reaches the standard (the maximum diameter of the bulge is 1.4~1.6 times the diameter of the steel bar, and the length is 1.2~1.4 times the diameter of the steel bar) and the joint red disappears, withdraw the pressure and remove the fixture.

After economic comparison, with pneumatic welding than with arc lap welding each joint savings about 3 RMB, each layer is completed 15 days earlier than the arc welding, showing good economic benefits.

At the same time to do: the operator is licensed to work; rebar welding parts by cleaning, the surface is smooth, clean, with no oil, impurities, etc.; welded joints in batches for appearance inspection and mechanical properties test.

3.3 *Reinforcement scheme of the connecting beam*

Due to the demand for building use function, it is necessary to shear in the core barrel. The window and door openings are set in the wall, and the part between the upper and lower openings forms a connecting beam. The span and height of the connecting beam have a great influence on the lateral stiffness of the shear wall, and the strength, ductility, and energy dissipation capacity of the connecting beam are very important to the seismic performance of the shear wall structure. In the seismic design of reinforced concrete shear walls, the principle of "strong wall and weak beam, strong shear and weak bending" should be followed to ensure that the connecting beam yields before the wall limbs and the shear bearing capacity of the connecting beam is higher than its bending capacity so that the structure has sufficient ductility. floor, its connecting beam span height ratio is often smaller. So far, many scholars have proposed a variety of different reinforcement schemes for small span-height ratios of connecting beams, as shown in Figure 1.

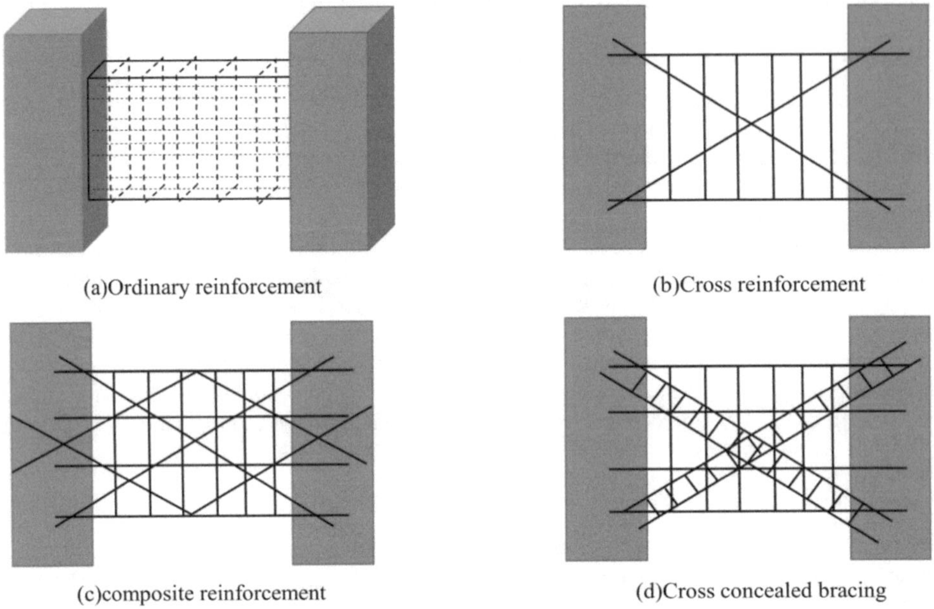

Figure 1. Reinforcement method of the connecting beam.

Tissios T P(Tissios,1996) compared the performance of beams with different types of reinforcement and showed that the shear resistance of specimens with bent-up reinforcement was better than that of conventional reinforcement, and the performance of beams with crossed concealed braces was significantly improved. Galan L and Vignoli A (Galan,2000) showed that the ductility of diamond-shaped reinforcement is the best, which can effectively prevent the sudden diagonal splitting shear failure, while the diagonal cross reinforcement is slightly lower. The energy dissipation capacity of diamond-shaped reinforcement is also better than that of crossed diagonal reinforcement.

4 CONSTRUCTION OF CONCRETE WORKS

In the construction of high-rise shear wall buildings, the amount of concrete placement is very large, the quality of concrete placement is significant to the whole project, and the requirements for construction technology management are also higher.

During the construction process, we should strictly follow the relevant operation specifications and process standards, strengthen the elimination of quality defects of concrete works, reduce the construction joints, treat the construction joints, and pay attention to concrete maintenance. Meanwhile, extra attention should be paid to the following problems to nip some quality defects in the bud.

(1) A detailed investigation and analysis should be done before the concrete construction. According to the degree of wall thickness, the types of vibrating bar and concrete varieties are chosen for rebar spacing; according to the construction requirements, its slump should be strict controlled to ensure the quality of concrete placement and design strength.

(2) If the pumping concrete is used in the wall concrete pouring, in the process of moving the pump pipe on the floor, do not dump the concrete in the pump pipe randomly on the floor, but to dump in the pouring wall, in order to prevent the formation of leakage phenomenon here when pouring the floor.
(3) When using the wall and floor slab as a whole for the pouring program, it is required to pour the concrete wall first, and after the wall concrete is initially settled (about 1~1.5 h), then it is needed to pour the floor slab or pour it at the same time, but the concrete at the intersection of the wall and floor slab is not initially set before the second pounding to prevent the concrete from cracking here due to the settlement of the wall.

5 CONCLUSION

Through an overview of the development process of shear walls, this paper discusses the process of high-rise shear walls, steel construction technology, and the multifaceted technical means of concrete construction technology. Based on these results and discussions, the following conclusions are drawn:

(1) With the wide application of shear wall structure, more and more problems will be exposed in the construction. Therefore, it is necessary to strengthen the optimization of the construction process of high-rise shear wall buildings in order to ensure smooth construction, save capital investment and reduce the project cost.
(2) In the construction of high-rise shear wall buildings, construction technology management is an important part of the construction management process, and the quality, cost, safety, and construction progress are closely related to technology management. In the process of construction operation, it is necessary to strictly control the quality and select the appropriate reinforcement method and welding process.
(3) There are many technical points in the construction of shear wall buildings of high-rise buildings. To ensure the quality of the construction of shear wall structures, various departments and professionals need to do a good job of cooperation, and only in this way can construction companies complete the construction within the specified period. Usually, the personnel concerned collect information after completion, which will lead to omissions and is not conducive to guiding later maintenance. Moreover, the information data recorded by the person concerned are also needed when the final account is made, and for this reason, the person concerned must do a good job of recording.
(4) In the future, shear wall structures in high-rise buildings will certainly be applied more and more, and more engineers and scholars will focus their attention on shear wall structure construction techniques, construction technologies, and management schemes. The integration of shear wall structure and assembled buildings applied to super high-rise buildings is also a mainstream trend today and even in the future. The construction technology and management scheme are also maturing in the process of continuous use and development. In the future, the shear wall structure is bound to shine in more fields.

REFERENCES

Adebar P, DeVall R, Mutrie J (2017) Evolution of High-rise Buildings in Vancouver, Canada. Structural Engineering International. 27(1):7–14.
Chity L (1974) On the cantilever composed of a series of parallel beams interconnected by cross-bars. PhilMag. 1947.
Coull A (1967) Analysis of coupled shear walls of variable thickness. Building Science,1967, 2(2) :181–188.
Fang E (1985) Influence of connecting beam on elastoplastic properties of twin shear wall. Earthquake Resistant Engineering and Retrofitting. 7(4) :5–10.(In Chinese)
Galan L, Vignoli A (2000) Seismic behavior of short coupling beams with different reinforcement layouts[J]. ACI Structural. 97(6) :876–885.

Huo T, Ye C, Zhang X (2021) Research on Construction Technology of External Hanging Scaffold for Prefabricated Concrete Shear Wall. Journal of Physics: Conference Series. 1952(3)

Paulay T (1997) Ductile behavior of shear walls subjected to reversed cyclic loading / /6th World Conference on Earthquake, India,1977.

Roseman R (1964) Approximate analysis of shear walls subjected to lateral loads. ACI Journal, 1964, 61(6):717–734.

Salonikios T N (2002) Shear strength and deformation patterns of R/C walls with aspect ratio 1. 0 and 1. 5 designed to Eurocode 8 (EC8). Engineering Structures.(24) :39–49.

Sozen M A (1975) Behavior of ten-story RC walls subjected to earthquake motions. The USA, the University of Lllinois at Urbana-Champaign.

Tissios T P (1996) On the behavior and ductility of reinforced concrete coupling beams of shear walls. ACI Structural Journal. 93(6) :711–720.

Zhao Ziwu, Li Shimin, Yu Ziqiang, et al (2021) Construction technology of super high rise ultra-thin shear wall bearing heavy tonnage tower crane under energy-saving and emission reduction[J]. IOP Conference Series: Earth and Environmental Science.804(2)

Advances in Urban Engineering and Management Science – Khalil & Yang (Eds)
© 2023 The Author(s), ISBN 978-1-032-30426-7

Study on technology of gangue backfilling and reclamation for construction land

Guang Sun*
China Coal Science and Technology Ecological Environment Technology Co., Ltd., Beijing, China

ABSTRACT: As a companion organism in the process of coal mining, coal gangue has been widely used in many fields, such as construction and the chemical industry, for many years. Using it as a backfill material for building foundations can not only reduce the environmental pollution caused by open-air stacking but also reduce the cost of gangue disposal. Under the premise of reasonable analysis of the feasibility of building land reclamation with gangue backfilling, this paper comprehensively analyzes the key technology of gangue backfilling and reclamation for construction land from multiple perspectives, such as planning site selection, design points, backfill elevation, and construction methods. Finally, the effect is introduced with an engineering example. The research results can provide references for related research and engineering design.

1 INTRODUCTION

Coal gangue backfilling and reclamation has been a research hot spot in various countries for many years. Since the 1970s, the UK has carried out a large number of experiments and researches on coal gangue backfilling to build highways, parking lots, trading yards, and coal preparation plants. Based on the experiment, the corresponding coal gangue backfill building reclamation standards were formulated. Countries such as the Federal Republic of Germany, the former Soviet Union, France, Poland, and the Netherlands have also done a lot of work in this area and achieved good results. Since the 1980s, many places in China have used coal gangue to backfill the collapsed area for construction reclamation, and accumulated a lot of successful experience (Li et al. 2008; Tang 2011; Wang et al. 2009).

The use of coal gangue for backfilling and mining subsidence as a building foundation has been applied in some parts of our country. Research on the physical and mechanical properties of coal gangue, engineering properties, backfill treatment technology, and construction technology has achieved more results (Liu et al. 2011; Xu et al. 2011; Zhang & Wang 2013), but there are still some deficiencies in in-depth theoretical research and specific implementation details. On the basis of summarizing the existing theories and technologies, this article conducts an in-depth analysis of the feasibility and key technologies of backfilling construction land with gangue in the coal mining subsidence area.

2 FEASIBILITY ANALYSIS OF BACKFILLING AND RECLAIMING CONSTRUCTION LAND

2.1 *Feasibility analysis of gangue backfill foundation construction*

China has conducted many field tests on coal gangue in engineering applications and accumulated a large amount of engineering experience and experimental data in the mining areas of Huaibei, Kailuan, Xuzhou, Tiefa, etc., and based on the same the research results have been vigorously

*Corresponding Author: 962314363@qq.com

DOI 10.1201/9781003305026-63

promoted. Especially in recent years, the rapid development of infrastructure has increased the use of coal gangue in roads, railways, and industrial foundations. Engineering practice has proved that it is fully feasible to select suitable coal gangue to backfill roads, railways, and building foundations under reasonable construction. In addition, backfill reclamation projects have been widely carried out in various mining areas across the country in recent years, and many successful experiences have been accumulated for reference. There are no problems in design, construction, surface water, and groundwater treatment, and the technology is relatively complete.

According to results of laboratory experiments on coal gangue, coal gangue generally has the following characteristics: (1) Coal gangue is mainly composed of sandstone, sandy phyllite and phyllite, and its main chemical composition is SiO_2; coal gangue has good gradation; (2) the indoor compaction test shows that compared with general sandy clay, coal gangue is easier to be compacted; (3) according to the indoor coal gangue compression test, the gangue is within the load range of 200 kpa and the compression modulus is 28.7 Mpa, indicating that the coal gangue when is close to or reaches the maximum bulk density after compaction, exhibits low compressibility; (4) coal gangue is prone to weathering and deliquescence when directly exposed to the atmosphere, and when the gangue is buried below the underground freezing line, weathering will hardly take place. At the same time, the backfilled coal gangue is compacted and the surface is covered with soil, which can effectively prevent the weathering and deliquescence of the gangue; (5) water immersion has a certain influence on the compression characteristics of the gangue, but has little effect on the compacted backfill foundation. The settlement is small, the settlement is uniform, and the building will not be damaged by the gangue soaking in water; (6) according to the indoor load test, when the dry bulk density of the gangue after compaction reaches more than 95% of the maximum dry bulk density, its gangue compacted foundation bearing capacity will reach 236 kpa or more, meeting the requirements of general civil building foundations; (7) comprehensive coal gangue indoor tests show that coal gangue has good compactness and low compressibility, which is ideal backfill materials for foundation works.

2.2 *Bearing capacity of coal gangue foundation*

The bearing capacity of the foundation is the key data in the design of the building foundation. The bearing capacity of the coal gangue foundation is suitable to be determined by static load test and heavy dynamic cone penetration test, and its deformation modulus can also be determined at the same time. Domestic and foreign data have confirmed that after the gangue foundation is treated, a relatively high bearing capacity can be achieved. The allowable bearing capacity and deformation modulus of some gangue backfill foundations at home and abroad are shown in the table below.

Table 1. Allowable bearing capacity and deformation modulus of some coal gangue backfill foundations at home and abroad.

Project		Allowable bearing capacity (kPa)	Deformation modulus (MPa)
Foreign	Uncompacted gangue foundation	150	
	Gangue foundation after vibrating pressure	250~350	>45
	Gangue foundation after rolling	200~300	>45
Domestic	Layered vibration compaction of gangue foundation in Huaibei	240~320	>45
	Layered vibration compaction gangue foundation in Xuzhou	220~320	>45
	Layered vibration compaction of gangue foundation in Northern Anhui	180~260	>45
	Layered vibration compaction of gangue foundation in Tiefa	250~320	>45

2.3 *Analysis of the impact of coal gangue backfilling and reclamation on the environment*

Coal gangue contains certain harmful ingredients. As a building foundation backfill material, the first thing to know is whether coal gangue harms the living environment, including the air, soil, and water quality around the living area, and the radiation of coal gangue is closely related to the health of residents. In recent years, many mining areas in our country have made useful attempts in this regard and accumulated successful experience.

By analyzing the impact of coal gangue backfilling on water, soil, atmosphere, and its radioactivity on the human settlement environment, it can be known that under the premise of effective early treatment (such as anti-seepage, spontaneous combustion, etc.), coal gangue backfilling construction land will affect the backfill area. The water, soil, and atmosphere showed no obvious negative effects.

3 ANALYSIS OF KEY TECHNOLOGIES

3.1 *Principles of site selection and planning for reclamation areas*

In site selection, the principle of no or less arable land shall be observed, and the subsided land of coal mining shall be fully utilized to reclaim construction land. When selecting a site, the construction land for reclamation needs to be arranged in a subsidence stable area to avoid the impact of mining subsidence on the village buildings (Duan et al. 2008; He 1991; Li 2014). The new village site is chosen as far as possible in the area with small subsidence and higher terrain. The purpose is to reduce the amount of backfilling and reduce the project investment. Second, starting from the principle of facilitating construction, the backfilling area is divided into several districts, and the backfilling of each district is considered as a whole engineering, and the technological process is designed that is conducive to the construction.

3.2 *Design of backfill reclamation project*

3.2.1 *Design points*

The backfill area is used for village relocation. To meet the requirements of the construction foundation, the water area, steep ridges, and cultivated soil backfilled in the area are treated with different treatment techniques according to the actual situation to ensure that the foundation after the backfill treatment does not leave any hidden dangers. Room security minimizes the secondary handling of earthwork. The artificial backfill boundary is likely to cause uneven settlement and damage to the building. The backfill project adopts the partition construction, and the partition boundary must be handled properly to prevent the backfill boundary from forming a weak zone of backfill and rolling.

3.2.2 *Determination of elevation*

The backfill elevation is an important technical and economic indicator. The determination of the backfill elevation is mainly to consider the surrounding surface elevation and drainage in the rainy season; it is in harmony with the surrounding topography, and drainage is convenient in the rainy season, and there is no accumulation of water in the village; for the fish pond, it is necessary to ensure that water level is not less than 1.5 m throughout the year.

3.2.3 *Construction method*

The division and layering of "backfilling, leveling, compaction, inspection, and acceptance" is a single cycle, striving to form a virtuous cycle of flow operation and optimized construction within one or several divisions. The thickness of layered filling is generally determined based on the comprehensive consideration of backfill materials and compaction machinery. The layer thickness is controlled by level measurement. The compaction is generally determined by an on-site rolling test. During construction, the number of compaction passes is used to control the final compaction. The number of passes is subject to the records of a dedicated person.

The layered thickness of gangue is generally 0.4~0.5 m, which can be smaller than this layer thickness, but it cannot be larger than the upper limit of the abovementioned layering; when layered backfilling, in principle, only one layer difference between adjacent partitions is required, and the boundary should be set to 1:4 transitional gentle slope; when a subarea has the condition to be backfilled to the design elevation first, and the backfilling work of its adjacent subarea is limited, the subarea backfill boundary should be specially treated, such as leaving a transition zone.

3.2.4 *Treatment method of coal gangue foundation*

Many documents at home and abroad prove that the natural backfill of gangue that has not been compacted has a large amount of sedimentation. Buildings built on the untreated gangue backfill layer will be damaged due to uneven settlement. Therefore, the natural backfill of gangue passed through compaction treatment can build buildings on it. At present, the treatment methods suitable for large-area gangue foundations are the right dynamic compaction method, vibration compaction method, and grouting method. The dynamic compaction method is very effective for the one-time backfilling of the thick gangue foundation. The vibration pressure method is economical and reasonable for the treatment of the large area layered gangue cushion, and the compaction effect is ideal. When the grouting method is applied in a large area, the cost is high, the speed is slow, and the encryption effect is poor. It can only be used when the dynamic compaction and vibration pressure method cannot be implemented.

Vibration compaction is a method of using vibration compaction machinery to apply vibration and impact force on the surface of the foundation to compact shallow loose soil with vibration. The excitation force can be twice the weight of the fuselage. The vibration compaction method is suitable for the treatment of layered backfill gangue foundations. The vibration effect of the gangue foundation depends on the engineering properties of the gangue, particle gradation, layer thickness, vibration pressure impact energy of the vibration press, the number of vibration pressures, etc.

The particle gradation of gangue has a greater influence on the foundation treatment effect. According to foreign data, the uneven coefficient of gangue is doubled, and the compaction effect can be increased by 1 to 1.5 times. Therefore, adding some fine-grained soil such as fly ash when backfilling the gangue can greatly improve the compactness of the gangue foundation and ensure that the gangue has sufficient bearing capacity and stability.

3.2.5 *Compaction test*

To ensure the effectiveness of the project, a compaction test is required.

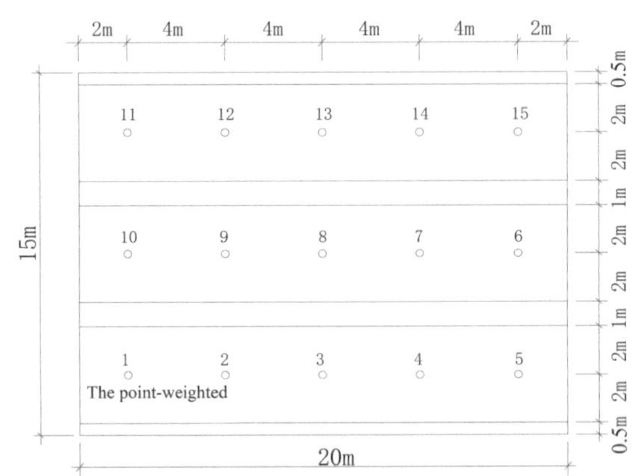

Figure 1. Schematic diagram of measuring point layout.

The specific steps are: (1) use a level to fix the point (fixed position of the measuring point) to accurately measure the thickness of the layer; (2) use the Rayleigh wavemeter to measure the compactness and bearing capacity of the layer; (3) the loose bulk density of the gangue belt can be measured by sand replacement method, and take one point for each of them, such as the positions of points 3, 8, and 13 (Figure 1); (4) fill the pits dug and use a road roller to vibrate each waste rock belt once and again. It is required to move forward by one gear, and adopt a strong vibration gear (the following vibration and pressure requirements are the same); (5) use a level to measure the thickness after compaction at the measuring point, and use a Rayleigh wave instrument to measure the compaction of the pavement and bearing capacity, measuring the bulk density of the gangue after compaction with the sand replacement method (to minimize the damage and compaction situation, each one is still digging a hole, such as the position of the 2, 7, and 12 measuring points, and then digging the hole later. Time delay, the error is corrected by the sinking value of the level measurement); (6) repeat the above (1)–(5) until the rolling pass reaches 10 times.

3.3 *Subarea backfill construction technology*

At present, there are mainly three types of coal gangue backfilling methods, with their technical standards: (1) Coal gangue backfilling. The burned coal gangue has a small particle size and is relatively uniform, with a small porosity, layered backfill, layered rolling, good compactness, and small uneven settlement. It is suitable for backfilling collapsed areas as construction land. (2) Mixed filling of coal gangue and fly ash. Coal gangue without spontaneous combustion has different particle sizes and small porosity. To solve the spontaneous combustion of coal gangue, coal gangue and fly ash are mixed, that is, a layer of coal gangue and a layer of fly ash are backfilled. After tests, the ratio of coal gangue to fly ash is 5:2. The coal gangue is backfilled with a 200 mm thick layer of fly ash, which not only prevents spontaneous combustion of the gangue but also reduces the porosity of the backfilled foundation, and at the same time improves the consolidation capacity of the gangue foundation and reduces the unevenness of the gangue foundation settlement. (3) The coal gangue is mixed with soil. To solve the spontaneous combustion of coal gangue, a mixture of coal gangue and soil is used, that is, a layer of coal gangue is backfilled, and a layer of soil is backfilled. After tests, the ratio of coal gangue to soil is 5:2. If a layer of coal gangue with a thickness of 500 mm is backfilled, the backfill is 200 mm. A thick layer of soil not only prevents spontaneous combustion of coal gangue but also reduces the porosity of the backfill foundation and effectively controls the uneven settlement of the gangue foundation.

Here we focus on the layered vibratory compaction method. The layered vibratory compaction method is economical and reasonable for the treatment of a large area of layered backfilled coal gangue cushion, and the compaction effect is ideal. It can improve the density of the coal gangue layer. Due to the layered compaction, the porosity is greatly reduced and the air is isolated, which can prevent the spontaneous combustion, weathering, and the precipitation of harmful substances in the coal gangue foundation. The process steps are as follows:

(1) Drainage of accumulated water in the subsidence area

To achieve layered backfilling of all the coal gangue (Figure 2) and layered vibration compaction, all the water in the gangue filling area should be drained, and no water gangue discharge operation should be implemented.

(2) Stripped topsoil and stockpiled in selected areas

Before coal gangue backfilling, the soil source of coal gangue pre-backfilling in non-water-filled areas shall be stripped by partitions and blocks, and stored in selected areas. The borrow area should choose a higher terrain area to keep the gangue filling area level and make the thickness of the gangue the same, to facilitate the gangue backfill construction in layers. If there is humus suitable for farming on the surface of the backfill area, it shall be stripped and piled in the soil field for surface covering the soil.

(3) Desilting process for water pit

After draining the water in the sump, it is required to use a mud pump or a hydraulic pond digging unit to remove the silt at the bottom of the pit to avoid the slippage of coal gangue and other solid wastes after backfilling and slushing during layered rolling, which will affect the quality of the project.

(4) Foundation treatment of coal gangue

The practice has proved that the natural backfill of coal gangue without compaction treatment has a large amount of settlement. Buildings built on the untreated coal gangue backfill layer will be damaged due to uneven settlement of the foundation, so the natural backfill of coal gangue must undergo compaction, after which buildings can be built on it.

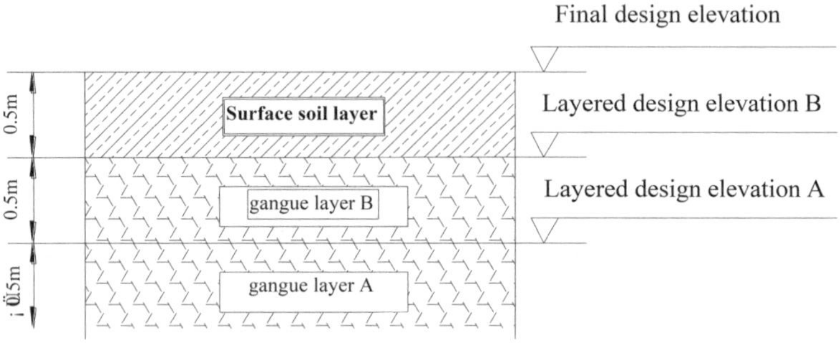

Figure 2. Schematic diagram of backfilling stratification in the reclamation area.

4 ENGINEERING PRACTICE

The reclaimed new village site area is located above the coal mining subsidence of a mine in Kailuan, and the original terrain is flat. Due to mining collapse, the ground is uneven, and there are many large and small water sinkholes. According to the actual measurement results of the current situation in this area, the current ground elevation is mostly +25.0~+27.0 m, the elevation of the bottom of the sump is +23.0~+24.0 m, and the elevation of the water surface of the sump is about +24.8 m. The collapsed puddles are mostly used for aquaculture, and crops are planted in the partially collapsed areas other than the puddles. The proposed new village site is 500 m long from north to south, 400 m wide from east to west, and covers an area of 200,000 m^2. After deducting the area of 36,667 m^2 of built villas, 163,333 m^2 of the new village site can be reclaimed.

Through the analysis of mining and subsidence in this area, it is believed that the surface movement and deformation in this area have been stabilized and will not affect the village buildings on the ground of the new village site. In the future, the nearest distance between the surrounding mining area and the new village site is about 800 m. Through the prediction and analysis of the impact of surrounding mining on the reclaimed area, it is obtained that future mining will not affect the reclaimed area of the new village site. The new village reclamation area receives small water area and has better conditions for flood discharge to the south. According to the location of the new village site and the surrounding ground elevation, the ground elevation of the new reclaimed village site is determined to be +27.0 m.

After the site selection of the new village site, the determination of the elevation, and the selection of backfill materials, according to the actual situation of the site of the new village site, to facilitate the construction and the combined reclamation of coal gangue backfill and soil, the planning area is divided into eight areas carry out the construction separately. The construction process design mainly includes: (1) construction of coal gangue dams; (2) drainage of accumulated water in the subsidence area; (3) removal of silt in the area of the subsidence pit; (4) recovering of soil; (5) leveling of subsidence; (6) layered backfilling of gangue; (7) densitometer detection; (8) analysis to determine the number of rolling passes.

Figure 3. Schematic diagram of the gangue backfill site.

Figure 4. Surface buildings in the demonstration area with gangue backfilling foundation.

According to the current topography of the new village site, determine a typical rolling test site (Figure 3). The test site is identified as Zone C. The surface soil of the rolling test site is stripped to +25.0 m, and the site is leveled, three blocks are divided. It is needed to backfill 300 mm, 500 mm, and 700 mm, respectively, and use a vibratory roller with a capacity of 120–200 kN for multiple passes. A level is used to measure until the difference in a settlement between two adjacent rollers is small. After analysis, the number of rolling passes is determined, and the subsequent construction is guided. In the process of layered backfilling, the density of the backfilled gangue is detected by layering with a densitometer. The coal gangue was backfilled to an elevation of +26.5 m, and the foundation load test was carried out to determine the bearing capacity of the foundation, which provided a basis for the design of the village buildings.

Since the implementation of the project, 600,000 m^3 of gangue has been processed, 31,800 m^2 of gangue area has been reduced, and 176,667 m^2 of coal mining subsidence has been reclaimed, which has significantly improved the local ecological environment and created huge social benefits for the local area (Figure 4). The relocation of the coal mining subsidence and reclamation site saves 10.6 million yuan compared with the relocation of normal villages; the Jiefang Village has lowered 4.54 million tons of coal, which has achieved great economic benefits.

5 CONCLUSION

(1) It should be pointed out that because coal gangue has larger sandstone and other hard particles, smaller and finely crushed mud, shale, and other soft particles, it has some properties of gravel soil and some clay. In nature, it is very necessary to study the engineering characteristics of this special foundation. Related experimental research must be carried out before engineering application.

(2) When using coal gangue as the foundation backfill material in the coal mining subsidence area for building reclamation, to avoid residual deformation from destroying the surface buildings, the site should be selected as far as possible in the surrounding stable settlement area where mining is no longer carried out, and the coal gangue is backfilled. The ground for reclamation construction should be backfilled in layers, and multiple vibration pressure treatment methods and construction should be carried out in divisions and blocks.

(3) Although coal gangue backfilling the building foundation has significant effects and strong practicability, the construction land is above the mined-out area, especially when the mined-out area of multiple coal seams. To ensure the safety of the building, while strictly implementing the foundation treatment indicators, surface buildings should also adopt a series of anti-deformation measures according to the actual situation. Observation stations should be established, when necessary, deformation observation should be carried out in time, and problems should be dealt with in time.

ACKNOWLEDGMENTS

The authors would like to thank the editors and the anonymous reviewers for their helpful and constructive comments. This research was funded by the China coal science and industry group science and technology innovation venture capital special key project (2018-2-ZD007). Beijing Science and technology project (Z181100005118012).

REFERENCES

D. D. Xu, G. L. Guo, S. Li. Metal Mine, 2011, 46(3):42–45.

G. Q. He. Xuzhou: China University of Mining and Technology Press,1991.

J. M. Duan, Y. J. Qian, X. T. Zeng. Coal Engineering, 2008, 55(8):44–46.

Q. F. Zhang, D. Q. Wang. Chinese Journal of Rock Mechanics and Engineering. 2013, 32(5):1049–1056.

S. Z. Li. Coal Science and Technology, 2014,42(1):93–97.

W. M. Li, H. R. Zhang, Z. H. Tan, et al. Journal of Shandong University of Science and Technology, 2008,27(6):24–27.

X. L. Liu, X. M. Wang, P. Wu. Metal Mine. 2011, 46(6):6–8+35.

X. S. Wang, P. W. Hao, G. Zhang. Subgrade Engineering.2009,27(2):44–45.

Z. X. Tang. Mine Surveying. 2011, (1):84–86.

Public infrastructure and urban traffic management

Advances in Urban Engineering and Management Science – Khalil & Yang (Eds)

Research on the needs of public space for old people sojourning in the background of "Internet Plus"—Take Southern Hunan as an example

Dandan Liu, Biao Tang*, Kangning Zhao*, Lei Chen, Canhua Qiu & Peng Fu
School of Architecture, University of South China, Hengyang, China

ABSTRACT: With the aging of the population in today's era, the model of Elderly Long-Stay Tourism has been gradually favored by the majority of the elderly. Under the educational background of internet plus, the characteristics of senior citizens' demand for public space are investigated. Taking the southern Hunan region as the research object, we select three sites for investigation and analysis, and use the computer-aided design software for drawing. We take the elderly in Southern Hunan as the main research object, and based on ERG theory, divide the public space needs for sojourning into the space for existence needs, relatedness needs, and growth needs. By means of sampling survey and interview, we construct a structural equation model about the degree of needs of the elderly for sojourning in public space, so as to explore the relationship among the three and analyze the influence degree on the needs of sojourners in public space. This result shows that there is a very obvious positive correlation among the three potential variables: the space for the existence needs of the elderly, the space for the relatedness needs and the space for the growth needs. Moreover, it has a positive impact on the public space where the elderly live. In addition, we also give some suggestions on the public space of sojourners' needs from the smart pension, and provide a theoretical basis for the research on the aging of sojourning in public spaces.

1 INTRODUCTION

With the deepening of China's aging population, the problem of providing for the aged in China has become a social concern (Cousins 2021). Therefore, creating a suitable living environment has an important impact on the physical and mental health of the elderly (Evans et al. 2002; Shan et al. 2020; Yang et al. 2020; Zhe & Mardelle 2018). People's living standards have been gradually promoted and improved, and the model of "Elderly Long-Stay Tourism" has been gradually favored by the elderly and developed rapidly. In China, "Elderly Long-Stay Tourism" was put forward by Yong Cheng in 2009 (Yong 2009). Since then, some related scholars such as Xiaoxiao Pang (et al. 2015) have made quantitative evaluation with the index of sojourning for the aged, and obtained the evaluation results such as the standard of the suitability of setting up a destination for Elderly Long-Stay Tourism in cities and towns. How to get the relevant configuration of public space and how to improve the utilization efficiency are one of the core issues in the design of aging-resistant buildings, and they are also the key indicators to measure the living quality of residents and affect their health (Yufeng 2016).

With the rapid development of Internet technology, 5G technology and other high-tech technologies, promoting the application of intelligent systems in old-age care has become an important measure to alleviate the pressure of old-age care. Our latest "Action Plan for Promoting the Development of Aged Care Services (2019–2022)" mentioned that it is necessary to further develop the

*Corresponding Authors: 191174636@qq.com and zhaokn@mail2.sysu.edu.cn

DOI 10.1201/9781003305026-64

Table 1. System structure composition diagram.

System category	Composition
Data acquisition subsystem	Vital signs data, monitoring data, decision data
Emergency call subsystem	Emergency calls, precise location systems, accident detection systems
Information interaction subsystem	Medical culture entertainment, housekeeping services, information query and release
The elderly suitable infrastructure	Activity center for the elderly, smart home for the elderly, barrier-free environment

Table 2. ERG model diagram.

Category	Intention	Example	Relationship
Existence needs	The most basic material needs for survival	Diet, bathing, sports, etc.	There are high and low levels of needs, and more than one kind of needs may work at the same time, and the three have coexistence
Relatedness needs	The needs to maintain important interpersonal relationships	Chatting, making friends, etc	
Growth needs	Pursuing the needs of personal self-development and perfection	Learn skills, acquire knowledge, etc.	

new format of "Providing for the aged in internet plus", understand the deep-seated social needs, and help promote the service reform of the community home-based aged care model. The model of smart pension is to use the technology of the Internet of Things to collect information, and efficiently and intelligently respond to the needs of various old-age services such as emergency assistance through the platform of old-age care and healthy public services (Li-hua H & Ya-hui C 2017). Its system structure composition diagram is shown in Table 1. Clayton Alderferr put forward an improved hierarchy of needs theory-erg theory. He believes that people have these three core needs: the needs for existence, the needs for relatedness and the needs for growth (Cheng-Liang et al. C 2011). See Table 2 for details.

Scholars at home and abroad have also carried out related research. Rodrigues, M. et al. proposed that the demand core of the elderly is "connectivity" when designing smart home (Rodrigues M et al. 2018); Lin Li et al. ranked the elderly care service demand system with analytic hierarchy process and life quality factor analysis, and concluded that the elderly care service platform was designed from medical aid service, housekeeping, meal service and cultural entertainment (Li & Huang 2019) and so on. However, there are few studies on the public space of intelligent elderly care nowadays.

The so-called southern Hunan mainly refers to three areas in the southern part of Hunan Province, namely Yongzhou, Chenzhou and Hengyang, which include 34 counties (county-level cities). The location here is very advantageous, and it also has the richest tourism resources in Hunan. By the end of 2017, there were 3,026,300 elderly people aged 60 or above in southern Hunan (including 1,275,300 in Hengyang, 975,900 in Yongzhou and 775,100 in Chenzhou). It is of great practical significance to make use of the existing tourism resources to create a public space for sojourning that meets the needs of the elderly.

According to ERG theory and Yanmin Zhou's detailed explanation of the architectural design of old-age facilities, this paper divides the public space for sojourning into the space for existence needs, relatedness needs, and growth needs according to the daily behavior needs of the elderly. Moreover, the questionnaire items are set up, and the analysis and research of the structural equation model are constructed through the way of issuing questionnaires and field interviews. From the perspective of "internet plus," the characteristics of the needs of the aged for residential public space are explored, which provides a theoretical support basis for optimizing the residential public space in southern Hunan.

2 DATA SOURCES AND RESEARCH METHODS

2.1 *Acquisition of data*

Take representative residential buildings in each hot tourist attraction of Hengyang, Yongzhou, and Chenzhou as the main object of study, which are Shouyue International Hotel, Huameishi, Yishanyishe. The statistical research on sojourning buildings in southern Hunan is carried out, and the architectural plan is drawn by AUTO CAD (see Figures 1–3) based on field investigation and visit in May of 2021. Questionnaires are distributed to the senior citizens about the demand level of public space for sojourning. Out of the total 340 questionnaires distributed, 306 valid questionnaires are collected, with an effective rate of 90.0%. Refer to Table 3 for the detailed composition of the sample, which is generally representative.

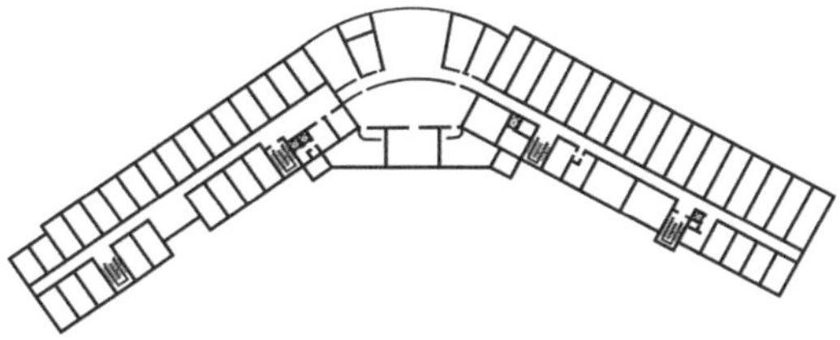

Figure 1. Standard floor plan of Shouyue International Hotel.

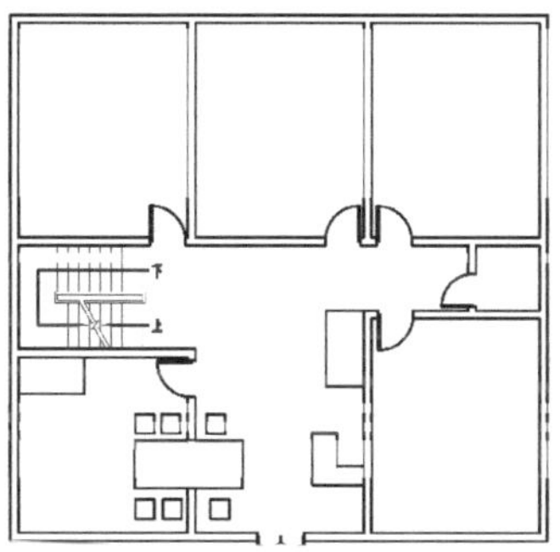

Figure 2. First floor plan of Yishanyishe.

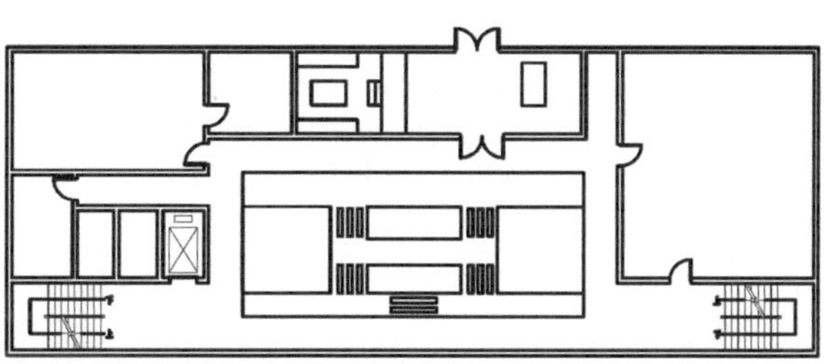

Figure 3. First floor plan of Huameishi.

Table 3. Composition of the survey sample table.

		Sample number	Proportion%
Gender	Man	150	49.20
	Woman	156	50.98
Age	60–65	156	50.98
	65–70	184	60.13
	70–75	80	26.14
	Over 75	26	8.50
Number of tours	Hardly	13	4.25
	1–3 times a year	50	16.34
	More than 3 times a year	220	71.90
Pocketbook	Pensions	35	11.44
	Free savings	258	84.31
	Children provide	33	10.78
	Other	10	3.27
The journey's destination	Natural Scenic Area	242	79.08
	Characteristic cultural scenic spot	43	14.05
	Artificial recuperation base	9	2.94
	Unique homestay style area	46	15.03
	Other	19	6.21
Attraction point of sojourn	Appropriate environment	218	71.24
	Experience the local folk customs	63	20.59
	Meet friends from different regions	10	3.27
	Other	14	4.58
Traveling objects	Independent	11	3.59
	With spouse	131	42.81
	With friends or children	165	53.92
Working background	Party and government offices	52	16.99
	Public institution	96	31.37
	Enterprise	114	37.25
	Army	10	3.27
	Other	29	9.48

The main design of the questionnaire mainly includes four parts, namely, the space for existence needs, relatedness needs, growth needs and the sojourners' needs for public space. Among them, there are ten specific items in the space for existence needs, relatedness needs, and growth needs, mainly involving public toilets, medical spaces and public restaurants (Yanmin Z 2019). The needs of public space for sojourning mainly includes three measured items, namely, "living distance with children," "living distance with children," and "living time" (Cohen-Mansfield J et al. 2016). The questionnaire is in the form of Linkert scale. What's more, the options and scores set according to the questionnaire answers are as follows. (Very important: 5 points, important: 4 points, generally important: 3 points, less important: 2 points, very unimportant: 1 point). The measurement items of travel time are assigned to 5–1 according to "starting at any time," "weekend," "cold and summer period," "depending on the situation," and "golden week or long holiday," respectively; The measurement items of the length of stay are set according to "within half a month," "half a month–1 month", "1–3 months", "3–6 months" and "more than six months", which are assigned 5–1 respectively; The measurement items of "living distance from children" are set according to "living together," "living in the same city," "living in neighboring provinces," "living in other provinces" and "others," and are assigned 5–1 respectively.

2.2 *Descriptive statistical analysis content*

The ratio of male to female of 306 elderly people who participated in the survey is basically in a balanced state, and the age is mainly concentrated in the younger elderly aged 60–70. About 85.61% of the elderly work in enterprises, institutions or party and government organs. The vast majority of the main economic sources come from pensions, accounting for about 84.31%. 71.9% of the elderly travel 1–3 times a year. Most of the travel destinations are natural scenic spots, accounting for about 79.08%. About 71.24% of the elderly believe that the most attractive place for them to live is a suitable environment. Most of them choose to travel together, and only 3.59% of them travel alone.

2.3 *Methods of research*

In this paper, SPSS 21.0 is used to analyze the needs of the elderly sojourning in public spaces except in southern Hunan by exploratory factor analysis. According to the analysis results, the structural equation model diagram of the elderly's degree of needs for sojourning in public space is drawn by AMOS24.0, then the model is tested and fitted, and finally, the results are drawn into a table by Microsoft Office 2010 software. This paper mainly discusses the relationship between each space for needs and the public space for elderly sojourners' needs based on ERG theory.

3 FACTORIAL ANALYSIS

Before there is any theoretical model with substantive content, exploratory factor analysis is used to obtain the SEM structural model. According to the reliability test of SPSS 21. 0, Cronbach α coefficients of spaces for existence needs, relatedness needs, growth needs and the sojourners' needs for public space are 0. 887, 0.869, 0.826 and 0.878 in turn, which shows that the questionnaire has high credibility. The significance value of the Bartlett spherical test of 13 variables in the questionnaire is 0.000, less than 0.05, and the KMO value is 0.866, greater than 0.7, which indicates that the data are well correlated and suitable for factor analysis. Through factor analysis, four main factors were extracted from 13 variables, with a cumulative contribution rate of 77.226%. See Table 4 for details.

Table 4. Exploratory factor analysis results of the needs of the elderly for sojourning in public space.

Latent variable	Measurement items	Factor loadings	Mean value	Contribution rate%
The space for relatedness needs	The public space of the traffic space in the elevator lobby of the building GX1	0.779	3.77	22.081
	Intergenerational communication space GX2	0.761	3.78	
	Chatting space GX3	0.852	3.87	
	Indoor cultural entertainment space GX4	0.820	3.82	
The sojourners' needs for public space	Living distance from children LJ1	0.859	2.26	19.042
	Travel time LJ2	0.832	2.33	
	Length of stay LJ3	0.890	2.13	
The space for existence needs	Public restaurant SC1	0.839	3.43	18.620
	Medical space SC2	0.842	3.51	
	Public toilet SC3	0.842	3.55	
The space for growth needs	Multifunctional hall FZ1	0.748	3.58	17.483
	Study classroom FZ2	0.839	3.61	
	Experience space with local characteristics FZ3	0.835	3.68	

4 CONSTRUCTION OF THE MODEL

According to the results of the above factor analysis, three exogenous potential variables are determined, which are the space for relatedness needs, the space for existence needs and the space for growth needs, and one endogenous potential variable is the sojourners' needs for public space, resulting in four ideas: 1) The space for existence needs has a significant positive impact on the sojourners' needs for public space; 2) The space for relatedness needs, existence needs has a significant positive impact on the sojourners' needs for public space; 3) The space of growth needs has a significant positive impact on the sojourners' needs for public space; 4) There is a correlation among the three exogenous latent variables. The structural equation model for the needs of the elderly sojourning in public space is shown in Figure 4.

Table 5. Test results of model goodness of fit.

Fitting index	Suggested value
CMIN	–
DF	–
CMIN/DF	<3
RMR	<0.05
GFI	>0.9
AGFI	>0.9
NFI	>0.9
RFI	>0.9
IFI	>0.9
TLI	>0.9
CFI	>0.9
RMSEA	<0.05

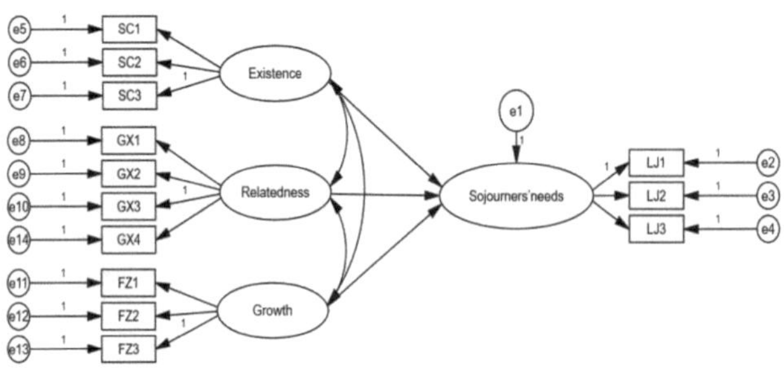

Figure 4. The structural equation model of the needs for the elderly sojourning in public space.

5 ANALYSIS OF MODEL RESULTS

5.1 *Model Fitting Evaluation*

Through the calculation of model fitting, obtaining that CMIN=83.488, DF=59, CMIN/DF=1.415, RMR=0.044, GFI=0.961, AGFI=0.939, NFI=0.964, RFI=0.953,

IFI=0.989, TLI=0.986, CFI=0.989, RMSEA=0.037. This index meets the requirements, and the model fits well.

5.2 *Explanation of relevant models*

In this model, the correlation coefficient of each path represents the relationship among variables, and the degree of mutual influence among variables is reflected by the size of the path coefficient.

The calculation result of the standardized path coefficient is shown in Figure 5. The fitting index of each item is statistically significant, the fitting degree of the model as a whole meets the known requirements, and the model is in an acceptable range.

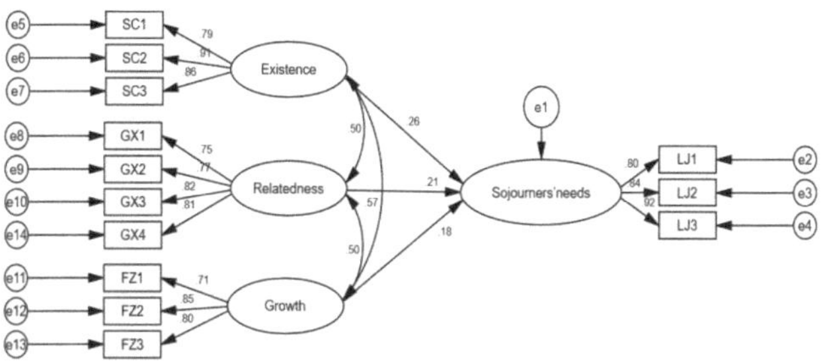

Figure 5. Model path parameters.

Table 6. The model paths.

Path	Relationship		
1	The sojourners' needs for public space	<—	The space for existence needs
2	The sojourners' needs for public space	<—	The space for relatedness needs
3	The sojourners' needs for public space	<—	The space for growth needs

Table 7. Standardized path coefficient (partial)

	Path 1	Path 2	Path 3
St. Estimate	0.262	0.21	0.181
Estimate	0.222	0.191	0.192
S.E.	0.064	0.066	0.083
C.R.	3.452	2.916	2.314
P	***	0.004	0.021
Establishment situation	found	Found	found

According to the specific ERG theory, we can know that the space for existence needs is to meet the most basic and lowest needs of human beings. From Tables 6 and 7, we can see that the path coefficient of the space needed for existence and the space needed for sojourning in public is 0.262, which shows that there is a significant positive correlation between the space needed for existence and the space needed by the elderly ($p<0.001$), so Hypothesis 1 can be established. The space for existence needs is the variable with the largest path coefficient among the three potential variables, which represents that it is a part of the elderly who have strong needs in the space for existence needs. The correlation coefficients of the public dining room, medical space and public toilet are all greater than 0.7, which indicates that the three variables have a great correlation with the space of existence needs. Among them, the correlation coefficient of medical space reaches 0.91, which indicates that medical space is the most concerned part of the sojourning space for the elderly.

The path coefficient between the spatial needs of relatedness and the spatial needs of sojourning in public is 0.21, showing an obvious positive correlation ($p<0.05$), which can prove that Hypothesis 2 holds. The correlation coefficients of public space, intergenerational communication space, chat space and indoor cultural entertainment space in the lobby traffic space of the elevator are all greater than 0.7, which indicates that the four variables have a great impact on the spatial relatedness needs. Among them, the maximum correlation coefficient of chat space is 0.82, which indicates that chat space is the most important space in the space for relatedness needs.

The path coefficient of the space for relatedness needs and the sojourners' needs for public space is 0.18, which shows that there is an obvious positive correlation between them ($p<0.05$), so hypothesis 2 holds. As the highest level of needs theory, its demand degree is lower than the first two spaces. The correlation coefficients of multi-function hall, study classroom and experience space with local characteristics are all greater than 0.7, which indicates that these three variables are all important parts of the space for relatedness needs. Among them, the maximum correlation coefficient of the experience space with local characteristics is 0.9, which indicates that it is also the most favored part of the space for the relatedness needs of the elderly, which can be built in the spatial design.

There are also obvious correlations among the three potential variables, namely, the space of existence needs, the space of relatedness needs and the space of growth needs, and the correlation among them is between 0.5 and 0.6, so hypothesis 4 can be established. The correlation coefficients of the three dominant variables, the distance from children's residence, travel time and sojourn time, are all greater than 0.8, which have a great influence on the needs of sojourn public space.

Based on the results of the structural equation model, we can know that the space for existence needs, the space for relatedness needs and the space for growth needs of the elderly influence each other and play a role of mutual encouragement. This is consistent with the ERG theory that "people may have more than one need at the same time, and all kinds of needs can have a stimulating effect at the same time." That is, the needs of the three are the same, and they meet the spatial needs of one level, and they all promote the spatial needs of the other two levels. According to the results of the model, we put forward corresponding suggestions for the three kinds of spaces for needs from the perspective of intelligent old-age care.

6 SUGGESTION

6.1 *Suggestions on smart pension in the public spaces for sojourners' existence needs*

According to the results of the model, the public dining room, medical space, and public toilet are all important parts of the concern of the elderly, especially the medical space. According to the actual situation of the interview when the questionnaire was actually distributed, most of the elderly mentioned that the medical space has efficient and safe nursing work. Therefore, we can create an information exchange subsystem to link and communicate efficiently at the hospital level in view of the urgent problems of the elderly during their sojourn. The subsystem of data collection can transmit the passenger's vital signs data to the nearest health service center through the Internet, and the medical institutions can monitor the health status of the elderly. The elderly can also publish information about their service needs through the information interaction subsystem, so that the operators and the elderly can realize effective information links (Yunhua X & Hong Y 2016)

6.2 *Suggestions on smart pension in the public space for sojourners' relatedness needs*

The results of the model show that, for the elderly, in the process of sojourning, the relatedness needs for public space is mainly manifested in the public space of the elevator lobby traffic space, intergenerational communication space, chat space and indoor cultural entertainment space. Among them, the chat space is an important public space for the elderly. According to the actual interview content, we can know that chatting with fellow travelers and other residents is an important part of the public space for existence needs and supporting the elderly. Some interviewees expressed their desire to create a variety of chat spaces. Therefore, some smart design contents can be created in the indoor space, which can bring certain convenience to the elderly passengers on the one hand, and enrich the experience of the aged passengers on the other hand. For example, to create an intelligent partition mode, some partitions are added to the intelligent sensing system, which makes the entry and exit of the elderly automatically open or close, thus achieving the purpose of reducing the inconvenience of the elderly (Xiaoyu C2021).

6.3 *Suggestions on smart pension in the public space for sojourners' growth needs*

In the structural model, it can be shown that multi-purpose halls and study classrooms, which can carry out meetings and other work activities, especially in the space for observing and experiencing with local characteristics, are the parts that the elderly pay more attention to. Therefore, sojourning in public space can integrate some local architectural features from materials, architectural forms, etc., and create an immersive and interactive virtual environment by means of VR technology according to local culture, so as to solve the problem that the elderly are still eager to know new things despite their weak learning ability, and also strengthen their attraction to all-age travelers (Krzysztof 2008).

7 CONCLUSION

In this paper, 306 valid questionnaires were sent to 3 residential buildings in southern Hunan, and the structural equation model of the demand for the elderly to live in public spaces was constructed under the perspective of "Internet+."

1. The results show that there is an obvious positive correlation among the three latent variables of the elder's living demand space, relationship demand space and development demand space, which have a positive influence on the elder's living in public space.
2. According to the analysis of the model results, this paper puts forward targeted suggestions from the perspective of intelligent elderly care, which provides a certain theoretical reference for future research on the intelligent system of living in public spaces.

The number of indicators selected in the questionnaire is limited, the demand type of living public space is not comprehensive, and the number of survey areas and samples is limited, which needs more empirical research in the later stage.

ACKNOWLEDGMENTS

This study was financially supported by the National Natural Science Foundation of China (31800366 to K. Z.)

REFERENCES

Cheng-Liang Y, Mark H and Ya-Chien C. An empirical study of the existence, relatedness, and growth (ERG) theory in consumer's selection of mobile value-added services[C]. African Journal of Business Management,2011,5(19). Public Administration(12th) & International Symposium on West African

Cohen-Mansfield J, Hazan H, Lerman Y, et al. Correlates and predictors of loneliness in older adults: a review of quantitative results informed by qualitative insights[J]. International Psychogeriatrics, 2016, 28(04):557–576.

Cousins M. The sustainability of China's Urban Employees' Pension Programme: A case of getting old before getting rich? [J]. International Social Security Review, 2021, 74(1).

Evans G W, Elyse K, Paul E. Housing quality and psychological well-being among the elderly population. [J]. J Gerontol B Psychol, Soc, 2002(4): P381.

Krzysztof W. Structured design of interactive VR applications[P]. 3D web technology,2008.

Li L, Huang YX.Research on the Smart Home-care services System of Chengdu Community based on "Internet plus "[C]. LOS ALAMITOS, IEEE COMPUTER SOC,2019:185–188.

Li-hua H, Ya-hui C. Study on the Service Mode of Smart Pension in Zhongshan City from the Perspective of the Supply-side Structural Reforms in China[C]. Proceedings of 2017 International Conference on Public Administration(12th) & International Symposium on West African Studies (1st) (Volume II).

Rodrigues M, Santos R, Queiros A, et al. Queiroz Smart Walk, Smart Cities for Active Seniors[C]. NEW YORK, IEEE,2018(7).

Shan W, Xiu C, Ji R. Creating a Healthy Environment for Elderly People in Urban Public Activity Space[J]. International Journal of Environmental Research and Public Health, 2020, 17(19):7301.

Xiaoxiao P, Fuhua B, Xiaohua L. Establishment and Application of Evaluation Index System for Residential and Elderly Towns[J]. Journal of Henan Commercial College, 2015, 28(01): 17–21.

Xiaoyu C. The feasibility study of the intelligent design of suitable old-style housing[J]. Architecture and Culture,2021(01):48–49.

Yang Z, Jiang Y, Wang M, et al. Current Status and Challenges of Community-Based Elderly Care Centers in Chongqing, China: A Cross-Sectional Study[J]. Risk Management and Healthcare Policy, 2020, Volume 13:2975–2983.

Yanmin Z. "Detailed Explanation of Architectural Design of Elderly Care Facilities" (Volumes 1~2) [J]. Architecture,2019(10):53.

Yong C. Talking about living for the elderly[A]. China Geriatric Health Association, Hainan Medical College. 2009 China Geriatric Healthcare and Industry Summit Forum [C]. China Geriatric Healthcare Association, Hainan Medical College: China Geriatric Healthcare Association, 2009(3).

Yufeng H, Xin W. Research on the Utilization Efficiency of Public Space in Urban Nursing Homes in my country: Investigation on the Utilization Status and Problems of Public Space in a Nursing Home in Hangzhou[J]. Architecture Journal, 2016(10): 103–107.

Yunhua X, Hong Y. Innovation of elderly care service system: local practice and countermeasures for smart elderly care[J]. Journal of Xi'an University of Finance and Economics,2016,29(06):110–114.

Zhe W, Mardelle M S. Can aging-in-place be promoted by the built environment near home for physical activity: a case study of non-Hispanic White elderly in Texas[J]. Journal of Housing and the Built Environment,2018,33(4).

Advances in Urban Engineering and Management Science – Khalil & Yang (Eds)
© 2023 The Author(s), ISBN 978-1-032-30426-7

Application and practice of intelligent infrastructure in urban transportation management

Yu Wang*
Intelligent Transportation Technology Department, CCCC Highway Consultants Co., Ltd., Beijing, China

Lu Pan*
Research Center for Detailed Planning, Beijing Tsinghua Tongheng Urban Planning & Design Institute, Beijing, China

ABSTRACT: The wide implementation of intelligent infrastructure promotes the continuous improvement of transportation management effect, improves the efficiency of resource utilization, enhances the level of traffic safety, and promotes the realization of energy-saving and environmental protection goals of urban transportation. The rapid development of urban intelligent transportation promotes the continuous implementation of urban intelligent infrastructure. This paper summarizes the application and practice of intelligent infrastructure in China on the basis of an analysis of urban transportation issues to better promote the development of an intelligent transportation system.

1 INTRODUCTION

As urban development gradually moves toward an era dominated by refined traffic design and management, the role of infrastructure combined with big data is becoming more and more prominent in the development of intelligent transportation. As an important part of smart cities, intelligent transportation system makes comprehensive management of transportation system planning, facility construction, demand management, operation regulation, transportation services and other aspects reliable through information collection, data communication, sensors, artificial intelligence and automatic control and other technical means. In 2012, the Ministry of Transportation released the 'Strategy for the Development of Intelligent Transportation in the Transportation Industry (2012–2020)', marking the rise of intelligent transportation as a national strategy. In 2014 and 2015, 'New Planning for National Urbanization (2014–2020)' 'Promoting transportation service and developing intelligent transportation with Internet +' were released continuously to promote the development of intelligent transportation. Besides, in 2019, 'Strategy for a strong transportation country' clarified the significance of intelligent transportation development, pointing out the four major tasks of intelligent transportation: alleviating traffic congestion and improving traffic safety, providing quality traffic services and modernizing the traffic management system. It is clear to see the great value of intelligent transportation for cities in China. The outline of the 2020 Digital Transportation Development Plan also proposes to promote the digitization of the whole element and cycle of transportation infrastructure and build an intelligent application system.

This paper illustrates future urban intelligent transportation solutions based on several problems of urban transportation and the application of intelligent transportation facilities in China, provides examples of high-quality development models for intelligent transportation development, builds a transportation innovation-driven system with perfect mechanisms and vitality, and explores a new model of transportation development driven by science and technology, system and talents. It attempts to offer some inspiration for planning ahead to construct new infrastructure, accelerate the application of 5G, Internet of Things, artificial intelligence, BIM and other technologies in traffic

*Corresponding Authors: wangyu@hpdi.com.cn and panlu@thupdi.com

DOI 10.1201/9781003305026-65

service management, provide strong support for the country's deep participation in the global governance of transportation, and provide Chinese experience for the world.

2 COMMON ISSUES FOR URBAN TRANSPORTATION

2.1 *Safety*

Traffic accidents are mainly influenced by human, vehicle, road, and environmental factors in cities, among which human factors account for a high proportion, even up to 95% of the total traffic accidents in some regions. Irregular driving behavior, failure to maintain a safe distance, inattentiveness of drivers, and lack of real-time accident releasing and advancing system warning are often the most significant factors leading to traffic safety accidents.

According to the 2018 China Highway Journal Report, the majority of traffic accidents in China occur on urban roads, rather than the highway in rural areas, with 68.31% of road fatalities and 31.69% of urban road fatalities. Among them, traffic accidents caused by human traffic violations accounted for 95.42% of the total number of traffic accidents and 95.42% of the total number of fatalities.

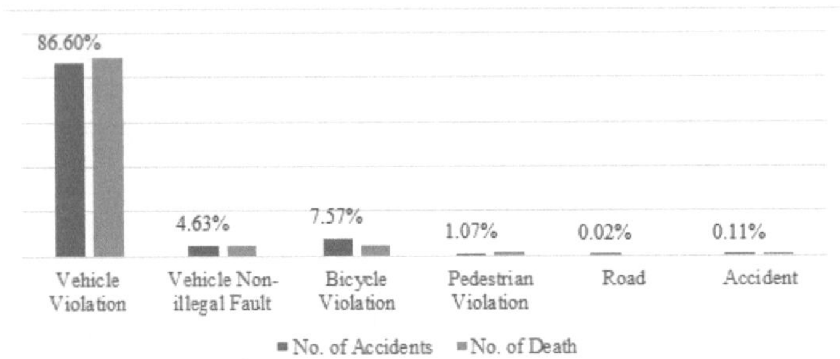

Figure 1. Main reasons for traffic accidents in China.

2.2 *Efficiency*

Due to the huge total number of private motor vehicles and relatively poor and uncomfortable urban public transportation services, travelers are highly dependent on motorized transportation, which leads to serious traffic congestion, and also results in high transportation travel time costs, high socioeconomic costs, and serious environmental pollution. The application of intelligent transportation infrastructure and Internet big data technology has a huge space to play in alleviating and improving traffic congestion.

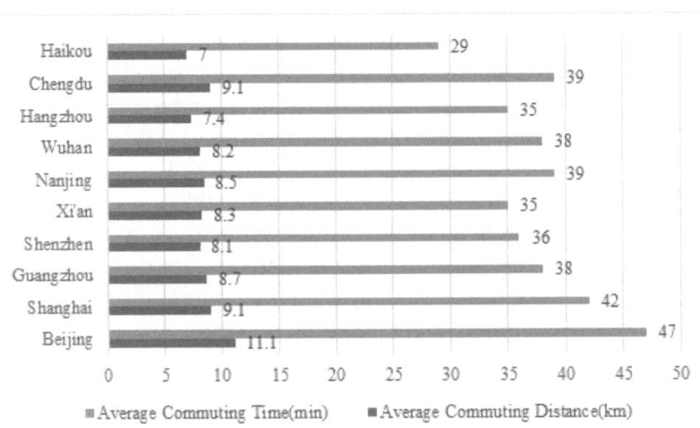

Figure 2. Average cost for commuting in main Chinese cities in 2020.

2.3 *Service*

Due to the lack of intelligent facilities and intelligent traffic management, there are often many inconvenient and unsuitable traffic services. For example, the setting of bus stops conflicts with pedestrian and non-motorized traffic flow, which affects the quality and safety of pedestrian travel. Meanwhile, it is difficult for bus services to accurately serve the different travel demands during peak and non-peak periods, which easily causes inefficient and wasteful resource allocation. The number of bicycle sharing, parking arrangements, and advance notice of motor vehicle congestion can be improved by intelligent transportation facilities to reduce environmental pollution and time costs caused by congestion.

3 URBAN INTELLIGENT TRANSPORTATION SOLUTIONS

According to the conventional understanding of integrated transport systems, urban transportation is usually divided into road transportation, public transportation, parking transportation and pedestrian and bicycling. Intelligent infrastructure in urban transportation is divided into on-road intelligent infrastructure and roadside intelligent infrastructure due to the different physical space distribution. This paper describes how intelligent infrastructure can improve urban transportation from the five aspects above. On-road intelligent infrastructure includes integrated sensing devices and traffic monitoring devices. The intelligent roadside infrastructure includes multifunctional information poles, variable information signs and roadside equipment for vehicle infrastructure cooperation systems (VICS). Intelligent infrastructure for public transportation mainly refers to intelligent connected buses. Intelligent infrastructure for parking transportation mainly refers to intelligent car parks. Intelligent infrastructure for pedestrian and bicycling includes smart pedestrian crossings and shared bicycle management equipment.

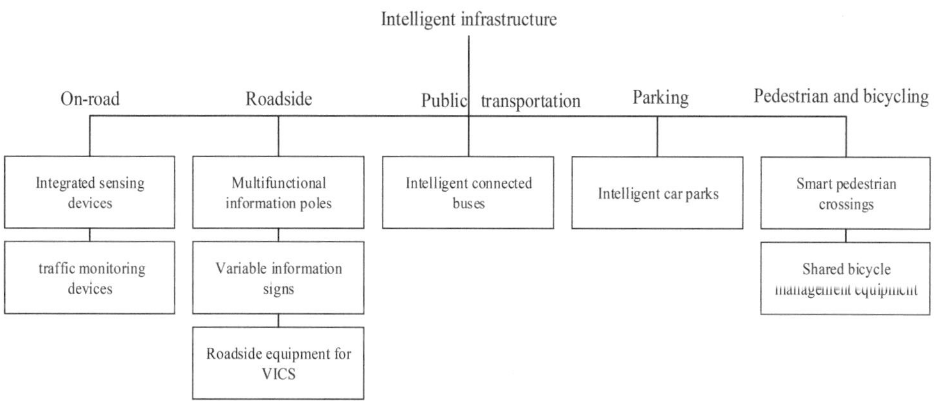

Figure 3. Intelligent infrastructure in urban transportation.

3.1 *On-road intelligent infrastructure*

3.1.1 *Integrated sensing devices*

Digitization of infrastructure has always been a hot topic pursued by the transport industry, and data collection using multiple sensors is an important part of digitization. Integrated sensing equipment for urban roads mainly includes key structure sensing equipment, environmental sensing equipment and manhole cover sensing equipment.

Key structure sensing equipment uses IoT technology and a variety of sensors to provide real-time monitoring of key structures such as roads, bridges, tunnels and high side slopes, with regular data analysis and report output. The main ones used in road monitoring are settlement sensors, road moisture content sensors and subgrade temperature sensors for monitoring subgrade settlement, excessive moisture content and subgrade temperature. The sensors used in bridge monitoring are used to monitor parameters such as deflection, displacement, vibration and structural strain. The data collected by the sensors is transmitted to the control center via a transmission system. When abnormal data occurs, early warnings are sent to management, and reports are issued to guide repairs and prevent more serious disasters from occurring.

The environmental sensing equipment collects data such as wind speed, wind direction, rainfall, temperature, humidity and PM2.5. The equipment collecting these data is integrated into a small weather station that allows real-time monitoring of the weather environment within 5 square kilometers of the surrounding area. When there are bad weather conditions such as heavy rain, strong winds and fog, the data can be released through information distribution terminals to provide timely warnings and alert passing vehicles to detour or slow down.

Manhole cover sensing equipment allows for a variety of sensing such as absence, displacement and tilt through sensors installed on the manhole cover. Through the integrated sensing equipment, city managers can obtain basic information in real time to ensure safe travel.

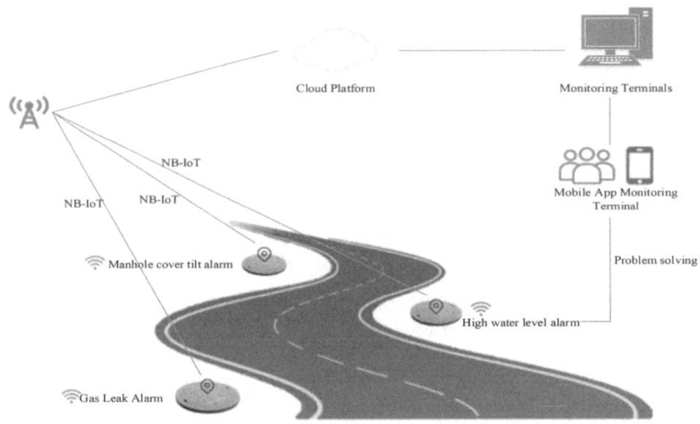

Figure 4. Manhole cover sensing equipment and systems.

3.1.2 *Traffic monitoring devices*

Traffic police are the direct participants and enforcers of urban transportation. However, due to factors such as manpower and deployment, not all traffic police officers are present where enforcement and supervision are required. This is where traffic monitoring devices come in handy. Traffic monitoring devices include cameras, radars and electronic license plates. The wide-area radar can collect traffic parameters such as traffic flow, speed, occupancy rate, vehicle type and vehicle length at intersections in real time, and supports lane separation statistics. It is suitable for important road sections with high traffic volume and accident rates and interconnected three-dimensional key nodes such as intersections, hubs, and car parks for encrypted deployment. Panoramic video monitoring cameras are generally arranged at large intersections, allowing for 360-degree blind spot-free monitoring. In addition, there are accident detection cameras, which can detect congestion, accidents, illegal lane changes, road collapse and throwing objects, and have video storage functions to facilitate later verification and assist in case processing. The main equipment of the electronic license plates includes the license plate reading and writing equipment and antenna, which is generally set up at the intersection. When vehicles carrying vehicle-mounted electronic signs pass by, the system can acquire vehicle identity information, real-time positioning information, operation status information, traveling track information, etc., which can assist city managers in signal timing, tidal lanes, etc., also play a great role in chasing fugitives.

3.2 *Roadside intelligent infrastructure*

3.2.1 *Multifunctional information poles*

For a long time, municipal construction has involved many departments, and there has not been a good overall plan for each department. The construction of various departments has resulted in many street-light poles, telephone poles, base station poles and sign poles in close proximity. It is both an obstruction to sightlines and an eyesore to the city. Collaborative design technology now allows these devices to be installed on the same pole, enabling multifunctional use of one pole. Multifunctional information poles are defined as poles used for communication, computing and information in areas such as roads, courtyards and venues, equipped with intelligent lighting, video security, mobile communication, intelligent traffic, information dissemination, citizen services, environmental weather monitoring and governance, green energy and data collection, depending on the needs of the scenario. Multifunctional information poles, also known as information poles and intelligent poles, are carriers of urban intelligent equipment for convenient portability, integrated transmission and centralized management.

Figure 5. A junction before and after integrating poles.

3.2.2 *Variable information signs*

At present, traditional signs on both sides of the road can only release fixed information such as speed limits, junctions and road names. In foggy weather, the information released is isolated and ineffective. Variable information signs are road traffic signs that are controlled by remote control equipment and display in real time the changes in the driving environment on the road ahead due to weather, natural disasters, traffic accidents and other reasons. This kind of sign improves the shortcomings of traditional signs, such as single information release and poor adaptability to foggy weather. It can release messages in time to guide lanes to avoid the accident site or to select a car park with remaining parking spaces.

3.2.3 *Roadside equipment for VICS*

There are two categories in the battle for the technological route of smart cars, namely single-car intelligence and vehicle infrastructure cooperation. Single car intelligence, represented by Western countries such as the United States, means that all sensors are installed on the vehicle, and the vehicles have rich sensing and computing capabilities that allow for driverless driving in unfamiliar traffic environments. In China, there are more challenges than in other countries in all aspects, be it sensors, chips, computing power or the complexity of road conditions. Therefore, China has chosen the strategy called vehicle infrastructure cooperation. In order to serve autonomous driving and future driverless driving, roadside equipment need to be built. This usually includes roadside units (RSU), edge computing units and 5G base stations. Information required by vehicles, such as accidents ahead, traffic lights ahead and vehicles overtaking from behind, is no longer obtained through its own sensing devices, but is provided by the RSU through V2V and V2X technologies. By building in this way, some of the sensing and computing power is transferred from the vehicle side to the roadside, reducing the high cost of the vehicle, making it more conducive to driverless industries and facilitating the coordinated development of a wider range of industries. In an intelligent networked environment, the previous logic will be broken, and more reliance will be placed on vehicle infrastructure cooperation, and the industrial ecology will be transformed from the tree-like structure of the past to a spatial network structure.

3.3 *Public transportation*

There is no denying that public transportation is one of the large-scale application scenarios most likely to be the first to see driverless buses on the ground. In fact, there are already a number of intelligent connected buses operating on the road. Subject to some legal and policy restrictions, the driver is still required to sit in the driving seat and take over the bus in an emergency, but is not required to perform any operations. A range of operations such as starting, accelerating, steering and braking can all be automated by the bus. What's even more surprising is that not only can the bus be intelligent, it can also obtain phase information for over-the-horizon signals, monitor the bus's data, monitor in-bus video information, provide active safety warnings and share traffic status, which is achieved by reforming the roads and buses along the route. At this point, the whole bus, as the receiving carrier of vehicle infrastructure cooperation information, has become the intelligent infrastructure. Through cooperation with traffic police, it is also possible to achieve bus priority. When a bus is about to reach an intersection, it is signal-controlled to enable it to cross the road without slowing down, greatly improving the efficiency of public transportation.

Figure 6. An intelligent connected bus.

3.4 *Parking*

In order to solve the problem of difficult parking caused by the rapid growth of the number of vehicles, combining technologies such as geomagnetic positioning and license plate recognition, the intelligent car park realizes real-time collection of parking space information and access to internal data of the car park through intelligent infrastructure such as geomagnetic IOT sensor terminals, reversing and finding terminals, and parking guidance screens. The intelligent car park guides the comprehensive use of parking facilities, reduces manual management costs and traffic congestion caused by vehicles searching for parking spaces, realizes the maximum use of parking resources and optimizes parking services for car owners. At the same time, the intelligent car park also has unattended and electronic payment functions.

3.5 *Pedestrian and bicycling*

3.5.1 *Smart pedestrian crossings*

A smart pedestrian crossing is an intelligent infrastructure used in conjunction with traffic lights. It is usually deployed at level crossings with high traffic flow, especially pedestrian traffic. It is designed to provide a strong visual reminder for motorists and pedestrians to prevent them from running red lights and other dangerous behavior due to inattention. The system usually consists of pedestrian warning posts and illuminated floor tiles, some of which are equipped with large exposure screens. The pedestrian warning posts are set up in the waiting area for pedestrians crossing the road, the illuminated floor tiles are arranged between zebra stripes, and the large exposure screens stand in a prominent position at the junction.

When the pedestrian signal in front of the pavement lights up green, the LED illuminated floor tiles on the pavement also flash green to remind pedestrians that they can pass. Accordingly, the stop line light strip in front of the motor vehicle lane lights up red to indicate that motor vehicles are prohibited from passing, complementing the original red signal for motor vehicles. At the same time, the pedestrian warning pillar emits a sharp beep or voice alert through the broadcast system to remind pedestrians to cross the street quickly.

When the pedestrian signal in front of the pavement lights up red, the pedestrian warning pillar activates the capture mode through the camera. If a pedestrian runs a red light to cross the road, it will be captured, recorded, and coded for display on the large exposure screen. The system can increase the penalty for the offense and strengthen the warning effect.

Figure 7. Smart pedestrian crossings.

3.5.2 *Shared bicycle management equipment*

The shared bicycle management equipment uses electronic fencing, Bluetooth positioning, and cloud technology to normalize urban shared bicycle parking areas and solve the chaos of shared bicycle parking and management. The shared bicycle management equipment can provide intelligent lock monitoring, return area control, data display, monitoring and early warning, intelligent scheduling and other functions. Bicycle return area control refers to the predefined bicycle parking area via an electronic fence. When returning a bicycle, the bicycle must be parked in this designated area to complete the return; otherwise, it cannot be locked, and billing cannot be stopped. For bicycle companies, bicycle sharing management equipment can help regulate operations and intelligently and accurately dispatch transport. For the community and government management, it can reduce the order management costs of shared bicycles, solve the parking chaos, improve the image of the city and help improve the quality of residents.

Figure 8. Shared bicycles parked on the electronic fence.

4 PRACTICE

According to public information released, in some large and medium-sized cities of China, some intelligent infrastructures have been built one after another, and they are playing an increasingly important role in improving traffic safety, convenience and the economy. For example, the Chuntianmen Bridge was completed and opened to traffic in Banan District, Chongqing City, in 2008. After 2016, many places in Chongqing experienced frequent earthquakes. The health monitoring of the Spring Gate Bridge began in September 2017, covering displacements and cracks, and in November of the same year, the sensing equipment warned of abnormal bearing displacements. After on-site investigation and analysis by experts, the traffic was interrupted along the line, and disease remediation work started, avoiding disasters.

In 2018, Shanghai carried out a comprehensive road improvement project, adopting an intensive approach and coordinating various factors to integrating the original components in the form of integrated poles as the main approach to comprehensive improvement, resulting in an estimated 72% and 65% reduction in the number of poles on city trunk roads and feeder roads respectively. While the number of poles is reduced, it also provides the best road space carrier for 5G micro-base stations.

In the same year, Shanghai's smart pedestrian crossings were put into operation at ten intersections, including the Bund. A traffic policeman, who has been on duty there for nearly five years, said that although the newly installed smart pedestrian crossings look complicated, they work well for places with huge traffic flow, such as the Bund, and have greatly eased traffic pressure in hot spots. The next step will be to extend the smart pedestrian crossing to other key intersections in Shanghai, making it a beautiful sight in the city.

The Wuxi Telematics pilot area has gone through three stages of technology testing and verification, scale verification application and scale commercial environment construction, and has built environmental sensing equipment, traffic monitoring equipment, roadside equipment for VICS and a series of intelligent infrastructures within a nearly 170 square kilometer area. It provides information services covering 12 categories and 26 application scenarios, including traffic light signal information push, traffic event alert, active safety warning and real-time access to surrounding traffic conditions for general users. The service scenario of priority access was tested and verified for social service vehicles such as emergency vehicles, fire engines and buses in the city. About 12,000 front and rear users, mobile phone APP users and industry users were developed.

In addition, the city of Changsha in Hunan Province launched an intelligent connected bus in April 2020. Based on the original traditional buses, the vehicles and road sections were intelligently transformed, loaded with smart onboard devices, equipped with smart traffic lights at intersections along the routes, and loaded with road-testing smart connected devices, which greatly improves safety and punctuality. An intelligent car park was built in a multi-purpose large urban complex in Zhanjiang, Guangdong Province, to achieve the benefits of fast, efficient, non-congested traffic and enhance the overall customer experience.

It is worth emphasizing that an integrated transportation system does not only contain one mode of transportation and that the above solutions are not applied singularly, but in an organic combination of many to improve transportation.

Table 1. Examples and effects of intelligent infrastructure applications.

City	Contents	Effects
Chongqing	key structure sensing equipment	timely warning of abnormal bearing displacements
Shanghai	multifunctional information poles	space saving and city beautification
Wuxi	roadside equipment for VICS	early over-the-horizon sensing and warning
Changsha	intelligent connected bus	improving access efficiency
Zhanjiang	intelligent car park	saving parking time and reducing the difficulty of finding a car

5 CONCLUSION

With the widespread dissemination and intensive application of technologies such as the Internet and artificial intelligence, the problems that have long plagued urban transport are expected to be solved. No matter how the city develops, safe, efficient and economical transport is the constant pursuit. China is in a phase of rapid urbanization, and considering intelligent transport infrastructure at the outset of city construction will reduce the cost of many unnecessary later renovations and upgrades. Through a series of new infrastructure policies and regulations promulgated in China, the transport industry is deeply applying technologies such as the Internet, big data and artificial intelligence to support the transformation and upgrading of traditional infrastructure, accelerate digital transformation and integration and innovation, and develop gradually. China should learn from the advanced experiences of Singapore, Japan and Hong Kong, China, and there should also be more exchanges between countries worldwide. Technologies such as digital twins, visualization and map infrastructure should be hot spots for research in the coming years. Along with the application of technology, we should pay more attention to the possible disturbance to human beings in terms of information security.

REFERENCES

Cattermole-Terzic. A human factors investigation into the effectiveness of traffic incident management. Sustainable Minerals Institute, 2017.

Editorial. Analysis of the main causes and characteristics of road traffic accidents in China. China Highway, 2018.

Wang Lei. (2020). Innovative practice of road improvement project in Shanghai. Urban Roads Bridges & Flood Control, 000(003), 110–114.

Wuxi Telematics Pioneering Application Demonstration White Paper (2019)

Zhao Pengjun, Zhu Junyi. (2020). Current condition and challenges for intelligent transportation, Contemporary Architecture, vol.12 (12), pp.46–48

Advances in Urban Engineering and Management Science – Khalil & Yang (Eds)
© 2023 The Author(s), ISBN 978-1-032-30426-7

Analysis of spatial characteristics of intercity traffic travel network based on mobile phone signaling datum—A case study of Fujian Province

Yi Liu*
Transport Planning and Research Institute, Laboratory for Traffic & Transport Planning Digitalization, Ministry of Transport, Beijing, China

Jun Ma* & Tong Liu*
Transport Planning and Research Institute, Ministry of Transport, Beijing, China

Di Wu*
Transport Planning and Research Institute, Laboratory for Traffic & Transport Planning Digitalization, Ministry of Transport, Beijing, China

ABSTRACT: Intercity traffic is the carrier of element flow and exchange of urban agglomeration. Based on the mobile phone signaling datum, this paper analyzed the characteristics of intercity traffic in Fujian Province, such as travel scale, connection strength, network agglomeration and sub-region. The results show that highway traffic is still the main mode of intercity travel in Fujian Province; the traffic volume of intercity highways decreases obviously with distance, but this characteristic of high-speed rail is not obvious. In the intercity travel network, the primacy and multi centrality are high, and two network subgroups of Xiamen-Quanzhou-Zhangzhou and Fuzhou-Putian-Ningde have been initially formed, indicating that the intercity travel network in Fujian Province is relatively balanced, showing a dual-core and multi-center pattern.

1 OVERVIEW

The basic attribute of traffic is moving. With the rapid development and wide application of mobile Internet, communication technology, navigation technology and satellite remote sensing technology, the technical bottleneck of traditional traffic information collection has been broken through, and the collection accuracy, frequency and sample volume of traffic travel elements have been significantly improved. Mobile signaling datum records the temporal and spatial changes of mobile objects, which makes it possible to track and study individual temporal and spatial movement characteristics for a long time, with high precision and high efficiency, and lays a foundation for the comprehensive analysis of the transformation from relatively independent and single-mode datum to multi-domain and multi-mode widely connected composite datum. Through the analysis and processing of massive mobile phone signaling datum, the mobile trajectory of mobile phone users can be obtained through datum mining and other methods, which can be further applied to the research of travel behavior such as residents' travel characteristic parameter extraction, travel spatial distribution characteristic analysis, travel mode discrimination and so on.

Since the 1990s, researchers in Europe and America have been committed to obtaining and mining traffic information through mobile signaling datum. With the rapid development of communication and network technology and the rapid popularization of smartphone devices, in recent years, more and more experts and scholars have used mobile signaling data to study and analyze traffic behavior. Ran bin (2013) and Niu Xinyi (2014) proposed to mine the spatial distribution

*Corresponding Authors: 6928224@qq.com, majun@tpri.org.cn, liutong@tpri.org.cn and wudi@tpri.org.cn

 DOI 10.1201/9781003305026-66

and activity patterns of population activities based on the combination of mobile phone signaling datum and geospatial location and restoring the real travel trajectory of users. Calabrese et al. (2011), Smorda (2013) et al. extracted the residents' mobile information from the location datum of several mobile phones and studied the residents' travel mobile mode in large-scale urban areas. Many scholars have conducted empirical research on population mobility (Aguilera et al. 2014), job-residence relationship (Luo et al. 2012; Xu et al. 2016), commuting behavior (Guo et al. 2018; Zhou et al. 2021) and traffic characteristics (Zhou et al. 2018) based on mobile phone signaling datum, respectively. Other scholars conduct travel path matching based on the auxiliary datum of the railway network, road network, highway network and aviation network, respectively, for mobile signaling datum to identify intercity travel modes such as automobile, railway, and aviation (Huang et al. 2020; Juan et al. 2010; Li et al. 2015). Mobile signaling datum analysis has become a new technical means to study the travel characteristics of intercity traffic, which will provide a reference for studying the travel behavior of intercity traffic and optimizing the planning, construction and management of intercity traffic networks.

2 INDEX AND METHOD OF ANALYSIS TO INTERCITY TRAFFIC TRAVEL BASED ON MOBILE SIGNALING DATUM

Based on the mobile signaling datum, through datum cleaning, processing, fusion and mining, the volume of intercity traffic between cities can be calculated. On this basis, the characteristics of intercity traffic such as travel scale, distance, travel mode, primacy and agglomeration of travel network can be deeply analyzed.

2.1 *Travel scale and connection intensity of intercity transportation*

Due to the migration of the population, the adjustment of industrial structure and the development of the economy and society, a large number of transportation needs will be generated between cities. The number of people and the income level of residents directly affect the size of intercity travel. With the improvement of residents' consumption level, the proportion of transportation in consumption will be greater, and the demand for off-site tourism, shopping and family visits will be greater. The stronger the city's economic strength, the closer the economic and industrial ties with surrounding cities, and the greater the amount of intercity travel. It is assumed that there are n cities in the region, which V_{ij} is the traffic volume between cities i and j, and the traffic distance between cities i and j, then $S=\sum_{i=1}^{n}\sum_{j=1}^{n}V_{ij}$ is the total amount of intercity travel and $T=\sum_{i=1}^{n}\sum_{j=1}^{n}V_{ij}\times d_{ij}$ is connection strength of the region.

2.2 *Network primacy and multi centrality*

Network primacy is generally expressed by the proportion of intercity traffic trips in central cities in the total regional trips. The higher the primacy value, the more significant the position of the central city in the intercity travel network of the whole urban agglomeration. The primacy contact intensity refers to the intercity connection with the largest travel volume to each city, and the primacy contact of each city constitutes the intercity travel backbone network (Li 2018). According to the intercity connection strength, the influence scope of each central city can also be determined. If the network connection degree between a city and the central city is the largest in the connection strength between the city and all cities, it means that the city belongs to the influence hinterland of the central city (Sun et al. 2015).

Network multi centrality mainly refers to the ratio of the average value from the second to the fourth city in the network to the first city. The higher the value, the more obvious the multi-centrality of the intercity travel network of the urban agglomeration. Generally speaking, the spatial distribution of intercity travel in a single-center area has a strong centripetality, and the surrounding cities gather in the central city; the spatial distribution of intercity travel in multi-center areas is more networked, and the gap in travel intensity among cities is relatively small.

2.3 *Analysis of attenuation characteristics of travel distance of intercity traffic*

According to the distance attenuation law of spatial effect, the radiation attraction effect of the central city to the surrounding area weakens with the increase of space-time distance. Weighted average travel distance between cities. Generally speaking, the intercity distance of 200 km is more suitable for road transportation. With the construction of high-speed railway and intercity railway, intercity travel convenience has been greatly improved, and intercity travel time has been greatly reduced, travel demand has further increased, and the suitable travel distance of the high-speed railway and intercity railway has been extended to less than 800 km.

2.4 *Travel network subgroup analysis*

In regional economic development, a sub-region refers to an urban agglomeration area with certain functions, which spans several administrative regions and is composed of many cities and their hinterland that are very close in space (Lao et al. 2016). By analyzing the interaction between cities, accessibility and contact strength of each node in the intercity travel network, and determining the subgroups in the travel network, we can further analyze the travel characteristics of the subregion. Using the k-plex in the agglomerative subgroup analysis method, a k-plex is an agglomerative subgroup that meets the following conditions. That is, in the subgroup, each node is directly connected to at least k points except k points. That is, when the scale of the agglomerative subgroup is n, each point is directly related to at least n-k points in the agglomerative subgroup; that is, the degree of each point is at least n-k. Using SPSS software, regional cities can be clustered according to the inter-city travel intensity to divide subregions.

3 ANALYSIS OF SPATIAL CHARACTERISTICS OF INTERCITY TRANSPORTATION IN FUJIAN PROVINCE

3.1 *Basic characteristics of intercity traffic development in Fujian Province*

Fujian Province is located on the southeast coast of mainland China, with a land area of 124 thousand square kilometers. In 2020, the GDP in Fujian province will reach 4.39 trillion yuan, with a total population of 39.22 million. The distribution of cities and towns and population will show the characteristics of dense in coastal areas and the southeast, sparse in inland areas and the northwest.

Firstly, the framework of a comprehensive transportation network has been basically formed, effectively supporting provincial economic development and urban system layout. The "three vertical and four horizontal" provincial channels composed of railways, expressways and ordinary national and provincial trunk lines have been basically connected, and the aim of "high railways reach every city and high-speed reaches every county" has been realized. The opening of the "ring Fujian" high railways line has greatly promoted the flow of people and logistics between cities. The journey between the central city of the province, the Yangtze River Delta, Guangdong, Hong Kong and Macao high-speed rail has been shortened to 3 hours, Fuzhou and Xiamen's position as a transportation hub has been effectively promoted to further drive the development of regional integration.

Secondly, the industries and population in coastal areas are highly concentrated, the transportation facilities network is developed, and the travel volume is large. There are nine cities in Fujian, of which six coastal cities account for about 45% of the land area of the province, carrying more than 80% of the total economic output and 75% of the total population of the province. The top

30 counties in the total economic output of the province are basically located in coastal cities. Cities, populations and industries are concentrated in long and narrow coastal areas, coupled with dense airports, and port hubs, the traffic in coastal areas presents the characteristics of high intensity, diversification, high frequency and strong timeliness, and operation pressure of road network is huge.

Thirdly, the agglomeration effect of the Fuzhou metropolitan area and Xiamen-Zhangzhou-Quanzhou city agglomeration is prominent, and the "one-hour" commuter circle has basically taken shape. Transportation integration is the leading field in the construction and development of the metropolitan area. At present, the large-capacity public transport system of the Fuzhou metropolitan area and Xiamen-Zhangzhou-Quanzhou urban agglomeration has been basically completed, the ability to serve large-scale and high-density business, commuting and other passenger trips has been continuously enhanced, the intra-city effect of the urban agglomeration has been highlighted, and a one-hour-commuter in the metropolitan area and two hours accessibility in the urban agglomeration has been basically formed, which provides favorable conditions for resource factor sharing and economic population agglomeration.

Lastly, the travel intensity in inland mountainous areas is low, and the traffic connection with coastal areas is still poor. Inland mountainous cities and towns are separated by mountains and distributed in spots. Trunk roads are the main mode of transportation, and the travel intensity is much lower than that in coastal areas; There is still no convenient channel between inland mountainous areas and some coastal cities, the detour distance is long, and the regional linkage effect has not yet been formed.

3.2 *Spatial characteristics of traffic travel in Fujian Province based on mobile signaling datum*

According to the characteristics of intercity traffic travel in Fujian Province, based on the mobile signaling datum in 2019, the traffic survey datum, highway toll datum, trunk highway operation monitoring datum, etc., are supplemented to calculate the traffic travel volume among cities in Fujian Province.

1) Analysis of the scale and intensity of intercity travel. In terms of travel scale, in 2019, the scale of intercity travel among cities in Fujian Province reached 1.089 million persons/day, of which about 116000 persons/day, accounting for about 10.6%, and about 973 thousand persons are based on the highway. There is only one intra-provincial route from Xiamen to Wuyishan in Fujian Province, and the average passenger flow on non-holidays is about 1000 persons/day. The intercity travel intensity in the province reached about 149.56 million persons · Km/day, including 24.25 million persons Km of cross-city travel relying on high-speed rail every day, accounting for about 16.2%. The intensity of cross-city travel relying on roads has reached 125.31 million persons Km every day.
2) Network Primacy and Multicentricity Analysis. The cities with the largest scale of intercity travel in the province are Xiamen, Quanzhou, Zhangzhou, and Fuzhou, of which the communication volume between Xiamen and cities in the province reached 598 thousand persons every day, the first degree reached 54.9%. The communication volume between Quanzhou, Zhangzhou and Fuzhou and cities in the province has reached 397 thousand persons every day, 360 thousand persons every day and 327 thousand persons every day. The intercity travel network presents a multi-center pattern, with a multi-center degree of 60.4%. In terms of the first contact strength, the agglomeration effect of central cities is obvious. Most of the first contact strength is the channel radiating the surrounding cities with the central city as the core. The intercity travel volume between Xiamen and Zhangzhou, the first contact strength, has reached 260000 persons every day, and the intercity travel volume between Xiamen and Quanzhou has reached 195 thousand persons every day. The intercity travel volume from the provincial capital city Fuzhou to the surrounding Ningde and Putian reached 61 thousand persons every day and 57 thousand persons every day, respectively.

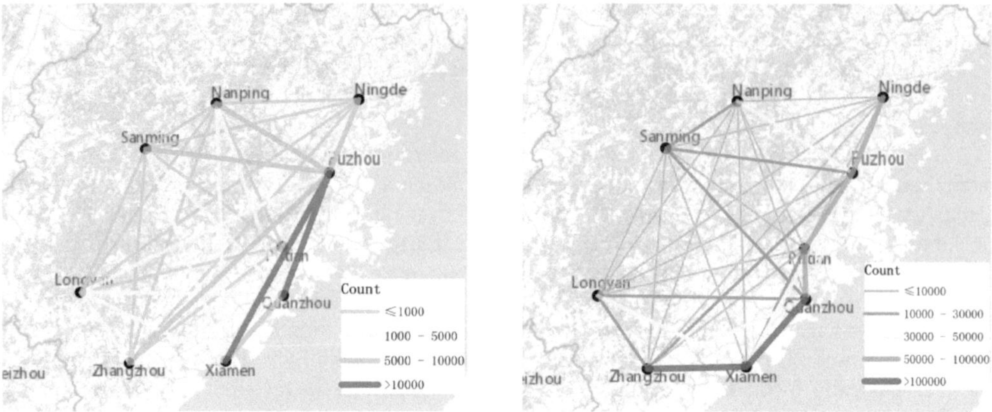

① High Speed Railway ② Highway

Figure 1. Schematic diagram of intercity travel scale in Fujian Province.

Table 1. Intercity travel scale of cities in Fujian Province.

	Travel Scale (1000 persons/day)	By High Speed Railway (1000 persons/day)	Proportion (%)	By Highway (1000 persons/day)	Proportion(%)
Xiamen	598.1	47.6	8.0%	550.5	92.0%
Quanzhou	397.2	28.7	7.2%	368.5	92.8%
Zhangzhou	360.0	16.4	4.6%	343.6	95.4%
Fuzhou	327.4	70.1	21.4%	257.2	78.6%
Putian	144.6	12.9	8.9%	131.7	91.1%
Longyan	95.6	12.7	13.3%	83.0	86.7%
Sanming	87.7	14.1	16.0%	73.6	84.0%
Ningde	86.1	14.9	17.3%	71.2	82.7%
Nanping	80.3	14.4	17.9%	66.0	82.1%

3) Analysis of attenuation characteristics of intercity traffic travel with distance. In terms of modes, the weighted average intercity travel distance of high-speed rail in the province is 209 km. Cities with high-speed rail intercity travel volumes of more than 10 thousand persons every day include Fuzhou-Xiamen and Fuzhou-Quanzhou, and cities with intercity travel volumes of 5–10 thousand persons every day include Fuzhou-Nanping, Fuzhou-Ningde, Fuzhou-Sanming, Quanzhou-Xiamen and Fuzhou-Zhangzhou. The weighted average intercity travel distance of highways in the province is 129 km. Cities with more than 10 thousand persons every day of highway intercity travel include Xiamen-Zhangzhou and Xiamen-Quanzhou. Cities with 50–100 thousand persons every day of intercity travel include Fuzhou-Putian, Fuzhou-Ningde and Quanzhou-Putian. Cities with 30–50 thousand persons every day of intercity travel include Quanzhou-Zhangzhou, Fuzhou-Quanzhou, Fuzhou-Xiamen, and Xiamen-Longyan. The travel distances between Fuzhou and Nanping are 72 km, 94 km and 189 km, respectively. The intercity highway travel volume decreases obviously with distance, while the cross-city travel volume of high-speed rail decreases obviously with distance, which may be related to the spatial scope of the study. The travel volume and distance of the intercity highway are fitted, which basically conforms to the negative logarithmic relationship.

Table 2. Intercity travel volume and travel distance.

High Speed Rail	Intercity travel volume	≥ 10000 persons/day	5000~10000 persons/day	1000~5000 persons/day	≤ 1000 persons/day
	Cities	Fuzhou-Xiamen Fuzhou-Quanzhou	Fuzhou-Nanping Fuzhou-Ningde Fuzhou-Sanming Quanzhou-Xiamen Fuzhou-Zhangzhou	15 groups	14 groups
	Average Distance	215 km	186 km	223	286 km
HighWay	Intercity travel volume	≥ 100000 persons/day	50000~100000 persons/day	30000~50000 persons /day	≤ 30000 persons/day
	Cities	Xiamen-Zhangzhou Xiamen-Quanzhou	Fuzhou-Putian Fuzhou-Ningde Quanzhou-Putian	5 groups	26 groups
	Average Distance	72 km	94 km	189 km	282 km

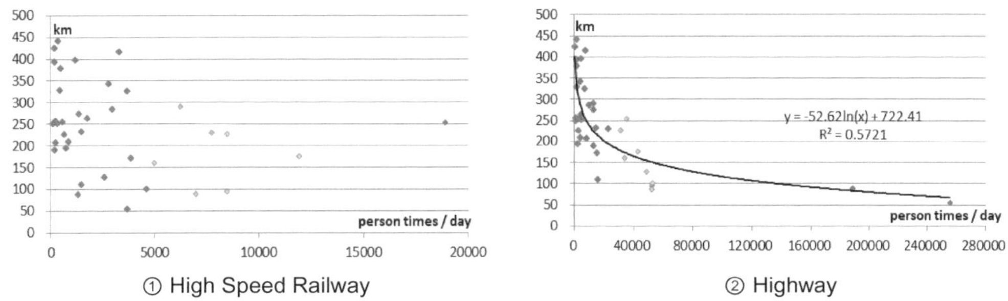

Figure 2. Scatter diagram of intercity travel volume and distance.

4) Subgroup analysis of provincial travel network. The subgroups in the travel network are divided based on the connection strength between cities, and each subgroup is a relatively closely connected urban group. SPSS software is used to divide subregions according to the intercity travel network, and the results are shown in Figure 3. From the perspective of the 2-plex (secondary level), there are two agglomerative subgroups in the economic network of urban agglomeration. From the perspective of the 3-plex, there are five agglomerative subgroups in the economic network of urban agglomeration. At the 3-plex level, Xiamen-Quanzhou-Zhangzhou and Fuzhou-Putian-Ningde have been formed, which are also the two urban agglomerations with

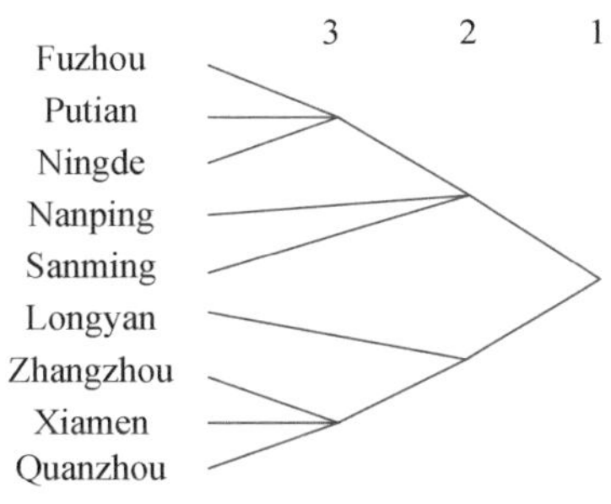

Figure 3. Provincial travel network subgroup analysis.

the most densely populated cities and towns and the most developed economy and industry in Fujian Province. At the 2-plex level, two subgroups of Xiamen-Quanzhou-Zhangzhou- Longyan and Fuzhou-Putian-Ningde-Nanping-Sanming have been formed, showing that Fujian Province has basically formed a pattern of radiating and driving the development of surrounding cities led by the double centers of Fuzhou and Xiamen.

4 MAIN CONCLUSIONS

Based on the mobile signaling datum, this paper calculates the intercity travel volume between cities in Fujian Province and makes a comparative study on the characteristics of travel scale, connection strength, network agglomeration and subregion. The results show that highway traffic is still the main mode of intercity travel in Fujian Province, accounting for about 90% of intercity travel. At the same time, the traffic volume of intercity highways decreases obviously with distance, but this characteristic of high-speed rail is not obvious. From the characteristics of the intercity travel network, the primacy degree of Xiamen has reached 54.9%, and the multi-center degree has also reached 60.4%. Two network subgroups of Xiamen-Quanzhou-Zhangzhou and Fuzhou-Putian-Ningde have been preliminarily formed, which is basically consistent with the current economic and urban distribution pattern of "strong coastal area and weak mountainous area" and "double core and multi-center" in Fujian Province.

REFERENCES

Aguilera V, Allio S, Benezech V, et al (2014). Using Cell Phone Data to Measure Quality of Service and Passenger Flows of Paris Transit System[J]. Transportation Research Part C: Emerging Technologies, 43: 198–211.

Calabrese F, Colonna M, Lovisolo P, et al (2011). Real-time Urban Monitoring Using Cell Phones: A Case Study in Rome [J] IEEE Transactions on Intelligent Transportation Systems,12(1):141–151.

Guo L, Bi Y F, Huang J Z, Zheng G Y, et al. (2018). Multi-scale Comparison and Analysis of Jobs-housing Spatiale Characteristics in Big Cities——Taking Wuhan as an Example[J]. Urban Planning Forum, 5: 88–97.

Huang J, Du D L, Wang J E et al. (2020). A Comparative Study on Network Structures of High-speed Train and Inter-city Coach Based on the Urban Agglomerations. Scientia Geographica Sinica, 40(12):1958–1966.

Juan C, Herrera A, Daniel B, et al (2010). Evaluation of Traffic Data Obtained via GPS-enabled Mobiles Phones: The Mobile Century Yield Experiment[J]. Transportation Research Part C: Emerging Technologies,18:568–583.

Lao X, Shen T Y, Yang Y, Zhang Y (2016). A Study on the Economics Network of the Urban Agglomeration in the Middle Reaches of the Yangtze River: Based on Social Network Analysis Method with Gravity Model[J]. Urban Development Studies23(7): 91–98.

Li H N (2018). Spatial Dimension Characteristics Analysis of Passenger Intercity Travel Demand in Different Urban Agglomeration Spatial Pattern in China[D]. Beijing: Beijing Jiaotong University.

Li T, Zhou R (2016). Urban Hinter World in Yangtze River Delta: Empirical Comparison of Two Network-Based Methods[J]. Acta Geagrophica Sinica, 71(2):236–250.

Li W, Cheng X, Yang D (2015). Mining Rail-transit Use Patterns from Mobile Phone Data: Case Study of Three Communities in Shanghai[C]. 94th Annual Meeting of the Transportation Research Board, Washington, D.C.

Luo Z D, He H M, Wei J L (2012). Intercity Relationship and Structure of Anhui Province from the Perspective of Passenger Flow of Coach Travel[J], Acta Geographica Sinica, 32(10): 1193–1199.

Niu X Y, Ding L, Song X D, et al (2014). Understanding Urban Spatial Structure of Shanghai Central City Based on Mobile Phone Data[J]. Urban Planning Forum,6:61–67.

Ran B (2013). Use of Cellphone Data in Travel Survey and Transportation Planning [J]. Urban Transport of China,1(11):72–83.

Smoreda Z, Olteanu-Raimond A M, Couronne T (2013). Spatiotemporal Data from Mobile Phones for Personal Mobility Assessment[J]. Transport Survey Methods: Best practice for decision making, 41:745–767.

Sun Q, Tang F, Tang Y (2015). An Economic Tie Network-structure Analysis of Urban Agglomeration in the Middle Reaches of Changjiang River based on SNA [J]. Journal of Geographical Sciences, 25(6):739–755.

Xu Y, Shaw S, Zhao Z, et al (2016). Another Tale of Two Cities: Understanding Human Activity Space Using Actively Tracked Cellphone Location Data[J]. Annals of the American Association of Geographers, 106(2): 489–502.
Zhou X G, Sun C C, Niu X Y (2021), Spatial Scale Problem of Jobs-Housing Relationship Based on Cellphone Data: Case Studies of Shanghai and Shenzhen [J]. Urban Planning International, 5(36):78–85.
Zhou X, Yeh A Go, Yue Y (2018). Spatial Variation of Self-containment and Jobs-housing Balance in Shenzhen Using Cellphone Big Data[J]. Journal of Transport Geography, 68: 102–108.

Advances in Urban Engineering and Management Science – Khalil & Yang (Eds)
© 2023 The Author(s), ISBN 978-1-032-30426-7

Short-term passenger flow prediction of urban rail transit based on SSA-GRU

E.Y. Chen* & T. Zhang*
College of Mechatronics and Control Engineering, Shenzhen University, Shenzhen, Guangdong, China

Y.X. He* & Q. Luo*
School of Urban Transportation and Logistics, Shenzhen Technology University, Shenzhen, Guangdong, China

ABSTRACT: In the context of the rapid development of urban rail transit, accurate and real-time short-term passenger flow forecasting is of great importance for the operation of urban rail transit. Considering stations with different passenger flow characteristics require different prediction models, this paper uses the K-means algorithm to cluster the stations with the characteristics of inbound and outbound passenger flow time distribution. After that, to improve the accuracy of the short-term passenger flow prediction results of urban rail transit, based on the passenger flow prediction using GRU alone, the SSA-GRU model combining GRU and sparrow search algorithm (SSA) is proposed for the problems of difficult adjustment of its model hyperparameters and slow convergence, and the hyperparameters of GRU are used as the parameter optimization target of SSA to complete the modeling and prediction. The experiment proves that the SSA-GRU model has good prediction performance and can be effectively applied to short-term passenger flow prediction of urban rail transit.

1 INTRODUCTION

With the rapid development of urban rail transportation, short-term passenger flow prediction has attracted the attention of researchers and scholars because of its important practical significance. If we can provide short-term metro passenger flow prediction service and accurately predict the passenger flow of metro stations in advance, the metro operation management can change the operation plan and take related control measures in advance, which can effectively prevent the occurrence of danger and maintain the order of metro, and improve the accuracy and effectiveness of dispatching. Decision-makers can plan the route direction, adjust the scale of stations and optimize the layout of station facilities based on the predicted passenger flow data. Travelers can change their travel plans according to their needs in time to avoid the risk of trampling caused by overcrowding. Travelers can change their travel plans according to their needs to avoid the risk of trampling caused by overcrowding. Therefore, it is necessary and urgent to research short-time passenger flow prediction.

As a branch of machine learning, deep learning models have become a mainstream approach for passenger flow prediction in recent years. Since deep belief networks were first proposed in 2006 (Hinton 2006), the application of deep neural networks in computer science has received much attention. With the rapid construction of transportation infrastructure, many researchers have applied deep neural networks to solve urban rail and road traffic problems due to their powerful capabilities. Park et al. (Park & Rilett 1999) used feed forward neural networks to accomplish

*Corresponding Authors: cchenenyi@163.com, ztdx2012@163.com, heyuxin@sztu.edu.cn and luoqin@sztu.edu.cn

 DOI 10.1201/9781003305026-67

the traffic flow prediction task. Lv et al. (Lv et al. 2015) used SAEs to predict traffic flow for the first time, and the results proved that it outperformed SVM and feedforward neural network (FFNN). Huang et al. (Huang et al. 2014) proposed a network structure consisting of a deep confidence network and a regression model and verified that the network could capture random features from traffic data of multiple datasets, and the model improved the accuracy of traffic prediction. In addition, since RNN and its variants long short-term memory network LSTM and gated recurrent neural network GRU can effectively use the self-loop mechanism, they can learn the time dependence well and obtain better prediction results, so they have gained wide attention and application. Ma (Ma et al. 2015) used LSTM for traffic speed prediction and demonstrated that the LSTM model is better than most nonparametric models and ARIMA (2, 1, 1) for better prediction. Fu et al. (Fu et al. 2016) first used GRU for traffic flow prediction in 2016.GRU is essentially a simplified version of LSTM, which simplifies the structure of LSTM and improves the training speed of the model, and is one of the common deep learning methods used to perform short-term passenger flow prediction for urban rail transit.

Deep learning models are essentially data-driven methods, which only need to learn from a large amount of data to learn the changing pattern of passenger flow and predict future passenger flow. However, these data-driven methods have difficulties, such as single models or difficulty adjusting model hyperparameters.

For hyperparameter tuning, the industry is currently using grid search, stochastic search and various optimization algorithms, respectively. In fact, grid search and stochastic are very common and generally effective methods, which are not necessarily better than personal experience in the case of limited computational resources. Therefore, optimization algorithms are introduced to tune the hyperparameters.

At present, the most used optimization algorithms include genetic algorithm, particle swarm algorithm, crow algorithm, whale algorithm and so on. Compared with the above algorithms, the sparrow search algorithm, a swarm intelligence optimization algorithm proposed by Xue (Xue & Shen 2020) of Donghua University in 2020 based on the behavior of sparrows foraging and fleeing from predators, has obvious advantages and simple structure in terms of convergence speed and optimization-seeking accuracy, and can accurately cope with complex problems.

Since different stations have different passenger flow characteristics and the requirements of prediction models are not consistent for stations with different passenger flow characteristics, the stations are first clustered and analyzed by using the K-means algorithm with the temporal distribution of incoming and outgoing passengers flows as the characteristics. After that, the SSA-GRU model is proposed for the long-term time dependence of urban rail transit passenger flow, in which the GRU model can effectively capture the time dependence and has a simple structure, while the sparrow search algorithm is used to solve the hyperparameter tuning problem of the GRU model. To verify the effectiveness of the model, different types of representative stations are selected as examples, and comparison experiments are conducted using simple GRU and SSA-GRU. The research framework of this paper is shown in Figure 1.

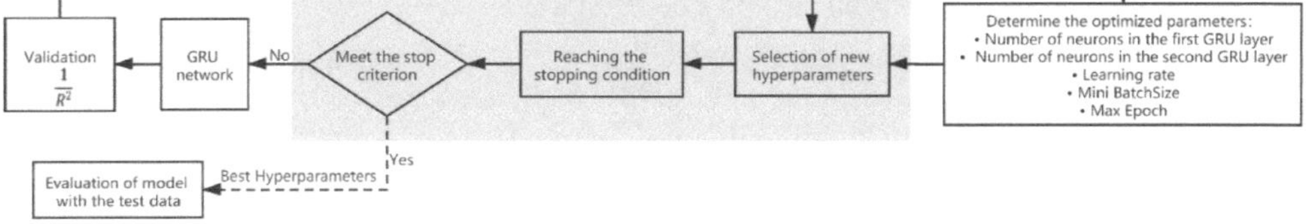

Figure 1. Proposed framework of this paper.

2 METHODOLOGY

2.1 *Station clustering based on K-means*

K-means is our most commonly used clustering algorithm based on Euclidean distance. It believes that the closer the distance between two targets, the greater the similarity.

Assuming that the total number of urban rail transit stations is N, the stations can be expressed as S_x, $x = 1, 2 \ldots N$. Suppose the time-sharing inbound passenger flow of the station S_x in one day is $I_{x,1}, I_{x,2}, \ldots, I_{x,t}$, where $x = 1, 2, \ldots, N$, the minutes in one day When the outbound passenger flow is $O_{x,1}, O_{x,2}, \ldots, O_{x,t}$, the inbound and outbound passenger flow of the station S_x can be expressed as: $\vec{S_x} = (I_{x,1}, I_{x,2}, \ldots, I_{x,t}; O_{x,2}, \ldots, O_{x,t})$, and set the number of cluster centers to K and the station cluster to G_y Then the passenger flow of y-type stations can be expressed as: $\vec{G_x} = (I_{x,1}, I_{x,2}, \ldots, I_{x,t}; O_{x,2}, \ldots, O_{x,t})$, where $y = 1, 2, \ldots, K$, the calculation process is as follows (Luo et al. 2021):

Step (1): Calculate the distance between the sample data and the original clustering center and make the objective function minimum, as shown in equation (1), i.e.

$$\sum_{i=1}^{N} \min_{j \in \{1,2,\ldots,k\}} \sqrt{\sum_{p=1}^{m} \left(\left(I_i^p - I_j^p\right)^2 + \left(O_i^p - O_j^p\right)^2 \right)} \tag{1}$$

Step (2): Set the set of stations included in category y as $L_{y,i}$, and calculate the cluster centers of each category. As shown in formula (2), it is:

$$\vec{G_J} = \frac{\sum_{\vec{S_l} \in L_i} \vec{S_l}}{|L_i|} \tag{2}$$

Step (3): Repeat steps (1) and (2). When the change of $\vec{G_y}$ is less than the threshold σ, the cluster analysis ends, as shown in equation (3).

$$\left| \frac{\sum_{\vec{S_l} \in L_i} \vec{S_l}}{|L_i|} - \vec{G_J} \right| < \sigma \tag{3}$$

2.2 *Gated recurrent neural networks*

Although the traditional RNN can handle the sequence data better, when the input sequence is beyond a certain range, the problem of gradient disappearance or gradient explosion easily occurs due to the back propagation algorithm and long time dependence. To solve the above problems, scholars have proposed the long short-term memory network LSTM and the gated recurrent unit network GRU.

GRU was proposed by Cho et al. (Cho et al. 2014), which is essentially a variant form of LSTM, but compared to LSTM, GRU has a simpler structure, possesses fewer parameters, is easy to compute, and can improve the training efficiency of the model to some extent.

The structure of GRU is the same as traditional RNN and LSTM. Based on the traditional RNN, GRU adds a reset gate and update gate, and the structure of GRU is shown in Figure 2.

2.3 *Sparrow search algorithm*

The Sparrow Search Algorithm (SSA) is a swarm intelligence optimization algorithm based on the behavior of sparrows foraging and escaping from predators. The algorithm mainly simulates the foraging process of sparrows, and also overlays a scouting and warning mechanism. The individual sparrow that finds the better food acts as the finder, and the other individuals act as followers, while

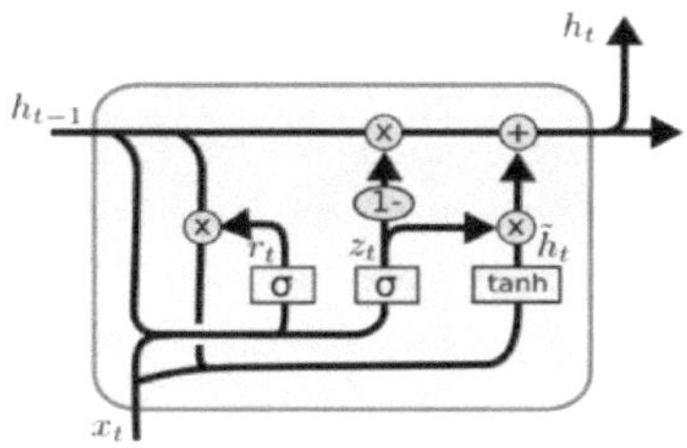

Figure 2. GRU structure diagram.

a certain percentage of the population is selected for scouting and warning, and if danger is found, the food is abandoned, and safety comes first. (Xue & Shen 2020)

The SSA algorithm flow chart is shown in Figure 3.

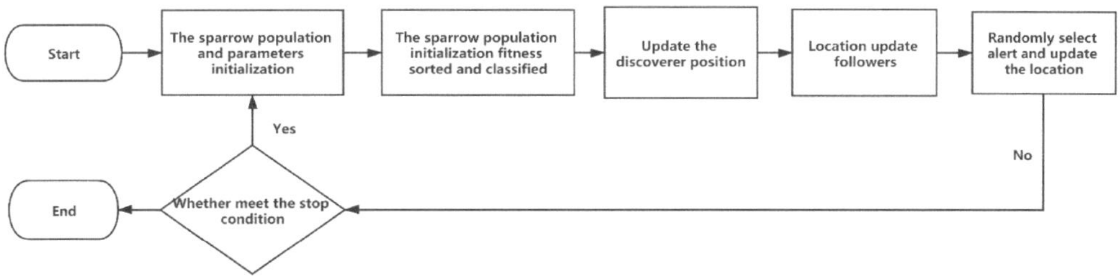

Figure 3. SSA flow chart.

3 MODEL DEVELOPMENT

In GRU, generally, hyperparameters such as learning rate, number of GRU hidden layer neurons, and epoch have a large impact on the accuracy of prediction results. In this paper, the number of GRU hidden layer layers set is 2. Therefore, the five hyperparameters of the first GRU hidden layer neurons, the second GRU hidden layer neurons, learning rate, MiniBatchSize and MaxEpoch in GRU are used as the optimization variables of SSA, and the SSA-GRU algorithm is proposed as follows.

Step 1: Sparrow population initialization and parameter determination.

Determine the number of sparrow populations, determine the sparrow location finding dimension according to the number of hyperparameter groups, and determine the proportion of the population occupied by the discoverer sparrow, which is generally taken as 20%. Determine the upper and lower boundaries of hyperparameters, and initialize the sparrow location, which consists of the superiority-seeking hyperparameter group. The initialization parameters include the best fitness p_{Fit}, the best sparrow position corresponding to p_{Fit}, the global optimal fitness f_{Min}, and the global best position $best_X$ corresponding to f_{Min}

Step 2: Initialization of sparrow population ranking and classification.

The initialized sparrow locations are evaluated and ranked according to the fitness function, the top 20% are found, and 10%–20% of sparrows are randomly selected to carry reconnaissance warning mechanism actions.

Step 3: Update the sparrow position by discoverer, follower and warning condition formulas.

Step 4: Constrain the hyperparameters with a boundary function, predict the hyperparameter pass-through parameters required for GRU, and return the result to evaluate the position by the fitness function. If the current sparrow fitness is better than the sparrow fitness at the optimal location, it is substituted; otherwise, it remains unchanged. If the optimal sparrow adaptation in this iteration is better than the global best adaptation, then substitute; otherwise, remain unchanged.

Step 5: Judge whether the set stopping conditions of reaching error and the maximum number of iterations are satisfied. If so, set the global optimal hyperparameter set as the parameters of GRU; otherwise, return to step 2.

4 CASE STUDY RESULTS AND DISCUSSION

4.1 *Dataset description*

The dataset was collected from the AFC system of Shenzhen Metro Cooperation in China, which covers a time span of one consecutive month from August 1 to August 31 (The opening hours are from 6:00 to 24:00) in the year 2019. By 2019, there will be eight metro lines and 172 stations in Shenzhen. Considering the difference in passenger flow characteristics between weekdays and weekends, in order to improve the prediction accuracy of the prediction model, this paper only uses weekday passenger flow data for prediction. The data of 21 working days were used in this experiment, and the specific data form and field descriptions of each record are listed in Tables 1 and 2. The AFC raw data need to be cleaned and processed to remove the wrong swipe data and employee self-entry and exit data and fill in the missing data. After that, the AFC raw data are aggregated into inbound passenger flow data with a time interval of 10 minutes. The first 16 days of inbound traffic were used as the training set, and the next five days of inbound traffic were used as the test set to verify the model's single-step prediction effect.

Table 1. Original AFC data record.

Card Number	Trading Date & Time	Transaction Types	Line Stations	Settlement Date
324312693	20190801022457	Bus	204	20190801
667285636	20190801023415	Bus	3	20190801
880023346	20190801043626	Subway Inbound	Baigelong	20190801
881300346	20190801045436	Subway Inbound	Shenzhen North	20190801
881301311	20190801050242	Subway Outbound	Baishilong	20190801

Table 2. Field description.

Fields	Field Description
Card Number	User Card Number
Trading Date & Time	yyyyMMddhhmmss
Transaction Types	Such as subway entry, subway exit, QR code bus
Line Stations	When the transaction type is subway inbound or subway outbound, the value is the name of the subway line; otherwise, the value is the name of the bus company
Settlement Date	yyyyMMddhhmmss

4.2 *K-means clustering results*

To eliminate the influence of single-day error and at the same time to avoid too many data dimensions and excessive complexity, the extracted 21-day time-sharing passenger flow data are taken as the average value of the original value of the single-day time-sharing passenger flow at each station. The inbound and outbound passenger flow statistics of each station from 6:00 to 24:00 are obtained.

Because of the large differences between the passenger flow levels of each station, it is easy to interfere with the analysis algorithm to make different stations with different levels of passenger flow comparable, so the data is based on passenger flow characteristics Z-score standardization. Finally, K-means is used to cluster the processed inbound and outbound passenger flows.

Due to the different nature of land development and location conditions around each station, stations of different nature are formed, and the corresponding passenger flow patterns and passenger flow composition are also different. According to the time distribution characteristics of passenger flow in and out of stations at different times of the day, the elbow method is used to determine the category number k. The experiment finds that the curvature is maximum when $k = 5$, so the 172 stations of Shenzhen Metro can be divided into five major categories, namely residential, office, mixed residential, mixed office, commercial and attractions. The station types and the representative stations are shown in Table 3. The time-distribution characteristics of inbound and outbound passenger flow of different stations are shown in Figure 4.

Table 3. Result of station cluster analysis.

Station Type	Representative Station
Residential Type	Liyumen, Pingzhou, Xixiang, Gushu, Airport East, Baishizhou, Buji, Longhua, Minle, Minzhi, Baishilong, Fanshen, Buxin etc.
Office Type	Shenda, Hi-Tech Park, Qiaocheng North, Chegongmiao, Xiangmihu, Convention & Exhibition Center Station, Guoao, Keyuan, Civic Center, Shopping Park Station, Futian, Gangxia North, Baohua, etc.
Mixed-Residential Type	Bao'an Stadium, Honglang North, Window of the World Station, Xiangmi, Shuanglong, Fumin, Jingtian, Lingzhi, etc.
Mixed-Office Type	Taoyuan, Bao'an Center, Zhuzilin, Gangxia, Chiwan, Shuiwan, Tongxinlin, Xili, Shangmeilin, etc.
Commercial and Attraction Type	Huaqiaocheng, Laojie, Science Museum, Luohu, Shenzhen North, Children's Palace, Futian Checkpoint, Houhai, etc.

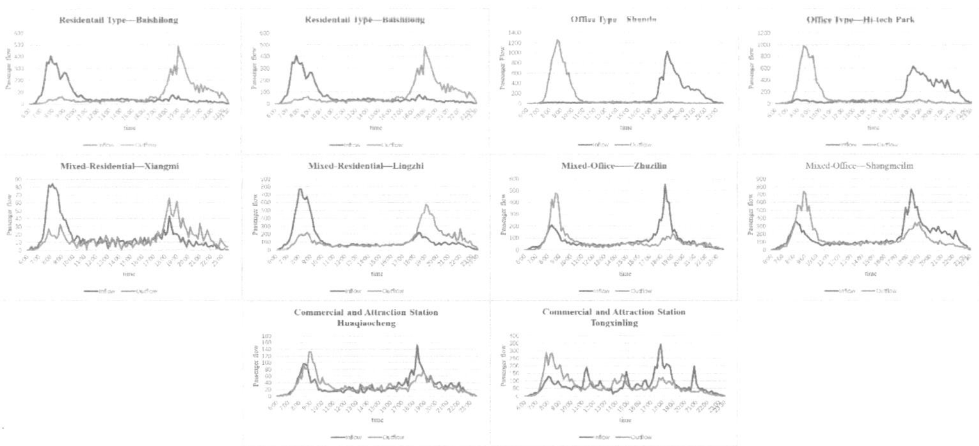

Figure 4. Time distribution curve of passenger flow in and out of stations of different types.

As shown in Figure 4, the inbound and outbound traffic of both residential and office stations have obvious single peaks, with residential stations having obvious morning peak inbound and evening peak outbound traffic and office stations having obvious morning peak outbound and evening peak inbound traffic. The mixed-residential and mixed-office sites are bimodal, but the peaks of the morning peak inbound and evening peak outbound traffic are more pronounced for the mixed residential sites than for the other two peaks, while the opposite is true for the mixed office sites. For the commercial and attraction type stations, the inbound and outbound traffic is more volatile than the other types of stations.

4.3 *Sparrow search algorithm parameter setting*

In this experiment, the parameters of the sparrow search algorithm are defined as follows: the number of sparrow populations is 10, the search dimension is 5, the number of iterations of the search algorithm is 2, the initialization range of sparrow positions (GRU hyperparameter initialization) are: learning rate [0.001, 0.01], the number of neurons in both GRU layers is [20, 100], MiniBatchSize is [10, 200], and MaxEpoch is [10, 50]. Contrast the definition of simple GRU hyperparameters: learning rate of 0.005, number of neurons in both GRU layers is 20, MiniBatchSize is 106, and MaxEpoch is 10.

4.4 *Evaluation metrics*

This experiment selects three evaluation indicators, namely R^2, RMSE and MAE.

$$R^2 = 1 - \frac{\sum_i (\hat{y}_i - y_i)^2}{\sum_i (\hat{y}_i - \bar{y})^2} \tag{4}$$

$$RMSE = \sqrt{\frac{1}{m}\sum_{i=1}^{m} (y_i - \hat{y}_i)^2} \tag{5}$$

$$MAE = \frac{1}{m}\sum_{i=1}^{m} |(y_i - \hat{y}_i)| \tag{6}$$

where m is the sample size, y_i is the true value, and $\hat{y}_i$ is the predicted value.

4.5 *Result analysis*

In this experiment, a residential station Xixiang, an office station Shenda, a mixed residential station Xiangmian, a mixed office station Zhuzilin and a commercial and attraction station Huaqiaocheng are selected for case studies to verify the validity of the SSA-GRU model. The SSA optimization algorithm finds the optimal hyperparameters, as shown in Table 4.

Table 4. Optimized hyperparameter results.

Station Type	Number of neurons in the 1st GRU layer	Number of neurons in the 2nd GRU layer	Learning rate	Mini-Batchsize	MaxEpoch
Residential Station Baishizhou	94	23	0.01	143	44
Office Station Shenda	100	90	0.01	113	50
Mixed Residential Station Xiangmi	96	100	0.01	200	50
Mixed Office Station Zhuzilin	50	89	0.003	16	50
Commercial and Attraction Station Huaqiaocheng	64	99	0.01	129	50

The optimal hyperparameters obtained by the SSA optimization algorithm are used to adjust the hyperparameters of the GRU model, and the experimental results of the SSA-GRU model and the GRU model are shown in Table 5. From Table 5, it can be seen that for different types of stations, the prediction effect of the SSA-GRU model is better than that of the simple GRU model. The RMSE and MAE of office type Shenda station, residential type Baisizhou station and mixed biased office type Zhuzilin station predicted by SSA-GRU all get a substantial decrease, and the R^2 of residential type Baisizhou, mixed biased office type - Zhuzilin, commercial and attraction type - Huaqiaocheng, mixed biased residential type - The average decrease in RMSE is about 24.5, the average decrease in MAE is about 14.4, and the average increase in R^2 is about 0.22. And as seen in Fig.5, the SSA-GRU prediction results are in good agreement with the actual values and significantly better than the simple GRU, indicating that the SSA algorithm can effectively tune the hyperparameters of GRU, thus verifying the validity.

Table 5. Comparison of experimental results between SSA-GRU model and GRU model.

Station	Evaluation Metrics	GRU	SSA-GRU	Optimization Results
Office Station Shenda	RMSE	91.1104	39.9032	51.2072
	MAE	54.9974	24.3738	30.6236
	R2	0.84985	0.96968	0.11983
Residential Station Baishizhou	RMSE	57.3866	20.9588	36.4278
	MAE	34.922	15.3549	19.5671
	R2	0.73633	0.96471	0.22838
Mixed-Office Station Zhuzilin	RMSE	58.5112	34.7724	23.7388
	MAE	34.5101	19.8785	14.6316
	R2	0.58809	0.85454	0.26645
Commercial and Attraction Station Huaqiaocheng	RMSE	19.2937	13.4492	5.8445
	MAE	13.3814	8.9927	4.3887
	R2	0.53573	0.77544	0.23971
Mixed-Residential Station Xiangmi	RMSE	12.8449	7.5895	5.2554
	MAE	8.3215	5.4824	2.8391
	R2	0.6243	0.86813	0.24383

5 CONCLUSION

In recent years, the emerging deep learning models have been widely used in urban rail transit short-term passenger flow prediction, but the problem of tuning hyperparameters comes along with it. Therefore, this paper innovatively uses the SSA algorithm to solve the hyperparameter tuning problem by combining GRU and SSA to form the SSA-GRU model for passenger flow prediction. Since the model requirements are different for stations with different passenger flow characteristics, the K-means clustering analysis is first performed on 172 stations of Shenzhen Metro with the time distribution of incoming and outgoing passenger flow as the characteristics. The stations are classified into five categories: residential, office, mixed residential, mixed office and commercial and attraction. The results show that compared with the GRU model, the prediction effect of the SSA-GRU model has improved significantly, with the average RMSE decreasing by about 24.5, the average MAE decreasing by about 14.4, and the average R^2 increasing by about 0.22. This proves that SSA is an effective super parametric optimization algorithm, and the SSA-GRU model can be effectively applied to urban rail transit short-term passenger flow forecasting. The real-time and accurate prediction results can help the operation management accurately grasp the trend of passenger flow, provide the basis for capacity allocation and passenger organization, and take early warning and evacuation measures for large events and emergencies; at the same time, it can also guide public travel to improve the speed, safety and comfort of travel. Thus, the SSA-GRU passenger flow prediction model proposed in this paper is of great practical significance.

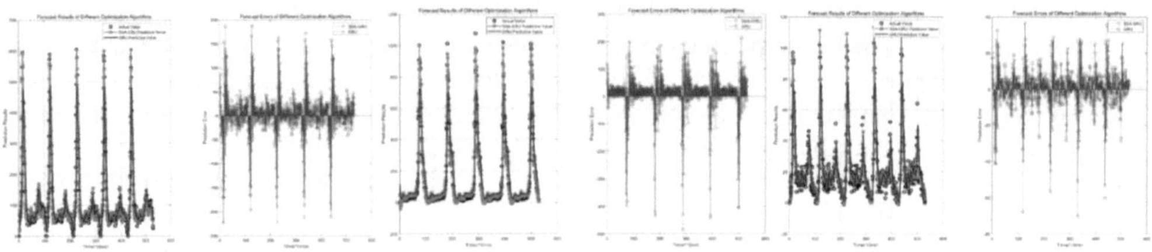

(a) Residential Station—Baishizhou (b) Office Station—Shenda (c) Mixed Residential Station—Xiangmi

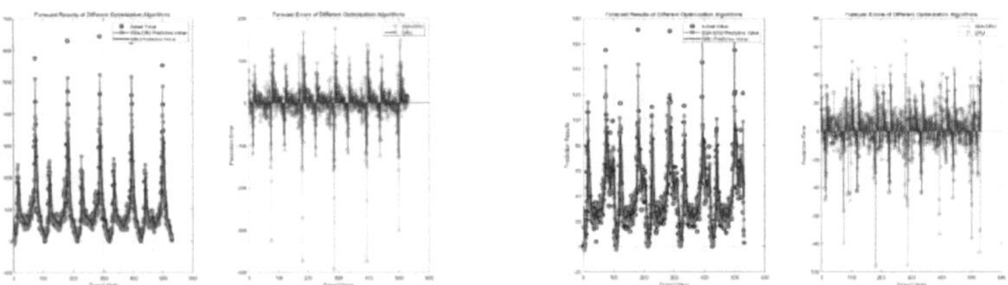

(d) Mixed Office Station—Zhuzilin (e) Commercial and Attraction Station—Huaqiaocheng

Figure 5. Experimental results of passenger flow prediction for different types of stations.

REFERENCES

Cho, K., B. V. Merrienboer, C. Gulcehre (2014). Learning Phrase Representations using RNN Encoder-Decoder for Statistical Machine Translation. J. Computer Science.

FU, R., Z. ZHANG, L. Li (2016). Using LSTM and GRU Neural Network Methods for Traffic Flow Prediction. C.//201631st Youth Academic Annual Conference of Chinese Association of Automation (YAC). Wuhan: IEEE, 2016: 324-328.

Hinton, G. E., S. Osindero, Y. W (2006). Teh, A fast learning algorithm for deep belief nets, Neural Comput, vol. 18, no.7, pp. 1527–1554.

Huang, W., G. Song, H. Hong (2014). Deep Architecture for Traffic Flow Prediction: Deep Belief Networks with Multitask Learning. J. IEEE Transactions on Intelligent Transportation Systems, 15(5): 2191-2201.

Luo, Q., Y. Cai, M. L. Xu, Y. Liang (2021). Passenger flow prediction model of urban rail transit network at origin and destination under new line access conditions. J. Urban Rail Transit Research, 24(08): 36-40+46.

Lv, Y., Y. Duan, W. Kang (2015). Traffic Flow Prediction with Big Data: A Deep Learning Approach. J. IEEE Transactions on Intelligent Transportation Systems, 16(2): 865-873.

Ma, X. L., Z. Tao, Y. Wang, H. Yu, Y. Wang (2015). Long short-term memory neural network for traffic speed prediction using remote microwave sensor data. J. Transportation Research Part C, 54.

Park, D. & L. R. Rilett (1999). Forecasting Freeway Link Travel Times with a Multilayer Feedforward Neural Network. J. Computer-Aided Civil and Infrastructure Engineering, 14(5): 357-367.

Xue, J., B. Shen (2020). A Novel Swarm Intelligence Optimization Approach: Sparrow Search Algorithm. J. Systems Science &Control Engineering, 8(1): 22-34.

Advances in Urban Engineering and Management Science – Khalil & Yang (Eds)
© 2023 The Author(s), ISBN 978-1-032-30426-7

Infrastructure construction of elderly community based on digital network intelligence

Yi Hou*
Consulting Department, GI Yingting Pty Ltd., Melbourne, Australia

ABSTRACT: With the intensification of population aging, pensions have become a hot issue of social concern. To improve the health monitoring and nursing service level of the elderly, this paper designs a smart elderly care system based on big data technology. The article first analyzes the problems of traditional community infrastructure in aging cities. Then we analyze how digital, networked and intelligent construction of new community infrastructure can improve the quality of life of the elderly. Finally, we discussed the potential problems of digital network intelligence in the infrastructure construction of elderly communities.

Keywords: digitization; networking; Internet of things; smart old-age care

1 INTRODUCTION

By 2020, China's population aging growth rate will have ranked first globally. It is estimated that by 2030, the proportion of China's population over 65 years old will exceed that of Japan, which has the most aging population in the world. Population aging has become an obstacle to China's economic development, and solving the problem of population aging has also become a global problem. The growth of the elderly population has greatly increased the pressure on employment and life for young people. Due to the increased pressure of life and work, children cannot take care of the elderly around the clock. Therefore, nursing homes have become the choice of many retirees (Sun & Qiao 2018). At present, most nursing homes in China are in the form of manual care, and there are problems such as insufficient management personnel, low professionalism, imperfect elderly care facilities, and untimely emergency measures (Figure 1 shows the aging of China's population).

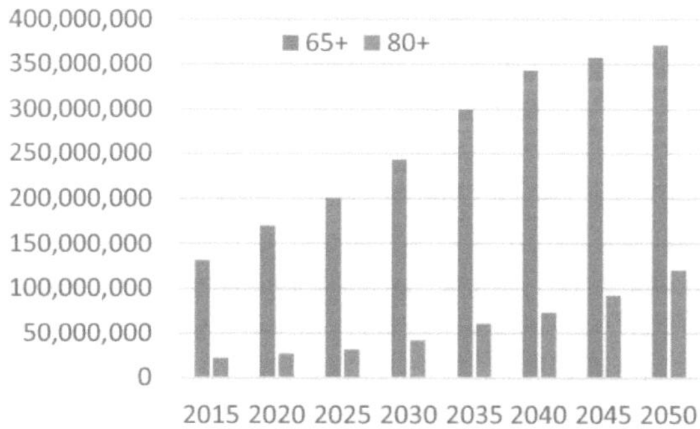

Figure 1. China's aging population.

With the superposition of intelligence, informatization and population aging, intelligent medical products have become the basic support and open interface of the social elderly care service system,

*Corresponding Author: houyihy2021@126.com

breaking through the time and space limitations of traditional medical care, expanding the scope of health management, and realizing the intelligent interconnection of medical products (Wu 2018). On the one hand, the user customization mode of smart medical products meets the individual needs of the elderly, and improves the experience of elderly users on the basis of functional services; on the other hand, the information interaction module of smart medical products can collect and process the health data of elderly users, track the health status of the elderly, and prevent the occurrence of major health hazards for the elderly. Intelligent medical products for the elderly closely follow the trend of technology and adapt to the current Chinese family structure and pension model. However, due to the complicated operation of the existing product program, the multi-level overlay of the interface, and the iterative update of the system, the characteristics of the elderly user group have not been taken into account, resulting in the difference between the design of the product and the actual needs of the elderly user. Aiming at the interaction problem of elderly users operating smart medical products, this paper proposes a barrier-free interaction strategy to achieve the experience goal of product interaction availability, ease of use, and enjoyment.

2 PROBLEMS REFLECTED IN TRADITIONAL COMMUNITY INFRASTRUCTURE IN AGING CITIES

At present, due to the constraints of the development level of information technology and the development of the elderly care industry, domestic research on smart elderly care is still in its infancy, and there is no unified definition and industry standard. Through reference research, we believe that smart old-age care refers to the use of modern information technology (such as the Internet of Things, cloud computing, etc.) The automatic monitoring, warning and alarming and autonomous processing of information related to the elderly by modern intelligent equipment can provide convenient, efficient, and flexible services and management for the life of the elderly, realize the intelligent interaction between modern technology and the elderly, and have a good understanding of the experience and wisdom of the elderly (Wang 2012). Carrying out redevelopment and reuse can promote the continuous improvement of the material life and spiritual life of the elderly while realizing the value created by the elderly for social development, so that the elderly can live a more valuable, dignified and happy life.

2.1 *Lack of unified layout planning for social elderly care facilities*

The scale of social pension facilities is too large. Rural nursing homes are limited by traffic, environment and facilities. The elderly in the suburbs is reluctant to stay, resulting in a high vacancy rate and low utilization rate. Most of the elderly want to build elderly facilities in the suburbs. Due to the unreasonable distribution, the elderly care facilities are in short supply on the one hand, and a large number of them are idle on the other hand.

2.2 *The smart pension model is still not understood and accepted by the public*

Smart old-age care has just started, the exploration and practice time has not exceeded five years, and there is no systematic theoretical guidance. Most theories are derived from the exploration of old-age care practices, and are often limited by their own conditions and the type of institution they are in, and they are only fragments (Chen et al. 2018). There is no complete systematic theory from upstream and downstream and horizontal linkages, and a complete theoretical system of smart old-age care is impossible to talk about. The field investigation of some community public facilities found that there are problems in the use of community public facilities by the elderly, such as information identification difficulties, operational obstacles, and unsatisfactory demands. Smart elderly care products have not been understood, accepted and used skillfully by most older people. Guiding smart elderly care services is still only a high-quality service that only a small number of elderly people can enjoy. At present, there is serious polarization in smart elderly care, especially for high-quality high-end elderly care services. The price is too high, leading to a wisdom pension often becoming an "aristocratic pension" The vast majority of the elderly are discouraged, and even more unacceptable or even enter it.

2.3 *The informatization and intelligence of smart old-age care are low*

The development and application of modern information technology and intelligent equipment are still in their infancy. In the application, integration and processing of old-related information and data, it is still impossible to fully collect, effectively analyze, integrate, process and share data and information to provide an intelligent elderly care service. At present, the products are too single. Generally there are only portable bracelets or smartphones, which cannot meet the diverse and personalized needs of the elderly. At the same time, the development of information data and smart devices often lags behind the development of elderly care services, and cannot achieve effective docking with elderly care services. So, it is difficult to meet the actual needs of the elderly.

3 DIGITAL, NETWORKED AND INTELLIGENT CONSTRUCTION OF NEW COMMUNITY INFRASTRUCTURE IMPROVES THE QUALITY OF LIFE OF THE ELDERLY

3.1 *Operational goals of smart community infrastructure for the elderly care service model*

The community home-based elderly care service model is based on the home, relying on community elderly care service institutions, and with the participation of the government, society and family. It aims to provide elderly care, medical care, spiritual comfort, emergency assistance, culture, education and entertainment services for the elderly model (Xiao 2016). Starting from the general operation mechanism of the community-based home-based care service model, this paper proposes the current operational goals of the community-based home-based care service model in China based on four perspectives: supply-demand connection, demand expression, service supply, and service feedback. The specific design process is shown in Figure 2 (the picture is quoted from Remote health monitoring of the elderly through wearable sensors).

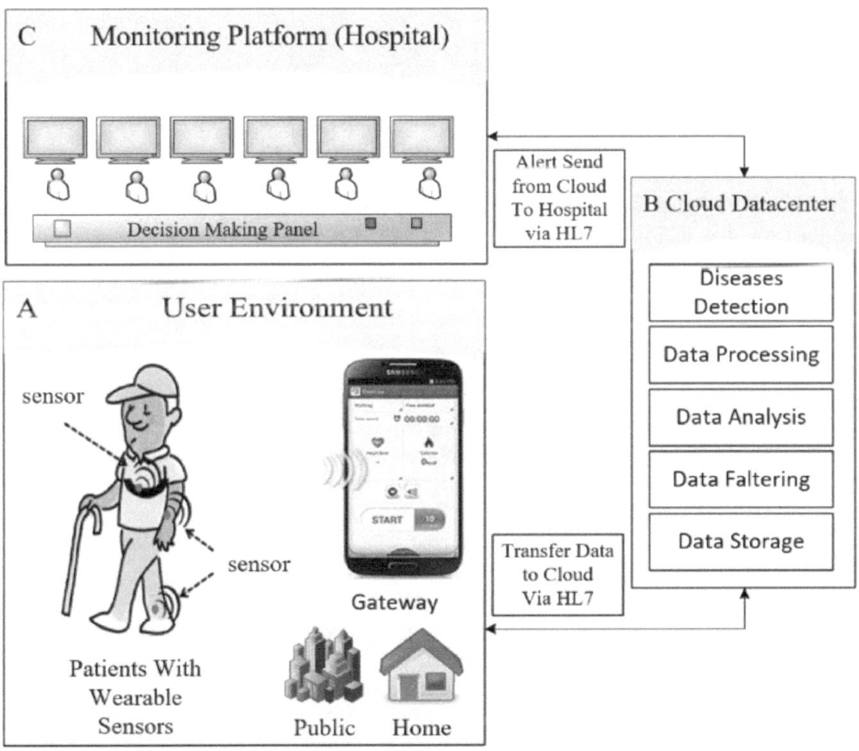

Figure 2. The overall architecture of the elderly monitoring system.

3.1.1 *Realize the diversity of information interaction channels*

With the development of the Chinese economy and society and the progress of medical care, the needs of the elderly for elderly care services are gradually escalating, showing a development trend of diversification, individualization, and specialization (Chen 2015). However, most of the services that can be effectively provided by most service providers are life care, simple medical care, culture, education and entertainment, etc. The rest of the services are difficult to provide effectively for various reasons. In other words, at this stage, most elderly care services can only meet the basic needs of the elderly, but cannot meet the further needs of the elderly, and there is a mismatch between service supply and service demand. The reason is that the current community home-based elderly care services in China have the problem of poor communication channels of multiple information. Specifically, first, the information exchange between the supply and demand sides is not smooth. On the one hand, the elderly lack a sufficient understanding of the elderly care services that service providers can provide; on the other hand, service providers also lack research on the needs of elderly care services. Second, the information exchange between service facilities is not smooth. On the other hand, the role that a single service facility can play is limited, and it is necessary to connect different service facilities to play the greatest role; on the other hand, the information interaction between service facilities is mainly realized by manpower. The interaction efficiency is extremely low, and it is difficult to verify the accuracy of the needs of elderly care services through the information generated by the service facilities. Therefore, from the perspective of supply and demand docking, the operational goal of realizing the diversity of information interaction channels is proposed.

3.1.2 *Improve the accuracy of service demand expression*

The elderly's demand for elderly care services exists objectively. However, the demand for elderly care services is affected by many factors, from its existence to its expression. Academic research has found that due to the influence of multiple factors such as traditional culture, family income, and their personality, the elderly in China like to support themselves and do not want to cause too much trouble to their families, and the expressed needs for elderly care services are often less than the actual needs. In addition, the needs expressed by the elderly cannot be directly used for multi-information interaction and need to be further analyzed and processed before they can be integrated into the actual service supply process (Liu et al. 2018). Therefore, improving the accuracy of service demand expression can narrow the difference between the actual needs and the expressed needs of the elderly and reduce the difference between the expressed needs and the service content provided by the service provider. Therefore, from the perspective of demand expression, the operational goal of improving the accuracy of service demand expression is put forward.

3.2 *"Four-dimensional model" smart city elderly care service model*

"Intelligent +" enables the continuous upgrading and development of elderly care services, "Inclusive Plus" makes elderly care services full of warmth, "Life Plus" makes elderly care services more convenient, and "Ecological Plus" makes elderly care service structures scientific and reasonable. In the fields of education and culture, medical care, social security, environmental protection, tourism and other public services, a digital social service system covering both urban and rural areas will be formed (Liu & Luo 2021). In terms of smart medical care, we will promote the integration of three databases of population information, residents' electronic health records and electronic medical records, establish a population health cloud platform, and coordinate six business application systems of public health, family planning, medical services, medical security, drug management, and comprehensive management, to promote the sharing of various medical information resources. In terms of smart social security, we will accelerate the construction of an "all-in-one card" for citizen cards, integrate existing social security cards, resident health cards, and city cards with integrated financial IC card functions, and promote multiple uses of one card in medical and health care, community services, civil affairs subsidies and other fields. We can also integrate social security information resources and develop online declaration, social security payment, and mobile client that integrates law enforcement and various personal information inquiries. In terms of smart community, relying on the community management system and comprehensive community database, combined with a geographic information system (GIS), to build and improve a comprehensive information system for community management and services, and to promote applications in many fields such as smart home, community medical care, property management, and home care. The terminal self-pickup network is based on smart mail express boxes (Figure 3 shows the smart community pension system). In terms of smart city management, the city's digital city management system has been comprehensively upgraded, with new information terminals as the main carrier, a perception network covering the urban area, a "universal urban management" system, and a digital supervision platform for domestic waste disposal. A spatial data platform is built for measurable real-world images to promote the establishment of an urban management model that is precise, precise, agile and efficient, full-time, all-around, and full-real-life coverage.

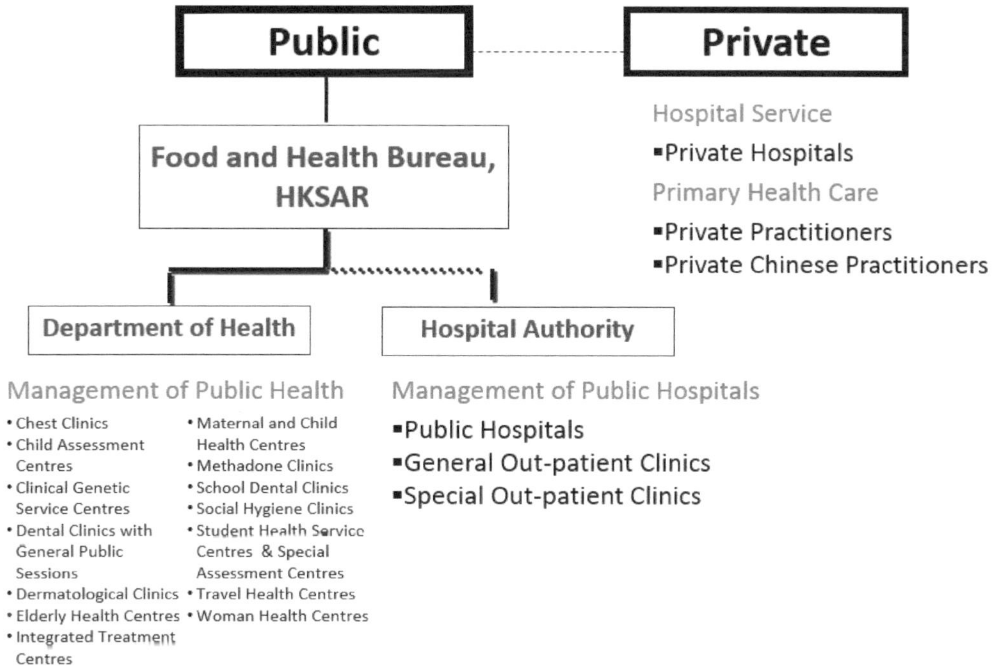

Figure 3. Smart community pension system.

3.3 *Building a community intelligent IoT*

The main reason for the low level of aging suitability of service facilities is that service facility producers lack research on the actual usage habits and usage environments of the elderly. To this end, artificial intelligence can collect and analyze the data generated when the elderly use service facilities through the "life model" "health model" and "happy model," form corresponding production indicators, and provide data support for service facility producers. After realizing the aging of service facilities, the next step is to use smart sensors, artificial intelligence algorithms, etc. to transform and produce service facilities, build a community intelligent Internet of Things to connect service facilities in the community, and then build a platform to realize the integration of these service facilities, information interaction between people, so as to realize the intelligent connection between people, people and things, things and things (Figure 4 is the smart IoT community pension model, the picture is quoted in). At the macro level, artificial intelligence can support the systematic construction and improvement of service standards, that is, the use of artificial intelligence technologies such as machine learning and knowledge graphs to summarize the pain points and difficulties faced by elderly care services from massive data with ultra-high efficiency, thus generating a relatively complete solution and providing valuable advice for the systematic construction of service standards. At the micro level, artificial intelligence can support the personalized construction and update of service standards, that is, based on the "life model," "health model" and "happiness model," combined with previous service records, build and update the personalized service of the elderly in real time. Service standards, so that service personnel can be more accurately guided and service facilities deployed. In short, artificial intelligence can give full play to the advantages of information acquisition, data analysis, computing speed, etc., support the construction and maintenance of service standards, and realize its systematization and personalization.

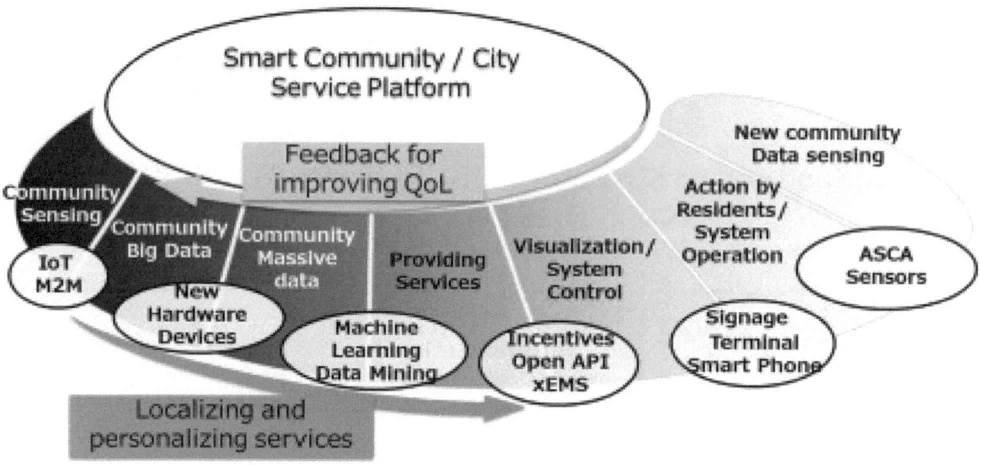

Figurc 4. Smart IoT community pension model.

3.4 *Survey and analysis of smart elderly care*

A control experiment is designed to verify the function of the smart community pension system. During the test, each functional parameter needs to set a threshold range scientifically. The test result is judged as "correct" if it does not exceed the set range and is judged as a "dangerous situation" if it exceeds the range. 1) Heart rate monitoring: The pulse sensor in the smart community pension system is used to detect the heart rate in different states, and at the same time, the measurement results of the commercial wristband (Xiaomi 2) are used as the standard for comparison. The difference is within 2 times/min. When judged as correct, the dangerous situation of heart rate out of range is simulated with the test results after strenuous exercise. 2) Fall monitoring: Volunteers wear the host computer around their waists, simulate normal behavior to record the number of false alarms, test abnormal behaviors such as falling forward, falling backward, and sideways, and count and record the number of alarms. 3) Location detection: Volunteers are positioned in different indoor and outdoor environments, observing and recording the positioning error and the number of successes. 4) Temperature and humidity detection: test the temperature and humidity of the environment under different conditions, and compare it with the measurement values of mercury thermometers and commercial hygrometers at the same time, and the difference is within 2°C and 3%. 5) Gas monitoring: use household gas stoves to measure, record the number of alarms, and compare with the data of household gas alarms. 6) Bed-wetting monitoring: touch the module with a cotton ball soaked in water, and record the number of alarms in multiple experiments. 7) Monitoring of falling out of bed: Volunteers simulate the elderly falling out of bed at night and record the number of falling out of bed and the number of alarms for comparison. The test results are shown in Table

Table 1. Multi-functional test results in the smart community pension system.

Monitoring function	Test sampling	Test accuracy rate/(%)	Hazardous Situation Test	Alarm accuracy rate/(%)
heart rate	100	96	50	98
Fall	100	95	50	95
Location	100	98	–	–
Temperature	100	100	50	100
Humidity	100	99	50	100
Gas	100	100	50	100
Wetting the bed	100	100	50	100
fall out of bed	100	100	50	100

All monitoring functions in the smart community pension system are fully realized, which can effectively monitor the parameters of the elderly physiology, location, environment, sleep, etc., with high practicability, accurate data, and high stability and reliability.

4 POTENTIAL PROBLEMS OF DIGITAL NETWORK INTELLIGENCE IN THE INFRASTRUCTURE CONSTRUCTION OF ELDERLY COMMUNITIES

The "digital divide" faced by the elderly needs to be addressed urgently. At the same time, the penetration of intelligent Internet technology into daily life is increasing, showing the characteristics of the full-scenario application and rapid iteration, which aggravates the difficulties and challenges for the elderly to integrate into digital life. The use of intelligent technology for the elderly to integrate into the digital society is an important issue related to future social operation and governance and sustainable economic development. Comprehensively improving the suitability of intelligent technology, products and services for the elderly has become a social hotspot and an industry consensus.

We suggest setting up intelligent care centers for the elderly in each community. The care center has at least ten beds based on the number of the aged in the community. The care center targets two types of the elderly, one is the elderly who need rehabilitation and conditioning, and the other is the healthy elderly. These aged learn and become familiar with the intelligent facilities in the center. The care center can set up an activity area so that the elderly can be interested in and master the intelligent technology in the activity to adapt to the intelligent products and services. The center's intelligent facilities are the same as those installed at the house of those aged. Therefore, after a while, those aged can use the intelligent facility at home.

5 CONCLUSION

Technological progress has changed the social pension model, but traditional communities, families and the elderly have not fully adapted to this change. Therefore, when designing smart community, home medical products in the context of smart elderly care, it is even more necessary to explore the physical and psychological needs of the elderly. Taking the elderly as the design object and using advanced science and technology as the means, the design is carried out according to the physical and psychological characteristics, lifestyle and emotional needs of the elderly, so as to meet the safety needs of the elderly and enable the elderly to better use a smart community facility, home medical care product. At the same time, the development and improvement of smart medical products will promote the development of smart elderly care and build a more comfortable elderly care environment for the elderly. As an important step in the development of community infrastructure, the construction of the Internet of Things connects the government public departments, service suppliers for the elderly and the living places of the elderly, thus improving the comfort, convenience and safety of the elderly life.

REFERENCES

Chen Feilong, Cai Chunxiao, Zhou Jianrong, et al (2018). Design and Implementation of a Health Monitoring System for the Elderly and Infants Based on the Internet of Things. Science and Technology Vision, vol. 7, pp. 50–52.

Chen Siqing (2015). The Application and Prospect of Internet of Things Technology at Home and Abroad in the Field of Elderly Care. Chinese Journal of Gerontology, vol. 4, pp. 54–59.

Liu Hang, Teng Qiaoying, Qu Xiaopeng, et al (2018). Research and Design of an Intelligent Monitoring System for Community Elderly Users Based on Big Data. Intelligent Computers and Applications, vol. 8, pp. 47–50.

Liu Yi, Luo Shuangyu (2021). Research on Remote Intelligent Monitoring System for the Elderly Based on Internet of Things Technology. Consumer Electronics, vol. 7, pp. 2–7.

Sun Juan, Qiao Pingan (2018). Design and Application of an Elderly Health Management System Based on the Internet of Things. China New Communication, vol. 20, pp. 27–33.

Wang Xiuping (2012). The Application of Internet of Things Technology in the Life and Health Services of Empty-nest Elderly. Chinese Journal of Gerontology, vol. 32, pp. 34–41.

Wu Weisheng (2018). Elderly Health Monitoring Scheme Based on Internet of Things Technology. Digital Users, vol. 70, pp. 71–78.

Xiao Hong (2016). Design of Application Monitoring Platform of Internet of Things Technology in Health Service for Empty-Nest Elderly. Industrial Design, vol. 5, pp. 2–8.

Advances in Urban Engineering and Management Science – Khalil & Yang (Eds)
© 2023 The Author(s), ISBN 978-1-032-30426-7

A dynamic scheduling method of satellites communication in an uncertain environment

Yue Liu*
School of Computer Science and Engineering, Tianjin University of Technology, Tianjin, P.R. China

Peng Yang*
Key Laboratory of Computer Vision and System, Ministry of Education, Tianjin University of Technology, Tianjin, P.R. China

ABSTRACT: Dynamic scheduling is crucial for the satellite communication system. The paper studies the problem of dynamic scheduling a group of communication satellites. A flexible resources assignment model and response method with the Multi Agent System (MAS) are proposed. The two models are integrated to meet the demands of dynamic and real time. Correspondingly, an improved hybrid genetic algorithm is developed to minimize the finish time with limited time. The effectiveness of the model was verified by a set of experiments, and the results show that the method can improve the utilization rate of the satellite repeater, strengthen the communication mission capable rate, and enhance the stability of the system.

1 INTRODUCTION

Communication satellites are traditionally scheduled statically; namely, the schedule is created off-line. In the process of satellite communication operations, there are often new demands for communication. The execution of the original schedule will be disrupted, and sometimes, the plan must be reworked, which will bring much communication load and computation load.

There are two main methods for communication scheduling satellites: ground control and onboard autonomous control. The complexity of the target requirements of communication tasks also significantly increases the number of communication satellites it contains. Inter-satellite collaboration leads to a more complex system structure. Satellite performance cannot be fully utilized by ground management alone, and fast response to highly time-sensitive tasks cannot be achieved. In the process of satellite communication, the determination of its autonomous collaborative operation structure becomes a key issue. In this context, the impact of random task requirements on mission scenarios is of great importance to the study of dynamic satellite scheduling optimization.

Traditionally, researchers often design a static scheduling model, and a heuristic algorithm is used to solve the model during the model-building phase to solve the problem of satellite scheduling for random tasks, such as an improved genetic algorithm for satellite data transmissions (Zhang & Xing 2022), such as using individual reconfiguration-based integer coding genetic algorithm for satellites scheduling (E. 2021) or proposing a hybrid heuristic including column generation and simulated annealing to solve linearized robust scheduling model (Wang et al. 2021). In addition, some scholars divide the pre-scheduling process into pre-allocation, coarse scheduling, and fine scheduling, and proposed a hierarchical scheduling algorithm based on the ant colony algorithm instead of dynamic scheduling (He 2019).

In the dynamic scheduling phase, the contract network protocol is widely used in the research of multi satellite and multi-task cooperation because of its clear process, flexibility and reliability. Yang designed a new mathematical model based on MAS, and proposed an improved contract

*Corresponding Authors: 344776209@qq.com and 29139475@qq.com

DOI 10.1201/9781003305026-69

network protocol (ICNP) and blackboard model (BM) negotiation mechanism to coordinate satellites (Yang et al. 2021). In addition, the Contract Net Protocol (CNP) based secondary allocation strategy and a new Interactively Re-planning Method (IRM) are developed to increase the observation benefit and reduce the loss by re-inserting tasks in case of failure (Du & Li 2019). Although task assignment can be achieved by using a contract network mechanism for negotiation, with the complexity of tasks and changes in the system environment, there are the following shortcomings: uneven computing and communication load, Insufficient stability. In summary, there are two main ways to generate dynamic tasks at present. One is to generate the initial schedule first, and then roll adjustments based on it. The other is the MAS-based distributed decision-making approach. In the dynamic environment, the emergence of new communication tasks often triggers the rolling adjustment of the first type of mode, and the system operation lacks flexibility. The second type uses distributed dynamic decision making, which has higher flexibility, but it has the problems of high communication operation load and low decision efficiency.

In summary, in view of the above problems. In this paper, the entire dispatching framework is divided into two stages of decision-making by combining the flexible dispatching model with the dynamic coordination strategy. In the first stage, the flexible model is used to generate the initial plan. In the second stage, the MAS-based dynamic coordination mechanism is used to adjust the initial scheme. The analysis of simulation results shows that the method used in this paper can effectively reduce the cost of task rescheduling, and the success rate of random task insertion is higher.

2 FLEXIBLE DECISION FRAMEWORK FOR SATELLITE DISPATCH

2.1 *Problem description*

Satellites can communicate with ground stations only when they travel around the earth through a specified area. The period of communication with ground stations is called the visible time window of a satellite. Usually, the window of a medium-low orbit satellite is between minutes and tens of minutes. The visible time window is as follows:

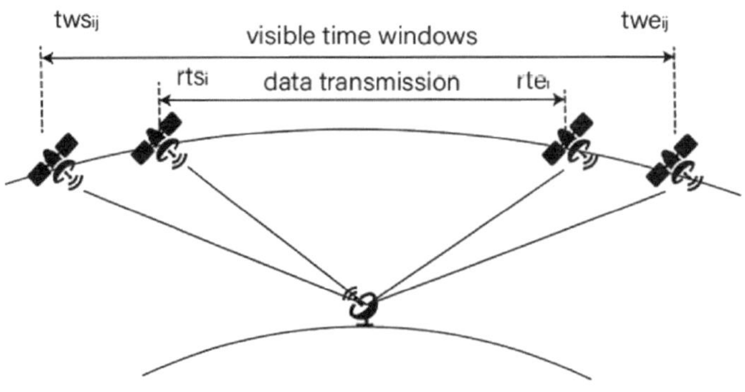

Figure 1. Visible time window of satellite.

Task planning of satellites means that under the condition of multiple satellites and multiple time windows, the execution time and execution of satellites are arranged to achieve certain goals to make full use of satellite resources. Usually, this goal is the rate at which a task is completed or the total benefit from the task. The tasks described in this paper are mainly for data transmission between satellites and ground stations.

There are two types of data transmission tasks. The first type is conventional tasks, which is a series of tasks that have already been acquired before a working cycle of a satellite system. The second type is random tasks, which occur outside the existing task planning scenario.

A satellite can emit electromagnetic waves in multiple bands. The transponder is the smallest unit of satellite frequency division, which constrains the bandwidth and power of the frequency. Generally, there are multiple transponders in a band. To make full use of the satellite resources, we have planned the missions for the units with transponders. The pre-scheduling model in this paper is designed based on setting up redundant transponders. The communication structure of the satellite is shown in the following figure. A bold transponder is a redundant transponder.

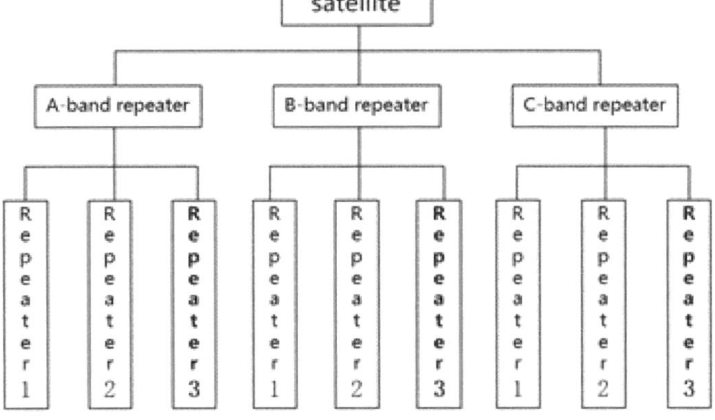

Figure 2. Satellite communications section structure diagram.

In a dynamic environment, the communication and data transmission tasks of a satellite can be divided into two types. The first is the initial task, which is a series of tasks already obtained before a working cycle of the satellite system. The second is new tasks, which occur during the execution of existing task planning scenarios.

The flexible scheduling framework designed in this paper is divided into two phases: the first phase completes the generation of the initial flexible scheduling scheme; Stage two uses MAS-based dynamic adjustment of communication tasks in the implementation stage.

2.2 *Pre-flexible schedule generation*

2.2.1 *Notations description*

Table 1. Notations.

Notation	Description
S	Set of satellites
G	Set of ground stations
$TASK$	Set of tasks
s_i	a satellite
$task_i$	a task
TW_i	Set of visible windows for satellites, $tw_{ij} = [tws_{ij}, twe_{ij}]$
tws_{ij}	The start time of visible windows of s_i
twe_{ij}	The end time of visible windows of s_i
rt_i	Execution time window of $task_i$, $rt_i = [rts_i, rte_i]$
rts_i	The start time of the time window to execute the $task_i$
rte_i	The start time of the time window to execute the $task_i$
w_i	Benefits of $task_i$
t_i	$task_i$ execution time
$tran$	Necessary work changeover time for satellites
$stor_i$	Storage capacity consumed by $task_i$
$stormax$	Satellite maximum storage capacity
$maxrec$	The maximum number of tasks that the ground station can receive at the same time
$exetimes_i$	$task_i$ execution times
rec	Number of ground stations receiving connections
$rec_{\max}$	Maximum value of ground station receiving connection

2.2.2 *Constraint condition*

$$\left(twe_{ij} - tws_{ij}\right) >= \left(rte_i - rts_i\right) \tag{1}$$

$$\mathrm{rte_i} + \mathrm{tran} <= \mathrm{rts_j} \tag{2}$$

$$exetimes_i \leq 1 \tag{3}$$

$$\sum_k \mathrm{stor}_k \leq stormax \tag{4}$$

$$rec < rec_{\max} \tag{5}$$

$$twe_{ij} < rts_n \tag{6}$$

$$twe_{ij} > rte_n \tag{7}$$

Equation (1) is the job length constraint; (2) is the task uniqueness constraint; (3) is the task execution times constraint; (4) is the storage capacity constraints; (5) is ground station acceptance capacity constraints; (6) and (7) are the visible window constraint.

Objective function

$$max \sum_{s \in S} \sum_{i=1}^{n} w_i \frac{\alpha}{t_i} \tag{8}$$

$$max \frac{\sum_{p=1}^{q} t_p}{\sum_{i=1}^{m} \sum_{re \in RE} \sum_{v=1}^{n} \left(twe_{ijv} - tws_{ijv}\right)} \tag{9}$$

Equation (8) is the revenue objective function combining the task revenue and the task execution time, w_i is the benefit of $task_i$, t_i is the execution duration of $task_i$, and a is a constant. Tasks with short execution time and high task income have higher overall income. Therefore, this objective function is similar to the traditional model to make the overall income as high as possible. (9) is a function of the occupancy rate of the time window to make the occupancy rate of the currently available repeaters as high as possible. Through the flexible task planning model of two-objective optimization, we can get an initial task scheme to be executed by the satellite.

2.3 *Dynamic scheduling strategy based on MAS*

Firstly, based on MAS (multi-agent system), this paper designs a dynamic coordination mechanism of task scheme with task intensity as the index. As shown in Figure 3, the model includes three types of agents: control center agent, dynamic unit agent and satellite agent. Under the dynamic scheduling framework, the satellite scheduling is completed through the interaction of these three types of agents to improve the overall income of executing tasks.

In the task pre-scheduling stage, a certain proportion of forwarders are reserved for dealing with the insertion of random tasks. In the dynamic scheduling stage, there are two task scheduling strategies. The scheduling strategy makes decisions through a task density index. There are two kinds of task density to be calculated in the scheduling process. One is to adjust the task scheme on a single satellite for a single time window. Second, for all time windows of the whole satellite, it is used to compare the task intensity of different satellites and adjust the task scheme among different satellites. The calculation formula of task intensity is as follows:

$$OR_i = \frac{\sum_{j=1}^{m} or_{ij}}{n} \tag{10}$$

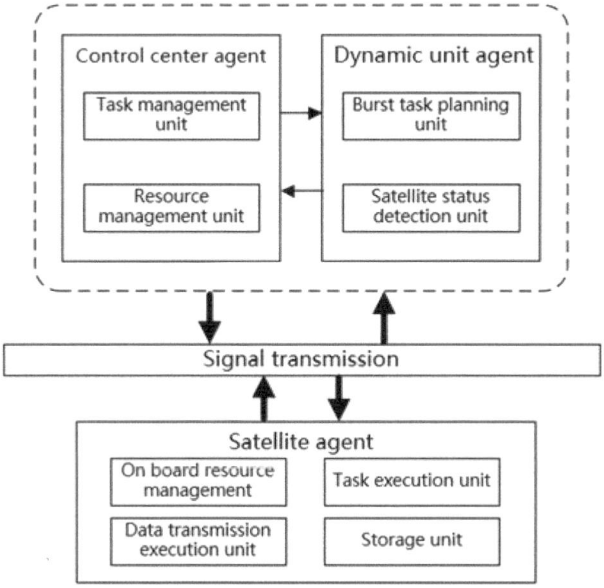

Figure 3. Schematic diagram of satellite MAS model.

$$or_{ij} = \frac{\sum_{k=1}^{n}(rte_k - rts_k)}{twe_{ij} - tws_{ij}} \tag{11}$$

OR_i indicates the mission density on the ith satellite, or_{ij} indicates the task intensity of the jth visible time window on the ith satellite.

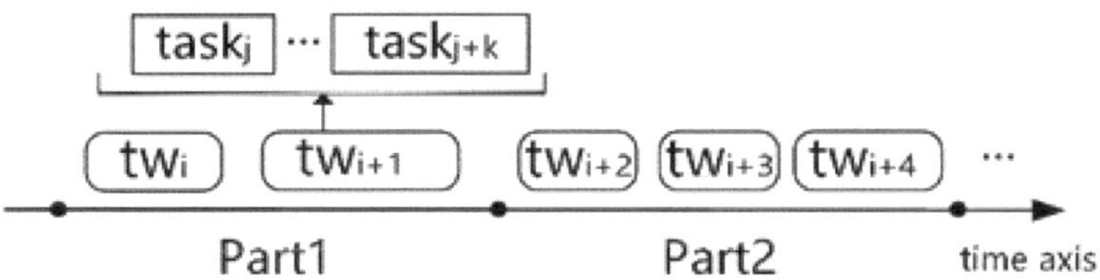

Figure 4. Example timeline for performing tasks.

As shown in the figure above, we divide the whole scheduling process into multiple intervals according to time, and each interval contains multiple visible time windows. Our adjustment strategy is based on intervals.

Inter partition insertion tasks are divided into task insertion in an unplanned task pool and random task insertion.

1) The specific steps for unplanned task insertion are as follows:
   Step 1: The pre-scheduling scheme starts to execute, and all reserved transponders are marked as available.
   Step 2: Calculate the task intensity of the current interval.
   Step 3: If the task density does not exceed the threshold, execute step 4; otherwise, it ends.
   Step 4: Search for executable tasks in the current interval time window from the unplanned task pool. If a task is found, perform step 5; otherwise, it ends.
   Step 5: Insert task, end.

2) The specific steps of random insertion are as follows:
   Step 1: The random task request arrives.
   Step 2: Search whether there is an available time window in the current interval. If not, proceed to step 3; otherwise, proceed to step 4.
   Step 3: Put the random task into the unplanned task pool and end.
   Step 4: Insert task, end.
   To further improve the success rate of task insertion, a task density adjustment algorithm is designed. That is, by calculating the density of tasks between partitions, adjust tasks across partitions to make the insertion success rate of tasks higher.
3) Task intensity adjustment steps:
   Step 1: The ith task execution interval starts.
   Step 2: Calculate the task intensity of all intervals and sort according to the task intensity.
   Step 3: Take a sequence from the beginning and end of the interval sequence and pair it in pairs.
   Step 4: Adjust the task from the interval with high task density to the interval with low task density and repeat this step until the difference between the two paired intervals is within a certain range, end.

3 SIMULATION EXPERIMENT ANALYSIS

3.1 *Experimental scene*

The experimental scenario consists of a ground station and nine satellites. There are 200 ordinary tasks and 50, 100, 150, and 200 random tasks. Random tasks arrive randomly in the whole task cycle. The experimental scene is simulated by simulation software. The task scheduling framework based entirely on MAS (CDSSB) uses the revenue-oriented model to generate the initial task execution scheme. In the dynamic adjustment stage, the negotiation of each agent adopts the contract network mechanism, and the decision strategy adopts the greedy strategy and the first come first service strategy. As it does not consider the insertion of random tasks in the initial stage, the scheme adopted in this paper (DSSBORR) is compared with CDSSB.

3.2 *Comparison between MAS task scheduling framework based on redundant repeater and task scheduling framework based on full MAS*

3.2.1 *Analysis of mission scheme intensity*

When the repeater redundancy is 30%, the whole task execution cycle is divided into five intervals. The task density data of 50, 100, 150 and 200 new tasks are compared. It can be seen from figure 5 that the dynamic scheduling framework proposed in this paper has an average task density in the whole task cycle, while the conventional scheduling framework has high task density in the early stage and low task density in the later stage. This is because the insertion of new tasks is not considered in the task pre-scheduling stage of the conventional scheduling framework, and all tasks are arranged in chronological order as far as possible. It leads to high congestion of previous tasks, which is not conducive to the insertion of new tasks. In the framework proposed in this paper, there is not particularly high or low task congestion in the whole task cycle, which will improve the success rate of new task insertion and reduce the waste of satellite resources.

3.2.2 *Task completion rate analysis*

The completion rate of new tasks is shown in the figure below. With the increase in the scale of new tasks and the limited satellite resources, the insertion success rate of new tasks shows a downward trend. The overall completion rate of the scheduling framework proposed in this paper is significantly higher than that of the conventional scheduling algorithm. This is because the scheduling algorithm based on redundant repeater has uniform task density in the whole task cycle, so it has an excellent ability to insert new tasks.

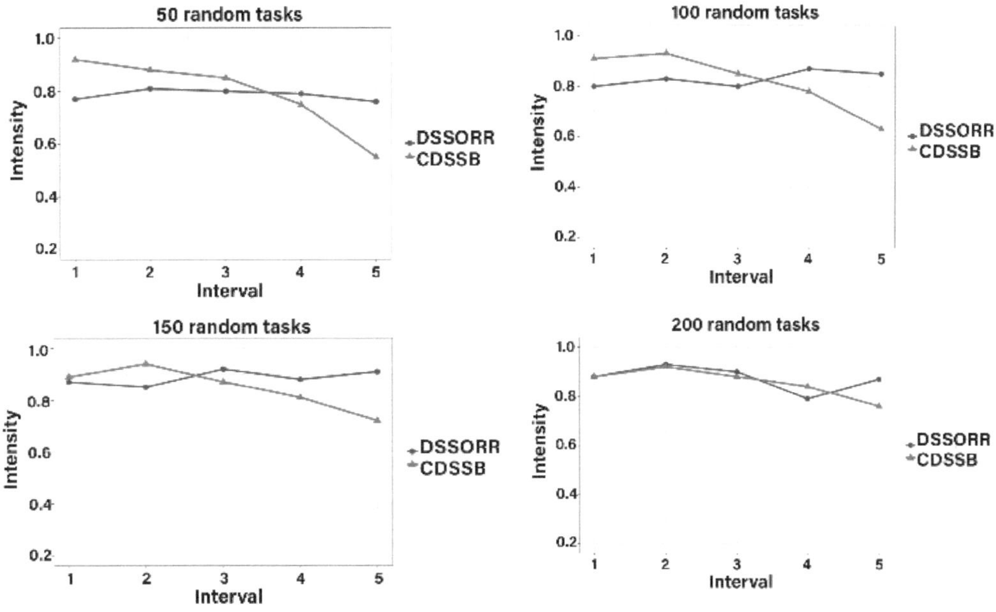

Figure 5. The density of new tasks in different intervals.

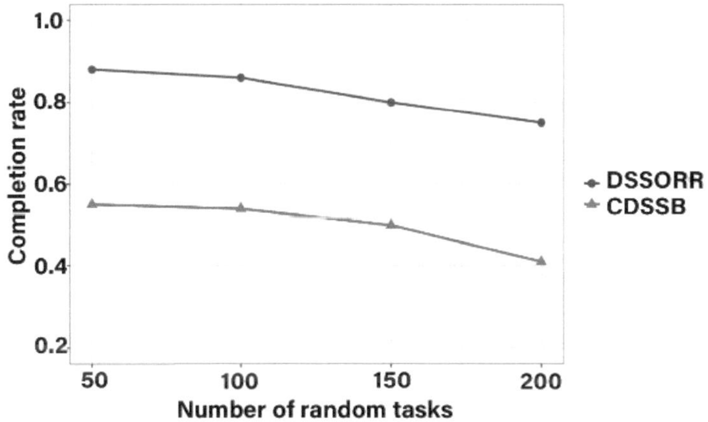

Figure 6. Completion rate of new tasks under different new task sizes.

4 CONCLUSION

A novel hybrid distributed planning framework is presented for multiple satellites systems. The framework consists of a generation of the flexible pre-scheduling scheme and a dynamic scheduling strategy based on MAS. First, the generation of a flexible pre-scheduling scheme uses an improved genetic algorithm as the search algorithm for solving the flexible local constraint satisfaction problems, while the MAS strategy is conducted to adjust the current plan for the newly emerged task. To verify the validity of the framework, numerical simulation is conducted, and the results show that the proposed method can effectively improve the insertion success rate of new tasks. The cost of task rescheduling is reduced. Our future research will focus on improving this method to face the sudden insertion request of a large number of new tasks.

ACKNOWLEDGMENTS

This work was supported by the Fundamental Research of Science and Technology on Complex Electronic System Simulation Laboratory under Grant DXZT-JC-ZZ-2019-010.

REFERENCES

Bin Du, Shuang Li (2019). A new multi-satellite autonomous mission allocation and planning method. Acta Astronautica. 163, 287–298.

Jiawei Zhang, Lining Xing (2022). An improved genetic algorithm for the integrated satellite imaging and data transmission scheduling problem. Computers & Operations Research. 139.

Lei He, Xiao-Lu Liu, Ying-Wu Chen, Li-Ning Xing, Ke Liu (2019). Hierarchical scheduling for real-time agile satellite task scheduling in a dynamic environment. Advances in Space Research. 63, 897–912.

Weiyi Yang, Lei He, Xiaolu Liu, Yingwu Chen (2021). Onboard coordination and scheduling of multiple autonomous satellites in an uncertain environment. Advances in Space Research. 68, 4505–4524.

Xinwei Wang, Yi Gu, Guohua Wu, John R. Woodward (2021). Robust scheduling for multiple agile Earth observation satellites under cloud coverage uncertainty. Computers & Industrial Engineering. 156.

Zhibo E., Renhe Shi, Lan Gan, Hexi Baoyin, Junfeng Li (2021). Multi-satellites imaging scheduling using individual reconfiguration-based integer coding genetic algorithm. Acta Astronautica. 178, 645–657.

Advances in Urban Engineering and Management Science – Khalil & Yang (Eds)

Multi-lanes detection in complex scenes based on Res2Net

Feng You, Zhihao Xiao*, Zhenjiang Wu, Jiaxia Yang & Jingshan Yang
School of Civil Engineering and Transportation, South China University of Technology, Guangzhou, Guangdong, China

ABSTRACT: In order to solve the problems of existing lane detection algorithms with low robustness and low real-time performance in complex road scenes, we propose a grid classification lane detection algorithm based on Res2Net. Lane detection is one of the most important parts of the autonomous driving perception module. Existing lane detection methods based on deep networks mainly treat lane detection as pixel segmentation, which is difficult to meet the real-time requirement as well as solve the problems of lane occlusion, lack, and detection at night or under the glare. We improve the grid classification by considering the lane feature point detection as a grid classification based on the image global features. Since Res2Net can extract multi-scale information, it is embedded in ResNet50 to obtain Res2Net50 as the backbone. By comparing the detection effects of several common feature extraction networks, we concluded that Res2Net50 could extract lane information under various complex road conditions. Extensive experiments on the CULane and TuSimple lane datasets show that our method achieves a 72.54% F1-Measure score on CULane and 95.73% accuracy on TuSimple, processing a 288*800 image in only 10.7 ms. Finally, the lane feature points are fitted using polynomials. Our method helps to improve the accuracy and speed of lane detection for the Advanced Driving Assistance System (ADAS) in complex road scenes.

1 INTRODUCTION

According to the statistics of the Ministry of Public Security in 2020, the number of motor vehicles and vehicle drivers reached 372 million and 456 million, respectively, in China. And the death toll from road traffic accidents per 10,000 vehicles was 1.66, of which 90% of traffic accidents were caused by human error. To reduce traffic accidents, self-driving cars came into being.

Lane detection has become a hot topic in the field of automatic driving perception. Lane is the main traffic sign on the road, and its detection technology is very important for lane-departure warning systems, lane-keeping systems, and trajectory planning decisions. Good detection methods can significantly improve the safety of driving. However, due to poor lighting conditions, the occlusion of other traffic participants, interference of irrelevant road markers and other reasons, lane detection in complex road scenes is challenging, which hinders the commercial application of autonomous driving technology. Therefore, it is very important to study lane detection in complex road scenes to improve road traffic safety and promote the development of autonomous driving.

2 RESEARCH STATUS

The difficulty of existing lane detection mainly lies in the need to face various complex road scenes, such as lane occlusion, lack, and detection at night or under glare (Liang 2018). At this time, the edge feature of the lane is not obvious, and false or missed detections are prone to occur.

The existing lane detection methods can be roughly divided into three categories: traditional image processing methods, the combination of traditional image processing and deep learning methods, and methods based on deep learning.

*Corresponding Author: xzh564238303@163.com

DOI 10.1201/9781003305026-70

The traditional lane detection methods are usually divided into three steps (Luo 2020): image preprocessing, local feature extraction and lane fitting. It mainly uses the edge, texture, and color features of the image to capture the local lane information. The disadvantage is that it is only for specific roads and is difficult to adapt to the complex scenes of the actual application of automatic driving.

The combination of traditional image processing and convolutional neural networks (CNN) makes use of the powerful nonlinear fitting ability of CNN. After the lane features are extracted by CNN, the traditional image processing algorithm can be used for post-processing. Or use the traditional method to extract the rough lane feature map as the input, and use the CNN to further extract and refine the lane (Chen et al. 2020).

The lane detection method based on deep learning usually adopts downsampling to realize the deep architecture. For example, VPGNet (Lee et al. 2017) aims to solve the detection of lanes and lane signs in rainy and dark scenes. It is a multi-task network guided by lane vanishing points. It has high robustness in various scenes. The disadvantage is poor real-time performance, only 20 frames per second (fps). In order to make more effective use of visual information, SCNN (Pan et al. X 2017) uses a special convolution operation in the segmentation module. The traditional depth layer by layer convolution is summarized as a slice-by-slice convolution in feature mapping to realize the information transmission between pixels across rows and columns in one layer. This structure is suitable for continuous lane detection on structured roads. The disadvantage is poor robustness in complex scenes. Some scholars try to explore lightweight methods that can be applied in real time, such as Enet (Paszke et al. 2016) and self attention distillation (SAD) (Hou et al. 2019). ENet uses a structure like ResNet (He et al. 2016), with a main branch and a convolutional extension branch, so as to perform large-scale computing faster and more effectively. SAD adopts an attention extraction mechanism. By adding the SAD module to the model, the shallow features learn the expression of high-level features, thereby strengthening the overall feature expression ability of the network. The above two networks need to be combined with other modules to consider the speed and accuracy of lane detection. PINet (Ko et al. 2020) is a lane detection algorithm based on keypoint estimation and instance segmentation. The key points are divided into a single instance. Several hourglass modules (Newell et al. 2016) are stacked in the network. The size of the model can be selected according to the computing power of the target environment. The disadvantage is that the network is too cumbersome. It is 25fps on GTX2080Ti and only 10fps after processing.

The research on lane detection has made remarkable progress and excellent achievements, but there are still some problems:

- The traditional lane detection methods are greatly affected by the environment. They are only suitable in specific scenes, and are poor in detecting complex scenes such as night and glare;
- The real-time performance of the algorithm is poor. A deep network involves a large number of convolution operations, especially pixel-level semantic segmentation. The deeper the network, the stronger the computing power of the equipment;
- The applicable scenes are limited. Most algorithms are only applicable to structured roads with clear lanes and are not robust to scenes such as congestion, night, and missing lanes.

To solve the above problems, we propose a lane detection algorithm based on Res2Net and grid classification. The main research contents and innovations are as follows:

- Replace semantic segmentation with grid classification, use the global features of the image to better detect the lanes with "semantics," and reduce the amount of calculation.
- ResNet50 is improved by Res2Net (Gao et al. 2021), and the network can extract multi-scale features of lanes. The algorithm not only achieves good results on the challenging CULane and TuSimple datasets, but also can be applied to most down-sampled feature extraction networks.
- Construct the fitting equation of lane feature points, and the discrete lane feature points are fitted into a continuous curve, which can correct the anomaly detection points.

Our method not only reduces the model parameters, but also enhances the robustness. It effectively improves the detection index of the algorithm on the dataset, considers the real-time and accuracy of lane detection in complex scenes, and provides a theoretical reference for the research and development of ADAS.

3 DESIGN OF LANE DETECTION ALGORITHM

The overall flow of the lane detection algorithm is shown in Figure 1. The image is input into the network, and the lane feature map is obtained through preprocessing and feature extraction. Then use the grid classification method to traverse the location of the lane feature points line by line and row by row. Finally, the third-order polynomial is used to fit the feature points. This section will introduce the idea of grid classification, location judgment of lane feature points, and lane fitting model design.

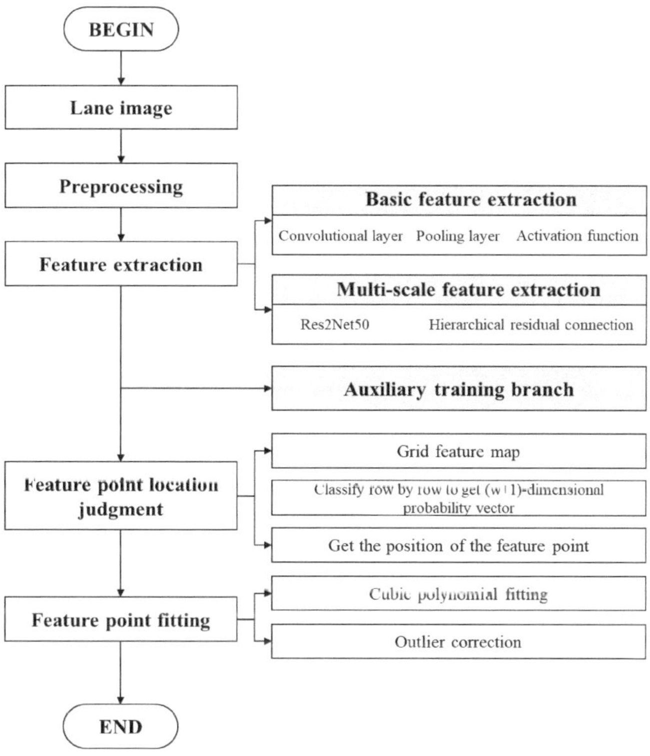

Figure 1. Flowchart of lane detection.

3.1 *Location judgment of lane feature points based on row*

In order to detect the lane feature points more simply and quickly, the grid classification is to divide the picture into several grids (Qin et al. 2020), and lane detection is regarded as looking for a collection of feature points in certain grid positions in the image; that is, based on the selection and classification of grid positions in the row direction. The detection principle is shown in Figure 2. Where, W and H are respectively the image width and height (pixel), h and w are respectively the number of grid rows and columns (piece), n is the number of lanes to be detected is set in advance. If the lane feature points are not detected in this row, the probability falls into the k column.

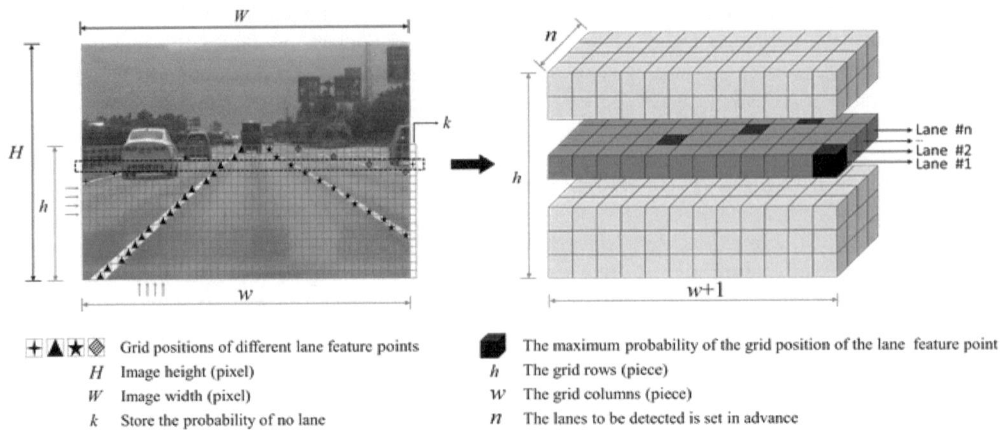

Figure 2. The principle of detecting feature points of lanes.

According to this idea, the lane detection process is regarded as line-by-line classification and grid-by-grid selection using the global features of the image. The search process is: divide the fixed area of the image into $h \times w$ grids, classify each grid on each line, and find the location of lane feature points. Take the row in the dashed box on the left side of Figure 2 as an example, respectively traverse n times to find the location of n lane feature points in the row. The dark part in the middle of the right picture represents the location of the lane feature points (as shown in the legend in Figure 2). Since there are n lanes, the probability matrix is n dimension.

For the same problem, the traditional semantic segmentation method needs $H \times W$ classification, while the grid classification method only needs to classify on h rows, and the classification on each row is w dimension. Therefore, the grid classification simplifies the $H \times W$ classification of semantic segmentation into h classification. And h can be set as needed. Generally speaking, h is far less than the image height H and far less than $H \times W$. Therefore, the grid classification reduces the computational complexity to a very small range and greatly improves the speed of the lane detection algorithm.

3.2 *Design of lane feature point detection network*

Based on grid classification, we propose a lane detection framework. The network extracts the feature map of the lane image by direct downsampling and then stitches it into a probability matrix of $h \times (w + 1) \times n$ size, and obtains the position of each lane feature point in each row by solving the expectation.

In the network training phase, activate the branch of auxiliary training. Specifically, the shallow feature layer and deep feature layer in the network are extracted and fused through upsampling to calculate the semantic loss. This branch is not active during prediction. By designing this branch, the network can not only increase the additional computing cost, but also improve the lane detection performance. The network structure is shown in Figure 3.

Backbone feature extraction network

The difficulty of lane detection is as shown in Figure 4. The scale of a lane changes obviously, near wide and far narrow. In the process of moving, the part close to the human eye is called the myopia field, and the part at the far end of the human eye is called the hyperopia field (Wang 2017). A lane appears in an image of different sizes. Besides, it is difficult to detect lanes in complex scenes such as lane occlusion, lack, or detection at night or under the glare. As shown in Figure 5, When the field of view is poor, it is necessary to rely on the information of adjacent vehicles or

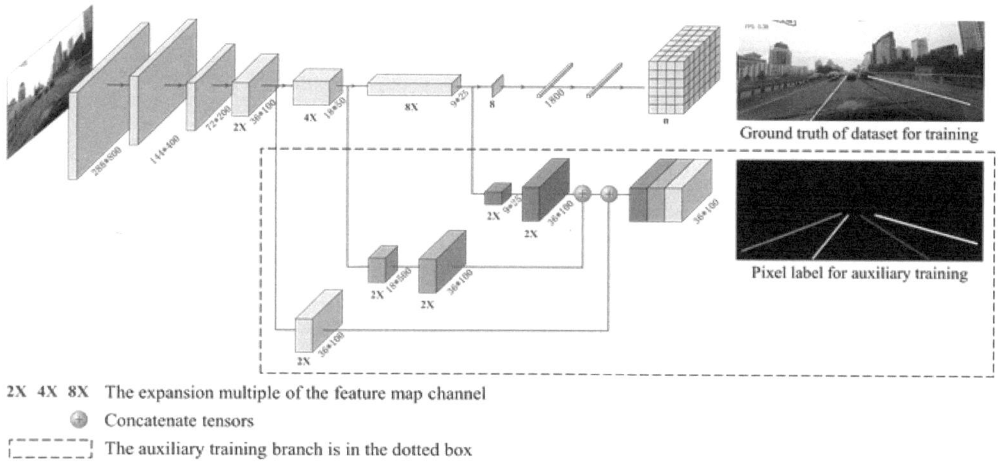

Figure 3. Network structure diagram.

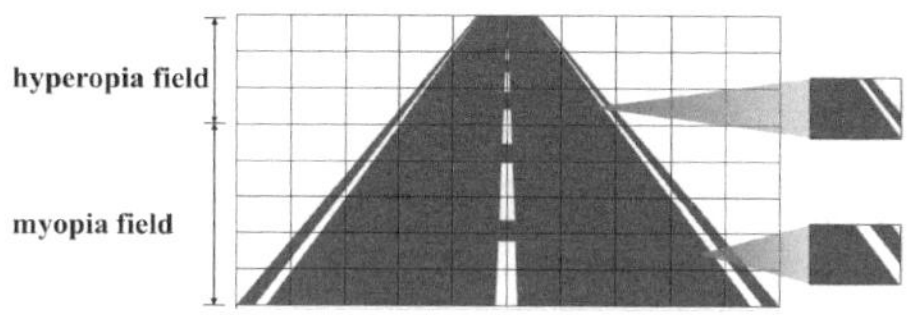

Figure 4. Multi-scale information of lanes.

Figure 5. Severely blocked lanes.

visible lanes to judge the lanes with "semantics," and the context information of lanes occupies a larger area than itself.

The multi-scale features of objects are of great significance for deep learning. Perceiving the target objects of different scales is conducive to understanding the local and overall lanes. Therefore, it is very important to design an excellent multi-scale feature extraction network for visual tasks.

We embedded the Res2Net module into ResNet50 to obtain Res2Net50. Res2net is a simple and effective multi-scale feature processing method. Unlike most existing methods to improve the hierarchical multi-scale representation of the neural network, it improves the multi-scale representation ability at a finer level. Res2Net constructs hierarchical residual connections within a single residual block. Specifically, Replace the n-channel convolution kernel with s w-channel convolution kernels ($n = s \times w$). It has a stronger multi-scale feature extraction ability when the amount of calculation is similar.

The depth network is stacked by modules. By building different module combinations, we can extract the depth features of the image. Figure 6 (a) is the basic bottleneck block in ResNet. The bottleneck layer reduces the channels through a convolutional layer of 1×1 convolution kernel. The advantage of using 1×1 convolution is that the amount of calculation can be greatly reduced, and the network structure can flexibly change the dimension. (b) is the Res2Net module, and its output can be written as:

$$y_i = \begin{cases} x_i & i = 1; \\ K_i(x_i) & i = 2; \\ K_i(x_i + y_{i-1}) & 2 < i \leq s. \end{cases} \quad (1)$$

After the first 1×1 convolution, the input feature map is divided into s subsets according to channels, which is defined as $x_i i \in \{1, 2, ..., s\}$. Each feature has the same scale size, but the channel is $1/s$ of the input feature. Except x_1, other sub-features have 3×3 convolution kernels, which are defined as $K_i(x)$, and their output is y_i. Sub-features x_i and $K_{i-1}(x)$ are added and input into $K_i(x)$.

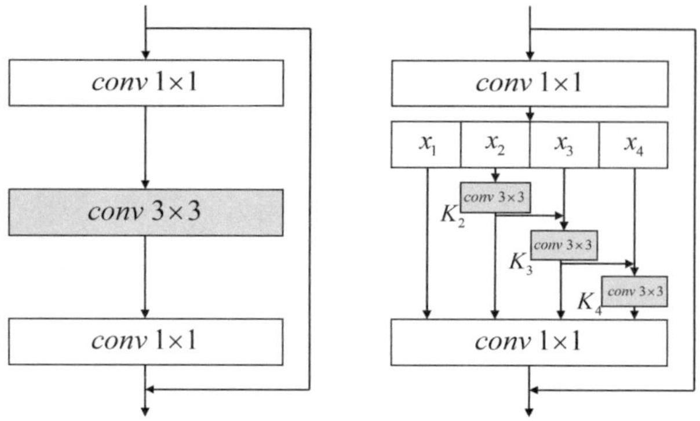

conv Convolutional layer
1×1, 3×3 Convolution kernel size

(a) bottleneck block (b) Res2Net module

Figure 6. Comparison of bottleneck and Res2Net.

Each 3 × 3 convolution operation of the Res2Net module can potentially accept all the feature information on its left, and each output can increase the receptive field. Therefore, the Res2net module can obtain a different number of feature combinations with different receptive field sizes. "*s*" is the number of feature groups in the Res2Net block, and the larger "*s*" means more features and richer receptive fields are learned. In the Res2Net module, hierarchical residual connections enable the receptive field to capture details and global characteristics at a finer-grained level. This method is a more fine-grained improvement, further improving network performance.

To distinguish, the Res2Net module structure is called bottle2neck. As shown in Figure 7, (a) and (b) are two types of bottle2neck structures. The two types correspond to two situations: the number of input and output channels is different (bottle2neck^A), and the number of input and output channels is the same (bottle2neck^B). In Figure 7, "*c*" is the number of input channels, "*width*" is

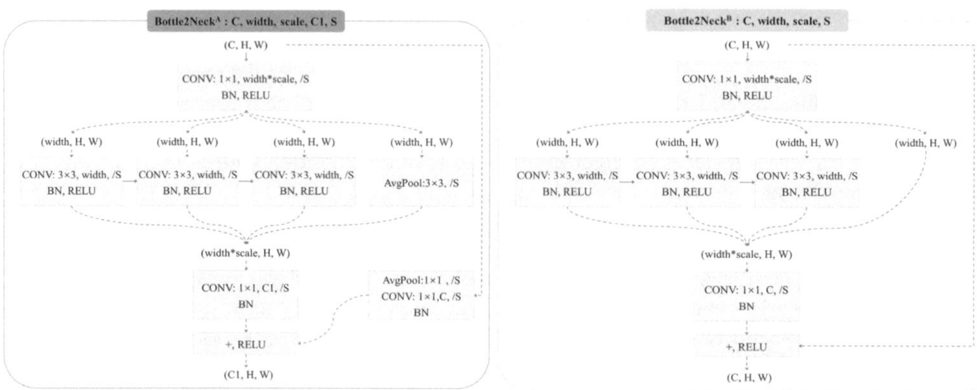

(a) Bottle2Neck^A (b) Bottle2Neck^B

C Image channels
H Image height
W Image width
width Channels after segmentation
scale Segmented subsets
C1 Output channels

S Stride
CONV Convolutional layer
AvgPool Average pooling layer
BN Batch Normalization algorithm processing
RELU ReLU activation function
3×3, 1×1 Convolution kernel size

Figure 7. Two different modules of Res2Net.

the number of channels after segmentation, "*scale*" is the number of segmented subsets, "$C1$" is the number of output channels and "S" is the stride.

We compared Res2Net50 with other backbone feature extraction networks. The overall structure of Res2Net50 is shown in Figure 8. Res2net50 is divided into five stages, in which STAGE 0 preprocesses the input, and the last four stages are composed of two kinds of bottle2necks, with similar structures.

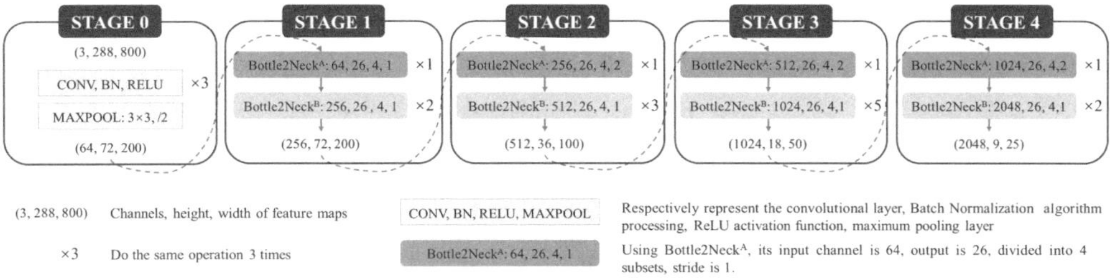

Figure 8. Res2Net50 backbone feature extraction network structure.

3.2.1 *Loss function*

The loss function of the network consists of two parts: the classification loss function of the main branch and the semantic loss function of the auxiliary branch.

Assume that "n" is the maximum number of lanes, "h" is the number of grid rows, "w" is the number of grid columns, and "X" is the global image feature. When the th lane and the lane position on the th row, the position prediction of the lane feature points can be written as:

$$P_{i,j,:} = f^{ij}(X), \quad s.t.\ i \in [1, n],\ j \in [1, h] \tag{2}$$

Where, $P_{i,j,:}$ is a $(w+1)$-dimensional vector, representing the probability that there are lane feature points on the grid of the ith lane and the jth row, which corresponds to the right part of Figure 2.

Assuming $T_{i,j,:}$ is the label at the correct position, the loss function for classification is defined as:

$$L_{cls} = \sum_{i=1}^{n} \sum_{j=1}^{h} L_{CE}(P_{i,j,:}, T_{i,j,:}) \tag{3}$$

The cross-entropy loss function is used as the auxiliary segmentation loss function. To sum up, the overall loss is defined as:

$$L_{total} = L_{cls} + \beta L_{seg} \tag{4}$$

Where, β is a constant and the value is in the range of [0,1].

3.2.2 *Lane fitting model*

For the area with small curvature of the lane, the lane information obtained by the line fitting model can basically meet the needs of lane detection, but it cannot adapt to the road environment with large curvature. Therefore, we adopt the cubic polynomial fitting model, and the specific formula is as follows:

$$f(x) = a_0 + a_1 x + a_2 x^2 + a_3 x^3 \tag{5}$$

Where, x and $f(x)$ are the horizontal and vertical coordinates of lane feature points on the image.

4 EXPERIMENTAL DESIGN AND ANALYSIS

4.1 *Lane dataset*

TuSimple dataset contains multi-lane images on expressways under different traffic conditions. The dataset only labels four lanes of the current lane and left and right lanes, and there are also a few five lanes. TuSimple dataset is the first large dataset with only lane annotation. It makes an important contribution to the development of lane detection algorithms and has become the benchmark dataset in this field.

The CULane dataset is a large-scale lane detection dataset released by the Multimedia Laboratory of the Chinese University of Hong Kong. The dataset is divided into general scenes and eight complex road scenes. The CULane dataset only focuses on the detection of 4 lanes. Because this scene is the most common in real life, other lanes are not marked. When the dataset labels the occluded lanes, it artificially uses a priori knowledge and context information to the label to make the trained model also can detect the occluded lanes. The distribution of scene images in the CULane test set is shown in Figure 9. The datasets are described in Table 1.

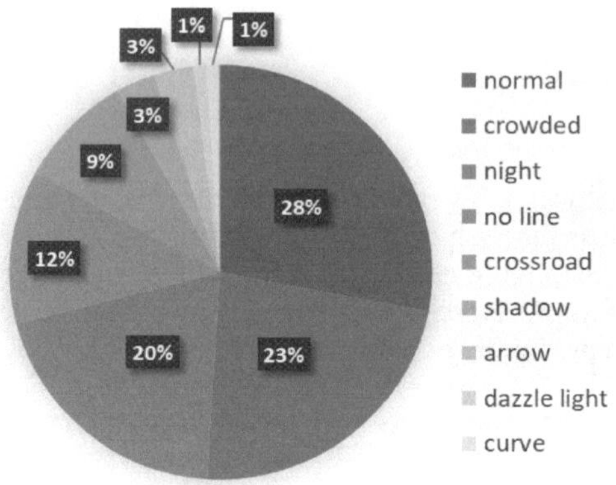

Figure 9. The distribution of images in each scene of the CULane.

Table 1. Dataset introduction.

Dataset	Frame	Train	Validation	Test	Resolution	Lane	Scenes
TuSimple	6408	3268	358	2782	1280×720	≤5	Highway
CULane	133235	88880	9675	34680	1640×590	≤4	urban and highway

4.2 *Valuation criterion*

4.2.1 *TuSimple dataset*

The calculation method of the official evaluation index of TuSimple is shown in (6).

$$Accuracy = \frac{\sum_{clip} C_{clip}}{\sum_{clip} S_{clip}} \tag{6}$$

Where, C_{clip} is the number of lane points correctly predicted, and S_{clip} is the total number of true points in each picture.

4.2.2 *CULane dataset*

In order to verify the effectiveness of our method, the official evaluation method is used when evaluating the CULane dataset. In order to make the evaluation more reasonable, [13] regards both the predicted lane and the real lane as lines with a width of 30 pixels, and the intersection over union (IOU) between them is calculated. In this paper, the IOU threshold is set to 0.5. In the auxiliary branch, if the existence value of the lane in the predicted probability diagram is greater than the IOU threshold, it is considered a true positive (TP). The F1-Measure score is used to quantify the effect of lane detection, and its definition is shown in (8).

$$Precision = \frac{TP}{TP + FP}, \ Recall = \frac{TP}{TP + FN} \tag{7}$$

$$F1 - measure = \frac{2 \times Precision \times Recall}{Precision + Recall} \tag{8}$$

Among them, *TP* are positive examples that are correctly classified, *FP* are negative examples that are incorrectly classified as positive examples, and *FN* are positive examples that are incorrectly classified as negative examples.

4.3 *Experimental design*

The backbones used in the experiment are ResNet18, ResNet34, darknet53 (Redmon & Farhadi 2018), ghost-ResNet18 (Han et al. 2020), ghost-ResNet34 (Chen et al. 2018), Res2Net50-26w-8s and Res2Net50-26w-4s, where the Res2Net50 suffix indicates the number of channels and subsets of the subset divided in the Res2Net50 module, such as Res2Net50-26w-8s divides the input into eight subsets of 26 channels. The above networks all use the pre-trained weight of ImageNet. Before training, adjust the image resolution to 288*800 pixels. All experiments were carried out under the PyTorch framework and trained with an NVIDIA 3080Ti GPU (10G).

4.4 *Experimental results and comparative analysis*

4.4.1 *Comparative experiment of different backbones*

We compared the effects of different feature extraction networks on the accuracy of lane detection. It can be seen from Table 2 that under the same experimental conditions and parameters, Res2Net50-26w-4s has higher scoring and accuracy.

Table 2. The impact of different backbones on model performance.

		CULane (test set)		TuSimple (test set)	
Backbone		$F1 - measure$/%	Single frame processing time/ ms	Accuracy /%	Single frame processing time/ ms
A	ResNet18	68.92	**2.69***	95.47	**2.82***
B	ResNet34	70.94	4.22	95.56	4.66
C	darknet53	70.95	5.17	95.42	5.81
D	ghost-ResNet18	69.71	5.32	95.42	6.20
E	ghost-ResNet34	69.75	9.79	95.43	10.81
F	Res2Net50-26w-8s	71.39	18.85	95.57	18.18
G	Res2Net50-26w-4s	**72.54***	10.78	**95.73***	11.30

In order to reflect the effectiveness of this algorithm, Table 3 and Table 4 compare the relevant indicators of this algorithm with other algorithms, respectively. When the backbone is Res2Net50-26w-4s, the algorithm achieves a 72.54% score on the CULane, and the single frame image processing time is only 10.78ms, which is better than other algorithms. On the TuSimple, the accuracy is 95.73%, and the data marked with bold and asterisk is the best result for each row.

Table 3. Comparison of F1-measure and running time of different networks on the CULane test set.

Scenes	Res50-Seg	SCNN	PINet(4H)	Res-34-SAD	Res-101-SAD	ResNet34-ultra	Res2Net50
Normal	87.4	90.6	90.3	89.9	90.7	**90.7***	90.4
Crowded	64.1	69.7	72.3	68.5	70.0	70.2	71.3
Night	60.6	66.1	67.7	64.6	66.3	**66.7***	66.3
No-lane	38.1	43.4	**49.8***	42.2	43.5	44.4	42.4
Shadow	60.7	66.9	68.4	67.7	67	69.3	**74.8***
Arrow	79.0	84.1	83.7	83.8	84.4	85.7	**85.7***
Dazzle light	54.1	58.5	**66.3***	59.9	59.9	59.5	61.8
Curve	59.8	64.4	65.6*	66.0	65.7	**69.5***	60.8
Crossroad	2505	1990	**1427***	1960	2052	2037	1574
Total	66.7	71.6	72.3	70.7	71.8	72.3	**72.54***
Single frame processing time/ms	–	133.5	40	50.5	171.2	5.7	10.78
multiple	–	1.3x	4.3x	3.4x	1x	30	15.9
FPS	–	7.5	25	19.8	5.8	175.4	93

Note: the data at the crossroad is FP. The smaller the value, the better the detection effect

Table 4. Comparison of accuracy and running time of different networks on the TuSimple test set.

Algorithms	Accuracy /%	Single frame processing time (ms)	Multiple
ResNet18-Seg	92.69	25.3	5.3x
ResNet34-Seg	92.84	50.5	2.6x
LaneNet (Wang 2018)	96.38	19	7.0x
PINet (4H)	96.75	40	3.3x
SCNN	96.53	133.5	1x
ENet-SAD	**96.64***	13.4	10.0x
ResNet34-ultra	95.56	12.65	10.6x
Res2Net50_26w_4s	95.73	**11.3***	**11.8x***

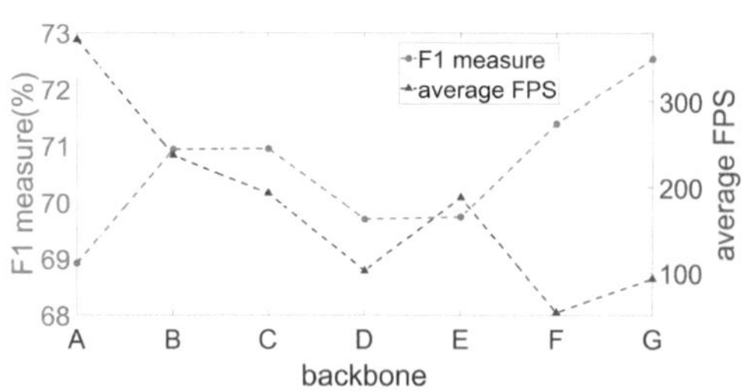

(a) CULane test set

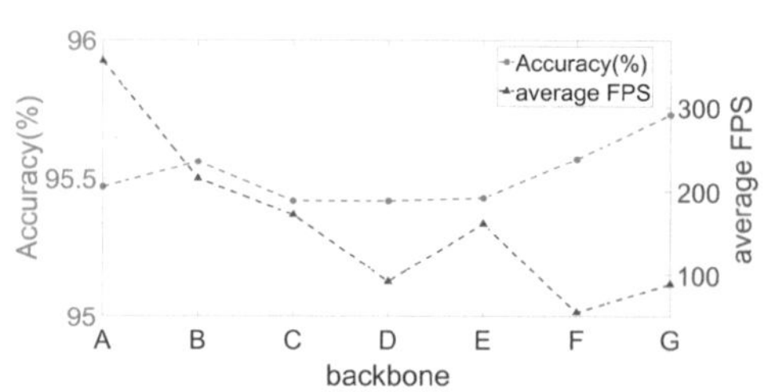

(b) TuSimple test set

Figure 10. The impact of different backbones on model performance.

4.4.2 *Comparative experiment of different number of grid elements*

As mentioned above, we use grid classification to determine the location of lane feature points. This section further explores the impact of a different number of grid cells on the Res2net50 effect. On the CULane dataset, images are segmented using grids of 150, 200, 250 and 300 columns, respectively. On the TuSimple dataset, images are segmented using grids of 50, 100, 150 and 200 columns, respectively. According to the annotation format of the dataset, the number of grid rows in the CULane and the TuSimple are set to 48 and 31, respectively. The final result is shown in Figure 11. With the increase of grid cells, the score increases on the CULane, and the classification accuracy decreases on the TuSimple. This is because more grid cells require finer and more difficult classification. Although a small number of grid cells means high classification accuracy, the positioning error will be large because the grid cells are too large to represent the location of lane feature points accurately. Therefore, we select 300 as the number of grid cells on the CULane and 50 as the number of grid cells on the TuSimple.

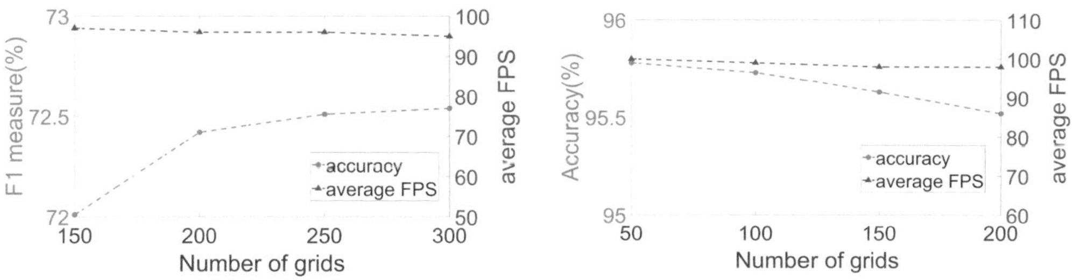

(a) The influence of the number of grids on the CULane test set on the results

(b) The effect of different grid numbers in the TuSimple test set on the results

Figure 11. The effect of different grid numbers on model performance.

4.5 *Lane feature point fitting*

The final output of the network is the position of the feature points of each lane in each row. Taking the center point coordinate of the grid as the feature point coordinate, the discrete feature points of each lane are obtained. The feature points of each lane are fitted by the cubic polynomial, and the results are represented as the green lane and yellow lane in Figure 12. At the same time, the feature point fitting method can modify some abnormal prediction points to make them closer to the real label.

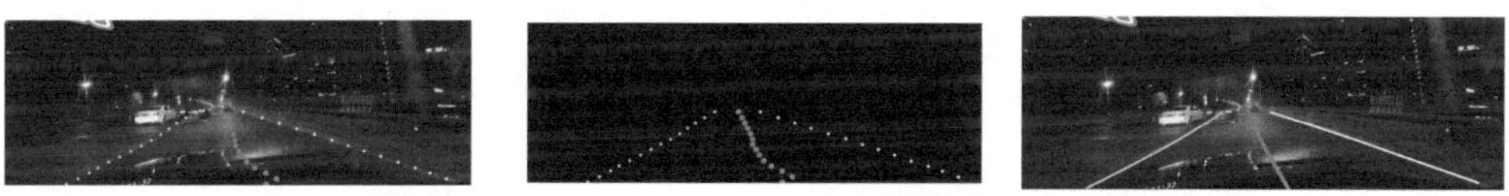

Figure 12. Feature point fitting of the lane line.

4.5.1 *Performance of the algorithm in data sets and real scenes*

Res2Net50-26w-4s has achieved good results in the complex scenes of the above two datasets. As shown in Figure 13, the first two lines are the results of the TuSimple dataset, and the rest are the results of the CULane dataset. From left to right are the original frame, the ground truth and the prediction result (in which the prediction result is marked in blue and the ground truth is marked in green), the ground truth (both datasets use a series of points to represent the ground truth) and the prediction result (different lanes are distinguished by different colors). The algorithm has achieved good detection results in curves, missing, congestion, glare, darkness, and night scenes.

The robustness of the algorithm needs to be reflected in the actual road scene. As shown in Figure 14, our algorithm is compared with the algorithm proposed in [12]. In the case of the night (1),

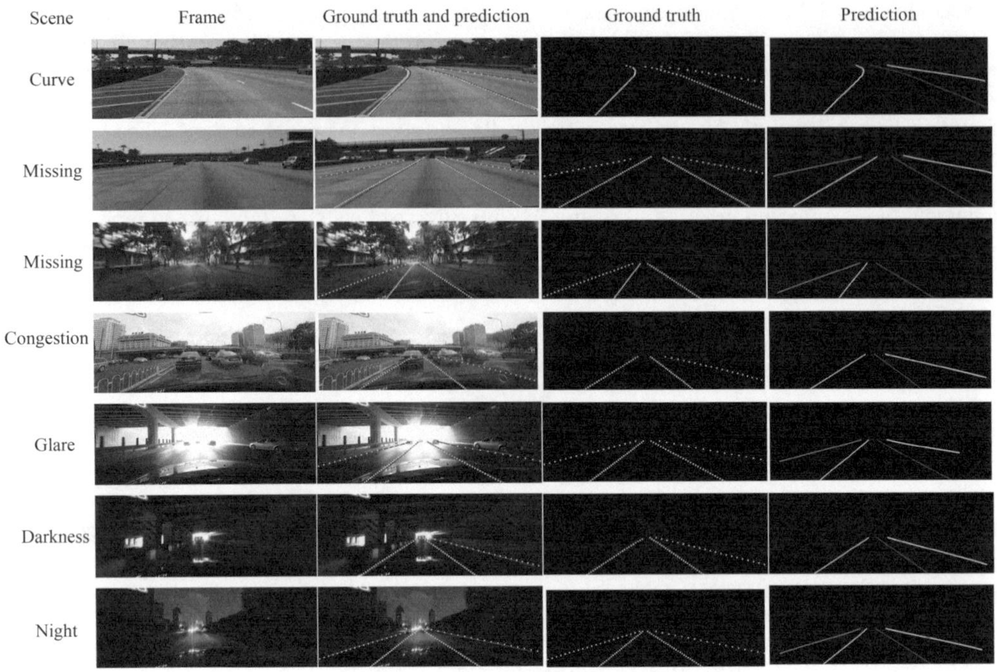

Figure 13. Visualization of Tusimple and CULane datasets.

Figure 14. Comparison of prediction results in actual road scenes.

occlusion, congestion and curve scenes, ResNet34-ultra missed detection, only identifies the innermost two lanes, and even cannot detect the lanes. At night (2), ResNet34-ultra has false detection. In contrast, our algorithm has better robustness for multi-lane detection in complex scenes.

5 EPILOGUE

This paper simplifies the extraction of lane feature points according to the idea of grid classification, and embeds Res2Net into ResNet50 with the help of its multi-scale feature extraction ability. A large number of experiments on datasets show that our method has excellent speed and accuracy. The main conclusions are as follows:

1) The semantic segmentation is transformed into a grid classification, which greatly simplifies the calculation of the depth of the network. At the same time, using the global features of the image, we can better obtain the semantic information of lanes in complex scenes.
2) Through the comparative experiments of different backbones, it is concluded that Res2Net50-26w-4s can effectively extract the multi-scale information of lanes. The F1-Measure score was 72.54% on CULane and 95.73% accuracy on TuSimple.
3) Through comparative experiments, it is concluded that on the CULane dataset, the effect is the best when the number of grid columns is 300, and on the TuSimple dataset, the effect is the best when the number of grid columns is 50.

REFERENCES

Baofeng Wang, Research on Vehicle and Lane Line Recognition Based on Machine Learning and Sensor Fusion. Beijing Institute of Technology, 2017.

Chen L. C., Papandreou G., and Kokkinos I. et al. DeepLab: Semantic Image Segmentation with Deep Convolutional Nets, Atrous Convolution, and Fully Connected CRFs. IEEE Transactions on Pattern Analysis and Machine Intelligence, Vol.40, No.4, 2018, pp.834–848.

Gao S., Cheng M., and Zhao K. et al. Res2Net: A New Multi-Scale Backbone Architecture. IEEE transactions on pattern analysis and machine intelligence, Vol.43, No.2, 2021, pp.652–662.

Han K., Wang Y., and Tian Q. et al. Ghostnet: More features from cheap operations. 2020, pp.1580–1589.

He K., Zhang X., and Ren S. et al. Deep residual learning for image recognition. 2016, pp.770–778.

Hou Y., Ma Z., and Liu C. et al. Learning Lightweight Lane Detection CNNs by Self Attention Distillation. 2019.

Ko Y., Lee Y., and Azam S. et al. Key Points Estimation and Point Instance Segmentation Approach for Lane Detection. 2020.

Lee S., Kim J., and Yoon J. S. et al. VPGNet: Vanishing Point Guided Network for Lane and Road Marking Detection and Recognition. 2017 IEEE International Conference on Computer Vision (ICCV), 2017.

Leyin Liang. Research on Lane Line Detection Algorithm Based on Deep Learning. Beijing Jiaotong University, 2018.

Lichao Chen, Xiuzhi Xu, Jianfang Cao, etc. Multi-scene lane line detection with auxiliary loss introduced. Journal of Image and Graphics", Issue 09, 2020, pages 1882–1893.

Newell A., Yang K. and Deng J. Stacked hourglass networks for human pose estimation. 2016, pp.483–499.

Pan X., Shi J., and Luo P. et al. Spatial as Deep: Spatial CNN for Traffic Scene Understanding. 2017.

Paszke A., Chaurasia A. and Kim S. et al. ENet: A Deep Neural Network Architecture for Real-Time Semantic Segmentation. 2016.

Qin Z., Wang H., and Li X. Ultra-Fast Structure-aware Deep Lane Detection. Springer, Cham, 2020.

Redmon J. and Farhadi A. Yolov3: An incremental improvement. arXiv preprint arXiv:1804.02767, 2018.

Wang Z., Ren W., and Qiu Q. LaneNet: Real-Time Lane Detection Networks for Autonomous Driving. 2018.

Yang Luo. Lane line detection in complex environment," University of Electronic Science and Technology, 2020.

Advances in Urban Engineering and Management Science – Khalil & Yang (Eds)

Optimization study of train stopping scheme of Guangzhou-Zhuhai intercity railway

S.Q. Chen*, W. Li*, Q. Luo* & J.T. Lu*
College of Urban Transportation and Logistics, Shenzhen Technology University, Shenzhen, Guangdong, China

ABSTRACT: Combining the social factors, passenger transport factors, and train factors of the stations themselves, this paper uses the information entropy method and the systematic clustering method to classify the railway stations. On this basis, a multi-objective model is established with the service frequency of passenger stations, the number of single train stops, station capacity and train transport capacity as constraints, and the minimum railway operation cost and the minimum total travel time of passengers as the objectives. The optimization model is solved by a genetic algorithm. The study shows that compared with the actual scheme, the average service frequency of stations of different levels in the optimization scheme is reduced by 2.05 trains, the average OD service frequency of stations of different levels is reduced by 2.4 trains, the total number of stops is reduced by 36, the total travel time of passengers is reduced by 2.27%, and the transportation cost of the intercity railway is reduced by 9.21%.

1 INTRODUCTION

In recent years, with the increasing size of urban agglomerations, the demand for passengers for short-distance transportation has been increasing. Although with the rapid development of high-speed railways in China, the radiation range and transport speed of intercity railway lines have been greatly enhanced, the competition from highways has made the need to improve the core competitiveness of intercity railway lines very urgent. The rational optimization of train stopping plans to meet passenger demand is conducive to attracting more passengers and improving the core competitiveness of intercity railways.

For the study of the train stopping scheme optimization, Li et al. (Li et al. 2013) established a nonlinear programming model with the station service frequency, the number of single train stops, the inter-station accessibility as constraints, and the minimum total number of station stops as the optimization objective. Lv et al. (Lv et al. 2014) established a 0-1 planning model for optimizing the stopping scheme with the maximization of the sum of service frequency and minimization of train running time for each station. Niu et al. (Niu et al. 2014) considered time-varying demands as well as minimizing passenger waiting times to optimize stopping schemes. Huang et al. (Huang et al. 2017) used grey correlation analysis to classify the passenger node classes of high-speed railways and construct a stopping scheme optimization model to minimize lost passenger time, the distance traveled by empty vehicles, and the total number of stops. Liu and Zhang (Liu et al. 2019, Zhang et al. 2021) established a 0-1 integer planning model for the optimization of the combination of the train stopping scheme and running map. Yan et al. (Yan & Li 2019) established a nonlinear 0-1 planning model for minimizing the occupancy time of train running lines. Huang et al. (Huang & Niu 2020) established a two-layer planning model, with the upper layer being a stopping plan optimization model to minimize operating costs and the lower layer being a passenger

*Corresponding Authors: 13049200228@163.com, liwei@sztu.edu.cn, luoqin@sztu.edu.cn and lujunting2020@email.szu.edu.cn

DOI 10.1201/9781003305026-71

flow allocation model based on the Waldrop user balance principle, which is solved using a genetic algorithm. Dong et al. (Dong et al. 2020) used a heuristic approach to integrate the optimization of commuter rail train stopping plans and schedules. The above studies mainly constructed single-objective planning models, multi-objective planning models and two-level planning models, and designed corresponding solution algorithms, but seldom considered station importance, which, as a quantitative indicator for studying the transport distribution capacity of stations in transport networks, has become the main basis for measuring the importance of cities or transport hubs in road network planning.

To sum up, this paper takes the Guangzhou-Zhuhai intercity railway as the research object and classifies the stations along the Guangzhou-Zhuhai intercity railway by constructing a station importance evaluation system. On this basis, an optimization model of the train stopping scheme is constructed to minimize the operating cost and total travel time of passengers and solved by designing a genetic algorithm.

The remainder of the paper is organized as follows. Section 2 establishes the station importance evaluation index system and combines the information entropy and systematic clustering methods to classify the station levels. In section 3, a stopping scheme optimization model of trains based on minimizing the operating cost and total passenger travel time is established, and then the model is solved by an agentic algorithm. In section 4, an example of the Guangzhou-Zhuhai intercity railway line is provided to optimize the train stopping scheme by using the model and algorithm above and evaluating it by the station service frequency, average service frequency of stations of various levels, and average OD service frequency of stations of various levels. The final section draws the conclusions.

2 STATION CLASSIFICATION

2.1 *Station importance evaluation index system construction*

Station level is an important factor in deciding the stopping scheme. For intercity railways serving urban passenger flows, the classification of station level should consider various factors such as passenger factors, train factors, and social factors. Therefore, the station importance evaluation index of the intercity railway is constructed as shown in Figure 1.

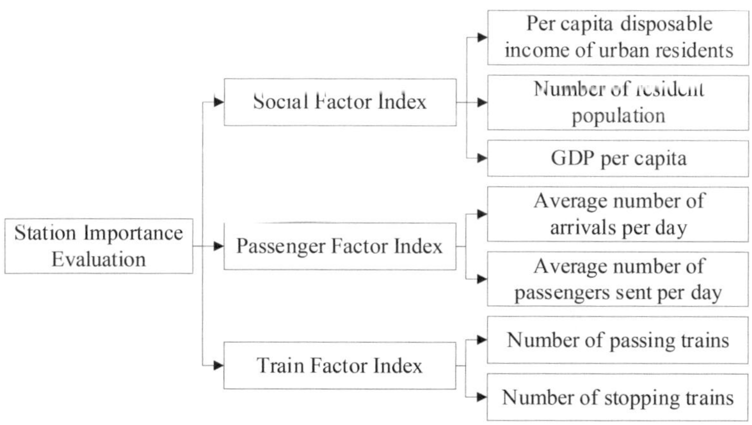

Figure 1. Station importance evaluation index system.

2.2 *Station importance calculation*

Using the information entropy method to calculate the importance of stations, the specific steps are as follows:

Step1: Standardize the index data. Since the measurement unit of each indicator is not uniform, the data needs to be standardized.

$$X'_{ij} = \frac{X_{ij}}{\sum_{i=1}^{n} X_{ij}} \tag{1}$$

where n is the number of stations; X_{ij} is the *i-th* station about the *j-th* indicator value; X'_{ij} is the normalized data value.

Step2: Calculate the differentiation of each indicator.

$$D_j = 1 + \frac{1}{\ln n} \sum_{i=1}^{n} X'_{ij} \ln X'_{ij} \tag{2}$$

where D_j is the discrimination of indicator j.

Step3: Calculate the weight of each indicator.

$$W_j = \frac{D_j}{\sum_{j=1}^{n} D_j} \tag{3}$$

where W_i is the weight of indicator j.

Step4: Calculate the importance of each station.

$$Y_i = \sum_{j=1}^{J} W_j X'_{ij} \tag{4}$$

where Y_i is the importance of station i.

2.3 *Station importance clustering*

The systematic clustering method is used to cluster the station importance by viewing each station as a class, specifying the distance between classes, selecting the pair with the smallest distance to merge into a new class, calculating the distance between the new class and the other classes, and merging the two classes with the closest distance until all stations are combined into one class. The systematic clustering method uses the similarity statistic between sets of objects as the basis for classification and generally uses the distance statistic between objects as the similarity statistic (Li 2014). In this paper, the distance between objects is defined as the Euclidean distance, and the distance between classes is the shortest distance method. Euclidean distance and shortest distance are calculated in Equations (5) and (6).

$$d_{ij} = \left[\sum_{m=1}^{M} |\mathrm{Y}_{im} - Y_{jm}|^2 \right]^{\frac{1}{2}} \tag{5}$$

$$d_{pd} = \min_{i \in G_p, j \in G_q} d_{ij} \tag{6}$$

3 OPTIMIZATION MODEL AND ALGORITHM

3.1 *Hypothesis*

First, although the line is divided into the upstream and downstream directions, the model only discusses the one-way situation where the passenger flow is larger. Second, the Guangzhou-Zhuhai intercity line, which this paper studies, can meet the demand for passenger retention. Third, passengers choose to travel on trains that stop at both the origin and destination stations. Fourth, passengers arriving can all be considered to follow the normal distribution.

3.2 *Modeling*

3.2.1 *Objective function*

The optimization of the intercity railway stopping scheme should consider the maximization of both passenger travel and intercity railway sector interests based on meeting passenger demand. Therefore, in this paper, the optimization model of the train stopping scheme is designed with the minimum railway operation cost and the minimum total passenger travel time as the optimization objectives. The decision variables are x_{ki}, $k = 1, 2, \ldots, K$, $i = 1, 2, \ldots, N$, $x_{ki} \in \{0,1\}$, $x_{ki} = 1$ means that train stops at station otherwise no stop. Clearly $x_{k1} = 1$, $x_{kn} = 1$. To facilitate the description of the model, the model symbols are listed in Table 1.

a) Minimal rail operating costs: Rail operating costs include variable costs and fixed costs. Variable costs include the cost of stopping intercity trains and the cost of running intercity trains in kilometers. The optimization of the stopping scheme in this paper is based on a fixed operating section and a fixed number of trains, so fixed costs and intercity train running time kilometer costs have no impact on the stopping scheme. Therefore, only the stopping costs of intercity trains are considered in this paper. The expression of the objective function is shown in Equation (7).

$$\min Z_1 = \sum_{k=1}^{K} \sum_{i=1}^{N} C_s x_{ki} \tag{7}$$

b) Minimal total passenger travel time: Passenger travel time consists of waiting time and onboard travel time. As the waiting time is not related to the train stopping scheme, only the travel time of the passenger onboard is considered. Passenger travel time onboard includes the running time, stopping time, and additional departure and stopping time of the train on which the passenger is traveling in the respective section. The expression of the objective function is shown in Equation (8).

$$\min Z_2 = \sum_{k=1}^{K} \sum_{i=1}^{N-1} \sum_{j=i+1}^{N} \min\left\{Q_{ij}\ ,\ Q_{ij}^{\max}\right\} P_{kij} T_{kij} \tag{8}$$

Table 1. Model symbols.

Symbols	Explanation
(i, j)	Travel OD pairs, where i is the up station and j is the down station
K	Total number of trains
N	Total number of stations
C_S	Train stopping costs
Q_{ij}	The number of passengers between (i, j)
Q_{ij}^{max}	The number of passengers that can be satisfied between (i, j) when all trains stop between (i, j)
P_{kij}	The passenger ratio choosing the train k to travel between (i, j)
T_{kij}	Travel time of train k between (i, j)
t_i^s	Stopping time of trains at station i
t_α	Additional train start/stop times
$t_{i-1,i}^r$	Train interval running time
θ	Scale parameter
$\phi_{min}(x)$	The minimum service frequency of station i, that is, the minimum number of trains stopping at the station
$\phi_{max}(x)$	The maximum service frequency of station i, that is, the maximum number of trains stopping at the station
φ_{min}	Minimum number of stops for a single train
φ_{max}	Maximum number of stops for a single train
C_k	The capacity of train k
w_i	Weights

$$T_{kij} = \sum_{n=i+1}^{j-1} (t_n^s + 2t^{\alpha})x_{kn} + t_{n-1,n}^r] + t_{j-1,j}^r \tag{9}$$

According to random utility theory (Zhao 2017), the behavior of passengers selecting trains can be analyzed by the Logit model like Equation (10).

$$P_{kij} = \begin{cases} 0 & x_{ki}x_{kj} = 0 \\ \frac{\exp(\theta/T_{kij})}{\sum_{k\in\{k|x_{ki}x_{kj}=1\}} \exp(\theta/T_{kij})} & x_{ki}x_{kj} = 1 \end{cases} \tag{10}$$

3.2.2 *Constraints*

According to the established objective function, the constraints of the train stopping scheme optimization model are established in terms of station service frequency, the number of the single train stop, station capacity and train capacity.

a) Station service frequency constraint: For a station, the number of train stops in a day cannot be lower than the service frequency requirement of the station level.

$$\sum_{k=1}^{K} x_{ki} \geq \phi_{\min}(i), \forall i = 1, 2, ..., N \tag{11}$$

b) Constraints on the number of single train stops: If the number of train stops is too low, the service frequency at intermediate stations will be reduced; if the number of train stops is too high, route delays for passengers will increase. Therefore, the number of stops for a single train should be limited.

$$\varphi_{\min} \leq \sum_{i=1}^{N} x_{ki} \leq \varphi_{\max}, \forall k = 1, 2, ..., K \tag{12}$$

c) Station capacity constraint: Station capacity means that the maximum number of trains allowed to originate and stop at that station in a day does not exceed the station's capacity.

$$\sum_{k=1}^{K} x_{ki} \leq \phi_{\max}(i), \forall i = 1, 2, ..., N \tag{13}$$

d) Train capacity constraint: The service capacity provided by the train between (*i, j*) satisfies the following Equation (14).

$$\sum_{k=1}^{K} C_k x_{ki} x_{kj} \geq \min\left\{Q_{ij}\ ,\ Q_{ij}^{\max}\right\} \tag{14}$$

3.3 *Multi-objective optimization*

To facilitate the calculation, this paper uses the weighted summation method to convert the multi-objective function into a single objective function. The objective functions in the constructed model are all minimized, but with different number outlines. Therefore, first, use Equation (15) for normalization, and then use Equation (16) for weighted summation.

$$Z_i^{'} = \frac{Z_i - Z_{i,\min}}{Z_{i,\max} - Z_{i,\min}}, i = 1, 2 \tag{15}$$

$$\min Z = \min\left(w_1 * Z_1^{'} + w_2 * Z_2^{'}\right) \tag{16}$$

3.4 *Solution algorithm*

In this paper, a genetic algorithm is designed to solve the above model. The fitness function is the inverse of the objective function, and the termination condition is the maximum number of iterations. The specific steps of the genetic algorithm are as follows.

Step 1: It generates N binary-coded individuals randomly, and the stopping scheme of the train is represented as a chromosome, where code 1 means that the train will stop at the corresponding station and 0 means no stopping. Besides, because the train must stop at the original and terminal stations, the encoding is 1. What is more, the initial population satisfies the constraints in Equations (11) to (14).

Step 2: Calculate the fitness of each chromosome in the current population. The number of iterations is noted as 0.

Step 3: The selection operation adopts the roulette method, the crossover operator is a uniform crossover, and the mutation operator is a single point mutation. After the selection, crossover and mutation are completed, it is determined whether the new individuals resulting from the reproduction satisfy the constraints in Equations (11) to (14). If this is the case, go to step 3; if not, start again at step 2.

Step 4: The number of iterations is increased by one, and the fitness value of the new individual is calculated. If the number of iterations reaches the upper limit, the individual with the best fitness is selected, and the algorithm is terminated; otherwise, go to step 2.

4 CASE ANALYSIS

4.1 *Line profile*

The Guangzhou-Zhuhai intercity railway opened for operation in January 2011, with trains operating at a speed of 200 km/h. The main line of the Guangzhou-Zhuhai intercity railway connects Guangzhou South and Zhuhai, and the branch line connects Guangzhou South and Xinhui (Figure 2). In this paper, the average daily passenger flow data of Guangzhou-Zhuhai intercity in March 2016 is selected as the initial passenger flow data for the optimization of the stopping scheme, and the passenger flow between some stations of the common line is divided according to the current train operation ratio (2:1) of the main and branch lines (Tables 2, 3). At the same time, according to the classification results of station levels, referring to the existing stop plan of the Guangzhou-Zhuhai Intercity Railway, the service frequency ranges corresponding to different levels of stations are customized. The data required in the example analysis is shown in Table 4. The length of the interval and pure runtime of trains are followed in Table 5.

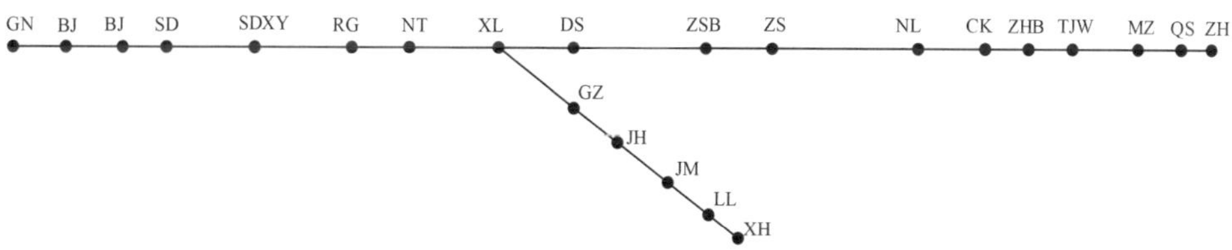

Figure 2. Guangzhou-Zhuhai intercity line.

4.2 *Classification of Guangzhou-Zhuhai intercity station*

In this paper, the 17 intermediate stations along the Guangzhou-Zhuhai intercity railway are used to classify the stations using the above-mentioned station importance classification method.

4.2.1 *Information entropy method for calculating weights*

According to Equations (1)(2)(3), the weights of the seven indicators of urban disposable income per capita, the number of permanent urban residents, the gross urban product per capita, the number

Table 2. The value of OD passenger flow of Guangzhou-Zhuhai intercity railway main line.

Station	GN	BJ	BJ	SD	SDXY	RG	NT	XL	DS	ZSB	ZS	NL	ZHB	TJW	MZ	QS	ZH
GN	–	46	86	1398	104	1449	545	2603	33	3954	364	155	228	545	1361	68	7063
BJ	28	–	0	0	0	4	3	11	0	8	3	1	1	4	0	0	14
BJ	57	0	–	0	0	3	1	9	0	4	0	0	0	1	0	0	6
SD	1274	1	1	–	0	32	8	54	1	124	6	5	12	41	69	3	520
SDXY	97	0	1	0	–	0	0	7	0	5	0	1	1	0	0	1	5
RG	1455	4	2	48	2	–	2	67	2	130	6	3	1	45	55	0	395
NT	662	1	3	18	2	10	–	20	1	11	8	0	4	4	8	0	40
XL	2407	8	7	80	6	109	23	–	1	471	47	45	27	87	136	9	806
DS	18	0	0	0	0	1	0	0	–	2	1	0	1	2	1	0	7
ZSB	4050	5	7	125	5	149	5	571	6	–	3	13	18	89	179	5	1130
ZS	160	1	0	4	0	3	3	31	0	1	–	0	3	15	10	4	82
NL	210	0	0	6	4	0	6	58	0	13	0	–	1	1	6	3	50
ZHB	168	0	1	7	0	6	2	22	1	22	3	0	–	0	0	2	8
TJW	647	4	0	55	2	45	3	113	2	136	8	7	0	–	18	4	258
MZ	1723	0	6	68	7	58	6	169	3	251	0	11	5	44	–	0	32
QS	273	0	1	7	3	6	5	29	0	24	5	4	6	9	1	–	1
ZH	7152	5	6	467	19	361	51	729	5	1197	37	64	15	236	61	27	–

Table 3. The value of OD passenger flow of Guangzhou-Zhuhai intercity railway branch line.

Station	GN	BJ	BJ	SD	SDXY	RG	NT	XL	GZ	JM	XH
GN	–	23	43	699	52	724	273	1302	2681	2145	1364
BJ	14	–	0	0	0	2	2	5	5	10	5
BJ	28	0	–	0	0	1	1	5	9	13	7
SD	637	0	0	–	0	16	4	27	35	54	40
SDXY	48	0	0	0	–	0	0	3	4	11	6
RG	728	2	1	24	1	–	1	33	77	95	61
NT	331	1	1	9	1	5	–	10	41	48	22
XL	1204	4	4	40	3	55	12	–	91	446	231
GZ	2573	6	7	81	2	108	45	168	–	73	88
JM	2137	9	7	93	3	122	53	542	102	–	20
XH	1317	5	3	73	2	78	24	266	104	34	–

Table 4. Values of case parameters.

	Value
The service frequency range of first-level stations (times)	30–38
The service frequency range of second-level stations (times)	8–28
The service frequency range of third-level stations (times)	1–10
Minimum number of stops for a single train (times)	2
Maximum number of stops for a single train (times)	9
Train capacity (person)	1000
Stopping time of train (minute)	2
Train stopping costs (yuan)	500
Weight	0.5
Additional train start/stop times (minute)	0.5
Scale parameter	600
population	100
Crossover probability	0.9
Mutant probability	0.1
Iterations	100

Table 5. The length of the interval and pure runtime of trains.

Interval	Length (km)	Runtime (min)	Interval	Length (km)	Runtime (min)
GN-BJ	5	5	ZS-NL	13	7
BJ-BJ	6	5	NL-ZHB	10	6
BJ-SD	4	4	ZHB-TJW	4	4
SD-SDXY	8	6	TJW-MZ	12	8
SDXY-RG	9	6	MZ-QS	4	7
RG-NT	5	5	QS-XL	3	9
NT-XL	8	5	XL-GZ	9	7
XL-DS	7	5	GZ-JM	11	7
DS-ZSB	12	7	JM-XH	7	7
ZSB-ZS	6	5			

of passengers sent, the number of passengers arrived, the number of passing trains and the number of stopping trains are 0.0027, 0.0798, 0.0101, 0.3203, 0.3348, 0.0260, 0.2263, respectively.

4.2.2 *Station importance calculation*

The importance of each station is calculated according to Equation (4) and is shown in Table 6.

Table 6. Station importance.

Railway station	Station importance	Railway station	Station importance
BJ	0.017339661	ZS	0.020456897
BJ	0.018217674	NL	0.015330448
SD	0.107167599	ZHB	0.016811855
SDXY	0.019016588	TJW	0.035850088
RG	0.113225156	MZ	0.059353402
NT	0.037569305	QS	0.012820376
XL	0.188641009	GZ	0.08545302
DS	0.00898061	JM	0.083692768
ZSB	0.160073545		

4.2.3 *Station classification results*

Using the principle of systematic clustering, with the help of SPSS software, the stations were systematically clustered for importance. The 17 intermediate stations along the Guangzhou-Zhuhai intercity line were divided into three levels, as shown in Table 7.

Table 7. Station classification results.

Station level	Station
First-level station	XL, ZSB
Second-level station	SD, RG, MZ, GZ, JM
Third-level station	BJ, BJ, SDXY, NT, DS, ZS, NL, ZHB, QS, TJW

4.3 *Optimization model solving*

According to the above-established model, algorithm and data, this paper uses MATLAB to solve the problem and obtain the optimized Guangzhou-Zhuhai intercity train stopping scheme results, as shown in Figure 3.

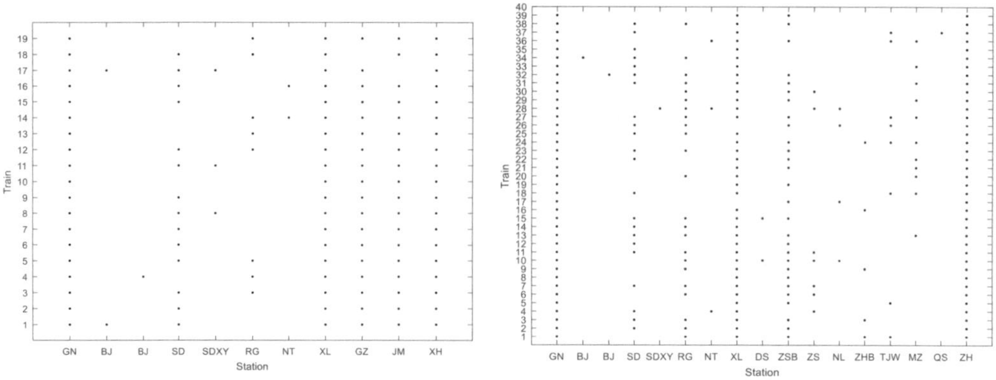

(a) Optimal stop plan of the main line.

(b) Optimal stop plan of the branch line.

Figure 3. Optimal stop plan of Guangzhou-Zhuhai intercity railway.

As can be seen in Figure 3, there are two main stopping modes for the optimized stopping scheme: large stop pattern and selective stop pattern, with a total of 31stopping modes for the main line and 11 stopping modes for the branch line.

4.4 *Result analysis*

4.4.1 *Station service frequency analysis*

Compared with the actual plan, the station service frequency of the optimal plan is significantly lower, as shown in Figure 4. This is mainly because the current Guangzhou-Zhuhai intercity passenger flow is not very high, and many trips are experiencing excess capacity.

4.4.2 *Analysis of average service frequency of stations of various levels*

Compared with the actual plan, the service frequency of stations of different levels in the optimal plan has generally decreased slightly, as shown in Figure 5. Assuming the weights of the first, second, and third-level stations are respectively 0.5, 0.3, and 0.2, it can be calculated that the weighted average service frequency of all stations in the optimization scheme is 28.66 trains.

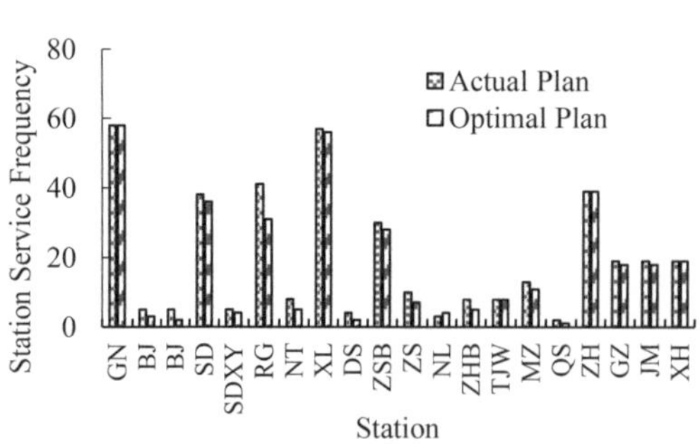

Figure 4. Station service frequency comparison.

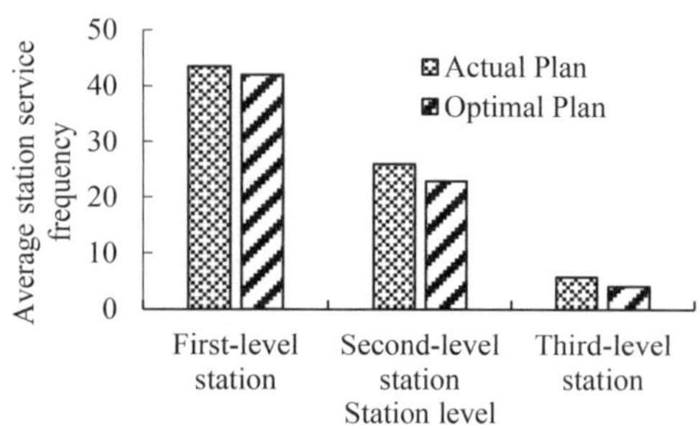

Figure 5. Comparison of average service frequency of stations of various levels.

4.4.3 *Analysis of average OD service frequency of stations of various levels*

As shown in Figure 6, there is a significant difference in the average OD service frequency between the first-level stations and other stations, with a more significant reduction than the other stations. This indicates that the current scheme focuses too much on the service frequency of the first-level

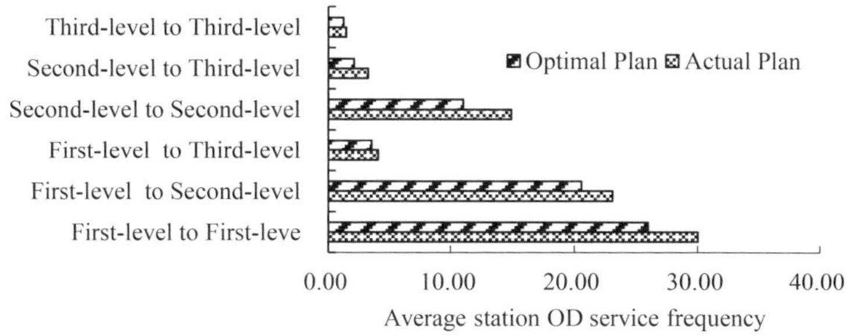

Figure 6. Comparison of average OD service frequency of stations of various levels.

stations and neglects the passenger demand between the other stations. By assigning weights of 0.3, 0.2, 0.15, 0.15, 0.1, 0.1 to the average OD service frequency of each level of the station from bottom to top, the weighted average OD service frequency between all stations of the optimized scheme can be calculated as 14.43 trains.

4.4.4 *Summary*

In summary, the optimal plan is better than the actual plan. As shown in Table 8, the average service frequency of stations of different levels is reduced by 2.05 trains, the average OD service frequency of stations of different levels is reduced by 2.4 trains, and the total number of stops is reduced by 36. The savings in stopping costs and passenger travel time, based on meeting passenger demand, include a 2.27% reduction in total passenger travel time and a 9.21% reduction in intercity rail transport costs.

Table 8. Comparison of actual and optimal plan indictors.

Indicator (unit)	Actual plan	Optimal plan	Benefit	Benefit rate
Average service frequency of each level station(trains)	30.71	28.66	2.05	6.68%
Average OD service frequency of each level station(trains)	16.93	14.43	2.4	14.18%
Total passenger travel time(minute)	1874143	1831640	42503	2.27%
Intercity railway transportation cost (yuan)	195500	177500	18000	9.21%
Total number of stops (times)	391	355	36	9.21%

5 CONCLUSION

This paper studies the optimization of the train stopping scheme of the Guangzhou-Zhuhai intercity railway. Firstly, a station importance evaluation index system is constructed, and the information entropy method and systematic clustering method are used to classify the station levels. On this basis, a multi-objective train stopping scheme optimization model is constructed with the objective function of minimizing railway operation cost and minimizing the total passenger travel time, considering the station service frequency, the number of single train stops, station capacity and train transport capacity constraints, and a genetic algorithm is designed for solving the model. The study shows that, compared with the actual stopping scheme, all the indicators of the optimized stopping scheme are reduced accordingly, shortening the passenger travel time, and saving the train stopping cost based on meeting the passenger flow demand.

REFERENCES

Dong, X., D. Li, Y. Yin, S. Ding, Z. Cao (2020) Integrated optimization of train stop planning and timetabling for commuter railways with an extended adaptive large neighborhood search metaheuristic approach. Transportation Research Part C Emerging Technologies, 117: 102681.

Huang, L.Y., H.X. Lv, L.X. Du, W.X. Wang (2017) Optimization of High-speed train stopping schemes based on node importance. Journal of Transport Engineering and Information, 15(03):49–57.

Huang, Z.P., H.M. Niu (2020) Optimizing train Skip-stop operations and frequencies with period-preference for high-speed rails. Transportation Information and Safety, 38(04):115–121+131.

Li, D.W., B.M. Han, X.J. Li, H.J. Zhang (2013) High-speed railway stopping schedule optimization model based on node service. Journal of Transportation, 35(06):1–5.

Li, T.Y. (2014) Research on Clustering Algorithms in Traffic Domain. Science Publishing, Beijing.

Liu, L., L.Y. Meng, X.Y. Li, G. Liu (2019) Integrated optimization of stopping pattern and train timetable for passenger demand. Journal of Railway Science and Engineering, 16(2):518–527.

Lv, H.X., Y.Y. Tu, S.P. Sun, C. X. Yang (2014) Study on setting-up of high-speed train stops based on station service frequency. Railway Transport and Economy, 36(08):32–37.

Niu, H., X. Zhou, R. Gao (2015) Train Scheduling for Minimizing Passenger Waiting Time with Time-Dependent Demand and Skip-Stop Patterns: Nonlinear Integer Programming Models with Linear Constraints. Transportation Research Part B. Methodological, 76(3):117–135.

Yan, H. F. & Y. U. Li (2019) Optimization of high-speed railway stop schedule plan. Journal of Transportation Engineering and Information, 17(2):8–16.

Zhang, T., P.J. Sun, Z.X. Qu (2021) Study of train stop scheme for inter-city railway during peak hours. Railway Transport and Economy, 43(02):7–15.

Zhao, K.H. (2017) Traveling habits and preference theories and model application. China Railway, (02):55–61

Advances in Urban Engineering and Management Science – Khalil & Yang (Eds)
© 2023 The Author(s), ISBN 978-1-032-30426-7

Evaluation of forward collision warning systems on driver behavior and EEG in heavy fog conditions: Based on a driving simulator

Ran Liu* & Xuedong Yan*
MOT Key Laboratory of Transport Industry of Big Data Application Technologies for Comprehensive Transport, School of Traffic and Transportation, Beijing Jiaotong University, Beijing, P.R. China

ABSTRACT: Forward Collision Warning (FCW) systems have been designed to enhance road safety by reducing the number of rear-end collisions. Nevertheless, little is known about how drivers adapt their behavior and physiology over time when using this kind of system. To investigate these effects, driving data were collected from the Beijing Jiaotong University (BJTU) driving simulator. This study seeks to investigate the impacts of the real-time FCW system on the behavior and EEG of drivers under poor visibility. Thirty-six drivers participated in this simulator study. Findings suggest that exposure to the FCW system has promising safety benefits manifested in improved driver behavior and response by influencing their brain electrical activity and shortening the brake reaction time. Nonetheless, the effectiveness of early warning decreases with the increase in emergency level, and the FCW system may have introduced little to cope with sufficiently urgent scenarios (TTC < 3s). Therefore, participants' physiological and behavioral responses to FCW were used to indicate their risk tendency in driving, which could be further exploited as a potential parameter of FCW system design.

1 INTRODUCTION

Rear-end collisions, in which a vehicle collides with the rear of a slower moving preceding vehicle, are among the most common vehicle-related accidents, especially in fog conditions. The main reason for the change in driving behavior in foggy conditions is blurred vision, which will affect drivers' ability to detect the surrounding road environment and traffic conditions, weaken their information processing ability and make their perception response time longer (Fors & Lundkvist 2009; Plainis & Murray 2002). Rear-end collision in foggy conditions has a high incidence, and the consequences are very serious once it happens. Forward Collision Warning (FCW) systems have been designed to enhance road safety by reducing the number of rear-end collisions. These systems are expected to induce positive and safer behaviors in driving, especially when providing assistance to drivers in critical situations.

Drivers can receive warning information, which mainly includes visual and auditory warnings. Different warning types stimulate drivers differently, thus bringing different warning effects. Wege et al. (2013) found that visual early warning technology could help drivers quickly detect potential risks in the traffic environment. However, it is necessary to consider the potential interference effect of a visual warning on drivers. Visual warning may occupy the visual resources needed by the driver for vehicle control. Since drivers wanted to know the location and content of visual warning information, their attention would shift from vehicle control to visual warning information, which may increase the possibility of accidents (Baumann et al. 2004). Compared with visual warnings, audio warnings transmitted danger information through voice signals and did not occupy drivers' visual resources (Mccallum et al. 2006).

*Corresponding Authors: 19120853@bjtu.edu.cn and xdyan@bjtu.edu.cn

DOI 10.1201/9781003305026-72

The vast majority of published studies (Hirst & Graham 1997; Lee et al. 2002) have shown the benefits of FCW in reducing the number and severity of rear-end collisions. Previous studies of the effects are mainly based on the driver's subjective evaluation and real driving behavior data (Itoh et al. 2013; Liu et al. 2013; Wu et al. 2018). In contrast, only a few studies have used physiological data such as electroencephalography (EEG) to identify driving behavior. EEG has been recognized as an effective, non-invasive technique for monitoring and assisting real-world driving (Kar et al. 2010; Papadelis et al. 2007; Wang et al. 2015). In Chuang's study (Chuang et al. 2015), EEG data were used to detect driving states, and warnings were provided to real-world drivers as feedback. Relative to traditional driving data, EEG data can provide extra information (physiological and emotional) in addition to kinematic vehicle indices. Thereby, a neural analysis of the brain electrical activity in drivers could offer novel information in assessing the effects of the FCW.

In this study, the driving simulation experiment platform and EEG acquisition equipment were used to collect and analyze the driving behavior data and EEG signal data of drivers in the early-warning scene. EEG was used as the physiological parameter to characterize drivers' perception, and the internal correlation mechanism between EEG and driving behavior was explored. As a result, participants' physiological and behavioral responses to FCW were used to indicate their risk tendency in driving, which could be further exploited as a potential parameter of FCW system design.

2 METHOD

2.1 *Experimental apparatus*

The Beijing Jiaotong University (BJTU) driving simulator was used to conduct the experiment and real-time data acquisition. The driving simulator is of high performance and fidelity with a linear motion base, capable of operating with one degree of freedom. Five screens provide the driver with 300° surrounding vision, and each of the screens has a resolution of 1400 × 1050 pixels.

A Neuroscan system was used to collect the EEG data, which was composed of an electrode cap, a SynAmps2TM EEG amplifier and a signal acquisition and analysis system. There were 64 channels in the electrode cap, and their locations followed the international 10–20 system. The driving simulator operates at a 60 Hz update rate while the Neuroscan system at a 1000 Hz rate.

Subjects

In this experiment, 46 participants (27 males and 19 females) aged from 30 to 50 years (M = 37, SD = 2.99) were recruited. Each participant held a valid Chinese driving license and had at least three years' driving experience and more than 10,000 km driving mileage per year. After arrival, each participant was briefed on the requirements of the experiment.

2.2 *Scenario design*

The designed road is a bidirectional circle road with a speed limit of 60 km/h. A double-yellow solid line was set in the middle of the road to avoid overtaking behavior. The simulated vehicle was stopped at the start point of 400 meters away from the lead vehicle. When the participant passed the curve section, the leading car would be found stopping in the driving lane and its brake lights on. Six deceleration points were designed on the driving lane. Once the leading vehicle passed the deceleration point, it would brake at a deceleration rate of 4 m/s2 or 6 m/s2 until it reached the predefined speed. The leading vehicle's driving speeds and decelerations are shown in Figure 1 (the dashed line indicates the starting point of the lead car's brake). Thus, each participant would experience six braking scenarios within one drive. In addition, to simulate the real driving environment, free traffic flow was designed in the opposite lane.

The FCW warning algorithm designed in this paper was a kinematics-based one. It can detect the real-time driving status of the leading vehicle and following vehicle. By doing so, how much

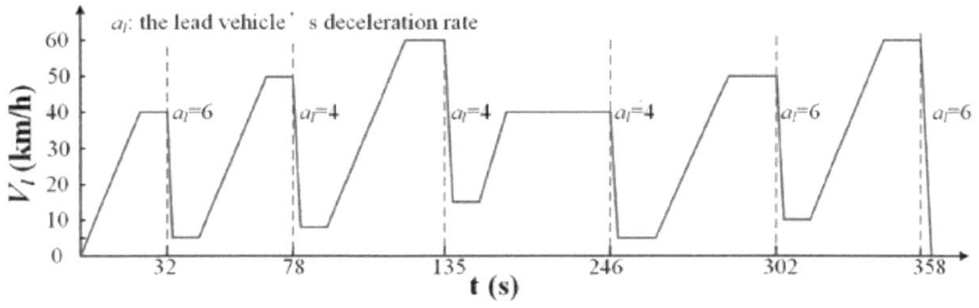

Figure 1. Driving speed profile of the lead vehicle.

time is available for the driver to respond can thus be estimated, represented by t_{max}. Once the lead vehicle's brake onset, the FCW system could detect the sudden brake of the lead vehicle, and the time that the FCW system spent to detect was denoted as t_s (1/60 s). The driver's expected response time was represented by t_r. The value of tr in this paper is 1.5s.

If the driver did not take any deceleration measures and $t_{max} \leq t_s + t_r$, the system would be forced to brake. Otherwise, if $t_{max} > t_s + t_r$, the system judged the warning level according to the specific value of t_{max}, as the following:

if $0 < t_{max} - (t_s + t_r) \leq t$ (1 s), the system sent a level-3 early warning;
if $t < t_{max} - (t_s + t_r) \leq 2t$, the system sent a level-2 early warning;
if $2t < t_{max} - (t_s + t_r) \leq 3t$, the system sent a level-1 early warning;
if $t_{max} - (t_s + t_r) > 3t$, the following vehicle was in a safe state, and the system did not give an early warning.

2.3 *Experimental procedure*

Before the experiment, each of them was required to complete a 10-min practice drive using the simulator. After the practice experiment, the subjects signed an informed consent form. The basic road scenario was explained, and participants were told that there would be a leading car that they should follow as usual and that overtaking was not allowed. The participants were not informed beforehand that the lead vehicle would be braking. For the formal experiment, participants had to drive six times and would rest for 5 min between the drives if they need. The order of the scenarios was randomly selected to minimize potential learning and adaptation effects. All participants received 600 RMB (around 85 USD) for their participation in the study.

3 RESULTS

3.1 *Brake reaction time*

As Table 1 shown, in fog weather conditions, the average BRT of drivers under visual warning the condition was the smallest (M = 1.44s, S.D. = 0.58), followed by audio warning (M = 1.49s, S.D. = 0.61), and the deceleration response time without warning was the longest (M = 1.62s, S.D. = 0.77).

A two-way ANOVA showed the main effect for warning type was significant (F = 16.43, $p < 0.05$). A post hoc Tukey HSD showed that the BRT of drivers without warning was significantly longer than that of audio warning ($p < 0.01$) and visual warning ($p < 0.01$), but no significant difference was found between the audio and visual warning conditions ($p = 0.675 > 0.05$). The main effect for warning level (F = 16.43, $p < 0.05$) and their interaction was also significant (F = 8.86, $p < 0.05$).

Table 1. Descriptive statistical results of BRT under different warning types and levels.

		BRT		MTH	
Warning type	Warning level	Mean	S.D.	Mean	S.D.
No warning	1	3.31	0.88	2.36	0.58
	2	1.81	0.54	2.46	0.42
	3	1.24	0.31	1.96	0.34
	total	1.62	0.77	2.15	0.46
Audio warning	1	2.78	0.87	2.64	0.71
	2	1.71	0.46	2.44	0.46
	3	1.17	0.31	1.97	0.27
	total	1.49	0.61	2.19	0.47
Visual Waring	1	2.03	0.49	3.23	0.42
	2	1.68	0.65	2.43	0.61
	3	1.27	0.47	1.84	0.41
	total	1.44	0.58	2.10	0.61

3.2 *Minimum time headway*

As Table 1 shown, the average MTH of drivers under audio warning conditions was the longest ($M = 2.19s$, $S.D. = 0.47$), followed by no warning ($M = 2.15s$, $S.D. = 0.46$, and the MTH with a visual warning was the shortest ($M = 2.19s$, $S.D. = 0.61$).

A two-way ANOVA showed the MTH of drivers was significantly different between different warning types for maximum deceleration ($F = 5.49$, $p < 0.05$. The main effect for warning level ($F = 103.09$, $p < 0.05$) and their interaction was also significant ($F = 6.50$, $p < 0.05$). A post hoc Tukey HSD showed that the lower the warning level, the longer the MTH. Especially when the warning level was level 1, the MTH of drivers under audio warning and visual warning was significantly higher than that without warning.

3.3 *EEG feature analysis*

As can be seen from Table 2, the relative power of the α wave in the central, partial, and temporal regions was significantly different between different warning types. In the audio warning scenario, drivers had the highest alpha wave relative power in these three brain regions, followed by the no warning scenario, and the relative power was the lowest under the condition of visual warning (See Figure 2).

Except for the temporal region, the relative power of α, β, δ waves in the central, partial, and occipital regions were significantly different between different warning types. With the increase of the warning level, the relative power of α and β waves in each brain region (except the temporal region) decreased, the relative power of δ wave increased, and the relative power of θ wave significantly decreased in the temporal region.

3.4 *Correlation analysis of behavior and EEG*

Through correlation analysis, the relationship between the relative power values of each brain region and the deceleration response time during the deceleration response period was tested. As shown in Table 3, the correlation test results showed that the BRT of drivers in the collision avoidance reaction stage was positively correlated with the relative power values of α, β and θ overall and negatively correlated with the relative power of δ. In contrast, there was no significant correlation between BRT and the relative power of alpha and beta in the temporal region and theta waves in the central region.

Table 2. MANOVA analysis results of EEG features by FCW(F-value).

Dependent variable	Independent variables	Frontal	Central	Parietal	Occipital	Temporal	All brain
relative_α	warning type	2.08	3.41*	4.70	1.73	4.13*	3.39*
	warning level	5.22*	5.67*	4.96*	3.26*	2.70	5.03*
	warning type * warning level	3.07*	3.04*	3.55*	3.77*	3.75 *	3.65*
relative_β	warning type	0.30	0.98	0.21	1.20	0.32	0.41
	warning level	3.50*	5.78*	6.65*	4.68*	2.03	3.80*
	warning type * warning level	0.88	1.74	0.74	1.39	1.32	1.23
relative_δ	warning type	1.51	2.63	2.39	2.39	2.60	1.70
	warning level	5.81*	4.70*	6.63*	4.24*	0.83	5.20*
	warning type * warning level	2.03	2.96*	3.07*	3.80*	2.97*	3.27*
relative_θ	warning type	0.19	0.30	0.28	0.49	0.15	0.27
	warning level	0.52	0.67	0.49	0.30	3.26*	0.46
	warning type * warning level	1.90	2.45*	1.62	2.55*	1.64	2.24

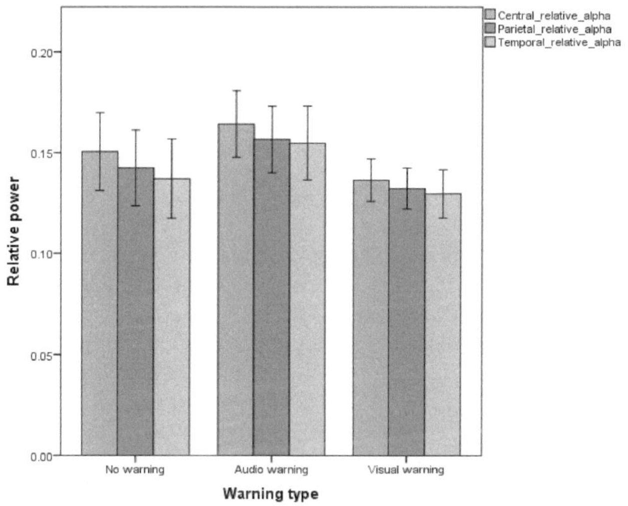

Figure 2. Alpha relative power significantly affected by FCW.

Table 3. Correlation analysis between BRT and brain wave relative power.

BRT	Pearson correlation	α	β	Δ	θ
	Frontal	0.151**	0.211**	−0.289**	0.170**
	Central	0.154**	0.261**	−0.280**	0.091
	Parietal	0.134**	0.270**	−0.299**	0.141**
	Occipital	0.145**	0.278**	−0.295**	0.137**
	Temporal	0.042	0.002	−0.129**	0.192**
	All brain	0.142**	0.241**	−0.293**	0.150**

4 DISCUSSION

4.1 *Effectiveness of FCW system*

The FCW system had a significant effect on shortening drivers' response time. With the assistance of early warning technology, drivers would recognize risks more quickly, take collision avoidance measures earlier, and thus increase the minimum headway in the collision avoidance process. In the case of a low warning level, when the leading car suddenly brakes, the distance between the leading car and the following car is large, and the driver's vision is not clear because of the fog. The early-warning information released in this case can make up for the driver's perception of the surrounding road traffic environment. Therefore, when the early warning level was low, the early warning technology had a very significant effect on shortening the driver's response time. The driver could take deceleration measures earlier, and the minimum headway was larger in the collision avoidance process than when there was no warning. The human response time had a certain limit, so in very urgent circumstances (TTC $< 3s$), the driver deceleration response time could not be shortened, and the minimum time headway was not significantly improved compared to the state without warning.

In general, α wave and β wave representing drivers' alertness and relative power values were larger under audio warning and visual warning conditions, while δ wave and θ wave relative power values representing drowsiness wave were lower than those without warning, which meant that the FCW system could improve their physiological alertness (Stevenson Brookes & Morris 2011). This also proved the driver's reaction process under the effect of early warning technology. Drivers receive early warning information through the auditory and visual sense organs, and the body's nervous system sends alert information to the brain; then, the brain synthesizes the information first and makes judgment decisions. Based on experience and attitude to collision risk, the brain helps drive people to take an emergency brake to avoid risk through the EEG change, then driving people to respond by taking risk reduction measures to avoid the collision.

4.2 *Correlation between behavior and EEG*

The BRT can reflect the speed of the driver's response in the emergency collision avoidance scenario. Under the condition of sudden braking of the leading vehicle, the state of the driver's brain in the collision avoidance reaction stage may affect his/her BRT, and the driver of a different BRT may lead to a change of collision risk in the following process. For example, if a driver responded quickly and took timely measures, the risk of collision would be greatly reduced. On the contrary, if the driver failed to respond timely, the risk of collision would increase. The results of this study showed that the higher the relative power values of α, β and θ waves, the shorter the BRT and the driver's response, while the delta waves were the opposite.

5 CONCLUSIONS

In this study, driving simulation experiments were conducted to design scenes of rear-end collision avoidance under different warning types. The driver's behavior and EEG characteristics during collision avoidance were analyzed, and the correlation between them was analyzed. Employment of the FCW system resulted in a reduction in BRT and a conditional increase in MTH. α wave and β wave representing drivers' alertness and alertness and relative power values were larger under audio warning and visual warning conditions, while relative power values representing drowsiness wave were lower than those without warning. The correlations between BRT and brain waves in different brain regions were also studied. Only warning type and warning level were considered in the analysis of FCW efficacy, while differences in warning contents were not taken into account. In the following research, different warning contents and warning times under the same warning type can be further discussed.

ACKNOWLEDGMENTS

This work was financially supported by the National Key R&D Program of China (2019YFF0301403).

REFERENCES

Baumann, M., Keinath, A., Krems, J. F., & Bengler, K. (2004). Evaluation of in-vehicle HMI using occlusion techniques: experimental results and practical implications. Applied Ergonomics, 35(3), 197–205.

Chuang, C., Huang, C., Ko, L., & Lin, C. (2015). An EEG-based perceptual function integration network for application to drowsy driving. Knowledge-Based Systems. 80,143–152.

Fors, C., & Lundkvist, S. O. (2009). Night-time traffic in urban areas: a literature review on road user aspects. Tillämpad Psykologi.

Hirst, S., Graham, R., (1997). The format and perception of collision warnings. In: Noy, Y.I. (Ed.), Ergonomics and Safety of Intelligent Driver Interfaces. Lawrence Erlbaum Associates, Mahwah, NJ, pp. 203–219.

Itoh, M., Horikome, T., & Inagaki, T. (2013). Effectiveness and driver acceptance of a semi-autonomous forward obstacle collision avoidance system. Applied Ergonomics, 44(5).

Kar, S., Bhagat, M., & Routray, A. (2010). EEG signal analysis for the assessment and quantification of driver's fatigue. Transportation Research Part of Traffic Psychology & Behaviour, 13(5), 297–306.

Lee, J.D., McGehee, D.V., Brown, T.L., Reyes, M.L. (2002). Collision warning timing, driver distraction, and driver response to imminent rear-end collisions in a high-fidelity driving simulator. Human Factors 44 (2), 314–334.

Liu, H., Wei, H., Yao, Z., Ai, Q., & Ren, H. (2013). Effects of collision avoidance system on driving patterns in curve road conflicts. Procedia - Social and Behavioral Sciences, 96, 2945–2952.

Mccallum, M., Campbell, J. L., Richard, C., & Brown, J. L. (2006). Integrated vehicle-based safety system heavy truck driver vehicle interface (DVI) literature review. University of Michigan Transportation Research Institute, Ann Arbor.

Papadelis, C., Zhe, C., Kourtidou-Papadeli, C., Bamidis, P. D., & Maglaveras, N. (2007). Monitoring sleepiness with onboard electrophysiological recordings for preventing sleep-deprived traffic accidents. Clinical Neurophysiology, 118(9), 1906–1922.

Plainis, S., & Murray, I. J. (2002). Reaction times as an index of visual conspicuity when driving at night. Ophthalmic and Physiological Optics, 22(5), 409–415.

Stevenson, C. M., Brookes, M. J., & Morris, P. G. (2011). β-band correlates of the fMRI bold response. Human Brain Mapping, 32(2), 182–197.

Wang, S., Zhang, Y., Wu, C., Darvas, F., & Chaovalitwongse, W. A. (2015). Online prediction of driver distraction based on brain activity patterns. IEEE Transactions on Intelligent Transportation Systems, 16(1), 136–150.

Wege, C., Will, S., & Victor, T. (2013). Eye movement and brake reactions to real-world brake-capacity forward collision warnings—a naturalistic driving study. Accid Anal Prev, 58(Sep.), 259–270.

Wu, Y., Abdel-Aty, M., Park, J., & Zhu, J. (2018). Effects of connected-vehicle warning systems on rear-end crash avoidance behavior under fog conditions. Transportation Research, 95(OCT.), 481–492.

Advances in Urban Engineering and Management Science – Khalil & Yang (Eds)
© 2023 The Author(s), ISBN 978-1-032-30426-7

Study on the influencing factors and implementation path of China's freight structure adjustment under the background of "Carbon Peak" and "Carbon Neutrality"

Yujian Gao & Jinxiang Cheng
Transport Planning and Research Institute, Ministry of Transport of the People's Republic of China, Beijing, China

Long Bai
School of Geographical Sciences, Hebei Normal University, Shijiazhuang, China

Gaoru Zhu, Shengqiang Liu* & Xiaowen Yang
Transport Planning and Research Institute, Ministry of Transport of the People's Republic of China, Beijing, China

ABSTRACT: Transportation structure is related to energy consumption, greenhouse gas emissions and the realization of "Carbon Peak" and "Carbon Neutrality" objectives in the field of transportation. This paper uses the structural analysis method to analyze the current situation and characteristics of China's freight structure in recent ten years from two aspects of freight volume and freight turnover and analyzes the transportation structure adjustment policies at the central, industry and local levels in recent years from the perspective of policy tools. The research shows that: (1) At present, China's freight structure is unreasonable, the development level of multi-modal transport is low, the connection of comprehensive transport is not smooth, the collection and distribution system is imperfection, the research and development of new technologies and new equipment are insufficient, and the imperfect system of transportation market mechanism. (2) It is proposed to deepen the construction of demonstration areas during the "14th Five-Year Plan" period, continuously improve support policies, strengthen the effective connection of comprehensive transportation infrastructure, promote information interconnection and sharing, increase the intensity of the railway market reform, and promote the construction of the benign market ecological environment. The implementation path provides a reference for optimizing the transportation structure.

1 INTRODUCTION

The goal of a carbon peak and the vision of carbon neutrality are the major strategies proposed by the CPC Central Committee to coordinate international and domestic situations. The adjustment of transport structure is a major decision made by the CPC Central Committee with Comrade Xi Jinping as the core. Recently issued by the Central Committee of the Communist Party of China and the State Council, "Opinions on Completely, Accurately and Comprehensively Implementing the New Development Concept and Doing Well the Carbon Neutralization Work in Carbon Peak" pointed out the development goals of optimizing the transportation structure, vigorously developing multi-modal transport, and improving the proportion of railway and waterway in comprehensive transportation, so as to continuously reduce the energy consumption and carbon dioxide emission

*Corresponding Author: liusq@tpri.org.cn

 DOI 10.1201/9781003305026-73

intensity of transportation (State Council Bulletin of the People's Republic of China 2021). Transportation is one of the four major areas of energy consumption and greenhouse gas emissions, and it is also one of the most difficult areas of carbon peak and the most obvious areas affected by climate change (Cao 2019; Zho & Zhao 2021). Therefore, reducing carbon emissions in transportation is of great significance to carbon peak and carbon neutrality, and has a far-reaching influence on global energy conservation and emission reduction, greenhouse gas control and global warming suppression. Among them, optimizing the freight structure has become an effective means to reduce carbon emissions and improve the environmental problems caused by the transport industry. Under the background of "Carbon Peak" and "Carbon Neutrality," it is urgent to clarify the basic situation of transportation structure and influencing factors and put forward feasible optimization paths. Based on this, this paper analyzes the change characteristics of freight structure in China in recent ten years by structural analysis method, analyzes the transportation structure adjustment policies in China in recent years from the perspective of policy tools, and puts forward the implementation path of optimizing the transportation structure in view of the problems existing in the process of transportation structure adjustment.

2 EVOLUTION TREND AND CHARACTERISTICS OF FREIGHT STRUCTURE IN RECENT TEN YEARS

Transportation structure includes not only the structural relationship between transportation system and socio-economic environment, but also the structural proportional relationship between railway, highway, waterway, aviation, and pipeline. The most direct external manifestation is the proportional relationship between transportation modes (Chen et al. 2018; Guo & Qiao 2017). This paper focuses on the proportion of various modes of transport.

2.1 *Freight volume structure analysis*

The proportion of pipeline freight increased gradually from 1.58% in 2011 to 2.13% in 2020. The proportion of railway freight volume decreased slowly from 10.64% to 7.06% from 2011 to 2016, and increased slowly from 2017 to 2020 (up to 9.59%). The proportion of water freight volume increased from 11.52% in 2011 to 16.05% in 2020. The proportion of road freight volume changed continuously from 2011 to 2020 (76.28%–77.76%–72.21%) and declined to 72.21% in 2020; the proportion of civil aviation freight volume is relatively small and stable. The freight volume structure of various transportation modes is shown in Figure 1.

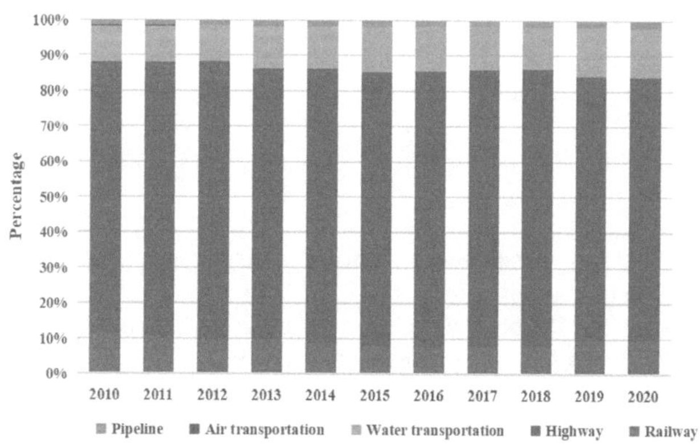

Figure 1. Diagram of freight volume structure of various transport modes in recent ten years.

2.2 *Freight turnover structure analysis*

The proportion of pipeline freight turnover is increasing from 1.81% in 2011 to 2.70% in 2020. After a slow decline from 2011 to 2016 (from 18.49% to 12.75%), the proportion of railway freight turnover began to rise slowly from 2017 to 2020 (rising to 15.09%); the proportion of freight turnover by water showed an overall increasing trend from 2011 to 2020 (from 47.34% to 52.35%); the proportion of highway freight turnover showed an overall downward trend from 2011 to 2020 (from 32.25% to 29.74%); the proportion of civil aviation freight turnover is relatively small and stable. Various modes of transport freight turnover structures are shown in Figure 2.

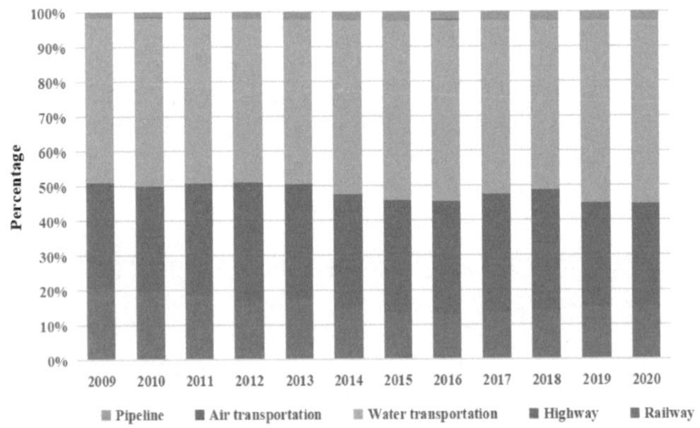

Figure 2. Diagram of freight turnover structure of various modes of transport in recent ten years.

The analysis of freight volume and freight turnover shows that China's freight transport structure has been continuously optimized. The proportion of road freight has declined but still dominated, the proportion of civil aviation and pipeline freight has increased slightly, the proportion of railway freight has increased tortuously, and the trend of waterway freight has been stable and progressive.

3 ANALYSIS OF CHINA'S TRANSPORT STRUCTURE POLICY IN RECENT YEARS

General Secretary Xi Jinping paid great attention to the adjustment of the transportation structure. At the 2017 Central Economic Work Conference, the first meeting of the Central Committee of Finance and Economics in 2018 and the symposium on comprehensively promoting the development of the Yangtze River Economic Belt in 2020, he repeatedly stressed the need to optimize and adjust the transportation structure, reduce highway freight volume, increase railway freight volume, build a unified, open, and orderly transportation market, and innovate the transportation organization model.

3.1 *Central level*

In September 2018, the three-year action plan for the restructuring of transport issued by the General Office of the State Council proposed a 30% increase in rail freight volume over 2017 (State Council Bulletin of the People's Republic of China 2018). In September 2019, the State Council of the Central Committee of the Communist Party of China issued the "Outline for the Construction of a Transport Power," emphasizing creating a green and efficient modern logistics system. It was proposed that we should optimize the transport structure, speed up the construction of "transit railway" key projects such as port collection and distribution railway, logistics park and railway special line of large industrial and mining enterprises, and promote the orderly transfer of bulk goods and long-distance goods to railway and waterway transport. In addition, we should also promote the development of rail, rail, water, air, and land intermodal transport, promote cross-modal rapid transshipment standardization facilities and equipment, and form unified multi-modal transport standards and rules. In February 2021, the guidance of the green low-carbon cycle development economic system issued by the State Council made it clear that we should actively adjust the transport structure, promote the multi-modal transport of hot metal, public rail, and public water, and speed up the construction of railway special lines (State Council Bulletin of the People's Republic of China 2021). In the same month, the Central Committee of the Communist Party of China and the State Council issued the national comprehensive three-dimensional transportation network planning outline, proposing to optimize and adjust the transportation structure, promote the construction of multi-modal logistics parks and railway special lines, and form a long-distance transportation pattern of bulk goods and containers dominated by railway and water transportation (State Council Bulletin of the People's Republic of China 2021).

3.2 *Industry level*

In October 2018, nine departments, such as the Ministry of Transport, issued a three-year action plan to promote the adjustment of transport structure, refined and decomposed the incremental transport tasks of railways, formulated the implementation plan for the construction of transport structure adjustment demonstration zones in Beijing-Tianjin-Hebei and surrounding areas, and proposed the construction of railway special lines and other "nine major projects" (State Council Bulletin of the People's Republic of China 2018). In September 2019, the Development and Reform Commission, the Ministry of Natural Resources, the Ministry of Transport, and other departments jointly issued guidance on speeding up railway special lines, emphasizing requirements such as simplifying access conditions and reducing processing time limits (Li 2019). In January 2021, the General Office of the Ministry of Transport issued a notice to improve the working mechanism to effectively promote the adjustment of transport structure, emphasizing the optimization of transport structure adjustment monitoring indicators, and establishing a monthly reporting and gradual assessment supervision mechanism. In April 2020, the Office of the Ministry of Transport forwarded the subsidy scheme to adjust the transport structure in Jiangsu Province, encouraging all regions to learn from the experience of Jiangsu Province and introduce relevant supporting policies in combination with the actual actualities local situation.

3.3 *Local level*

All provinces in China have issued the implementation plan to adjust transport structure, clarifying the working objectives and implementation plan. Jiangsu Province has issued a transportation structure adjustment subsidy program to develop a subsidy program, clarify the subsidy object, subsidy, and capital declaration, and review the process to promote bulk cargo, container transport "transit iron," and "transit water." Sichuan Province has issued a notice to do a good job in the transport structure and multi-party efforts to accelerate the adjustment of the transport structure. Jinan has issued a transport structure adjustment and domestic train development subsidy program to subsidize mining, processing, and manufacturing enterprises that shift bulk cargo from road transport to railway vehicle transport and to fund operators engaged in domestic railway freight trains.

3.4 *Effect analysis of optimization adjustment*

From 2017 to 2020, the national railway freight volume increased by about 5%, and the road freight volume decreased by 5%; the proportion of railway water freight turnover increased by about 5%, and the proportion of highway freight turnover decreased by about 5%. In 2018 and 2019, the national port container rail-water intermodal volume increased by about 48 percent, and by about 30 percent in 2020. In 2020, the average overload rate of the expressway is less than 0.5%.

4 MAIN PROBLEMS OF FREIGHT STRUCTURE ADJUSTMENT IN CHINA

4.1 *Unreasonable freight structure*

Judging from the proportion of transportation volume completed by different modes of transportation in China, such as railways, highways, water transport and aviation, highways undertake too much long-distance transportation of bulk goods such as coal, steel and grain, and the ore transportation in some ports is still dominated by highways. The advantages of low cost and low energy consumption of railway and water transport have not been fully exploited. The transportation structure is unreasonable, and the proportion of highway freight is high. The transportation structure dominated by highway freight has not fundamentally changed (Feng 2019).

4.2 *Integrated transportation connection is not smooth*

The overall layout of the railway shows the structural characteristics of eastern denseness and western sparseness, and the double-track rate is low. The comprehensive transportation hub is not coordinated with urban functional institutions and does not match the industrial spatial layout. The lagging construction progress of railway special lines and the idleness of existing special lines have caused low railway accessibility in some regions. According to statistics, about one-third of the major coastal port hub port areas have not yet been connected to railways, the overall planning and construction coordination of cross-modal transport infrastructure is insufficient, the efficiency of hub station collection and distribution is low, and the problems such as "last kilometer" remain prominent (Gao 2020; Yang et al. 2019).

4.3 *Application of new technologies and equipment lags behind*

Insufficient innovation, research and development and application of new technologies and equipment in freight transport and low standardization, assembly and compartmentalization of transport equipment result in low transport efficiency and comprehensive benefits. It is mainly manifested in the lagging development of specialized transport equipment such as cold chains, commodity vehicles and dangerous goods, and the multi-modal transport equipment system such as inland containers is in its infancy. The railway container type is single, and the scale of special carrier and transport equipment is small. The technological innovation system in multi-modal transport has not been established, the basic research investment is insufficient, and the key technical standards are lacking.

4.4 *Market environment needs to be improved*

The transformation and upgrading of highway freight are difficult. There are still excessive competition and low-price competition in road freight (Li 2018; Tang 2019). Railway freight transport capacity and service level are insufficient, and there are still transfer investments, high costs, forced operation and maintenance and other phenomena. The uncertainty of freight transport time, inflexible pricing mechanism and low transport convenience reduces the competitiveness of railway transportation, resulting in the comparative advantage of railway transportation is not fully exploited (Cai et al. 2019).

4.5 *Low-level of information sharing*

Due to technical barriers and market segmentation, the compatibility of various modes of transport in technology, equipment, transport, insurance rules, policies and legal systems is not enough, the exchange and sharing of information between departments, modes, regions and enterprises are insufficient, and there is a lack of a unified information sharing platform for multi-modal transport, which cannot truly realize the interconnection, sharing and win-win of multi-modal transport information.

5 IMPLEMENTATION PATH OF OPTIMIZING FREIGHT STRUCTURE IN "14TH FIVE-YEAR"

5.1 *Further promoting the construction of demonstration zones*

Continue to deepen the Beijing-Tianjin-Hebei and surrounding areas of transport structure demonstration area construction, research to build Shanxi, Shaanxi, western Inner Mongolia, and other major coal production areas of transport structure adjustment demonstration area, promote long-distance coal transport using railway transport. Explore to promote the Yangtze River Delta, Pearl River Delta area of rail-water intermodal demonstration area construction, improve the proportion of container rail-water intermodal transport in coastal ports.

5.2 *Strengthening infrastructure connectivity*

Strengthen the effective convergence of industrial structure and transportation structure. Further complement the logistics channel shortboard, improve the trunk channel multi-modal transport service capacity. Strengthen the planning and construction of integrated transport hub station cohesion, improve the level of hub transport services, solve the "first" and "last" kilometer problem, smooth "big cycle" and "microcirculation."

5.3 *Information connectivity construction*

Strengthen the information interconnection and sharing, promote the electronic of multi-modal transport documents, and promote the interconnection of information resources between railway, highway, waterway, aviation and customs, market supervision and other departments, so as to realize the information sharing of transport capacity arrangement, transport organization, goods on-way information, inspection and quarantine, customs clearance inspection and credit evaluation.

5.4 *Create a healthy market environment*

Promote the formation of a reasonable transport price relationship, to create a fair market environment. Continue to strengthen overload control, reduce road unreasonable competition, promote the reform of railway marketization, improve the railway market mechanism, and promote the modernization of the railway transport governance system and governance capacity. Deepen railway freight organization reform, improve service quality, rational use of price policy, vigorously develop modern logistics industry.

6 CONCLUSION

Based on the data on freight volume and freight turnover in the past ten years, this paper studies the evolution trend and characteristics of China's freight structure in the past ten years by using the structural analysis method and analyzes the adjustment policies of transportation structure in recent years from the perspective of policy tools. The study finds that: (1) China's transportation structure is unreasonable, the transportation structure dominated by road freight has not changed fundamentally, the level of multi-modal transport is still not high, the information exchange and sharing are not smooth, and the support of land and funds is insufficient. (2) It is proposed to deepen the construction of transportation structure adjustment demonstration area, promote the planning and effective connection of comprehensive transportation infrastructure, the construction of collection and distribution system, the interconnection of information, and the construction of benign market ecological environment, so as to provide a reference for optimizing and adjusting the transportation structure. However, this paper has shallow research on improving the convergence of comprehensive transportation infrastructure and the construction of collection and distribution systems, and the subsequent research focuses on the technical methods to improve the convergence of comprehensive transportation infrastructure to reduce carbon emissions in the transportation industry.

FOUNDATION

NSFC-Funded Project (42071288); Transportation Industry Key Science and Technology Projects (2020-MS4-113); Green Intelligent Inland Ship Innovation Programme; Henan Provincial Department of Transportation Science and Technology Project (2020G-2-15).)

REFERENCES

Cai Jing, Liu Ying, Zhang Minghui, Ma Hao, Zhu Yuting. "Influencing Factors and Strategies of Transportation Structure Adjustment in Beijing Area." Railway Transportation and Economy, vol.41, pp. 59–63, June 2019.

Cao Zhan. "Research on the optimization of freight structure in Sichuan Province from the perspective of carbon emissions." Southwest Jiaotong University, 2019.

Central Committee of the Communist Party of China, The State Council of the People's Republic of China. "Opinions on the complete, accurate and comprehensive implementation of the new development concept to do a good job of carbon neutralization of carbon peak." State Council Bulletin of the People's Republic of China, vol.31, pp. 33–38, 2021.

Central Committee of the Communist Party of China, The State Council of the People's Republic of China. "National Comprehensive Three-dimensional Traffic Network Planning Outline." State Council Bulletin of the People's Republic of China, vol.31, pp. 25–27, 2021.

Chen Shuling, Kang Zhaoxia, Wu Jianhong. "International comparison and enlightenment of transportation structure adjustment policies." Railway transportation and economy, vol. 40, pp. 33–37, February 2018.

Feng Wenbo. "Study on Influencing Factors and Strategies of Transportation Structure Optimization and Adjustment in China." Railway Transportation and Economy, vol.41, pp. 18–23, September 2019.

Gao Fei. "China's railway transportation structure adjustment present situation analysis and countermeasure." Rail-freight, vol.38, pp. 12–16, June 2020.

Guo Zhongqing, Qiao Rui. "Trend Analysis of Transportation Structure Evolution in China." Transportation Energy Conservation and Environmental Protection, vol. 13, pp. 77–80, January 2017.

Li Minghui. "Solve the development problem of railway special line." China investment. pp. 58–59, November 2019.

Li Yunhan. "Discussion on Policies to Promote Adjustment of Freight Transport Structure in China." Rail-freight, vol.36, pp. 1–5, August 2018.

Tang Jinsong. "Research on optimization countermeasures of transportation service of railway special line." Rail-freight, vol.37, pp. 9–13, April 2019.

The State Council of the People's Republic of China. "Circular of the General Office of the State Council on the issuance of the three-year action plan (2018–2020) to promote the adjustment of transport structure". State Council Bulletin of the People's Republic of China, vol. 30, pp. 61–66, 2018.

The State Council of the People's Republic of China. "Notice of the General Office of the State Council on the issuance of the three-year action plan (2018–2020) to promote the adjustment of transport structure". State Council Bulletin of the People's Republic of China, vol.28, pp. 61–66, 2018.

The State Council of the People's Republic of China. "State Council on accelerating the establishment of a sound green low-carbon cycle of economic development system guidance." State Council Bulletin of the People's Republic of China, vol.31, pp. 39–43, 2021.

Yang Hualong, Jian Lingxiang, Sun Guangqi. "Analysis of Transportation Structure Adjustment in China." Transportation Systems Engineering and Information, pp. 11–13+18, February 2019.

Zhou Wei, Zhao Hongxue. "Thoughts on the development of multi-power road transport equipment in China under the background of carbon peak and carbon neutralization strategy." Traffic energy saving and environmental protection. vol. 17, pp. 1–4, March 2021.

Advances in Urban Engineering and Management Science – Khalil & Yang (Eds)

Quantitative risk analysis of major accidents in dangerous cargo consolidation areas at port

Pan Shao* & Majing Lan*
China Waterborne Transport Research Institute, Beijing, China

ABSTRACT: The safety of the port, an important hub for logistics trade, is the focus of attention. Dangerous cargo consolidation areas at the port are prone to major safety risks, and the quantitative risk analysis of major accidents is an effective method of controlling and reducing the risks in the consolidation areas. The computational simulation and personal and social risk analysis for the accident consequences of typical containers can provide scientific guidance and a basis for the planning of dangerous cargo consolidation areas at the port, as well as the risk control and emergency rescue work.

1 INTRODUCTION

Port is the hub for sea-land transportation, international trade, commodity distribution and logistics node, and also serves as a strategic resource and important support for economic and social development. With the global economic integration, the cargo throughput at port presents a year-by-year growth trend in China, accompanied by the ever-increasing throughput of dangerous cargoes. A dangerous cargo port undertakes the loading/unloading, storage and transshipment of dangerous cargoes, where its storage tank area, storage yard, warehouse and wharf can easily become dangerous cargo consolidation areas. Due to the inflammable, explosive and toxic features of dangerous cargoes, the dangerous cargo consolidation areas are susceptible to the high risk of major and extraordinarily serious accidents. In this study, the quantitative evaluation method was used to analytically calculate the accident consequences of dangerous cargo consolidation areas, expected to provide a scientific basis for the administrative departments at the port. It also can be used to effectively control and reduce the major safety risks in the consolidation areas and further ensure the safe operation of dangerous cargo ports.

With dangerous cargo consolidation areas at one port taken as an example, foreign regional risk indexes were taken for reference in this study, and the personal risk value and social risk value were regarded as the risk indexes for the consolidation areas [1]. A quantitative risk analysis of major accidents in these areas was carried out. Their risk levels and risk acceptance levels were comparatively analyzed, and the results could provide a scientific and effective basis for the safety management of dangerous port consolidation areas at this port.

2 INTRODUCTION OF MAJOR ACCIDENT RISK ANALYSIS METHOD FOR DANGEROUS CARGO CONSOLIDATION AREAS AT PORT

The procedure chart of major accident risk analysis for the dangerous cargo consolidation areas at the port is shown in Figure 1.

2.1 *Calculation of major accident occurrence probability*

The occurrence probability of major accidents is calculated through the accumulation and statistics of multi-year accident occurrence data. In this study, the accident occurrence probability is given in

*Corresponding Authors: shaopan@wti.ac.cn and lanmajing@wti.ac.cn

DOI 10.1201/9781003305026-74

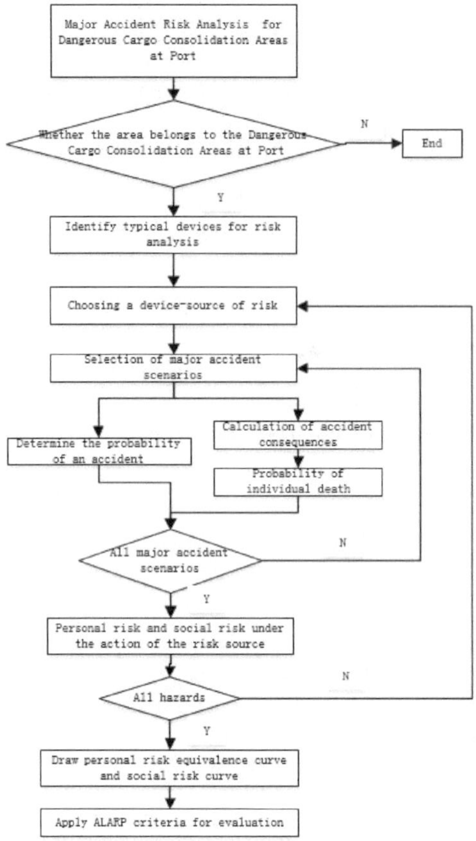

Figure 1. Procedure chart of major accident risk analysis for cargo consolidation areas at port.

Risk Criteria for Hazardous Chemicals Production Unit and Storage Installations (GB36894-2018) was taken as the accident occurrence probability in the analytical risk study.

2.2 *Consequence calculation of major accidents*

In this study, the possible accident consequences of each risk factor were calculated through the EFFECT software recommended by The Netherlands Organization for Applied Scientific Research (TNO).

The EFFECT software provides a large number of computational models, including the leakage model, evaporation model, fire model, explosion model, diffusion model and damage model. Besides, this software owns complete chemicals databases, which contain the toxicity and inflammability-related parameters and thermodynamic properties.

The accident consequences of dangerous chemicals storage devices in the dangerous cargo consolidation areas at the port were calculated via EFFECT, as shown in Figure 2.

2.3 *Personal and social risk analysis*

In this study, the personal and social risks in dangerous cargo consolidation areas at the port were quantitatively simulated and calculated via the RISKCURVES software of TNO. The RISKCURVES software includes multiple scenes, and each production or storage facility has its specific scenes (e.g., catastrophic actual effect and leakage) [2]. Based on each scene, the analog computation of cumulative risks in the surrounding environment was implemented under

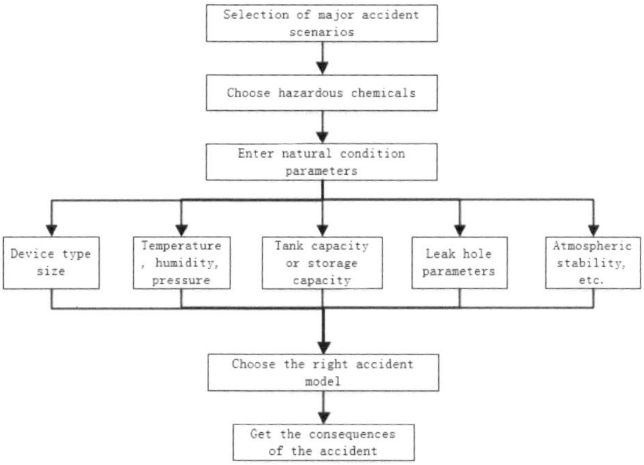

Figure 2. Calculation flowchart of accident consequences.

six representative climatic conditions, in an effort to provide technical support for the development planning of dangerous cargo consolidation areas at the port and analyzing the risk mitigation strategies.

The quantitative risk evaluation procedures via RISKCURVES software are displayed in Figure 3.

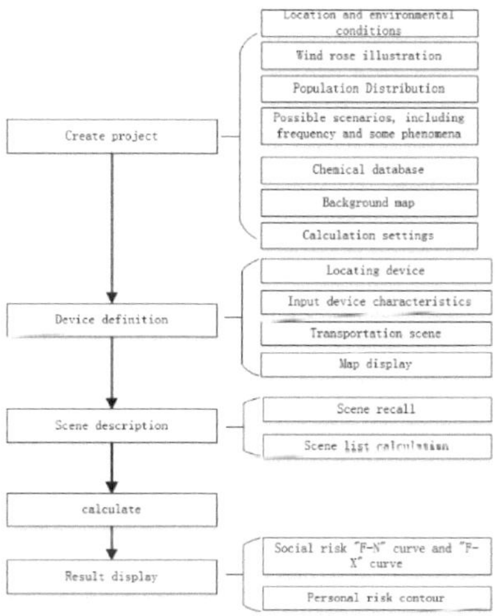

Figure 3. Calculation flowchart of personal and social risks.

3 CASE STUDY OF MAJOR ACCIDENT RISK ANALYSIS FOR DANGEROUS CARGO CONSOLIDATION AREAS AT PORT

Hereby the major accident risk analysis process was explained by taking the dangerous cargo consolidation area at a port as an example. This consolidation area included one dangerous cargo storage tank area, one liquid chemicals wharf and one oil product wharf, which covered an area of about 460,000 m2. A lot of dangerous cargoes were stored in this area. Once taking place, any accident might impose enormous threats to personal health and property safety in the production area and its surrounding area. The emphasis in this study was laid on the study of the accident scenes with a large scope of influence and serious harm, like intoxication, fireball and steam cloud explosion, etc., which were induced by leakage accidents.

3.1 *Meteorological conditions of dangerous cargo consolidation area at the port*

The main meteorological parameters in this port area are listed in Table 1, and the wind-rose diagram is shown in Figure 4.

Table 1. Meteorological parameters of dangerous cargo consolidation area at port.

NO.	Meteorological elements	Numerical value
1	The annual average temperature/°C	22.7
2	Average air pressure over many years/mbar	1015.8
3	Dominant wind direction throughout the year	Northeast, North
4	Annual average wind speed/(m/s)	3.7

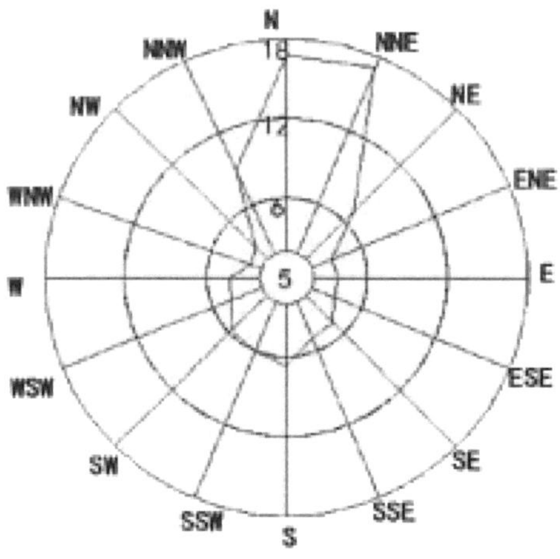

Figure 4. Annual wind-rose diagram.

3.2 *Leakage calculation of typical equipment*

3.2.1 *Selection of typical equipment and leakage scenes*

Considering the possible fire, explosion and intoxication accidents after the leakage of substances stored, the typical accident simulation equipment was chosen according to the most unfavorable principle. Accordingly, the typical equipment selected in the dangerous cargo consolidation area at the port was a V1308 gasoline storage tank in the storage tank area, with a storage capacity of 21,000 m^3.

The atmospheric storage tank was analytically calculated by selecting the large hole leakage as the typical failure mode. Given that the related enterprises had not defined the exact emergency response time to large hole leakage like gasoline leakage in their contingency plans in the current phase, the leakage time was set as 10 min based on the configuration of the reservoir self-control device (high and low liquid level monitoring, inflammable gas monitoring and early warning, etc.) and other equipment and facilities.

3.2.2 *Analog calculation of leakage*

As for the target gasoline storage tank, the diameter of the medium leakage hole was set as 25 mm, and that of the large leakage hole was 100 mm. The release rate at the leakage hole was obtained through the formula of liquid leakage, and the 10 min mass release was further calculated.

Table 2. Analog calculation results of mass release at leakage hole.

Location	Leak source	Type of leakage (Equivalent leakage caliber) (mm)	Leak rate (kg/s)	Leakage (kg)
Storage tank area	Gasoline leak	25	5.051	3045
		100	80.742	48498

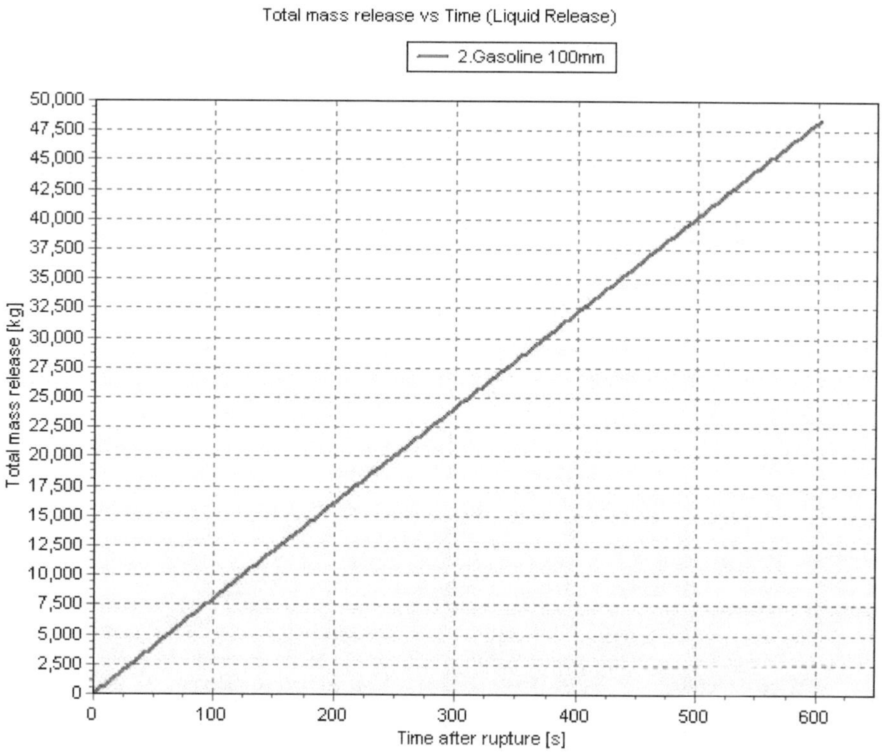

Figure 5. Relationship between mass release and leakage time at gasoline leakage hole (Diameter: 100 mm).

3.2.3 *Analog calculation of evaporation and diffusion*

The scope of pool liquid diffusion of mass release generated by different leakage hole diameters was simulated through the "Pool Evaporation" model in TNO EFFECTS software. Furthermore, the volatilization and diffusion scenes of leaking cargo type in the atmosphere with time were simulated via "Dense Gas Dispersion" in the software, and the simulation results are as seen in Table 3:

Table 3. Analog calculation results of evaporation and diffusion after gasoline leakage through medium and large holes.

Leak source	Type of leakage (mm)	Leak rate (kg/s)	Leakage (kg)	Pool area (m^2)	Maximum length of explosive cloud concentration (m)	Maximum explosion mass (kg)	Explosion cloud area (m^2)
Gasoline	25	5.051	3045	507.83	12.79	8.0703	29.296
	100	80.742	48498	8026.1	71.33	750.99	637.46

Figure 6. Simulation results of diffusion scope of steam cloud after gasoline leakage through 100 mm hole.

3.3 *Simulation of fire and explosion accident consequences of typical equipment*

3.3.1 *Selection of accident scenes*

The meteorological conditions play a leading role in the fire and explosion accident, including wind velocity, wind direction, air temperature, relative humidity and atmospheric stability. According to the natural conditions in this project area, the main environmental parameters chosen in the analog calculation are listed in Table 4.

Table 4. Selection of typical meteorological conditions.

Meteorological factors	Selected result	Remark
Wind direction	N	Normal wind direction
Wind speed	5 m/s	Average wind speed in normal wind direction
Temperature	20°C	Yearly average temperature
Relative humidity	82%	Yearly average relative humidity
Atmospheric stability	D	

The simulation analysis of pool fire accident consequences was implemented for the V1308# gasoline storage tank in the consolidation area, and the porthole leakage occurred to the gasoline storage tank. A large quantity of leaked gasoline formed a liquid pool in the fire dike, and pool fire could be generated when encountering any ignition source. The accident consequences generated by such pool fire accidents are displayed in the following Table 5:

Table 5. Scope of influence of thermal radiation generated by pool fire after gasoline leakage through hole.

				Downwind heat radiation distance (m)		
Leak source	Leakage aperture (mm)	Flame length (m)	Flame temperature (°C)	4.0 kw/m2 (Green line)	12.5 kw/m2 (Red line)	37.5 kw/m2 (Yellow line)
Gasoline	25	22.86	714.95	52.7	37.259	21.396
	100	63.043	765.28	179.56	125.14	74.876

Figure 7. Simulation results of pool fire accident consequences induced by gasoline leakage through large hole.

Once a leakage accident occurs in the gasoline and diesel storage tank, the leaking liquid is mixed with the air through evaporation and diffusion, thus forming explosive steam clouds within a certain scope. If the steam clouds are located in an area densely distributed with equipment, buildings and structures, and encounter any ignition source with strong energy, the steam cloud explosion may occur. In this study, the harms of steam cloud explosion accidents that might take place at the porthole of the V1308 gasoline storage tank within the consolidation area were simulated and evaluated, and the overpressure distance of explosive shock wave of steam clouds is listed in Table 6.

Table 6. Overpressure distance of steam cloud explosion after gasoline leakage at porthole.

Leak source	Leakage aperture (mm)	Influence distance from the center of the vapor cloud (m)		
		0.017 MPa	0.044 MPa	0.14 MPa
V1308 Gasoline storage tank	25	166.83	25.238	10.947
	100	756.01	114.37	49.608
Remarks: degree of hazard		Typical pressure for glass breakage	The house is partially destroyed and cannot be inhabited	Low limit of severe result damage

3.4 *Simulation analysis of personal and social risks*

In accordance with *Risk Criteria for Hazardous Chemicals Production Unit and Storage Installations* (GB36894-2018), the personal and social risks of the V1308 gasoline storage tank in the dangerous cargo consolidation area were simulated via TNO RISKCURVES[3].

According to No. 13 announcement—*Personally and Socially Acceptable Risk Criteria for Production and Storage Device of Dangerous Chemicals (Trial)*—released by the Ministry of Emergency Management (former State Administration of Work Safety), if a storage device is in service, the personally acceptable risk criterion is not higher than 3×10^{-5} in places with low-density personnel (number of people < 30), that in densely populated residential places and public

Figure 8. Scope of influence of steam cloud explosion 600 s after Gasoline Leakage at 100 mm Hole (shock wave overpressure: 0.044 MPa).

Figure 9. Simulation curves of personal risks in dangerous cargo consolidation area.

gathering places ($30 \leq$ number of people < 100) is not higher than 1×10^{-5}, and that in highly sensitive places, important targets and special densely populated places is not higher than 3×10^{-6}. Based on No.274 document—*Measures for the Supervision and Administration of Major Hazard Installations of Dangerous Goods at Port (for Trial Implementation)*—issued by the Ministry of Transport in 2013, the permissible risk level is 3×10^{-7} in highly sensitive places, important targets and special densely populated places and that in densely populated residential places and public gathering places is 1×10^{-6}. It could be seen from Figure 9 that no contour line of personal risk probability value exceeding 1×10^{-5} was generated in the consolidation area, and the contour line of personal risk probability value 3×10^{-6} was basically in the storage area or around the quay berth, while highly sensitive places, important targets and special densely populated places not included. Hence, the personal risks of the existing facilities inside this dangerous cargo consolidation area (only the enterprises operating dangerous cargoes at the port within the scope of evaluation were considered) were acceptable, and the social risk curves of the dangerous cargo consolidation area are presented in Figure 10.

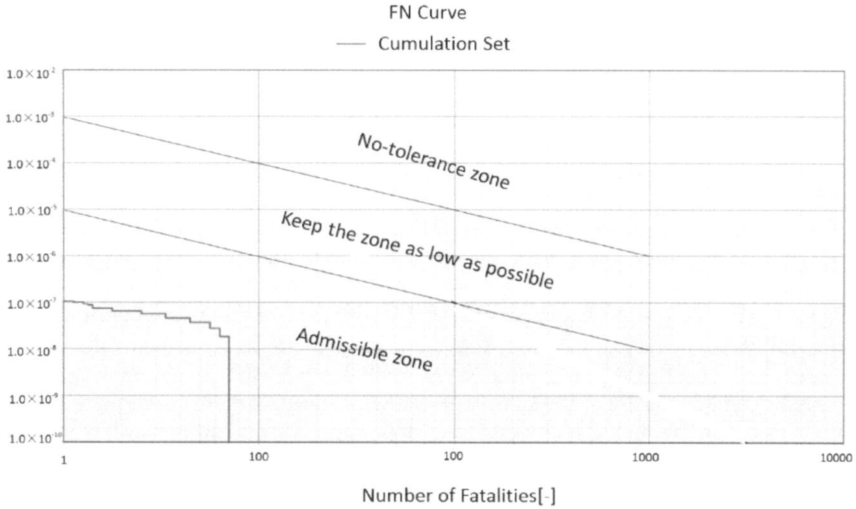

Figure 10. Social risk curve chart of dangerous cargo consolidation area.

4 CONCLUSION

A large number of dangerous cargoes or devices are gathered in the storage area, tank area and wharf at dangerous cargo ports, which are special high-risk areas with a dense arrangement of major hazard sources. The risk evaluation of a dangerous cargo consolidation area is a very effective solution for better mastering the safety status in the dangerous cargo consolidation area, doing its safety management work very well, enhancing its ability to prevent and resolve the accident accidents and promoting healthy development of the port. This study analyzed the possible accident types at each dangerous source and their scope of influence in the dangerous cargo consolidation area. Next, the personal risk values of multiple major accident risk sources were quantitatively calculated and superposed, and then the social risks in the area were calculated and analyzed. Therefore, scientific guidance and basis can be provided for the planning of the dangerous cargo consolidation area at the port as well as its risk control and emergency rescue work.

REFERENCES

Chen G H, Zhang H, et al. (2006). Study on regional risk assessment methodology[J]. China Safety Science Journal,16(6); 112–117.

Healthy and Safety Executive. Guidance on "as low as Reasonably practicable" decisions in control of major accidents hazards.SPC/Permissioning/12[EB/OL]. [2009-11-09].

Xu Anqi. Qian Yixiao, Qian Yu. (2018) Study on optimal layout of barriers against domino effects in chemical enterprise [J]. China Safety Science Journal, 28(2): 175–180.

Advances in Urban Engineering and Management Science – Khalil & Yang (Eds)
© 2023 The Author(s), ISBN 978-1-032-30426-7

Comprehensive importance evaluation of urban rail transit network based on complex network

Shiyan Han*
Suzhou Rail Transit Group Co., Ltd., Suzhou, China

Junlan Chen* & Jumei Zhu*
School of Transportation, Southeast University, Nanjing, China

Fanghui Tang* & Xing Yan*
Suzhou Rail Transit Group Co., Ltd., Suzhou, China

ABSTRACT: The increasing degree of urban rail transit network greatly increases the complexity of train operation organization and connection optimization. In order to ensure that passengers can get a good transfer experience, it is necessary to coordinate and optimize the connection of trains in all directions according to the transfer rules of rail passenger flow in each period. Combined with Suzhou rail video statistics data and AFC data, through the analysis and calculation of the transfer waiting number, this paper quantitatively evaluates the transfer direction importance of a single transfer station, the line direction importance of multiple transfer stations and the line importance, so as to calculate the comprehensive importance of the rail network, and to directly reflect the functional level of each transfer station and line in the rail network. Under the complex network, there are often conflicts and contradictions in the connection of different transfer stations. Therefore, according to the comprehensive importance of the network, the operation of urban rail transit network lines can be more orderly organized.

1 INTRODUCTION

Under the condition of the formation of an urban rail transit network, the line form and passenger flow scale show diversified and different characteristics, which greatly affects the function and positioning of different stations and lines in the line network, especially the junction transfer station, where multiple rail lines converge (Xu & Li 2012). Coordination of connection relations between transfer stations and lines can effectively optimize the operation organization of rail transit under network conditions (Kang 2016). In order to organize the line operation of rail transit in an orderly way and improve the transport efficiency of the rail network, this paper analyzes the importance of transfer connection between transfer stations and lines in the rail network.

Aiming to optimize train operation organization, many scholars have carried out studies on the evaluation of the importance of urban rail transit transfer (Bai et al. 2013; Kang et al. 2104). By analyzing thc importance of transfer stations and lines, Zhang and Jiang (2008) established a multi-direction transfer connection model to minimize passengers' transfer waiting time. Huang (2014), based on the bus topological network, measured the importance of each transfer point with transfer intensity and transfer passenger flow as indicators, and adjusted the departure time and frequency of rail lines on this basis. Bian (2015) analyzed the characteristics of rail transit passenger flow under network conditions by defining transfer nodes and the importance of lines, and studied the characteristics of passenger flow at transfer stations and the spatial and temporal distribution of passenger flow in the line networks based on the difference of passenger flow of a single line

*Corresponding Authors: 18452766620@163.com, 937365671@qq.com, 982228244@qq.com, tangfanghui@sz-mtr.com and yanxing@sz-mtr.com

 DOI 10.1201/9781003305026-75

and the entire line network. Kou Chunge (2016) weighted the sum of the internode number and transfer passenger flow and proposed a quantitative evaluation method for the importance of transfer direction, which can determine the importance of a certain transfer direction from two aspects of passenger transfer demand and node location. By analyzing the departure time of the first and last trains and the AFC passenger flow data of urban rail transit, Guo Xin (2018) found that the first and last passengers had regular transfer travel needs. Based on this, he studied the transfer lines and station importance of the first and last trains. Qin (2019) quantified the importance of lines and transfer stations by considering the region where transfer stations are in the importance evaluation of rail transit transfer.

This paper takes the Suzhou rail network as the research object, analyzes, and calculates the number of transfer waiting in a certain transfer station infrastructure and passenger transportation organization conditions. Combined with the video statistics data and AFC data in September 2019, the importance of transfer direction within a single transfer station, the importance of multiple transfer stations by line direction and the importance of line are quantitatively evaluated, providing data basis and theoretical basis for the coordination and optimization of urban rail transit operation organization.

2 SURVEY OF TRANSFER PASSENGER FLOW

This paper selects four main transfer stations of Suzhou rail transit as the research objects, which are Guangjinan Road Station, Leqiao Station, Suzhou Railway Station and Shihu East Road Station. The field survey adopted the method of setting up instruments and recording videos at the stairway of the transfer route, and then manually counting the number of transfers. Among them, the video recording time of Guangzhou Jinan Road station is 8:06–8:43 (morning peak), Leqiao Station is 17:46–18:15 (evening peak), Suzhou Railway Station is 8:18–8:51 (morning peak), and Shihu East Road Station is 17:50–18:19 (evening peak).

2.1 *Survey of transfer numbers at transfer stations*

According to the way of platform setting, these four transfer stations can be divided into two types: first, the second underground floor is the side platform, and the third underground floor is the one island side platform transfer station, including Guangjinan Road station and Leqiao Station, the second is a double-island platform transfer station with island platforms on the second and third underground floors, including Suzhou Railway Station and Shihu East Road Station. The statistical methods of transfer number of the two platforms are as follows:

2.1.1 *One island side platform interchange station*

Location of instruments: set up four instruments to record video at two transfer staircases on the second floor and two transfer staircases on the third floor.

Number statistics of transfer: connect the same two instruments to record video, up and down the stairs records vehicle arrival time and swarms of respectively the number of transfers to the crowd, arrive or to transfer passengers from different directions, respectively, according to the vehicle's arrival time will be two videos count the same change to the match, the crowd finally had counted the change to the number of the transfer line.

2.1.2 *Double island platform transfer station*

Location of instrument erection: since there is only one transfer stairway on the third underground floor of this kind of transfer station, and about two stairways on the second underground floor are bifurcated from it, three instruments are set up at the three stairways to record video.

Transfer statistics: In connection with the videos recorded by the two instruments at the upper and lower levels of the same staircase, the number of transfers from trains at the upper (lower) level to trains in the two directions at the lower (upper) level, and the number of transfers from trains at the two directions at the lower (upper) level to trains at the upper (lower) level are calculated from the video at the stairway of the lower (upper) level. Record the beginning moment when the crowd from different directions appears in the video on the (upper) layer; then, according to the arrival time of the upper and lower floors, the same group of transfer crowd in the upper and lower floors videos are matched, and the number of transfer passengers of each transfer route is finally counted.

2.1.3 *Transfer number statistics at transfer stations*
When transfer passengers arrive at the platform connecting lines train station, wait on the platform until the next shift of cohesion. The train transfer waiting number indicates that the connection lines stand for transfer to the number of passengers, this study defines it as bridging the trains on a line cut adjacent to the next period of trains from the inner platform gathered on the number of transfer passengers. The number of transfer passengers is the number of transfer passengers waiting for train q on connecting line ls, which can be expressed as:

$$n_{U,r,s}^{wait} = \rho \cdot \int_{d_{q-1}}^{d_q} f(t)dt \tag{1}$$

Let the random variable obey the three-parameter Weibull distribution; write it as $t \sim (\beta, \lambda, \eta)$, where β is the shape parameter, λ is the scale parameter, η is the position parameter, and the probability density function of t is as follows:

$$f(t) = \begin{cases} \frac{\beta}{\lambda}\left(\frac{t-\eta}{\lambda}\right)^{\beta-1} \cdot \exp\left[-\left(\frac{t-\eta}{\lambda}\right)^{\beta}\right] & t \geq \eta, \beta > 0, \lambda > 0 \\ 0 & t < \eta \end{cases} \tag{2}$$

3 IMPORTANCE EVALUATION OF TRANSFER

3.1 *Importance evaluation of transfer direction within a single station*

The proportion of passenger flow distribution in each direction of the transfer station is calculated using the number of transfers from the video statistics, and compared comprehensively with the AFC data of Suzhou Railway in September 2019. The spatial and temporal distribution of passenger flow in the network can be quantitatively analyzed from the micro perspective, and then the transfer passenger flow of the line network (divided into transfer directions) can be obtained. Taking Shihu East Road Station as an example, Table 1 is the statistical table of transfer passenger flow (transfer direction) and the proportion of passenger flow distribution in a flat peak period.

Table 1. Statistics of passenger flow (by transfer direction) at Shihu East Road Station during peak hours.

Transfer Station	Transfer relationship		Transfer flow distribution proportion	Transfer flow (peo)	Order
Shihu East Road	From Line 2 to Line 4	From Qihe to Longdaobin	0. 032	32	8
		From Qihe to Tongli	0. 183	184	2
		From Sangtiandao to Longdaobin	0. 144	145	3
		From Sangtiandao to Tongli	0. 129	130	5
	From Line 4 to Line 2	From Longdaobin to sangtiandao	0. 135	136	4
		From Longdaobin to Qihe	0. 037	37	7
		From Tongli to sangtiandao	0. 123	123	6
		From Tongli to Qihe	0. 216	217	1

In order to display the passenger flow distribution of each transfer relationship in the transfer station more directly, the passenger flow distribution map of the transfer relationship is drawn.

Based on Table 1 and Figure 1, the source and direction of transfer passenger flow are analyzed. In the flat peak period, the transfer passenger flow of Shihu East Road station mainly comes from the upward direction of Line 4 and concentrates on Line 2. Among them, the transfer relationship with the largest passenger flow is Line 4 ascending to Line 2 ascending. From the perspective of urban form and spatial distribution of the urban rail transit line network, Shihu East Road Station is in the center of Suzhou city and its line network. The upward direction of Line 4 is mainly from the suburbs to the downtown area. The transfer direction with a large transfer passenger flow reflects that more passengers travel from the suburbs to the downtown area during this period. This kind of travel rule is consistent with the rule that the daytime passenger goes from the suburb to the city center to participate in daily work leisure activities.

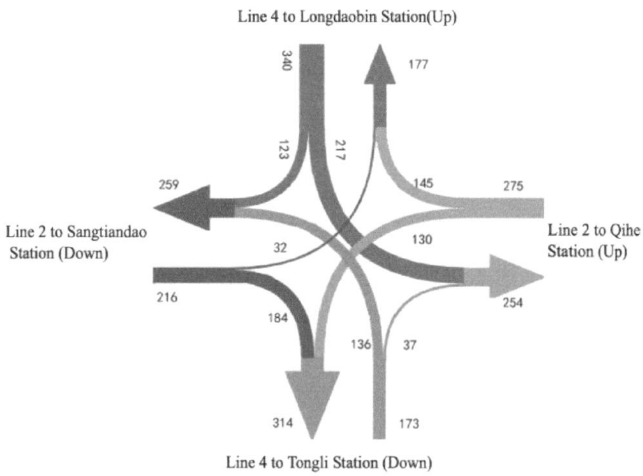

Figure 1. Distribution of the direction of the change of passengers during the peak hour at Shihu East Road Station.

In order to analyze the characteristics of passenger flow in the flat peak period in a more detailed way, the distribution of passenger flow in each transfer station was counted in the unit of 15 minutes, and the distribution of passenger flow in the transfer relationship was drawn in Figures 2–5.

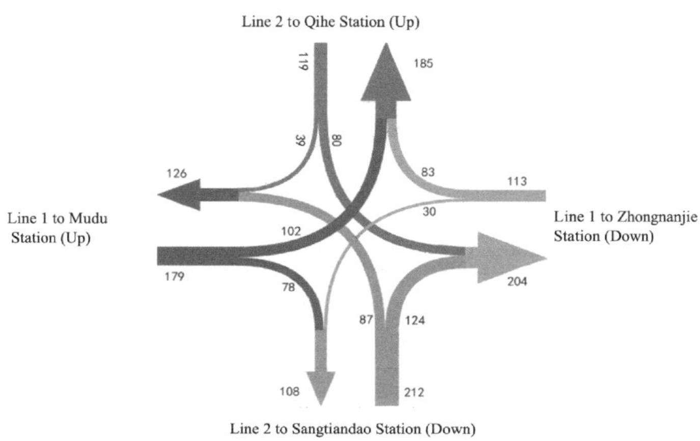

Figure 2. Distribution of changing passenger flow direction during peak hours at Guangjin South Road Station.

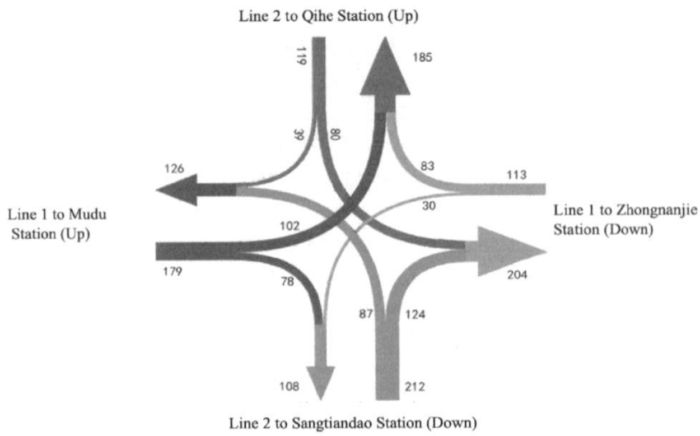

Figure 3. Distribution of changing passenger flow direction during normal peak hours at Leqiao Station.

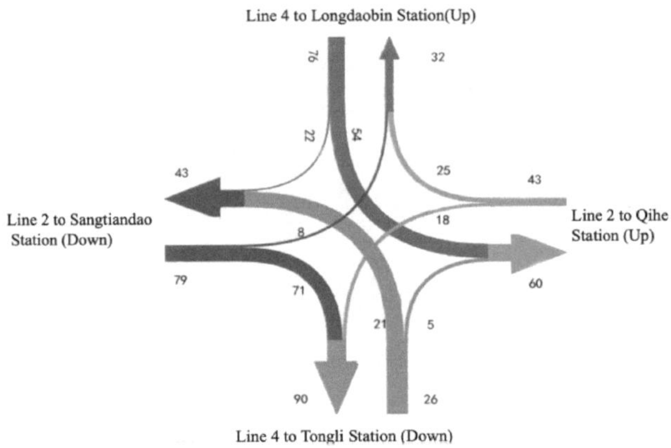

Figure 4. The distribution of the direction of the change of passenger flow during peak hours at Suzhou Railway Station.

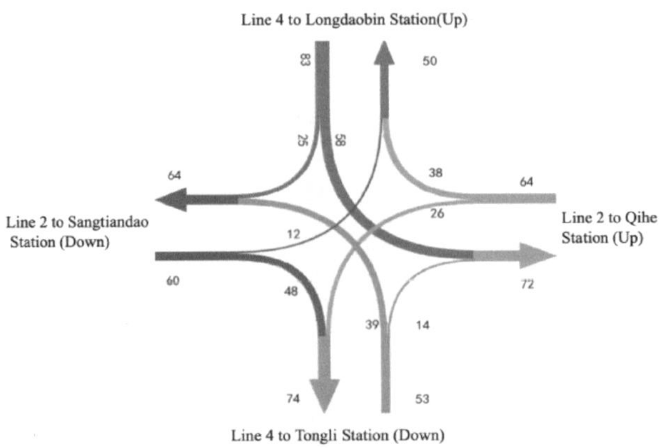

Figure 5. Direction distribution of changing passenger flow during peak hours at Shihu East Road Station.

3.2 *Importance evaluation of multi-station sub-line direction*

The structure of the rail transit line network is complex, so it is necessary to use complex network theory to analyze the importance of transfer nodes, and measure the importance of each transfer relationship by considering the transfer passenger flow in different directions, so as to provide a reference for optimizing the configuration scheme of train running density in each direction in the rail transit network.

3.2.1 *Importance of network transfer nodes*

In the analysis of a complex network, centrality is an index to measure the importance of nodes in the network, which can effectively reflect the value of nodes in the network. The importance of network nodes is described in the table:

Table 2. Network midpoint importance index.

Index	Concept	Characteristic
Point centrality	How many lines are there at this point	Emphasize the value of a point
Near centrality	The reciprocal of the sum of the distance between this point and other points in the network. The larger it is, the more it is in the center, the faster it can reach other points	The greater the value of the network, the more central it is
Intermediate centrality	The more “shortest path” passes through this point, the more important this point is.	It emphasizes the regulation ability between other points, control ability index and intermediary regulation effect
Centrality of eigenvector	Measure the value of adjacent points according to their importance. Firstly, the adjacency matrix is calculated, and then the eigenvector of the adjacency matrix is calculated.	It emphasizes the value of points in the network, and compared with the proximity to centrality, the point value is determined according to its adjacent points.

In the rail transit network structure, the passengers’ route choice is usually based on the shortest path. The center of the middle degree index is the comprehensive comparative network center degree that can generate the shortest path of the node in the network, measure transfer station in the whole railway network connected potential, and obtain line net transfer station connected ability of concentration. The higher the center degree is, the more times the shortest path passes through this transfer station in the rail transfer network. In addition, the number of lines connected to transfer stations also reflects the importance of transfer stations, so point centrality and intermediate centrality are selected to comprehensively measure the importance of transfer nodes in the network.

Assuming a directed graph $G=(V,E)$, V is the set of the subway station, E is the line edge set for the subway station, the weight of the edge is l, a path. $P=\{s,(s,v_1),v_1,(v_1,v_2)\dots(v_n,t),t\}$ is the path between the start $s\in V$ and destination $t\in V$, which is a sequence of vertices and edges. The length of the path *dist*(*P*) is the sum l of the weights of the edges contained by the path. If there is a shortest path between s and t, it can be represented as dist (s, t). $SP(s,t)$ is the shortest path between s and t. Assume that $\sigma(s,t)$ is the number of shortest paths between s and t. When $v\in(s,t)$,$\sigma(s,t|v)=0$; When $s=t$, $\sigma(s,t)=1$. Then, the intermediate centrality of vertices in G, $v\in V$is:

$$B(v_i)=\sum_{s,t\in V}\frac{\sigma(s,t|v_i)}{\sigma(s,t)} \tag{3}$$

0–1 standardization of centrality and intermediate centrality:

$$x^*=\frac{x-x_{\min}}{x_{\max}-x_{\min}} \tag{4}$$

The importance of the network transfer node is obtained:

$$I(v_i)=B(v_i)^*+C(v_i)^* \tag{5}$$

Table 3. Centrality and intermediate centrality of Suzhou rail transfer station.

Number	Transfer station	Lines	Centrality $C(v_i)$	Betweenness centrality $B(v_i)$
1	Suzhou Railway Station	Line 2 and Line 4	4	1440
2	Guangjinan Road Station	Line 1 and Line 2	4	1081
3	Leqiao Station	Line 1 and Line 4	4	2148
4	Shihudong Station	Line 2 and Line 4	4	2253
5	Hongzhuang Station	Line 4 and Branch Line 4	3	1531

3.2.2 *Proportion of relative transfer flow distribution*

Transfer passenger flow is one of the factors affecting the importance of the transfer relationship. In this study, the relative transfer passenger flow distribution ratio of R_i^v in the whole network is defined as $Pd(R_i^v)$, which can be expressed as:

$$Pd(R_i^v) = \frac{Q^{R_i^v} - Q_{\min}^R}{Q_{\max}^R - Q_{\min}^R} \tag{6}$$

$Q^{R_i^S}$ is the transfer passenger flow corresponding to different transfer relationships in the transfer station, $Q_{\min}^R$ and $Q_{\max}^R$ are the minimum and maximum of transfer passenger flow under different transfer directions, respectively. The higher the $Pd(R_i^v)$of the transfer relationship is, the higher the proportion of passenger flow it undertakes in the online network, which reflects the importance of the transfer relationship to a certain extent.

Table 4. The distribution ratio of relative passenger transfers before and after the morning rush hour at Suzhou interchange station on working days in September 2019.

Transfer Station	Suzhou Railway Station		Guangjinan Road Station		Leqiao Station		Shihudong Station	
The direction	Line 4 to Line 2	Line 2 to Linc 4	Line 1 to Line 2	Line 2 to Line 1	Line 1 to Line 4	Line 4 to Linc 1	Line 2 to Line 4	Line 4 to Line 2
6:00:00	0. 05	0. 23	0. 23	0. 59	1. 00	0. 00	0. 33	0. 74
6:15:00	0. 00	0. 32	0. 34	0. 81	0. 28	1. 00	0. 31	0. 94
6:30:00	0. 00	0. 19	0. 69	0. 67	0. 33	1. 00	0. 33	0. 54
6:45:00	0. 00	0. 18	1. 00	0. 71	0. 59	0. 63	0. 26	0. 54
7:00:00	0. 00	0. 04	0. 92	0. 82	0. 59	1. 00	0. 32	0. 56
7:15:00	0. 00	0. 01	0. 60	0. 77	0. 88	1. 00	0. 20	0. 44
7:30:00	0. 00	0. 04	0. 60	0. 74	0. 49	1. 00	0. 28	0. 40
7:45:00	0. 00	0. 10	0. 74	1. 00	0. 44	1. 00	0. 26	0. 52
8:00:00	0. 00	0. 07	0. 56	0. 80	0. 27	1. 00	0. 17	0. 39
8:15:00	0. 00	0. 08	0. 68	0. 98	0. 37	1. 00	0. 19	0. 44
8:30:00	0. 00	0. 02	0. 57	0. 88	0. 33	1. 00	0. 11	0. 34
8:45:00	0. 00	0. 01	0. 61	0. 83	0. 42	1. 00	0. 18	0. 21
9:00:00	0. 00	0. 04	0. 68	0. 82	0. 42	1. 00	0. 15	0. 20

It can be seen from the above table that the transfer relationship between Line 4 and Line 1 of Leqiao Station and Line 2 and Line 1 of Guangjinan Road Station is of high importance before 9:00. The importance of the transfer relationship between Line 4 and Line 2 and Line 2 to Line 4 of Suzhou Railway Station and Shihu East Road is low. It is worth noting that Shihu East Road has high importance of transfer relationship when the train transfers from Line 4 to Line 2 earlier.

Table 5. The distribution ratio of Pingfeng relative passenger transfer flow at Suzhou transfer station on working days in September 2019.

Transfer Station	Suzhou Railway Station		Guangjinan Road Station		Leqiao Station		Shihudong Station	
The direction	Line 4 to Line 2	Line 2 to Line 4	Line 1 to Line 2	Line 2 to Line 1	Line 1 to Line 4	Line 4 to Line 1	Line 2 to Line 4	Line 4 to Line 2
11:00:00	0. 12	0. 00	0. 84	0. 91	0. 73	1. 00	0. 05	0. 10
11:15:00	0. 13	0. 03	0. 69	1. 00	0. 74	0. 82	0. 00	0. 18
11:30:00	0. 01	0. 00	0. 75	1. 00	0. 79	0. 95	0. 05	0. 11
11:45:00	0. 10	0. 00	0. 85	0. 93	0. 80	1. 00	0. 09	0. 20
12:00:00	0. 07	0. 00	0. 81	1. 00	0. 70	0. 86	0. 09	0. 22
12:15:00	0. 07	0. 00	0. 82	1. 00	0. 68	0. 96	0. 08	0. 17
12:30:00	0. 03	0. 00	0. 78	0. 99	0. 81	1. 00	0. 07	0. 21
12:45:00	0. 10	0. 00	0. 70	1. 00	0. 72	0. 96	0. 15	0. 19
13:00:00	0. 10	0. 00	0. 89	1. 00	0. 77	0. 93	0. 18	0. 16
13:15:00	0. 05	0. 00	0. 77	0. 96	0. 67	1. 00	0. 03	0. 08
13:30:00	0. 06	0. 00	0. 69	1. 00	0. 62	0. 82	0. 04	0. 12
13:45:00	0. 04	0. 00	0. 77	1. 00	0. 78	0. 91	0. 03	0. 16
14:00:00	0. 10	0. 00	0. 82	1. 00	0. 81	0. 95	0. 04	0. 11
14:15:00	0. 00	0. 03	0. 99	1. 00	0. 84	0. 86	0. 06	0. 09
14:30:00	0. 03	0. 00	0. 76	1. 00	0. 78	0. 90	0. 09	0. 05
14:45:00	0. 08	0. 00	0. 84	1. 00	0. 73	0. 87	0. 12	0. 16
15:00:00	0. 08	0. 00	0. 97	1. 00	0. 92	0. 88	0. 10	0. 17

It can be seen from the above table that from 11:00 to 15:00, Line 2 of Guangzhou Jinan Road Station transfers to Line 1, Line 4 of Leqiao Station transfers to Line 1, and then Line 1 of Guangzhou Jinan Road Station transfers to Line 2, and Line 1 of Leqiao Station transfers to Line 4. The importance of the transfer relationship between Suzhou Railway Station and Shihu East Road is the lowest.

As can be seen from the table above, the transfer relationship between Line 1 of Leqiao Station and Line 4, Line 2 of Guangjinan Road Station and Line 1 to Line 2 is of high importance before and after the evening rush hour. The transfer relationship of Line 4 to Line 2 and Line 2 to Line 4 of Suzhou Railway Station and Shihu East Road is of low importance.

In this study, both physical network structure and passenger flow index are considered, and a comprehensive importance index of transfer relationship is proposed based on the vulnerability of rail network nodes and the proportion of relative transfer passenger flow distribution.

The comprehensive importance index of transfer relationship in line network structure was defined as $I_R(R_i^S)$.

$$I_R(R_i^v) = \alpha \times I(\mathrm{v}_i) + \beta \times P_d(R_i^v) \tag{7}$$

Here, $I_S(S)$ is the importance of node S where the transfer relationship is located, $P_d(R_i^v)$ is the relative transfer passenger flow distribution ratio of the transfer relationship, α and β are the weight coefficients of the importance of transfer nodes and the relative transfer passenger flow distribution ratio of the line network respectively, satisfying $\alpha + \beta = 1$ ($\alpha = 0.6$ and $\beta = 0.4$ in the application of Suzhou case). In the measurement process, α and β reflect the tendency of the importance of transfer nodes and the relative proportion of transfer passenger flow. The values of α and β in the coordination process are affected by the degree of inclination of the coordinator or the operation manager to the two factors, and the calculation needs to combine the relevant practical production experience. In complex networks, the greater the value of the comprehensive importance of the

Table 6. The distribution ratio of relative passenger flow before and after the evening peak at the Suzhou interchange station on a working day in September 2019.

Transfer Station	Suzhou Railway Station		Guangjinan Road Station		Leqiao Station		Shihudong Station	
The direction	Line 4 to Line 2	Line 2 to Line 4	Line 1 to Line 2	Line 2 to Line 1	Line 1 to Line 4	Line 4 to Line 1	Line 2 to Line 4	Line 4 to Line 2
17:30:00	0.00	0.01	1.00	0.80	0.89	0.87	0.24	0.16
17:45:00	0.03	0.00	0.95	0.63	1.00	0.59	0.27	0.16
18:00:00	0.01	0.00	1.00	0.48	0.88	0.39	0.30	0.12
18:15:00	0.01	0.00	0.97	0.59	1.00	0.45	0.32	0.12
18:30:00	0.03	0.00	1.00	0.56	0.98	0.44	0.37	0.07
18:45:00	0.03	0.00	0.92	0.75	1.00	0.43	0.47	0.12
19:00:00	0.10	0.00	1.00	0.82	0.93	0.52	0.44	0.11
19:15:00	0.04	0.00	1.00	0.78	0.96	0.47	0.33	0.05
19:30:00	0.08	0.00	1.00	0.94	1.00	0.54	0.28	0.13
19:45:00	0.14	0.00	0.79	0.76	1.00	0.53	0.23	0.02
20:00:00	0.03	0.00	0.99	0.86	1.00	0.50	0.16	0.06
20:15:00	0.21	0.00	1.00	0.87	0.99	0.41	0.21	0.09
20:30:00	0.13	0.00	0.81	0.76	1.00	0.29	0.20	0.10
20:45:00	0.12	0.00	0.89	0.87	1.00	0.40	0.20	0.10
21:00:00	0.10	0.02	0.82	0.84	1.00	0.30	0.14	0.00

transfer relationship is, the greater the comprehensive impact of the transfer relationship on the network efficiency and transfer passenger flow. In this study, the comprehensive importance of the transfer relationship is taken as an index to measure the coordination and optimization of rail transit. The operation organization, the distribution of transfer passenger flow and the influence of transfer stations on network structure are considered.

3.3 *Importance evaluation of lines*

There may be multiple lines in the same transfer station in the rail line network, so it is necessary to coordinate the connection relationship between the lines, and clarify the importance of each rail transit line in the system in the process of operation organization optimization.

3.3.1 *Rail transit loops are more important than rail transit rays*

The rail transit loop usually surrounds the central area of the city. Its purpose is to alleviate the fast, high density, large concentration of passenger flow in the downtown, and large transfer hub stations are usually located on the loop. Because there is no terminal in the loop line, the train operation will be difficult to reset if the trip is delayed. Therefore, the loop line has stricter requirements on the accuracy of the running time, which is of higher importance than the ray in the system.

3.3.2 *Rail transit rays with long lines and passing through key areas are of higher importance than ordinary rays*

The purpose of the long line ray is usually to facilitate commuters in the suburbs and alleviate the problems of connecting the main city with the suburbs or satellite cities. Therefore, the line length ray plays an important role in the rail transit system under the condition of network formation, especially the lines passing through key areas (important business districts) bear a great demand for passenger flow and play an important role in dredging. Compared with other rays, these rays are of higher importance in the rail transit system.

All lines of Suzhou rail transit are linear. Line 1 runs east-west. Line 4 runs from south to north in the Y shape of the main branch line. Line 2 is a C-shaped line that connects the downtown area to the northern and eastern directions, partially coinciding with Line 4. Line 3 is an east-west U-shaped structure, carrying passenger flows from the south of the central city to the northwest and northeast, respectively. The order of line importance is Line 1, Line 4 and Line 2.

4 CONCLUSION

The urban rail transit system under the condition of network is complex and huge. It is composed of numerous rail transit line units, which are subsystems of the system. A transfer station is the main passenger flow distribution point in the rail transit line network, which has the important function of passenger flow exchange and train connection between different lines. In combination with the video statistics data and AFC data of Suzhou Railway in September 2019, the paper conducts a quantitative evaluation and analysis of the importance of transfer direction within a single transfer station, the importance of multiple transfer stations by line direction and the importance of line, so as to reflect the function level of each transfer station and line more intuitively in the rail network. This study has reference significance for optimizing and adjusting the train operation scheme, which is beneficial to strengthen the urban rail network operation and further realize the fine management of rail passenger flow.

REFERENCES

C. G. Kou, "Research on the connection optimization of the last train schedule of urban rail transit under the network condition," Beijing Jiaotong University, 2016.

G. Z. Bai, J. Guo, H. G. Shi, Y. Yang and T. H. Zhang, "Research on Optimization of Train Connection in Metro Interchange Station Based on Coordination of Departure Time," Transportation System Engineering and Information,2013,13(05):134–139+201.

L. J. Kang, J. J. Wu, H. J. Sun, X. N. Zhu, and Z. Y. Gao, "A case study on the coordination of last trains for the Beijing subway network," Transportation Research Part B,2014.

L. Kang, "Collaborative organization and optimization of the last train operation plan of urban rail transit," Diss. Beijing Jiaotong University, 2016.

Q. Tian, "Research on the optimization of the connection between the first and last trains of urban rail transit under networked conditions," Southwest Jiaotong University, 2019.

R. H. Xu and X. Li, "Comprehensive optimization of the last train connection plan of the urban rail transit network," Journal of Tongji University: Natural Science Edition, 2012.

R. H. Xu, M. Zhang and Z. B. Jiang, "Research on the Departure Time Domain of the First and Last Trains of Urban Rail Transit Based on the Coordination of Line Network Operation," Journal of the China Railway Society,2008(02):7–11.

X. Guo, "Collaborative optimization of train schedules in urban rail transit network by stages," Beijing Jiaotong University, 2018.

Y. Huang, "Research on regional bus schedule model considering the importance of stops," Southwest Jiaotong University, 2014.

Z. Y. Bian, "Coordination and optimization of urban rail transit network operation organization," Southwest Jiaotong University, 2015.

Advances in Urban Engineering and Management Science – Khalil & Yang (Eds)
© 2023 The Author(s), ISBN 978-1-032-30426-7

Comprehensive evaluation of highway development based on DEA and gray correlation

Jianyou Yu* & Ding Wu*
Hebei Province Expressway Yanchong Management Center, Hebei, China

Shuangshuang Qin*
School of Economics and Management, Shijiazhuang Tiedao University, Hebei, China

ABSTRACT: The development of the expressway plays a very important role in regional economic development. In this paper, comprehensive evaluation indicators of expressway development are constructed from the perspectives of investment, infrastructure construction, transportation capacity, economy and safety. Based on the relevant data on the expressway in Hebei Province from 2010 to 2019, DEA-BCC and the grey correlation model are applied. The comprehensive development of the expressway in Hebei province is evaluated and analyzed, and the key indicators affecting the development of the expressway are further explored. The results show that the overall development of the expressway in Hebei province is rapid, and there is a strong correlation between fixed assets investment, gross regional product, and high-speed development.

1 INTRODUCTION

China's expressway construction has entered a comprehensive standardization, and high-quality development stage, and expressways are the link between cities for communication and drive the development of the inter-regional economy. At present, China is moving towards a strong transportation country, and the rapid development of expressways also affects people's daily life, so it is crucial to conduct a comprehensive evaluation of expressway development.

The development of highways is an important indicator of the level of regional economic development. Foreign scholars Pelayo Arbués et al. (2015) and Galina Tokunova (2018) found that transportation infrastructure affects economic development. Domestic scholars Li Jianlong (2021) and Jia Ruining (2017), respectively, using the benchmark Doberman regression method and space model, the highway has a promoting effect on economic growth.

At present, the highway evaluation methods used by scholars at home and abroad mainly include fuzzy comprehensive evaluation method, analytic hierarchy process, neural network, and DEA method, among which DEA can analyze the development of decision-making units from two perspectives of input and output. The basic models include CCR and BCC, among which BCC is for variable scale. Therefore, this paper uses the DEA-BCC model to comprehensively evaluate the Hebei expressway from 2010 to 2019 by selecting six indicators related to expressway development from different aspects and further uses the grey correlation model to analyze the impact of evaluation indicators on expressway development

2 SELECTION OF INDICATORS FOR COMPREHENSIVE EVALUATION

In order to better reflect the development level of expressways, this paper selects input-output indicators from the perspectives of investment, infrastructure construction, safety, transportation capacity and regional economic development. The selected indicators are shown in Table 1.

*Corresponding Authors: Y565656yk@163.com, 104868006@qq.com and qinshuang456@163.com

 DOI 10.1201/9781003305026-76

Table 1. Indicators of highway inputs and outputs.

Indicator	Variable	Name	Unit	Indicator Description
Input index	X1	Transportation Infrastructure Investment	Billion	Reflect highway development and construction
	X2	Integrated highway density		Represents the average mileage of expressways per 10,000 people per 100 square kilometers
	X3	Highway mileage	Km	The total length of the expressway, reflecting the scale of its development
	X4	Mortality per 10,000 vehicles		It can reflect highway safety. The lower the index, the better. Forward processing is required
Output Indicator	Y1	Comprehensive turnover of highway	Million ton-km	The flow of personnel and materials driven by the development of expressways
	Y2	GDP	Billion	Can reflect the highway development, the impact on the regional economy

3 COMPREHENSIVE EVALUATION OF HIGHWAY DEVELOPMENT

3.1 *Comprehensive evaluation model and data source of highway development*

3.1.1 *DEA-BCC*

Charnes et al. proposed the first DEA model-CCR model, which was a method to evaluate multiple inputs and multiple outputs. Banker (1984) proposed the BCC model on this basis, which further considered the change of return to scale, as shown below.

$$\begin{cases} min\theta \\ s.t\ \ \sum_{j=1}^{n}\lambda_j X_j \leq \theta X_0 \\ \sum_{j=1}^{n}\lambda_j Y_j \leq Y_0 \\ \sum_{j=1}^{n}\lambda_j = 1 \\ \lambda_j \geq 0 \\ j = 1, 2, \ldots, n \end{cases} \tag{1}$$

In the formula: n indicates the number of decisionmaking units x is the input variable; y is the output variable; θ denotes the efficiency value; λ_j is the coefficient of the decision unit.

The BCC model decomposes Technical Efficiency (TE) into pure Technical Efficiency (PTE) and Scale Efficiency (SE), establishing the following relationship.

$$\mathrm{TE} = \mathrm{PTE} \times \mathrm{SE} \tag{2}$$

3.1.2 *Grey correlation degree model*

To better reflect the influence degree of the index of highway development, based on DEA is calculated, the grey correlation method is used to calculate the efficiency value and indicators. The higher the relevance, the index weight, and the more influence on the development of highways. Specific steps are as follows.

The comparison sequence and reference sequence were determined and marked as X_{ij} and X_{0j} $i = 1, 2, 3, \ldots, m; j = 1, 2, 3, \ldots, n$ respectively. X_{ij} was the number of indicators, and X_{0j} was the number of comparison sequences

Finally, calculate the grey correlation

$$r_i = \frac{1}{n}\sum_{j=1}^{n} \frac{\min_i \min_j |X'_{0j} - X'_{ij}| + \rho \max_i \max_j |X'_{0J} - X'_{ij}|}{|X'_{0J} - X'_{ij}| + \rho \max_i \max_j |X'_{0J} - X'_{ij}|} \tag{3}$$

Where: r_i is the gray correlation; $\min_i \min_j |X'_{0J} - X'_{ij}|$ indicates the extreme difference minimal value; $\max_i \max_j |X'_{0J} - X'_{ij}|$ indicates the extreme difference maximal value; ρ is the resolution coefficient, which is taken as 0.5.

3.1.3 *Data source*

Hebei Provincial Expressway was selected as the research object. The data were mainly obtained from the Hebei Economic Yearbook, China Statistical Yearbook, China Logistics Yearbook, and Statistical Bulletin of National Economic and Social Development Hebei Province.

3.2 *Efficiency analysis and projection analysis based on DEA-BCC highway development*

3.2.1 *Efficiency analysis:*

DEA-BCC model is used to comprehensively evaluate the Hebei expressway, and the evaluation results are shown in Table 2.

(1) Comprehensive technical efficiency is a comprehensive evaluation index for resource allocation and use efficiency of decision-making units. In this paper, it represents the development status of expressways comprehensively mapped by several factors such as the economy. According to Table 3, it can be found that from 2010 to 2019, the average comprehensive technical efficiency was 0.973, indicating that expressways in Hebei Province had good overall development. During this period, the comprehensive technical efficiency of 7 years was higher than the average, among which, in 2012, 2015, 2017 and 2018, the comprehensive technical efficiency of these four years was 1, indicating that DEA was effective in these four years, and the scale and structure of expressways in Hebei Province had reached the optimal state and developed rapidly.
(2) The mean values of pure technical efficiency and scale efficiency were 0.987 and 0.986, respectively. In terms of quantity, pure technical efficiency was higher than average in 6 years, and scale efficiency was higher than average in 8 years. In non-DEA effective years, in 2013, 2014, 2016 and 2019, pure technical efficiency is less than scale efficiency. Therefore, it can be seen that pure technical efficiency is the main reason for the inefficiency of comprehensive efficiency. In 2010 and 2011, the scale efficiency is less than the pure technical efficiency, but the value of the pure technical efficiency is 1, and the return to scale is increasing, indicating that the Hebei expressway resources input in these two years is effective, but there is a difference between the scale and the optimal scale.

Table 2. DEA evaluation results.

Year	TE	PTE	SE	Compensation for size	Year	TE	PTE	SE	Compensation for size
2010	0.906	1	0.906	irs	2015	1	1	1	
2011	0.988	1	0.988	irs	2016	0.975	0.976	0.999	irs
2012	1	1	1		2017	1	1	1	
2013	0.968	0.974	0.994	irs	2018	1	1	1	
2014	0.983	0.984	0.999	drs	2019	0.906	0.931	0.973	drs
Average value	0.973	0.987	0.986	–	–	–	–	–	–

3.2.2 *Projection analysis*

From the perspective of input, projection analysis is conducted on the non-DEA effective years for each element to find out the gap between the target value and the actual value and analyze the causes of input redundancy or output insufficiency, as shown in Table 3.

From the perspective of investment, transportation infrastructure investment, the investment redundancy value is the largest in 2019, which needs to reduce the investment by 6.632 billion yuan. The maximum redundancy value of expressway network comprehensive density in four years is 1.36. The maximum redundancy value of expressway mileage is 536.23 km; The minimum difference between the target value and the actual value of the death rate per 10,000 vehicles in 2013 was 0.02, which increased to 0.15 over time, indicating that the number of traffic accidents in Hebei province was increasing.

From the perspective of output, the comprehensive turnover in 2016 and 2019 has the situation of insufficient output, which is 91.59 and 228.52, respectively. The GDP of Hebei Province is less than 837.12 in 2014, and there is the phenomenon of input redundancy in 2014, 2016 and 2019. This indicates that over-input did not bring the comprehensive turnover and increase in GDP, so the investment should be curtained to an appropriate amount to achieve the effective DEA.

Table 3. Analysis of input-output slack in Hebei Province during 2010–2019.

Variable		Year	Input redundancy	Insufficient output	Variable	Year	Input redundancy	Insufficient output
Input index	X1	2013	−23.94	−13.32	X2	2013	−0.39	−0.02
		2014	−14.73	−43.23		2014	−0.26	−0.03
		2016	−21.92	−38.25		2016	−0.42	−0.01
		2019	−62.58	−3.74		2019	−1.36	
	X3	2013	−144.97		X4	2013	−0.02	
		2014	−95.35			2014	−0.01	−0.02
		2016	−155.76			2016	−0.02	
		2019	−515.27	−20.96		2019	−0.06	−0.09
Output Metrics	Y1	2013			Y2	2013		
		2014				2014		837.12
		2016		91.59		2016		
		2019		228.52		2019		

3.3 *Correlation analysis based on gray correlation highway evaluation index*

The grey correlation model was established based on Matlab. Further analysis of indicators for the degree of influence on the development of the highway is shown in Table 4.

Table 4. Correlation of input-output indicators of Hebei expressway.

Indicators	Relevance	Relevance Ranking	Indicators	Relevance	Relevance Ranking
X1	0.7076	1	X4	0.6500	4
X2	0.6843	2	Y1	0.5079	6
X3	0.6708	3	Y2	0.6136	5

As shown in Table 4, the selection of evaluation indexes of the expressway and the efficiency value of the correlation degree is greater than 0.5, illustrating the selection of indicators has a great influence on the development of the expressway in Hebei province. From the perspective of the investment index, the biggest impact is brought by traffic investment in fixed assets. From the point of view of the output index, the GDP of Hebei province has the greatest influence on the development of the Hebei expressway.

4 CONCLUSION

Based on the DEA-BCC model and grey relational degree model, the comprehensive evaluation of expressways in Hebei Province from 2010 to 2019 was carried out by establishing the evaluation index system of comprehensive expressway development, and the following conclusions were drawn. (1) The average comprehensive technical efficiency of highway development in Hebei province is 0.973. The pure technical efficiency and scale efficiency affect the comprehensive technical efficiency simultaneously; (2) From the perspective of years, only 2012, 2015, 2017 and 2018 in Hebei Province expressway input-output reached the optimal state; (3) By using the grey correlation model, it is found that fixed-asset investment and GDP of Hebei province are the key indicators affecting the development of expressway.

The construction of the expressway improves regional transportation and drives the social and economic development of the city. Combining the analysis results, the following suggestions are made. (1) Propose a scientific and reasonable plan of highway construction and ensure that it matches the regional economic development level; (2) Combined with the current situation, do overall planning, optimizing traffic construction investment plan to reduce unnecessary investment; (3) Strengthen the urban traffic control and take measures to mobilize the people, attention to the traffic rules, to reduce the number of traffic accidents.

REFERENCES

Banker R D&Charnes A&Cooper W W. (1984), Some models for estimating technical and scale inefficiencies in data envelopment analysis[J]. Management Science, (9):1078–1092.

Cândida Ferreira. (2020), Evaluating European Bank Efficiency Using Data Envelopment Analysis: Evidence in the Aftermath of the Recent Financial Crisis[J]. International Advances in Economic Research, (pre-publish).

Charnes A & Cooper W W & Rhodes E. (1978), Measuring the efficiency of decision-making units[J]. European Journal of Operational Research, (2):429–444.

Galina Tokunova. (2018), Assessment of the transport infrastructure influence on urban agglomerations development[J]. Transportation Research Procedia, 36: 754–758.

Jianlong Li. (2021), Study on Highway Construction and Regional Economic Development. [J]. China Circulation Economy,06:71–75.

Kamran Rashidi. (2019), Evaluating the sustainability of national logistics performance using Data Envelopment Analysis[J]. Transport Policy, (9):35–46.

Pelayo Arbués & José F. Baños & Matías Mayor. The spatial productivity of transportation infrastructure[J]. Transportation Research Part A, 2015, 75: 166–177.

Ruining, J & Haicheng, X. (2017) Adaptation of highway scale and regional economic development-an empirical analysis based on provincial panel data in China from 2001–2015. China Distribution Economy,07:20–29.

Wei, W. (2020) DEA-based evaluation of highway construction and socio-economic adaptability in Guizhou. Marketing Community,51:98–99.

Xiaoqing, W. (2017) Analysis of Spatial Differences in Economic Benefits of Construction Land in Shenzhen and Optimization Countermeasures-Based on DEA's CCR Model and BCC Model. Homeland Resources Science and Technology Management,01:73–80.

Xuxiao, W. (2012) Study on the efficiency of economic development in urban areas of the Central Plains city cluster. Journal of Henan University of Technology (Social Science Edition),04:38–44.

Yanmei, D & Yingming, Z. (2015) Evaluation of innovation efficiency in China's high-tech industries – based on two-stage dynamic network DEA model[J]. Science and Technology Progress and Countermeasures,32(24):106–11

Advances in Urban Engineering and Management Science – Khalil & Yang (Eds)

A framework for intelligent network architecture-based transportation infrastructure resilience assessment index system

XuanHe Yu* & GuiXiong Chang*
Business School, Sichuan University, Chengdu, Sichuan, China

Wei Wei*
School of Architecture and Environment, Sichuan University, Chengdu, Sichuan, China

ZiQiang Zeng*
Business School, Sichuan University, Chengdu, Sichuan, China

ABSTRACT: As the internal components of the transport infrastructure system and the external associated infrastructure system are more correlated to the intelligent network architecture, the transport infrastructure system is more vulnerable to cascading failures under emergencies, and its vulnerability is different from that of the traditional transport infrastructure system. There is little existing research, and new systematic vulnerability assessment methods are lacking. This paper uses the pressure-state-response model to put forward the vulnerability evaluation index system framework of transportation infrastructure systems under intelligent network architecture. Compared with traditional vulnerability assessment methods, this model classifies correlation indicators, considering the interaction and influence on "objecting" and "environment," which has the advantages of systematization, operability, and accessibility. This paper puts forward the framework of the transportation infrastructure vulnerability assessment index system under the intelligent network architecture, comprehensively considering the impact of internal and external cascading failures and system correlation indicators on system vulnerability, which has certain enlightening significance for subsequent transportation infrastructure vulnerability assessment studies under the intelligent network architecture.

1 INTRODUCTION

1.1 *Research background*

Major emergencies, such as earthquakes, hurricanes, tsunamis, and terrorist attacks, often cause huge economic losses and casualties to society and damage to important infrastructure systems. On January 12, 2010, at 16:53 (local time), in the capital of Haiti, Port-Au-Prince, a magnitude 7.3 earthquake occurred, resulting in Above 100,000 deaths. On February 27, 2010, at 14:34 Beijing time, in Chile's second-largest city Concepcion, a magnitude 8.8 earthquake occurred, killing 799 people; On the afternoon of September 30, 2009, a magnitude 7.9 earthquake on the Richter scale struck the Indonesian island of Sumatra, with the final death toll likely to exceed 5,000. All the cities affected by the earthquakes have suffered damage to their transport infrastructure, resulting in the inability of rescue teams and supplies to arrive and the inability of the affected people to move. Compared with floods and other natural disasters, earthquakes damage to transportation infrastructure systems is permanent and intense, having a more obvious impact on transportation infrastructure systems and post-disaster emergency rescue efficiency. How to reduce the resilience of the transportation infrastructure system and speed up the recovery of the transportation infrastructure system has become an important issue in the research of various countries.

*Corresponding Authors: yuxuanhe@stu.scu.edu.cn, changguixiong@stu.scu.edu.cn, qqbsqq@163.com and zengziqiang@scu.edu.cn

DOI 10.1201/9781003305026-77

1.2 *Problem statement*

The resilience of the traditional transportation infrastructure system is internal and depends only on the results of the internal functions of the system. However, the resilience of transportation infrastructure systems under intelligent network architecture has externality besides internality. The transportation infrastructure system under the intelligent network architecture forms an association relationship with multiple types of infrastructure systems such as communication infrastructure and emergency rescue infrastructure through new and high technologies such as the Internet of Things and 5 G. Therefore, when dealing with emergencies, interference from other external infrastructure systems will also affect the resilience of transportation infrastructure system. However, the correlation between systems and the resilience of transportation infrastructure systems is not a simple negative correlation. For example, under the intelligent network architecture, the emergency rescue infrastructure system can effectively predict and deal with emergencies, reduce the damage to emergencies to the transportation infrastructure system, and reduce the resilience of the transportation infrastructure system. However, when the damage degree exceeds the threshold, the function of the associated infrastructure system is damaged, which may directly interfere with the normal operation of the transportation infrastructure system, and the resilience of the transportation infrastructure system is significantly increased. For example, after the communication infrastructure system is damaged, the central control and command system in the traffic infrastructure system may fail to issue scheduling information in time due to poor communication and other reasons, resulting in traffic congestion on some main roads.

According to the results of the literature review, there are five common methods of resilience analysis. The research on resilience assessment methods mainly includes commonly used assessment methods and assessment models, the improvement of assessment models or methods, the construction of new models and empirical tests, and the analysis of assessment results, etc., which are the focus of current research. Many related papers for resilience assessment methods have been discussed, including the analytic hierarchy process, Delphi method, fuzzy mathematics method, regression analysis and principal component analysis, the EFI method and grey system theory, the DEA data envelopment method, the triangle method, etc. Although these methods have been studied and applied, there are generally some problems such as lack of depth, insufficient comprehensive application of various methods, and failure to consider the impact of intelligent network architecture on the resilience of infrastructure systems. Hence the research on the resilience assessment method of infrastructure systems under intelligent network architecture is still the focus of future research in resilience assessment.

1.3 *Intelligent network architecture*

Under the guidance of "Intelligent Transportation and Green Transportation" development, intelligent network architecture has become an important trend and vane in China. The intelligent network architecture selected in this paper mainly consists of UAV mobile base station technology and road capacity control technology. The traffic network's capacity control system mainly includes variable lane control device "intelligent road stick," portable communication base station "I-Node," and related communication protocols. Intelligent road sticks can be laid along the road surface and control the traffic direction of different lanes in the road section through LED signal lights, to realize the dynamic configuration of road passage capacity. The function of I-Node is to realize Communication connection with intelligent road patches through Dedicated-Short-Range-Communication (DSRC) technology and integrate it into a real-time traffic dynamic management system. I-Node can remotely control LED signal lights of intelligent road stickers by receiving instructions from a real-time traffic dynamic management system, to realize the dynamic configuration of traffic capacity of road sections. In the event of an emergency, damaged I-Nodes can be temporarily restored using drone mobile base station technology. Intelligent network architecture connects multiple infrastructure systems, such as emergency rescue, communication, power grid, and transportation, to form a unified intelligent network architecture (Bachmann et al. 2013).

2 RESEARCH METHODS

To simplify the research object, this paper takes the transportation infrastructure system as the basic assessment object based on studying the new characteristics of the resilience of the transportation infrastructure system under the intelligent network architecture. On the basis of the study of historical accident statistics of critical transportation infrastructure and literature review results, this paper constructed a PSR resilience assessment model based on pressure (P), state (S), and response (R). The following will explain the meanings of P, S, and R, respectively.

P represents the pressure index, which is divided into exogenous pressure and endogenous pressure. Exogenous pressure represents the probability or frequency that the evaluation unit is vulnerable to damage or loss in the region, which mainly refers to the influence and pressure of external factors such as natural disasters on the evaluation unit. This paper takes earthquake disasters as an example to study. Endogenous pressure refers to the impact on the transportation infrastructure system after the damage to the associated infrastructure system. This paper measures the system's relevance.

S represents the status index, representing the ability of the evaluation unit to resist pressure P. The status index is derived from system functions and device running status. Through literature review, this paper selects the prediction and early warning ability of transportation infrastructure systems for emergencies and the operating years of the equipment itself as the basis for measuring index S.

R represents the response index, which represents the resilience of the transportation infrastructure system after damage under specific situational pressure. It mainly refers to the ability of the transportation infrastructure system to respond to emergencies and recover after disasters. In the intelligent network architecture, the response index is mainly represented by the correlation between transportation infrastructure systems and associated infrastructure systems.

Finally, the calculation method of resilience indexes of transportation infrastructure systems under intelligent network architecture and the classification standard and basis of resilience indexes of transportation infrastructure systems under intelligent network architecture is further studied by referring to the evaluation method.

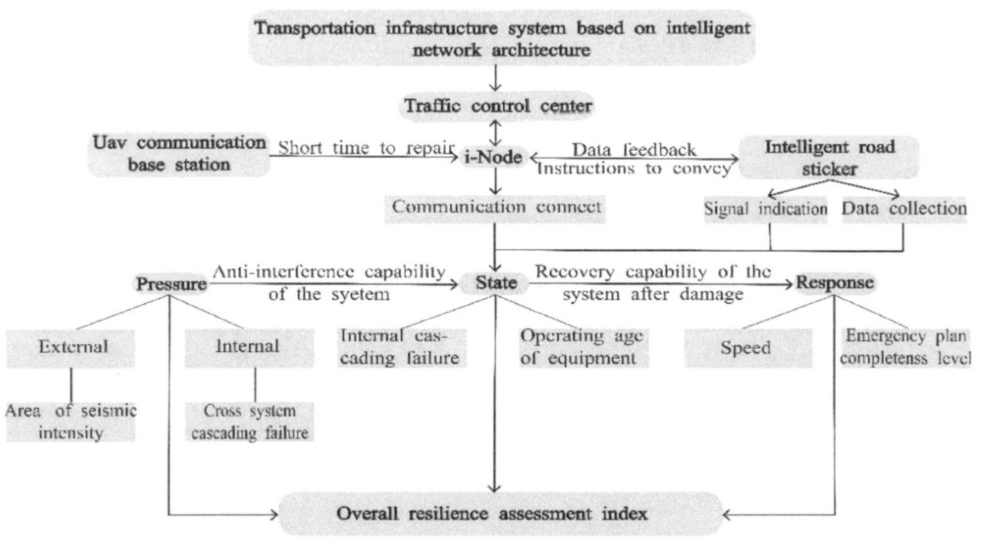

Figure 1. Framework of the assessment index system.

3 EMPIRICAL RESULTS AND DISCUSSIONS

3.1 *Pressure index*

This paper refers to the basic idea of the risk index evaluation method, assigns hierarchical values to each index, and determines that the value range of each index is 0–1. Taking earthquake disasters as an example, the evaluation criteria of each index are explained in detail. Based on the seismic intensity zoning map and seismic grade distribution map of China, this paper determines the basic intensity region of the transport infrastructure system. According to the Chinese Earthquake intensity Table (GB/T 17742-2008), the influence and damage degree of different earthquake intensities on the ground and the buildings are calculated.

Table 1. Pressure parameter A_{11} value table (Men 2015).

Area of seismic intensity	A_{11} Value range
Below V degree	$0 \leq A_{11} < 0.1$
VI degree	$0.1 \leq A_{11} < 0.2$
VII degree	$0.2 \leq A_{11} < 0.4$
VIII degree	$0.4 \leq A_{11} < 0.8$
Above IX degree	$0.8 \leq A_{11} < 1$

In this paper, the cross-system cascading failure probability of the associated infrastructure system is selected as the main evaluation index of endogenous pressure of transportation infrastructure system under intelligent network architecture. Large-scale cascading failures of associated infrastructure will directly interfere with the normal operation of the transportation infrastructure system, like what happens when communications infrastructure systems are damaged. Abnormal communication directly affects the regional multimodal transport and traffic signal control functions of the transportation infrastructure system and indirectly leads to the collapse of emergency rescue functions such as emergency vehicles and goods transportation. Referring to the power grid infrastructure system, the new impact of cascading failure probability on the resilience of the associated infrastructure system is shown in Table 2:

Table 2. Pressure parameter A_{12} value table.

Cross-system cascading failure	A_{12} Value range
Below 10%	$0 \leq A_{12} < 0.3$
10%–30%	$0.3 \leq A_{12} < 0.5$
30%–40%	$0.5 \leq A_{12} < 0.7$
Above 50%	$0.7 \leq A_{12} < 1$

As for the exogenous and endogenous factors that affect the pressure index, as much as possible to avoid subjective decision-making in the evaluation process, this paper adopts the weighted average method to obtain the seismic disaster resilience pressure index. Therefore, the maximum value of the resilience pressure index(P) of the transportation infrastructure system based on the pressure index is 1.

$$P = \frac{1}{2}(A11 + A12) \tag{1}$$

3.2 *State index*

The status index is related to the probability of internal cascading failure and the operating years of the equipment. The probability of internal cascading failure is negatively correlated with the operating status. The higher the probability of internal cascading failure is, the stronger the system's ability to resist external interference is, and the higher its resilience is, the lower the value of this parameter is. The operating life of the equipment is positively correlated with the operating state. The longer the operating life of the equipment is, the worse the anti-interference capability is, and the higher the value of this item is. Specific values were obtained through a literature review, referring to the power grid infrastructure system. Operating status values are shown in Table 3:

Table 3. State parameters A_{21} and A_{22} value table.

Internal cascading failure	A_{21} Value range	Operating age of equipment	A_{22} Value range
Below 10%	$0 \leq A_{21} < 0.3$	0–1 year	$0 \leq A_{22} < 0.1$
10%–30%	$0.3 \leq A_{21} < 0.5$	1–6 year	$0.1 \leq A_{22} < 0.6$
30%–40%	$0.5 \leq A_{21} < 0.7$	6–10 year	$0.6 \leq A_{22} < 0.9$
Above 50%	$0.7 \leq A_{21} < 1$	Above a decade	$0.9 \leq A_{22} < 1$

The probability of emergency prediction and early warning and operating years were given a weight of 0.5, respectively, and the status index was obtained by the weighted sum. Therefore, the maximum value of the resilience state index (S) based on earthquake disaster species is 1.

$$S = (\frac{1}{2}A_{21} + \frac{1}{2}A_{22}) \tag{2}$$

3.3 *Response index*

Similar to the pressure index and state index, each index in the response index system is graded and assigned, and the value range of each index is 0–1.

The efficiency of transportation infrastructure systems to respond to emergencies under the intelligent network architecture mainly depends on the response speed of relevant departments in post-disaster reconstruction and the completeness of emergency response plans. The response speed of post-disaster reconstruction is negatively correlated with the response index. The faster the response speed is, the higher the efficiency of post-disaster reconstruction is, and the weaker the vulnerability of the transportation infrastructure system is. The completeness of the emergency response plan is negatively correlated with the response index. The more complete the emergency response plans are, the faster the system can be repaired, and the lower the resilience is. Specific values are obtained through literature review and refer to the power grid infrastructure system. Values of operating status are shown in Table 4:

Table 4. Response parameters A_{31} and A_{32} value table (Men 2015).

Response speed	A_{31} Value range	Emergency plan completeness level	A_{32} Value range
0–18 h	$0 \leq A_{31} < 0.3$	Level 1	$0 \leq A_{32} < 0.3$
18–36 h	$0.3 \leq A_{31} < 0.5$	Level 2	$0.3 \leq A_{32} < 0.5$
36–54 h	$0.5 \leq A_{31} < 0.7$	Level 3	$0.5 \leq A_{32} < 0.7$
54–72 h	$0.7 \leq A_{31} < 1$	Level 4	$0.7 \leq A_{32} < 1$

To avoid subjective decision-making in the evaluation process, the resilience response index (R) of the transportation infrastructure system is calculated by a weighted average of the index of the response speed of relevant departments in post-disaster reconstruction and the completeness of the emergency response plan. Therefore, the maximum value of the resilience response index is 1.

$$R = (\frac{1}{2}A_{31} + \frac{1}{2}A_{32}) \tag{3}$$

3.4 *Overall resilience assessment index*

The overall resilience index of the transportation infrastructure system under the intelligent network architecture can be obtained by a weighted sum of the pressure index, state index, and response index obtained above. The overall vulnerability index grade refers to the construction risk grade and is divided into five grades: slightly resilience, average resilience, significantly resilience, high resilience, and extremely resilience. The classification standards of the overall resilience index are shown in Table 5:

$$Resilience = \frac{1}{3}(VP + VS + VR) \tag{4}$$

Table 5. Classification standard of an overall resilience index.

Resilience index	Resilience level
~0.2	slightly resilience
0.2~0.4	average resilience
0.4~0.6	significantly resilience
0.6~0.8	high resilience
0.8~1.	extremely resilience

4 CONCLUSIONS

In this paper, the new characteristics of the resilience of transportation infrastructure systems under intelligent network architecture are clarified by the literature review. Based on the PSR model, this paper establishes the resilience analysis framework of transportation infrastructure systems under the intelligent network architecture, which provides a certain theoretical basis for the subsequent resilience analysis of associated infrastructure systems under the intelligent network architecture. Based on the above research results, the overall resilience index and resilience assessment level of transportation infrastructure systems under intelligent network architecture are preliminarily established in this paper, which provides some ideas and guidance for emergency management departments to reduce the resilience construction of transportation infrastructure.

ACKNOWLEDGMENT

This research was supported by the National Natural Science Foundation of China (Grant No. 71971150).

REFERENCES

Bachmann, C.; Roorda, M.J.; Abdulhai, B.;(2013) Moshiri, B. Fusing a bluetooth traffic monitoring system with loop detector data for improved freeway traffic speed estimation. J. Intell. Transp. Syst. 17, 152–164.

Chen J, Gong X Y. (2017) Study on the community of shared future in cyberspace along the Belt and Road [J]. International Inspection and Inspection, (05) :60–73.

Men Y S.(2015) Vulnerability Assessment of Shanghai power grid's important infrastructure [D]. University of Science and Technology Beijing.

Nikoo N, Babaei M, Mohaymany A. (2018) Emergency transportation network design problem: Identification and evaluation of disaster response routes. International Journal of Disaster Risk Reduction, 27: 7–20.

Pánek J, Marek L, Pászto V, Valuch J. (2017) The crisis map of the Czech Republic: the nationwide deployment of a Ushahidi application for disasters. Disasters, 41: 649–671.

Qiu T.(2019) Study on the effectiveness evaluation of network political inquiry in Yantai [D]. Harbin: Harbin Engineering University.

Voosen P. (2017) Hurricane Harvey provides a lab for US forecast experiments. Science, 357(6354): 854–854

Yi X P, Yang Z Y, Wu Wende. (2020) Exploration and application of image processing and Internet of Things technology in intelligent transportation system [J]. Computer Products and Circulation, (11):52.

Zou M Q. (2020) Application of Internet of Things Technology in Intelligent Transportation System [J]. Traffic World, (26):20–21.

Advances in Urban Engineering and Management Science – Khalil & Yang (Eds)
© 2023 The Author(s), ISBN 978-1-032-30426-7

Research and application of safety performance model in port enterprises based on fuzzy theory

Chen Huang* & Yu Bai*
Tianjin Dongfang Tairui Technology Co., Ltd., Tianjin, China

Linlin Lu* & Hang Jin*
Tianjin Research Institute for Water Transport Engineering, Tianjin, China

ABSTRACT: According to the characteristics of safety production management in port enterprises, the safety performance of port enterprises is quantitatively assessed by taking ordinary cargo terminal enterprises as an example, and the safety performance indicator system is established. Fuzzy Borda number analysis is used to establish a model. The Delphi method clarifies the factors concerned by industry regulators and enterprise managers, and the relative weight and absolute weight of each indicator are determined. Finally, the model is used to assess and evaluate the safety performance of enterprises quantitatively and comprehensively. The example shows that the model is scientific and feasible and can directly reflect the advantages and weak links of safety production in port enterprises to achieve the purpose of continuously improving the safety production level of port enterprises.

1 INTRODUCTION

The goal of safe production in port enterprises is to protect the life and health of workers, guarantee the integrity of machinery and equipment, and ensure stable, safe production of enterprises (Zhan et al. 2012). Most port enterprises regard the limitation of safety production as a branch of operational performance or human resource performance assessment. The assessment mostly takes safety production accidents and hidden dangers rectification as the main indicators and adopts qualitative methods. Due to the different footholds and starting points of the assessment subject, the qualitative assessment method is subjective, which leads to one-sidedness in the assessment process of safety production and differences in assessment results. With the deepening of the reform of the port management system in China, the safety performance evaluation of port enterprises has become particularly important, and it is particularly critical to build a set of independent safety performance evaluation models, which is suitable for the evaluation and audit of competent units and the internal assessment of enterprises.

2 CONNOTATION AND RESEARCH STATUS OF SAFETY PERFORMANCE

2.1 *Connotation of safety performance*

The definition of Safety Performance can be roughly divided into three types: The first is to directly define the safety performance by virtue of the occurrence and consequences of safety

*Corresponding Authors: huangchen@tk-aq.com, baiyu@tk-aq.com, lulinlin@tk-aq.com and jinhang@tk-aq.com

 DOI 10.1201/9781003305026-78

accidents; the second is to consider the operation effect of safety production with the actual performance of enterprises; the third is to define safety performance by accident occurrence and on-site performance of enterprises (Liu et al. 2010). Some scholars put forward to define enterprise safety performance by using accident data and the actual performance of enterprises' safety production: enterprises' safety performance is the synthesis of accident occurrence, occupational injury, and the overall operation status of enterprises' safety production system in a period of time (Luo 2011).

On the basis of summarizing the definitions of safety performance by all parties, combined with the explanation of "performance" in GB/T 28001 Specification for Occupational Health and Safety Management System (Bagheri 2009), the safety performance of port enterprises can be defined as the measurable results obtained by port enterprises in managing safety production risks based on safety production policies and safety production work objectives, which are related to their own safety production risk control.

2.2 *Research status of safety performance*

Safety performance is an important indicator that affects the overall performance of enterprises. In its regular business recollection, safety performance was firstly discussed by DuPont Company of the United States (Caliendo & Guglielmo 2016). Amid the evaluation of safety performance, the traditional accident evaluation methods such as accident frequency and severity rate are used to evaluate the safety status of companies, departments, or equipment, which cannot make clear whether the system is effective, whether the diagnosis is correct and whether the system is under control (Petersen 2003). It is found that if the evaluation of safety performance only depends on accidents, injuries, and other aspects, the statistical data lacks sensitivity, accuracy, and timeliness, ignoring the exposure of risks (Glendon et al. 2001), so it is inappropriate to measure safety performance only by accident data.

At present, domestic scholars are more inclined to take the enterprise's own safety production objectives as the basis of the indicator system. The department objectives, team objectives, and post objectives should be stripped, with the safety production objectives as the basis for formulating the safety production indicator system, which has better practicability and adaptability.

3 DETERMINATION OF INDICATOR SYSTEM OF SAFETY PERFORMANCE ASSESSMENT MODEL

The safety performance assessment model of port enterprises needs to determine its assessment indicators first. Because there are different types of cargo in port enterprises, the indicator setting of specific safety performance evaluation may reflect their operating characteristics. This paper takes ordinary cargo terminal enterprises (such as container terminals, bulk cargo terminals, etc.) as examples for the focus of the research.

On the basis of aggregating and summarizing the daily safety production work of an ordinary cargo terminal enterprise, combined with the requirements of relevant laws and regulations of national safety production and regulations of industry authorities, this paper puts forward a core indicator system of the safety performance of port enterprises, which is composed of six first-class core elements (A) and 16 second-class elements (B).

Daily management, safety guidance, accident emergency, supervision and management, performance management, and standardization construction amid the setting of the first-class core elements are the main work contents that port enterprises need to carry out safety production management according to laws and regulations, the setting of which has wide universality and applicability.

The three core tasks of daily safety management of enterprises when implementing laws and regulations, formulating documents and coordinating and cooperating in the setting of second-class elements; at the same time, work plan, safe operation, training and publicity, and safety documents are important measures for enterprise managers to guide personnel and ensure safety conditions; emergency plan, emergency duty, emergency rescue, and accident treatment are the main work contents of enterprises' emergency management; risk prevention and control, and hidden danger investigation and management are the core requirements for the construction of enterprises' safety management system in China at present; the implementation and improvement of performance assessment is an important measure for enterprises to guide safety performance to play a role in the continuous improvement of safety management; the construction of port enterprise standardization is not only a necessary condition for our port enterprises to carry out legal production and business activities, but also objectively reflects the level of enterprise safety production management.

4 CONSTRUCTION OF SAFETY PERFORMANCE ASSESSMENT MODEL

4.1 *Using the fuzzy Borda method to solve weight distribution*

This paper evaluates six first-class indicators and 17 second-class indicators in the assessment model and considers factors, such as personal experience, subjective emotion, field reality, and objective environmental influence. There will be great differences in judging the importance of indicators, so this paper adopts the fuzzy Borda number analysis method to determine the weight distribution of the first-class indicator system. The specific operation steps are as follows:

1. Establishment of the hierarchical relationship
   Establishing the hierarchical relationship means decomposing complex problems into various components of indicators, continuing to decompose them until they can be evaluated, and finally forming a hierarchical relationship with a dominant relationship from top to bottom.
2. Creation of the grid
   In this paper, the importance evaluation of the first-class indicators is divided into the following five grades, which can be expressed by a linear scale as shown in Figure 1.

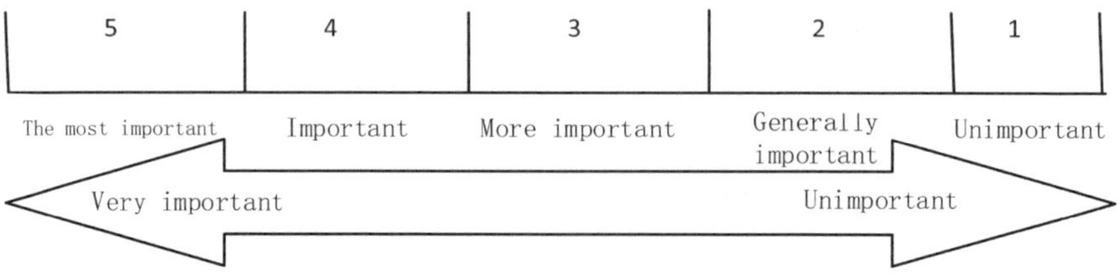

Figure 1. Scoring scale diagram of the element in the grid.

   The selected industry experts use a linear scale to judge and score the importance of the first-class indicators, respectively, thus obtaining the completed grid data.
3. Analysis of the grid
   The relative weight under a single criterion is obtained by analyzing the grid. Let the m-th attribute of indicator DP be divided into Bm (DP) (where m=1, 2, …, M; p=1, 2 …, N).

1) Determining the membership degree

In the m-th attribute evaluation, the membership degree Ump that each evaluated indicator belongs to the "the most important" is obtained, and its plan formula is:

$$U_{mp} = \frac{B_m(D_p)}{max[B_m(D_p)]} = (0 \leq U_{mp} \leq 1) \tag{1}$$

2) Making a statistical table of fuzzy frequency

$$R_p = \sum_{m=1}^{m} f_{hp}, f_{hp} = \sum_{m=1}^{m} \delta_m^h(D_p) U_{mp} \tag{2}$$

3) Calculating the fuzzy Borda number FB (Dp)

If it is stipulated that the weight of the evaluated indicator ranking at the h-th position in the Dp pecking order relationship is Qh, there are:

$$\Phi BD_p \sum_m \frac{f_{hp}}{R_p} \times Q_h = \sum_m W_{hp} Q_h \tag{3}$$

$$W_{hp} = \frac{f_{hp}}{R_p}, Q_h = \frac{(N-h)(N-h+1)}{2} \tag{4}$$

4) Calculating the relative weight

The weight value is normalized to get the relative weight under a single criterion, that is: $W_p' = FB(D_p)/\sum^N FB(D_p)$.

4.2 *Method for determining the weight of the underlying indicators of the model*

Using the same calculation method, the relative weights of the first and second-class indicators are obtained, and the final combined weights are obtained.

The calculation formula is as follows:

$$W_{ij} = W_i \times W_{ij}^i (i - 1, 2, 3, 4, 5, 6; j - 1, 2, 3 \ldots)$$

Where, W_{ij} is the combination weight of the second-level indicator; W_i is the weight of the first-level indicator; W_{ij}^i is the relative weight of the second-level indicator.

4.3 *The method of setting the bottom indicator of the model*

Because the indicators in this model are all "non-statistical" indicators, the qualitative explanation is adopted first, and the quantitative description method is adopted. That is to say, by using the method of comment set, each indicator is divided into five grades: excellent, good, medium, slightly poor, and very poor, and then the scores of each grade are set as 100, 80, 60, 40 and 20. The performance assessment personnel score the indicator, on the basis of a comprehensive score, the final score of each indicator can be obtained. The formula is as follows:

$$G_{ij} = 100 f_{ijA} + 80 f_{ijB} + 60 f_{ijC} + 40 f_{ijD} + 20 f_{ijE} \tag{5}$$

Where, fijk=Rk/S; Rk is the number of experts selected on k standard; S is the total number of experts; Gij is the final score of second-level indicator ij.

5 EXAMPLE APPLICATION

5.1 *Distribution of indicator weights*

Using the above summary formulas 1 ~ 4 and combined with the Delphi method, the weight value of the safety performance model indicator is obtained, and finally, the weight distribution table of the safety performance assessment model is obtained.

Table 1. Weight distribution of production safety performance model indicators.

Serial number	First-level indicators	Weight	Second-level indicators	Combination weight
1	Daily management	0.17	Implement of laws and regulations	0.0765
2			Documentation	0.0595
3			Coordination and cooperation	0.034
4	Safety guidance	0.2	Three Simultaneities of Construction Projects	0.04
5			Work plan	0.03
6			Safe operation	0.038
7			Training and publicity	0.048
8			Safety culture	0.044
9	Accident emergency	0.22	Emergency plan	0.0616
10			Emergency duty	0.0352
11			Emergency rescue	0.0726
12			Accident treatment	0.0506
13	Supervision and management	0.13	Risk management and control	0.078
14			Investigation and management of hidden dangers	0.052
15	Performance management	0.08	Implementation of performance assessment	0.032
16			Performance improvement	0.048
17	Standardization construction	0.2	Standardization standards	0.2

5.2 *Comprehensive evaluation*

When using the above model to assess the safety performance model of port enterprises, the assessors only need to give quantitative scores according to the qualitative comments of each indicator, and finally use the linear weighting method to synthesize the scores of each indicator, and get the total evaluation score of the safety performance assessment of the enterprise or a certain department. The following formula is used:

$$\sum_{i=1}^{6}\sum_{j=1}^{n} GijWij \tag{6}$$

Where, F is the comprehensive overall evaluation score of port safety performance; Gij is the final score of second-level indicator ij; Wij is the combined weight of the ij-th indicator; N is the number of underlying indicators contained in the corresponding indicator i.

A port enterprise uses the safety performance model constructed in this paper to assess and uses Formulas 5 and 6 to calculate the assessment results, as shown in Table 2.

Table 2. Scoring results of the model application.

First-level indicators	Second-level indicators	Score	First-level indicators	Second-level indicators	Score
Daily management	Implement of laws and regulations	6.81	Accident emergency	Emergency plan	4.31
	Documentation	4.46		Emergency duty	2.64
	Coordination and cooperation	1.97		Emergency rescue	6.10
Safety guidance	Three Simultaneities of Construction Projects	3.80		Accident treatment	4.45
	Work plan	2.43	Supervision and management	Risk management and control	6.71
	Safe operation	3.19		Investigation and management of hidden dangers	3.9
	Training and publicity	3.74	Performance management	Implementation of performance assessment	2.34
	Safety culture	2.86		Performance improvement	3.12
Standardization construction	Standardization construction	17.00	Total score	/	73.02

Finally, according to the safety performance classification standard preset by the enterprise or the competent unit of the industry, the safety performance level of the port enterprise is qualitatively evaluated. The grading standards of port enterprises for safety performance are shown in Table 3.

Table 3. Grading standards of port enterprise's safety performance.

Grade	Excellent	Good	Medium	General	Poor
Total score	90 ~ 100	80 ~ 89	70 ~ 79	60 ~ 69	Below 60

Combined with Tables 2 and 3, it can be analyzed that port enterprises use the safety performance assessment model designed in this paper to assess, and the final score is 73.02. The total assessment result is "medium." Specific analysis: Among the 17 second-level indicators of the model, although the assessment results of 8 indicators are "good" or above, the assessment results of 9 indicators are "medium" or below, so enterprises should strengthen the corresponding work of these second-level indicators in the next step. Among the indicators with lower scores, two indicators with "average" assessment results are safety culture and performance improvement. This shows that enterprises should further strengthen the creation of the safety culture, create a safe working atmosphere with enterprise characteristics, and implement safety performance evaluation to improve the on-duty performance of personnel, highlighting the status of safety performance. Finally, the assessment result of "coordination and cooperation" is "very poor," which shows that safety management personnel have not effectively organized and coordinated enterprise resources and fully mobilized the enthusiasm of all employees. It is necessary to put forward rectification requirements as soon as possible in this work, complete rectification within a time limit, and improve the coordination and cooperation level of enterprise safety management.

In a word, it can be seen from the above table that only when enterprises actively maintain the indicators with "good" and above assessment results and carry out targeted rectification and promotion on the work corresponding to the indicators with "medium" and below, can the safety performance assessment of enterprises be maintained or improved.

6 CONCLUSION

Firstly, taking ordinary cargo terminal enterprises as examples, the author puts forward a core indicator system of the safety performance in port enterprises, which is composed of 6 core first-class elements and 17 second-class elements. The fuzzy Borda number analysis method and Delphi method are used to solve the weight distribution problem of all levels of indicators in the model, which makes the safety performance model proposed in this paper have universal applicability. Secondly, the safety performance assessment model and the adopted method of the port enterprise can be used for both internal self-assessment of enterprises and annual assessment of enterprises by the competent unit of the industry, which can intuitively reflect the strengths and weak links of safety production management of enterprises, and point out the direction for the next stage of key work of enterprises, so as to achieve the purpose of continuously improving the safety production level of port enterprises.

Of course, due to the different management modes and types of goods operated by port enterprises, there are also obvious differences in the level, method, and content of safety production management. It is important to refine the second-level indicators further, expand the third-level indicators, especially further refine the "measurable" standard of each second-level indicator, thus improving the adaptability and availability of the model, and gradually extending the safety performance model designed in this paper to the port enterprises which operate dangerous goods, and this will be the main direction of the next research.

ACKNOWLEDGMENTS

Fund support: Fundamental Research Funds for the Central Public Welfare Research Institutes (Grant no. TKS20200316, TKS20210308).

REFERENCES

Bagheri M (2009). Risk analysis of stationary dangerous goods railway cars: a case study. Journal of Transportation Security; 2(3): 77–89.

Caliendo C, Guglielmo M L D (2016). Quantitative Risk Analysis on the Transport of Dangerous Goods through a Bi-Directional Road Tunnel. Risk Analysis an Official Publication of the Society for Risk Analysis; 37(1): 116.

Glendon A I, Litherl and D K (2001). Safety Climate Factors, Group Differences and Safety Behavior in Road Construction. Safety Science, 39 (3): 157–188.

Liu S X, Mei Q, Shen B, et al (2010). Literature Review on Safety Performance. China Safety Science Journal, 20 (5): 131–139.

Luo Y (2011). Evaluation of Safety Performance. Beijing: Coal Industry Press.

Petersen D (2003). Techniques of Safety Management: A System Approach. Des Plaines, Illinois: American Society of Safety Engineers.

Zhan S F, Huang C, Liu Y, et al. (2012). Guidelines for Safety Production Standardization Construction of Port General Cargo Terminal Enterprises. Beijing: People's Communications Publishing House.

Advances in Urban Engineering and Management Science – Khalil & Yang (Eds)
© 2023 The Author(s), ISBN 978-1-032-30426-7

Research on urban road safety early warning system

Sun Ying*
Party School of Liaoning Provincial Party Committee, Shenyang, P.R. China

ABSTRACT: Based on AI, big data, the Internet of things, edge computing and other technologies, providing perfect, rich, and professional software and hardware solutions can further promote the development of the transportation industry and improve the informatization and automation level of intelligent transportation. LDWS is a system that assists drivers in reducing traffic accidents caused by lane departure by means of alarm. When the vehicle exceeds a certain speed, it can send out an alarm before unconsciously (the driver does not turn on the turn signal) departing from the original lane, providing more reaction time for the driver, and greatly reducing the collision accidents caused by lane departure. It detects the lane line in real-time and estimates the departure time according to the steering wheel direction, vehicle speed and the angle between the vehicle and the lane. If the departure time is less than a threshold, it will send an alarm to remind the driver. The lane departure warning method based on computer vision technology uses a monocular camera to calibrate the lane line and obtain the holography matrix of the world object plane and image plane; Then detect, track, and calculate the slope of the lane line in the obtained lane image, and judge the lateral displacement of the vehicle and the change of the lane line slope. Through optical fiber 5g, AI vision chip and artificial intelligence algorithm, artificial intelligence network calculation, logic operation and embedded system operation are realized, and applied to vehicle deviation early warning system, intelligent speed measurement early warning system, curve passing early warning system, traffic accident early warning system, intersection early warning system, intelligent traffic video perception system and traffic accident early warning system. The early warning system supports the import of a control database, can generate data reports, and provide edge operation support for the solution through traffic flow thermal map distribution imaging and traffic flow distributed counting. It can accurately capture and identify the machine, non-machine, human and other elements, and master the road traffic information in real time. Use the data to analyze the road conditions, provide accurate video image data for the traffic management department, and guide the traffic system to divert and adjust the traffic information in time in combination with the flow data. At the same time, real-time detection of road status, number of pedestrians, moving track and other information, prediction of risk coefficient, intelligent reminder, active broadcast, and early warning, and update the early warning content at any time.

1 GENERAL INTRODUCTIONS

The road safety early warning system includes a cloud server and an early warning device arranged in the vehicle. The early warning device receives the video information photographed by the first camera device and obtains the detection signal sent by the safety sensor. The early warning device judges whether there is an abnormal condition on the road where the vehicle is located, and sends early warning information to other early warning devices within the predetermined range of the vehicle location. The early warning device sends the alarm signal of the abnormal state, the location information of the vehicle and the video information within the time period corresponding to the location of the abnormal state to the cloud server (Bai Jing & Zhao Qing 2021). When the cloud server confirms that there is an abnormal condition on the road according to the video information,

*Corresponding Author: 12017366@qq.com

DOI 10.1201/9781003305026-79

it sends early warning information to other early warning devices within the predetermined range of the vehicle location, so as to form a complete early warning process.

Table 1. Schematic diagram of urban road safety early warning system.

User layer	Command center	Traffic control scene police	Patrol Officer	Public
Application layer	Equipment management service platform	Intelligent transportation system	Construction management information system	Equipment management WeChat official account
Transport layer	Cloud data service center			
Equipment layer	Warning light		Early warning equipment	Mobile terminal

2 ROAD SAFETY EARLY WARNING ALGORITHM

Lane departure early warning algorithm, the risk assessment of lane departure, refers to using the information obtained from the sensing part, such as the motion state of the current vehicle and the geometric structure of the road ahead, to judge whether the vehicle is in danger of deviating from the lane. Whether there is danger is usually described by early warning time. An appropriate warning time should not only ensure that frequent false alarms will not cause unnecessary interference to the driver, but also ensure that an appropriate response time is reserved for the driver to take corrective measures (Chen Zhenning 2021). This is because different drivers have different physical and psychological qualities and different driving styles between people, so their perception characteristics of vehicle deviation risk are also different. Even for the same driver, his driving behavior characteristics will change with time, space, and mental state. Therefore, different types of drivers have different requirements and effects on the early warning system.

2.1 *Evaluation algorithm based on lateral velocity*

The early warning algorithm takes the lateral speed Vy of the vehicle as the evaluation index. If the vehicle deviates from the road boundary line at a relatively large speed, the system will give an early warning. The formula is as follows: $v_y > v_{th} \rightarrow$ alerting. v_{th} is the given speed threshold. This method will also lead to a false alarm, because some drivers do not closely follow the road lane line, but swing left and right on the road. At this time, the lateral speed of the vehicle will be large. For this kind of driver, this method will also lead to frequent false alarms, and the driver will not recognize the early warning function of the system (Chen Xin 2021). In addition, if the driver finds that the car deviates from the road centerline by a large distance, he will respond quickly and turn the steering wheel to make the car return to the vicinity of the road centerline. At this time, the lateral speed of the car is very high. If the alarm is given at this time, it will interfere with the driver's correction and annoy the driver.

2.2 *Key technologies of related data acquisition*

The early warning algorithm takes the lateral speed Vy of the vehicle as the evaluation index. If the vehicle deviates from the road boundary line at a relatively large speed, the system will give an early warning. The formula is as follows: $\varepsilon = x_m - x - \tau v \varphi$. Where x is the lateral position of the vehicle centroid at the current time, Xm is the lateral position of the vehicle centroid after τ seconds, which represents the vehicle yaw angle, and V represents the vehicle speed. The algorithm generally assumes that the driver can better follow the change of road curvature, so the target trajectory is usually the centerline of the driving road. At the same time, this algorithm assumes that the yaw angle of the vehicle is constant, and the predicted trajectory is a straight line.

2.3 *Early warning decision algorithm based on TLC*

The TLC method is a very popular decision algorithm in all kinds of lane departure early warning systems globally. It is the method adopted by most institutions and universities that study lane departure early warning. TLC refers to the movement time required from the current position of the vehicle to the beginning of contact between the vehicle and the lane line, which can also be referred to as the remaining time from the current time of the vehicle to the time before the vehicle deviates from the lane. The TLC method aims to identify the possible trajectory deviation in the future as quickly as possible. This method generally makes effective assumptions about the vehicle dynamics model at a specific time in the future, and finally calculates the time when the vehicle is about to cross the road boundary according to the established vehicle motion model and the correct identification of the road model ahead. The early warning algorithm based on TLC was originally proposed by GodHelp. The basic principle is that if TLC is less than the given time threshold Tth, i.e., TLC < Tth. We believe that the vehicle will deviate from the lane and trigger the system alarm.TLC evaluation algorithm can be divided into horizontal TLC algorithm and vertical TLC algorithm, which is distinguished by the different lane departure directions considered. Aurora system (Carnegie Mellon University, USA) adopts a horizontal TLC algorithm, and its formula is as follows:

$$TLC = \frac{L_P}{v_y}$$

In the formula, L_P represents the lateral position of the vehicle, that is, the lateral distance between the longitudinal axis of the vehicle and the road centerline, vy represents the lateral speed of the vehicle. The time is calculated by calculating the distance that the vehicle sign line moves relative to the vehicle in the last half-second. In this method, the road width is known and constant. The algorithm uses the information of lateral displacement and lateral speed, and considers the driving track of the vehicle, which can ensure that the alarm is given to the driver within a certain time range and reserve a certain reaction time for the driver. However, this algorithm assumes that the lateral speed of the vehicle remains unchanged in a short time interval and the heading angle of the vehicle remains constant, but this assumption is incorrect in some cases. We know that when the steering wheel angle is a fixed value, the car will travel along the arc track, so the lateral speed of the car on the road is constantly changing, and the heading angle of the car is also constantly changing.

Table 2. Decision algorithm based on TLC.

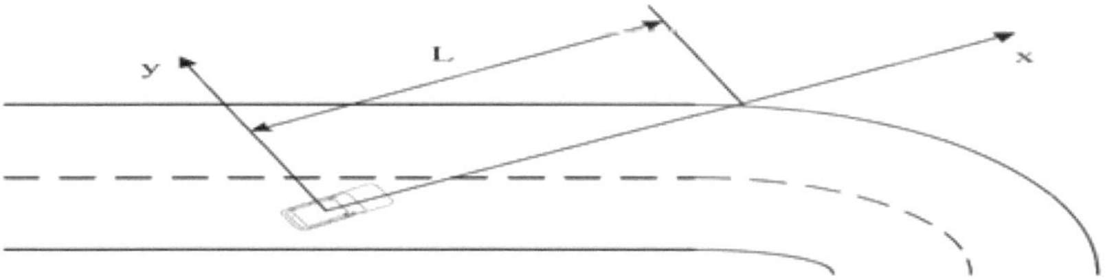

2.4 *Early warning algorithm based on instantaneous lateral displacement*

This algorithm uses the instantaneous lateral displacement L1of the vehicle center from the lane center as the evaluation index. It is a relatively simple lane departure early warning algorithm. When LP is greater than the preset threshold, the system will give an alarm. This algorithm is relatively simple and easy to implement in practical application; However, it ignores the motion track of the vehicle, especially when the motion track of the vehicle deviates from the center of the road by a distance and runs parallel to the lane, it will send out an error alarm. Usually, drivers cannot drive strictly along the road centerline, but deviate from the road centerline by a specific distance. This value can reach 10 cm at most. For drivers with such driving behavior, the early warning algorithm can lead to frequent false alarms. When the vehicle deviates from the current driving lane at a large angle, the response time left to the driver after the system sends the warning signal is too short, and the driver generally has no time to correct the lane departure behavior, so the warning sent by the system will lose its function. Both the decision algorithm based on TLC and the decision algorithm based on predicted track deviation need to extract the information of lane sign line from the road image obtained by the on-board camera, so as to accurately locate the position of the vehicle in the lane, calculate the transverse distance between vehicle and lane sign line, and then determine an optimal alarm time according to the vehicle speed information. The decision algorithm has been proved to be effective and accurate. The disadvantage is that it needs to establish the geometric model of the camera, road, and vehicle system, which is easily affected by camera selection, optical lens, camera installation position and vehicle model. In addition, these two decision algorithms need to calibrate the camera, and the results are easy to be disturbed and have poor robustness in the process of vibration and impact caused by vehicle bumps. It cannot meet the complex and changeable road traffic environment. The threshold setting does not consider the driver's characteristics and driving habits and the current driver's driving behavior. In the decision algorithm based on the system yaw angle and heading and the decision algorithm based on the sum of the left and right lane sign line angles, the angle between the lane line and the image plane coordinate axis will not change greatly, there is no need to calibrate the camera, and it is not sensitive to impact, vibration and other interference. It has strong environmental adaptability and is more suitable for use in the vehicle environment. However, there will also be false positives due to the complex external environment. From the actual driving situation, high accuracy is not required in vehicle departure warning, and drivers with rich driving experience do not need accurate lane departure values when judging lane departure. The disadvantages are a high false alarm rate, less consideration of driver characteristics and driving behavior, and low accuracy in a complex road traffic environment.

3 THE MEASUREMENT OF THE DISTRIBUTION AND EFFECT OF URBAN INFRASTRUCTURE

3.1 *Video vehicle detection technology*

At present, video vehicle detectors can be roughly divided into three types: point detector, line detector and surface detector. The first is the point vehicle detector. By setting certain detection points on the image and detecting the gray changes of these detection points, it can infer whether there are vehicles passing and the speed, density, and flow of traffic flow. Its disadvantage is that it is easy to be disturbed by the change of environmental illumination and the shadow of the vehicle itself. The second is the linear vehicle detector. It mainly includes a transverse linear detector and longitudinal linear detector. A transverse linear detector refers to a detector that draws a line perpendicular to the road direction at a specific position of the image, judges the vehicle tree and the width of the vehicle passing through the detection line by checking the gray change on the detection line, judges the vehicle type according to the width of the vehicle (Ke Wenlu 2021). If two detection lines are drawn, the vehicle speed can be calculated. The longitudinal linear detector draws a detection line parallel to the road, and then judges the length of the vehicle according to the gray change on the line. The longitudinal linear detector is often used together with the transverse linear detector, which can improve the accuracy of vehicle classification. The linear detector is easy to misjudge when the vehicle changes lanes, and it is easy to miss judgment when the vehicle distance is too short. However, because of its small amount of calculation and easy access to vehicle model, speed and other information, the video vehicle detector commonly used abroad is the linear detector. The third is the area detector. It extracts some features of vehicles in the detection area, such as area, edge, and other information, by performing operations such as edge detection on the captured image. It is easier to use this information for vehicle classification, and the accuracy can be greatly improved. Because the area detector often extracts the shaving gradient information of image gray, compared with the point detector and line detector, the influence of ambient lighting on the detection accuracy can be greatly reduced. Therefore, although the amount of computation of area detector increases greatly, due to its inherent advantages, with the continuous improvement of the operation speed of the microprocessor, area detector has become the main research direction of video vehicle detector.

3.2 *Key technologies of related data analysis*

Fuzzy control technology can solve many problems that are difficult to be solved by classical control theory. Its unique simplicity and robustness have won the general recognition of designers. However, when solving the multi-input problem, the fuzzy control technology faces the bottleneck problem of expanding the number of rules. The urban traffic early warning system is a typical multi-input monitoring system. If the traffic flow, traffic flow speed, saturation and the actual input analog value and its change of main models are taken as the input of the system, there are eight input variables, and four language values taken for each state variable. The total number of rules required in theory is 48, and the expansion of the number of rules and the difficulty of obtaining knowledge make the direct design of the fuzzy controller very difficult. However, it has a great feature that the input information can be fused. The multi-input fuzzy controller method based on information fusion can be used as a solution. Information fusion, also known as data fusion, is to make full use of multi-source information to obtain a reliable understanding and interpretation of the consistency of monitoring objects, so as to facilitate the system to make a correct response, decision-making and control. Firstly, the "fusion function" is used to reduce the dimension of input variables, and then the "fuzzy action function" is designed by using the reduced dimension input variables to realize the control of the system, which can greatly simplify the design process of the multi-input controller. The system uses BP neural network method to reduce the order of the system, uses a fuzzy neural network for overall decision-making, makes full use of the powerful nonlinear operation ability of the neural network, can combine fuzzy rules, accurately judge the environmental characteristics and their change trend based on limited measured values, finally achieve the early warning function.

Table 3. Fuzzy control technology system.

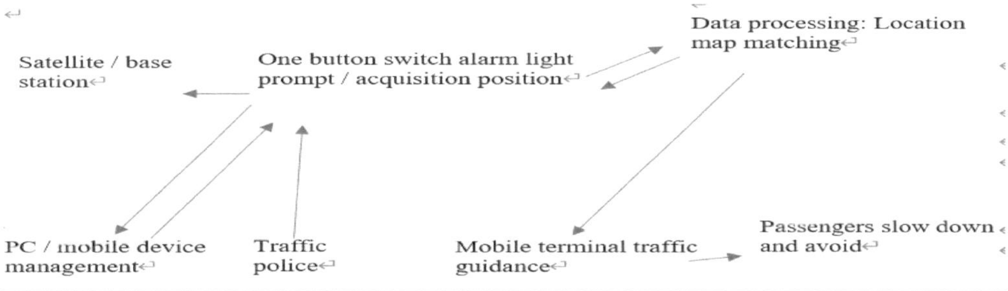

3.3 *Key technologies of related data acquisition*

The urban traffic early warning system based on machine vision technology is an extremely important link in the advanced traffic management system. Whether the early warning performance of the system is accurate or not, the key point is to accurately and timely collect all kinds of traffic information, including traffic flow, traffic flow speed, a saturation of each section, vehicle length and width, etc. At present, the commonly used vehicle detector is the ring coil detector, which can monitor the presence of vehicles and the saturation of road sections (Wei Yan 2021). By installing multiple coils on a single lane, the ring coil detector can also monitor the speed and length of vehicles. Up to now, the research and development history of this technology is long, the technology is relatively mature and stable, the detection accuracy is high, and it can work in bad weather environment, but the ring coil detector also has its inherent disadvantages. First, the ring coil detector cannot monitor multiple lanes simultaneously. If you want to know the traffic flow of multiple lanes, you can only install a set of coil detectors in each lane on the multi-lane road section. Second, when installing or maintaining the coil vehicle detector, it is necessary to block the traffic flow. Third, when two vehicles are too close, the ring coil detector will be mistaken for one vehicle. Fourth, for some special sections, such as the surface of the overpass, coils cannot be buried due to thickness constraints.

4 EARLY WARNING SYSTEM

4.1 *Lane departure warning system*

Lane Departure Warning System is a system that assists the driver in reducing traffic accidents due to lane departure by means of alarm. The lane departure early warning system provides intelligent lane departure early warning. When the driver unconsciously (the driver does not turn on the turn signal) deviates from the original lane, it can give an alarm 0.5 seconds before departing from the lane, or the steering wheel begins to vibrate to remind the driver of the current vehicle deviation, provide more reaction time for the driver, greatly reduce the collision accidents caused by lane departure. According to statistics, about 50% of automobile traffic accidents are caused by vehicles deviating from the normal driving lane. The main reasons are driver's upset, inattention or driving fatigue. The working principle of a lane departure warning system is like this: the lane departure warning system is mainly composed of a HUD head-up display, camera, controller, and sensor. When the lane departure system is turned on, the camera (usually placed on the side of the vehicle body or the rear-view mirror) will collect the identification line of the driving lane at all times and obtain the position parameters of the vehicle in the current lane through image processing. When the vehicle deviates from the lane, the sensor will collect the vehicle data and the driver's operation status in time. Then the controller will send an alarm signal. The whole process is completed in about 0.5 seconds, providing more reaction time for the driver. If the driver turns on the turn signal and changes the line normally, the lane departure warning system will not give any prompt. At present, the lane departure warning systems equipped by various manufacturers are developed based on the data collected by vision (camera), but their accuracy in collecting lane identification lines will decline in rainy and snowy weather or on roads with low visibility. In order to solve this problem, smart technical engineers have developed the acquisition method of infrared sensors, which are generally arranged on both sides of the front bumper, to analyze the road condition through infrared collection signals. Even on the road in a bad environment, they can also identify the lane sign line, so that they can timely remind the driver of the deviation state of the vehicle under any environmental road condition.

4.2 *Intelligent speed measurement and early warning system*

Intelligent speed measurement and early warning system is an Intelligent speed measurement and early warning system. It is a driving safety speed measurement and early warning software. The software system is mature and stable, and the user experience value is very high. It can quickly process the received signals according to the actual situation of the current road section, and screen the interference signals or data that does not exceed the early warning standard. The intelligent speed measurement and early warning system are mainly used on the full series of high-end driving safety speed measurement and early warning equipment. Intelligent speed measurement and early warning is the fifth-generation product. It is a high-tech driving safety early warning instrument integrating a global satellite positioning system, South Korea's top radar detection technology and intelligent speed measurement and early warning system. It has five intelligent characteristics: Intelligent filtering, intelligent retrieval, intelligent processing, intelligent upgrading, and intelligent broadcasting, which are not possessed by ordinary speed measurement and early warning.

Intelligent speed measurement and early warning can not only accurately detect the radar signal without missing or false alarms, but also filter out the interference signal. It adopts South Korea's top radar detection technology (SKS), which can retrieve radar signals 360 degrees in an all-round way without leaving a dead angle; Its intelligent speed measurement and early warning system (ISW system) can quickly process the received signals according to the actual situation of the current road section, and screen the interference signals or data that does not exceed the early warning standard. Its built-in speed measuring point data is updated in real time, and different intelligent upgrading schemes can be selected according to the needs of vehicle owners to cooperate with the global positioning system (GPS) for accurate positioning. Finally, it provides a variety of humanized and intelligent broadcasting modes for users to use in different driving environments.

4.3 *Intersection warning system*

At present, the intersection early warning equipment generally adopts the radar safety early warning system, which combines the radar monitoring system with the LED display screen. This system can be set at the intersections of urban roads, national and provincial roads, county, and township roads, at the intersections of shopping malls, communities, and hospitals, or at the level intersection of main roads and branches with visual blind areas. Among them, the radar monitoring system can adopt two radar detection and reminder methods, the first is to detect vehicles, and the second is to detect pedestrians. The working principles of the two intersection early warning systems are as follows: First, when the radar detects that the vehicle with continuous acceleration (exceeding) speed is driving towards the intersection or zebra crossing, the detection results are transmitted to the system control chip, which sends instructions to the corresponding LED display screen or screen display voice broadcaster to inform the driver of the vehicle about to enter the intersection in the warning area through jumping / eye-catching subtitles. Or pedestrians who have not yet entered the zebra crossing can even tell vehicle drivers or pedestrians at the zebra crossing to avoid vehicles through voice broadcasting. Second, when the radar detects that the pedestrian is about to enter or pass through the zebra crossing, the detection results are transmitted to the system control chip, which sends instructions to the corresponding roadside facilities such as luminous signs, LED display screens, led warning posts, led ground lights, send long-distance strong warning information to the vehicle driver by means of stroboscopic and color change. The driver shall take corresponding yield measures in advance. Third, pedestrian detection areas are generally set at pedestrian intersections or road zebra crossings and non-motor vehicle road intersections with visual blind areas. Similarly, pedestrian detection can also be combined with the text reminder of LED display screens, audible and visual alarm systems, led warning posts, led ground lamps and other roadside facilities for intersection early warning to remind vehicle drivers to "pay attention to pedestrians." Adopting the new generation millimeter-wave MMIC integrated RF scheme, combined with FMCW, MIMO and other technologies, it has the advantages of high ranging and velocity measurement accuracy, high angle measurement resolution and low false alarm rate. It can accurately monitor, accurately locate and stably track the pedestrians crossing the street in the detection area. In addition, wtr-820-millimeter-wave radar with innate anti-jamming capability also has its own startup automatic calibration and temperature compensation functions. The center frequency does not drift, which further ensures that the radar detection function is not disturbed by wind, light, temperature changes, smoke, fog, dust, and other external environmental factors.

4.4 *Curve passing warning system*

The curve early warning system is composed of a front radar detector and a rear integrated vehicle early warning device. The rear integrated vehicle early warning device is also composed of a power module, system control module, voice warning module, text warning module, speed display module and yellow flash warning module. Through the radar equipment installed at both ends of the entering curve, the driving data such as the speed of passing vehicles are collected in real time and sent to the display screen of the integrated vehicle early warning device beside the opposite vehicle road. Through the display screen, passing drivers can understand the information of opposite vehicles ahead before entering the curve. When a vehicle comes from the opposite lane of the front, a red character prompt will appear on the display screen, and a warning light will flash to remind the driver of the coming vehicle in front, display the instantaneous speed when the vehicle passes, and remind the driver to slow down. The equipment composition of the curve early warning system is installed in pairs. When a vehicle is detected passing through the radar detector installed on the equipment, the radar detector will output two groups of signals: the first group is the incoming vehicle trigger signal, which will be sent to the control card receiver of the opposite vehicle detection and warning device. After the receiver receives the signal of the radar detector, two sets of instructions are also output. The first group is the control card that sends the command warning signal to the LED screen. The LED control card displays the preset content on the LED display screen, and sends the command to the voice control card. The voice control module broadcasts the warning voice according to the preset content. The second group is the data information of the incoming vehicle speed, which is sent to the control card receiver of the opposite vehicle detection and early warning device. The receiver receives the vehicle speed data information sent by the radar detector, and sends the speed information data of the incoming vehicle to the LED screen control card after analysis and judgment. The LED control card displays the received speed information on the LED display above the vehicle detection and warning device. When there is no radar detector signal, the LED screen below the front of the vehicle detection and warning device displays a green "curve section, slow down," the LED display in the middle, the red "slow" word and "red blue flash" are displayed alternately with the upper light panel "yellow flash"; When receiving the signal from the radar detector, the LED screen at the bottom of the front of the vehicle detection and warning device displays green "incoming vehicle," the LED screen in the middle displays the speed of the incoming vehicle, the yellow light on the upper light panel flashes.

4.5 *Intelligent traffic video perception system*

Through intelligent video analysis technology, we can sort out some key information worthy of attention from the video, obtain traffic condition information through real-time analysis and processing, and make real-time responses to the road conditions on the implementation site. Intelligent video analysis technology comes from computer vision technology. Computer vision technology is one of the branches of artificial intelligence. It can establish a mapping relationship between image and image description, so that computing can understand the content of video through digital image processing and analysis. The intelligent video technology mentioned in video surveillance mainly refers to "automatic analysis and extraction of key information from video sources." If the camera is regarded as human eyes, the intelligent video system or device can be regarded as the human brain. With the help of the powerful data processing function of the computer, intelligent video technology carries out high-speed analysis of the data in the video picture, filters out the information that users do not care about, and provides users with key information useful for monitoring and early warning [5]. Using video camera equipment for traffic flow measurement and accident detection belongs to the category of vehicle monitoring. Video monitoring systems have great advantages over other sensing technologies. They do not need to deploy any equipment on the road or subgrade, so they are also called "non-implantable" traffic monitoring. When a vehicle passes by, the video captured by the black-and-white or color camera will be input into the processor for analysis to find out the changes in video image characteristics. Cameras are usually fixed to buildings or columns near the driveway. Most video monitoring systems need some initial configuration to "teach" the processor the basic background image of the current road environment. This process usually includes entering known measurement data, such as lane line spacing and camera to road height. According to different product models, a single video monitoring processor can process the video data of $1 \sim 8$ cameras at the same time. The typical output of the video monitoring system is the vehicle speed, number of vehicles and lane occupancy of each lane. Some systems also provide some additional outputs, including stop vehicle detection, wrong driving vehicle alarm, etc. In the case of complex background, it can also better complete the segmentation of moving area and can effectively overcome the dynamic changes of the background image, such as weather, illumination shadow and chaos interference. Fusion of shape information and motion characteristics: combine the shape information and motion characteristics of moving targets, distinguish people, vehicles and chaotic disturbances using dispersion and area information. Compared with the motion of the rigid object, the motion of a non-rigid object has a higher average residual optical flow, and it also presents the characteristics of periodic motion.

Table 4. Intelligent traffic video perception system.

			Start			
			Capture video			
			Loading parameters			
Extracting feature points of projection sequence	Vehicle target extraction	Detect abnormal area		Focus extraction matching tracking		Target segmentation and morphological analysis
Look up the mapping table for trajectory fitting	Target matching	Bidirectional tracking		Trajectory direction analysis	Trajectory velocity analysis	Target tracking and trajectory analysis
Calculate vehicle speed	Statistical traffic flow	parking	Dropped object	Retrograde	traffic jam	pedestrian
			Send event and parameter information			
			End			

5 CONCLUSION

At present, the integrated application and unified management platforms of intelligent transportation multi-sensor data fusion mostly focus on B / S architecture and take the portal, large screens, and other terminals as display carriers. It is a professional monitoring and management platform integrating intelligence, networking, functionalization, and integration, which provides powerful management functions for massive monitoring of video and audio data based on an IP network. The system adopts a modular architecture design, and has special functional modules for system core data and services, storage and forwarding, alarm linkage, decoding display, and video analysis. Using the unified database, an interface service platform is established between the decentralized system and users to complete the unified management, alarm linkage and centralized control of all monitoring equipment and security subsystems in the system. The system has a comprehensive design instability, integration, maneuverability, data security, storage efficiency and reliability, and can be flexibly combined according to needs to meet the needs of projects of different scales.

REFERENCES

Bai Jing, Zhao Qing. (2021). Research on traffic congestion control in Qinhuangdao city from the perspective of intelligent transportation. Smart city. 117–118.

Chen Zhenning. (2021). Smart city concept and future urban development. Data.73–74.

Chen Xin. (2021). Design and use of intelligent traffic management system. Transportation world. 125–126.

Ke Wenlu. (2021).Coven road. Problems and solutions in the construction and development of intelligent transportation in Quanzhou. Transportation world.8–9 + 20.

Wei Yan. (2021).Preliminary discussion on key points of intelligent transportation design. Building electrical. 7–11.

Advances in Urban Engineering and Management Science – Khalil & Yang (Eds)
© 2023 The Author(s), ISBN 978-1-032-30426-7

A review and insight of vosviewer-based China community renewal-related studies

Senyu Lou*
Faculty of Humanities and Arts, Macau University of Science and Technology, Faculty of Construction and Engineering, Jinhua Polytechnic, Zhejiang, China

ABSTRACT: Community regeneration is an important topic of current urban renewal in China, yet there is not enough current experience to deal with it. Therefore, this paper uses Vosviewer software to sort out and analyze the domestic literature related to community renewal in CNKI in the past ten years through the bibliometric method. Firstly, we review the concept, development history, and current status of community planning; then, we organize and summarize the community renewal models under the new perspective; finally, we analyze the current mechanism of community renewal. Based on the bibliometric approach, it is found that the research on community renewal in China suffers from a single research perspective, outdated renewal mechanisms, and a lack of quantitative research. This paper aims to provide an effective reference for future research on community renewal in China by sorting out and summarizing the research content of the literature in the past decade.

1 INTRODUCTION

After rapid expansion, Chinese cities have experienced a significant increase in urbanization rate and a large improvement in the urban environment, but the mismatch between community conditions and living needs has led to a series of social problems such as livelihood and environment. In addition, as China's urban and rural planning enters the era of stock planning and renewal planning, urban built-up areas, including communities, have become a new research hotspot (Journal of Urban Planning 2016). In the face of this transformation, cities return to the community at a basic level, leave behind the grand narrative of physical space planning, and begin to follow a new human-centered planning concept (Yu & Cao 2013). However, urban planning has long focused on public facilities and physical environment. In the design of urban settlements, the objects are often residential areas or neighborhoods rather than "communities," and there is little interaction with community members, which can be said to be "seeing things but not people." This situation is related to the immaturity of China's community organizations and the lack of a sound public participation system. In the face of the diversification of community subjects and the complexity of related interests, the existing urban planning field lacks sufficient experience to deal with the problem. Therefore, this paper reviews the literature on community renewal in China in the past ten years, summarizes the current status of research in this field, and outlooks research trends, to provide a reference for future research in the field of community renewal in China.

*Corresponding Author: 2109853zat30001@student.must.edu.mo

 DOI 10.1201/9781003305026-80

2 ANALYSIS METHODS AND DATA SOURCES

Vosviewer is one of the most common knowledge mapping tools, and it has been applied in the field of urban renewal in recent years. In this paper, we searched "community regeneration" as the subject term in China Zhiwang Journal Database from 2011 to 2021, and obtained 659 valid documents by reading the titles and abstracts.

3 LITERATURE CHARACTERIZATION

3.1 *Analysis of the number of articles issued over the years*

In terms of the time and quantity of literature development (Figure 1), the number of publications in the field of community renewal in China shows an overall upward trend [1]. 2011-2013 is the initial development stage, with annual publications below 20. 2014-2016 is the slow development stage, with overall fluctuating growth, with annual publications reaching 33 in 2014 and 23 and 24 in 2015 and 2016, respectively. This represents that as China's urban and rural planning enters the era of stock planning and renewal planning, urban built-up areas, including communities, become a new research hotspot. 2017-2022 is a phase of sustained growth, with a rapid increase in the annual number of articles compared with the previous phase, and the average annual number of articles reaches more than 100; this shift indicates that cities return to the basic level of communities, leave behind the grand narrative of material. This shift indicates that cities are returning to the basic level of community, leaving behind the grand narrative of physical planning and starting to follow a new concept of human-centered planning (Yu & Cao 2013).

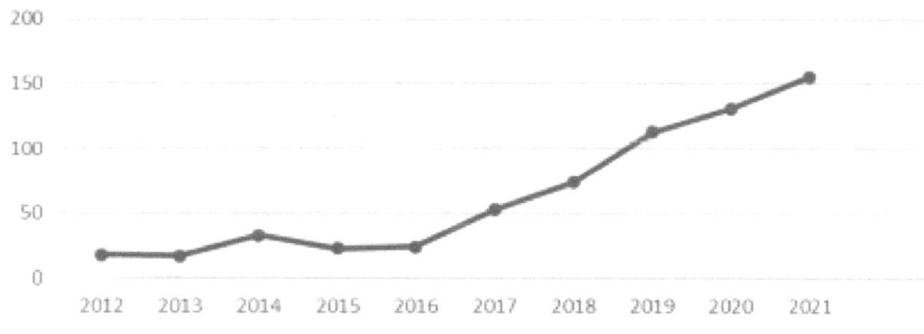

Figure 1. Analysis of the number of articles published in community renewal-related papers over the years.

3.2 *Literature distribution and authorship analysis*

As shown in Figure 2, Chinese community renewal research is mainly concentrated in the discipline of building science and engineering (80.49%) and is involved in the fields of Chinese politics and international politics, macroeconomic management, sustainable development and sociology and statistics, but there is less interdisciplinary integration. From the source of journals, the most published article in domestic general academic journals is Urban Architecture, and the core journal is Planner. Through the statistics of all authors, the author with the most publications is Professor Huang Ling from Chongqing University, who has published ten papers related to community renewal, but the analysis of the authors' cooperation network reveals that the degree of cooperation among the authors is low, and their respective research fields are relatively independent, without forming a core group of authors (Figure 3).

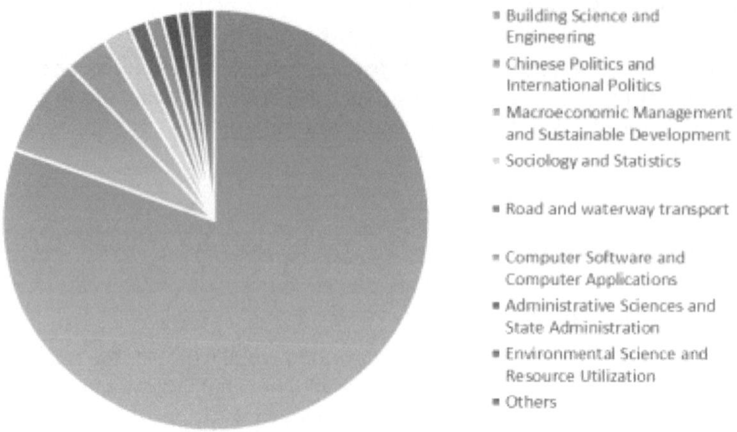

Figure 2. Community renewal literature subject distribution.

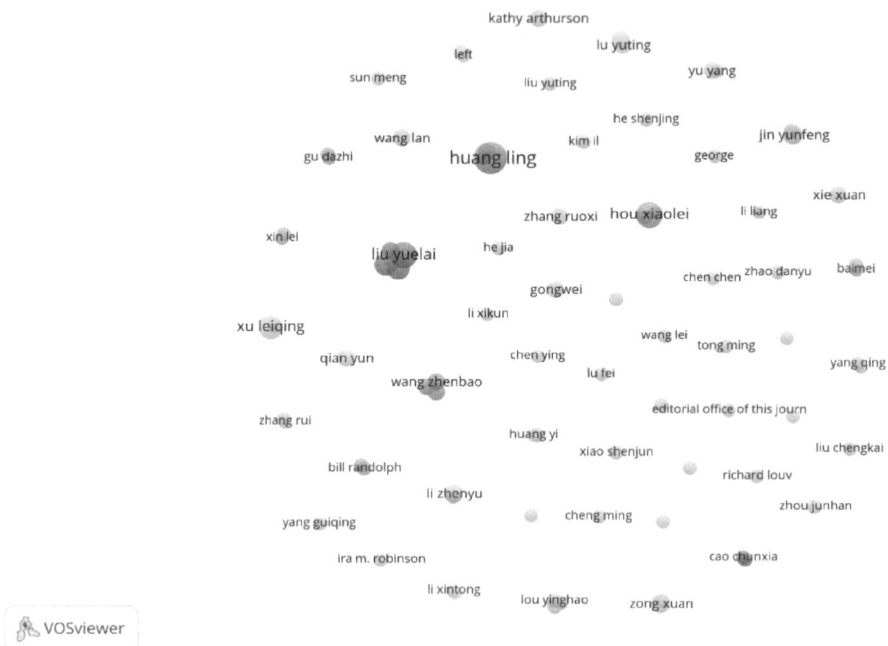

Figure 3. Author collaboration chart.

3.3 *Analysis of high-frequency keywords and research hotspots*

Based on Vosviewer software to analyze the keywords, from the co-occurrence density view (Figure 4), it can be intuitively seen that the research hotspots in the field of community renewal in China are urban renewal, old communities, community renewal, micro-renewal, and public participation. Following that, keywords with word frequency greater than and equal to 3 times were counted, and the data were cleaned and merged based on preliminary statistics to obtain a total of 153 keywords, which were clustered and analyzed based on Pajek software to obtain the keyword co-occurrence label view (Figure 5).

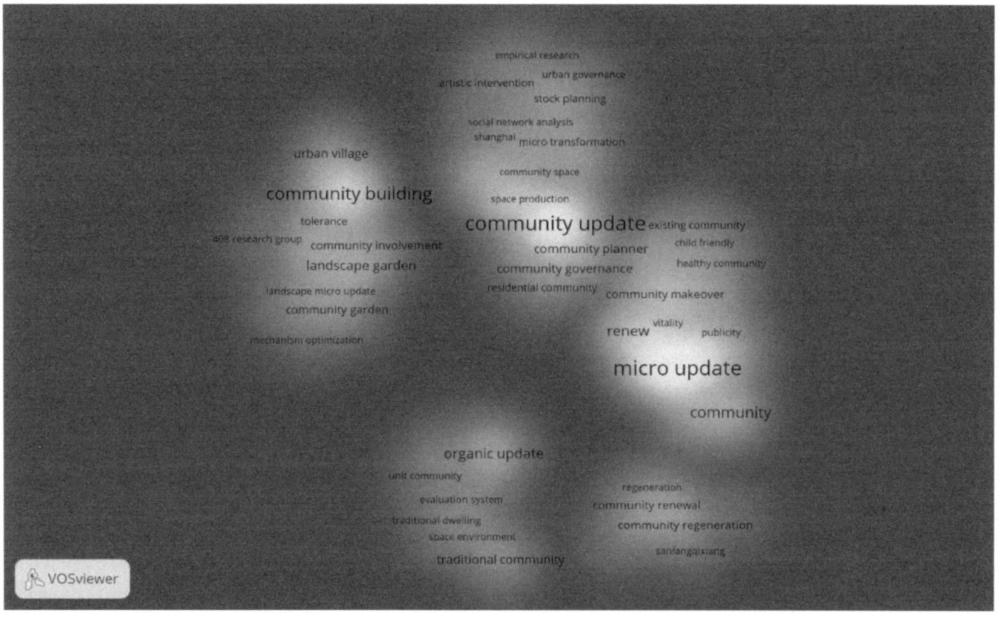

Figure 4. Keyword co-occurrence density view.

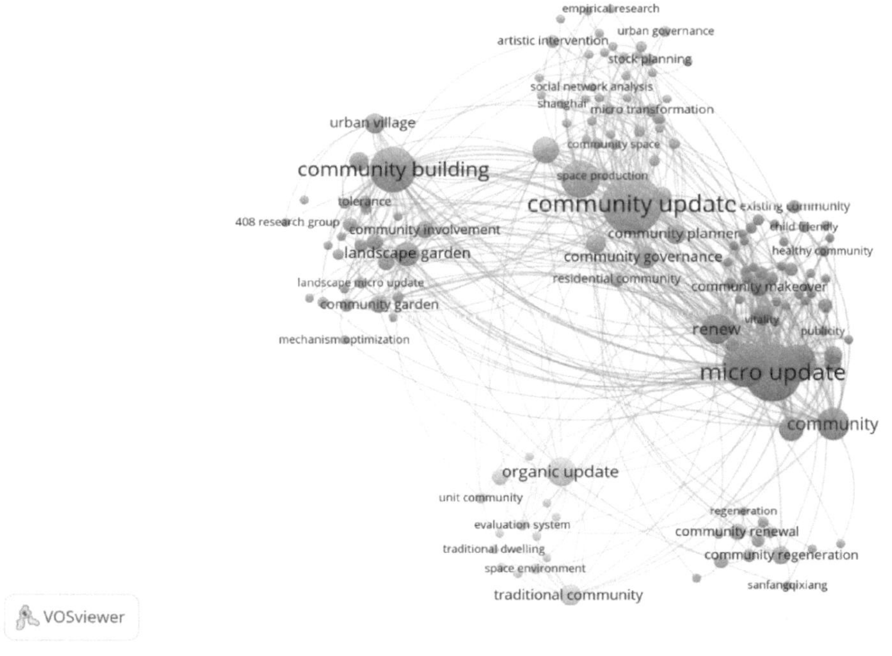

Figure 5. Keyword co-occurrence tag view.

In the keyword co-occurrence label view, the larger the circle area represents, the higher the word frequency of the keyword, and the circles of the same color belong to the same research cluster. The thickness of the connecting lines represents their association strength, and it can be seen from the figure that urban renewal, community renewal, and micro-renewal nodes are the largest and have the highest word frequency. In the following section, we will interpret the research content in the context of the literature.

4 RESEARCH CONTENT ANALYSIS

4.1 *Community*

The concept of community has its origins in sociology and is defined in a complex way. The British scholar Mayne first considered villages as communities, and in 1887, the German sociologist Ferdinand Tennis proposed in his book "Community and Society" that community is an abstract conceptual description of a group of people established through blood, neighbors, and friends, and is a type of relationship (Tennessee & Lin 1999). The concept of "community" in China was first proposed by Professor Wu Wenzao, and his student, Mr. Fei Xiaotong, defined community in his Introduction to Sociology as a number of social groups (families, ethnic groups) or social organizations (institutions, groups) gathered in a geographical area to form a large interrelated collective in life. Modern urban planning theory considers a community to be a spatial complexity of spatial, demographic, structural, and psychosocial elements with clear territorial boundaries and a living, economic, and public activity environment centered on housing (Li et al. 2012), i.e., a community is a social unit based on a certain territory and value identity. Although the community has many definitions, its three essential attributes are territoriality, commonality, and interaction (Pang 2017), they are all places with some spatial and social characteristics in the suburbs of cities or towns, and there are differences between communities (Lowe 2011).

4.1.1 *Community development*

In China, during the period 1952–1979, the concept of "community" was replaced by the "unit system" due to the constraints of the administrative and planned economy, and the unit settlements formed based on employment ties had inherent attributes of a community, such as a sense of place and social identity. In 1986, when communities were established in China, social functions previously undertaken by the government and state-owned enterprises, such as elderly care, health care, and education, were returned to communities, which gradually became the name of urban and rural grassroots management units and were linked to certain territorial jurisdictions and organizations. A community is defined as a community of social life formed by people who live together within a certain geographical area. Yu Ying and Cao Chunxia classify the types of urban communities in China into six types: communities of "agricultural conversion," communities of "urban villages," communities of foreign populations, communities of old city transformation, communities of unitary transformation and newly built residential communities.

4.1.2 *Community planning*

Although large-scale urban renewal has had some positive effects, such as concentrating scattered land uses, alleviating traffic problems, and remediating slums (Lowe 2011), there are significant limitations to solving urban problems solely through the improvement of physical space. William Lowe reviews a century of community planning in the United States and suggests that its insights lie in making planners aware of the importance of community networks and local relationships, the need to institutionalize the participation of local residents, and how to relate economic, environmental, and other related topics on the basis of planning. The current community planning in China is at a specific point in time, when the development of urban renewal in China has entered a period of concentration and urban governance and national discourse of remediation, and has the necessary conditions to move toward a flourishing social time (Huang 2020).

At the beginning of the 21st century, community planning was mainly residential planning, which was in the incremental stage. Then it was oriented to both new community development and old community renewal, the latter providing improvement solutions for existing communities; current community planning faces new challenges, such as age-appropriate, all-age, low-carbon renewal, historical conservation, and other new perspectives also drive new community renewal models.

4.2 *Community renewal from different perspectives*

4.2.1 *Community renewal under the influence of intergenerational conflict*

In recent years, with the emergence of "intergenerational conflicts" in the use of community public space, community planning and construction have begun to reflect on the differences in the demands of age groups brought about by changes in age structures. Some scholars and organizations have proposed the goal of building an "all-age community" and strengthening the renewal of public space in all-age communities through inclusive design theory in terms of rational allocation of spatial resources, creation of micro spaces, weaving of pedestrian networks, and improvement of spatial accessibility (Hu et al. 2016). Based on this reality, physical renovation and soft services are combined to clarify the focus of all-age renovation in terms of physical and mental care, travel convenience, interaction and companionship, and participation promotion (Xu & Hu 2017).

4.2.2 *Community renewal approach in historical and cultural context*

Although the commercial and tourism-led regeneration of historic communities has played a positive role in economic revitalization and landscape restoration, the spirit of the community is declining, and the community relationship is gradually alienated. Therefore, the primary goal of regenerating historic communities is to inherit the community culture and promote the return of community spirit (Zhou & Liu 2013). Yuan Yuan and Xiao Dawei proposed the concept of "community activation," which not only improves the living space, but also reprograms the social activity pattern, and finally achieves the purpose of revitalizing the surrounding areas (Xia et al. 2020). Under the guidance of rural revitalization strategy, the general rule of traditional village historical and cultural preservation and community revitalization is to focus on the community and people based on the development of the overall regional value framework, to stimulate the motivation and potential of the community, and to use the catalytic effect to continuously and gradually promote the large-scale change of the community landscape (Chen et al. 2021).

4.2.3 *Eco-health-led community renewal*

Low carbon and low energy consumption represent the fundamental concept of contemporary habitat construction, and communities have received extensive attention as the basic unit for implementing healthy city strategies. Wang Yi proposed that community planning under a healthy city orientation should focus on meeting activity needs and protecting the ecological environment, and on this basis, he proposed evaluation and planning strategies for healthy communities. The "Sustainable Low Carbon Demonstration Community" is a large scale practical exploration in the field of community low-carbon renewal in Taiwan Province, targeting completed communities. The main content is to develop low-carbon improvement projects through community carbon emission measurement, conduct investment estimation and benefit analysis, and develop phased implementation guidance. Lin Jian and Ye Zijun proposed some suggestions for green urban renewal in the 14th Five-Year Plan period, including the exploration of green zero-carbon community renewal. It is also a new research proposition to build a healthy community governance model and create a resilient community service space in response to the epidemic problem.

4.3 *Community update mechanism*

The public participation mechanism emphasizing "bottom-up," which is dominated by the field of sociology, has brought new challenges to the traditional planning field in China. Some new planning concepts have emerged, such as advocacy planning and communicative planning, which are all types of bottom-up planning (Yu & Cao 2013). Guided by the theory of communicative planning, L. W. Zhang and W. Y. Yang proposed to fully expose the contradictions and conflicting interests of all parties in practice, to form a negotiated plan that meets the requirements of all parties and is implementable. Li Bin and Xu Xinyan argue that formal public participation with legal constraints is the foundation of urban renewal, while informal public participation plays a complementary and corrective role; through a comparison of Chinese and German community renewal cases, they propose that public participation can enhance the rationality of renewal policies, effectively promote the implementation of renewal, and realize public demands to assist the government in implementing renewal activities (Li et al. 2012). Jian Zhuo and Yuanduo Sun analyzed the reasons for the current dilemmas faced by community renewal from three aspects: power structure, renewal model, and public spirit, and proposed strategies for the participation of multiple parties in community co-governance.

5 COMMUNITY MICRO UPDATES

Community micro-renewal has entered the academic field in recent years, which aims to improve spatial quality through the improvement and maintenance of buildings, reuse of fragmented land and low-utility land, and cultivation of community culture. Wang Chenghui divided community micro-renewal into two circles: the micro-renewal project action circle and the micro-renewal institutional environment circle, and classified community micro-renewal. Huang Tiling and Shen Mo took the Zhangjiayuan community in the Yuzhong District of Chongqing as an example. Following the concept of "asset-based" community development, they sorted out the characteristics of four types of assets: physical, social, human, and cultural in the linear space of mountainous urban communities, and explored the micro-renewal path of community linear space asset extraction and activation and community power cultivation from both spatial and governance levels. Li Hao, starting from the evolution and reality of inner-city street communities, reveals the problematic representation of the loss of publicness in community public spaces, and proposes a collaborative and inclusive regeneration governance paradigm and micro-renewal strategies for community public spaces based on the Chinese transformation of the connotation of "publicness."

6 CONCLUSION AND OUTLOOK

Based on the above review, the current research on community renewal has three main shortcomings: a single research perspective, a renewal mechanism to be innovated, and a lack of quantitative research. This paper first reviews the concept of community, its development history, and the current status of planning, then organizes and summarizes the community renewal models under the new perspective, and finally analyzes the current mechanism of community renewal. It is expected to provide an effective reference for future research on community renewal in China.

ACKNOWLEDGMENT

The authors appreciate the funding support from "Point axis village update policies based on industry-oriented research and practice of" 2020 Annual Housing and Construction Office, Zhejiang Province, the construction of research projects (No. 2020K135) and the 2020 Zhejiang Provincial Department of Education's scientific research project "Research on the Practice of Vigorous Country Construction from the Perspective of Adaptive Renewal" (No. Y202043245).

REFERENCES

Chen Ye, Zhang Shangwu, Shi Yu, et al. A study on the regeneration and governance of Shanghai's old neighborhoods from the perspective of aging: A case study of the workers' new village community in Lane 396, Changning Road, Shanghai [J]. Urban Development Research, 2021, 28(01): 39–44.

Ferdinand Tennessee, Lin Rong Yuan. Community and society [M]. The Commercial Press. Beijing, 1999.
Hu GY, Huang JC, Niu Qiang. A review and insights of research on community service facilities in the context of aging [J]. Urban Development Research, 2016, 23(02): 78–83.
Huang Yi, Wu Changfu. An analysis of community planning in China based on urban renewal and governance: an example of community planning in Jin Yang Xincun Street, Pudong New Area, Shanghai [J]. Urban Development Research, 2020, 27(04): 110–8.
Journal Editorial Board. Are planners ready to go into the community? –A Colloquium on Community Development and Planning Change [J]. Journal of Urban Planning, 2016, 05): 1–8.
Li B, Xu Xinyan, Shao Yi, et al. Study of public participation models in urban renewal [J]. Journal of Architecture, 2012, S2): 134–7.
Pang Guoyu. Research on the model of non-governmental organizations' involvement in urban community planning [D]; Zhejiang University, 2017.
William Lowe, Chun Zhang. From local to global: 100 years of community planning in the United States [J]. International Urban Planning, 2011, 26(02): 85–98+115.
Xia Dawei, Lin Yuyun, Chen Yi'an, et al. Study on the renewal model of old urban communities in the context of empty nesting [J]. Urban Development Research, 2020, 27(05): 87–93.
Xu Zhiqiu, Hu Huiqin. Exploration of methods to transform old houses into community elderly facilities [J]. Journal of Architecture, 2017, 03): 95–101.
Yu Ying, Cao Chunxia. Innovation in urban community planning and management [J]. Planner, 2013, 29(03): 5–10.
Zhao Min. Research on "community creation" and "community orientation" of urban planning [J]. Planner, 2013, 29(09): 5–10.
Zhou Yanmin, Liu Jiayan. Aging-friendly design of outdoor environment in residential areas [J]. Journal of Architecture, 2013, 03): 60–4.

Advances in Urban Engineering and Management Science – Khalil & Yang (Eds)

Modeling cascading failures in wireless mesh network technology-based communication networks

Xinyu Wang*
Business School, Sichuan University, Sichuan, China

Wei Wei*
School of Architecture and Environment, Sichuan University, Sichuan, China

Mochi Shi*
Computer Science School, Sichuan University, Sichuan, China

Ziqiang Zeng*
Business School, Sichuan University, Sichuan, China

ABSTRACT: Emergencies often disrupt communications systems and disrupt other rescue and recovery efforts. When the communication system of certain signaling nodes is attacked by emergencies, it can lead to a base station fault, and connection and communication will avoid the failure of the base station, then incorporated into the adjacent base stations, leading to the adjacent base station capacity surged, more than its maximum capacity, which causes the fails, the failure of dynamic transmission cascading failure can lead to communication systems. Based on the bidirectional transmission of a communication, the communication system has dual requirements of receiving and feedback, and traditional fault modeling of communication system is difficult to meet the dynamic failure of the communication system completely. Therefore, based on the wireless mesh network technology, this paper conducts a simulation analysis of the cascading failure process of communication systems. It can be found that the cascading failure propagation speed is faster when the key nodes are attacked than when the signal path is attacked.

1 INSTRUCTIONS

1.1 *Research background*

Earthquakes, tsunamis, and other emergencies often cause serious damage to the infrastructure system, resulting in significant economic losses and casualties. In 2004, the aquifer and groundwater regime was affected by the earthquake and tsunami of December 26, particularly on the islands and coastal regions of India. Thailand's flood of 2011 was devastating for the communities and inhabitants of the country, affecting approximately 13 million people and causing damages totaling THB 1.43 trillion (46.5 billion USD).

In particular, in disaster emergency rescue work, the communication system plays a very important role (Del Re et al. 2013). Wireless telecommunications represent an important asset for public protection and disaster relief (PPDR) organizations as they improve the coordination and the distribution of information among first responders in the field.

It can be said that an emergency information system is an important foundation of disaster emergency system construction; it plays a vital role in emergency management and emergency response process.

*Corresponding Authors: scu_nwangxinyu@163.com, qqbsqq@163.com, aatroxshi@outlook.com and zengziqiang@scu.edu.cn

DOI 10.1201/9781003305026-81

1.2 *Literature review*

Cascading failure, as a common dynamic phenomenon of complex networks, was first proposed by Newman et al. in the study of the power network. Many scholars have done a lot of research on cascading failures of infrastructure networks, including communication, power grid, transportation, water supply, and other fields. In the study of the cascading failure of the communication network, most scholars adopted the secondary control to study the cascading failure process of the communication system and proposed that the cascading failure develops dynamically. Holme and Kim constructed a node cascading failure model based on an evolutionary network. Motter et al. proposed an ML model when a node in the system is attacked by the outside world, or the load of the node exceeds its maximum carrying capacity, or the node fails, it will be removed from the network.

However, in an emergency, with the failure of a node, the functions of the node will usually be taken over by other nodes, resulting in changes in the functions of relevant nodes. It is not possible to simply remove the failed node, and the overall network structure can be described. Therefore, this paper assumes that the occurrence of sudden disasters will lead to a surge in the communication path demand of the affected nodes. In this paper, the cascading failure caused by the demand surge is simulated by changing communication path pairs. At the same time, it is assumed that the impedance of the adjacent links at the accident site becomes very large, indicating that the communication will avoid choosing the failure path.

1.3 *Wireless mesh network technology*

A wireless mesh network, also known as a "multi-hop" network, is a new wireless network technology completely different from the traditional wireless network. The idea is to let each node in the network send and receive signals, solve the traditional WLAN technology application with low scalability and poor robustness, and many other problems. WMN consists of multiple distributed network nodes that are connected wirelessly. The device deployment must be planned, and the device remains stationary for a long time after the deployment. The change of WMN logical topology is usually caused by the interference of its wireless links; that is, wireless interference can lead to the dynamic behavior of WMN logical topology.

Meanwhile, data in a wireless mesh network can jump from node to node, so that nodes too far apart to receive each other's data directly can still communicate and exchange data. Therefore, there is no single point of failure in its network. As long as there is an effective path, no matter how far they are, each wireless node can communicate with other wireless nodes on the network; that is, the network can be formed and restored automatically.

2 RESEARCH METHODS

The cascading failure response in a communication network system is slightly different from that in the traffic network system. Communication network system is composed of communication base stations as nodes connected through wireless signals. This study mainly considers wireless communication base stations, including 5G and 4G communication base stations. The coverage area of a 5G base station is about 250 meters, while the coverage area of a 4G base station is about one kilometer. The specific coverage area is affected by the power of the base station, signal quality, geographical location, and other factors. The communication base station has a limited load capacity. Usually, the maximum load capacity of a 5G communication base station is about 250 communication devices; that is, about 250 mobile phones can be allowed to talk at the same time. If the number of communication equipment loaded by the base station is lower than its maximum load capacity, the probability of failure is very low. However, if the number of communication equipment it is loaded exceeds the maximum load capacity, the probability of failure will be greatly increased. This is why, after emergencies such as the Wenchuan earthquake in 2008, the communication network system in Chengdu was paralyzed for a time as most of the communication base stations in some densely populated areas were overloaded due to the surge in communication demand.

To solve this problem, it is necessary to expand the capacity of the communication network system. In emergency rescue, mobile communication base station technology, such as UAV communication base station, and vehicle-mounted communication base station, can be used to expand the communication capacity. The advantages of a UAV communication base station are fast flight speed, not being limited by ground traffic congestion, flexibility, high mobility, and wide signal coverage. China's research and development of Wing Loong II UAV communication base station coverage radius can reach about 4 kilometers. The problem is that the communication capacity that drones can carry is very limited and of short duration. The Wing Loong II UAV communication base station can only operate continuously for 35 hours. Vehicle-mounted communication base stations can carry relatively large communication capacity. They can work continuously for a long time, but their mobility is affected by traffic congestion, and their flexibility and mobility are weak. The signal coverage is not as good as that of UAV communication base stations. Another way to solve this problem is to use wireless mesh network (WMN) or multi-hop mesh network (MHMN) technology. In the case of emergency rescue, the equipment supported by this technology can be set up quickly and can transfer the excess load of the overloaded base station to the adjacent base station, to realize the allocation of the communication surge demand across the base station, and reduce the risk of the base station failure to a certain extent. But this approach is also of limited use when communication demand surges are particularly large.

Based on the above discussion, the cascading failure model of the communication network system is constructed as follows. This study assumes that the excess load of overloaded base stations can be equally transferred to adjacent base stations so that the excess load can be spread as widely as possible among as many base stations by wireless mesh network technology. If the excess load of the overloaded base station can only be transferred to the adjacent base station that is not overloaded, the excess load may not be transferred. For example, all the adjacent base stations of a certain base station are overloaded, which will greatly weaken the effect of the amortization of excess load.

$L_i(t)$ said the communication load generated by base station i in its area at time t, $L_i^s(t)$ represents the communication load generated by base station i in its region at time t, $\sum_{k \in A_i} (L_k^s(t) - C_k)/D_k$ represents the sum of apportioned load received by base station i from its adjacent overload base stations at time t, where, D_k represents the output degree of the adjacent base station k as a node, that is, the number of adjacent base stations of base station k, C_k represents the maximum liability capacity of base station k, A_i represents the collection of adjacent overload base stations of base station i, that is, only when $L_k^s(t) - C_k > 0, k \in A_i$.

$$L_i(t) = L_i^s(t) + \sum_{k \in A_i} (L_k^s(t) - C_k)/D_k \tag{1}$$

At this point, if the base station I is not overloaded, the load of t+1 at the next moment remains unchanged. If base station I is overloaded, it will fail with A high probability. If it, fortunately, does not fail, its load at the next moment T +1 will become C_i, and the excess load will be allocated to adjacent base stations at t+1. Therefore, there is the following state transition equation:

$$L_i(t+1) = \begin{cases} L_i(t), \text{if} L_i(t) \le C_i \\ C_i, \mathit{if} L_i(t) > C_i \end{cases} \tag{2}$$

Different from traditional methods, a probability measure is used in this study to describe base station failure. It is assumed that the load of base station i is $L_i(t)$, which will affect the failure probability of the base station P_i. Construct the base station i failure probability function $f_i : R^+ \to [0, 1]$, which is used to describe the base station i failure probability $P_i = f_i(L_i(t))$, and set the failure function to meet the following basic assumptions:

- In the normal working state of each base station, i.e., its load capacity does not exceed the maximum load capacity, there will be a low prior failure probability, which is mainly affected by the service life of equipment, manual operation failure, and other factors.

- The maximum load capacity of base station i can be regarded as a threshold value. A load of the base station has no influence on the failure probability under the condition that the threshold is not exceeded, but the failure probability will increase sharply once the threshold is exceeded.
- When $L_i(t) > C_i$, the failure probability function is assumed to be A monotone function that grows as an S-shaped curve.
- The upper bound of the failure probability function f_i is 1, and $\lim_{L_i(t)\to\infty} f_i(L_i(t)) = 1$.

Based on the above basic assumptions, the sigmoid function was used to construct the failure probability function as follows:

$$f_i(L_i(t)) = \begin{cases} P_i^0, \text{ if } L_i(t) \le C \\ \frac{1}{1+(1/P_i^0-1)e^{-(L_i(t)-C_i)}}, \text{ if } L_i(t) > C_i \end{cases} \quad (3)$$

Figure 1 shows a graph of the above failure probability function. In practice, the parameters of the failure probability function corresponding to different base stations may be different, so the failure probability may be different under the same load level.

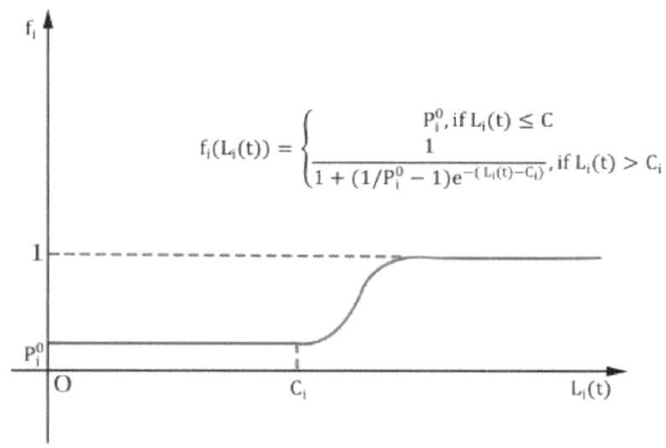

Figure 1. Failure probability function of communication base station.

3 EMPIRICAL RESULTS AND DISCUSSIONS

To simulate the cascading failure response of the communication network system, the communication base station simulation analysis network, as shown in Figure 2, is constructed, with a

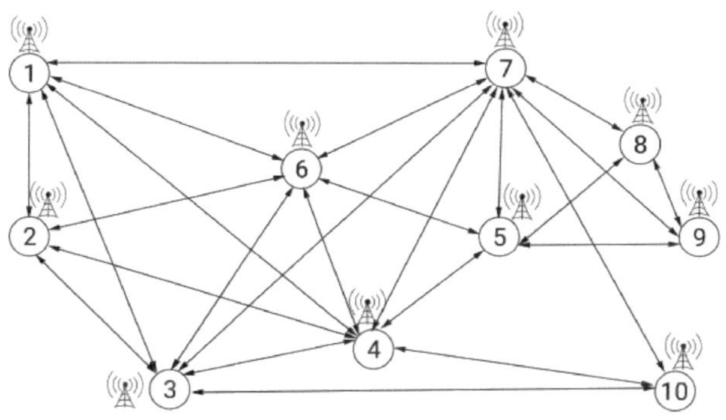

Figure 2. Communication base station simulation analysis network.

total of 15 communication base stations and 35 adjacent relationships. If there is an adjacency relationship between two communication base stations, it means that in the case of communication demand overload, the excess load of the overloaded base station can be transferred to the adjacent base station through wireless mesh network technology, to realize the allocation of communication demand surge across base stations.

Table 1 shows the basic parameters of the communication base station, including the maximum load capacity (unit: station/period), the population of the covered area (unit: ten thousand people), and the prior failure probability of the base station equipment (unit: %).

Table 1. Basic parameters of communication base station.

Base station number	Maximum load C_i (Station/time period)	The population of the covered area (ten thousand)	Prefailure probability P_i^0(%)	Thenodes D_k
1	300	8	5	4
2	350	12	10	4
3	350	11	8	5
4	300	7	9	6
5	350	15	6	5
6	450	20	5	6
7	250	4	10	8
8	350	10	2	3
9	350	13	4	3
10	300	9	1	3

In this study, the simulation process is divided into ten periods. Assuming that an emergency occurs in the third period, the communication demand of each communication base station in each period is shown in Table 2.

Table 2. The communication demand of each communication base station in each period.

Base station number	Communication demand of each period (station/period)							
	1	2	3	4	5	6	7	8
1	60	60	300	800	800	700	600	700
2	80	120	360	750	950	850	800	700
3	60	70	200	350	500	800	900	700
4	50	50	250	400	600	700	900	800
5	80	80	300	400	700	900	1000	900
6	90	120	500	900	1500	1500	900	900
7	30	40	90	200	300	400	400	300
8	60	70	250	350	600	800	800	700
9	65	80	70	500	700	800	650	550
10	60	50	50	350	500	700	600	500

Based on the above data, the increment of communication demand in each period can be obtained, as shown in Table 3.

Table 3. The increment of communication demand in each period.

Base station number	Incremental communication demand of each period (station/period)							
	1	2	3	4	5	6	7	8
1	\	0	240	500	0	−100	−100	100
2	\	40	240	390	200	−100	−50	−100
3	\	10	130	150	150	300	100	−200
4	\	0	200	150	200	100	200	−100
5	\	0	220	100	300	200	100	−100
6	\	30	380	400	600	0	−600	0
7	\	10	50	110	100	100	0	−100
8	\	10	180	100	250	200	0	−100
9	\	20	−10	430	200	100	−150	−100
10	\	−10	0	300	150	200	−100	−100

The process of base-station cascading failure in the communication network system can be simulated and analyzed using the POLARIS platform (Figure 3). The cascading failure process described in the figure starts from period 3 and lasts to period 5, a total of 3 periods, until all the base stations fail.

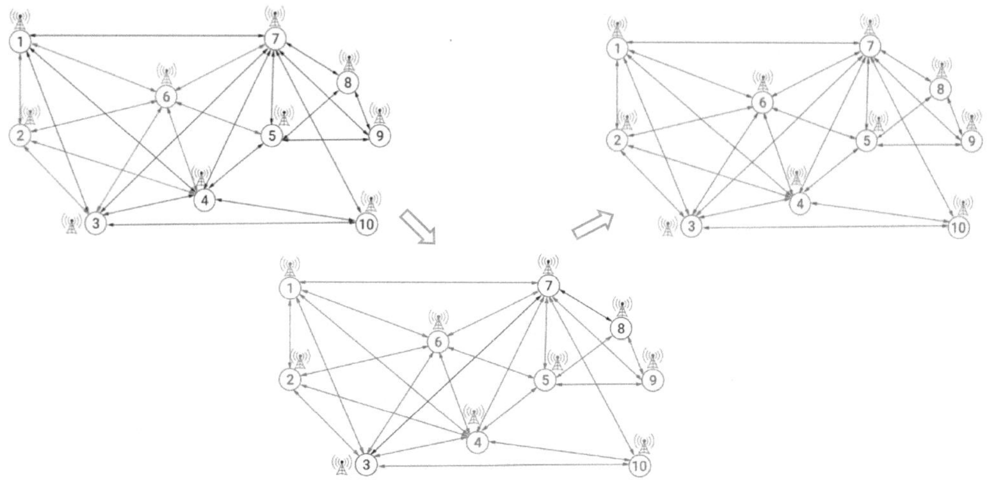

Figure 3. Simulation results of the dynamic process of communication base station cascading failure.

Figure 4 shows the number of cascading failures of communication base stations in each period. It can be seen that about 50% of the base stations fail in time period 4. The main reason is that the surge of communication demand needs a process after the occurrence of an emergency. The surge of communication demand broke out in 4 periods, resulting in the breakdown of a large area of communication base stations.

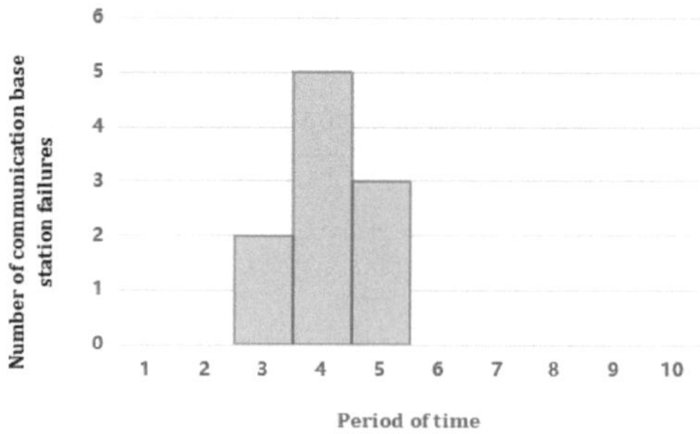

Figure 4. Communication base station failure time distribution.

4 CONCLUSIONS

The cascading failure response of the communication network system mainly depends on the dynamic communication demand and the maximum load capacity of the communication base station. When the load of the communication base station exceeds its maximum capacity limit, it will fail with a high probability, and this failure has a very fast conduction speed, and can quickly spread to a large range of communication base station networks. The main reason is the use of wireless mesh network technology to transfer excess load from overloaded base stations to adjacent base stations. In this way, the allocation of the communication surge demand across base stations can be realized, but when the communication surge demand leads to the group overload of base stations, this technology will further accelerate the breakdown of the communication base station network system. Based on the intelligent network architecture mainly composed of UAV mobile base station technology and communication network capacity control technology, the cascading failure model of communication system is constructed in this paper. The characteristics of cascading failure of communication infrastructure systems based on wireless mesh network technology are defined by using POLARIS platform simulation analysis and other methods. It provides a theoretical basis for subsequent research on cascading failure of associated infrastructure systems and provides some ideas for emergency management departments to alleviate cascading failure construction of communication infrastructure.

REFERENCES

Baldini, G, Sturman, T, Dalode, A, Kropp, A, Sacchi, C. (2014). An emergency communication system based on software-defined radio. [J]. EURASIP Journal on Wireless Communications and Networking,1–16.

Del Re, E., Jayousi, S., Morosi, S., Ronga, L.S., Sanctis, M.De., Cianca, E., Ruggieri, M., Falletti, E., Iera, A., Araniti, G., Sacchi, C. (2013). SALICE project: Satellite-Assisted Localization and Communication Systems for Emergency Services [J]. IEEE Aerospace and Electronic Systems Magazine,4–15

Famh, S.C., Wu, S.Y., Qiao, Y.G., Chen, C.G., Sun, Y.G. (2020). Principle and application of wireless mesh network technology. Equipment for Geophysical Prospecting.30(1):5–8

HOLME, P., KIM, BJ. (2002). Vertex overload breakdown in evolving networks – art. no. 066109[J]. Physical review. E. Statistical physics, plasmas, fluids, and related interdisciplinary topics, 6109-0.

Newman, M.E., Watts, D.J. Scaling and percolation in the small-world network model. [J]. (1999). Physical Review, Statistical physics, plasmas, fluids, and related interdisciplinary topics. 60(6):7332–7342.

Phibbs, S, Kenney, C, Rivera-Munoz, G, Huggins, TJ, Severinsen, C, Curtis, B. (2018). The Inverse Response Law: Theory and Relevance to the Aftermath of Disasters [J]. International Journal of Environmental Research and Public Health.15(5):916.

Shahid, M. U., Muhammad, M. K., Hashmi, K., Habib, S., Jiang, H., & Tang, H. (2018). A control methodology for load sharing system restoration in islanded D.C. micro grid with faulty communication links. Electronics, 7(6):90

Singh, V.S. (2007). Impact of the Earthquake and Tsunami of December 26, 2004, on the groundwater regime at Neill Island (south Andaman) [J]. Journal of Environmental Management,.89(1), 58–62.

Advances in Urban Engineering and Management Science – Khalil & Yang (Eds)
© 2023 The Author(s), ISBN 978-1-032-30426-7

Analysis of freeway traffic crash severity considering time differences

Li Jiang, Chenggang Duan & Zhongqiu Li
Hebei Freeway Yanchong Preparation Office, Zhangjiakou, China

Xianyu Wu*
School of Traffic and Transportation Beijing Jiaotong University, Beijing, China

Yanli Jiao
Hebei Provincial Communications Planning and Design Institute Co., Ltd, Shijiazhuang, Hebei, China

Yusen Chen
Beijing Waystone Information Technology Co., Ltd, TNO (The Netherlands Organization for Applied Scientific Research), Beijing, China

ABSTRACT: In this paper, the study area was divided into plains, mountainous areas and hilly areas, and the time was further divided into working days and rest days. The cause and severity of freeway traffic crash under different space-time conditions were studied, respectively. From the selection of vehicle, road and environment, three dimensions involved in vehicles, roadside protective facilities type, lighting conditions, a total of 13 variables for working day and rest day crash in plains, respectively applied sequential Logit model and the multinomial Logit model to construct the freeway traffic crash severity model, the calibration model using maximum likelihood estimation method, Likelihood ratio test, Pearson statistics, deviation statistics and information criterion statistics were used to test the fitting effect of the models, and the accuracy of prediction of the two models was compared. The results show that there are different risk factors for freeway traffic crash in different time conditions, and different models have their own advantages and disadvantages for analyzing crash in different time conditions and different severity levels. The results show that the ordered Logit model is more suitable for the analysis of the severity of traffic crashes on the plain freeway, and can reveal the differences in the causes and severity of the crash in different time conditions.

1 INSTRUCTION

According to statistics, one-tenth of road traffic crash deaths in China are caused by freeway crashes, which only account for 5% of the total number of road crashes (World Health Organization 2018). In 2015, about one-third of the major crash (more than ten deaths) in China took place on freeways (Osorio & Pedraza 2020). In recent years, China's freeways have developed rapidly, with the total mileage of freeways open to traffic reaching 149,600 kilometers by the end of 2019 (Bucsuházya et al. 2020), ranking first in the world. Freeways have the characteristics of large capacity, fast speed, and relatively perfect infrastructure, taking on the main task of road traffic (Yang et al. 2019). With the rapid growth of mileage and traffic flow, freeway traffic safety pressure is increasing day by day. Therefore, in order to realize the sustainable development of traffic, it is necessary to analyze the current situation of freeway traffic safety and put forward the corresponding crash prevention countermeasures.

*Corresponding Author: wuxy@bjtu.edu.cn

DOI 10.1201/9781003305026-82

In the analysis of road traffic crash, the existing research pays little attention to the temporal and spatial correlation of crash data, and ignore the possible influence of time and space factors on the occurrence frequency and severity of the crash. From the perspective of space, Yang Yang (Yang et al. 2017, 2019) considered the differences between regional types and divided the study area into urban areas, mountain areas and rural areas. He mined association rules for freeway traffic crashes by region and concluded that different association rules could be obtained from traffic crash mining in different regional types. From the perspective of time, Mussone et al. (Mussone et al. 2017) analyzed the influencing factors of traffic crash severity by taking the intersections in Turin, Italy, as the research object, and considered the possible influence of different lighting conditions at different times on crash severity, taking day and night as an independent variable. Ji Xiaofeng et al. (Ji & Zhu 2020) explored the factors influencing the severity of road traffic crash in megacities represented by Shenzhen, and made a day-night cause comparison. It is found that the severity of the crash is significantly affected by different factors during the daytime and at night. For example, fatigued driving only has a significant impact on the severity of crashes at night. From the perspective of space, Chen et al. (Chen & Chen 2013) studied the severity of interstate freeway crash in the United States from the perspective of spatial division between mountainous and non-mountainous areas, and the results showed that the crash mechanism of mountainous and non-mountainous roads was significantly different. Huang et al. (Huang et al. 2018) explored the correlation between the severity of crashes on mountain freeways including tunnels and four risk factors including driver behavior, vehicle characteristics, geometry, and environmental factors, and found that driving behavior is the core factor determining the severity of the crash. Both in the study of freeway traffic safety, usually for crash risk factors and crash severity for general analysis, ignoring the different time and space under the condition of the differences of the mechanism of a freeway traffic crash, such as working day and day off crash peak time of different sections, plains, and mountains of a high incidence of the crash and the potential impact. Some studies consider the time factor (day and night, working day and rest day) as an independent variable, and the results may not include the time factor, ignoring the influence of other factors on crashes under the condition of the specified time.

In this paper, freeway as the research object, the use of time division, considering the actual crash data volume of plain freeway traffic crash severity analysis. The ordered Logit model and multinomic Logit model of the discrete choice model are selected to establish the freeway traffic crash severity model, respectively, and the performance of the two models under different time conditions is compared and analyzed, and the safety improvement suggestions are made according to the model results.

2 DATA PROCESSING

2.1 *Data source*

This paper takes the Hebei freeway as the research object. The terrain conditions of Hebei province are complex and diverse, mainly showing the plain (North China plain), mountain (Yanshan and Taihang Mountains) and hill (south Yanshan and east Taihang Mountains).

2.2 *Independent variable correlation test*

A correlation coefficient is used to measure the correlation degree of variables, and there are three common types: Pearson correlation coefficient, Kendall correlation coefficient and Spearman correlation coefficient (Figueira et al. 2017). Spearman correlation coefficient is used for linear correlation analysis with the rank of the two variables, which has nothing to do with the overall distribution or numerical size of the variables. It is applicable to continuous variables and ordered classification variables and has a wider application range. Therefore, Spearman's

Table 1. Table type styles.

NO.	Attribute name	NO.	Attribute name
1	Administrative division of province	19	Scene situation
2	Prefectural administrative divisions	20	Weather
3	Road NO.	21	Visibility
4	Road name	22	Road condition
5	KM of distance	23	Road surface
6	Meters of distance	24	Traffic control
7	Week	25	Light
8	Month	26	Road type
9	Year	27	Freeway administrative class
10	Crash_date	28	Terrain
11	Crash_hour	29	Road alignment
12	Fatal in 24 hours	30	Section type
13	Injuries in 24 hours	31	Physical road isolation
14	Location	32	Pavement structure
15	Property damage	33	Type of central isolation facility
16	Crash type	34	Type of roadside protection facilities
17	Reason	35	Number of motor vehicles involved
18	Crash form		

correlation coefficient was adopted in this paper to test the correlation of independent variables. The specific calculation formula is as follows:

$$\rho = 1 - \frac{6\sum_{i=1}^{n} d_i^2}{n\left(n^2 - 1\right)} \tag{1}$$

Where, n is the number of observed values and di is the rank difference of variables.

2.3 *Variable processing*

Based on the crash data of Hebei Province, 13 candidate independent variables were selected to analyze the severity of traffic crashes from the perspectives of vehicle, road and environment. Firstly, the value of the candidate independent variable is simplified, such as interval discretization of a continuous variable. After processing, all independent variables were transformed into categorical variables.

To intuitively show the correlation between variables, a correlation heat map is drawn, as shown in Figure 1. When there is multicollinearity between variables, i.e., the correlation coefficient is greater than 0.7, the stability of prediction results is poor. It was observed that the correlation coefficient between variables X11 (lighting condition) and X12 (crash time period) is 0.89, so one of them was removed, and only X11 (lighting condition) was retained.

3 MODELING AND RESULTS

Traffic crash severity analysis is to reveal the correlation between crash influence factors and crash consequences by analyzing specific crash data, and identify the significant factors that affect crash severity. The study found that in the previous chapter different time under the condition of the freeway traffic crash severity causes and differences, thus divided into working day and day off. This chapter will time were built based on the ordered Logit model and multinomial Logit model

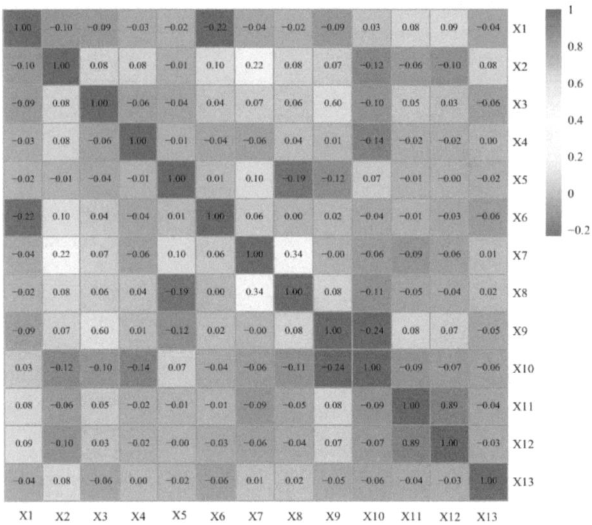

Figure 1. Heatmap of variable correlation.

of freeway crash severity, Hebei province freeway traffic crash for instance analysis, compare the merits and demerits of the two models, explore the methodology suitable for this kind of problem, and put forward the crash prevention countermeasures.

Hebei province has the longest freeway mileage and the highest crash frequency in plain area. The amount of plain crash data collected under the background of large cardinality is the most sufficient, which is enough to support the refinement time for in-depth study. However, the amount of existing crash data in mountainous areas and hills is only about 100, and the refinement time will further divide and reduce the amount of data, so significant results cannot be obtained. Therefore, this chapter takes the plain freeway traffic crash as the research object and analyzes the crash severity from working days and rest days, respectively.

3.1 *Comparison of incident severity analysis results of ordered Logit with multiple Logit models on weekdays*

The prediction accuracy of the two models is shown in Table 2.

Table 2. Comparison of classification prediction accuracy of two types of models.

	Prediction accuracy	
Severity	Ordered Logit model	Multinomial Logit model
Property damage crash	0.0%	100.0%
	43.0%	43.0%
Casualty crash	84.5%	84.5%
Average	69.1%	69.5%

Through the comparison of calculation results, it can be found that:

(1) from the significant influencing factors of crash analysis, two kinds of model analysis results: whether ordered Logit model or a multinomial Logit model, in the days of a plain freeway

traffic crash, all sections type found in the ordinary roads and programmed can increase the crash severity, which occurs more often than any other special sections deaths. At the same time, the model test proves that the fitting effect of the two models is good.

(2) From the analysis of different crash severity, there are similarities and differences between the two models: no matter the ordered Logit model or multiple Logit model, the prediction results of the plain crash on weekdays are similar, and the prediction accuracy of an injury crash and death crash is the same, which is close to 70%. The ordered Logit model has the strongest ability to predict death crash, and the multinomial Logit model can predict property damage crash more accurately.

3.2 *Comparison of crash severity analysis results of ordered Logit and multinomial Logit models on weekends*

The prediction accuracy of the two models was statistically analyzed, and the results are shown in Table 3.

Table 3. Comparison of classification prediction accuracy of two types of models.

	Prediction accuracy	
Severity	Ordered Logit model	Multinomial Logit model
Property damage	33.3%	66.7%
crash	67.6%	64.9%
Casualty crash	91.1%	85.7%
Average	80.2%	77.1%

Through comparison, it can be found:

(1) From the analysis of significant factors affecting the crash, there is a significant gap between the analysis results of the two types of models: the ordered Logit model obtained 4 categories of 7 significant variables, such as motor vehicle lane, metal guardrail, sunny, and less than 100 meters, while the multiple Logit models did not find significant variables affecting the severity of the crash. At the same time, the model test shows that the ordered Logit model fits well, while the multinomial Logit model fits poorly.

(2) From the analysis of different crash severity, there are similarities and differences between the two models: in plain crash on rest days, the prediction accuracy of the two models for death crash is the highest, followed by injury crash; Considering the average prediction accuracy, the ordered Logit model is superior to multiple Logit models, but the prediction ability of property loss crash is weak.

4 RASH SEVERITY ANALYSIS AND ACCIDENT PREVENTION COUNTERMEASURES UNDER DIFFERENT TIME CONDITIONS

4.1 *Analysis and comparison of accident severity on weekdays and weekends*

(1) For working days, the two significant variables of the ordinary road segment and ramp will increase the severity of accidents; that is, with the increase of the value, death accidents are more likely to occur than injury accidents.

(2) For rest days, metal guardrail, waveform guardrail, sunny days, cloudy days, rainy days and visibility less than 100 meters will increase the severity of the accident. The cross section of

the road for motor vehicles can reduce the severity of accidents and is more likely to cause injury accidents.

(3) Both the ordered Logit model and multiple Logit model have high fatality prediction rates in plain areas, generally reaching about 85%, regardless of working days or rest days.

4.2 *Study on prevention countermeasures of the freeway traffic crash*

According to the modeling and analysis results of accident severity, the hidden danger of freeway traffic safety is discovered, and the corresponding prevention and control measures are put forward from the Angle of accident prevention.

(1) Ramp is a section of plain freeway with a high incidence of traffic accidents. An on-ramp is usually composed of two lanes, and the road alignment is more radian or slope, that is, a curve or ramp, which is easy to cause traffic accidents due to temporary lane change, exceeding the speed limit, parking or reversing. The suggestions are as follows:
   1) Add road safety facilities: add dynamic information signs reasonably, set speed limit signs and warning signs upstream of the ramp to remind drivers to change lanes and slow down in advance; Reserve a safety area at the ramp, and set up anti-collision MATS with guiding function.
   2) Strengthen traffic safety education and management: strengthen the publicity of highway safety knowledge, strictly prohibit overtaking, reversing and parking near the ramp, and strictly prohibit driving along the pressure diversion line; Know the driving route and exit location in advance to avoid missing ramps or temporary lane changes.

(2) Traffic accidents on plain freeways mostly occur in rainy weather, and the dark environment and wet road surface caused by adverse weather pose higher challenges to safe driving. The suggestions are as follows:
   1) Popularize severe weather warnings: actively cooperate with transportation, meteorological, mobile, and other departments to remind drivers to drive carefully through mobile phone messages, vehicle-mounted traffic broadcasts, and dynamic information signs on the side of the road or in service areas.
   2) Improve lighting facilities along the highway: regularly check the use of lighting facilities along the whole highway, add reasonable lighting facilities in accident-prone areas, and ensure good lighting conditions for drivers when driving.

(3) For plain freeways, the type of roadside protection facilities has a significant impact on the severity of traffic accidents. The suggestions are as follows:
   1) Strengthen material research: Carry out special research for different types of roadside protection facilities, select construction materials or forms with higher safety, absorb or transfer vehicle collision energy as much as possible, and then reduce the severity of accidents.
   2) Enhance daily road maintenance: regularly maintain and inspect roadside protection facilities to ensure good conditions and maintain high safety performance.

5 CONCLUSIONS

In this paper, the ordered Logit model and multinomial Logit model are used to construct the traffic crash severity model of the plain freeway, which is divided into working days and rest days. The maximum likelihood estimation principle was used to estimate the parameters of the model, the logistic regression models under different time conditions were obtained, and the significant influencing factors of the crash on working days and rest days were found. The feasibility and fit degree of the model were proved by parameter test and goodness of fit test, and the accuracy of prediction of the two models was compared. By analyzing the solution results of the model, the main conclusions are as follows:

(1) Consistent significant variables can be obtained by using different models: Whether using an ordered Logit model or multiple Logit models, road section type (common road section, ramp entrance) has a significant influence on the plain crash severity under working day conditions.
(2) Under different time conditions, the significant variables that increase the severity of the crash are different: ordinary road section and ramp on weekdays, metal guardrail or corrugated guardrail on the roadside on rest days, sunny days, cloudy days, rainy days, and visibility less than 100 meters.
(3) For different crash severity, the prediction performance of the two models has its advantages and disadvantages: the ordered Logit model has the best prediction effect on death crashes; Multinomial Logit models have better performance in predicting property damage crashes. Overall, the ordered Logit model is more applicable to the analysis of traffic crash severity of plain freeway.

This study shows obvious differences in the significant factors and severity of crashes on the plain freeway under the time division of working days and rest days. By comparison, it is found that the ordered Logit model is more suitable for the study of the plain crash, and can reveal the risk factors and severity differences of freeway crashes under different time conditions. The model results are analyzed, and corresponding safety improvement schemes are put forward to provide theoretical guidance for traffic safety management personnel.

REFERENCES

Bucsuházya, K. & Matuchováa, E. & Zůvala, R. et al. (2020). Human Factors Contributing to the Road Traffic Accident Occurrence[J]. Transportation Research Procedia, 45: 555–561.

Chen, F. & Chen, S. (2013). Differences in Injury Severity of Accidents on Mountainous Highways and Non-mountainous Highways[J]. Procedia – Social and Behavioral Sciences, 96: 1868–1879.

Figueira, A.C. & Pitombo, C.S. & Oliveira, P.T.M.S. et al. (2017). Identification of Rules induced through Decision Tree Algorithm for Detection of Traffic Accidents with Victims: A Study Case from Brazil[J]. Case Studies on Transport Policy, 5(2): 200–207.

Huang, H. & Peng, Y. & Wang, J. et al. (2018). Interactive Risk Analysis on Crash Injury Severity at a Mountainous Freeway with Tunnel Groups in China[J]. Accident Analysis & Prevention, 111(Feb.):56–62.

Ji, X.F. & Zhu, G. (2020). Comparative analysis on causes of the severity of traffic accidents at day and night in megacities: taking Shenzhen as an example[J]. Journal of Safety Science and Technology, 16(2): 142–148.

Mussone, L. & Bassani, M. & Masci, P. (2017). Analysis of Factors Affecting the Severity of Crashes in Urban Road Intersections[J]. Accident Analysis & Prevention, 103: 112–122.

Osorio, C. & Pedraza, C. (2020). Modern Data Sources and Techniques for Analysis and Forecast of Road Accidents: A Review[J]. Journal of Traffic and Transportation Engineering (English Edition), 7(4): 432–446.

World Health Organization. Global Status Report on Road Safety (2018). [R].

Yang, Y. & Yuan, Z.Z. & Chen, J. &Guo, M. (2017). Assessment of osculating value method based on entropy weight to transportation energy conservation and emission reduction[J]. Environmental Engineering & Management Journal (EEMJ). 2017 Oct 1;16(10).

Yang, Y. & Yuan, Z. Z. & Sun, D.Y. et al. (2019). Analysis of the Factors Influencing Highway Crash Risk in Different Regional Types based on Improved Apriori Algorithm[J]. Advances in Transportation Studies, 2019(49): 165–178.

Advances in Urban Engineering and Management Science – Khalil & Yang (Eds)
© 2023 The Author(s), ISBN 978-1-032-30426-7

Exploring the factors affecting cyclists' detour decisions in the curbside bus station area

Jiajia Liang*
School of Civil Engineering, Xi'an University of Architecture & Technology Xi'an, Shaanxi Province, China

Lingyun Qiu*
China Aviation Planning and Design Institute Group Co., Ltd., Beijing, China

ABSTRACT: At the curbside bus station on the mixed road, buses stop at the station will take up the bicycle lane. Many cyclists decided to detour to the motor lane to reduce delays. Cycling detours could disturb the traffic order of the motor lane and increase the danger of cycling. To explore the factors affecting cyclists' detour decisions in the curbside bus station area, a video survey was conducted in Xi'an to collect data about bicycle traffic. Based on the analysis of cyclists' psychology and cycling habits, cyclists are classified as adventurous, neutral, and conservative types according to their position on the road when entering the bus station area. It is found that the riding tracks of adventurous and conservative cyclists are relatively fixed, while that of neutral cyclists are uncertain. The Binomial logistic model was used for exploring the relationship between the neutral cyclists' detour decisions in the curbside bus station area and the gender of the cyclist, cyclists' perception of the bus state, the number of passengers getting on and off the bus, the distance between the bus and the platform, the number of buses parked in the station, the number of vehicles in the adjacent lane, and the number of non-vehicles in the adjacent lane. The results show that cyclists' perception of bus state, the number of passengers getting on and off the bus, and the distance between the bus and the platform are significant variables. Among them, the number of passengers getting on and off the bus is positively correlated with the neutral cyclist's detour decision, and the cyclist's perception of the bus state and the distance between the bus and the platform are negatively correlated with the neutral cyclist' detour decision. This study provides important guidance for studying bicycle micro-traffic behavior and optimizing the design of curbside bus stations, and it is beneficial to improve the safety of bicycle riding.

1 INTRODUCTION

Recently, with the advocacy of green transportation travel and the popularity of sharing bicycles in China, bicycles have returned to the daily life of urban residents as one of the green travel modes of transportation. However, in some cities, the road space resources allocated to bicycle traffic can hardly meet the demands. On many urban roads, bus lanes are mixed with non-motor vehicles lanes. Therefore, conflicts between bicycles and buses occur frequently, especially in the bus station areas. The bus station is mainly divided into the bay bus station and the curbside bus station. Curbside bus stations, with the low construction cost, can make full use of roadside space, and buses are not restricted by road facilities when entering and leaving the station. However, compared with the bay bus station, buses occupy the non-motor vehicles lane when they arrive at the curbside bus station, which leads to frequent traffic conflicts. Many cyclists detour motorized lanes, which increases the risk of bicycles and disturbs the traffic order of motorized lanes. Cyclists are complex traffic participants, and their decisions and willingness in the process of cycling will be affected by the

*Corresponding Authors: 17319988360@163.com and qiuly007@avic.com

 DOI 10.1201/9781003305026-83

combined affecting of psychological and physiological factors caused by the environment (Li et al. 2018). The cycling track is the most direct reflection of cyclists' traffic behavior (Sun 2012). Therefore, it is necessary to analyze the cycling tracks and explore the factors that affect cyclists' detour track decisions in the curbside bus station area.

Previous studies have studied bicycle traffic conflict in the bus station area. Gregory R. Krykewycz, taking Philadelphia, USA, as an example, summarized the solutions of the same type of cities for conflicts with bicycles when buses arrive at the curbside bus station, including two main categories: changing the form and building facilities (Krykewycz 2009). Chen studied the interference mechanism of buses entering and leaving the station with other traffic modes and the interference mechanism of bus passengers getting on and off with non-motorized vehicles and constructed the simulation model for traffic flow in the bus station area (Chen 2014). Scholars have studied delays of non-motorized vehicles near the bus station when the buses arrive at the station. X. Y. Kuang assumed that non-motor vehicles would pass through the channel between the bus and the platform, and the distance between the bus and the platform was fixed. It was concluded that the arrival rate and evacuation rate of vehicles would be the important indicators of delay (Kuang et al. 2016). X.B. Yang assumed that the bicycle would detour to the motor lane and cause a delay in traffic flow, and established a delay model based on the queuing theory (Yang et al. 2011). For cyclists, A.H. FAN divided the cyclists into opportunists, risk-preferring, and risk-averse types based on the non-motor vehicle riding track at the bus station. Analyzed the data of the speed, longitudinal distance from the bus when changing lanes, lateral distance from the machine-non-divided line, and delayed waiting time of the three types of cyclists. Then, a survival analysis method was established based on the delay in the passage of small cars at the mixed bus stops model (Fan 2015). H.P. Shao studied how two-wheelers passed both sides of the bus in the curbside bus station area and concluded that the instantaneous headway of the left lane and the traffic environment has a significant impact on the left interpenetration of bicycles (Shao et al. 2017).

It is evident from previous studies that the research on the influence of cyclists' characteristics on the cycling track and the analysis of cyclists' psychology are limited. Studies exploring the factors affecting cyclists' detour decisions in curbside bus station areas are rare. This study aims to explore significant factors affecting cyclists' detour decisions in the curbside bus station area. The cyclists' cycling data were collected in Xi'an city, China. The binomial logistic model was applied to explore the affecting of cyclists' characteristics and road environment on the cyclists' detour decisions in the curbside bus station area. The results could enrich the research on the micro-characteristics of cyclists and provide the theoretical basis for optimizing the design of bus stations.

2 DATA COLLECTION

2.1 *Type video survey*

The video survey was conducted in Xi'an, China, in June 2020. 2nd Ring South Road and north of Yanta interchange station (west of road) was used as the survey site. The location of the bus station is shown in Figure 1.

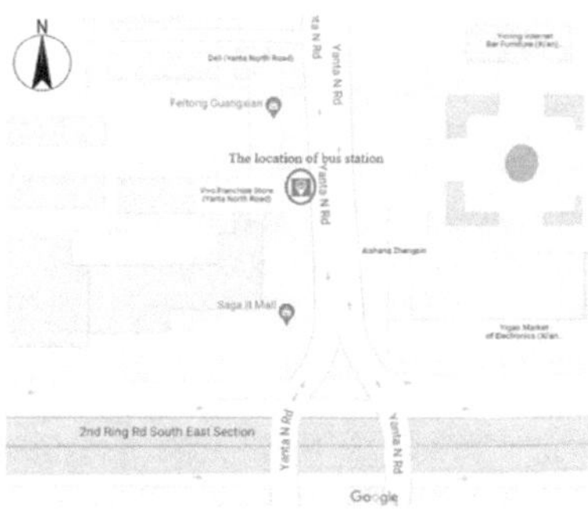

Figure 1. Location of the bus station.

It is a curbside bus station located on the main road of the city. There were 15 bus lines in this bus station, and the arrival rate of buses and bicycles was about 32 vehicles/min and 2.7 vehicles/min in the morning peak hours (7:45~8:45) and evening peak hours (17:30~18:30) on weekdays. The observation site was located on the pedestrian bridge near the station, and the positions of bus shelters, station signs, communication poles, and parking piles were recorded as the reference points for sample data extraction. The length of the bus station area is 41m, with one dedicated parking space and two potential parking spaces.

2.2 *Data description*

To accurately analyze the cyclist's riding track and riding characteristics when passing through the bus station area, we established the axes as shown in Figure 2. The shaded part of Figure 2 shows the area of the bus station, and the axis scale was divided according to the actual measured distance between the markers.

Figure 2. Coordinates of the bus stop area.

Recorded the moment when the front wheel of the bicycle arrived and the rear wheel left as the moment when the bicycle entered and left the bus station area, and calculated the time taken by the bicycle to pass the bus station area. A certain visible part of the cyclist was taken as the reference point to extract the track. During the whole process of the cyclist passing through the bus station area, the position of the point within the captured unit time (1s) was recorded as the track point in the cycling process. Then, found the position marker of each point in the graph with axes to get the diagram of cyclists' riding tracks.

A total of 360 cycling sample records of bicycles passing through the bus station area when buses stop at the station were screened out. There was a large number of bicycle samples obtained from the survey. To represent the characteristics of cyclists' riding behavior during their pass through the bus station area specifically, the samples were classified according to their cycling track. We analyzed the psychology and cycling habits of cyclists and classified them into three types: adventurous, neutral, and conservative, according to the position of bicycles on the road when they entered the bus stop area. Among them, adventure cyclists are more than 4m away from the platform, and neutral cyclists are 1-4 m away from the platform. Conservative cyclists are at a distance less than 1 m from the platform. As shown in Figure 3, the red track represents the track of adventurous cyclists, who pass through the bus station area quickly along the motor vehicle lane. The yellow track represents the track of neutral cyclists, who are cautious at the beginning of entering the bus station area and ride at a slower speed than adventurous cyclists. The green track represents the track of conservative cyclists, who always ride on the inside and slow.

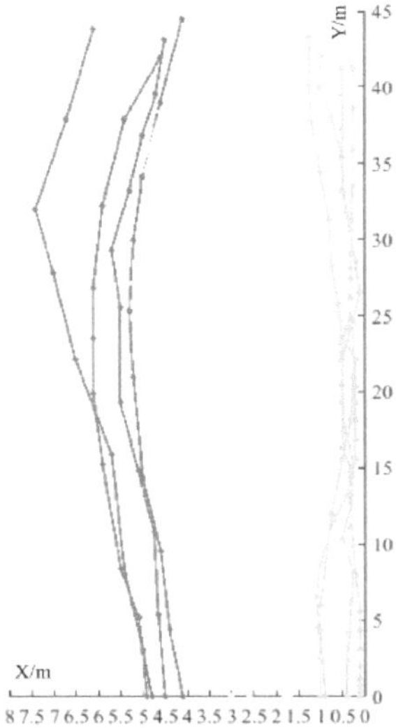

Figure 3. The caption of a typical figure.

The proportion of cyclist types and the distribution of passing time are shown in Figure 4 and Figure 5.

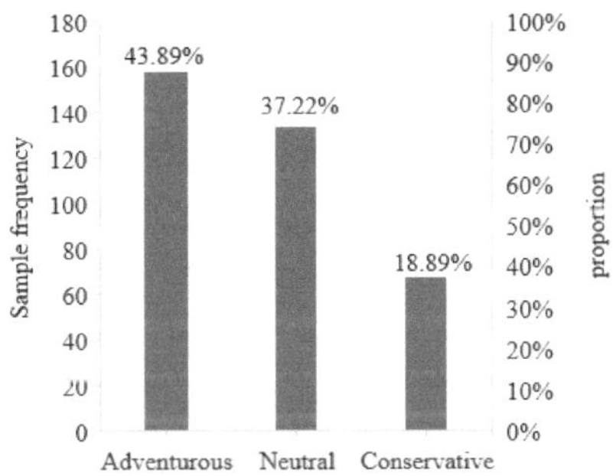

Figure 4. The proportion of cyclist types.

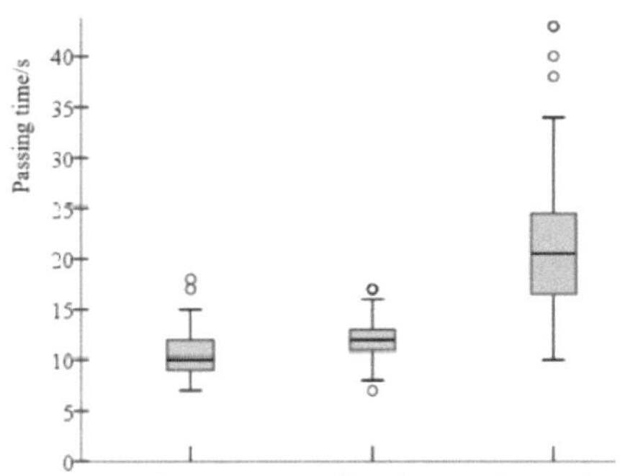

Figure 5. Distribution of passing time.

According to analyzing the cycling tracks of cyclists through the bus station area, adventurous and conservative cyclists have a relatively fixed path through the curbside bus station area. This is due to their cycling habits and psychology. Adventurous cyclists are more likely to detour to the motor lane when passing through the curbside bus station area, even when there are no bus stops at the station. Conservative cyclists always pass close to the inside when passing through bus station areas. However, the neutral rider's cycling track is uncertain when passing through the curbside bus station area. This is because neutral cyclists are affected not only by their habits and psychology but also by objective traffic factors such as the riding environment, the current state of buses, motor vehicles in the adjacent lane, and other non-motor vehicles. Cyclists need to make judgments and decisions about cycling tracks in a few seconds. Therefore, it is likely to conflict with buses. Cycling safety risks are higher. If cyclists are hesitant, they are more likely to have conflicts with buses, resulting in greater safety risks. Therefore, the factors affecting the detour decisions of neutral cyclists in the curbside bus area are selected for study and discussion.

2.3 *Potential factors*

Through field investigation and video analysis, the factors affecting cyclists' detour decisions were classified into three categories: the affecting of cyclists' characteristics, the affecting of bus traffic, and the affecting of other traffic participants. The affecting cyclists' characteristics include the gender of cyclists and cyclists' perception of the bus state. The affecting factors of bus traffic include the number of passengers getting on and off the bus, the distance between the bus and the platform, and the number of buses parked in the station. The affecting factors of other traffic participants include the number of vehicles in the adjacent lane and the number of non-vehicles in the adjacent lane. The calculation method is as follows:

2.3.1 *Gender of cyclist*

Man and woman differ in physical strength, patience, endurance, and have different mental activities when making decisions. Based on the statistical analysis of the above samples, the cyclist type distribution frequency of different genders is obtained, as shown in Figure 6. Gender has no significant difference in the classification of cyclist types. Therefore, when screening neutral cyclists, there is no need to select the gender of samples in a specific proportion.

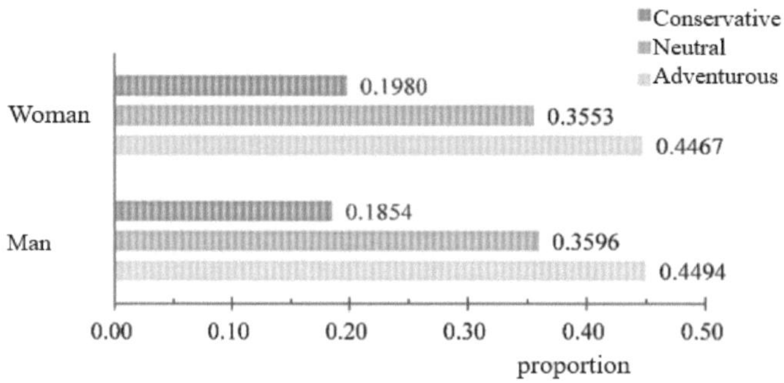

Figure 6. Distribution of different gender cyclist types.

2.3.2 *Cyclists' perception of the bus state*

Cyclists' perception of the bus state is related to the moment when the bus enters, stops, starts, and leaves. And once the bicycle reaches the rear of the bus, it is considered that the cyclist has decided on the way to pass. K_{re} is taken as the cyclists' perception of the bus state in this study. It quantifies the relationship between the moment the bicycle enters the bus station area and the moment of the bus stops, and the distance between the bicycle entering the bus station area and the stopping bus. For the same parked bus, bicycles entering the bus stop area at different times will have different judgments about their state. The calculation formula is as follows:

$$K_{re} = \begin{cases} \dfrac{t_{pass} - t_{bst}}{t_{Bst} - t_{bst}}, & t_{Bsp} < t_{bst} \\ \dfrac{t_{pass} - t_{Bsp}}{t_{Bst} - t_{Bsp}}, & t_{Bsp} \geq t_{bst} \end{cases} \tag{1}$$

t_{Bsp} is the moment when the bus stops; t_{Bst} is the moment when the bus starts again after it finishes service at the station; t_{bst} is the moment when the bicycle enters the bus station area; t_{pass} is the moment when the bicycle rides to the rear of the bus.

Among them,$t_{Bsp} - t_{bst} \geq 3s$, the time required to open and close the doors of a bus is generally considered to be 1~2s (Ge 2006). It is believed that the judgment of cyclists will be affected by the bus entering and leaving the station when the time is less than 3s.

2.3.3 *The number of passengers getting on and off the bus*

The number of passengers getting on and off the bus not only has a direct impact on the bus stop time at the platform but also has an impact on the level of congestion in the inside lane of the bus. Therefore, read from the video the number of passengers who do not get on or off the bus when the bicycle enters the bus station area. The maximum value is $N_{\max}$.

2.3.4 *The distance between the bus and the platform*

The greater the distance between the bus and the platform, the less likely it is that conflicts will occur between bicycles and passengers. The distance between the bus and the platform affects the decision of the cyclist, and it is specified as x. The axes established in Figure 2 are used to read the x values from the video samples.

2.3.5 *The number of buses parked in the station*

When the bicycle enters the bus station area, more buses stop at the station, the greater possibility of uncertainty in the cycling process, and affect the decision of the cyclists. Record the number of buses parked in the station at the moment the bicycle enters the bus station area.

2.3.6 *The volume of traffic in the adjacent lanes*

When the distance between a bicycle and a motor vehicle is close, the driver will feel the lateral pressure (Jia & Ma 2010), and vice versa is also true for cyclists. Although the collision avoidance force between the two has no physical source, it will affect the cycling track of the bicycle. Record the number of motor vehicles q_m and the number of non-motor vehicles q_{nm} that overtake the bicycle when the bicycle passes the bus station area. These vehicles are the number of vehicles visible to cyclists when they observe adjacent lanes and will affect their decision.

2.4 *Independent variables*

After analyzing the factors affecting cyclists' detour decisions in the curbside bus station area, the factors that are appropriate to the curbside bus station area environment were selected as independent variables in conjunction with the actual data survey. The independent variable codes are assigned in Table 1.

Table 1. The factors affecting cyclists' detour decisions in the curbside bus station area.

Serial number	Independent variables	Variables	Unit
1	The number of passengers getting on and off the bus	$N_{\max}$	People
2	Cyclist' perception of bus state	K_{re}	/
3	The distance between the bus and the platform	x	m
4	The number of buses parked in the station	n_{stp}	Vehicle
5	The number of non-vehicles in the adjacent lane.	q_{nm}	Vehicle
6	The number of vehicles in the adjacent lane	q_m	Vehicle
7	Gender of cyclist	G	/

According to the valid sample principle, when more than one bicycle enters the bus stop area at the same time and chooses the same way to pass, only one valid sample is counted. The second screening was carried out among the neutral cyclists, and 71 valid samples were obtained, including 41 male and 30 female cyclists. The statistical results of selected samples of cyclists are shown in Table 2.

Table 2. Descriptive statistics of independent variables.

Variables	Min	Max	Mean	S.D.
K_{re}	0.10	0.94	0.425	0.199
$N_{\max}$	0	13	4.65	2.914
x	1	2	1.20	0.401
n_{stp}	0.45	1.90	0.98	0.332
q_{nm}	0	5	1.37	1.427
q_m	0	3	0.73	0.736

3 METHODOLOGY

3.1 *Binomial logistic model*

Binomial logistic regression models are developed to explore the significant factors affecting the neutral cyclists' detour decisions in the curbside bus station area. The advantage of logistic regression models is that the dependent variable is a dichotomous variable or the probability of an event occurring. If the categorical variables or probabilities are fitted directly to the influencing factors, a regression equation will be obtained. The equation (2) is as follows:

$$\hat{P} = \alpha + \beta_1 x_1 + \beta_2 x_2 + \ldots + \beta_m x_m \tag{2}$$

However, in (2), there are some problems, such as inconsistent value intervals on both sides of the equation and nonlinear correlation, which are difficult to be established as the premise assumptions. Therefore, the Logit transformation is performed. First, take the logarithm of the ratio of the probability of an outcome to the probability of no outcome, and solve for logit(P). Then, take it as the dependent variable component of the logistic regression model containing M independent variables (Zhang & Dong 2013).

$$\text{logit}(P) = \beta_0 + \beta_1 x_1 + \beta_2 x_2 + \ldots + \beta_p x_p \tag{3}$$

$$P = \frac{\exp(\beta_0 + \beta_1 x_1 + \beta_2 x_2 \ldots + \beta_p x_p)}{1 + \exp(\beta_0 + \beta_1 x_1 + \beta_2 x_2 \ldots + \beta_p x_p)} \tag{4}$$

This model solves the problem that the values cannot correspond in the analysis of binary dependent variables, so the Logistic regression model is often the most common multivariate quantitative analysis method used in the regression of this type of dependent variable.

3.2 *Modeling approach*

The detour decision of the neutral cyclist passing the curbside bus station area with a bus stopping at the station is studied and defined as the dependent variable y. The value of the decision is as follows: when the cyclist passes through the inside of the bus, $y = 0$. When the cyclist detours the motor lane, $y = 1$. When the variable y is 0 or 1, the corresponding probabilities are P and 1-P respectively. The maximum likelihood method is used to estimate and test the regression parameters.

The independent variables are the gender of the cyclist, cyclists' perception of the bus state, the number of passengers getting on and off the bus, the distance between the bus and the platform, the number of buses parked in the station, the number of non-vehicles in the adjacent lane, and the number of vehicles in the adjacent lane. They are defined as $x_1, x_2, \ldots, x_7$. According to the logistic regression model, the probability of neutral cyclists' detour decision is:

$$P(y = 1 \, | x_1, x_2, \ldots, \; x_7) = \frac{\exp(\beta_0 + \beta_1 x_1 + \beta_2 x_2 + \ldots + \beta_7 x_7}{1 + (\beta_0 + \beta_1 x_1 + \beta_2 x_2 + \ldots + \beta_7 x_7} \tag{5}$$

3.3 *Factors analysis*

With the help of SPSS, a reverse stepwise selection method is adopted. The gender of cyclist G, cyclists' perception of the bus state K_{re}, the number of passengers getting on and off the bus $N_{\max}$, the distance between the bus and the platform x, the number of buses parked in the station n_{stp}, the number of non-vehicles in the adjacent lane q_{nm}, and the number of vehicles in the adjacent lane q_m were used as independent variables to conduct stepwise regression screening and set 95% confidence interval. The seven independent variables were brought into the model, and the results of the partial regression coefficients and significant level *sig*. values of each of the independent variables are shown in Table 3. When the calculated *sig*.value is not greater than the given α=0.05, this factor is considered to have a significant affecting on the value of the dependent variable.

Table 3. Correlation analysis results.

Independent variable	G	K_{re}	$N_{\max}$	x	n_{stp}	q_{nm}	q_m
sig.	0.899	0.002	0.031	0.025	0.153	0.943	0.945

From Table 3, it can be seen that the calculated sig. values for the four variables of the gender of cyclist G, the number of buses parked in the station n_{stp}, the number of non-vehicles in the adjacent lane q_{nm}, the number of vehicles in the adjacent lane q_m, are greater than 0.05 and do not meet the significant requirements, and could be removed. And the calculated *sig*. values of cyclists' perception of the bus state K_{re}, the number of passengers getting on and off the bus $N_{\max}$, and the distance between the bus and the platform x meet the significance test requirements, which can be performed to construct the model. The calculation results are shown in Table 4.

Table 4. Parameter estimate results.

Independent variable	*B*	DF	*sig*.	95% confidence interval
$N_{\max}$	0.395	1	0.042	[1.015,2.170]
K_{re}	10.744	1	0.001	[0.000,0.010]
x	2.513	1	0.048	[0.007,0.976]

Combining the results of the calculations in Table 4 and the constants from the data fit analysis, the probability model for the decision of the neutral cyclist detouring to the motor lane in the curbside bus station area is:

$$P(y=1)=\frac{\exp(0.395N_{\max}-10.744K_{re}-2.513x+6.534)}{1+\exp(0.395N_{\max}-10.744K_{re}-2.513x+6.534)} \tag{6}$$

After considering the index size of each factor, it is obtained that the cyclist will choose to ride through the outer motor lane with this probability.

B is the calculation of regression coefficients for each variable. It can be found that the number of passengers getting on and off the bus is positively correlated with the neutral cyclist's detour decision. The uncertainty of the cyclists' detour decision comes from the fact that there is no way to know what is happening on the side of the platform when the bus stops. Thus, the number of passengers getting on and off the bus when the bicycle enters the station is an important reference that affects the neutral cyclist's decision. Cyclists' perception of the bus state is negatively correlated with the neutral cyclist's detour decision. The distance between the bus and the platform is negatively correlated with the neutral cyclist's detour decision. The smaller the stopping distance between the bus and the station, the more likely the cyclist will detour to the motor lane. In addition, the test results of the model are consistent with the conclusion that the gender of the cyclist does not have a significant effect on the choice of the passing method (Fan 2015).

4 RESULTS AND DISCUSSION

1) This study investigated the bicycle traffic data at a curbside bus station in Xi'an, China, in June 2020. Based on the psychology and cycling habits of cyclists, cyclists were classified into three types according to the position of bicycles on the road when they entered the bus stop area: adventurous cyclist, neutral cyclist, and conservative cyclist. It was concluded that adventure and conservative cyclists have more fixed riding tracks, while neutral cyclists are more affected by psychological influences and objective traffic factors, with wavering tracks and great safety risks.
2) This study analyzes cyclists' psychological perception of bus state and creates a formula for cyclists' perception of bus state coefficients to quantify cyclists' psychological states when passing through the bus station area.
3) Combined with the survey video, the analysis yielded seven factors affecting neutral cyclists' detour decisions in the curbside bus station area: gender of cyclist, cyclists' perception of bus state, the number of passengers getting on and off the bus, the distance between the bus and the platform, the number of buses parked in the station, the number of vehicles in the adjacent lane, and the number of non-vehicles in the adjacent lane. Apply a binomial logistic regression model to analyze the significance of each affecting factor and establish a probability model of neutral cyclists' detour decisions in the curbside bus station area. The results show that cyclists' perception of the bus state, the distance between the bus and the platform, and the number of passengers getting on and off the bus are significant influencing factors. Among them, cyclists' perception of the bus state has the most significant influence. It indicates that the cyclists' psychological perceptions are more important in affecting cyclists' decisions than actual objective traffic factors.

This study analyzed cyclists' psychology and the affecting factors of their detour decision when passing through the curbside bus stop area. It is beneficial to optimize the cycling environment and reduce cycling risks. It also provides a theoretical basis for the optimal design of bus stop locations and passenger waiting points for the curbside bus station.

However, in this study, the classification of cyclist types is not combined with the physiological change characteristics of cyclists. The interaction between each factor is not analyzed. The model of cyclists' detour decision applies to general curbside bus stations, and the level of service of the road where the station is located should be between level 1 and level 3. With the traffic flow increasing, it will affect the decision of cyclists. The model can be further revised and improved by introducing relevant coefficients subsequently.

REFERENCES

Chen, Z. J. (2014). Simulation on Traffic Behaviors in Bus Stop Area. D. Beijing Jiaotong University.

Fan, A. H. (2015). Unsafe behavior and impact analysis of non-motor vehicles near the bus stop with mixed traffic flow. D. Beijing Jiaotong University.

Ge, H. W. (2006) Traffic Effect Analysis and Optimization Techniques of Bus Stop. D. Southeast University.

Jia, N. & Ma, S. F. (2010). Cellular automaton model for the motor-vehicle flow under the interference of bicycles. J. Xitong Gongcheng Lilun yu Shijian/System Engineering Theory and Practice. 30 (7), 1333-1339.

Krykewycz, G. R. (2009). Bicycle-Bus Conflict Area Study. J. Bicycling.

Kuang, X. Y. & Ba, J. & Sun, S. (2016). Model of Non-Motor Vehicle Traffic Flow Delay Caused by Public Traffic Vehicle. J. Transport Research. 2 (2), 12-18.

Li, C. Y. & Yang, Z. Z. & Shao, Z. Z. et al. (2018). Characteristics of Urban Cyclist Perception of Fatigue. J. China Journal of Highway and Transport. 31(6), 291-298.

Shao, H. P. & Li, J. X. & Xie, P. et al. (2017). Research on crossing behavior of two-wheelers at bus station based on Bayesian network. J. Journal of Wuhan University of Technology (Transportation Science & Engineering). 41(06), 969-973.

Sun, D. (2012). Analysis and Modeling of Pedestrian Crossing Behavior. D. Jilin University.

Yang, X. B. & Gao, Z. Y. & Gao, S. L. (2011). Car capacity near bus stops with mixed traffic derived by additive-conflict-flows procedure. J. SCIENCE CHINA Technological Sciences. 54(3), 733-740.

Zhang, W. T. & Dong, W. (2013). Advanced Tutorial on SPSS Statistical Analysis. M.

Advances in Urban Engineering and Management Science – Khalil & Yang (Eds)
© 2023 The Author(s), ISBN 978-1-032-30426-7

Verification of improved social force model based on vehicle trajectory data

Churan Lin*, Yanya Huang, Guowei Su, Xuexue Yan, Chaodang Long & Zhenyou Xie
Guilin University of Electronic Technology, Guilin, China

ABSTRACT: Car following behavior is one of the basic behaviors that affect the micro traffic state. Therefore, the research on the car following behavior model is also a focus in the field of traffic engineering. This paper validates Delpiano's improved car following model based on the social force model. The validation indexes are absolute error, and root mean square error and Pearson coefficient. The data are derived from the vehicle trajectory video data obtained from the aerial photography of the Nanjing elevated expressway. The results show that the absolute error of the model is concentrated in [0.5, 0.7], and the root mean square error is concentrated in [0.6, 0.8], which are relatively large. The correlation coefficient is about 0.5, and the overall fitting effect is poor.

1 INTRODUCTION

Real vehicle speed change is becoming an increasingly important demand for traffic flow models, thanks to the increasingly micro acquisition means of traffic flow data. Simple models cannot adapt to extreme scenes (Newell 2002), while complex models can only be solved numerically (Treiber et al. 2000). After referring to the driver's behavior characteristics, some scholars turn their attention to the social force model.

In the social force model (Helbing & Tilch 1998), the vehicle is set as a passive body, which is affected by the virtual force from the driver's behavior. Since then, it has been improved by some scholars (Fellendorf et al. 2012; Schönauer et al. 2012). Among them, the Delpiano model has improved the walking and stopping wave under the condition of finite acceleration and deceleration (Rafael et al. 2015) and deduced this improved model to the two-dimensional condition (Rafael et al. 2020).

However, Delpiano's improved model is derived theoretically, and its formula has not been verified by actual data. In this paper, it is verified by the fast-track vehicle trajectory video data obtained from Nanjing elevated aerial photography. Assuming that the model is applicable, the acceleration is calculated and compared with the actual acceleration and tested by absolute error, root mean square error and Pearson coefficient.

2 SOCIAL FORCE MODEL

2.1 *Helbing's original model*

Helbing proposed a social force model in 1998 to simulate the interaction between vehicles. He believes that with the change of speed and spacing, the driver will be stimulated accordingly and

*Corresponding Author: 806558961@qq.com

DOI 10.1201/9781003305026-84

thus produce a corresponding response, which is reflected in the effect of a virtual force on the vehicle. The model formula is as follows:

$$f = f_a + f_r \tag{1}$$

Where f is the social force, fa is the acceleration force, fr is the repulsive force.

Helbing believes that the driver will tend to reach his desired speed under the action of the expected force, which is generally the maximum speed in the free flow state. Due to the repulsive force of other vehicles in the lane, drivers will follow and change lanes according to the changes of surrounding vehicles in order to prevent accidents.

The acceleration formula of the expected force is as follows:

$$f_a = c_1(V - v_1) \tag{2}$$

Where, V is the maximum expected speed, usually the road design speed, v1 is the current speed of the rear vehicle, and c1 is the parameter.

Helbing proposes a formulation for the repulsive force that obscures the nature of acceleration force by subtracting the desired speed. As shown in the figure below, when no other vehicles are in the lane, the driver tends to reach the maximum speed. When the current vehicle is too close, the repulsion force is greater than the expected force, resulting in the deceleration of the rear vehicle. When the current vehicle is far away, the expected force is greater than the repulsion force, resulting in the acceleration of the rear vehicle.

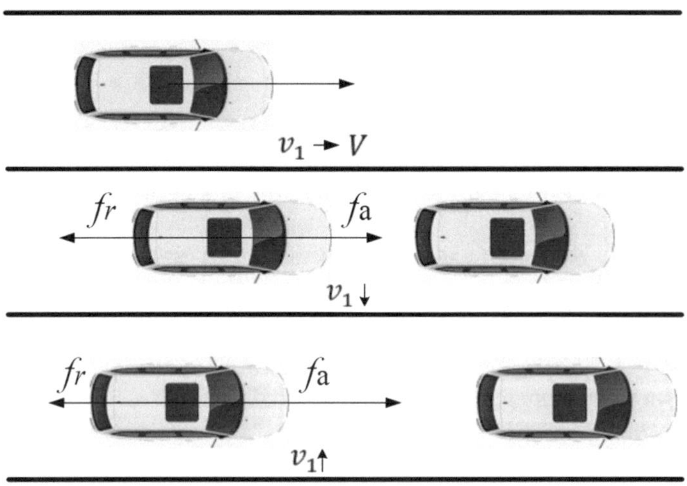

Figure 1. Function diagram of social force model.

2.2 *Delpiano's improved model*

Delpiano improved the social force model, followed the formula of expected force, and put forward the calculation conditions of parameter c1. the specific formula is as follows:

$$c_1 = \frac{a_m}{v_m} \tag{3}$$

Where a_m is the maximum acceleration and v_m is the maximum speed.

Delpiano modified the acceleration formula of repulsion force into the speed difference and spacing of following vehicles.

$$f_r(t) = \{\Delta v \cdot c_2 + \Delta y \cdot c_3\}^- \tag{4}$$

$$\Delta y = y_{i-1}(t) - y_i(t) - \tau_r v_i(t) + S_r \tag{5}$$

$\{\cdot\}^-$ refers to min $\{., 0\}^-$, c_2, c_3 are dimensional parameters, and the units are s^{-1} and s^{-2}, Δv is the speed difference between the front and rear vehicles, $y_{i-1}(t) - y_i(t)$ is the distance between two vehicles, τ_r is the parameter of time and S_r distance respectively.

In order to prove that the interaction force of the social force model is two-dimensional, Delpiano extended the formula to the multi-lane environment. He believes that the vehicle is not affected by the lateral force in the following state because the lateral forces of the vehicles on both sides of the isolation belt offset each other, and the longitudinal formula of the two-dimensional model is still the social force model.

However, Delpiano only explained from the perspective of formula proof, and did not use data for verification. Moreover, the model is aimed at the transition from free flow to congestion and does not mention the change in traffic flow under the continuous state of congestion. Therefore, this paper uses trajectory data to verify the model.

3 TRACK DATA ACQUISITION

The vehicle trajectory data used in this study comes from the bottleneck section of the confluence area near the Yuhua shopping center section in the viaduct of Yingtian Street Expressway on the South Inner Ring Road in the south of the main city of Nanjing. The specific data of the section are as follows:

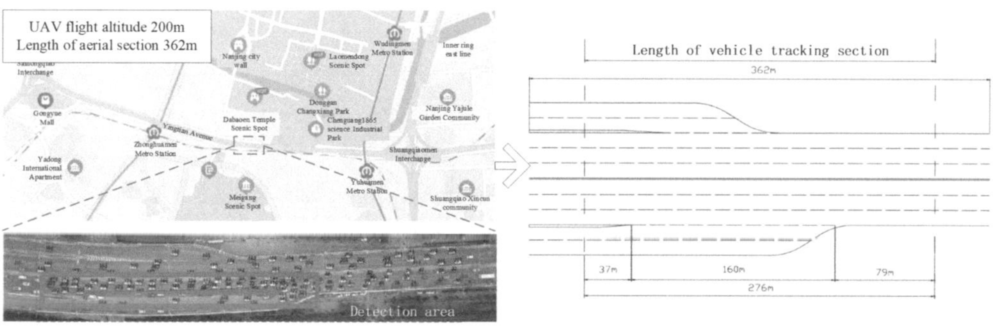

Figure 2. Data acquisition section.

The aerial photographing data is obtained by continuously photographing at an altitude of 200m in the elevated area of Yingtian Street Expressway in Nanjing, Jiangsu Province, China, for 9 minutes in the morning peak. The length of the aerial photographing section is 362M, and 278m is finally intercepted as the length of the vehicle tracking section. The traffic flow video is obtained by UAV aerial photography, the time and vehicle displacement data are obtained by an image recognition algorithm, and finally, the final vehicle trajectory data is generated by data cleaning technology.

In the obtained track data, the congestion period of 545 seconds is selected for research. At this time, the track data of lanes 1-3 are shown in the figure below.

As can be seen from the figure below, the weaving area of on-ramps 4 and 5 is concentrated at 50-200 m (red box area), and the impact of the on-ramp vehicle lane change behavior on the three lanes is gradually delayed and weakened. Considering that this paper mainly focuses on the car following behavior, the selection of car following data is mainly focused on Lane 1. In this paper, the front and rear vehicles without vehicle lane change insertion in the track map are selected as a car following track groups. The time length of each group of data is more than 45s. Finally, 70 track data are selected.

In the black dotted box of the three lanes, it can be found that under the same traffic shock, the process of eliminating the impact and returning to the equilibrium state of the front and rear vehicles with different average speeds, speed differences and the spacing difference is different. It is worth noting that the limit distance that drivers of the three lanes can bear decreases after experiencing congestion (i.e., relaxation), and the density of vehicles increases on the same road cross section after digesting the shock.

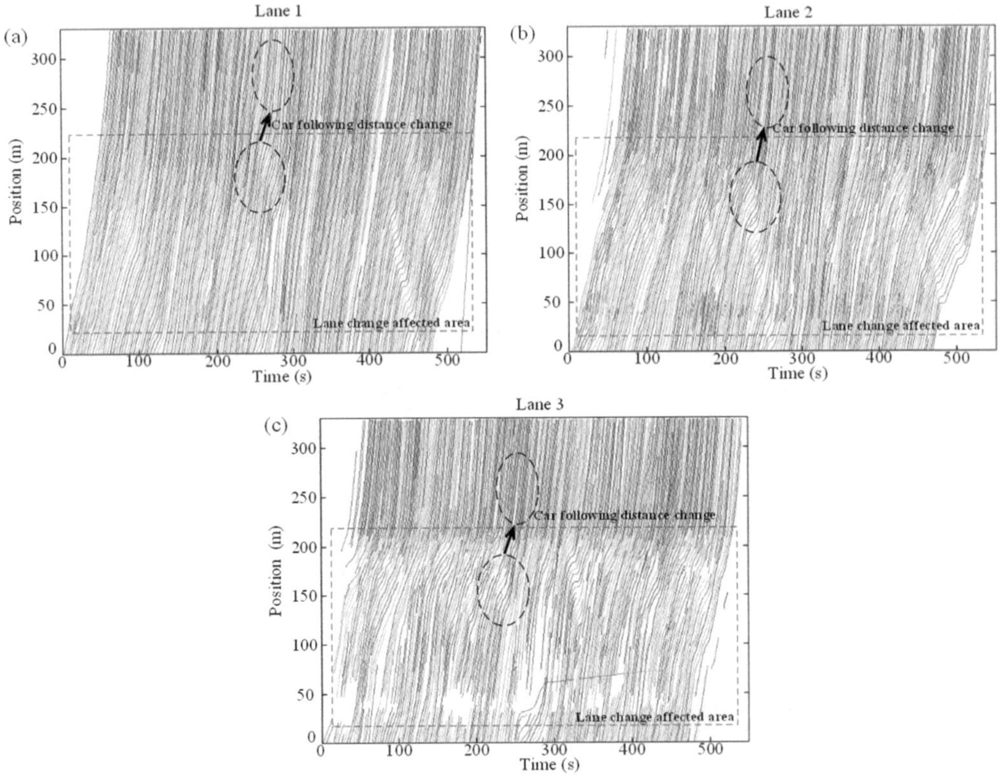

Figure 3. Lane trajectory data diagram.

4 DATA ANALYSIS

Based on Delpiano's improved social force model, this study uses the displacement and velocity data of trajectory data to calculate the calculated acceleration of 70 groups of data, and then compares it with the actual acceleration calculated by the acceleration formula to study the fitting effect of the improved model.

In this paper, absolute error (MSE), root mean square error (RMSE) and Pearson coefficient are used to test the fit degree of data fitting. MSE reflects the deviation degree between the car following model and the measured data, RMSE reflects the change degree of error difference, mainly characterizes the stability of the model, and the Pearson coefficient represents the fitting degree between the model data and the measured data. The calculation results of 70 groups of trajectory data are as follows:

As can be seen from the above figure, the absolute error is mainly concentrated between [0.5, 0.7], with a median of 0.59, while the root mean square error is mainly concentrated between [0.6,

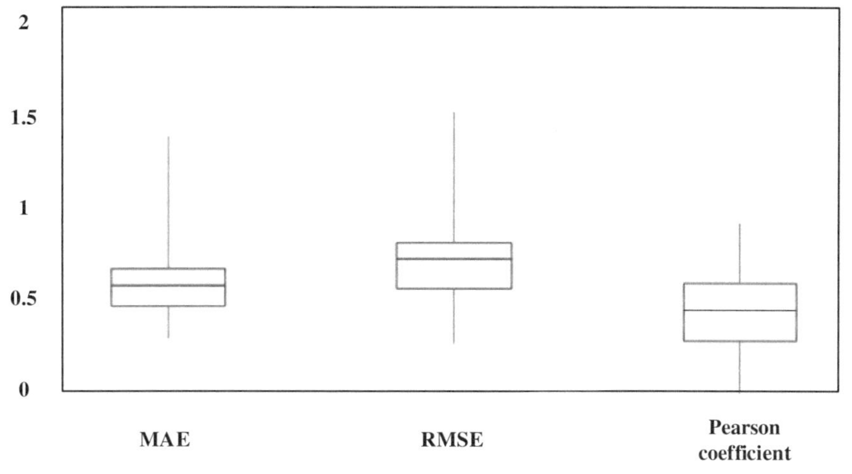

Figure 4. Data calibration box diagram.

0.8], with a median of 0.7, and 75% of the data is greater than 0.5. Therefore, this paper believes a large error exists between the improved model and the real data.

The correlation coefficient is about 0.5. Generally, the interval of the Pearson coefficient is [−1,1]. The closer it is to 1, the stronger the correlation is. 0.5 can be considered to have a certain correlation, but the correlation is weak. Therefore, it can be considered that Delpiano's model is not completely convincing to explain the interaction between vehicles in congestion.

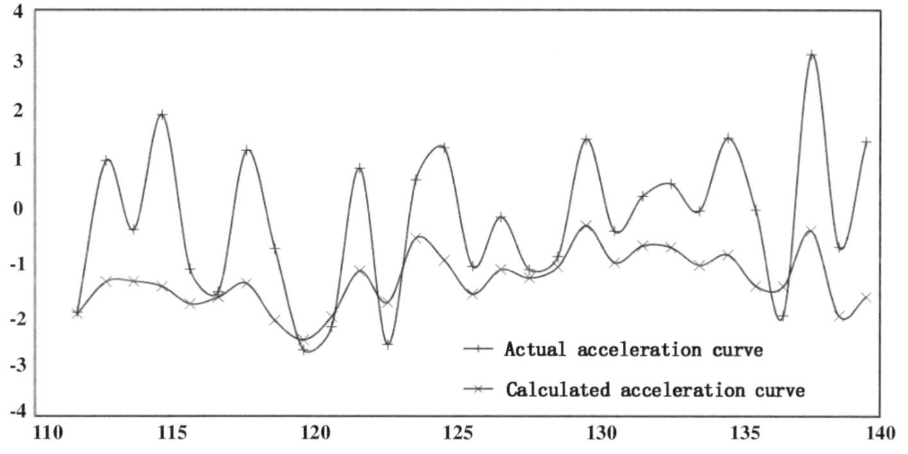

Figure 5. Fitting effect drawing.

It can be seen from the fitting diagram that the fluctuation of the curve calculated by the model tends to be consistent with the real acceleration curve. The difference is that the oscillation amplitude is much smaller than the real curve, which is also the main reason for the error. From the average of the data, the calculated curve is also inconsistent with the real curve. This paper believes that the reason for this phenomenon is that the drivers in the congested area no longer expect to reach the ideal speed of free flow on the premise of understanding their own state, but change to pass through the congested area as soon as possible. Therefore, the tolerance limit spacing is reduced, and the expected speed is changed.

5 CONCLUSION

In this paper, the trajectory data obtained by the Nanjing Yingtian viaduct are used to verify Delpiano's improved social force model. The fitting degree of the data is characterized by absolute error, root mean square error and correlation coefficient. It is found that the model has some defects in the state of congested traffic flow, with large error and low correlation. When analyzing the specific fitting data, it is found that the fluctuation range of the improved model is not enough, and the average value is lower than the real data. It is speculated that the reason is that the improved model does not consider the fluctuation caused by driving characteristics and the assumption of expected force.

ACKNOWLEDGMENT

The authors appreciate the funding support from the scientific research project of the National Ministry of Housing and Urban-Rural Construction (2019-K- 151), the Guangxi Key Project of Research and Development (Grant No. Guike AB17292087), and the innovation project of GUET graduate education (2020YCXS124, 2021YCXS179).

REFERENCES

Fellendorf, M., Schönauer, R., Huang, W., 2012. Social force-based vehicle model for two-dimensional spaces. In: *Transportation Research Board 91st Annual Meeting*, pp. 12–1347.

Helbing, D., Tilch, B., 1998. Generalized force model of traffic dynamics. *Physical Review E* 58, 133–138.

Newell, G., 2002. A simplified car-following theory: a lower order model. *Transportation Research Part B* 36, 195–205.

Rafael D., Jorge L., Juan C., Juan H., 2015. The kinematic wave model with finite decelerations: A social force car-following model approximation. *Transportation Research Part B*, 71 182–193.

Rafael D., Juan H., Jorge L., Juan C., 2020. A two-dimensional car-following model for two-dimensional traffic flow problems. *Transportation Research Part* 504–516.

Schönauer, R., Stubenschrott, M., Huang, W., Rudloff, C., Fellendorf, M., 2012. Modeling concepts for mixed traffic. Transportation Research Record: *Journal of the Transportation Research Board* 2316, 114–121

Treiber, M., Hennecke, A., Helbing, D., 2000. Congested traffic states in empirical observations and microscopic simulations. *Physical Review E* 62, 1805–1824.

Advances in Urban Engineering and Management Science – Khalil & Yang (Eds)
© 2023 The Author(s), ISBN 978-1-032-30426-7

One-way traffic organization optimization method of microcirculation road network in Handan City

Shunlin Li*
Xi'an University of Architecture and Technology, Xi'an, China

ABSTRACT: Through the investigation of the traffic status in Handan city, in order to alleviate the traffic congestion problem, the one-way traffic organization optimization design is carried out in this area from two aspects of road organization adaptability and concrete implementation scheme Based on OD data, one-way traffic organization design was evaluated by Transad, and the service level of Han Shan Street xiaoguangming Street was upgraded from GRADE C to Grade B, and the service level of Heping Road was upgraded from grade D to grade C In this paper, the applicability of microcirculation one-way traffic organization in the old city is analyzed, and the optimization scheme is put forward, which improves the problem of regional traffic congestion and complements the method of urban traffic planning.

1 INTRODUCTION

As the old city of Handan city, the traffic congestion in Handan District gradually develops from the line to the surface, and presents a further worsening trend, which seriously affects the travel of residents in Handan District and the quality of life of residents. "One-way traffic" refers to the way in which only certain vehicles are allowed to pass in one direction. According to the traffic congestion problem in the old city, a measure is proposed to change part of the road from two-way to one-way (Gao 2021). In China, Chengdu began to try out a one-way traffic organization in 2009, and now it has set up fixed one-way traffic and reversible one-way traffic. Chengdu has improved the driving speed by nearly 20% through the implementation of a one-way traffic organization (Hu 2020). Recently, Shanghai has set the planning target of improving the one-way network in the central area in the 13th Five-Year Plan of Comprehensive Transportation. The plan is to expand one-way roads to 186 (Wang 2019).

In foreign countries, the UK, and the narrow streets of 17 began to implement one-way traffic rules, Japan's main special measures in 5.5 m below the right road, road network density is higher, Manhattan in New York Manhattan Road divided into a number of the rectangular areas, equipped with dual carriageway, on the one hand, to ensure that the clear on the road, on the other hand, can make travel has many choices (Xi et al. 2019).

From the perspective of the implementation of one-way traffic at home and abroad, the organizational forms of one-way traffic are also very different because of the different traffic environments. Based on the experience of one-way traffic on urban roads at home and abroad, this paper puts forward a one-way traffic organization optimization method of the micro-circulation network in the old city, which can alleviate the traffic congestion in Han Shan district to a certain extent.

*Corresponding Author: 1132103582@qq.com

DOI 10.1201/9781003305026-85

2 BASIC PRINCIPLE OF UNIDIRECTIONAL TRAFFIC ORGANIZATION OF MICROCIRCULATION NETWORK IN OLD CITY

2.1 *Overview of the old city*

An old city generally refers to a city with a long history of human habitation. Handan city is one of the old cities. It is in the central location of Hebei Province, the basic road system was built in the 1980s, and the city updates slow due to historical reasons.

2.2 *Definition of the microcirculation road network*

The term "circulation" was first described in 2005 in the Beijing municipal government work report. The report explains its function and status: a perfect urban road microcirculation system can provide a reasonable distribution of traffic flow, ease traffic pressure, improve traffic speed, and is an indispensable part of city roads.

2.3 *Organization of one-way traffic*

For one-way traffic, its role is mainly reflected in diverting the short distance travel volume on the road network during peak hours, and alleviating traffic congestion on the trunk road. In order to improve the utilization rate of branch roads in the road network, traffic organization or traffic control measures are generally adopted.

3 FIELD INVESTIGATION AND ANALYSIS IN THE HANDAN MOUNTAIN AREA

3.1 *Regional land development and use analysis*

3.1.1 *The traffic demand in Handan District is great and the branch road network is systematic*

The district was built earlier in Handan District, which failed to fully consider the future traffic composition, resulting in a large population of residents. Residents mostly choose motor vehicles to travel, and the land is mostly residential, commercial land, historical site parks, and narrow roads will often occur traffic congestion. Maintaining the Integrity of the Specifications.

3.1.2 *Narrow width of road red line*

The district of Handan District was built earlier, and the future traffic composition was not fully considered in the development process. Meanwhile, due to the historical problems in the district, the road red line was narrow, and the road could not be widened now.

3.2 *Investigation and analysis of road capacity in Handan Districts*

Most of the land in the study area is commercial and residential land, while there are lands for education, administrative office and hospital, refer to HCM2010, Calculate the road capacity, traffic flow within the region, as shown below.

Table 1. Road network traffic index table of the north area of Handan District.

Road name	Possible capacity (pcu/h)	Traffic flow (pcu)	Density (pcu/km)	Service level
Yuxin South Street	640	575	9.6	E
Lingyuan road	437	600	15	F
Ling Southwest Street	1316	940	23.5	C
Heping road	722	590	14.8	D
Huo Mo Street	1600	150	5.0	C
Xin Shichang Street	1600	120	1.7	C
Xiao Guangming Street	1600	135	1.5	C
Han Shan Street	1600	155	1.8	B

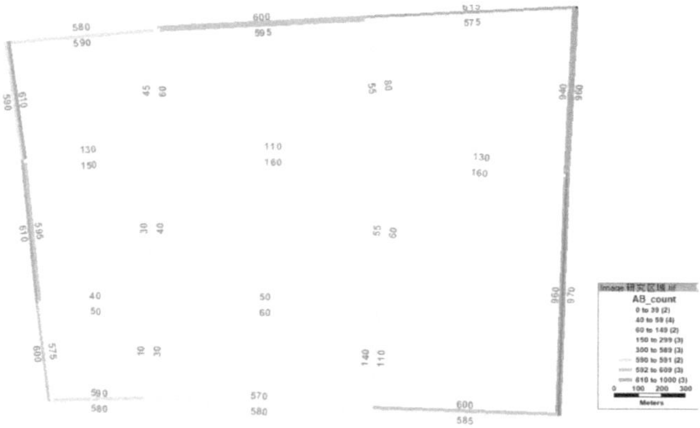

Figure 1. Map of traffic flow in the northern part of Handan District.

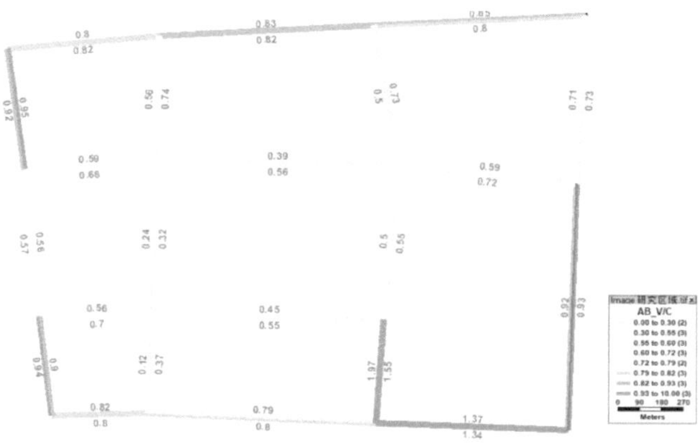

Figure 2. Map of V/C in the northern part of Handan District.

4 ONE-WAY TRAFFIC ORGANIZATION OPTIMIZATION DESIGN

Coordination of the relationship between regional road traffic system and regional road network form determines the effectiveness of the urban road traffic system to implement a one-way traffic organization scheme. In order to optimize the scheme and adapt to the traffic characteristics of the research region, it is necessary to analyze the road network form within the region. It is also necessary to analyze whether the road condition of the region can adapt to a one-way traffic organization form based on the traffic environment of the region (Zhao & Zhang 2021).

4.1 *Han Shan street*

Hanshan Street has a total length of about 436m, which is divided into the north section, middle section and south section by Huomo Street and Xinmarket Street, with Heping Road in the north and Lingyuan Road in the south, as shown in Figure 3-5. Among them, the north section is about 160 m long, the middle section is about 163 m long, and the south section is about 110 m long. The road section is shown below.

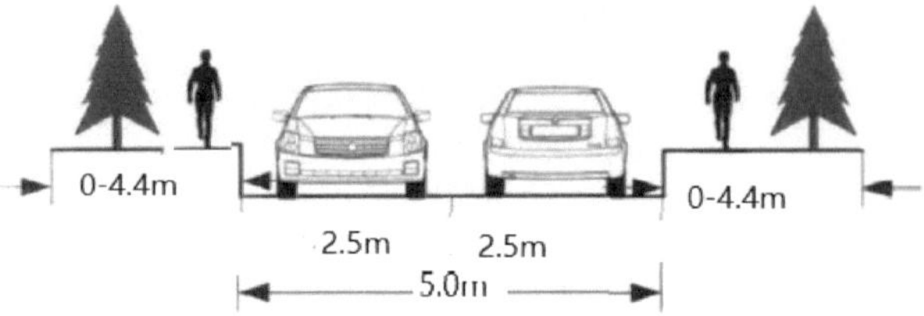

Figure 3. Schematic diagram of Hanshan Street section.

4.2 *Xiao Guangming street*

Xiao guangming Street has a total length of about 410 m, with Heping Road in the north and Lingyuan Road in the south. It is divided into north, middle, and south sections by Huomo Street and Xinmarket Street, as shown in Figures 3–10. The north section is about 135m long, the middle section is about 165 m long, and the south section is about 112 m long. The road section is shown below.

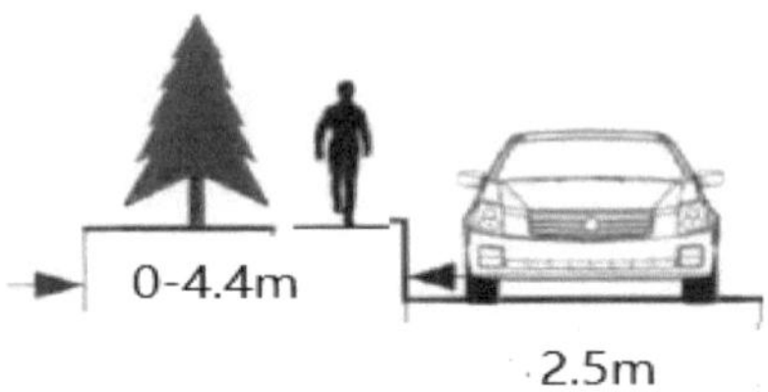

Figure 4. Road section of the north section of Xiao Guangming Street.

4.3 *Huo Mo street*

Huo Mo Street has a total length of about 586 m, connecting Yuxin South Street in the west and Ling Southwest Street in the east. It is divided into east, middle, and west, from Xiaoguangming Street and Han Shan Street. The eastern section is about 185m long, the middle section is about 257 m long, and the western section is about 144 m long. Figure 5 shows the road sections of the middle and east sections.

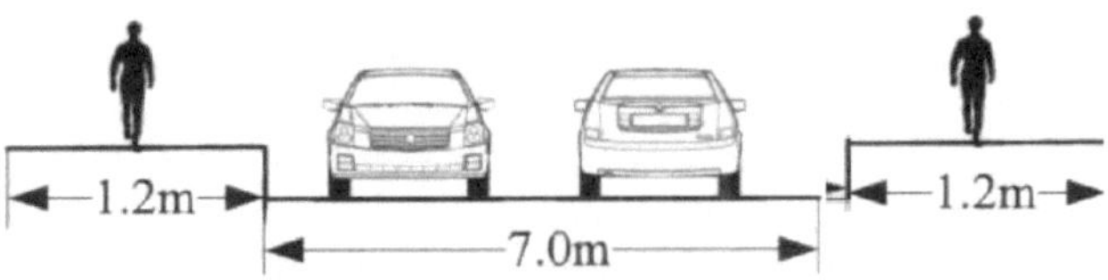

Figure 5. Road section of the north section of Xiao Guangming Street.

4.4 *Xin Shichang street*

Xinmarket Street is 365m long, with Yu Xinnan Street in the west and Han Shan Street in the east. It is divided into two sections: the west section is 125 m long, and the east section is 240 m long.

On the north side of the west section of Xinmarket Street is Yijia Express Hotel, and on the north side of the middle section is Yashi Mingdi Residential Area. On one side of the road, vehicles occupy the road and parking, as shown in Figures 3-4, so the one-way road is taken from east to west.

Figure 6. Parking on New Market Street.

5 EVALUATION OF ONE-WAY TRAFFIC WEAVE EFFECT IN HANDAN DISTRICT

In this paper, TranCAD software is used to simulate and analyze the road section after the implementation of the one-way traffic organization and compare it with the road section before the implementation to judge whether the implementation plan is reasonable.

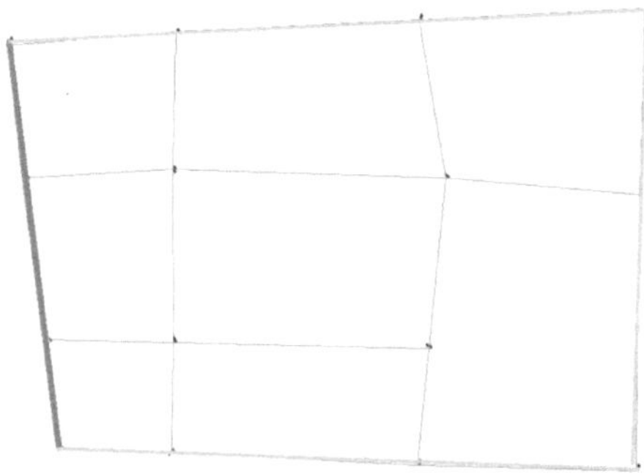

Figure 7. Study the simplified map of the road network in the northern part of Handan District.

5.1 *Existing OD matrix and prediction OD matrix*

The study area can be divided into eight traffic areas by TransCAD software, and The USER balance method is used to record OD backward.

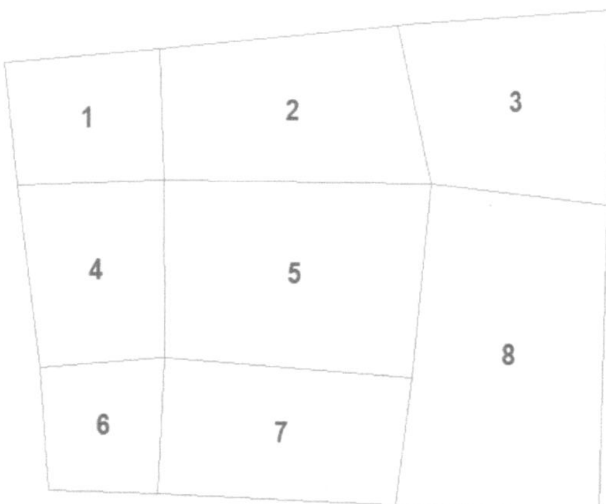

Figure 8. Schematic diagram of traffic district division.

	1	2	3	4	5	6	7	8
1	0.00	92.59	40.00	28.22	28.95	63.33	63.33	63.33
2	80.01	0.00	20.31	17.56	21.16	21.16	10.86	10.86
3	36.11	18.54	0.00	15.27	21.60	21.60	15.00	13.20
4	31.80	21.54	25.09	0.00	31.29	31.29	31.29	31.29
5	31.59	24.26	27.87	31.03	0.00	1.00	1.00	1.00
6	70.00	24.26	27.87	31.03	1.00	0.00	1.00	1.00
7	70.00	16.11	40.00	31.03	1.00	1.00	0.00	1.00
8	70.00	16.11	17.75	31.03	1.00	1.00	1.00	0.00

Figure 9. Inverse deduction of O-D matrix.

5.2 *Comparison of traffic parameters between current road network and one-way optimized road network*

After Huomo Street, Xinshiji Street, Han Shan Street and Xiaoguangming Street are set as one-way roads in this scheme. The road capacity will increase to a certain extent due to the increase in lane width and reduction of interference factors during driving. The traffic volume, saturation and service level of the optimized road section are shown below.

Table 2. Comparison of traffic volume saturation service level of road sections before and after optimization.

	Traffic Flow			Service level	
Road name	Before	After	Rate of change	Before	After
Huo Mo Street	150	212	41.33%	C	C
Xin Shichang Street	50	73	46%	C	C
Xiao Guangming Street	45	53	17.78%	C	B
Han Shan Street	55	97	76.36%	C	A
Yu Xinnan Street	575	513	−10.78%	E	D
Ling Yuan Road	600	529	−11.83%	F	F
Ling Xinan Street	940	830	11.7%	C	C
He Ping Road	590	537	8.98%	D	C

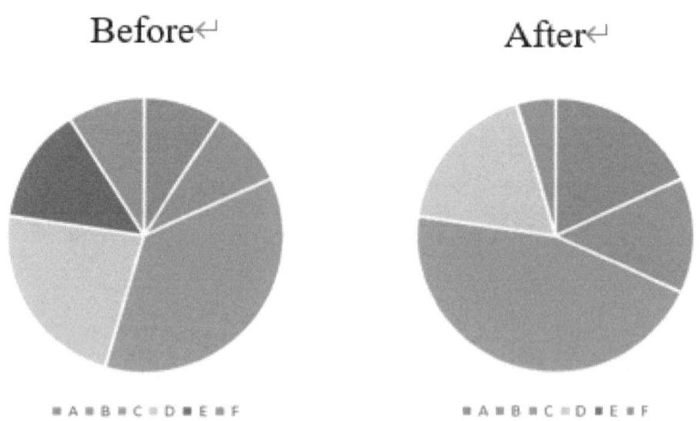

Figure 10. Comparison chart of road service level before and after optimization.

As shown in Figure10, by implementing one-way traffic organization, the road service level is higher, the saturation is reduced, and the traffic pressure is relieved. The optimized design scheme is reasonable.

6 CONCLUSION

The main achievements of this paper are as follows:

1 The traffic characteristics of the old city are analyzed, and the one-way traffic organization can be implemented in Handan District.
2 Analyze the road network form, road conditions, traffic demand and public transport in Han Shan District, and obtain an effective one-way traffic organization optimization scheme.
3 Using TransCAD software to evaluate the effect of one-way traffic organization, the results prove that the scheme alleviates the congestion situation in the study area and improves the service level in the study area.

This paper provides a new idea for solving the problem of traffic congestion in the old city, and explains the applicability of the single traffic organization method in the old city. TranCAD is used to analyze the optimization plan before and after, and the analysis results prove the rationality of this point of view, which can serve as a reference for future traffic planning in the old city.

REFERENCES

Gao Binqi (2021). Research on Urban Microcirculation Bus Network Planning Method. D. Guilin University of Electronic Science and Technology.

Hu Songtao, Zhang Kexin, Zhou Yili, Zhu Chengcheng (2020). Improvement Strategy of Traffic Microcirculation in YanTai Old City. J. Traffic and transportation, 36(05):96–100.

Wang Yanning (2019). Research on Optimization of Urban Traffic Microcycle Organization. D. Lanzhou Jiaotong University.

Xi Kuanxiang, Zha Weixiong, Ji Wenqian (2019). Research and Application of Microcirculation Theory in Small Cities. J. IOP Conference Series: Materials Science and Engineering, 688(4).

Zhao Yanfeng, Zhang Jianfeng (2021). Research on Construction Strategy of Urban Traffic Microcirculation System: A case study of Zhengzhou. J. Integrated Transportation, 43(09):132–138.

Advances in Urban Engineering and Management Science – Khalil & Yang (Eds)
© 2023 The Author(s), ISBN 978-1-032-30426-7

The impact of the opening of high-speed railways on corporate financing constraints based on the spatial effect

Zanlu Tian*
Beijing Jiaotong University, Beijing, China

ABSTRACT: The opening of the high-speed railway plays a positive role in promoting urban economic development. Based on the sample of listed companies registered in prefecture-level cities from 2009 to 2020, we use the financing constraint variable kz index as the Financing Constraint Index. The time-varying difference in differences method is employed to analyze the impact of the high-speed railway on corporate financing constraints, and a spatial econometric model is constructed to verify the network effect of the high-speed railway. It has been proved that high-speed rail has a significant network effect between high-speed rail and urban enterprises, and the scale of enterprises is heterogeneous in the financing effect. Our findings carry implications for optimizing the high-speed railway network, promoting the urban economy, and improving digital transformation and traffic efficiency.

1 INTRODUCTION

The construction of the high-speed railway deepens the openness between the cities. Through the high-speed railway connecting different cities, the time distance will be shortened, and the accessibility between cities will be improved. The rapid flow of factor resources brought by the opening of the high-speed railway will accelerate the spread of financial resources from the central city to the cities along the line. The opening of a high-speed railway may directly relieve the financing constraints of local enterprises and indirectly ease the financing constraints of enterprises in surrounding cities, thus promoting the spatial pattern. Based on this, we utilize listed enterprises as the research object, and investigate the impact of the high-speed railway on the financing constraints of enterprises. Our study contributes to promoting the network effect of high-speed railways and provides a reference for the economic development of urban enterprises.

Most of the existing literature focuses on the impact of high-speed rail on economic effects. These researches can be divided into two aspects: macro-effect and micro-effect. On the macro level, Liu and Hu (2010) argue that transportation infrastructure has a significant positive effect on China's economic growth. According to Liu and Li (2017), high-speed rail construction has boosted not only local economic growth, but also the economic growth of neighboring cities. Yang et al. (2015) argue that transport infrastructure has a significant spatial spillover effect on urban processes and has implications for policymaking. On the micro-level, high-speed rail has optimized the allocation of capital elements among enterprises, thus improving the production efficiency of enterprises (Li 2017; Tan 2019).

In summary, although the impact of the high-speed railway on microeconomics is significant, there is little research on the relationship between the high-speed railway and corporate financing constraints, which is not enough to build a theoretical system. Based on this, this paper empirically analyzes the role of high-speed railways in alleviating corporate financing by using the multi-time-point double-difference method and constructs a spatial econometric model to verify the network effect of high-speed railways.

*Corresponding Author: 1029063331@qq.com

 DOI 10.1201/9781003305026-86

2 SAMPLE AND RESEARCH METHODS

2.1 *Sample selection*

In this paper, the A-share listed companies are selected as the research object, and the sample data are processed as follows: Remove the financial listed companies, the ST/*ST listed companies, the missing data and use Winsorize method (1% and 99%). After screening, 3393 effective observation data samples were obtained. Data on the opening of the high-speed rail line are collected by hand from the China Railways Corporation, local news, etc. At the same time, to ensure the endogenies of high-speed rail, only the data of the local cities are retained. Other financial data comes from CSMAR and Wind databases.

2.2 *Time-varying differences in differences model design*

The premise of the double difference model is that there is an exogenous policy shock, which only exists in the treatment group and does not exist in the control group. The time of high-speed railway opening in different cities is not the same. It controls the time effect and city effect by the following model:

$$loan_{it} = \alpha + \beta_1 Treated_{it} + \beta_2 X_{it} + \delta_t + \gamma_c + \varepsilon_{it} \tag{1}$$

Among the equation, *loan* represents the corporate financing constraints, δ_t is a time-fixed effect, γ_c denotes the city-fixed effect, and ε_{it} is a random error.

2.3 *Spatial autoregression model design*

The indirect effect of transportation construction is realized by strengthening the effect of the inter-city transportation network and accelerating factor circulation. Based on the method of Hu and Liu (2009), this paper constructs a spatial autoregression model to verify the network effect model of high-speed railway opening to alleviate the financing constraints of enterprises:

$$loan_{it} = \alpha + \beta_1 Treated_{it} + \beta_2 X_{it} + \rho W \times loan_{it} + \delta_t + \gamma_c + \varepsilon_{it} \tag{2}$$

In the equation, W represents spatial weight between cities, which reflects the spatial connection between the cities, $W \times loan_{it}$ is the spatial lag variable, ρ is the spatial lag coefficient that reflects the spatial dependence of the observed value, which was used to test the network effect of high-speed railway on corporate financing.

2.4 *Variables selection*

The core variable of this paper is the corporate finance constraint. According to the KZ index calculation method proposed by Kaplan (1997), this paper constructs the KZ index based on the financial data of Chinese listed companies to measure the degree of financing constraint.

The central explanatory variable of this paper is Treated. If the enterprise establishment is in the opening of high-speed rail in the year and after, the value of 1 will be taken; otherwise, the value of 0 will be taken. If no high-speed rail is opened during the inspection period, 0 will be taken for all years.

Corporate-level control variables include firm size, asset-liability ratio, return on assets, and time since listing. The city-level control variable is introduced as per capita GDP measured in logarithms. The specific variable definitions are shown in Table 1.

3 EMPIRICAL RESULTS

3.1 *Basic regression result*

The first column in Table 2 shows the basic regression results for the impact of the high-speed railway on corporate financing constraints, as compared to the regression results not shown in the table, which are not based on the urban fixed effect and the time fixed effect, whether control fixed effect is added, Treated and kz is significantly negative at 5% or 1% level, which indicates that high-speed railway can significantly ease the financing constraint of listed companies. It demonstrates that better regional transport infrastructure will strengthen the economic radiation and factor flow space of the central city, resulting in the diffusion of financing resources from the central city to the surrounding cities, which alleviates the financing restriction of the city-level enterprises along the line.

Table 1. Definition of variables.

Variables	Symbol	Name	Notes
Explained variable	*loan*	Corporate financing constraints	KZ indicator (Kplan,1997)
Explanatory variable	*Treated*	Opening of high-speed railway	The year HSR is opened at the company location and in the following years = 1, not opened = 0
Control variables	*Size*	Company size	The total assets are in natural logarithms
	Lev	The ratio of assets to liabilities	Total liabilities/total assets
	Roe	The ratio of net assets to liabilities	Net profit/net assets
	Age	Time since listing	Time since listing
	PGDP	Per capita GDP	Per capita GDP measured in logarithms

3.2 *Selection of spatial effect model and result analysis*

The methods of selecting the spatial effect model in this paper are as follows: Based on the preparation of the previous spatial data, firstly, calculate the Global Moran'I to carry out spatial autocorrelation measure; secondly, the selected spatial autocorrelation model is determined by LM test, The significance test was carried out by Hausman. Finally, we choose the double fixed-effect model according to the regress result. The results of the spatial effect are listed in the second column of Table 2. According to the results, has passed the test of significance, which shows that high-speed railway has a positive network effect on alleviating the financing of enterprises.

Table 2. HSR opening and corporate financing constraints.

Variables	Basic regress	Spatial
Treated	−0.2896*** (−3.02)	−0.2758*** (−3.27)
ρ	–	−0.024* (−2.29)
Other Controls	Yes	Yes
City Fixed Effects	Yes	Yes
Year Fixed Effects	Yes	Yes
Adjusted R^2	0.348	0.350
Observations	3393	3393

3.3 *Heterogeneity tests*

The size of enterprises can have a direct impact on their financing capacity. According to the existing literature, large-scale enterprises have better financing ability and obtain more financial support than smaller enterprises. To analyze the impact of the high-speed railway on financing constraints of enterprises on a different scale, we classify enterprises above the median value of total assets as a large-scale group and enterprises below the median value of total assets as a small-scale Group. The results are shown in Table 3. The regression coefficients are significantly negative at the levels of 5% and 1%, indicating that the high-speed railway can help alleviate the financing constraints faced by enterprises of all sizes. Furthermore, the comparative coefficient shows that the impact of high-speed rail on its opening for smaller enterprises is greater.

3.4 *Robustness tests*

To guarantee the reliability of the results, the following robustness tests are carried out: (1) KZ index measurement. We use the method in Wei et al. (2014) to replace Tobin's Q with sales Growth. The empirical results are consistent. (2) Change the timing of policy implementation. The results show that the coefficient of high-speed railway opening is not significant, so the interference of other factors can be eliminated.

Table 3. The influence of scale heterogeneity.

Variables	Large-scale	Small-scale
Treated	−0.232**	−0.353*
	(−1.99)	(−1.90)
Other controls	Yes	Yes
City fixed effects	Yes	Yes
Year fixed effects	Yes	Yes
Adjusted	0.348	0.350
Observations	1389	2004

4 CONCLUSIONS

Based on the data from 2009 to 2020 of China's A-share listed companies registered in prefecture-level cities, we empirically test the impact of high-speed rail on corporate financing constraints by using Time-varying DID. The main conclusions show that: (1) When the high-speed railway is opened in the registered place, the financing constraints faced by enterprises are relieved remarkably. (2) There are positive network effects between the high-speed railway and enterprises through spatial overflow and accelerating inter-regional factor flow. (3) There is heterogeneity in the scale of enterprises in the financing mitigation effect brought by the high-speed railway. In contrast, for smaller enterprises, the opening of high-speed rail on corporate financing constraints to ease the greater role.

Based on the conclusions indicated previously, we suggest the following policy implications: (1) Rely on high-speed rail to drive the growth of enterprises, thus helping the city's economic development. In the future, the construction of the main passage should be promoted nationwide, and the density of the high-speed railway network should be strengthened to shorten the space-time distance between regions. (2) Based on the high-speed railway, implement the regional driving strategy of differentiation. In the eastern and central areas where the high-speed railway network is formed, the strategy of "network to the entirety" will be carried out. In the West and Northeast regions, where the high-speed railway does not present the network pattern, the strategy of "network line" will be carried out. (3) Promote the digital and intelligent transformation of transportation, and improve the convenience of transportation.

REFERENCES

Hu, A. & S. Liu (2009). Transportation, economic growth, and spillover effects: a spatial econometric analysis based on China's provincial data. *China's Industrial Economy* 05: 5–14.

Ju, X., Lu, D. & Y. Yu (2013). Financing Constraints, working capital management and sustainability of enterprise innovation. *Economic Research* 48: 4–16.

Kaplan, S. N. & L. Zingales (1997). Do investment-cash flow sensitivities provide useful measures of financing constraints? *The quarterly journal of economics* 112: 169–215.

Liu, S. & A. Hu (2010). Transportation Infrastructure and economic growth: A Perspective of regional disparity in China. *China's industrial economy* 04: 14–23.

Liu, Y. & Y. Li (2017). High-speed railway construction and urban economic growth in China. *Financial Research* 11: 18–33.

Wei, Z., Zeng, A. & B. Li (2014). Financial ecology and firm financing constraints: An empirical study of Chinese listed companies. *Accounting Research* 05: 73–80+95.

Yang, C. & Q. Han (2015). The impact of Transportation Infrastructure and spatial spillover on urbanization: An analysis based on provincial panel data. *Urban issues* 12: 62–68.

Advances in Urban Engineering and Management Science – Khalil & Yang (Eds)
© 2023 The Author(s), ISBN 978-1-032-30426-7

PM2.5 spatial distribution characteristics analysis in subway station

Liyang Liu, Hui Liu* & Pan Xu
School of Architecture and Urban Planning Huazhong University of Science and Technology Wuhan, China

Yiming Ma
School of Electrical and Electronic Engineering Huazhong University of Science and Technology, Wuhan, China

ABSTRACT: The air quality of subway stations has attracted much more attention. The study of the spatial distribution characteristics of PM2.5 is important in both pollutant control and subway station design. In this paper, combining pedestrian flow analysis (PFA) and Computational Fluid Dynamics (CFD) simulation, a novel method for characterizing the spatial concentration distribution of PM2.5 in subway stations is proposed, in which PFA is used to analyze the spatial-temporal movement characteristics of pedestrians to consider the values of pedestrians as the one of main PM2.5 source, and the CFD model is used to analyze the PM2.5 concentration distribution under different airflow organizations to determine the spatial average concentrations. A double-layer underground station with a platform screen door system is studied as an example. When the outdoor PM2.5 concentration is 61 $\mu g/m^3$, the simulated average concentration for the concourse floor and platform floor is 53.2 $\mu g/m^3$ and 64.5 $\mu g/m^3$, respectively.

1 INSTRUCTION

Owing to the rapid economic construction and the increase in population density, subways have become one of the most important transportation modes. However, with the dramatic increase in passenger flow and urban air pollution problems, the air quality problems caused by various pollutants in underground subway stations deserve extensive attention. Various types of particulate matter have been identified as one of the major air pollutants in the subways (Cao et al. 2017). Therefore, research on the distribution characteristics of fine particulate matter in subway stations is of great importance in the construction of subway stations and pollutant control.

Research related to air pollution in subway stations initially began with the study of the composition of pollutants (Colombi et al. 2013; Passi et al. 2021) gaseous, particulate matter, and microorganisms are three main sources of pollution. Then, some scholars set out to study the effects of various facilities in subway stations on the distribution of pollutant concentrations. For example, Kim et al. investigated the effect of Platform Screen Door (PSD) systems on airflow organization and spatial distribution of particulate matter in metro stations (Kim et al. 2014); Guan et al. analyzed the variation of pollutant concentrations under the combined effect of entrance/exit air infiltration and ventilation systems (Guan et al. 2020). However, relevant studies rely too much on the actual measurement data of pollutant concentrations, and obtaining the concentration data requires a lot of labor and material resources, thus making it hard to take pollutant control into account in the design of subway stations. Therefore, some scholars have tried to apply numerical simulation methods, such as Computational Fluid Dynamics (CFD), to the study of pollutant

*Corresponding Author: liuhui@hust.edu.cn

DOI 10.1201/9781003305026-87

distribution characteristics (Bolourchi et al. 2018; Peng et al. 2021; Teodosiu et al. 2016), but pedestrians, which are the most important source of pollutant generation, had not been properly considered (Shakya et al. 2020).

To solve the above problems and facilitate the positive consideration of pollutant distribution characteristics in the design of subway stations, this paper proposes a simulation analysis method for PM2.5 pollutants in subway stations based on CFD theory and pedestrian flow analysis (PFA), which is carried out for a typical double-layer PSD underground subway station. PFA is used to determine the size and location of pedestrian pollutant generating sources, which in turn gives the boundary conditions for particle spatial concentration distribution simulation.

2 MATERIALS AND METHOD

2.1 *Experimental site description*

This paper takes a double-layer PSD underground subway station as an example to illustrate the proposed methods. This station is divided into a concourse floor and a platform floor, with a total area of 13,940 m^2, the total length is 204 m, and the width of the platform floor is 13 m. The station has four entrances/exits A, B, C, D, and two sets of wind pavilions. The floor plan of the concourse floor and platform floor are shown in Figure 1.

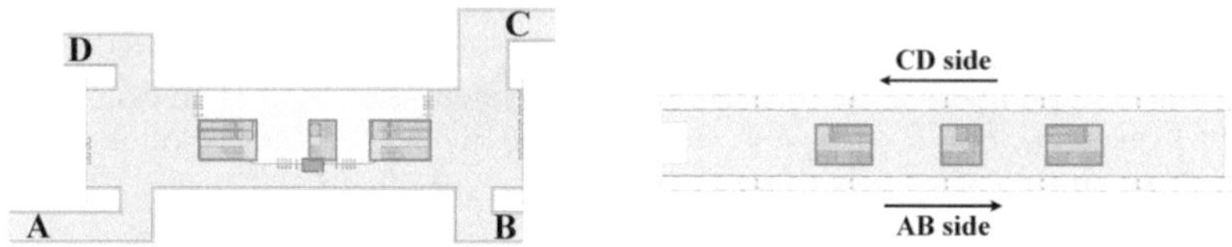

(a) Concourse floor. (b) Platform floor.

Figure 1. The floor plan of the concourse floor and platform floor.

2.2 *PFA model*

PFA is performed with the help of the Massmotion software based on the social force model; the effectiveness of this software has been demonstrated (Rivers 2014). According to the floor plan of this station, the PFA model was built. Figure 2 shows the PFA model of the concourse floor.

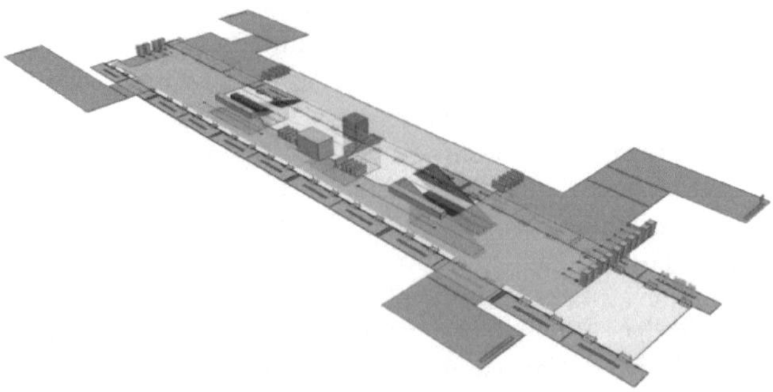

Figure 2. The floor plan of the concourse floor and platform floor.

For the boundary condition setting, the pedestrian step speed was set to 1.25 m/s. 20% of the incoming pedestrians went to the ticketing machine to purchase tickets, and 80% of the pedestrians entered the station directly. The software also needs to give the number of pedestrians getting on and off each train in different running directions. Table 1 shows the statistics of pedestrians' information during one-hour weekday morning off-peak hours. The interval between trains is 6 minutes. In addition, according to the statistical results, the ratio of pedestrian choice for the four entrances/exits, A, B, C, D, is set to 4:1:1:4.

Pedestrians' choice of action path in the MassMotion simulation considers action path length, queuing time, and facility usage. The path is chosen based on the perceived cost of all available routes so that pedestrians reach their final goal without backtracking, as shown in Table 1,

Table 1. Statistic of pedestrians' information during one hour weekday morning peak hours.

	Getting on	Getting off
Up direction	2113	951
Down direction	708	918

$$Cost = \left(\mathrm{W}_D \times \left(\frac{D_G}{V} \right) \right) + (\mathrm{W}_q \times Q) + (\mathrm{W}_L \times L) \tag{1}$$

where total route cost ($Cost$) denotes the total travel time along the route (s); W_D is the "distance" weight; D_G is the distance from the person's position to the final target (m); V is the person's speed (m/s); W_q is the "queue" weight; Q is the expected time in the queue before reaching the link entrance (s); W_L is the "geometric component" weight; L is the geometric component type cost (s), the geometric component type can be specifically divided into escalators and stairs.

2.3 *CFD simulation model*

Airpak3.0 and ANSYS Fluent19.0 together form a CFD simulation model, in which the former is used to build a subway station model and mesh division while solving the continuous-phase airflow field; while the latter is used to simulate the particle concentration field based on the Airpak model, and the particles in the simulation are selected as PM2.5, which is the most harmful to the human body.

In the choice of the numerical simulation model, the standard $\kappa - \varepsilon$ turbulence model is used to simulate airflows and turbulence in this semi-closed building (Song et al. 2018). Meanwhile, considering that the volume fraction of PM2.5 is less than 10% in air, the Discrete Phase Model (DPM) based on the Euler-Lagrange equation is used to perform gas-solid two-phase flow simulation. Finally, the following assumptions are given,

- Air is isotropic;
- The filtration efficiency of PM2.5 by the primary filter of the ventilation system is 40%;
- Consider the effect of continuous relative to discrete phase particle population and ignore the effect of discrete relative to continuous phase;
- Disregarding the adsorption of PM2.5 by pedestrians;
- When the platform screen door is closed, the amount of air leakage from the screen door is negligible

The modeling dimensions of the public area are 90.4 m×20 m×3.5 m (the concourse floor) and 110 m×12.8 m×3 m (the platform floor). The arrangements of supply and return air outlets on the two floors are shown in Figure 3. The air outlet sizes of the concourse floor and the platform floor are 300 mm×300 mm and 600 mm×500 mm, respectively.

Due to the existence of PSDs, there are two main airflow organization states in this station. One is PSD closed state when the train does not stop. The PSD is well sealed, and the air inside the subway flows to the outdoor through the exits. The other is PSD opened state. Due to the low air pressure in the tunnel, the platform air flows to the tunnel, and the outdoor air flows from the exits to the concourse floor. When PSDs are closed, the only sources of particulate are air supply outlets, equipment, and pedestrians, but when the PSDs are opened, outdoor air enters so that the exits are also particulate sources; therefore, the two airflow organization states need to be simulated separately.

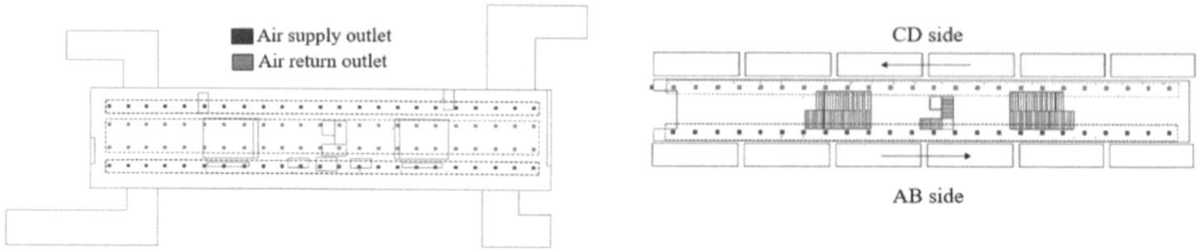

(a) Concourse floor. (b) Platform floor.

Figure 3. Arrangements of supply and return air outlets on the two floors.

In the CFD simulation model, the boundary conditions are given as follows: the concourse/platform floor air supply velocity are set to 2.3 m/s and 1.9 m/s; the particulate density and diameter are set to 1050 kg/m^3 and 2.5e-6 m; the outdoor PM2.5 concentration and air supply outlets particulate matter concentration is 61 μg/m^3 and 24.4 μg/m^3, the latter is 40% of the former. Meanwhile, particle generation rates per pedestrian and equipment are 10 mg/(person·h) and 3.78e-9 kg/s (Whyte 2001). The particulate concentration at the outlet is equal to the outdoor particulate concentration.

3 RESULTS AND ANALYSIS

3.1 *PFA result and validation*

The spatial average density characterizes the average number of pedestrians per unit time and unit area and is also used as the basis for setting the location and size of pedestrian particulate generation sources in the CFD model, which is mainly determined by the location and density of pedestrians gathered in the space. The Massmotion offers the service level of the area to represent the average density of pedestrians. It is based on the International Air Transport Association (IATA) waiting service level table (Oasys Software Limited 2021), as shown in Table 2.

The simulation results of pedestrian spatial density at the station concourse floor and platform floor are shown in Figure 4, respectively, while the average number of pedestrians in the area below service level B is also marked in the figure.

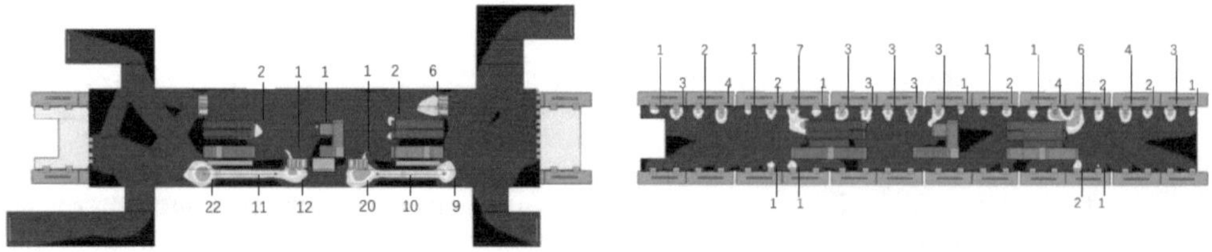

(a) Concourse floor. (b) Platform floor.

Figure 4. Simulation results of pedestrian spatial density on two floors.

Table 2. IATA waiting level table.

Service level	Density (person/ m^2)	Color Representation
A	$x \leq 0.370$	
B	$0.370 < x \leq 0.435$	
C	$0.435 < x \leq 0.526$	
D	$0.526 < x \leq 0.667$	
E	$0.667 < x \leq 1$	
F	$1 < x$	

3.2 *CFD result*

Figures 5–6 show the simulation results of the spatial concentration distribution of PM2.5 at the breathing plane (Height = 1.5 m) with different airflow organizations.

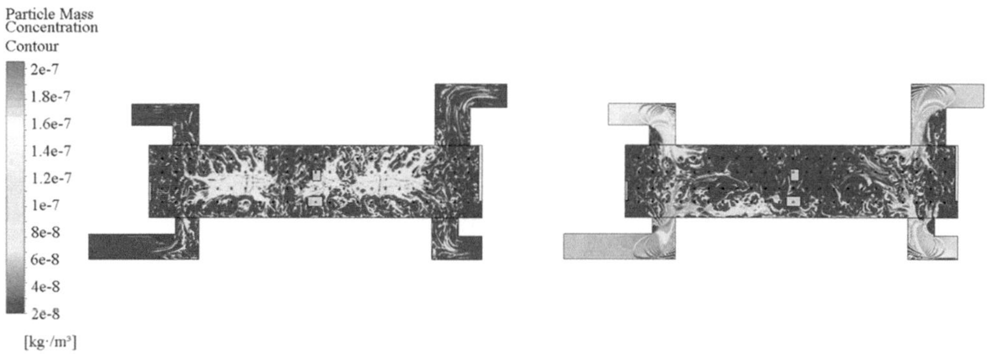

(a) PSDs closed. (b) PSDs opened.

Figure 5. Simulation results of particle mass concentration contour at station concourse floor under different airflow organization.

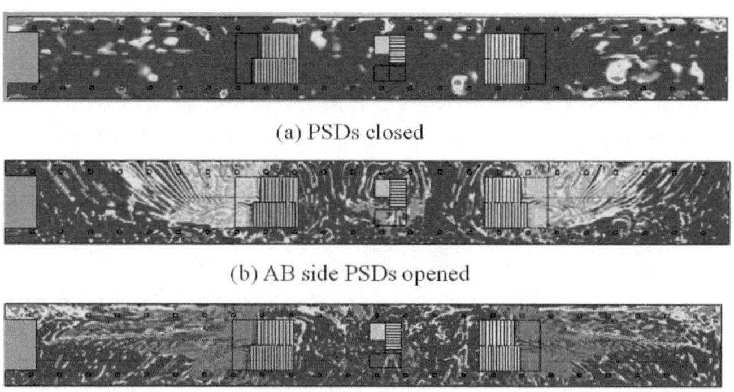

(a) PSDs closed

(b) AB side PSDs opened

(c) CD side PSDs opened

Figure 6. Simulation results of particle mass concentration contour at station platform floor under different airflow organization.

On the concourse floor, the concentration of PM2.5 is mainly concentrated at the inbound security checkpoint, inbound vending machines, outbound vending machines, the stairway entrance, and the return air outlet where the crowd gathers. Since pedestrians are PM2.5 sources, therefore the concentration of PM2.5 is larger where pedestrians gather. Secondly, PM2.5 from the platform floor can move to the concourse floor with the airflow, so the PM2.5 pollution near the stairway is also more serious. Meanwhile, the PM2.5 is easy to gather under the return air outlet due to the suction effect. The average concentration of PM2.5 on the concourse floor when the PSDs are closed is 59.73 $\mu g/m^3$; the average concentration of PM2.5 when the PSDs are opened is 46.67 $\mu g/m^3$. The total mean value of the two states is 53.2 $\mu g/m^3$. When the PSDs are opened, outdoor air enters the concourse floor through the exits, so the concentration of PM2.5 in the four passages is greater, reaching 57.23 $\mu g/m^3$.

On the platform floor, the concentration of PM2.5 mainly gathers at the PSDs on both sides; the position of the air supply and return outlet is one side air supply and the other side air return. Due to the suction effect, the concentration of PM2.5 under the return outlet is greater. The average PM2.5 concentrations on the platform floor are 60.03 $\mu g/m^3$, 53.59 $\mu g/m^3$, and 79.88 $\mu g/m^3$ for PSDs closed, AB side PSDs opened, and CD side PSDs opened, respectively. The total mean value of the three states is 64.5 $\mu g/m^3$. The platform floor is smaller than the concourse floor and has a higher pedestrian density, so the average concentration of PM2.5 is larger. When CD side PSDs are opened, the return air outlets are on the opposite side from the CD side PSDs, and the particulate PM2.5 tends to settle in the slower airflow area, so the concentration is higher.

It is important to emphasize that the numerical simulation method in this paper can significantly save labor and material resources compared with the actual measurement methods (Guan et al. 2020; Kim et al. 2012). The boundary conditions for the number of pedestrians in the PFA model can be determined based on the passenger flow analysis at the location of the subway station, and the boundary conditions for the concentration of PM2.5 generating sources in the CFD model can also be determined based on the environmental information near the landing point. Therefore, this method can efficiently and accurately simulate the spatial concentration distribution of PM2.5 when the facilities and layout of the subway station are changed, which is helpful for the comprehensive design and optimization during the construction of subway stations.

4 CONCLUSION

Combining PFA and CFD analysis, this paper presents a new method to analyze the spatial concentration distribution characteristics of PM2.5 in subway stations. Based on the PFA results, the areas where pedestrians gather in subway stations and the density of pedestrians in these areas can be obtained, so that the location and size of pedestrian particulate generation sources in CFD models are determined. The concentration of PM2.5 is higher in the pedestrian gathering area and under the return air outlet. The simulated average concentration for the concourse floor and platform floor is 53.2 $\mu g/m^3$ and 64.5 $\mu g/m^3$, respectively. Overall, this method only relies on outdoor environmental data to obtain the PM2.5 concentration distribution at the station concourse floor and platform floor under different airflow organizations, without the need for a large number of measurements, saving labor and material resources required for measurement, and helping subway designers to positively consider the pollutant concentration distribution in subway stations at the early stage of design.

REFERENCES

Bolourchi, A., Atabi, F., Moattar, F. & M. Ehyaei (2018). Experimental and numerical analyses of particulate matter concentrations in underground subway station. *Int. J. Environ. Sci. Technol.* 15: 2569–2580.

Cao, S.J., Kong, X.R., Li, L.Y., Zhang, W.R., Ye, Z.P. & Y.L. Deng (2017). An investigation of the PM2.5 and NO2 concentrations and their human health impacts in the metro subway system of Suzhou, China. *Environ. Sci.: Processes Impacts*. 19: 666–675.

Colombi, C., Angius, S., Gianelle, V. & M. Lazzarini (2013). Particulate matter concentrations, physical characteristics and elemental composition in the Milan underground transport system. *Atmospheric Environ.* 70: 166–178.
Guan, B., Zhang, T., Liu, X.H., Wang, J. & C.M. Gao (2020). Performance of mechanical ventilation system on particulate matter control in the public area of a subway station. *Chinese J. Environ. Eng.* 14(8): 2270–2276.
Kim, K.H., Ho, D.X., Jeon, J.S. & J.C. Kim (2012), A noticeable shift in particulate matter levels after platform screen door installation in a Korean subway station. *Atmospheric Environ.* 49: 219–233.
Oasys Software Limited (2021). *Massmotion Help Guide*. London, the United Kingdom: Oasys Software Limited.
Passi, A., Nagendra, S.M. & M.P. Maiya (2021). Characteristics of indoor air quality in underground metro stations: A critical review. *Build. Environ.* 198, no. 107907.
Peng, Y.L., Gao, Z., Ding, W.W., Zhang, J.S., Xu, J.F. & Y.L. Wei (2021). Application of computational fluid dynamics in subway environment without fire and smoke—Literature review. *Build. Environ.* 206, no. 108408.
River, E., Jaynes, C., Kimball, A. & E. Morrow (2014). Using case study data to validate a 3D agent-based pedestrian simulation tool for building egress modeling. *Transport. Res. Procedia*. 2: 123–131.
Shakya, K.M., Saad, A. & A. Aharonian (2020). Commuter exposure to particulate matter at underground subway stations in Philadelphia. *Build. Environ.* 186, no 107322.
Song, X.Y., Chang, L.Q. & Z.R. Peng (2018). Spatial distribution of fine particulate matter in underground passageways. *Int. J. Environ. Res. Public Health*. 8: 1574–1594.
Teodosiu, C.I., Ilie, V., Dumitru, R.G. & R.S. Teodosiu (2016). Assessment of ventilation efficiency for emergency situations in subway systems by CFD modeling. *Build. Simul.* 9: 319–334.
Whyte, W. (2010). *Cleanroom Technology: Fundamentals of Design, Testing and Operation*. New Jersey, the United States: Johnson Wiley & Sons Inc.

Advances in Urban Engineering and Management Science – Khalil & Yang (Eds)
© 2023 The Author(s), ISBN 978-1-032-30426-7

Information systems architecture development of smart expressway for large scale road network across China

Kun Chen*
Transport Planning and Research Institute, Ministry of Transport, China

Jie Ding*
Intelligent Transportation Division, Tencent Beijing Headquarters, Beijing, China

ABSTRACT: Since 2017, smart expressway pilot projects have increased rapidly in China. However, independent single project construction is therefore difficult to bring large-scale benefits of the smart expressway. It is imperative to develop the information system architecture of the smart expressway from the perspective of the countrywide large-scale road network. A general framework for smart expressway is developed in this study by using the Open Group Architecture Framework (TOGAF) method. The framework includes an application architecture as well as a business architecture. Built on the statistical analysis of more than 70 smart highway project proposals, the business architecture is designed. It composes four business domains and their sub-domains, which include highway construction, maintenance, operation, and service. The application architecture is further designed as a three-level architecture composed of top-level comprehensive applications, middle-level professional applications and bottom-level supporting applications. Core technology directions in smart expressway construction are proposed to guide the forthcoming construction of smart expressways. The general information systems architecture developed in this study is greatly significant to avoid the fragmentation of single projects of smart expressway. Meanwhile it can shed light on unified and homogeneous services for the national smart expressway network across China.

1 INTRODUCTION

The new infrastructure investment is not only a short-term strategy for economic recovery, but also will play a leading and basic role in developing the smart society and achieving the "two centenary goals" (Li 2020). By the end of 2020, the total mileage of the expressway in China was 161000 kilometers (Ministry of Transport 2021a). With the increasing shortage of resources and environment, it is required to improve the traffic efficiency, safety, and service quality of existing expressways, by taking advantage of advanced sensing, computing, communication, artificial intelligence, and other technologies. The smart expressway has become a key investment field of new infrastructure. Jiangsu, Zhejiang, Guangdong, Sichuan, Shandong, and other regions are active in building smart expressways. It is clearly proposed to build smart expressways on national highways and intercity expressways in The Action plan for the new transportation infrastructure Construction (2021-2025) (Ministry of Transport 2021b).

Smart expressway projects have been developing in China for several years. However, previous studies have focused on a single project. The systematic study of the information system architecture from the angle of the whole enterprise or the whole road network is few. The Ministry of Transport requires designing the entire framework and major directions of smart expressways

*Corresponding Authors: chenkun17@163.com and dingjie.tpri@foxmail.com

 DOI 10.1201/9781003305026-88

to conduct the forthcoming construction of smart expressway projects, and more importantly, to provide unified and homogeneous services for the national expressway network. By referencing the construction contents of more than 70 previous or ongoing smart expressway projects, a general framework of the smart expressway for expressway companies is developed in this paper. It defines the business domains responding to the requirements of users at different levels, realizes their application functions, and ensures effective information sharing and business collaboration among diverse users.

2 DEVELOPMENT PROCESS AND CONCEPT OF SMART EXPRESSWAY

2.1 *Development process of smart expressway*

As the reflection of time from the deep integration of new-generation information technology and transportation system, smart highways are changing human travel mode and becoming the focus of competition among countries (Wu & Song 2020). In the 1980s, Federal Highway Administration (FHWA) studied Automated Highway Systems (AHS) (Horowitz 1997). Magnetic nails in the dedicated lane have been used to guide vehicles, which was the first interconnection between vehicles and road infrastructures (Editorial Department of China Journal of Highway and Transport 2017). In 2005, the U.S. Department of Transportation (DOT) initiated a program to develop and test a 5.9GHz-based Vehicle Infrastructure Integration (VII) proof of concept in Michigan (Kandarpa et al. 2009). In 2009, the IntelliDrive initiative was proposed, which established an interconnected traffic environment among vehicles, road infrastructure and passenger portable devices by using wireless communication. IntelliDrive can reduce the perception-reaction (P-R) time in the drive to improve the flow of multilane highway systems (Mohammad et al. 2011). In 2016, the smart corridor project was promoted. I-80 smart project in California integrated a variety of electronic identification, information interaction and dynamic management technologies to enhance security and travel time reliability (Pande 2018). In Europe, the EasyWay projects were brought forward to deploy the core Intelligent Transportation Systems (ITS) services on the European road network, such as traveler information, traffic management, and freight and logistic services (Stocchi 2012). In Japan, "the next generation ITS: Smartway" and ITS spots were proposed, which integrated the Vehicle Information and Communication System (VICS), Electronic Toll Collection (ETC) and Advanced Safety Vehicle (ASV) (Fujimoto et al. 2008). ITS spot services have been installed at approximately 1,600 locations, primarily on expressways in Japan by August 2011, which were deployed every 10-15 kilometers on suburban expressways, and every 4 kilometers on urban expressways (Kanazawa et al. 2012).

In 2017, smart highway pilot projects were proposed in China. Six pilot directions were mainly promoted, including digital infrastructure, integration of highway and transportation services, comprehensive application of the Beidou navigation system, road network operation based on big data, travel services based on "Internet plus", and the new generation of national traffic control system (Chen 2020). Yanqing-Chongli expressway, Beijing-Xiongan expressway and Hangzhou-Shaoxing-Ningbo expressway established the digital infrastructure and carried out the cooperative vehicle-infrastructure system (CVIS) and autonomous vehicles tests. Hunchun-Ulanhot expressway in Jilin province promoted the infrastructure performance monitoring, maintenance, and emergency command in seasonal frozen regions. Jiangxi province explored the innovative application of Beidou high-precision technology in a free flow electronic toll collection system.

The construction of international smart highways is mainly oriented to the needs of intelligent connected vehicles (ICVs), the interaction between roads and vehicles, using the sensing facilities installed on the roadsides or vehicles, to improve road traffic safety and efficiency. The construction of smart expressways in China has applied a variety of technologies. There are even differences in the definition, construction standards, functional and technical requirements of smart expressways, with the insufficiency of sustainable construction and business operation models (Zhou et al. 2021). There influences the large-scale application of smart expressways in China.

2.2 *Connotation of smart expressway*

A smart expressway is a road infrastructure system integrating holographic sensors, digital twins, intelligent decision-making, information interaction and other technologies, providing a lot of real-time information about highway facilities, traffic operation and the environment to drivers or ICVs, using the intelligent traffic control cloud platform to realize better coordination between the central unit and roadside units, so as to improve the traffic safety, efficiency and service quality.

At present, the construction of smart expressways in China mainly relies on new expressway projects or reconstruction projects. However, existing expressways should be the main objects of smart expressway construction, because of their long mileages, heavy traffic flow and fixed environmental constraints. Therefore, it is very necessary and meaningful to design the architecture of a smart expressway by considering the overall situation of the expressway network.

3 ARCHITECTURE DEVELOPMENT METHOD (ADM) BASED ON TOGAF

Information systems architecture of smart expressway can adopt the structural analysis method to design the system functions of all components and their physical models, according to an analysis of the business requirements and the relationship for data sharing (Cen et al. 2020). The mainstream ADMs include the Zachman framework, the Federal Enterprise Architecture (FEA) and TOGAF. TOGAF is a proven enterprise architecture methodology and framework used by the world's leading organizations to increase business efficiency. The TOGAF-ADM starts with an architecture vision, analyzes the situation of information systems, and develops a target architecture to implement and migrate plan (The Open Group 2018). ADM for smart expressway is as shown in Figure 1.

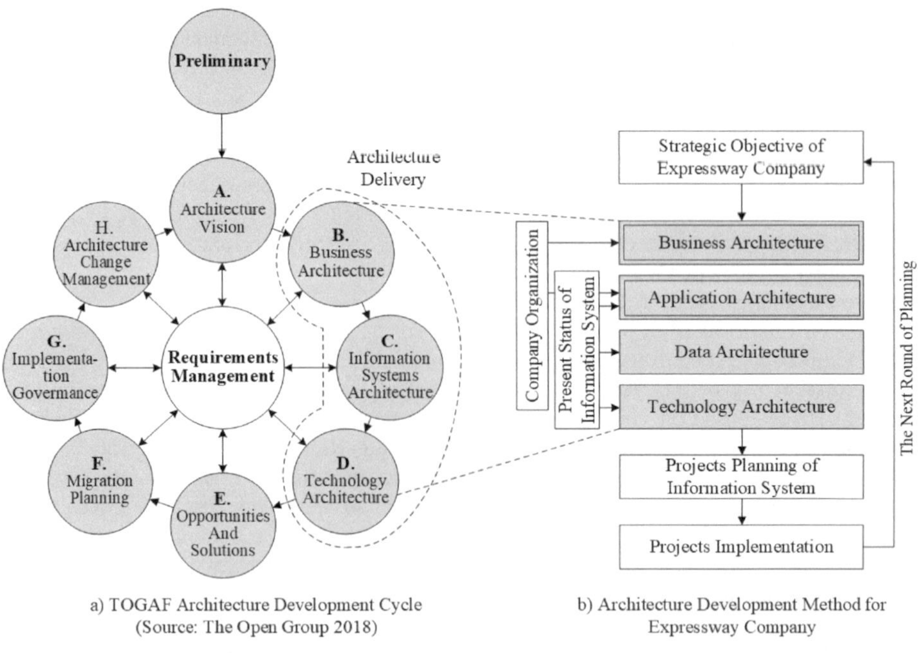

Figure 1. Architecture development method based on TOGAF.

The first stage is developing the business architecture by analyzing the business principles, goals, requirements, and organizational model. The business architecture is the basis of information systems architecture, which is expected to be long-lived and stable. It describes the objects and elements involved in the enterprise management and operation, as well as their features, behaviors and relationships, in a systematic way.

The second stage is developing the information systems architecture, which involves some combination of data and application architecture, in either order. This paper takes an application-driven approach, by analyzing certain core applications for the mission-critical business processes, and taking the implementation and integration of certain core applications as the primary focus of the architecture effort. The application architecture should consider the requirements balance of diverse users, and the maintainability of information systems.

4 INFORMATION SYSTEMS ARCHITECTURE DEVELOPMENT OF SMART EXPRESSWAY

4.1 *Business architecture development*

Xu, Z. G. et al. presented the directions of the intelligent road system, and its system architecture, which consists of a management layer, a network layer, and an application layer (Xu et al. 2019). Zhang, J. S. et al. designed the seven user service fields of the smart expressway, including traffic management and control, electronic toll collection, travel service, safety driving assist, transportation management and service, emergency and safety, energy management and service (Zhang et al. 2018). Du, Y. C. et al. clarified the core components and technical sections of the new generation of the smart highway, including wide-area holographic perception, digital twins, vehicle-road collaborative applications, and real-time communication media, and designed its logical architecture, physical architecture, and application architecture (Du et al. 2021). Although the technical framework of the smart expressway has been studied, it is still necessary to further analyze the service functions and their implementation architecture from the perspective of users.

Considering the business requirements of the expressway company and its subsidiaries, the business architecture is divided into four core business areas: expressway construction, maintenance, operation and service, shown in Figure 2.

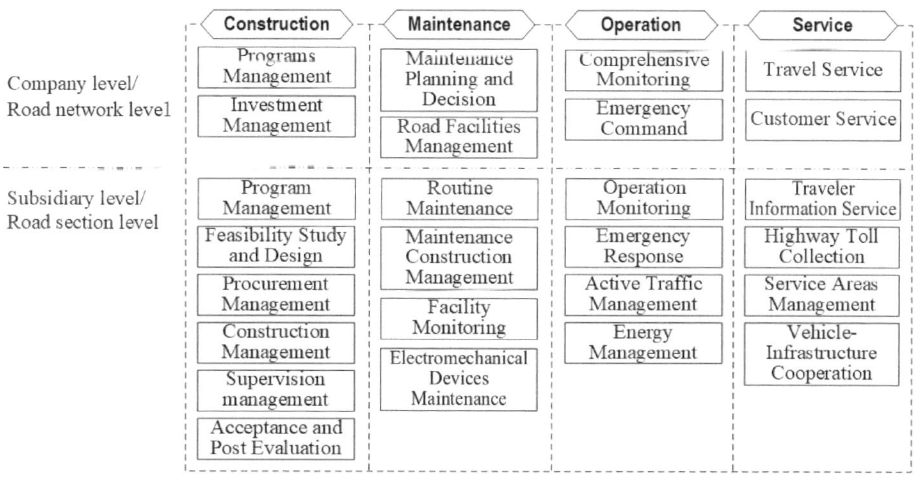

Figure 2. Business architecture of smart expressway.

4.1.1 *Construction area*

The life cycle management of expressway construction projects is provided, including programs management, feasibility study and design, procurement management, construction management, supervision management, acceptance, and post evaluation, etc. Considering the whole company, the programs management and investment management of all projects are more important.

4.1.2 *Maintenance area*

With the deep integration of information infrastructures and transportation infrastructures, maintenance business includes road facilities maintenance and information infrastructures maintenance, and pays more attention to the real-time monitoring and remote maintenance of infrastructures. The company will make maintenance plans and decisions on the basis of road facilities management and technical condition evaluation.

4.1.3 *Operation area*

Expressway operation is focused on the real-time monitoring and early warning analysis of traffic performance, meteorological environment, and emergencies, to ensure traffic safety and efficiency. Active traffic management is provided in sections with frequent accidents. Distributed generation and charging stations along the highway are new operations.

4.1.4 *Service area*

The traveler information service, highway toll collection, service areas management and vehicle-infrastructure cooperation should be provided. Considering the whole company, the comprehensive travel service is important, which establishes a complete human-based customer service system for various users of expressways.

4.2 *Application architecture development*

Application architecture is developing the target architecture of application systems that enable the achievement of business objectives. It is designed from multi-dimensional aspects of business, system users and data development process, shown in Figure 3.

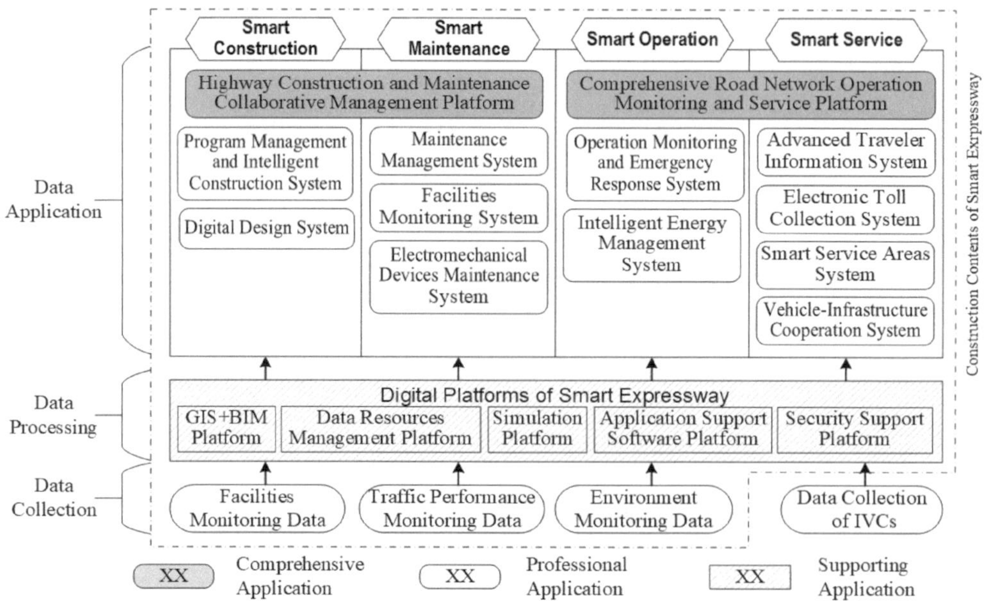

Figure 3. Application architecture of smart expressway.

4.2.1 *From the aspect of business*

The application architecture is decomposed into smart construction, smart maintenance, smart operation and smart service.

4.2.2 *From the aspect of system users*

The application architecture comprises top-level comprehensive applications, middle-level professional applications and bottom-level supporting applications. The comprehensive application system is mainly developed for the comprehensive management and macro decision-making business for the entire company or road network, such as highway construction and maintenance collaborative management platform. The professional application system mainly realizes special application functions for a department or a subsidiary, such as the maintenance management system. The supporting application system is developed as a cloud platform based on digital twins, which consist of a series of basic software and tools, such as the data resources management platform.

4.2.3 *From the aspect of data*

Various data are gathered, managed, analyzed, and encapsulated as data services provided to various comprehensive and professional applications through the digital platform.

5 TECHNOLOGY DIRECTIONS OF SMART EXPRESSWAY

In 2021, the National Highway Administration of China organized an investigation on new highway infrastructures and digital highways. The investigation results are demonstrated that 80% of the total 76 previous or ongoing smart highway projects are built on expressways. Various digital and intelligent applications have been encouraged on these expressway projects, as shown in Figure 4.

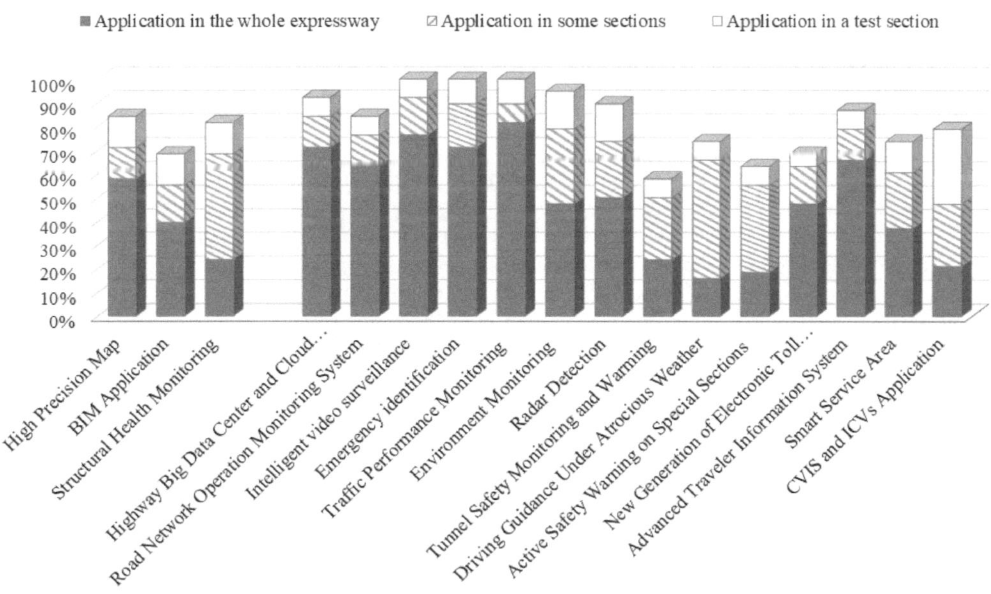

Figure 4. Statistical analysis of core technology directions of the smart expressway.

5.1 *Technology directions of smart construction and maintenance*

Digitization of highway infrastructure has received widespread attention. The digital road infrastructure management system is developed to manage the pavements, bridges, tunnels, slopes, electromechanical systems, and other online facilities. The high precision map and building information modeling (BIM) is implemented in the engineering design of some projects, especially used for modeling critical bridges or tunnels. Due to the high cost and long period, BIM is a supplement to the traditional paper documents or CAD drawings. It is quite necessary to study the application of high precision maps and BIM in the whole cycle of engineering design, construction, and maintenance. The structural health monitoring of bridges, tunnels and slopes is sometimes difficult to be applied on a large scale, because of the prohibitive cost and low utilization.

5.2 *Technology directions of smart operation and service*

Intelligent monitoring technology has been widely applied. The highway big data center and cloud platform are established to improve traffic efficiency and service quality, which integrates the functions of road network monitoring, dispatching, management and control, emergency command and service. According to statistics results of this investigation, more than 60% of these projects have applied intelligent video surveillance, traffic performance monitoring, emergency identification and the highway big data center and cloud platform in the whole expressway, more than 40% of these projects have applied the environment monitoring and radar detection in the whole expressway. Tunnel safety monitoring and warming, driving guidance under atrocious weather, active safety warning on special sections, and the CVIS and ICVs applications have been applied cautiously in some sections or a test section. Therefore, it is proved that using the new generation of monitoring, big data analysis and artificial intelligence technologies to improve the intelligence level of traditional electromechanical systems is the main direction of the smart expressway construction at the present stage.

6 CONCLUSION

A smart expressway is a road infrastructure system integrating holographic sensors, digital twins, intelligent decision-making, information interaction and other technologies. This study mainly develops a general architecture of smart expressways for large-scale road networks based on TOGAF. The information systems architecture of the smart expressway is designed as a 3-level architecture composed of comprehensive applications, professional applications and supporting applications. Meanwhile, the architecture is further decomposed into four core applications, including smart construction, smart maintenance, smart operation, and smart service. The enterprise architecture development of smart expressway will promote the integration of new infrastructures and traditional infrastructures, and contribute to the digital transformation of expressway companies.

This paper summarizes the main technology directions of the smart expressway. The conclusion has certain guiding significance for the future construction of smart expressway projects. With the continuous maturity of relevant technologies and standards, the construction of smart expressway will be more standardized and has the possibility of countrywide promotion.

ACKNOWLEDGMENT

This study was funded by National Key Technologies R & D Program (2018YFB1600304), Transportation Standard Budget Project (062110-037) and Transportation Strategic Planning & Policy Research Project (2021-18-1). The authors thank Shandong Hi-Speed Group for its help in field tests.

REFERENCES

Cen, Y. Q., Song, X. H., Wang, D. Z., Sun, L. and Liu, N. (2020). Establishment of technology system of smart expressway. *Journal of Highway and Transportation Research and Development,* 37(7), 111–121.

Chen, X. B. (2020). The development path and ideas on the smart highway. *China Transportation Review,* 2(11), 82–86.

Du, Y. C., Liu, C. L., Wu, D. F. and Zhao, C. (2021). Framework of the new generation of smart highways. *China Journal of Highway and Transport,* in press.

Editorial Department of China Journal of Highway and Transport. (2017). Review on China's automotive engineering research progress: 2017. *China Journal of Highway and Transport,* 30(6), 1–197.

Fujimoto, A., Sakai, K. and Ogawa, M. (2008). Toward realization of Smartway in Japan. *Proceedings of 15th World congress on intelligent transport systems and ITS America's 2008 annual meeting*, Washington D.C., ITS America, 1–9.

Horowitz, R. (1997). Automated Highway Systems: The smart way to go. *IFAC Proceedings Volumes,* 30(8), 449–460.

Kanazawa, F., Sakai, K., Suzuki, K. and Nakamura, S. (2012). Effectiveness evaluation of ITS Spot services in full-Scale operation throughout Japan. *Proceedings of 19th ITS World Congress.*

Kandarpa, R., Chenzaie, M., Dorfman, M., Anderson, J., Marousek, J. et al. (2009). *Final Report: Vehicle Infrastructure Integration (VII) Proof of Concept (POC) test-executive summary.* Research and Innovative Technology Administration, U.S. Department of Transportation.

Li, X. H. (2020). New infrastructure construction and policy orientation for a smart society. *Reform,* 315(5), 34–48.

Ministry of Transport. (2021a). Statistical bulletin on the development of transportation industry in 2020. <https://xxgk.mot.gov.cn/2020/jigou/zhghs/02105/t20210517_3593412. html>, May 19, 2021.

Ministry of Transport. (2021b). The action plan for the new transportation infrastructure Construction (2021–2025). Beijing.

Mohammad, N., Hossein, P.N. and Ni, D. H. (2011). The effect of IntelliDrive on the efficiency of highway transportation systems. *International Journal of Vehicular Technology,* 2011(3).

Pande, A. (2018). *Evaluation of adaptive ramp metering on I-80 in the San Francisco Bay area.* Research and Innovative Technology Administration, U.S. Department of Transportation.

Stocchi, E. (2012). The EasyWay project for a harmonized deployment of its systems in Europe. *Tollways,* 9, 58–67.

The Open Group. (2018). The TOGAF®standard, version 9.2, a standard of The Open Group (C182). <www.opengroup.org/library/c182>, April 2018.

Wu, J.Q. & Song, X. G. (2020). Review on smart highways critical technology. *Journal of Shangdong University (Engineering Science),* 50(4), 1–18.

Xu, Z. G., Li, J. L., Zhao, X. M., Li, L., Wang, Z. R. et al. (2019). A review on intelligent road and its related key technologies. *China Journal of Highway and Transport,* 32(8), 1–24.

Zhang, J. S., Li, B., Wang, X. J., Zhang, F. and Sun, X. L. (2018). Design of architecture and development roadmap of smart expressways. *Journal of Highway and Transportation Research and Development*, 35(1), 88–94.

Zhou, W., Zhou, H. T., Zhi, G. L., Cheng, P. and Xu, R. Q. (2021). Investigation of construction situation of smart highway pilot projects. *Special Report of Transportation Think Tank.*

Advances in Urban Engineering and Management Science – Khalil & Yang (Eds)

Research on the coupling and coordination of transportation and socio-economic system in southwestern minority areas

Juan Liu, Yuan Zheng, Xuesong Han* & Xuehong Zhou
National Territory and Spatial Planning Research Institute of Sichuan Province, Chengdu, Sichuan, China

ABSTRACT: In order to study the coupling and coordination law of the two major systems of transportation and social economy in the southwestern ethnic area, this paper uses the data of Aba Prefecture from 2009 to 2017 as an example to construct two system index systems, establish a coupling coordination degree model, and calculate the corresponding coordination degree index. Quantitatively measure the degree of coupling and coordination between systems. During the study period, the degree of coupling between transportation and social economy in Aba Prefecture is relatively high, while the degree of coordination fluctuates. The results of the study indicate that Aba Prefecture is a region with lagging traffic, and investment in transportation should be increased in the future.

1 INTRODUCTION

As the basic construction of social development and social economy, transportation has promoted the expansion of economic activities to a certain extent. At the same time, economic development has also improved the construction of the transportation industry, and the two are very closely related (Bai et al. 2019). Taking Aba Prefecture in Sichuan Province as an example, this paper constructs a system index system and establishes a coupling coordination degree model to quantitatively measure the coupling coordination degree between the transportation system and the socio-economic system in different periods in the southwestern ethnic region, providing a scientific evaluation of the coordination between the two refer to.

2 INTERACTION BETWEEN SOCIAL ECONOMY AND TRANSPORTATION

2.1 *Transportation advances the process of economic development*

From a macro perspective, a perfect transportation system plays an important role in supporting the flow of social resources. The social economy must realize its production flow, exchange and consumption process through transportation. Transportation realizes the connection between products and consumers. Efficient transportation reduces ineffective residence time, improves the efficiency of product flow, and promotes economic expansion and flow. In addition, the development of the transportation system can promote the coordinated and balanced development of the region.

2.2 *Economic development drives the development of transportation*

The rapid development of the social economy requires that transportation must effectively integrate the links between various production areas, so that goods and people can flow more safely and at a higher speed. At the same time, a large amount of transportation infrastructure construction

*Corresponding Author: songshu183@163.com

 DOI 10.1201/9781003305026-89

requires a certain level of economic support. In the era when social economy and transportation are integrated and developed, improving transportation levels has become a kind of economic development investment.

3 RESEARCH ON COUPLING AND COORDINATION OF TRANSPORTATION AND SOCIO-ECONOMIC SYSTEM IN ABA PREFECTURE

3.1 *Overview of Aba Prefecture*

The Aba Tibetan and Qiang Autonomous Prefecture is in the upper reaches of the Yangtze River and the Yellow River. By 2017, the total mileage of highways in the prefecture reached 13,500 kilometers, 8954 kilometers more than in 1978, 51 kilometers of expressways, and 13,000 kilometers of graded highways. The annual passenger turnover of highways was 2,045.9 million kilometers, and the annual freight turnover was 6,095 million tons. The total annual flight volume of Jiuzhai Huanglong Airport and Hongyuan Airport reached 541,164 passengers.

3.2 *Analysis of coupling degree of transportation and economic system*

3.2.1 *Construct an analysis and evaluation index system*

There are many relevant indicators to measure social and economic development and the transportation system. When selecting indicators, the changing characteristics and functional relationships of the two systems should be reflected as much as possible (Jiao et al. 2016). The level of social and economic development can be reflected by economic scale, economic structure, economic benefits, etc. This article selects three subsystems of the economic scale, economic structure, and economic benefits, each of which covers specific target values, a total of 11 (Table 1). The evolution of the transportation system is mainly reflected in the changes in transportation scale, transportation capacity, and transportation service level. This paper selects three subsystems: transportation scale, transportation capacity, and transportation service level. Each subsystem covers a specific target value. 10 (Table 2).

Table 1. The social and economic system index system and weight of Aba Prefecture.

Main system	Sub system	Index layer	unit	+ or	Weights
Urban socio-economic system	Economic scale A1	a1 GDP	100 million yuan	+	0.089
		a2 Permanent population	Ten thousand people	+	0.093
		a3 GDP per capita	Yuan	+	0.087
	Economic structure A2	a4 Unemployment rate	%	-	0.066
		a5 The proportion of tertiary industry in GDP	%	+	0.098
		a6 Total sales of social retail products	100 million yuan	+	0.121
		a7 Total social investment in fixed assets	100 million yuan	+	0.081
		a8 Total import and export	Ten thousand U.S. dollars	+	0.071
	Economic Benefit A3	a10 Per capita disposable income of rural residents	Yuan	+	0.118806
		a11 Per capita disposable income of urban residents	Yuan	+	0.093255
		a9 Fiscal revenue	100 million yuan	+	0.07827

Table 2. The index system and weights of transportation system in Aba Prefecture.

Main system	Sub system	Index layer	unit	+ or −	Weights
Transportation system	Transportation scale B1	b1 Total mileage of highway	kilometer	+	0.055
		b2 Transportation fixed-asset investment	100 million yuan	+	0.083
		b3 Total highway passenger transport	Ten thousand people	+	0.115
	Transportation capacity B2	b4 Total road freight	Ten thousand tons	+	0.063
		b5 Highway passenger turnover	Ten thousand people/km	+	0.112
		b6 road freight turnover	10,000 tons/km	+	0.087
		b7 passenger vehicles	Vehicles	+	0.167
		b8 Air passenger throughput	Person times	+	0.100
	Traffic service level B3	b9 Number of deaths in traffic accidents	Person times	-	0.074
		b10 Number of traffic accidents	rise	-	0.141

3.2.2 *Use the entropy method to process data*

The information entropy is calculated by the entropy method, and the evaluation value is calculated by combining the data of various indicators in each year, and then the degree of coupling is used to judge the degree of mutual coordination between the socio-economic system and the transportation system of Aba Prefecture.

Data matrix establishment

$$A = \begin{pmatrix} X_{11} & \cdots & X_{1n} \\ \vdots & \vdots & \vdots \\ X_{m1} & \ldots & X_{mn} \end{pmatrix}_{m \cdot n} \tag{1}$$

Among them, Xij is the value of the jth index in the i-th year. The sample number in this article is 9, which is the indicator data of Aba Prefecture from 2009 to 2017, m=9; for the socio-economic system, n=11; for the transportation system, n=10.

Entropy method to process data

Data normalization processing. The entropy method calculates the comprehensive ratio of a certain index of each sample to the same index. Since each index has its dimensions and attributes, it needs to be standardized.

Positive indicators:

$$X'_{ij} = \frac{X_{ij} - \min\left(X_{1j} \ldots X_{mj}\right)}{\max\left(X_{1j} \ldots X_{mj}\right) - \min\left(X_{1j} \ldots X_{mj}\right)} \tag{2}$$

Negative indicators:

$$X'_{ij} = \frac{\max\left(X_{1j} \ldots X_{mj}\right) - X_{ij}}{\max\left(X_{1j\ldots} \ldots X_{mj}\right) - \min\left(X_{1j} \ldots X_{mj}\right)} \tag{3}$$

Calculate the proportion P of the index in the i-th year under the j-th index.

$$P_{ij} = \frac{x'_{ij}}{\sum_{i=1}^{m} x'_{ij}} \tag{4}$$

Calculate the information entropy value e of the jth index.

$$e_j = -k * \sum_{i=1}^{m} (P_{ij} \ln P_{ij}) \tag{5}$$

Where k is a constant, $k = \frac{1}{lnm}$

Calculate the difference coefficient g of the j-th index. For the j-th index, the greater the difference in the index value Xij, the greater the effect on sample evaluation, and the smaller the entropy value.

$$g_j = 1 - e_j \tag{6}$$

The greater the coefficient of difference, the greater the weight. Calculate the information weight W of index j.

$$W_i = \frac{g_j}{\sum_{j=1}^{n} g_j} \tag{7}$$

The index information weight calculated by the entropy method essentially uses the degree of dispersion of the index information. The greater the degree of dispersion, the greater the uncertainty; the greater the impact on the comprehensive evaluation of the system, and the greater the weight occupied by the index (Chen & Zhou 2018).

Data matrix establishment

Calculate the comprehensive evaluation value of the index sample.

Using the weighted sum formula to obtain the comprehensive evaluation value S of the transportation system and socio-economic system samples of Aba Prefecture,

$$S_i = \sum_{j=1}^{n} (W_j * X'_{ij}) \tag{8}$$

3.2.3 *Evaluation of coupling degree and coupling coordination degree*

Coupling coordination degree refers to the degree of benign coupling in the interaction, which reflects the quality of coordination. It is an in-depth study of the coupling coordination relationship and reflects the dynamic evolution process of the two systems. According to the comprehensive evaluation value of each index, the coupling degree between the transportation system of Aba Prefecture and the socio-economic system is calculated.

$$C = \frac{2 * \sqrt{(S_T * S_E)}}{S_T + S_E} \tag{9}$$

$$T = \alpha S_T + \beta S_E \tag{10}$$

$$D = \sqrt{C * T} \tag{11}$$

In the formula, C is the coupling degree, ST is the comprehensive evaluation value of the transportation system index, SE is the comprehensive evaluation value of the socio-economic system index, T is the comprehensive evaluation index, which mainly reflects the comprehensive evaluation index value between systems, α β is to be determined, and the coefficient needs to satisfy $\alpha + \beta$=1. According to the weights of the transportation system and socio-economic system of Aba Prefecture, α=0.525 and β=0.475 are calculated; D is the degree of coupling and coordination.

The coupling degree C value is at (0,1]. Generally speaking, the closer the C value is to 1, the higher the coupling degree between systems. The closer the C value is to 0, the smaller the interaction between the systems. Coupling coordination degree D range is also at (0,1]; the closer D is to 1, the better the coordination between the transportation system and the socio-economic system indicators, and vice versa, it indicates the contradiction between related elements.

With reference to the results of the related coupling coordination research (Yuan 2014; Wang & Zhu 2013), the system coupling degree is divided into four stages (0,0.3], (0.3,0.5], (0.5,0.8], (0.8,1), and the coupling coordination degree Divided into (0,0.4], (0.4,0.5], (0.5,0.6], (0.6,0.7], (0.7,0.8], (0.8,0.9], (0.9,1) seven intervals, see the specific classification Table 3.

Table 3. Coupling degree and coupling coordination degree classification standard.

Coupling degree C	Coupling phase	Coupling coordination degree D	Coordination level
0	Disorderly development period	$0.0<D\leq0.4$	Severe imbalance
$0.0<C\leq0.3$	Low-level coupling	$0.4<D\leq0.5$	Mild disorder
$0.3<C\leq0.5$	Antagonistic period	$0.5<D\leq0.6$	Barely coordinated
$0.5<C\leq0.8$	Run-in period	$0.6<D\leq0.7$	Low coordination
$0.8<C\leq1.0$	High level of coupling	$0.7<D\leq0.8$	Moderate coordination
		$0.8<D\leq0.9$	Well-coordinated
		$0.9<D\leq1.0$	Quality coordination

3.3 *Analysis of coupling and coordination results of transportation system and socio-economic system in aba prefecture*

3.3.1 *Comprehensive evaluation value of Aba Prefecture's transportation system and socio-economic system indicators*

Table 4 shows the comprehensive evaluation values of the indicators of the transportation system and socio-economic system in Aba Prefecture. Among them, the coordination type whose comprehensive evaluation value of the transportation system is lower than the comprehensive evaluation value of the socio-economic system is the traffic lag type, and the coordination type whose comprehensive evaluation value of the socio-economic system is lower than the comprehensive evaluation value of the transportation system is the economic profit and loss type.

Table 4. Comprehensive evaluation value of Aba Prefecture's transportation system and socio-economic system indicators.

years	S_E	S_T	Coordination type
2009	0.112727	0.3398	Economic profit and loss
2010	0.18977	0.37898	Economic profit and loss
2011	0.40579	0.44252	Economic profit and loss
2012	0.493149	0.59809	Economic profit and loss
2013	0.54279	0.70947	Economic profit and loss
2014	0.62437	0.65947	Economic profit and loss
2015	0.74231	0.59063	Lagging traffic
2016	0.846812	0.491248	Lagging traffic
2017	0.781730	0.488679	Lagging traffic

It can be seen that the comprehensive evaluation value of the socio-economic system in Aba Prefecture has generally been increasing year by year, while the comprehensive evaluation value of the transportation system has fluctuated greatly decline. This is mainly due to the fact that there were basically no new transportation infrastructures in 2013~2017, and both passenger and freight volumes have decreased. Among them, air passenger throughput has decreased significantly.

3.3.2 *Result analysis of coordination degree between transportation system and economic system in Aba Prefecture*

According to the judging criteria, the results of the coordination and coupling between the socio-economic system and the transportation system in Aba Prefecture from 2009 to 2017 are obtained, as shown in Table 5.

Table 5. Comprehensive evaluation value of Aba Prefecture's transportation system and socio-economic system indicators.

years	S_E	S_T	*Coordination type*
2009	0.112727	0.3398	Economic profit and loss
2010	0.18977	0.37898	Economic profit and loss
2011	0.40579	0.44252	Economic profit and loss
2012	0.493149	0.59809	Economic profit and loss
2013	0.54279	0.70947	Economic profit and loss
2014	0.62437	0.65947	Economic profit and loss
2015	0.74231	0.59063	Lagging traffic
2016	0.846812	0.491248	Lagging traffic
2017	0.781730	0.488679	Lagging traffic

The coupling degree between the socio-economic system and the transportation system in Aba Prefecture was the lowest in 2009, which was a running-in period. The coupling degree in other research years was all higher than 0.9, which is a high-level coupling. In particular, from 2011 to 2015, the coupling degree was close to 1.

In 2009, the coupling coordination degree was less than 0.5, and it was in a state of slight imbalance. The development of the social and economic system lags behind the development of the transportation system.

In 2010, the degree of coupling coordination was improved compared with that in 2009. According to Figure 1, compared with 2009, the comprehensive evaluation value of the socio-economic and transportation system in 2010 has improved. Although the development of social economy still lags the development of transportation, the difference between the evaluation values of the two systems has decreased, which reflects the mutual promotion between social economy and transportation, and shows that the transportation system still plays a leading role in this period. The development of transportation has driven the improvement of social and economic levels to a certain extent.

From 2011 to 2013, the degree of coupling coordination rose to a moderately coordinated state. During this period, the development of the social economy and transportation continued to improve. The development of the transportation system promoted the development of the socio-economic system to a large extent, and promoted the optimization of industrial structure and the re-planning of land use. In addition, the national and Sichuan Province The support for the development of tertiary industries such as tourism in Aba has promoted the sustained economic growth of Aba.

From 2014 to 2016, the coupling coordination state of the two reached a good coordination state. In 2014, the comprehensive development level of the social economy and transportation in Aba Prefecture was highly similar. Compared with 2013, the social and economic evaluation value increased, while the transportation evaluation value decreased year by year. During this period, the economic system played a leading role.

In 2017, the coordination between the two dropped to moderate coordination. During this period, due to the continuous decrease in the evaluation value of the transportation system in Aba Prefecture, which affected the development of the social and economic system, the evaluation value of the social and economic system in Aba Prefecture declined, resulting in a decrease in the degree of coupling and coordination.

In summary, at different stages of the development of Aba Prefecture, there are differences in the coupling and coordination relationship between transportation and social economy. Under the long-term interaction between the two systems, the overall development of the coordination level is improving.

4 CONCLUSION

Due to its special topography, the stage of the transportation system and socio-economic system in the southwestern ethnic area is far from that of the plain area. Research and analysis of the coupling coordination between the two, identify the development stage, and explore the appropriate coordinated development path of transportation and economy.

This paper has achieved preliminary results on the research practice in Aba Prefecture. The research results show that the coupling coordination model constructed in this paper is in line with the development of Aba Prefecture. Through the analysis, the following four conclusions can be drawn:

The socio-economic system and the transportation system are long-term mutual influence systems, and the dominant systems are different in different periods. Moreover, the two systems are not completely developing in an upward spiral trend, which shows that they are influenced by many other factors besides mutual development.

Under the long-term interaction, the social economy and transportation system of Aba Prefecture are generally developing in the direction of improving the level of coordination.

This paper analyzes and reveals the coupling and coordination relationship between transportation and social economy in Aba Prefecture from the macro level. In the future, I hope to study the coordination mechanism between the two systems from the micro level, adjust the key development direction in a targeted manner, and create a transportation system. The development trend of spiral interaction between infrastructure construction and economic development.

REFERENCES

Bai Lifei, Li Yanjia, Zhang Jiaqiang. Research on the relationship between economic development and transportation infrastructure construction—Taking the five provinces along the "Belt and Road" in northwest China as an example. Comprehensive Transportation, 2019.

Chen Lu, Zhou Guohua. Evaluation of the coordination degree of human-land relationship in Wangcheng District, Changsha City. Journal of Western Human Settlements and Environment,2018,33(01):54–58.

Jiao Huafu, Yang Xianming. Coupling of industrial structure and spatial morphology evolution of coal resource-based cities: Taking Huainan City, Anhui Province as an example. Acta Geographica Sinica, 2016, 71(6): 998–1009.

Wang Lankai, Zhu Meng. Research on the coordinated development of human settlements and economy in the Chongqing section of the Three Gorges Reservoir area. Interior Design, 2013, (2): 20–24. DOI: 10.3969/j.issn.1006-2181.2013. 02.005.

Yuan Dingning. Research on the coordinated development of urban transportation system and economic system—Taking Changsha as an example. Changsha University of Science and Technology, 2014.

Advances in Urban Engineering and Management Science – Khalil & Yang (Eds)
© 2023 The Author(s), ISBN 978-1-032-30426-7

Evaluation of community grid management effect based on gray scale correlation model takes Jiang'an District of Wuhan, Hubei Province as an example

Yan Wang* & Feiyang Guo*
School of Business, Wuhan Huaxia Institute of Technology, Wuhan, Hubei, China

ABSTRACT: In the context of normalization of epidemic prevention and control, this paper uses the entropy method and the gray scale correlation analysis model to evaluate the grid management effect of 148 communities in Jiang'an District of Wuhan, and evaluates the grid management level of communities by establishing a comprehensive evaluation index system. The results show that the current community grid management system in Jiang'an District of Wuhan is not mature enough, the information resource platform is relatively lacking, the management personnel are weak, the needs of residents cannot be satisfied, and the management effect is not very significant. On this basis, it is proposed to strengthen the community affairs management level and the residents' service level, accelerate the construction of the information management platform, and improve the emergency linkage and rapid response mechanism. This paper reflects the shortcoming of community grid management and related solutions, and opens up new ideas for the community grid management mode.

1 INSTRUCTIONS

In February 2020, the Communist Party of China Central Political and Legal Affairs Committee issued the Notice on further playing the role of grassroots comprehensive governance center and grid members to build the first line of defense of epidemic prevention and control, pointing out that we should make full use of the "comprehensive governance center + grid + information" system to serve the overall work of epidemic prevention and control, and require the extension of prevention and control forces to sink to the community grid (Lv 2021). The Fourth Plenary Session of the 19th CPC Central Committee further put forward the specific requirements of "improving the community management and service mechanism and implementing the grid management and service." As a concrete embodiment of the modern technology and means of China's social governance capacity, community grid management can not only improve the comprehensive service efficiency, but also meet the needs of residents reasonably and appropriately (Shao 2021). With the normalization of epidemic prevention and control, community information statistics, safety monitoring, market supervision, sanitation management, government information publicity, social service guarantee, community grid management is more important in data sharing, public health event early warning and other aspects (Liu 2020). Based on the focus of improving the scientific operation management system, the construction of a high-quality and professional grid member team, and the guarantee of improving the grid way to serve the masses, and the construction of "co-construction, co-governance, sharing" as the goal, we will accelerate the modernization of China's grassroots social governance system and governance capacity (Liu 2020).

At present in the theory of community grid management technology, means, process, mode, etc., has made certain results, but the research of grid management model, the construction of relevant

*Corresponding Authors: mandy9905303@126.com and 917392439@qq.com

DOI 10.1201/9781003305026-90

indicators and quantitative evaluation has not been involved, and this is the measure of community governance accuracy, the important factor of grid management effect. This paper uses entropy and multi-level gray scale correlation analysis to quantify the community grid management index and determine its weight to provide a scientific and reliable basis for reasonable evaluation.

2 CONSTRUCTION OF THE EVALUATION INDICATORS

2.1 *Questionnaire design*

Based on the community grid management theory, fine management theory, and seamless government theory, this study builds an evaluation index system. Take the community grid management effect as the target layer; the management team and the construction (B1), information management (B2), community grid management efficiency (B3), public service (B4), community grid management effect (B5), and the social benefit of community grid management (B6) as the standard layer; the other 23 specific evaluation indicators as the index layer to design the questionnaire. The 27 multiple-choice questions of the questionnaire were divided into three parts: basic situation investigation, evaluation of community grid management effect, and suggestions on grid management. The Likert five-level scale is divided into very dissatisfaction to very satisfaction, indifference to active degree, with 1, 2, 3, 4 and 5 points.

2.2 *Reliability and validity test of the questionnaire*

2.2.1 *Confidence test*

Using the SPSSAU platform, the questionnaire was tested. Excluding the options of four basic situation surveys, there were 23 small items of the five-level scale problems involved in this questionnaire. The calculation showed that the coefficient was 0.839 > 0.8, indicating a high degree of credibility of the questionnaire and a high internal consistency of the scale.

Table 1. Coefficients of standard dimension variables.

Code layer dimension	Clone of bach alpha	Number of terms
Management team and construction	0.880	7
Information management	0.796	3
Community grid management efficiency	0.716	2
Public service	0.754	2
Community grid management results	0.920	4
Social grid management of social benefits	0.970	5
Scale as a whole	0.839	23

2.2.2 *Validity test*

The KMO test was 0.948> 0.80, and the Bartlett sphere test showed that the approximate chi-square value was 9299.465, with large value and significance probability of 0.000 <0.005. Therefore, the null hypothesis of the Bartlett sphere test was rejected, believing that the validity of this survey scale is good and suitable for factor analysis.

Table 2. KMO and Bartlett tests.

Number of KMO sampling suitability quantities		0.948
Bartlett's Test of Sphericity	Approximate chi-square	9229.465
	Degree of freedom	231
	Significance	0.000

2.2.3 *Exploratory factor analysis*

The results of the principal component analysis method (Principal Factor Analysis) show that the factor load of each option is> 0.50, so they should be retained; the total variance interpretation rate of the four factors is 78.675%> 60%, so the validity of this survey scale is considered good.

Table 3. Factor analysis statistics.

	Initial eigenvalue			Extract the sum of square of the load			Sum of square of the rotating load		
project	characteristic root	variance percentage	cumulative contribution %	characteristic root	variance percentage	cumulative contribution %	characteristic root	variance percentage	cumulative contribution %
1	11.941	54.278	54.278	11.941	54.278	54.278	7.231	32.869	32.869
2	2.755	12.524	66.802	2.755	12.524	66.802	4.09	18.589	51.458
3	1.421	6.458	73.26	1.421	6.458	73.26	3.168	14.401	65.859
4	1.191	5.415	78.675	1.191	5.415	78.675	2.82	12.816	78.675

2.3 *Determination of the evaluation index weight*

The index weights in this question mainly adopt the entropy method. According to the weight information theory, as a psychometric unit, the smaller the entropy value, the greater the degree of variation of the index, the greater the information that can be provided, the greater the utility achieved in the evaluation of the index system, and the greater the weight of the corresponding index. Combining the information obtained from the questionnaire survey, the weight relationship of various indexes was obtained using the entropy method.

3 COMMUNITY GRID MANAGEMENT EVALUATION BASED ON THE HIERARCHICAL GRAY ASSOCIATION ANALYSIS

3.1 *Gray association analysis method*

Gray association method (Grey Relational Analysis) is a branch based on the development of gray system theory, which refines the factors affecting things. In the development of things, study the future development trend of things through the development direction. In this process, the gray correlation analysis provides a very suitable analysis of the dynamic process of things development. The groups investigated in this paper do not understand the construction of the community grid system, and the final results of the community grid management are only in the subjective perception. Therefore, there are gray areas for all the indicators investigated in these areas. Therefore, grey system theory can be fully practiced here to obtain a comprehensive analysis based on a small amount of information.

3.2 *Acquisition and processing of the data*

This questionnaire survey mainly adopts the field survey method and sampling survey in 148 communities of Jiang'an District, Wuhan. A total of 420 questionnaires were distributed, and 370 were effectively recovered, with an effective recovery rate of 88.1%. According to the evaluation index of the questionnaire scale, the attitude of the respondent was divided into five levels: very dissatisfied, dissatisfied, generally satisfied, more satisfied, very satisfied, corresponding to the scale 1,2,3,4,5. After the completion of the questionnaire data information collection, the questionnaire was divided into two categories based on whether the community has a big data information management platform as the standard (1 means no, 2 means yes). Based on the relevant sample data, the final score of the corresponding index was obtained.

Table 4. Weight relationship of various indicators.

Evaluation index system	Code layer weight	Index layer weight
Management team and build (B1)	22.33%	
The number of grid worker visits (C11)		3.27%
Service attitude of the community staff (C12)		3.01%
Grid member's responsibility for the work (C13)		3.26%
Professional level for community staff (C14)		3.51%
Service efficiency for community workers (C15)		3.18%
Quality of service for Community staff (C16)		3.07%
Volunteer team service satisfaction (C17)		3.03%
Information management (B2)	18.99%	
Knowledge degree of the information management platform is (C21)		3.96%
The situation has been reported to the community through the information platform (C22)		6.96%
Is there an increase in the number of information provided (C23)		8.07%
Community grid management efficiency (B3)	13.29%	
Work efficiency (C31)		7.91%
The degree of improvement of community governance at the grassroots level by technical means (C32)		5.38
Public service (B4)	10.74%	
Infrastructure coverage (C41)		6.73%
The ability to help vulnerable groups and special groups (C42)		4.01%
Community grid management effect is (B5)	19.27%	
Community affairs management level (C51)		6.06%
Community resident service level (C52)		5.98%
Grassroots visit effect (C53)		3.62%
Overall work evaluation (C54)		3.61%
Community grid management of social benefits (B6)	15.73%	
Security monitoring capability improves (C61)		3.14%
Community environment improvement (C62)		3.23%
Community security capability (C63)		3.21%
Community residents live in harmony with the (C64)		3.09%
Community public health improvement in the (C65)		3.06%

Table 5. Results of community grid management in Jiang'an district, Wuhan city.

Construction of the evaluation index system for community grid management effectiveness in Jiang'an District, Wuhan City			appraise	
Target layer	Code layer	Index layer	1	2
Grid management effect of Wuhan Jiang'an District	Management team and construction	Number of visits made by the grid workers	3.32	3.86
		Service attitude of the community staff	3.74	3.11
		Grid member's sense of responsibility for the work	3.75	3.09
		Professional level of the community staff	3.59	2.99
		Service efficiency of the community staff	3.65	3.04
		Quality of service for the community staff	3.71	3.06
		Service satisfaction of the volunteer team	3.63	3.07

(*continued*)

Table 5. Continued.

Construction of the evaluation index system for community grid management effectiveness in Jiang'an District, Wuhan City			appraise	
Target layer	Code layer	Index layer	1	2
	Information management	Knowledge degree of the information management platform	2.29	2.89
		Reflect the situation to the community through the information platform	2.25	2.81
		Is there an increase in the number of relevant information provided	2.39	2.83
	Community grid management efficiency	Work efficiency	2.11	2.36
		The improvement degree of technical means to grassroots community governance	1.87	2.17
	public service	Infrastructure coverage level	2.29	3.00
		The ability to help vulnerable groups and special groups	3.00	3.56
	Community grid management results	Community affairs management level	1.97	2.20
		Service level of community residents	2.00	2.19
		Grassroots visit effect	3.56	3.04
		Overall work evaluation	3.66	3.07
	Social grid management of social benefits	Improve the safety monitoring ability	3.57	2.87
		Community environment improvement	3.49	2.79
		Community security guarantee capability	3.57	2.91
		Community residents live in harmony	3.53	2.91
		Community public health improvement	3.65	2.91

3.3 *Comprehensive evaluation*

From the above table, in the evaluation indicators of the grid management effect of Wuhan Jiang'an District, the three dimensions of information management are close to 3, indicating that the community grid management in the information construction is in a general satisfactory state. The two

Table 6. Means and standard deviation of the evaluation indicators.

Evaluation index system	Sample capacity N	Average value	Standard error	Variance
Management team and build (B1)				
The number of grid worker visits (C11)	370	3.78	1.147	1.315
Service attitude of the community staff (C12)	370	3.73	1.099	1.207
Grid member's responsibility for the work (C13)	370	3.71	1.137	1.292
Professional level for community staff (C14)	370	3.60	1.149	1.320
Service efficiency for community staff (C15)	370	3.67	1.124	1.262
Quality of service for community staff (C16)	370	3.68	1.105	1.221
Volunteer Team service satisfaction (C17)	370	3.67	1.099	1.207
Information management (B2)				
Knowledge degree of the information management platform is (C21)	370	2.93	1.027	1.055
The situation has been reported to the community through the information platform(C22)	370	2.86	1.335	1.782
Is there an increase in the number of information provided (C23)	370	2.82	1.448	2.096

(continued)

Table 6. Continued.

Evaluation index system	Sample capacity N	Average value	Standard error	Variance
Community grid management efficiency is (B3)				
Work efficiency (C31)	370	2.42	1.254	1.572
The degree of improvement of community governance at the grassroots level by technical means (C32)	370	2.33	0.980	0.960
Public service (B4)				
Infrastructure coverage (C41)	370	2.83	1.300	1.690
The ability to help vulnerable groups and special groups (C42)	370	3.49	1.201	1.443
Community grid management effect is (B5)				
Community Affairs Management Level (C51)	370	2.39	1.074	1.154
Community resident service level (C52)	370	2.38	1.066	1.136
Grassroots visit effect (C53)	370	3.58	1.169	1.366
Overall work evaluation (C54)	370	3.66	1.110	1.232
Community grid management of social benefits (B6)				
Security monitoring capability improves (C61)	370	3.56	1.096	1.200
Community environment improvement (C62)	370	3.51	1.090	1.188
Community security capability (C63)	370	3.56	1.098	1.206
Community residents live in harmony with the (C64)	370	3.55	1.074	1.153
Community public health improvement in the (C65)	370	3.61	1.084	1.175

dimensions of the efficiency of community grid management are both below 2.5, indicating that residents have a low evaluation on the efficiency of community grid management. In the dimension of public service, the score of the infrastructure coverage index is 2.83, indicating that the coverage of community infrastructure is not complete. In the effectiveness dimension of community grid management, the average value of the two indexes of community affairs management and residents' service level is close to 2.4, which is not very satisfactory.

4 CONCLUSION

According to the above evaluation model, in view of the weak level of community informatization management, the government needs to further strengthen the construction of an informatization management platform, including making full use of and integrating existing information and network resources, strengthening the co-construction, and sharing of information resources, and establishing a more effective mechanism of information resource sharing and exchange (Wu 2021). In view of the low efficiency of community grid management, it is necessary for the government to establish and perfect emergency linkage and rapid response mechanism. For example, each district can establish an emergency linkage center to assist communities or all departments in unified coordination and command, so that relevant departments can carry out the joint emergency response. In view of the low level of community affairs management and residents' service, it is necessary for relevant government departments to further emphasize the responsibility consciousness of relevant staff, and at the same time, pay attention to the establishment of corresponding working mechanisms and relevant system guarantees, such as standardizing the management system of grid staff, strengthening the training of grid staff, and improving their salary and treatment.

ACKNOWLEDGMENT

This work was supported by the General Humanities and Social Sciences Project of Hubei Provincial Education Department: Research on the Interactive Mechanism of science and technology resource allocation and regional independent innovation capability (No. 18G140)

REFERENCES

Lv Tong. Discussion on the Optimization Path of Grid Governance Structure—From the Perspective of Structural Functionalism[J]. Social Sciences of Beijing. 2021, (04), 106–115.

Liu Chuncheng. Community Grid Management against the Background of Corona Virus Disease 2019 Prevention and Control[J]. Journal of Jiangnan University (Humanities&Social Science), 2020, 19(02), 5–13.

Liu Runqiu. Exploration, Reflection and Optimization of Community Governance Mechanism for Epidemic Prevention and Control[J]. Peoples Tribune. 2020, (15), 20–22.

Shao Xinzhe, JiGuojun. Research on Collaborative Operation Mechanism of Urban Grid Management and Intelligent Community——Illustrated by the Case of S City, Sichuan Province[J]. Soft Science. 2021, 35(02), 137–144.

WuXiaolong, Kang Xuhui. Participation Logic and Practical Dilemma of Multiple Subjects in Grid Governance: An Analysis Perspective of the Empowerment Theory[J]. Social Work and Management. 2021, 21(05), 83–89.

Advances in Urban Engineering and Management Science – Khalil & Yang (Eds)
© 2023 The Author(s), ISBN 978-1-032-30426-7

An improvement of car retrieve process and system configuration of compact robotic automated parking system

Yunfeng Ma
Service Science and Engineering Research Center of WUST, Wuhan, China

Junhao Fan*
Wuhan University of Science and Technology, Wuhan, China

Liang Ren
Service Science and Engineering Research Center of WUST, Wuhan, China

ABSTRACT: Compact robotic automated parking system is a new system with higher storage utilization. The existing retrieve process causes the time of the car waiting for the lift to be longer than the time of the lift moving to the target floor from the first floor. The system always chooses the lift that is closest to the car, which may increase the expected retrieve time. Firstly, we made a new retrieve process to improve by sending commands simultaneously to the car storage floor and the lift, allowing the lift rising before the car completes the movement on the target floor. Secondly, two strategies for selecting the lift, the shortest distance strategy, and the shortest time strategy, are equipped with the process improved, which is used to compare with the process not being improved. Meanwhile, the system configuration design is discussed. Finally, simulation results show that the process improved using the shortest time strategy reduces the average retrieve time by 20%-50% compared with the process not being improved. Under the experimental conditions, the maximum space utilization rate of the system was 60.5%. Furthermore, considering to retrieve time and space utilization rate, the circular fixed platform radius is 5.85 meters, the system radius is 17.85 meters, the number of floors is 14, and the number of lifts is 6. The experimental results show that the process improved with the shortest distance strategy is better, and the system configuration design is feasible.

1 INSTRUCTIONS

Compact robotic automatic parking is a cylindrical automatic parking system with high storage density. Based on a Compact Automatic Parking (CAP) system, Wu put forward a CRAP, in which the centralized lift in the center is replaced by a distributed lift in the outer ring and the shuttle on the rotatable inner ring is removed to allow car storage (Wu et al. 2020). Provide each storage tier and each lift with AGV (Automatic Guided Cars).

The first floor of the CRAP system allows people to drive their cars in or out of the system (Figure 1). From the second floor to the top floor (Figure. 2), each storage floor is composed of a circular fixed platform with Floor AGV (F-AGV), the rotatable inner ring and the fixed outer ring for parking. The outer ring of the system is equipped with lifts and Lift AGV (LAGV) for the movement of cars between the inner ring and the lifts, or the first floor and the lifts. The LAGV loads cars on the lifts and moves up or down with them.

Research on CRAP systems is inspired by references to other theories. Fan Yuan et al. proposed a genetic algorithm to solve the cooperative scheduling problem of multiple mobile robots (Fan

*Corresponding Author: wustfanjunhao@163.com

 DOI 10.1201/9781003305026-91

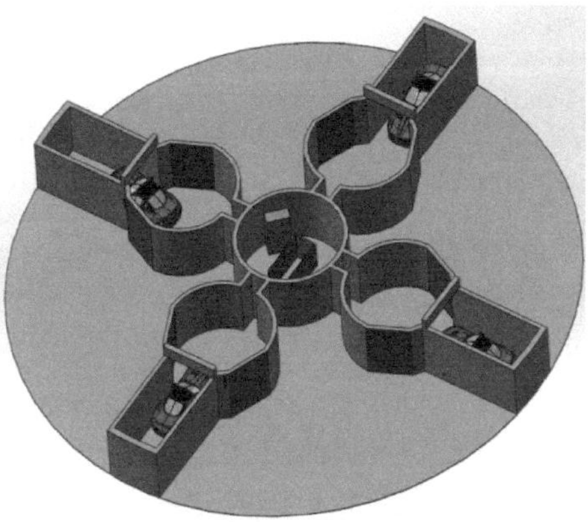

Figure 1. I/O floor.

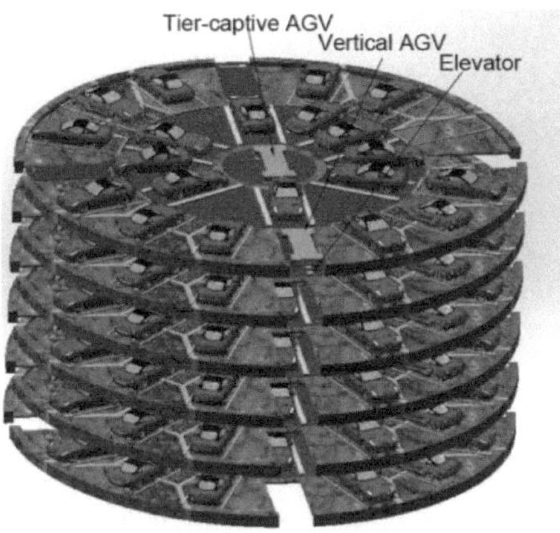

Figure 2. Storage floor.

et al. 2019). Zhou Xin et al. established a double depth cargo position optimization model for the stacker to execute cross access instructions, and proposed a Close & Open algorithm and assignment algorithm to solve the model (Zhou et al. 2017). Han Dongya et al. transformed the integrated optimization problem of task order and location selection in an automatic three-dimensional warehouse into a mixed-integer programming model and designed a two-stage heuristic algorithm to solve it (Han et al. 2020). Guo Di et al. constructed a discrete event simulation model based on queuing theory for the configuration of storage robots in an intelligent warehouse (Guo et al. 2020). Heragu et al. build an open queuing network to estimate the performance of tier-captive AVS/RS, and make a comparison between the AVS/RS and the traditional automated storage and retrieval system (AS/RS) (Heragu et al. 2011). Marchet et al. also formulate an open queuing network to get the throughput time of the tier-captive AVS/RS system, considering the acceleration/ deceleration of shuttles and lifts (Marchet et al. 2012). Zou et al. propose a parallel policy under which the lift and the vehicle can simultaneously move. They build fork-join queuing network models to analyze the system performance (Zou et al. 2016). Based on the development of artificial intelligence, some cutting-edge theories have also been put forward (Abubakar et al. 2019; Duan et al. 2016; Mahrook 2019).

CRAP System is designed based on CAP (Compact Automated Parking System) system. Wu studies the CAP system, in which the storage floor is composed of a fixed outer ring for car storage and an inner ring that can be rotating but cannot be stored; the inner ring is equipped with a shuttle

(Wu et al. 2019). System lifts are located in the center of the system. The comparison is made with a movable cube parking system studied (Zaerpour et al. 2016). On this basis, Wu further studied the problem of continuous multiple lifts in the CAP system and used two operating strategies of shuttles, focus strategy, and random strategy (Wu et al. 2019). The corresponding analytical algorithm was proposed and verified by the simulation method, and it was concluded that continuous multiple lifts were better than single lifts in the same configuration. The results show that the focus strategy is better than the random strategy. Wu established the CRAP system based on the CAP system and gave the analytical algorithm of system retrieve time. Through simulation experiments, the optimal number of CRAP floors is 8-10. In comparison with the CAP system, it is shown that the CRAP system has significant advantages in reducing the time of cars waiting for lifts and the time of cars moving in the system. However, the CRAP system can still be improved, further reducing the retrieve time of the system. Based on Wu (2020), the main contents of the research are as follows:

(1) Improving the retrieve process and proposed two lift selection strategies, through sending command simultaneously to the storage floor and the lift to improve the retrieve process, allowing the lift to rise from the first floor before the car complete the movement of the target floor, reducing the time of car waiting for the lift; Meanwhile, a way to explore the optimal configuration of CRAP is put forward.
(2) CRAP simulation model is established to evaluate its various values under different conditions, such as the time of car waiting for the lift, the whole time of retrieve, the average transport time of lift for each car, etc.
(3) The simulation experiment is used to compare the origin process with the improved process, which adopts two different lift selection strategies. It is concluded that the process improved which adopting the shortest time strategy is the best retrieve process under general conditions. The optimal configuration of the system is obtained by using the simulation model under experimental conditions, which verifies the feasibility of the system configuration design

2 PROBLEM DEFINITION

The car retrieve process of Wu is shown in Figure 3. The significance of the original process is to construct the basic logic of the CRAP system for car retrieving, which is good for experimental analysis. But it increases the expected retrieve time of the system. The original process of lift behavior is when the car arrives at the lift's position on the target floor, the lift begins to rise from the first floor if the lift is idle, or it will rise after finishing the last job. In the original process, the time of the car waiting for the lift is always greater than or equal to the time of the lift raise from the first floor to the target floor. Only one inner ring equipped with F-AGV can rotate on every floor when a car stays in the inner ring and waiting for the lift. The inner ring can only stay and wait for the lift rather than continue to retrieve the next car, thus prolonging the time of the car waiting for the inner ring of the next task, resulting in a longer retrieve time in the next retrieve task. Therefore, the key to shortening the retrieve time is to improve the retrieve process and rationally plan the lift's behavior to reduce waiting time.

In addition, the original process always chooses the lift nearest to the car, namely the shortest distance strategy. When multiple cars are waiting in a queue for the same nearest lift and other lifts are idle, the choice of the same lift may result in a longer retrieve time because the time of waiting for the lift is significantly increased. The shortest time strategy is to select the lift that can achieve the shortest time for the current retrieve task. However, if multiple tasks are considered as a whole, the nature of the shortest time strategy is a greedy algorithm, so further experiments are needed to test whether it is better than the shortest distance strategy under general conditions.

Finally, one of the core advantages of CRAP is the higher utilization of space compared to traditional parking systems. Wu et al. (2020) adopts several different configurations to explore the performance of the CRAP system, which takes retrieve time as the main indicator. However, the value of system configurations it adopts makes the system space utilization rate too low. For example, the value of one of the configurations is an outer ring diameter of 107 meters, an inner ring diameter of 53 meters, seven floors, an outer ring parking of 125 spaces, an inner ring parking of 42 spaces, and six lifts. The problem with the configuration is that the whole floor area of each is about 9551 square meters, with a parking area of about 2408 square meters; therefore, the space utilization rate is only 25%, even lower than the utilization of much common traditional parking, and cannot reflect CRAP system's advantages, space utilization. So, it is necessary for the system to design better configurations.

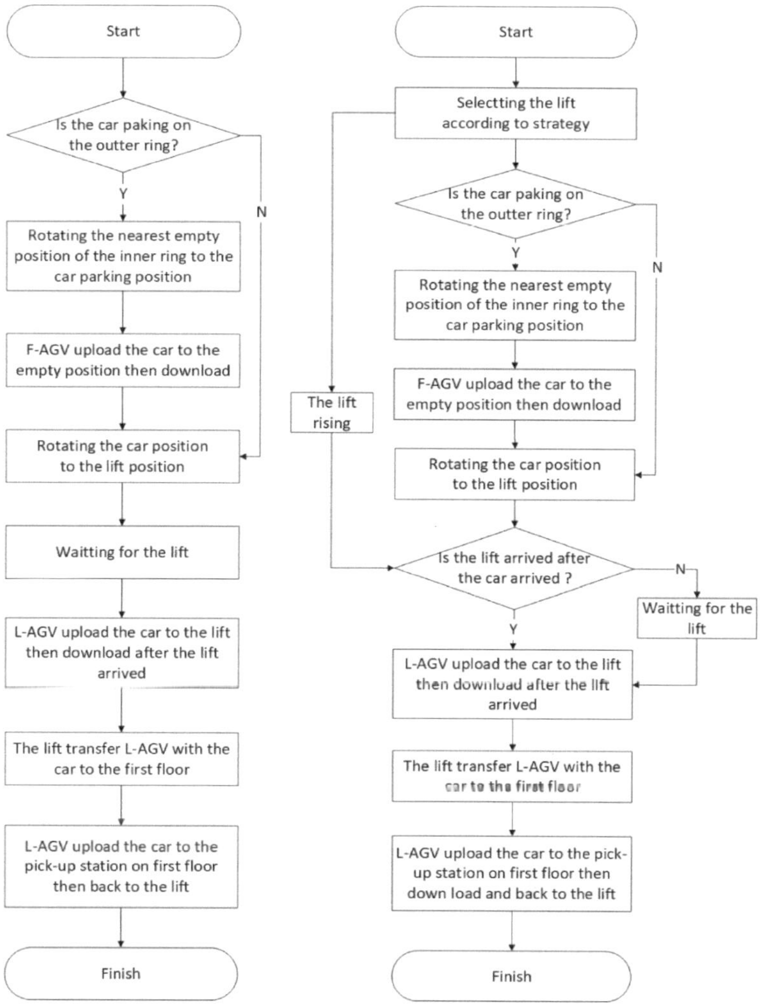

Figure 3. Original process (left) & process improved (right).

3 CONFIGURATIONS DESIGN

3.1 *Configurations design of retrieve process*

The key to reducing retrieve time is to reduce the time of the car waiting for the lift. That is, allowing the inner ring and the lift to move at the same time and planning their behavior. The behavior of the inner ring is relatively simple. It takes the car immediately began to rotate after receiving the order or after the end of the last task. The behavior planning of lifts shall meet the following requirements:

Case 1: When $L_F + T_L \leqq C$, $L_U = C - T_L$

Case 2: When $L_F + T_L > C$, $L_U = L_F$

The time at which the selected lift begins to rise is L_U; The time spent for the selected lift to reach the target floor from the first floor is T_F; L_F is the time after the selected lift finishes the previous retrieve task; the time when the car arrives, the lift's position on the target floor is C.

The process improved after the car arrived at the lift's position on the target floor: in case 1, the car retrieved did not need to wait for the lift, since the lift had been there. The F-AGV loads the car onto the lift and moves to the first floor, then the F-AGV unloads the car to the first floor, and the F-AGV returns to the lift. In case 2, the target car needs to wait for the lift to arrive.

Using the symbols defined above to show the lifting behavior of the original process meets the following:

Case 3: When $L_F \leqq C$, $L_U = C$;

Case 4: When $L_F > C$, $L_U = L_F$

To sum up, in the single retrieve task with the same other conditions, if $L_F \leqq C$, the retrieve time of the improved process will be reduced by $C - L_U$ compared with the original process, and the maximum time reduction is T_L.

3.2 *Configurations design of the system*

When designing system configuration, consider the retrieve time and the number of parking positions based on the maximum space utilization of the system. Configuration variables that need to be adjusted mainly include CRAP system diameter, number of lifts, and the total number of floors.

System diameter is the key to space utilization rate and the number of parking positions on a single parking floor. A too large diameter will reduce the space utilization rate, while too small a diameter may reduce the space utilization rate due to too few parking positions. The number of lifts and the total number of floors need to be adjusted together because the number of floors determines the number of inner rings, affecting the retrieve time. Meanwhile, when the system diameter is determined, the number of floors is positively correlated with the total number of parking positions, while the number of lifts is negatively correlated with it.

Diameter: In the system, when a certain diameter, outer ring width will affect the inner diameter, which determines the inner ring number of parking positions, and a very wide inner ring will increase energy when it needs to rotate, so the inner and outer ring width (Figure 4) should be on the premise of secure access to get the minimum. The determining factor of the diameter of the system is the diameter of the circular fixed platform. The number of parking positions is a geometric problem "given the radius of the circle and the side length of the polygon, figure out what polygon is the outer tangent regular polygon of the circle" (Figure 5), w_{sp} is the width of each parking position, and the number of cars on rings n_c can be expressed as the formula below:

$$n_c = 360 \div (2 \times arctan(w_{sp} \div 2 \div r)) \tag{1}$$

Calculate the number of parking positions on the inner and outer ring as a geometric problem, multiply the parking area, calculate the system space utilization ratio δ by Formula (2):

$$\delta = ((n\_ci + n\_co) \times w\_sp \times L\_"sp")/(\pi("12" + r) \wedge 2) \times "100\%" \tag{2}$$

n_{ci} is the number of parking positions on the inner ring, n_{co} is the number of parking positions on the outer ring, and L_{sp} is the length of each parking position.

The total number of floors and lifts number design method: When other conditions are determined, the number of floors and the number of lifts will affect the total number of parking positions.

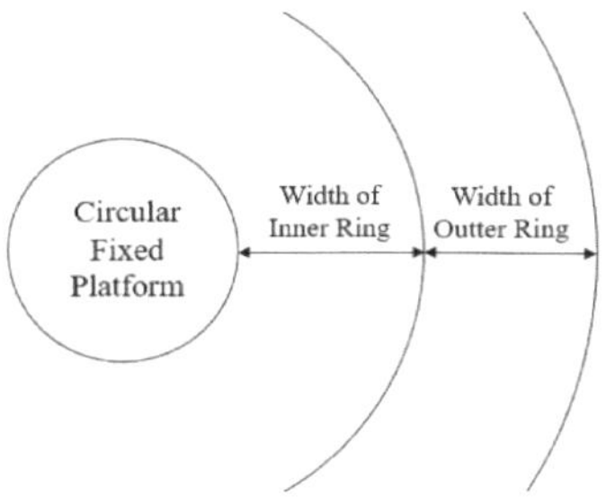

Figure 4. Width of rings.

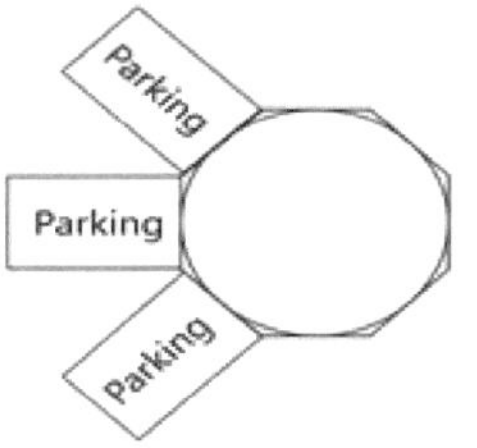

Figure 5. The geometric problem.

The relationship between the total number of parking spaces N_C and the number of lifts, the number of system floors is shown in Formula (3):

$$N_c = (n_{ir} + n_{or} - N_L) \times (N_T - 1) \tag{3}$$

4 SIMULATION MODEL BUILDING

The stopping point strategy of the inner ring and the lift follow the "service completion point is the stopping point" (POSC), and the lift and the inner ring follow the first-come, first-served (FCFS) principle. Assuming every AGV loads and unloads car costs same time. The car retrieved be generated in any parking position with equal probability. From the perspective of service, retrieve time is the main factor affecting user experience, so only the retrieve task is considered.

Table 1. Symbol definition.

N_T	Total floor number
N_L	Total lift number
SC_n	The time of "n"th task command issuing
SR_n	The time of "n"th car being retrieved
SL_n	The time of "n"th car being unload to the lift
FR_n	The time of "n"th car arrived lift position on inner ring
FL_n	The time of L-AGV back to the lift
P_n	The position which the target car in
T_n	The floor which the target car in
Policy	1: Shortest distant strategy; 2: Shortest time strategy
TT	Total time spent of simulation
t	Current time of simulation
λ	Rate of issuing retrieve command
h	Height of every floor
t_{agv}	Time spent of AGV loading/unloading a car
a_e	Accelerated speed of lift
v_e	Max speed of lift
a_{ir}	Accelerated speed of inner ring
v_{ir}	Max speed of inner ring
d_{ir}	Diameter of inner ring
n_{or}	Parking number of outter ring
n_{ir}	Parking number of inner ring

Table 1 shows the symbol definitions. The symbol with subscript "n," the main meaning unchanged after the subscript is replaced. For example, T_n stands for "the floor of the car of number n," and T_a stands for "the floor of the car of number a."

The main part of the simulation model is composed of a car information matrix, two types of service station matrices, and multiple service station queue matrices, which update the information by one step per second in the time-step method. Meanwhile, the logger matrix is set to record the relevant information per second.

The car information matrix is randomly generated in the initial stage of the model, which generates finite commands to retrieve the car within simulated periods (FIG. 6), which is equivalent to the "ID card" of each car identified and recorded in the experiment.

1	2		n
SC_1	SC_2	......	SC_n
T_1	T_2		T_n
P_1	P_2		P_n

Figure 6. Matrix of car retrieved.

Each lift is regarded as a lift service station, and all the lift service stations combine a lift service station matrix with columns equal to the numbers N_L. Each inner ring of the storage floor is regarded as an inner ring service station, and all the inner ring service stations combine an inner ring service station matrix with columns equal to the numbers of $N_T - 1$. The last behavior logical value, 1 means that the service station is busy, 0 means that the service station is idle. Others are related to the information of cars currently in the service station (Figure 7).

SC_a	SC_b		SC_c
SR_a / SL_a	SR_b / SL_b		SR_c / SL_c
FR_a / FL_a	FR_b / FL_b	......	FR_c / FL_c
T_a	T_b		T_c
P_a	P_b		P_c
1 / 0	1 / 0		1 / 0

Figure 7. Matrix of service station.

Each service station has its own service station queue matrix. The number of queue matrix of the inner ring service station is equal to N_T, and the number of queue matrix of the lift service station is equal to N_L. Each column of the queue matrix is a car waiting for service. The inner ring service station queue matrix is three rows (Figure 8), and the lift service station queue matrix is five rows (Figure 9).

SC_a	SC_b		SC_c
T_a	T_b	......	T_c
P_a	P_b		P_c

Figure 8. Matrix of service station queue of inner rings.

SC_a	SC_b		SC_c
SL_a	SL_b		SL_c
FL_a	FL_b		FL_c
T_a	T_a	......	T_a
P_a	P_b		P_c

Figure 9. Matrix of service station queue of lift.

Based on the above matrix, the code flow chart of the simulation model is shown in Figure 10. At the beginning of the program, the car retrieved matrix is generated through initialization, and the main loop is updated by the time step method. In the loop, the inner ring service station queue matrix and inner ring service station matrix of each floor are updated first, and then the lift service station queue matrix and lift service station matrix are updated. Among them, when the matrix of the inner ring service station is updated, the service station of the chosen lift is updated at the same time, because the choice of the lift is related to the time when the cars leave the floor, which will affect the time when the inner ring ends its busy state. The program ends After the end of the main loop.

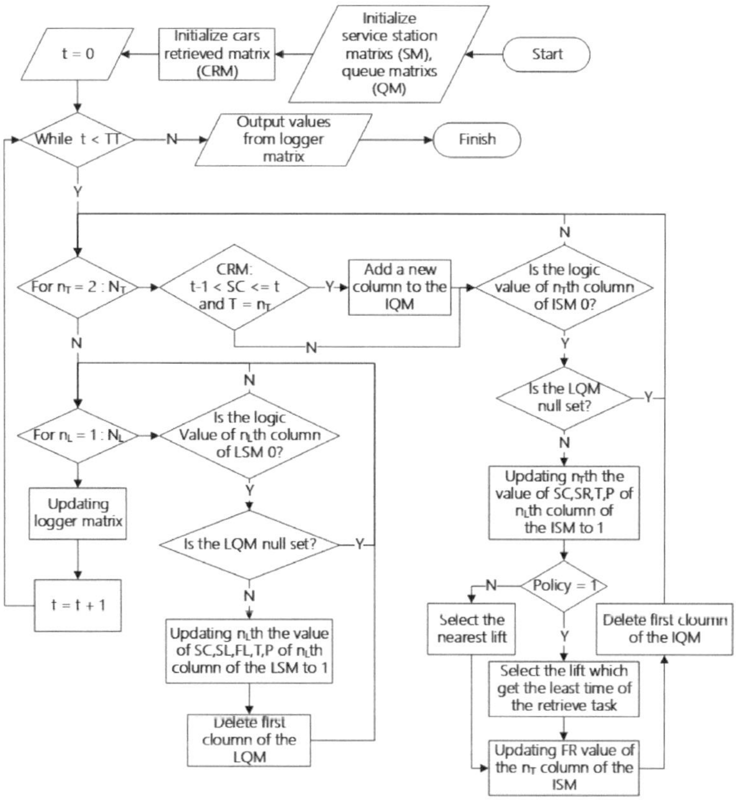

Figure 10. The code flow chart of the simulation model.

5 SIMULATION EXPERIMENT

5.1 *Experimental design*

5.1.1 *Comparative experiment on the improvement of retrieve process*

MATLAB was used to complete the simulation modeling. Firstly, in the original process, the process improved with the shortest distance strategy, and the process improved with the shortest time strategy were simulated and compared.

Set the five research indicators of "average retrieve time, the average time of cars waiting for the lift, the average time of cars waiting for inner rings, the average time of lifts moves the car, and the average time of inner rings moves the car." Although the main index to evaluate the performance of the system is the average retrieve time, other indexes are also helpful for the analysis of the retrieve process. Wu (2020) sets two indexes: the average retrieves time and the average time of cars waiting for the lift. Therefore, only two indexes include the comparison of the three results of the original process, while other indexes only include the comparison of the process improved with two different strategies.

Table 2. Value of model.

h	t_{agv}	a_e	v_e	a_{ir}	v_{ir}	w_c	w_{sp}
2.13 (m)	12 (s)	0.5 (m/s2)	1.6 (m/s)	0.6 (m/s2)	1.8 (m/s)	2 (m)	2.67 (m)

After the fitting experiment with the original process, the same preset values (Table 2) and eight configurations (Table 3) are used to conduct 100 simulation hours of trial operation for each scene and 100 times of simulation 1000 hours of an experiment for each scene respectively with two strategies.

Table 3. Values of configuration.

Scenario	N_T	n_{or}	λ	d_{ir}	n_{ir}	n_L
1	7	75	210	32	25	6
2	8	65	190	28	22	6
3	9	57	175	25	19	6
4	10	50	165	22	17	6
5	7	125	210	53	42	6
6	8	108	185	46	36	6
7	9	94	180	40	32	6
8	10	84	160	36	28	6

5.1.2 *Verify the feasibility of configuration improvement*

According to Table 2, the parking space length is 5.4 meters, and the width of the inner and outer rings is 6 meters. Combined with Wu (2020), the width of each parking position is set as 2.67 meters. It can fully guarantee the safe movement of the car during movement. The circular fixed platform is used to store F-AGVS. We can also set the diameter as 6 meters.

Setting λ varies in correlation with total car positions, because normally, the more parking cars are, the higher rate of retrieve rate. λ_c is the rate of a single car will be retrieved, the sum of being retrieved rate of all parking positions is λ, their relationship as shown in Formula (4):

$$\lambda = \lambda_c * N_c \tag{4}$$

A simulation model was used to take the radius of the circular fixed platform from 3 to 9 meters, according to the step length of 0.05 meters, and Formulas (1) and (2) were used to make a full enumeration calculation to obtain the radius of the circular fixed platform with the maximum space utilization. The radius of the circular fixed platform plus the width of the inner ring and the width of the outer ring were the radii of the system.

After obtaining the optimal diameter that can maximize the space utilization rate of each floor, set $\lambda_s = 0.4$, the probability of each parking position issuing a car retrieve command every hour is 0.4. This value is close to the value of λ in the scenario in Table 3 and is also convenient to calculate. Full enumeration simulation experiments were carried out with the number of floors from 5 to 20 and the number of lifts from 3 to 6, and the experimental results were analyzed to verify the effectiveness of the configuration improvement design.

5.2 *The comparative experimental results of the improvement of the retrieve process*

After the comparative experiment of improving the retrieve process, the experimental results were obtained, as shown in Table 4.

Table 4. The result of the experiment.

	Scenario	1	2	3	4	5	6	7	8
Avenrage retrieving time	Oringinal	122.963	122.693	120.617	126.804	134.912	129.278	143.508	133.111
	Shortest distance	97.9289	84.1557	78.5477	74.5906	129.2098	99.8601	90.2431	110.1141
	Difference	19.14%	16.39%	15.91%	15.60%	15.90%	14.34%	13.55%	8.32%
Avenrage time spent of car waitting for lift	Oringinal	28.078	30.582	30.046	37.483	23.942	23.470	41.304	33.024
	Shortest distance	11.6485	11.0605	10.9688	10.8425	13.2619	12.1291	11.9211	13.334
	Shortest time	0.2885	0.2124	0.1776	0.1572	0.6439	0.437	0.3579	0.7578
	Difference	97.52%	98.08%	98.38%	98.55%	95.14%	96.40%	97.00%	94.32%
Avenrage time spent of car waitting for Inner	Shortest distance	28.4819	16.3176	11.2195	8.2285	49.6939	23.6311	16.0044	20.6004
	Shortest time	16.8351	9.7845	6.54	4.6059	35.0859	15.999	10.8218	16.6085
	Difference	40.89%	40.04%	41.71%	44.03%	29.40%	32.30%	32.38%	19.38%
Avenrage time spent of lift moving car	Shortest distance	39.5828	40.9285	42.279	43.5873	39.565	40.9175	42.2718	43.6037
	Shortest time	39.5675	40.9348	42.3009	43.6293	39.5546	40.926	42.2883	43.6311
	Difference	0.04%	0.02%	0.05%	0.10%	0.03%	0.02%	0.04%	0.06%
Avenrage time spent of inner ring moving car	Shortest distance	26.0071	24.3134	23.2198	21.726	34.4715	31.6412	29.1816	42.3779
	Shortest time	30.2762	27.8981	26.18	24.3776	41.1576	36.642	33.6926	49.7665
	Difference	16.42%	14.74%	12.75%	12.20%	19.40%	15.80%	15.46%	17.44%

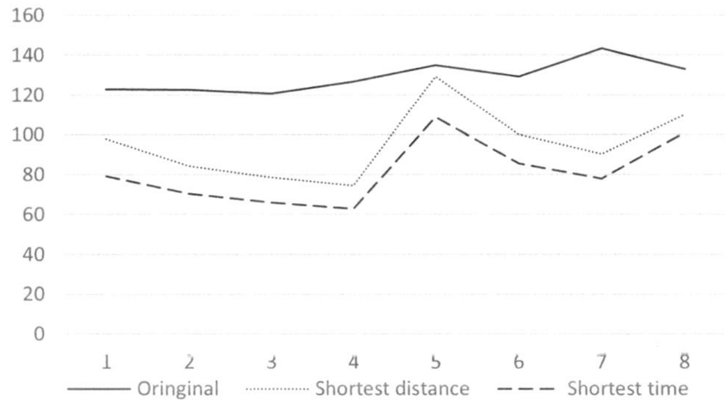

Figure 11. Average time of retrieve.

The process improved with any strategy. Under the eight kinds of scenarios, the average retrieve time has a significant reduction compared with the original process (Figure 11), and by comparing the index data in Table 4, the main reason is less time of car waiting for the lift.

Compared with the original process, in different scenarios, the average retrieve time is reduced by 10% to 40%, and the average time of cars waiting for a lift is reduced by 50% to 70%. Compared with the original process, the process improved using the shortest time strategy has reduced the retrieve time by 20% to 50% in different scenarios, and the average time of cars waiting for a lift has been reduced by more than 97%. Therefore, the process improved is better.

Compared to the average retrieve time of the two strategies, the time of the shortest time strategy is less than the shortest distance, which decreased by 8% to 20%. Through the analysis of other indicators, getting the result of average to retrieve time reduction is the average time of car waiting for the lift reduced, and the time of car waiting for the inner ring reduced. In the process improved, using the shortest time strategy, the time of car waiting for the lift is reduced to almost zero. There was also a significant reduction in the time of car waiting for the inner ring, about 19% to 45%.

Compared with the original process, the process improved with the shortest distance strategy or the shortest time strategy can significantly improve the retrieve performance. In the comparison of the two strategies, the CRAP System performs best with the shortest time strategy.

5.3 *The experimental results of configurational improvement*

After the full enumeration calculation with a radius from 3 to 9 meters with a step size of 0.05 meters for the circular fixed platform, the change result of space utilization rate was obtained as shown in Figure 12.

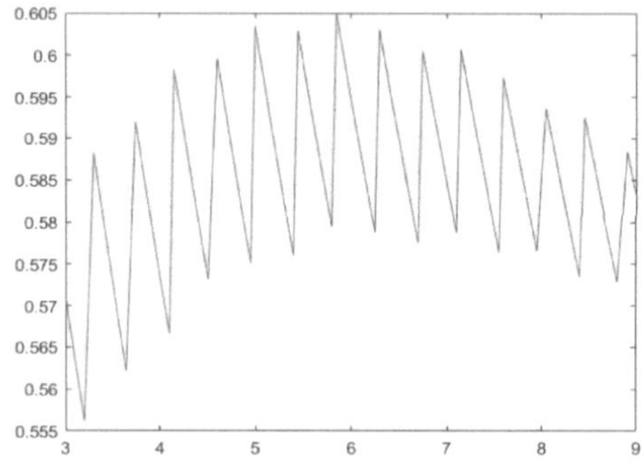

Figure 12. Result of the experiment on the utilization rate.

When the radius gradually increases from 3, due to the insufficient value-added of the radius, the parking positions cannot be increased. The area of a single parking floor becomes larger, but the parking area remains the same, and the space utilization rate decreases. When the radius is increased enough to increase the number of parking positions, the space utilization rate will be improved significantly. When the radius of the circular fixed platform increases to a certain extent, the increase of the non-parking area begins to be greater than the parking area. When the space utilization reaches 60.5%, which is the maximum rate, the circular fixed platform radius is 5.85 meters. We can get the optimal diameter of the CRAP system is 35.7 meters. And the inner ring can accommodate a maximum of 14 parking positions, and the outer ring can accommodate a maximum of 28 parking positions.

After determining the radius of the circular fixed platform as 5.85 meters, the experiment will be continued to explore the combination of the number of lifts and the number of floors.

According to Formula (4), the average retrieve time under different combinations of lift numbers and floors can be obtained (Table 5). The actual number of parking positions can be obtained according to Formula (3) (Table 6). Under the same number of floors, the increase in the number of lifts will reduce the average retrieve time, and the number of positions on a single parking floor will decrease. Under the same number of lifts, the increase in the number of floors will prolong the average retrieve time and increase the total number of parking positions.

Six lifts under the same floor number condition get the absolute advantage for retrieve time when the floor number is less than 13, making up for a parking position relatively less of a disadvantage. If we increase one floor for six lifts system, compared with other quantity lifts system, the parking position number dominant and retrieve time is generally superior to other. When the system is added to 14 floors, however, an additional floor does not make six lift system keeping the advantage of retrieve time and parking positions. Therefore, under the experimental condition, it can be preliminary thought that the radius of the circular fixed platform is 5.85, namely, the system radius of 17.85 meters, 14 floors and six lifts is the best. CRAP has an optimal configuration based on achieving maximum space utilization. At this point, its inner ring can accommodate 14 parking positions, and its outer ring can accommodate 22 parking positions. The total number of parking positions is 468, and the average retrieve time can reach 89.09 seconds.

Table 5. Average retrieve time of configurations.

Lifts Floors	3	4	5	6
5	65.2	62.0	60.4	58.6
6	71.5	64.4	61.9	60.2
7	77.1	68.9	64.8	62.2
8	84.8	72.7	67.9	64.6
9	98.2	77.5	71.4	67.1
10	115.0	86.3	75.0	70.3
11	147.3	94.6	79.9	73.9
12	248.3	111.9	86.9	78.9
13	2340.7	133.3	98.2	84.4
14	13618.4	169.7	109.1	89.1
15	29683.2	254.8	126.9	97.4
16	38385.1	716.3	159.5	108.4
17	52200.1	9973.6	207.5	127.5
18	61454.8	18649.4	366.9	146.5
19	70415.4	31631.3	1355.4	187.5
20	78557.5	38653.5	6726.9	283.7

Table 6. Configurations.

Lifts Fioors	3	4	5	6
5	156	152	148	144
6	195	190	185	180
7	234	228	222	216
8	273	266	259	252
9	312	304	296	288
10	351	342	333	324
11	390	380	370	360
12	429	418	407	396
13	468	456	444	432
14	507	494	481	468
15	546	532	518	504
16	585	570	555	540
17	624	608	592	576
18	663	646	629	612

Different definitions can produce different optimal configuration combinations, and when economic factors and other constraints are introduced, the optimal system configuration will also change greatly. This experiment is only a preliminary study and provides a feasible idea.

6 CONCLUSIONS

CRAP system and its related systems have been widely used in reality, but there are few papers in the literature. To further improve the efficiency of the CRAP system, this paper improves the CRAP system's retrieve process based on relevant research. Combining two lift selection strategies, through a simulation experiment, the retrieve process improved with the shortest time strategy is the best and reduces the average retrieve time by 20%-50% compared with the original process.

To improve CRAP system configuration, this paper puts forward a design of system configuration. Under experimental conditions, the maximum space utilization rate of the system is 60.5%. Based on this, the optimal configuration of the system with a circular fixed platform radius of 5.85 meters, 14 floors and six lifts and a series of expected retrieve times under different configurations are obtained. The feasibility of system configuration design is proved.

CRAP also has many research directions: First, in practical application, if the cost control problem is added to the configuration research, then the configuration may be adjusted. The main cost control is reflected in the control of the number of lifts or to reduce energy consumption, timely stop part of the operation of lifts, and then start them in the peak period.

Second, the automated parking system has a big drawback that car owners need to wait for a while at the point of picking, although from a practical perspective, even the traditional parking systems, car owners also need to find the car on foot, it may have an even longer time to take a car, but from the point of psychological factors, waiting at the point of picking still can let a person more easily produce an impatient mood. To improve user experience, therefore, the design of retrieve reservations in actual scenes is also an important research direction;

Thirdly, the CRAP system may have a new adjustment in terms of detailed layout. For example, the parking space is not oriented toward the center of the circle, but around the center of the circle. The minimum difference between the inner and outer radius is no longer determined by the length of the parking position, but by the width of the parking position.

REFERENCES

Abubakar A. & M. Behravesh & E. Rezapouraghdam & H. Yildiz (2019) Applying artificial intelligence technique to predict knowledge hiding behavior. J. International Journal of Information Management. 49, 45–57.

Duan Y & Edwards J & S. Dwivedi (2016). Artificial intelligence for decision making in the era of big data–evolution, challenges and research agenda. J. International Journal of Information Management. 48, 63–71.

Fan Yuan & Li Wenfeng & and He Lijun (2019). Collaborative scheduling of intelligent warehouse multi-mobile robots based on improved Genetic Algorithm. J. Journal of Wuhan University of Technology. 41(3), 298-311.

Guangmei Wu & Bipan Zou & Xianhao Xu (2019). Shuttle-based operating policies for multiple-lift compact automated parking systems based on queuing networks. J. Cluster Computing. 22, S6449–S6460.

Guangmei Wu & Xianhao Xu & Xinyuan Lu (2020). Considering the infiuence of queue length on performance improvement for a new compact robotic automated parking system. J. International Journal of Information Management. 50, 487–497.

Guangmei Wu & Xianhao Xu & Yeming Gong (2019). Optimal design and planning for compact automated parking systems. J. European Journal of Operational Research. 273, 948-967.

Guo Di & Xie Danlan & Ji Yuan (2020). Research on Configuration Simulation optimization of intelligent storage Robot under multiple constraints. J. Journal of System Simulation. 32(10), 2066-2072.

Han Dongya & Chen Ran & Yu Yugang (2020). Research on integrated optimization of task order and location selection in automated three-dimensional warehouse. J. Chinese Management Science. 28(10), 156-164, 2020.

Heragu Cai & X. Krishnumarthy & A. Malmborg (2011). Analytical models for the analysis of automated warehouse material handling systems. J. International Journal of Production Research. 49, 6833–6861.

Mahroof, K. (2019). A human-centric perspective exploring the readiness towards smart warehousing: The case of a large retail distribution warehouse. J. International Journal of Information Management. 45, 176–190.

Marchet G & Melacini M & Perotti S (2012). An analytical model to estimate performance of autonomous vehicle storage and retrieval systems for product totes. J. International Journal of Production Research. 50, 7134–7148.

Zaerpour & Yu Gong & Koster (2016). A framework for evaluating and optimizing live-cube compact storage systems. J. Transportation Science. 51, 34-51.

Zhou Xin & Ma Yunfeng & Ren Liang (2017). Double depth cargo position optimization based on double instruction. J. Journal of Wuhan University of Technology. 39(6), 47-54.

Zou B. & Xu X. & De Koster (2016). Modeling parallel movement of lifts and vehicles in tier-captive vehicle-based warehousing systems. J. European Journal of Operational Research. 254(1), 51–67.

Advances in Urban Engineering and Management Science – Khalil & Yang (Eds)
© 2023 The Author(s), ISBN 978-1-032-30426-7

Research and practice of digital protection of historic buildings relying on information models—Take the historical buildings of Taizhou Prefectural City as an example

Yukun Xin*
School of Civil Engineering and Architecture, Wuhan University of Technology, Wuhan, China

Yusu Xu
Instructor, School of Civil Engineering and Architecture, Wuhan University of Technology, Wuhan, China

ABSTRACT: With the acceleration of urbanization, many ancient buildings face the inevitable situation of being destroyed or even demolished, making the digital protection of historical buildings more and more practical. At the present stage, the digital protection methods of buildings are mainly three-dimensional modeling and image data recording, whose results and data are archived by local institutions. The inter-regional building materials are poorly correlated and lack systematic management. This research takes the Ancient city of Taizhou in Zhejiang Province as the research object, conducts digital protection and analysis of historical buildings with high protection value, and regulates the protection results in the form of a building space database. This research not only provides information for the restoration of ancient buildings after Taizhou prefectural city, but also has reference significance for the development of its cultural industry. It also provides templates for the protection of other historical areas.

1 INTRODUCTION

In 2015, President Xi said, "to remember nostalgia, we must promote the protection of fine traditional Chinese culture, the continuation of the city's historical context and preserve Chinese culture gene" in the middle of the city work conference, which pointed out that to deal with cultural heritage protection a high priority. In 2019, the document Preservation Planning Standards for Historical and Cultural Cities was promulgated. The overall planning and protection of ancient buildings based on districts once again ushered in a new situation of leapfrog development. The protection method has changed from simple physical space protection to architecture, culture, society, and economic space integrating protection and changes in the development model, but how to achieve rational and standardized protection templates requires continuous experimentation and exploration.

2 DEVELOPMENT OF DIGITAL PROTECTION TECHNOLOGY FOR HISTORICAL BUILDINGS

The international community has always maintained a more important attitude towards the digital protection of ancient buildings. The overall protection stage can be summarized as the unified protection of regional historical buildings with historical blocks as a unit-the architectural heritage as a combination of historical significance–Emphasis on protection and partial sacrifice-the reuse

*Corresponding Author: 287067@whut.edu.cn

DOI 10.1201/9781003305026-92

of reconstructed historical buildings. In 1980, the concept of "Building Description System," BIM was first time be proposed. In 1992, UNESCO put forward a related item, "Memory of the World," in order to promote the historic preservation of cultural heritage; since the 21 century, with the Three-dimensional modeling technology is constantly improving, foreign organizations have successively carried out digital protection research on famous historical buildings such as Basilica church and Roman villa (Cai 2020). Using three-dimensional scanning modeling and virtual reality technologies as tools, they have carried out a series of digital protection of historical buildings.

China first clearly stipulated relevant regulations on the protection of historical buildings in the Cultural Relics Protection Law of 1930. The prototype of digital protection appeared in the History of Chinese Architecture compiled by Mr. Liang Sicheng in 1994. The preserved rare architectural images and various drawings provided valuable data for later research on ancient buildings. In 1990, "Digital Forbidden City" and "Digital Dunhuang" were launched. Since 2006, the use of BIM technology has been vigorously promoted, and various scanning equipment and 3D modeling software have been put into use one after another.

Comparing the international trend with the digital protection technology of historical buildings in China, it can be found that the domestic investigation and management methods are relatively backward, and because the protection standards and protection methods of each protection section are different, there is a lack of systematic templates, and the protection of science is insufficient. The real-time protection monitoring and update mechanism are missing. Some ancient building protection practices are limited to the internal professions of the industry, are out of touch with society, and do not integrate the coordinated development of related cultural industries, resulting in low public participation and difficult promotion.

3 PRACTICE OF DIGITAL PROTECTION OF HISTORICAL BUILDINGS

This paper selects case-the Taizhou prefectural city in Linha, Zhejiang Province, as the practical object of research. Through surveying and mapping, we have mastered first-hand building materials (physical information, related data, video recordings, etc.) to model and recover key protected buildings in this area. The establishment of restoration and related information databases will lay the foundation for subsequent protection and cultural promotion.

3.1 *Select survey objects*

Linhai City is located on the east coast of central Zhejiang Province and the southern wing of the Yangtze River Delta Economic Belt. It is a county-level city under the jurisdiction of Taizhou City and the seat of Taizhou's capital city. Linhai takes Taizhou Ancient City as an important space carrier of historical and cultural value and has now been included in the third batch of national historical and cultural cities. Taizhou city is surrounded by two mountains and two rivers which are rich in cultural heritage resources. The entire city takes the Ming and Qing ancient city wall sites as its core cultural heritage. The checkerboard layout of the city pools and streets together constitutes the historical context of the ancient city, and the southwestern wall belt plays an important role in the internal spatial pattern of the city (He 2017).

The author has participated in the surveying and mapping of some historical buildings in Taizhou in the early stage, conducted a preliminary investigation on the parameters of local historical buildings and the overall spatial relationship, and summarized the current problems in the protection of ancient buildings in the city:

1) The city wall of Taizhou Prefecture has been included in the World Cultural Heritage Preliminary List as the China Ming and Qing Dynasty Wall. Its protection requirements conflict with other historical protection elements (such as buildings, streets, etc.). For example, in the southwest area, buildings built with footing and external streets and alley spaces had to be selected and dismantled to restore and protect the city wall.

2) The historical buildings in the city and its traditional streets retain the scale and features of the Ming and Qing Dynasties, and the facades along the street and the internal courtyard space show real-life scenes in the old days. In the process of protection, attention should be paid to the integrity of the environment, and valuable historic environments should be protected.
3) The connection between the new buildings in the city and the historical buildings needs to be guided by the overall architectural style of the historical area, and at the same time, provide data to support the function positioning and development forms of the new buildings.

In response to the key issues in the above protection process, this article will rely on digital modeling and databases to provide solutions one by one.

3.2 *Two-dimensional data collection and processing*

1) In this stage of information collection, on-site surveying and mapping are mainly used: rough surveying of data using surveying and mapping instruments, hand-painted building facade and plane data recording, information collation and verification and supplementary testing; supplemented by electronic images, interviews, etc. form on-site record.
2) Use image processing, computer graphics and other information development technologies to realize data collection of existing historical buildings, such as taking photos or audio and video recordings, and using traditional surveying and mapping tools for preliminary measurements (such as laser rangefinders, tape measures, and radar surveys, etc.), combined with the computer graphics software AutoCAD to draw the measurement data.

3.3 *Three-dimensional data acquisition and processing*

Assign ancient building surveying and mapping information to the three-dimensional model, so that the three-dimensional digital model contains more information than the two-dimensional graphic. It can accurately and vividly record the appearance, internal structure, and data size of the building, and it is also easier to understand and read the information.

1) Provide historical building information in a three-dimensional space. Using the previous two-dimensional data, the CAD drawing data is established in the three-dimensional model, the missing data is verified to fill in the missing data, and the damaged part of the building is speculatively restored. Some part of the modeling steps is shown in Figure 1.
2) In addition to using two-dimensional data to reproduce three-dimensional models, there are many spatial scanning tools that can directly realize the basic restoration of virtual models. The simplest method is a mobile device with its own radar scanning function, such as 3D Scanner (based on the Apple store platform), which can perform small-scale three-dimensional spatial data measurement and real-time modeling on the mobile terminal equipped with a radar scanner to facilitate future related data. The software also supports the export of 3D models in multiple formats, supports the reprocessing of the model, and can also export the original point cloud data. In addition, some professional methods can perform fine scanning and data collection of the entire ancient building. Commonly used are 3D laser scanners (laser point cloud modeling technology) and panoramic cameras, etc., capturing the key point position information in the building and establishing the corresponding three-dimensional coordinates in the virtual space. Then the contours of the building and each part of the component are obtained through post-processing optimization.

3.4 *The establishment of the database*

For historic buildings that must be demolished or destroyed, it is necessary to take the form of three-dimensional modeling to build a database for the preservation of virtual images of the building. The space information and the attribute information of ancient buildings are organically combined to

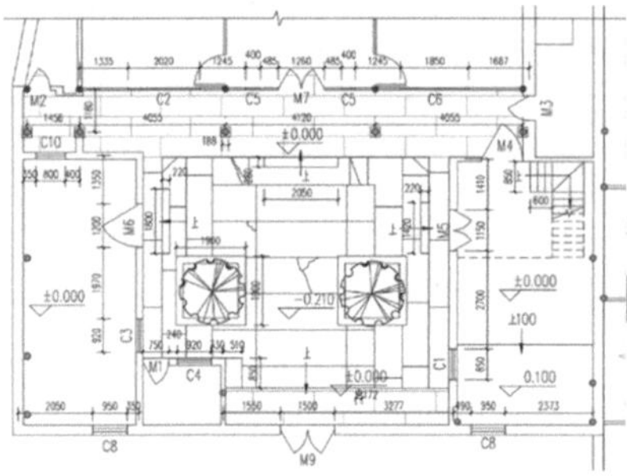

(a)Mark on 2D drawings.

(b)Mapping against photos.

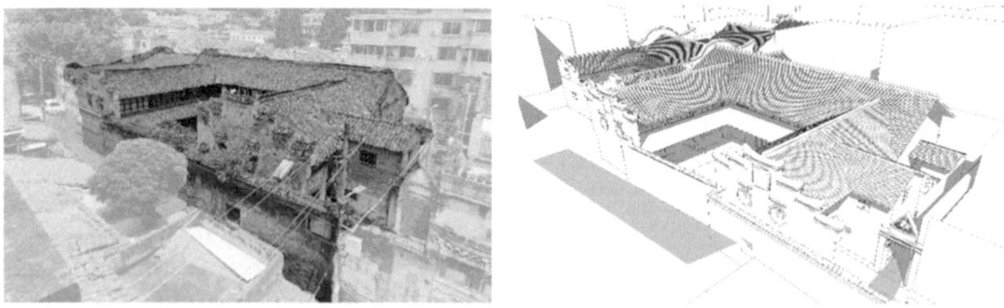

(c)Refer to external space scale.

Figure 1. Part of the modeling steps.

Figure 2. Mobile radar scanning model.

Figure 3. Laser point cloud modelling.

form a complete database of ancient buildings. The former is the spatial data obtained during the 3D modeling process, and the latter is the physical attributes corresponding to the spatial data, including historical evolution, humanistic intentions, building environment and building materials, component sizes, types, etc.

3.4.1 *Spatial information database of ancient buildings*

Spatial data is stored in a three-dimensional space, managed, and read by a computer, and is integrated into a convenient form for electronic storage and processing by using graphics, images, models, etc., prepared in the early stage. The laser scan pictures are stored in the form of a raster, and the building primitive information (including the information that divides the building boundary such as points, lines, and surfaces) and the three-dimensional model are stored in vector data. Some are shown in Figure 4.

In addition, after the preparation of the overall building space information database, the important historical building components in the three-dimensional model will be screened, and

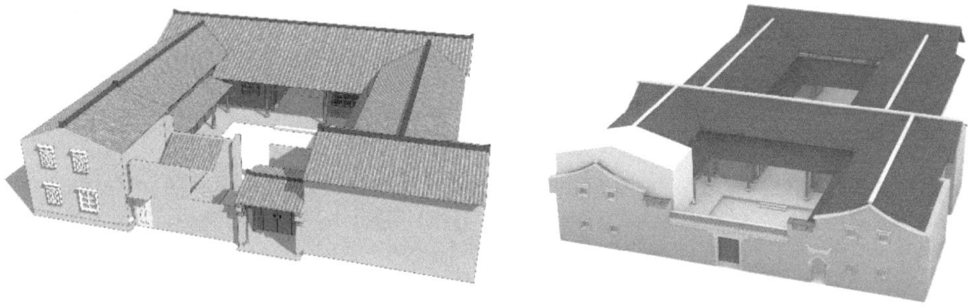

(a) No. 21 Sanjing Alley. (b) No. 8 Santai Alley.

Figure 4. Part of the 3Dl model in the spatial information database.

the BIM ancient building component library will be separately established for refining the connection relationship of historical building components and for the subsequent reconstruction of the building. And provide a reference for research work. Table 1 shows the content.

First, the building components are divided according to their usage and location, load-bearing components, such as beams, columns, walls, buckets, and roofs; detailed components, such as doors, windows, ceilings, foundations, other small wooden works, and other parts, such as rafters, tiles, carved, etc. Afterward, BIM parameter settings are carried out for the key dimensions of each component, and a three-dimensional component family library will be established. When each part of the component is built separately, the related components should be merged into a linked group to facilitate the modification of the overall relationship.

Table 1. Summary of family components.

Entity name	Component name	Family name	Location	Key parameter
		Tabia	Entrance, hall	Pavement size
ground	Pavement	Bluestone	Platform, aisle	Pavement size
Beam frame	column	Long column, short column	Courtyard gate, hall	Diameter, height
		Pillar foundation	Courtyard gate, hall	Diameter, decoration
	Beam	Beam	Courtyard gate, hall	Length, height
roof	roof	Pantile, Vadang	Courtyard gate, hall	Thickness, radian
		Rafters and ridges	Courtyard gate, hall	Width, thickness
		Dripping	Courtyard gate, hall	Width, thickness
Wall	Gable	Gable	Hall, splitting wall	Thickness, height
	partition	Partition	Hall	Thickness, height
	Fence	Low wall	Courtyard gate	Thickness, height
door	door	Doors, entrance doors, compartment doors	Courtyard gate, hall	Width, Height
window	window	Slope windows, willow windows	Hall	Width, Height
other	Decorative components	Carved flowers, queti, etc.	Throughout	Specific pattern

In addition, point cloud data can also be used to identify component types and summarize them automatically. Depending on the characteristics of ancient building data that mostly exist in the form of building groups, the PointNet network and improved network can be used to compare and experiment with ancient building data and perform partial experiments to visualize and analyze results. Different components are displayed in different colors and textures, data tags are added to establish a data set, and local details are subjected to secondary noise reduction processing and detail optimization (Cui 2021).

3.4.2 *Ancient building attribute information database*

In the actual protection of ancient buildings, in addition to the architectural model data, it also includes the attribute information corresponding to each spatial data, such as building materials, geographic location, historical background, humanistic features, etc. In the preliminary surveying stage, the literature data, sampling specimens retained and interview records, etc., in the preliminary surveying stage, can all be used as building attribute resources to establish a unified database. This not only provides a reliable source for the preservation and analysis of building intangible historical information, but also provides an intuitive display of the external space scale of the building, which is convenient for the overall protection of the entire historical area. Some of its specific applications include: extracting local details in the photos to form a building material texture library, adding material information to the previously produced information components; attaching scene photos to the 3D model to visually show the spatial scale and overall of the building and streets Scenery; analysis and summary of the overall environment in the historic building area.

3.5 *Practical application of database information*

1) Retain building attributes and spatial information to provide data support for historic building restoration and damage cause analysis. The aggregated building information can be used to analyze the building and the surrounding environment, such as the long-term impact of local climate and geological conditions on the building, the overall style and spatial texture evaluation of the building group, the protection value of historical buildings and the evolution of historical forms, etc., and it can also be used for new construction architectural style, distance and height controlling reference.
2) Conduct a holistic analysis of local conditions to provide a basis for the protection and renewal strategies of historical buildings. Understand the internal relevance of the building, provide streamlined optimization for future tourism development and reliable resources for the coordinated development of cultural industries, and improve information sharing.

4 CONCLUSIONS

This research provides a standardized ancient building digital protection template for the long-term weak protection of ancient buildings and poor systematization. Through the establishment of an ancient building database, it organically combines ancient building information and converts it into computer vector data and raster data. Each of them is not only related to each other, but also independently stored and modified, providing data storage for the restoration and protection of ancient buildings. The BIM building components in the database can also be used for modeling and restoring other ancient buildings to reduce repetitive tasks.

More importantly, ancient buildings not only have high historical value and artistic value, but are also an important source of reference for new architectural design. Ancient buildings are often important places for cultural tourism. The analysis of the ancient buildings' background information can effectively promote the surrounding area, develop cultural industry, improve the sharing of information, and create economic benefits.

This article belongs to National Innovation Training Program for College Students. Project name: Research on the construction of historical and cultural city architecture and humanistic image information. Project number: S202110497186.

REFERENCES

2017. Regulations on the Conservation of Historic City, Town, and Village.

Cai Qiang. (2020) Research and practice of digital protection technology for historical buildings: Taking general Xujiaqiao Mansion as an example. pp.1-8.

Cui Zhe. (2021). Automatic classification of ancient building components based on deep learning (Master's thesis, Beijing University of Civil Engineering and Architecture). pp. 26-40.

He Sining. (2017) The Study of Conservation and Design Guidelines Based on Southwest Historic Area in Taizhou Prefectural City (Master's thesis, Tsinghua University). pp.1-12.

Huang Yonglin, Tan Guoxin. (2012) Research on the Digital Protection and Development of China's Intangible Cultural Heritage. Journal of Central China Normal University (Humanities and Social Sciences Edition), vol. 51(02): 49-55.

Hu Zhangjie, Li Xiang. (2019). Research and application of the protection of historical and cultural cities and towns based on spatial information. Bulletin of Surveying and Mapping (S2), 267-270.

Lin Yinan. (2020). The public participation in the conservation of cultural heritage in the digital age: An observation based on Fuyuan road, Fuzhou (11), 100-104.

Wang Liming, Feng Shangrong. (2019) Research on the Protection of Taizhou prefectural city in Linhai City, Zhejiang. Famous Chinese Cities (03), 68-74.

Advances in Urban Engineering and Management Science – Khalil & Yang (Eds)
© 2023 The Author(s), ISBN 978-1-032-30426-7

The analysis of high-speed railway opening and enterprise performance—Empirical test based on difference-in-difference method

Lixue Sun*
Beijing Jiaotong University, Beijing, China

ABSTRACT: This paper takes the listed companies registered in prefecture-level cities from 2008 to 2014 as samples and uses the difference in difference method (DID) to explore the relationship between the opening of high-speed railways and enterprise performance level. The research finds that the opening of high-speed railways promotes the improvement of enterprise performance levels. The group test found that for companies with higher financing degrees, the opening of high-speed railways has a more significant effect on improving corporate performance. This paper enriches the research on the microeconomic consequences of high-speed railway opening and provides valuable suggestions for improving the enterprise management system.

1 INTRODUCTION

With the rapid development of the economy, transportation has become the artery of the sustainable development of the national economy and the deepening of reform. The state continues to increase investment in transportation infrastructure, and the impact of transportation infrastructure construction on economic development is self-evident. Studies have shown that after the opening of the high-speed railway, the GDP level of cities along the route increases significantly, and the wage level also increases correspondingly (Wang Yufei & Ni Pengfei 2016). In addition, the construction of a high-speed railway has brought a series of positive effects to the enterprise operation, prevented the information asymmetry of the company, and attracted more venture capital. The company can obtain more bank credit funds and further improve investment efficiency.

So, what effect will the construction of high-speed rail have on corporate performance? What is the mechanism of action? The research on these issues will help to understand the micro mechanism of high-speed rail opening promoting economic development and provide empirical evidence on the driving factors of enterprise performance.

2 THEORETICAL ANALYSIS AND RESEARCH HYPOTHESIS

This paper believes that after the opening of high-speed rail in the location of the company, the performance of the company will be significantly improved. The reasons are as follows:

First, the opening of a high-speed railway shortens the geographical distance between enterprises and their customers, suppliers, and other partners, reduces the trade cost of production factors and products, and enables enterprises and partners to carry out closer cooperation. The study found that the improvement of transportation infrastructure can encourage enterprises to expand their market scope and increase the transaction volume between companies and customers. Liu (2011) found that transportation infrastructure helps manufacturing enterprises reduce inventory costs by increasing order frequency. The expansion of the product market and the reduction of operating costs will significantly improve enterprise performance.

*Corresponding Author: sunlx2962@163.com

 DOI 10.1201/9781003305026-93

Secondly, with the improvement of the transportation infrastructure, transport costs decline in regional trade in goods becomes more frequent, and enterprises expand the sales range of products and services. In addition, due to the improvement of the transportation and distribution, logistics chain, enterprises expand the size of the market also, better for the enterprise market segment and differentiation, leading to economic agglomeration and market expansion. Thus, the scale efficiency of enterprises can be improved. The rapid development of high-speed railway has improved the level of division of labor, specialization and operation scale, which is conducive to enterprise knowledge creation and business communication, and thus further improves the labor productivity of the company by giving play to economies of scale effect (Elpman & Krugman 1985; Krugman 1980), and thus improved performance.

Finally, the opening of high-speed rail will help ease the financing constraints of enterprises. Chu and Fang (2019) found that after the opening of the high-speed railway, the degree of information asymmetry between enterprises and banks was reduced, which led to a larger scale, longer-term and lower cost of bank credit, which means that enterprises could obtain external financing at a lower cost. The research of Long Yu et al. (2017) shows that the opening of a high-speed railway helps enterprises to obtain more financial support from venture capital. Sufficient fundraising helps enterprises grasp good investment opportunities, and further improve investment efficiency. Enterprises will also increase investment in innovation, and corporate performance will be significantly improved.

Based on the above analysis, this paper proposes hypotheses:

H: The opening of the high-speed railway is significantly positively correlated with enterprise performance.

3 RESEARCH DESIGN

3.1 *Sample selection*

To verify the research hypothesis mentioned above, listed companies registered in prefecture-level cities from 2008 to 2020 were selected as samples. In this paper, the data on high-speed railway opening are collected manually from Chinese railway companies and local news. At the same time, in order to avoid the endogeneity of high-speed railway opening, the data of municipalities directly under the central government, provincial capital cities and vice-provincial capital cities are excluded. Other data came from the CSMAR database. Finally, 367 valid samples and cities along the route were obtained, with a total of 3393 samples.

In this paper, Winsorize method (1% and 99%) was used for tail reduction of extreme values to ensure the robustness of results, and STATA was used for data processing and regression analysis.

3.2 *Definition of variables*

3.2.1 *Business performance*

This article uses a yield valve to measure corporate performance, specifically, ROE = net margin/net asset

3.2.2 *Opening of high-speed railway*

The Train value is 1 if the company's location has experienced a high-speed rail opening during the sample period; otherwise, the Train value is 0; the TrainPost value is 1 if the company's location has experienced a high-speed rail opening during the sample period; the value is 0 if the *company's* location has experienced a high-speed rail opening during the sample period.

3.2.3 *Control variable*

As shown in Table 1, this paper introduces the size of the enterprise, the ratio of assets and liabilities, the growth rate of sales revenue, the structure of the board of directors and the proportion of independent directors as control variables.

3.3 *Model design*

To explore the impact of high-speed rail on corporate performance, the following empirical models were developed:

$$ROE = \beta 0 + \beta 1Train + \beta 2TrainPost + \beta 3\ Controlst + Year + Industry + \varepsilon t \quad (1)$$

In Model (1), ROE is used to measure business performance. The larger the Roe, the better the company's performance and the more profitable it is. In the expected formula (1), $\beta 2$ is significantly positive, indicating that the opening of high-speed rail has improved corporate performance.

4 ANALYSIS OF EMPIRICAL RESULTS

4.1 *Descriptive statistics*

Table 2 lists the descriptive statistical results of the main study variables. As can be seen from Table 2, the mean value of Train is 0.925, indicating that by the end of 2018, about 92% of the company's offices had opened high-speed rail, indicating that China's high-speed rail construction is developing rapidly. The average size of the company is 21.7, and the average asset-liability ratio is 0.328, which is consistent with previous studies.

Table 1. Definition of variables.

Variable	Variable name	Variable symbol	Variable explanation
Explained variable	Corporate performance	ROE	ROE=net profit/net asset
Explanatory variables	Whether high-speed rail is open	Train	The value is 1 if the office of the company experienced the opening of the high-speed railway during the sample period; otherwise, the value is 0
	Before and after the opening of high-speed rail	TrainPost	The value is 1 for the year sample where the company's office is located after the opening of the high-speed railway, and 0 for the year before
Control variables	The company size	SIZE	The natural log of total assets
	Asset-liability ratio	LEV	liability/asset
	Sales revenue growth rate	GROWTH	(Current period operating income - previous period operating income)/previous period operating income
	Board structure	BOARD	Number of board members
	Years	YEAR	Annual dummy variable
	Industry	INDUSTRY	Industry dummy variable

4.2 *Multiple regression analysis*

In order to verify the relationship between the opening of high-speed railways and enterprise performance, this paper adopts Equation (1) for regression analysis. The regression results are shown in Table 3, column (1). The results showed that ROE was significantly positively correlated with TrainPost at the level of 10%, with a regression coefficient of 0.011 and t value of 1.826. The results show that the opening of the high-speed railway is positively correlated with enterprise performance. That is, after the opening of the high-speed railway, the performance level of enterprises has significantly improved. The regression results verify hypothesis H in this paper.

Analysis based on financing constraints. Referring to Hadlock and Pierce (2010), this paper uses the SA index to measure the degree of financing constraints. SA index =-0.737* Company

Table 2. Descriptive statistics of important variables.

variable	N	mean	sd	min	max	p50
ROE	3388	0.0657	0.116	−0.537	0.388	0.0684
Train	3393	0.925	0.264	0	1	1
TrainPost	3393	0.727	0.445	0	1	1
LEV	3393	0.328	0.168	0.0495	0.759	0.309
SIZE	3393	21.70	0.934	19.90	24.70	21.60
BOARD	3392	8.350	1.410	5	12	9
GROWTH	3070	0.0694	0.344	−10.20	0.988	0.101

Table 3. Multiple linear regression results.

	Explained variable ROE		
		Group by degree of financing constraint	
	Fundamental regression	Large financing constraints	Less financing constraints
Variables	(1)	(2)	(3)
Train	−0.005*	−0.003	−0.002
	(−1.670)	(−0.887)	(−0.453)
TrainPost	0.011	0.004*	0.002
	(1.826)	(1.683)	(0.595)
Intercept item	−0.377***	−0.330***	−0.433***
	(−18.159)	(−9.985)	(−15.558)
Controls		Control	
Year		Control	
Industry		Control	
Observations	3388	1690	1698
Adjusted-R^2	0.135	0.132	0.141

size +0.043* Company size 2-0.04* Listing years. The greater the absolute value of the SA index is, the greater the financing constraint is. The grouped regression results are shown in columns (2) and (3) of Table 3.

In Table 3, it can be seen that when the degree of enterprise financing constraint is large, TrainPost's coefficient is significantly positive. When the degree of enterprise financing constraint is small, TrainPost's coefficient is not significant. The above results show that the greater the degree of financing constraint, the more obvious the promotion effect of high-speed railway opening on performance improvement. In addition, industry fixed effects and annual fixed effects are controlled.

5 CONCLUSION

In view of the quasi-natural experiment of high-speed railway opening, this paper adopts the difference in difference method (DID) to study the relationship between high-speed railway opening and enterprise performance level. The research finds that the opening of high-speed railways promotes the improvement of enterprise performance levels. The group test found that for companies with higher financing degrees, the opening of a high-speed railway has a more significant effect on improving corporate performance.

The conclusion of this paper has important reference value in both theory and practice. At the theoretical level, starting from the exogenous event of the opening of a high-speed railway, this paper analyzes the relationship between the opening of a high-speed railway and enterprise performance level, enriching the existing literature on enterprise performance from the company level or macro-environment level. At the same time, this paper provides a new perspective for the study of the micro-economic consequences of high-speed railway opening. At the practical level, the conclusion of this paper expands the theoretical basis of promoting the development of the capital market through transportation infrastructure construction in China and provides a useful reference for investors and policymakers.

REFERENCES

B.F. Liu, Y.H. Liu (2011) Transportation infrastructure construction and Inventory cost reduction of Chinese manufacturing enterprises. China Industrial Economics (05), 69–79.

Hadlock C.J., Pierce J. R. New. (2010) Evidence on Measuring Financial Constraints: Moving Beyond the KZ Index[J]. Review of Financial Studies,23 (5), 1909–1940.

J. Zhu, J.X. Fang (2019) Transportation infrastructure improvement and bank credit decision: Based on the quasi-natural experiment of high-speed railway opening. Accounting and Economic Research (05), 38–60.

M.W. Wang, W.J. Sun and J.H. Ye (2020) High-speed railway opening and analyst prediction accuracy. Journal of Capital University of Economics and Business (01), 35–47.

X.Q. Du, F. Hou and S.J. Lai (2018) Does transportation infrastructure improvement inhibit auditor's "geographical preference"? Empirical evidence based on the natural experimental background of high-speed trains in China. Audit Research (01), 103–110.

Y.F. Wang, P.F. Ni (2016) Spillover of economic growth and regional spatial optimization under the influence of high-speed railway. China Industrial Economics (02), 21–36.

Y. Long, H.L. Zhao and X.D. Zhang (2017) Venture capital under spatio-temporal compression: High-speed rail traffic and regional changes in venture capital. Journal of Economic Research (2004), 195–208.

Advances in Urban Engineering and Management Science – Khalil & Yang (Eds)

An evaluation index system based on the supply-demand of urban parks from the perspective of "Park City"

Jianren Shi & Qiaoyun Guo*
School of Tourism and Urban-Rural Planning, Zhejiang Gongshang University, Hangzhou, China

ABSTRACT: The concept of "Park City" shows that China attaches importance to the ecological environment in the construction of human settlements, but there is a phenomenon of uneven spatial distribution in the process of urban construction. Coordinating the construction of urban parks on the supply side and the demand side has far-reaching significance for the sustainable development of the city. This paper constructs an evaluation index system for urban parks' social equity in three dimensions: the supply of urban parks, the link between people and the land, and the needs of urban residents. The results show that the overall distribution is unbalanced, presenting a spatial pattern of "high on both sides, low in the middle," and social equity is poorly balanced. It provides a certain reference value for the development of urban parks and the construction of an ecologically livable park city in Hangzhou.

1 INTRODUCTION

Urban parks are very important for increasing the opportunities for enjoying nature and engaging in recreational activities. Its function is to improve the aesthetics and environmental quality of the urban environment, provide public space for recreation, and improve the overall quality of life of urban residents. In recent years, more attention has been paid to the research on the factors affecting the accessibility of parks (Cao et al. 2021), the fairness and efficiency of park use (Liu et al. 2017), and the exploration of urban parks (Wen et al. 2020) from the perspective of age. However, there is a great contradiction between the supply-demand of urban parks in terms of recreation, so urban parks do not give full play to their due role in improving and enhancing residents' happiness index and well-being. Therefore, it is very important to control the social equity between the supply-demand of urban parks and green spaces (Tang & Gu 2015). Most studies in this regard are based on the social performance evaluation of park green space distribution under the concept of social equity proposed by Tang (Tang & Gu 2015). For example, Wang analyzed the impact of urban park green space spatial allocation on the effective supply capacity of social services (Wang et al. 2019). Liu constructed the urban green space spatial evaluation system, and the corresponding planning method for urban green space is put forward (Liu et al. 2012). Some scholars have proposed to apply the coupling degree model to the research on the supply and demand measurement of urban parks. This model is widely used in urban development, social economy, and other fields, but it is rarely used in the field of urban parks. Li constructed the evaluation system of the supply-demand system and the quantitative model of synergy to analyze and evaluate the spatial differences between urban parks in administrative units (Li 2017).

"Social equity" considers demand, fairness and justice when distributing public service facilities, and focuses on the service level difference between public facilities in a region. The service function of urban parks and residents' equitable enjoyment of urban park space can be used as important

*Corresponding Author: 1143539624@qq.com

DOI 10.1201/9781003305026-94

indicators to measure residents' quality of life. Therefore, the spatial layout of urban parks should aim to support fair and efficient supply-demand and maximize limited space resources. However, the traditional evaluation indexes of urban parks seldom consider the quality of parks and the fairness of spatial layout, and the balance between the supply of urban parks and the demand of residents is still difficult to measure. Moreover, through the research on the traditional coupling relationship between supply-demand, it is found that these studies usually establish the evaluation index system from the two dimensions of supply-demand, pay little attention to the path connection between urban park supply and residents' demand, and pay little attention to the path connection factors (actual traffic origin and destination, urban road construction degreeless.) It can directly affect residents' willingness to go to urban parks, resulting in the lack of scientificity in the social equity evaluation of urban parks. Based on the perspective of supply and demand, this research puts forward the evaluation factor of "human-land linkage" and incorporates it into the context of the full coverage of urban parks in the "15 min community life circle" and the connection and link relationship between urban residents and urban parks. In the study of social equity evaluation of urban parks, 15 minutes of walking is selected as the time threshold to conduct social equity evaluation analysis of urban parks, and provide a reference basis for the social equity evaluation and planning layout of urban parks.

2 STUDY AREA AND DATA

2.1 *Study area*

This study area is the administrative division of the Yuanjianggan District in the eastern city of Hangzhou. There are a total of 98 urban parks in the study area, scattered in various areas (Figure 1); According to the standard, it can be divided into three types, a total of 8 comprehensive parks ($S \geq 10$), 37 community parks ($1 < S < 10$), and 53 amusement parks ($S \leq 1$). As the sub-city center of Hangzhou, it is an important transportation hub center in Hangzhou. It is a new city center, a new paradise for innovation, an international window, and a new smart city. Its economic development is extremely advantageous. However, the spatial distribution and planning of urban parks within the scope of this study have hindered urban development, and selecting the eastern urban area of Hangzhou as the research object has certain research value.

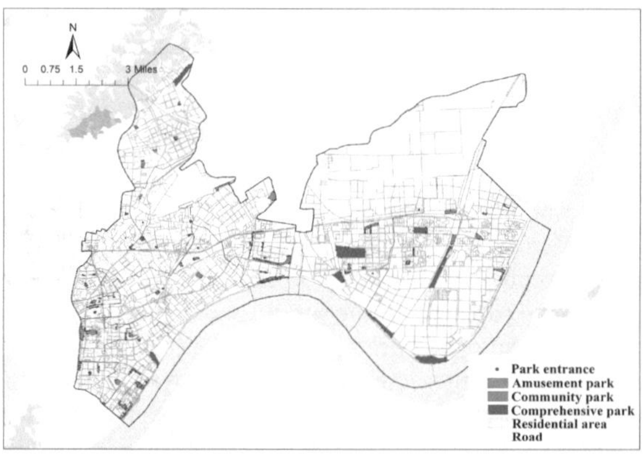

Figure 1. Distribution map of urban parks.

2.2 *Data*

Demographic data: The total population data of each street is collected from the sixth population census data of the Hangzhou Bureau of statistics and the 2019 statistical yearbook of Jianggan

District. Since the data cannot directly reflect the residential population of each residential area, in order to ensure the accuracy of the data, the AOI data of the residential area is compared with the population data of the administrative unit assigned a value, take "the area of the residential area / the area of the residential area in the spatial unit" as the weighted field, obtain the population data of each residential unit. The residential area and urban park data: AOI and POI data (including gate location, scope, area, etc.) were crawled through the AutoNavi API interface using Python language in June 2021. The housing price of the residential area was obtained from Anjuke in August 2021, while the university housing area, employee housing, etc., did not have a price tag, but it did not have a significant impact on this research, so it was set to 0 during processing. Road network data: The vector data of urban trunk roads, secondary trunk roads and urban expressways in the study area are extracted by Open Stree Map. In order to build the traffic network for network analysis, the road network data is preprocessed, including road merging and clipping, so as to facilitate the subsequent analysis in ArcGIS. Other data: Mainly related to POI data, using Python language through AutoNavi API interface to crawl all kinds of POI data in the study area in August 2021, including commercial service facilities, school education facilities, hospital medical and health facilities, etc. After obtaining the above data, sampling investigation and deduplication processing are carried out, and screening and correction are carried out in combination with the Google map to ensure the timeliness and accuracy of the data. At the same time, the corresponding spatial database is established in ArcGIS for later use of ArcGIS for data analysis.

3 METHODOLOGY

3.1 *Construction of evaluation variable system*

Table 1. Equity performance evaluation index system.

Target layer	Constraint layer	Index parameter	Index nature	Index weight
Supply	Quality of supply	Number of parks	+	10.11%
		Park area	+	33.09%
		Number of park entrances	+	9.31%
		Park area per capita	+	13.30%
		Park type	+	9.91%
	Supply efficiency	15-minute service radius coverage rate	+	4.13%
		15-minute service radius overlap rate	–	13.33%
		Park service coverage rate	+	6.81%
Human-Land linkage Demand	Accessibility	Density of bus stops	+	33.85%
		Road network density	+	41.83%
		Average distance between starting and ending points	–	24.33%
	Social demand	Population density	+	20.39%
		Proportion of population aged 0-14	+	3.90%
		Proportion of population aged 65 and over	+	7.47%
	Material demand	Average house price	+	8.08%
		Number of residential quarters	+	10.66%
		Residential area	+	10.06%
		Number of commercial service facilities	+	11.78%
		Number of school education facilities	+	16.20%
		Number of medical and health facilities	+	11.46%

This study considers the supply-demand relationship between urban residents and urban parks, and constructs three comprehensive evaluation index systems: urban park supply (supply side), human-land linkage (connection path) and urban residents' demand (demand side) based on the concept of coordinated development (Table 1).

3.2 *Construction of social equity evaluation model*

3.2.1 *Determine index weight*

Since the data indicators of each group are not unified, they cannot be compared directly. In order to eliminate the interaction between the indicators, the original data must be normalized first. The specific weight of each index is determined by SPSS and entropy method (Table 1), and the membership value of each index is calculated by the fuzzy membership function, and then the weighted average method is used to calculate the comprehensive evaluation indexes of urban park supply, human-land linkage, and urban park demand respectively. The specific formula is as follows:

$$S_i = \sum_{i=1}^{3} x_{ij} \times s_{ij} \tag{1}$$

$$L_i = \sum_{i=1}^{3} z_{ij} \times l_{ij} \tag{2}$$

$$Q_i = \sum_{i=1}^{3} y_{ij} \times q_{ij} \tag{3}$$

Where: Si, Li and Qi are the comprehensive evaluation index of supply, human-land linkage and demand of the ith city park, and xij, zij and yij are the membership values of the j index of the ith administrative unit, respectively, sij, lij and qij are the entropy weighting coefficients of the j index of the ith administrative unit. i represents the number of administrative units, and j represents the number of specific indicators.

3.2.2 *Constructing social equity evaluation model*

The degree of coupling mainly describes the intensity of the interaction between the elements of the system. The greater the degree of coupling, the stronger the interaction relationship. Therefore, the process of the interaction between the system and elements can be judged, but the degree of coupling cannot be judged more scientifically according to the complexity and imbalance of the relationship between different systems or elements. The social equity evaluation model in this study mainly includes two parts: the coupling degree model and the coordination degree model. The specific formula is as follows:

$$D = \sqrt{C \times T} \tag{4}$$

$$C = \left[\frac{Si \times Li \times Qi}{\left(\frac{(Si+Li+Qi)}{3}\right)^3} \right] \tag{5}$$

$$T = \alpha S_i + \beta L_i + \gamma Q_i \tag{6}$$

Where: D is the coordination degree, C is the coupling degree, and T is the comprehensive evaluation index of the supply-demand level of urban parks; K is the adjustment coefficient; α, β and γ are undetermined coefficients. Since the supply-demand of urban parks is equally important, the values of α and β are both 0.4, and γ is 0.2.

4 ANALYSIS OF SOCIAL EQUITY OF URBAN PARKS

4.1 *Spatial analysis of supply-demand*

According to the corresponding calculation method, the individual indicators in Tab.1 are spatially analyzed and visualized in ArcGIS through the classification of natural breakpoints, and the distribution diagrams of the individual indicators are obtained. Then, through ArcGIS, the supply, human-land linkage, and demand indicators are spatially weighted and superimposed to obtain the spatial distribution map of the three comprehensive indicators of urban parks in eastern Hangzhou. The comprehensive evaluation index of supply, human-land linkage, and demand of urban parks are 0.32, 0.5 and 0.39. The overall performance is "supply exceeds demand," with obvious spatial differentiation. The results show that the comprehensive level of supply in the eastern urban area of Hangzhou presents the characteristics of "high in the east and low in the west"; the comprehensive level of human-land linkages shows the characteristics of "high in the west and low in the east." The comprehensive comparison shows that the supply-demand relationship in some streets is not balanced. For example, the comprehensive supply evaluation index and the comprehensive demand evaluation index of Baiyang Street are relatively high, but the two indexes are quite different, and the balance between supply-demand is seriously unbalanced. The supply of the comprehensive evaluation index is low in Jianqiao Street and Kaixuan Street, but its demand comprehensive evaluation index is relatively high, and the spatial distribution of supply-demand is seriously unbalanced. The demand comprehensive evaluation index of other streets is relatively balanced. The comprehensive index of human-land linkage of Kaixuan Street, Sijiqing Street, Zhalongkou Street, Caihe Street, Pengbu Street and Jiubao Street all reached 0.5, indicating that most areas of the eastern city of Hangzhou have good traffic accessibility and the construction of transportation facilities is relatively complete, which can meet the travel requirements of residents to the park. However, the comprehensive evaluation index of human-land linkage in Xiasha Street and Baiyang Street is low, and their traffic accessibility severely restricts residents' travel needs in the streets.

4.2 *Spatial analysis of coupling degree*

Table 2. Calculation results and coupling coordinated development types.

Name	D value	C value	T value	Coupling coordination	Coordination
Dinglan Street	0.33	0.821	0.13	Mild dysregulation	Supply-demand damaged
Jianqiao Street	0.37	0.586	0.23	Mild dysregulation	Supply profit-loss
kaixuan Street	0.24	0.262	0.23	Moderate dysregulation	Supply profit-loss
Sijiqing Street	0.64	0.866	0.48	Primary coordination	Supply lag
Zhalongkou Street	0.40	0.613	0.27	On the verge of dysregulation	Supply lag
Baiyang Street	0.74	0.679	0.81	Intermediate coordination	Demand lag
Xiasha Street	0.30	0.531	0.17	Mild dysregulation	Supply profit-loss
Caihe Street	0.49	0.812	0.30	On the verge of dysregulation	Demand lag
PengBu Street	0.29	0.417	0.20	Moderate dysregulation	Supply-demand damaged
Jiubao Street	0.36	0.559	0.24	Mild dysregulation	Supply-demand damaged

The degree of coupling can reflect the degree of interaction and dependence between various systems. When the coupling degree is 0, the supply-demand is independent, and the system will tend to develop disorderly. When $0<C\leq 0.4$, the supply-demand coupling degree of urban parks in the main urban area of Hangzhou is in a tight growth stage, which is mainly reflected in the strong demand of urban residents for urban parks, and there is also a state of the supply side of urban parks and low demand of urban residents. Because Kaixuan street is close to the main urban area of Hangzhou, with a high urbanization level, dense residential land, superior urban construction and development conditions and fast development speed, the residents in the street have a great demand for urban parks, but the scale and number of urban parks in the street are very small,

resulting in uneven distribution and insufficient supply of urban parks, showing a state of low supply and high demand. When $0.4 < C \leq 0.8$, the coupling degree of supply-demand of urban parks in the main urban area of Hangzhou is in the running stage. At this stage, the supply of urban parks and the demand of urban residents do not reach the coupling coordination state, but the trend and state of benign coupling begin to appear. Jianqiao street, Zhalangkou street, Baiyang Street, Xiasha street, Pengbu street and Jiubao Street are all in this range. The streets in the eastern urban area of Hangzhou are affected by the economic radiation of the main urban area and the economic pull of Metro Line 1. Although urban construction starts late, the urban development and construction are relatively fast, the needs of urban parks and urban residents can reach a balance to a certain extent, and urban construction has quite stable development potential. When $0.8 < C \leq 1.0$, the supply-demand coupling degree of urban parks in the main urban area of Hangzhou is in the coordination stage, the development level of urban parks is stable, and the supply can meet the needs of urban residents. The population density of Dinglan street is small, the urban construction is relatively slow compared with other streets, and the allocation of urban parks can relatively meet the needs of residents. There is a large-scale Jianqiao airport in Jianqiao street, which covers a large scale of land, resulting in less distribution of urban parks and small population density, and the supply-demand can basically reach a coordinated state. The construction scale of urban parks in Sijiqing street and Caihe street is relatively large, which can accommodate more residents for leisure activities, and the allocation of supply-demand can reach a balance.

4.3 *Spatial analysis of coordinated development degree*

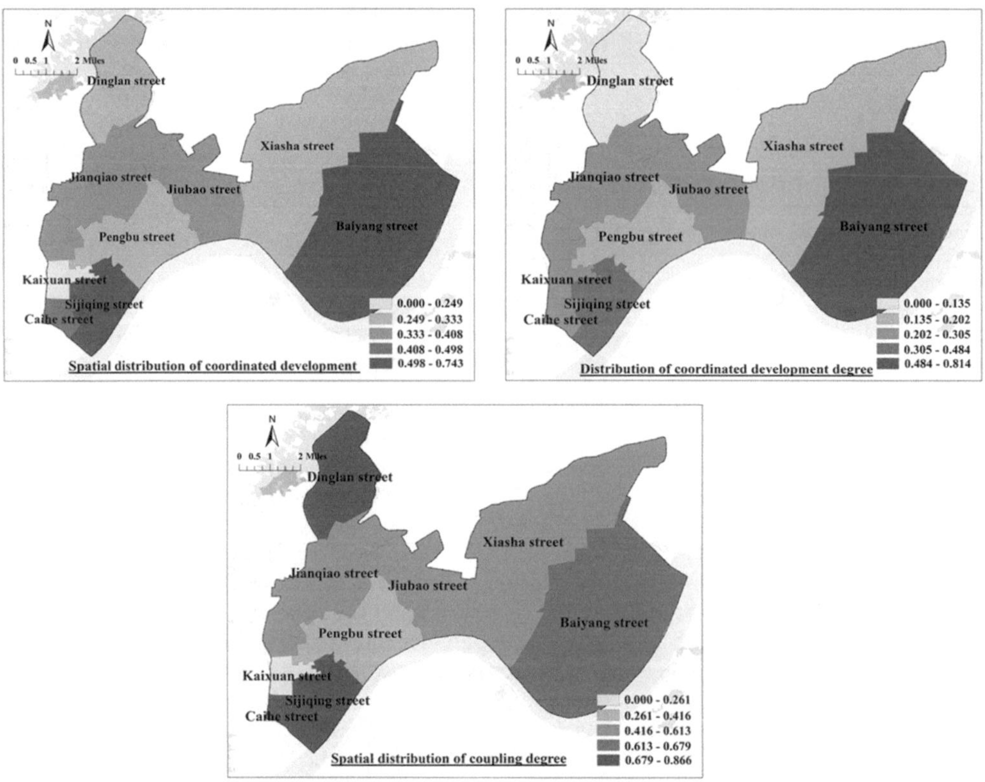

Figure 2. Spatial distribution analysis of coupling degree.

According to the idea of fuzzy mathematics, the level of coordinated development is divided into ten different coupling coordination degrees, high-quality coordination, good coordination and intermediate coordination, and the level of coordinated development in the eastern urban area of Hangzhou can be divided into five categories: primary coordination, near imbalance, intermediate coordination, mild imbalance, and moderate imbalance. The results show that the degree of coordination and coupling of urban parks is poor, ranging from 0.249 to 0.743, with an average of 0.42, the highest is Baiyang street, and the lowest is Kaixuan street. Among them, Sijiqing street and Baiyang street have a high degree of coordination and coupling, indicating that the spatial distribution of urban parks in these two streets is relatively fair. However, overall, the proportion of mild imbalance is 40%, which is concentrated in the north of the administrative region, which can directly reflect the imbalance between supply-demand in the administrative region, indicating that the coordinated development of urban parks in the eastern urban area of Hangzhou is poor, and the social fairness and evenness of urban parks are seriously unbalanced. Based on the degree of coordinated development, combined with the comprehensive evaluation index of supply-demand of urban parks, the types of coordinated development of each street are divided into four parts: supply profit-loss type, supply lag type, demand lag type and supply-demand damaged type (Table 2). The main types of coordinated development in the eastern urban area of Hangzhou are supply profit-loss and supply-demand damage. From the perspective of the supply-demand relationship, there are two demand lag types, five supply profit-loss types and supply lag types, and three supply-demand damaged types. The data clearly reflects that the supply of urban parks in the eastern urban area of Hangzhou is far less than the demand, the contradiction between supply-demand is prominent, and the development and construction of urban parks and social equity need to be improved.

5 CONCLUSION

This paper uses the coupling coordination degree to connect the three dimensions of supply-side, human-land linkage and demand side, constructs the evaluation index system, and forms the social equity evaluation model of urban parks. The results show that, overall, the distribution of urban parks in each street is unbalanced, showing a spatial pattern of "high school and low school on both sides." Except that Baiyang street and Sijiqing Street are in the coordination stage, other streets are in the imbalance stage, and the balance of social equity of urban parks is poor. In the following urban planning of the eastern urban area of Hangzhou, the coordinated development of urban parks in various regions should be considered to determine the supply according to the demand and ensure the balance between supply-demand. It is necessary to provide sufficient urban parks for residents and ensure the rational utilization and maximization of urban internal resources.

ACKNOWLEDGMENTS

Project supported by Natural Science Foundation of Zhejiang Province, China (Grant No. LY19E080004), Graduate teaching research and teaching reform project of Zhejiang Gongshang University, China (Grant No. YJG2021216) and Research project of the urban-rural planning department of Zhejiang Gongshang University, China. (Grant No.202104).

REFERENCES

Cao, M. Yao, H. & Xia, J. et al. (2021). Accessibility-Based Equity Assessment of Urban Parks in Beijing. Journal of Urban Planning and Development, 147(3):05021018.

Li, Y. (2017). Research on Fairness Evaluation of urban park layout based on supply-demand balance. Chongqing: Southwest University.

Liu, B. Y. He, W. & Liu, S. (2012). Study of the evaluation and planning of urban green space based on the coupling theory of green space and city space. Chinese Landscape Architecture, 28 (05):42–46.

Liu, W. Chen, W. & Dong. C. (2017). Spatial decay of recreational services of urban parks: Characteristics and influencing factors. Urban Forestry & Urban Greening, 25:130–138.

Tang, Z. L. Gu, S. (2015). An evaluation of social performance in the distribution of urban parks in the central city of Shanghai: from spatial equity to social equity. Journal of urban planning, (02):48–56.

Wang, M. Zhu, A. & Wang, J. Q. et al. (2019). Supply and demand analysis of urban park distribution based on social equity and justice: A case study of Xuhui District, Shanghai. Acta Ecologica Sinica, 39 (19):7035–7046.

Wen, C. Albert, C. & Von, C. (2020). Equality in access to urban green spaces: A case study in Hannover, Germany, with a focus on the elderly population. Urban Forestry & Urban Greening,55:126820.

Advances in Urban Engineering and Management Science – Khalil & Yang (Eds)
 ISBN 978-1-032-30426-7

Research on BIPV design strategy and technology of green public buildings based on data analysis—Taking Xiong'an high-speed railway station as an example

Bin Xu*
Land-Based Rationalism D.R.C China Architecture Design & Research Group, Beijing, China

Yuhao He
Hua Qiao University, Xiamen, China

ABSTRACT: This paper examines and analyzes the data of the demonstration project from the aspects of site optimization, roof organization, daylighting in the optics valley, window-to-wall ratio of the waiting room, air conditioning efficiency, lighting power and energy consumption of building. The combination of solar photovoltaic system technology with the roof of 42,000 km^2 greatly improves the energy efficiency of the building in an all-round manner and provides a reference for the follow-up application and promotion of BIPV.

1 INTRODUCTION

With the acceleration of global integration (Guo Lu 2010), the greenhouse effect becomes increasingly obvious. Relevant research data show that the carbon dioxide emissions from construction, transportation and production account for more than 80% of the total urban emissions, while more than 30% of the total urban emissions come from the construction industry (Cai Xiaoshuang 2011). Currently, the total area of urban and rural buildings in China has exceeded 60 billion square meters, and the total building energy consumption in China has exceeded 1 billion tons of standard coal, among which the area covered by public building area is about 12 billion square meters, and the annual energy consumption of public buildings reaches 270 million tons of standard coal. The public building area accounts for less than 20% of the total building area. However, the energy consumption of public buildings takes up more than 30% of the total building energy consumption. Public buildings feature high energy consumption, so there is a great potential for energy conservation and emission reduction in public buildings (Qin Wenjun 2019), which makes public buildings an important field to promote peak carbon dioxide emissions and carbon neutrality in the construction field.

As a technology of integrating solar power generation products into buildings, building-integrated photovoltaics (BIPV) has many strengths, e.g., wide distribution of solar energy resources, nearby power supply, simple power generation process and noiseless operation. Since solar energy is recognized as one of the most promising renewable energy sources in the world, BIPV has a promising market prospect, which has aroused enthusiastic investment from governments and industrial enterprises in various countries. Considering climate adaptability in cold regions, based on architectural design strategies in terms of site, form, function, space and interface, this paper combines green building technologies such as BIPV to develop targeted design methods and strategies, thereby strengthening the comprehensive efficiency of buildings.

*Corresponding Author: xubin@cadg.cn

DOI 10.1201/9781003305026-95

2 METHODS

2.1 *Technical routes*

Based on the design features of BIPV, this paper firstly analyzes the advantages of the BIPV technology, determines the development trend of BIPV in public buildings, and sorts out the design system to achieve the integration of solar cell modules and buildings. The BIPV design strategy is implemented throughout the whole process and all dimensions of green design, so as to build a green design system involving architectural layout, architectural form, spatial organization, maintenance system and equipment system.

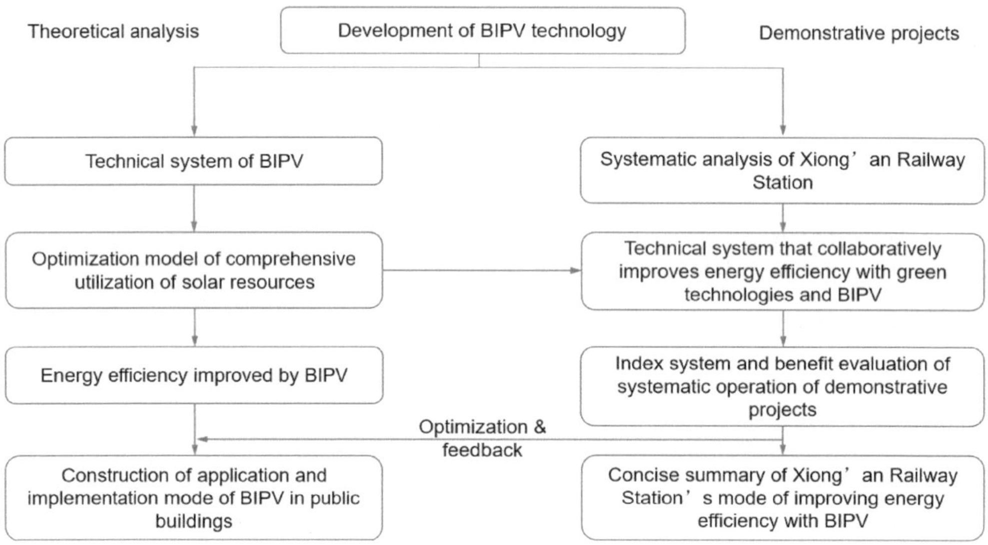

Figure 1. Technical routes.

Secondly, combined with the solar photovoltaic system and energy monitoring system used in the project of Xiong'an Railway Station, through analyzing the operation data, this paper investigates the comprehensive benefits of BIPV in improving the energy efficiency of this project, and verifies the feasibility of popularizing and implementing the solar photovoltaic integration strategy in the design of public buildings. Finally, this paper proposes the BIPV strategy for further application and promotion (Figure 1).

2.2 *Climatic analysis and comprehensive strategy*

As one of the five climatic regions divided by the Code for Thermal Design of Civil Building (GB50176-2016), the "cold region" is located in the south of the Great Wall and north of the Qinling Mountains.

Cold regions enjoy strong solar radiation, with an annual total solar irradiance of 50–190W/m2, a sunshine duration of 2000-2800 hours and an annual sunshine percentage of 40–60% (Lu Huimin 2007). For most buildings in cold regions, both heat preservation in winter and thermal insulation in summer should be considered in the architectural design. Besides, the strong sunshine and limited rainfall in this region serve as favorable conditions for BIPV.

The Xiong'an New Area, where the demonstrative project is located, receives the strongest solar radiation (mainly direct radiation) from the southwest direction all year round, with the solar radiation scattering ratio being relatively low. Therefore, when it comes to the architectural design, consideration should be given to thermal insulation and shading south by the west in summer to avoid excessive heat radiation from entering the room from the southwest direction. Secondly, based on the functional and physical requirements of architectural design, we ought to consider the change of photovoltaic equipment arrangement, and orderly lay of the solar photovoltaic panels on the roof or outer wall according to the overall design effect of the buildings. For some semi-outdoor or indoor-outdoor transitional spaces, natural lighting can be reasonably combined with BIPV.

3 TEST AND TYPICAL CASE

To apply the BIPV design strategy in cold regions, it is necessary to determine the form, quantity and arrangement of photovoltaic modules based on the collected local climate parameters and factors such as the layout, form, function, and space and interface of buildings. Then, the photovoltaic system and the building are designed simultaneously.

3.1 *Architectural Layout*

In the architectural layout, the local meteorological data, e.g., solar radiation and temperature change, is collected first, and the actual amount of radiation received in this region is calculated according to the theoretical and computational models. Then, the overall layout is designed with the aim of better receiving the illumination. Moreover, the relationship between photovoltaic power generation and the municipal power grid should be taken into comprehensive consideration. When it comes to the detailed layout design, site selection, architectural layout, volume division and function arrangement should be carried out based on the electricity demand and photovoltaic power generation situation of the buildings and the influence of architectural design on the grid-connected mode of photovoltaic power generation should be considered (Wang Zhe 2021). Eventually, the power generation efficiency of photovoltaic buildings can be optimized, with the investment and income being effectively balanced.

In the project of Xiong'an Railway Station, importance is attached to the close integration of planning layout and urban space. The overall layout design featuring "station-city integration" gives full play to the role of the railway station as an urban hub, and achieves an overall regional layout design by closely connecting with the form, space and streamlining of the city. With "green travel" as the overall goal, this project sorts out the transportation system, constructs a travel system integrating various modes of transportation, and optimizes the site design in combination with the design concept of low-impact development, so that 88% of the total runoff of the site is controlled (see examples in Table 1) and the roof greening area ratio reaches 63.5%. In terms of compact layout, the ratio of the underground building area to the total land area reaches 46.7%. However, thanks to the application of a passive design strategy, the proportion of the area where the lighting coefficient meets the requirements reaches 82.6%. The photovoltaic integrated design of Xiong'an Railway Station enables the station to generate solar electrical energy for its own consumption and transfer the surplus electricity onto the power grid to be consumed by nearby users. The annual amount of electricity generated by the photovoltaic power generation system reaches 12.7% of the total annual electricity consumption of the project, which reflects the green, efficient and convenient planning concept of this project.

3.2 *Architectural form*

The architectural form of BIPV design directly determines the amount of solar radiation the buildings can accept. Hence, while designing the architectural form, we should first select the appropriate direction of the building and then optimize the plane shape of each floor according to the functional layout. Then, the optimal inclination angle of the roof of the building and the sunshine incident angle of the light-receiving surface is calculated based on the meteorological data to realize the highest power generation efficiency of the photovoltaic system. Moreover, with the traditional Chinese design elements combined, the design elements such as function, space and interface are fully considered, and the architectural form design for increasing the light-receiving area is added. Apart from satisfying the requirements of design and power generation, the architectural form design

Table 1. Optimization analysis table for low-impact development site.

Type	Area	Green infrastructure area (m^2)	Total area of the area (m^2)	Percentage (%)
Rainfall infrastructure	Sunken green space Between 1-2 axis Between 37-38 axis	9700	9700	100
Plaza	Northside of West Plaza	7600	54400	61
	Southside of West Plaza	7600		
	Northside of East Plaza	2700		
	Southside of East Plaza	2700		
	Area between 1-2 axles	6200		
	Area between 37-38 axles	6200		
	Total	33000		
Calculated	The annual comprehensive runoff coefficient of the site is 0.30 The total annual runoff control rate of the site is 88%			

adapting to BIPV can improve the architectural form and space by optimizing the shape of the roof, adding lighting balconies and changing the facade material of the sunny-side surface, so as to create a suitable green space. Through the integrated design, the architectural form usually has a characteristic roof, obtains more photovoltaic application area, and features facade richness and architectural texture that perfectly matches the roof form.

The architectural form of Xiong'an Railway Station is completely geared to the local climate conditions. BIPV and adaptation to the cold region are realized through a number of design strategies. First of all, the architectural form of Xiong'an Railway Station adopts the compact design concept by applying the multi-story space superposition arrangement mode while reasonably organizing the passenger flow and functions. The waiting hall, the city corridor and many transportation facilities are set up in the space below the rail transit area, and a plurality of large spaces such as the ground entrance floor, the elevated entrance floor, and the underground transportation floor are set up superposed. What's more, comprehensively considering the factors of illumination and functional space, this project organically organizes large-area solar photovoltaic panels, daylighting skylights, platforms for human activities and roof greening. Parametric design tools are used to conduct topological modeling and optimization of the shape of the roof (see examples in Table 2). On the premise of guaranteeing the height of a tall space, this project takes into account multiple objectives such as solar radiation acquisition, reduction of wind load on the roof and decrease of surface area coefficient, to create a convex roof form in the shape of a water drop. In addition, soilless cultivation devices are properly set up in the space between the shading facades of Xiong'an Railway Station to plant green plants, thereby creating a green environment.

Table 2. Organic organization analysis table of the roof.

Type	Number	Position	Number	Percentage
Equivalent solar photovoltaic panel area	35000 m^2	Roof surface area	127200 m^2	21.8%
Green area of the roof	8409 m^2	Area of the roof that can be afforested	13293 m^2	63.5%
Transparent part of the roof	23093.4 m^2	Roof area	192445 m^2	12%
Average annual solar power generation	583,0000 kwh	Total annual power consumption of the project	5600,0000 kwh	10.4%

3.3 *Spatial organization*

In terms of spatial organization, it is necessary to learn the requirements of the functional layout of buildings, the orientation, size and spacing of different functional spaces, and the requirements of surrounding road network layout, square green space and ventilation and lighting. Although the BIPV design will affect the indoor lighting effect in some cases, a rationally designed photovoltaic system can achieve both light transmission and shading. Therefore, the lighting surface should be scientifically designed according to specific demands in spatial organization. Atrium space can not only strengthen the lighting, improve the ecology and enhance the visual comfort, but also serve as a climate buffer to adjust the comfort of the microenvironment.

Xiong'an Railway Station is an architectural massing. To improve the quality of spatial organization, the design strategy of cavity adjustment is adopted in spatial design. When it comes to the light environment, an ecological "optical valley" horizontally running through the east and west is ingeniously introduced, which combines commercial functions and a green landscape to provide passengers with a horizontal transportation space with ecological functions. A vertical cavity is formed based on the vertical traffic space between the ticket hall and the east-west urban corridor on the ground to improve the indoor light environment by introducing light into the indoor space. In terms of optimizing the performance of the light environment, the daylight simulation tool is used to analyze the light environment of the waiting hall equipped with the "Optical Valley" so as to determine the optimal form and size of the valley. This project also guides the light into the corridor through the opening of the exit stairs on the platform floor, which improves the lighting and ventilation conditions of the corridor (See examples in Table 3).

Besides, for the need for ventilation, an openable skylight is set at the top of the Optical Valley, which adjusts the indoor air distribution and improves environmental comfort by achieving thermal pressure ventilation in the high space during transition seasons and horizontally guiding the wind pressure ventilation in the corridor of the Optical Valley. Secondly, a double-height shared hall is vertically embedded in the project as a vertical atrium cavity to adjust the microclimate of the multi-story space. There is a variable sunshade device and an openable skylight on the interface of the cavity to guide natural ventilation, which avoids overheating and glare caused by sun exposure in summer and introduces solar radiation in winter to form a greenhouse effect.

Through the optical valley, green landscape and atrium cavity, this project has comprehensively improved the indoor microclimate environment and formed a "green valley" with ecological benefits.

Table 3. Analysis table of daylighting up to standard.

Analyzed area	Main space area (m^2)	Area up to standard (m^2)	Proportion of daylighting area up to standard (%)
Above-ground inner space	77661.13	56168.17	72.32
Underground space (First floor of the underground)	111568.42	12726.56	11.41
Main above-ground functional space	160873.46	132948.78	82.6

3.4 *Interface System*

In BIPV projects, the green and efficient interface system designed through modularization and integration in a performance-based way can effectively improve the green performance of buildings.

The modular design and fabricated construction technology extensively applied in the design of curtain walls and photovoltaic components have improved the efficiency of the design and construction of Xiong'an Railway Station and realized its goal of energy saving and emission reduction. To minimize the adverse effects of light radiation on the interior of the building, the first floor and the east and west outward windows of the interlayer are designed to be concave to form a fixed self-shading structure. By analyzing the natural lighting and natural ventilation in the elevated waiting hall and optimizing the area ratio of window-to-wall in the waiting hall (see examples in Table 4), this project realizes the lighting and shading design based on comfort control. The project adopts the integrated technology to improve the comprehensive green performance of the envelope enclosure through composite construction, and adopts a large number of low-carbon, environmentally friendly and recyclable materials to reduce carbon emissions.

Through a series of interface optimization design, the thermal performance of Xiong'an Railway Station in various aspects (heat transfer coefficient of the transparent parts of roof, outer wall, outer window and curtain wall, and solar heat gain coefficient of the transparent parts of outer window, curtain wall and roof) has been reduced by more than 10% as compared to the corresponding indicators stipulated in the Standard for Energy Consumption of Building (GB/T 51161-2016).

Table 4. Analysis table of the window-to-wall ratio of the waiting room.

Orientation	Façade	Area of window (m^2)	Area of wall (m^2)	Window-to-wall ratio	Limit	Conclusion
South	South – default facade	11422.21	25959.24	0.44	0.70	Suitable
North	North – default facade	12021.05	23973.70	0.50	0.70	Suitable
East	East – default facade	9731.98	24651.77	0.39	0.70	Suitable
West	West – default facade	10599.24	23321.37	0.45	0.70	Suitable
Basis of "Standard"		China, "Energy-saving Design Standard for Public Buildings in Hebei Province" (DB13(J)81-2016), 3.2.2				
Standard requirement		In cold areas, the window-to-wall area ratio of each facade of Class A public buildings (including translucent curtain walls) should not be greater than 0.70.				
Conclusion		Suitable				

3.5 *Equipment system*

In the equipment system, this project applies the green energy system, high-efficiency electromechanical equipment and green intelligent lighting system to improve the utilization efficiency of photovoltaic power generation and meet building electricity consumption with on-site power generation, thereby achieving the purpose of energy-saving and carbon reduction.

In Xiong'an Railway Station, a green energy system integrated with a solar photovoltaic system is installed on the roof and the canopy of the station building. Consisting of polysilicon components, the photovoltaic system has a high light energy conversion efficiency. The total capacity of the system is about 6 MW, with the laying area reaching about 42000 m2, which can reduce the coal consumption by 1800 tons and the carbon dioxide emission by 4500 tons. The photovoltaic system can provide about 20% of the power needed by the station, providing a steady flow of clean energy for the building. What's more, this project uses efficient electromechanical equipment such as the air conditioning system (see examples in Table 5), heating system and fresh air system, and the intelligent lighting control system realizes the switch control (see examples in Table 6), induction control, remote control and time control, which comprehensively improves the overall energy efficiency strategies by organically combining strategies like improving energy efficiency, saving energy and replacing traditional energy with green electric energy.

4 TEST RESULTS AND DISCUSSIONS

Through the comprehensive application of design strategies in terms of architectural layout, architectural form, spatial organization, interface system and equipment system, the demonstrative project of Xiong'an Railway Station combines the application of solar photovoltaic system technology to improve the comprehensive energy efficiency of buildings and facilitate the energy conservation and emission reduction in public buildings (see examples in Tables 7 and 8).

Table 5. Analysis table of air conditioning efficiency.

No.	Equipment type	Rated cooling capacity (KW)	Performance parameter(W/W) Actual equipment	Standard requirements	Whether satisfy
L-1	Centrifugal chiller unit	3868	8.206	6.10	Satisfy
L-2	Centrifugal chiller unit	2637	8.206	6.10	Satisfy
L-5	Screw chiller unit	1060	7.7448	5.85	Satisfy
ZX-DLW5-450	Multi-connected air conditioning (heat pump) unit	45	4.47	3.85	Satisfy
ZX-DLW1-400		40	4.47	3.85	Satisfy
ZX-DLW3-335		33.5	4.47	3.85	Satisfy
ZX-DLW10-615		61.5	4.47	3.85	Satisfy
ZX-DLW6-900		90	4.47	3.75	Satisfy
ZX-DLW7-960		96	4.53	3.75	Satisfy
ZX-DLW8-850		90	4.35	3.75	Satisfy

Table 6. Analysis table of intelligent lighting design.

Room type		Design illuminance value (Lx)		Lighting power density (W/m^2)	
		Actual value	Standard value	Actual value	Target value
Main functional rooms	Front hall at east entrance	206	200	3.80	8
	Front hall at the west entrance	233	200	6.20	8
	Elevated waiting area	219	200	4.47	8
	Ground floor waiting area	244	200	3.68	8
	Basic platform canopy	159	150	2.58	6
	Platform canopy	112	75	1.77	4.5
	City corridor	160	150	2.98	6
	Toilet of the front hall	130	75	2.70	3
	Hallway at the west entrance	155	150	3.25	4.5
	VIP lounge	317	300	9.25	10
	Ticket office	234	200	6.37	8
Other rooms	Toilet of the elevated waiting area	82	75	2.97	3
	Commercial corridor	193	150	3.58	6
	Maternal a waiting room	330	200	5.34	6
	Parking lot	79	75	1.62	3.5

Table 7. Building energy consumption value of design.

Energy consumption type	Building energy consumption value of design Annual power consumption (MWh/a)	Converted (kWh/a)	Annual energy consumption per unit area (kWh/m^2·a)
Cooling	9792.64	9792640	20.61
Heating	24174.56	24174560	50.87
Fan	2448.16	2448160	5.15
Water pump	3626.184	3626184	7.63
Lighting	7025.96	7025960	14.78
Equipment	4717.375	4717375	9.93
Air exhaust	898.54	898540	1.89
Elevator	5897.23	5897232	12.41
Total	52683.419	52683419	110.86

Table 8. Building energy consumption value of reference.

Energy consumption type	Building energy consumption value of reference		
	Annual power consumption (MWh/a)	Converted kWh/a	Annual energy consumption per unit area (kWh/m^2·a)
Cooling	11389.59	11389590	23.97
Heating	28904.36	28904360	60.82
Fan	2847.40	2847398	5.99
Water pump	4335.654	4335654	9.12
Lighting	9485.05	9485050	19.96
Equipment	4717.375	4717375	9.93
Air exhaust	898.54	898540	1.89
Elevator	5897.23	5897232	12.41
Total	68475.1965	68475196.5	144.09

5 CONCLUSION

The BIPV strategy should integrate the layout, form, function, space, and interface of architectural design, combine the green building technology, and make rational use of the passive design concept, so as to make public buildings greener, more ecological, efficient and energy-saving. The following conclusions are gained in this paper:

- BIPV can easily obtain clean and renewable energy and realize nearby power supply.
- The BIPV strategy is applicable to large-scale public building projects for it can make full use of the space, such as the roof and facade of buildings, without occupying extra land.
- The components of the photovoltaic power generation system have stable long-term service performance and the energy recovery period is comparatively short.
- The photovoltaic power generation system produces no noise, pollutants or light pollution during its operation.
- The BIPV strategy is worthy of further promotion for its wide application value and promising market prospect in the design of green buildings.

ACKNOWLEDGMENT

This research was supported by The Research on High Quality Development of Architectural Design Industry Under the Concept of Green and Low-carbon (Y2021152) of the scientific and technological innovation research project of China Architecture Design & Research Group.

REFERENCES

Cai Xiaoshuang, Research on low carbon building evaluation based on LCA. Jiangnan University, 2011.

Guo Lu, Qiu Mingzhu, Research on internal control defects and improvement measures of Listed Companies in China. Modern commerce and industry, 2010, pp. 66–67.

Lu Huimin, Research on energy efficiency design of residential buildings in cold areas. Tongji University, 2007.

Qin Wenjun, Li Xiang, Overview and development prospect of China's photovoltaic construction integration industry. Journal of architecture, 2019 (S2), pp. 6–9.

Wang Zhe, Yang Jinpeng, Green railway passenger station with sustainable development – Thoughts on the design and implementation of Xiong'an station of intercity railway from Beijing to Xiong'an New Area. Architectural skills, 2021, pp. 82–84.

Advances in Urban Engineering and Management Science – Khalil & Yang (Eds)
© 2023 The Author(s), ISBN 978-1-032-30426-7

Study on multi-level analysis method of highway slope maintenance decision during operation period

Zhiyuan Cheng, Yimin Wang*, Kunbiao Zhang & Qi Li
School of Civil Engineering and Transportation, South China University of Technology, Guang Zhou China

ABSTRACT: Implementing the characteristics of fuzzy classification standards for slope safety status in existing highway slope maintenance guidelines to scientifically formulate slope maintenance countermeasures and improve the maintenance quality of highway slope, as per the characteristics of highway slope during the operation period, considering the comprehensiveness, scientificity, and operability of the index system, we established a multi-level evaluation model of slope safety, which contains 13 evaluation factors and four kinds of sub-evaluation objects, using improved multi-level analysis method to determine the weight parameters of each level evaluation index. The improved multi-level analysis method can greatly reduce the subjectivity of the traditional analytic hierarchy method, and establish the safety classification standard and corresponding maintenance countermeasures of highway slope during the operation period. The model method was used to evaluate and grade the safety of slope of three section units on S356 in Guangdong province, and accordingly, the maintenance plan of the slope of each section was made. The slope on the S356 highway has been stable up to now after undertaking maintenance measures based on the analysis and decision of this model. This method can be applied to the decision of highway slope maintenance during the operation period.

1 INTRODUCTION

Stability and safety of highway slopes have always been a major problem faced by highways in the operation period (Han et al. 2003). After years of exploration by scholars at home and abroad, a slope stability research principle of "qualitative analysis as the premise and quantitative analysis as the means" has been formed (Zhang & Shen 2011). Among these, the analytic hierarchy process (AHP) has great advantages in dealing with the problems of slope discreteness, fuzziness, and uncertainty (Xie & Xia 2003), and can be better applied in the study of highway and subgrade slopes. It includes highway risk assessment (Zayed et al. 2008), slope stability assessment (Huang et al. 2007), optimal decision-making of slope treatment scheme (Xie & Xia 2002), and more.

After the highway traffic enters the operation period, a set of quantitative grading evaluation standards is formed in the roadbed engineering (Ministry of Transport of the People's Republic of China 2018; Wang et al. 2017; Zhang et al. 2015). As an important part of the subgrade, as per the guiding principles of the "Maintenance Guide for Highway Cutting Slope," a slope should be maintained and managed at different levels (Guangdong Hualu Traffic Technology Co. Ltd. 2014). However, the classification standard of the slope is vague in the guide, and how to establish a set of effective quantitative evaluation standards and maintenance countermeasures for slope safety is an important problem to be solved urgently during the highway operation period. Currently, a large number of scholars have conducted research on the stability classification and risk assessment system of highway slopes during the construction period (Lin et al. 2019; Qin et al. 2016; Zhang et al.

*Corresponding Author: ctymwang@scut.edu.cn

DOI 10.1201/9781003305026-96

2010). For the highway slope in the operation period, its disease develops gradually, and the safety of the slope is affected by many factors. There are few studies on the maintenance decision based on the slope disease condition, geological environment, and engineering design scheme. Therefore, this paper, based on an improved analytic hierarchy process (AHP), establishes the problem combined with slope, slope shape, slope engineering geology, and hydrogeology conditions, such as multi-level and multi-factor operation period of highway slope stability evaluation model, and the territory of Guangdong province S356 highway as a background for the field application engineering, the three sections of the slope safety evaluation and grading unit. Based on this, the maintenance plan of each section slope is formulated, and the applicability and reliability of the model are verified. The research results can provide a reference for the safety evaluation and maintenance decision-making of other highway slopes during the operation period.

2 FUNDAMENTAL PRINCIPLE OF MULTI-LEVEL ANALYSIS METHOD FOR SAFETY OF HIGHWAY SLOPE IN OPERATION PERIOD

2.1 *Establishment of multi-level structure model*

The key to the research of AHP is to analyze the core target and decompose it into various impact factors. The impact factors of the next layer are controlled by the factors of the upper layer, and the impact factors of the upper layer are associated with the impact factors of the next layer. Therefore, the target is hierarchized, and a multi-level structure evaluation model is established (Li & Wang 2012). Thus, a slope safety evaluation model including 13 indexes, such as fragmentation and slope scour, is established, as shown in Figure 1.

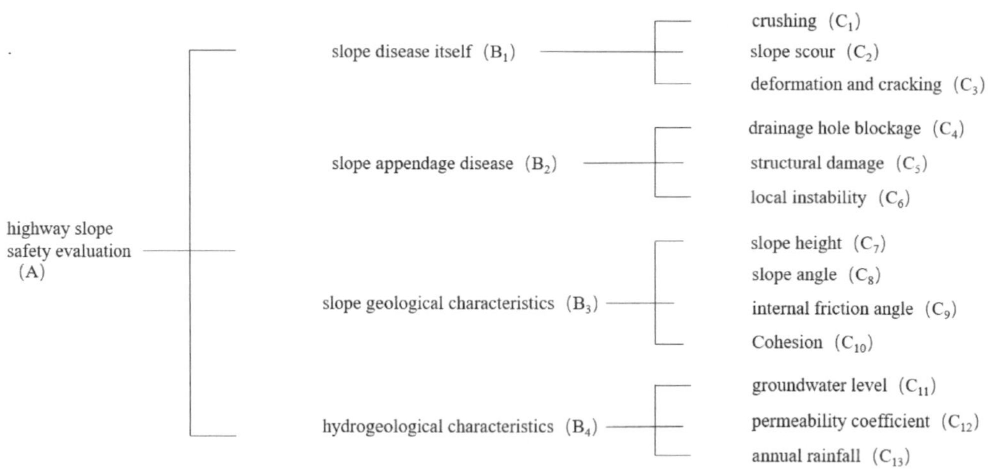

Figure 1. Multilevel model of highway slope safety evaluation during the operation period.

2.2 *Evaluation index weight determination and consistency verification*

The traditional analytic hierarchy process is very subjective. In this paper, the improved analytic hierarchy process is used to evaluate the inter-element importance on a scale of 0.1~0.9, which can reflect the relation between elements at all levels more objectively.

(1) Construction of fuzzy complementary judgment matrix

According to the hierarchical structure shown in Figure 1, by comparing the importance of two factors, the fuzzy complementary judgment matrix of the upper and lower sub-targets is constructed.

The general form of the judgment matrix is shown in Equation (1).

$$A = \begin{bmatrix} a_{11} & \cdots & a_{1n} \\ \vdots & \ddots & \vdots \\ a_{n1} & \cdots & a_{nn} \end{bmatrix} \tag{1}$$

Based on previous research results (Wang et al. 2017; Zhang 2019) and expert evaluation, five judgment matrices are obtained, as shown in Tables 1 and 2.

Table 1. Fuzzy complementary judgment matrix of A-B and B_3-C.

Judgment matrix (A-B)				
A	B_1	B_2	B_3	B_4
B_1	0.5	0.3	0.5	0.4
B_2	0.7	0.5	0.6	0.6
B_3	0.5	0.4	0.5	0.4
B_4	0.6	0.4	0.6	0.5
Judgment matrix (B_3-C)				
B_3	C_7	C_8	C_9	C_{10}
C_7	0.5	0.4	0.5	0.5
C_8	0.6	0.5	0.6	0.6
C_9	0.5	0.4	0.5	0.5
C_{10}	0.5	0.4	0.5	0.5

Table 2. Fuzzy complementary judgment matrix of B_1-C B_2-C and B_4-C.

Judgment matrix (B_1-C)			
B_1	C_1	C_2	C_3
C_1	0.5	0.4	0.4
C_2	0.6	0.5	0.5
C_3	0.6	0.5	0.5
Judgment matrix (B_2-C)			
B_2	C_4	C_5	C_6
C_4	0.5	0.7	0.4
C_5	0.3	0.5	0.1
C_6	0.6	0.9	0.5
Judgment matrix (B_4-C)			
B_4	C_{11}	C_{12}	C_{13}
C_{11}	0.5	0.5	0.7
C_{12}	0.5	0.5	0.7
C_{13}	0.3	0.3	0.5

(2) Calculating the weight of each index

Calculate the sum of each element in the fuzzy complementary judgment matrix:

$$b_i = \sum_{j=1}^{n} a_{ij} = a_{i1} + a_{i2} + \cdots\cdots + a_{in} \tag{2}$$

Based on $r_{ij} = 0.5 + {}^{b_i - b_j}\!/_{2(n-1)}$, produce the fuzzy consistency judgment matrix R$=(r_{ij})_{n\times n}$, normalize the matrix R to obtain the weight vector W of the judgment matrix, where W$_i$ satisfies Equation (3)

$$W_i = (n(n-1))^{-2} \sum_{i=1}^{n} r_{ij} + \frac{n}{2} - 1 \tag{3}$$

(3) Consistency test

For the judgment matrix weight vector W $W_{ij} = {}^{w_i}\!/_{w_i + w_j}$ W*=$(w_{ij})_{n\times n}$ is the eigenmatrix of fuzzy complementary judgment matrix A Calculate the norm A-W* according to Equation (4)

$$A - W^* = \rho(A, W^*) = \sum_{i=1}^{n}\sum_{j=1}^{n} \left| a_{ij} - w_{ij} \right| \tag{4}$$

Calculate consistency ratio I (A, W*):

$$I(A, W^*) = \frac{1}{n^2}\rho(A, W^*) \tag{5}$$

If the I (A, W*) is lower than 0.1, the matrix will be considered consistent, and the generated weight vector from this matrix will be accepted.

Taking judgment matrix A-B as an example, the fuzzy consistency judgment matrix R is calculated, as shown in Table 3. The weight vector W = $(0.2332,0.2723,0.2389,0.2556)^T$ of the judgment matrix can be obtained from Equation (3). The eigenmatrix W* is calculated according to this vector, as shown in Table 3. According to Equations (4) and (5), I (A, W*)=0.0545 < 0.1, the judgment matrix A-B is considered consistent.

Table 3. Fuzzy consistency matrix R and eigenmatrix W* of the matrix A-B.

Fuzzy consistency matrix R			
0.5	0.383	0.483	0.433
0.617	0.5	0.6	0.55
0.517	0.4	0.5	0.45
0.567	0.45	0.55	0.5
eigenmatrix W*			
0.5	0.461	0.494	0.477
0.539	0.5	0.533	0.558
0.506	0.467	0.5	0.483
0.523	0.442	0.517	0.5

According to Equations (2)–(5), the weights of other judgment matrices and consistency verification are calculated. The weight vector is as follows: $W_{B_1-C}=(0.3084,0.3458,0.3458)^T$, $W_{B_2-C}=(0.3458,0.2667,0.3958)^T$, $W_{B_3-C}=(0.2444,0.2668,0.2444,0.2444)^T$, $W_{B_4-C}=(0.3583, 0.3583,0.2834)^T$, and the I(A,W*) are all lower than 0.1.

2.3 *Safety evaluation and classification of highway slope*

In this paper, the method of scoring and evaluation is adopted to calculate the safety score of the evaluated section slope through the formula, and the safety classification of the slope is carried out. Different treatment measures are taken for the slope with different safety grades, which can meet the requirements of the maintenance department for the graded management and governance of the slope. The specific steps are as follows.

First, according to local and national norms and guidelines as well as references (Ministry of Transport of the People's Republic of China 2014; Zhang et al. 2015), the classification table of index layer evaluation factors and the classification table of highway slope safety categories are formulated, as shown in Tables 4–6.

Then, the evaluation scores of indicator layer $C_1 \sim C_{13}$ are obtained from Tables 4 and 5, and the evaluation scores of criterion layer B1~B4 are calculated according to Equations (6)–(9):

$$B_1 = C_1 W_{B_1-C} + C_2 W_{B_1-C} + C_3 W_{B_1-C} \tag{6}$$

$$B_2 = C_4 W_{B_2-C} + C_5 W_{B_2-C} + C_6 W_{B_2-C} \tag{7}$$

$$B_3 = C_7 W_{B_3-C} + C_8 W_{B_3-C} + C_9 W_{B_3-C} + C_{10} W_{B_3-C} \tag{8}$$

$$B_4 = C_{11} W_{B_4-C} + C_{12} W_{B_4-C} + C_{13} W_{B_4-C} \tag{9}$$

Finally, the overall objective "highway slope safety evaluation score" A is calculated by Equation (10).

$$A = B_1 W_{A-B} + B_2 W_{A-B} + B_3 W_{A-B} + B_4 W_{A-B} \tag{10}$$

According to the slope safety classification principle in Table 6, the safety of the slope is evaluated, and the slope maintenance countermeasures are further formulated to ensure the safety of the highway during the operation period.

Table 4. The slope itself and its appendages disease values[a].

Disease classes	Grade and number	Score	Disease classes	Grade	Score
Crushing	none	5	Drainage hole blockage	none	5
C_1	1	25	C_4	1	35
	2	45		2	65
	3 or more	65		3 or more	95
Slope scour	none	5	Structural damage	none	5
C_2	2	30	C_5	1	25
	4	55		2	45
	5 or more	80		3 or more	65
Deformation and cracking	none	5	Local instability	none	5
C_3	1	30	C_6	1	55
	2	55		2 or more	100
	3 or more	80			

[a]Each score can be added or reduced according to the severity of the disease. The number of diseases is a statistical unit of 1 km section.

3 THE ENGINEERING APPLICATION

3.1 *Engineering background*

Provincial Highway S356 is a secondary highway, located in the territory of Huizhou City, Guangdong Province. The section of this highway in Gotham Town is located in hilly mountains, and a large number of deep-cut slopes have been designed and built along the highway, without too much engineering reinforcement treatment. Influenced by rainstorms in recent years, the slope

Table 5. Slope engineering and hydrogeological condition values.

Parameter	Grade	Score	Parameter	Grade	Score
Slope height (m) C_7	$0 \leq H \leq 20$	10	Groundwater level (m) C_{11}	$h \leq 0.25H$	20
	$20 < H \leq 40$	30		$0.25H < h \leq 0.5H$	40
	$40 < H \leq 60$	50		$0.5H < h \leq 0.75H$	60
	$60 < H \leq 80$	70		$0.75H < h \leq H$	80
	$H \geq 80$	90	Permeability coefficient (cm/s) C_{12}	$k < 1.0 \times 10^{-6}$	10
Slope angle C_8	$\theta \leq 4°$	5		$(1.0\sim2.0)\times10^{-6}$	30
	$40° < \theta \leq 5°$	25		$(2.0\sim3.0)\times10^{-6}$	50
	$50° < \theta \leq 6°$	45		$(3.0\sim4.0)\times10^{-6}$	70
	$60° < \theta \leq 70°$	70		$k > 4.0 \times 10^{-6}$	90
	$\theta \geq 70°$	100	Annual rainfall (mm) C_{13}	< 1000	5
Internal friction angle C_9	$\varphi > 18°$	5		100~1300	20
	18° ~22°	25		1300~1600	35
	15° ~18°	45		1600~2000	50
	12° ~15°	65		>2000	80
	10° ~12°	85			
	$\varphi < 10°$	100			
Cohesion (kPa) C_{10}	c>25	5	Cohesion (kPa) C_{10}	12~16	65
	20~25	25		10~12	85
	16~20	45		c < 10	100

has been damaged by water, resulting in interruption of highway traffic, as shown in Figure 2. How to prevent the slope disease in time and scientifically maintain the subgrade slope is an urgent problem to be solved by the management department. According to the road site situation, taking 1 km as a unit, three sections with serious diseases (pile numbers: K5+000~K6+000, K6+000~K7+000, K9+000~K10+000) are selected to carry out field investigation and highway slope safety evaluation.

Figure 2. Slope water disaster of the S356 test section.

Table 6. Safety classification table of highway slope.

Grade	A	Safety condition of the slope	Maintenance policy
1	A < 40	safe	Daily maintenance, including cleaning slope, drainage pipe mud debris, removing loose boulders, repairing cracks, pruning vegetation, etc.
2	40≤A < 50	Basically safe	Regular maintenance and reinforcement of lattice beams, retaining walls, anchor anti-slide piles and other supporting structures, regular washing, dredging, and renovation of drainage facilities to prevent the possibility of gradually decreasing the safety of the slope.
3	50≤A < 60	Partly dangerous	Appropriate monitoring should be carried out, including surface, subsurface, structural deformation, structural and geotechnical stress, rainfall and groundwater. If the diseases in Table 4 are found, the ground and underground interception drainage facilities should be improved in time, and grouting reinforcement and necessary anchoring and retaining works should be carried out.
4	A≥60	Highly dangerous	For all sections of the earth's surface and underground, the deformation of the structures, structure, rock mass stress, rainfall and groundwater monitoring, the monitoring data of slope stability should be considered for detailed analysis combined with site condition and geological conditions such as drainage, paring back pressure, fence, emergency measures including micropile for the design of the permanent maintenance measures.

3.2 *Condition of subgrade slope*

The slope height of this section is 8~40 m, and the slope angle is about 40~55 degrees. The rock and soil body of the slope is mainly strongly weathered ~ fully weathered argyle sandstone, which is easy to soften when it meets water. The cohesion is 8.2 kPa~20.1 kPa, the internal friction angle is 12.2° ~21.4°, and the permeability coefficient of the overlying rock is 1.24×10^{-6}cm/s~4.55×10^{-6}cm/s. Under the influence of subtropical monsoon climate, the rain in the region is always huge, with the annual average rainfall of up to 1918.2 mm. Mainly for unconsolidated rock pore water and groundwater type bedrock fissure, water and underground water level change according to the season and terrain. Atmospheric precipitation during the rainy season causes the underground water level to rise significantly, as the slope runoff velocity is fast due to the steepness of the slope. Easy-to-form runoff scour along the slope affects the safety of the slope. The method of manual investigation is adopted to investigate the slope diseases of the sections of the three units mentioned above. The statistical results of slope diseases are shown in Table 7.

Table 7. Statistics of slope diseases in the S356 test section.

Diseases	Quantity		
	K5+000~K6+000	K6+000~K7+000	K9+000~K10+000
Crushing	3	2	3
Slope scour	5	3	3
Deformation and cracking	2	0	1
Drainage hole blockage	3	1	2
Structural damage	1	1	1
Local instability	0	0	0

3.3 *Slope safety evaluation classification*

Based on the multi-level analysis model method constructed in this paper, the slope safety evaluation results of the sections of provincial highway S356 K5+000~K6+000, K6+000~K7+000, and K9+000~K10+000 are calculated as shown in Table 8.

Table 8. Evaluation results of slope safety.

Test section	B_1	B_2	B_3
K5+000~K6+000	79	36.09	49
K6+000~K7+000	43	25.18	42
K9+000~K10+000	58	29.22	44
	B_4	A	Security grade
K5+000~K6+000	62.66	51.58	3
K6+000~K7+000	59.11	39.52	1
K9+000~K10+000	60.42	44.20	2

According to the calculation results, the slope of the K5 +000~K6+000 section belongs to the third grade "partly dangerous slopes." The slope of the K6+000~K7+000 section belongs to the first-grade "safety slope." The slope of the K9+000~K10+000 road section belongs to the second-grade "basic safety slopes." According to Table 6, appropriate maintenance measures have been implemented. S356 highway slope has remained stable so far with good effect.

4 CONCLUSION

(1) In summarizing the operational phase of the highway and slope maintenance and slope safety classification research, based on the multi-level analysis method to establish the operational phase of the highway slope safety evaluation model combined with the slope situation of Guangdong province that put forward the model of quantitative scores of all the evaluation factors, we established the operation period of highway slope safety evaluation standards based on the standard slope maintenance countermeasures. The research results can provide a reference for the safety evaluation and maintenance decision of highway slopes in the operation period.

(2) The multi-level structure evaluation model established in this paper includes 13 influencing factors, such as crushing, slope scour, and so on, which effectively integrate the disease of highway slope and regional geological and hydrogeological characteristics into the safety evaluation of slope.

(3) Taking the slope of three sections on the provincial highway S356 as an example, the slope safety evaluation and classification were carried out, and the maintenance plan was formulated. The slope safety of the K5+000~K6+000 section is partly dangerous, the slope safety of the K6+000~K7+000 section is safe, and the slope safety of the K9+000~K10+000 section is basically safe. Based on the research results of the maintenance decision in this paper, appropriate maintenance measures have been undertaken for slopes of each section of the S356 highway, which have remained stable until now.

ACKNOWLEDGMENTS

Supported by the Traffic Science and Technology Project of Guangdong Province (2014-02-041)

REFERENCES

Guangdong Hualu Traffic Technology Co. Ltd. (2014) Technical specification for expressway slope maintenance: GDCG J03-2014. Guangdong Communications Group Co. Ltd, Guangzhou.

Han L, Xiao B, Xu S.C. (2003) Importance of dynamic information about slope control along road. Chinese Journal of Rock Mechanics and Engineering., S2:2725-2730.

Huang J.W, Li J.L, Zhou Y.H. (2007) Application of fuzzy analysis based on AHP to slope stability evaluation. Chinese Journal of Rock Mechanics and Engineering, S1:2627-2632.

Li Z.G., Wang Y.P. (2012) Application of analytic hierarchy process in slope landslide prevention and control. Coal Engineering, 08:117-119.

LIN D.M, CHEN P, MA J.L, et al. (2019) Assessment of slope construction risk uncertainty based on index importance ranking. Bulletin of Engineering Geology and the Environment, 78:4217-4228.

Ministry of Transport of the People's Republic of China. (2014) Guide for safety risk assessment of highway cutting high slope engineering construction. China Academy of Transportation Science, Beijing.

Ministry of Transport of the People's Republic of China. (2018) Highway Performance Assessment Standards: JTG 5210-2018. China Academy of Transportation Science, Beijing.

Qin S.W, Cao R.G, Li G.J, et al. (2016) Stability classification of rock slope based on FAIIP efficiency coefficient method. Journal of Catastrophology, 31:27-32.

Wang Y.M, Guo J.X, Li L.S, et al. (2017) Condition evaluation of subgrade in highway operation period based on analytical hierarchy process. Journal of South China University of Technology (Natural Science), 45:37-43 and 51.

Xie Q.M, Xia Y.Y. (2002) Multihierarchy and multiobjective optimization decision methods of slope treatment schemes and its application. Journal of Wuhan University of Technology, 10:21-24.

Xie Q.M, Xia Y.Y. (2003) Fuzzy hierarchy analysis on decision making of rockmass slope treatment based on entropy weight. Chinese Journal of Rock Mechanics and Engineering, 07:1117-1120.

Zayed T, Amer M, Pan J. (2008) Assessing risk and uncertainty inherent in Chinese highway projects using AHP. International Journal of Project Management, 26:408-419.

Zhang J.L, Shen M.R. (2011) A new approach to expressway slope stability assessment. Rock and Soil Mechanics, 32:3623-3629+3636.

Zhang J.W, Liao X.P, Wei T.R, et al. (2015) Study on safety risk assessment system for high slope of highway in operation. Subgrade Engineering, 06:197-203.

Zhang L, Wang Y.M, Li L.S, et al. (2015) Analysis on the state-of-art of highway subgrade disaster prevention and maintenance based on expert questionnaire. Highway Engineering, 40:210-213.

Zhang X.J. (2019) Study on safety evaluation of highway slope based on online monitoring system. Chang'an University.

Zhang Y.H, Li H.X, Sheng Q, et al. (2010) Study of stability gradation of highway rock slopes based on fuzzy comprehensive evaluation. Rock and Soil Mechanics, 31:3151-3156.

Advances in Urban Engineering and Management Science – Khalil & Yang (Eds)
© 2023 The Author(s), ISBN 978-1-032-30426-7

Study on the construction of China's international logistics system under the international public health crisis

Hongli Hu* & Luning Yan*
School of Business Administration, Zhejiang Gongshang University, Hangzhou, Zhejiang, China

ABSTRACT: The global outbreak of COVID-19 at the end of 2019 has brought a huge impact on the stable operation of the international logistics system, which poses great risks to the smooth transportation of epidemic prevention materials and various goods. Therefore, under the international public health crisis, it is of great significance to accelerate the construction of the international logistics system of land, sea and air controlled by China all over the world, and to realize the effective connection and interaction of commodity inspection, transfer, unloading, transportation, storage, circulation processing and information in the international logistics system. Firstly, this study verifies that China must accelerate the construction of an international logistics system under the international public health crisis from three aspects: international trade, digital transformation of enterprises and value chain. Secondly, based on the framework of the international logistics system, the paper studies the system structure of China's international logistics affected by the international public health crisis. Finally, it analyzes the strategic mode that is beneficial to developing China's international logistics system under the international public health crisis, which provides a theoretical basis for China's construction of an international logistics system.

1 INTRODUCTION

The global COVID-19 outbreak at the end of 2019 has brought a huge impact on the good coordination and stable operation of all links in the international logistics system, putting the smooth transportation of epidemic prevention materials, raw materials, spare parts and related commodities in the international community at great risk. Since the "closure" of Wuhan on January 23, 2020, the Cainiao network has successively delivered millions of face masks donated by Chinese in Europe and ventilate machines donated by the Belgian government from Liege Airport to China. On March 16, the Jack Ma Foundation and Alibaba Foundation delivered a batch of face masks and testing reagents to the E-WTP hub of Liege Airport via the Cainiao network chartered plane and donated them to Belgium and Italy. During this period, China has experienced many "turnbacks," which can reflect the shortcomings of China's international logistics system construction. In addition, the low efficiency of logistics service during the epidemic period caused by the dock warehouse bursting, the lack of cabinets, the increase of epidemic prevention procedures caused by the decline of international logistics on-time rate, the rise of the international logistics service price and other problems also show that China has a long way to go in the international public health crisis in the construction of international logistics system.

A Chinese express association industry expert pointed out: "It is the responsibility of every real manufacturing country to have a logistics company with global service capability." In his view, the special test brought by the epidemic is ringing a wake-up call to China, "China should speed up the construction of its own international logistics system." In the last century, the United States, Germany, Japan and other big manufacturing countries finally formed a number of logistics giants

*Corresponding Authors: 805641926@qq.com and 2570942140@qq.com

 DOI 10.1201/9781003305026-97

with international or regional service capabilities through the construction of efficient international logistics systems, such as FedEx, DHL and Yamato Group. While companies such as China Post, Cainiao and SF Express have also tried to build their own international logistics systems in recent years, progress has been slow.

The current international logistics system is a collection of information, technology and intelligence in one of the logistics systems, customer demand from the homogenization to customized development, service mode from a single operation to the sharing and win-win logistics ecosystem development change. Countries need to optimize all aspects of the chain on a global scale, such as procurement, production and logistics, and rely on technology and model innovation to link capital flow, logistics and information flow, so as to achieve global product interconnection. Once an international public health crisis breaks out, it will have a huge impact on China's international trade and supply chain. Therefore, during the crisis, it is necessary to accelerate the construction of an international logistics system controlled by China that covers the world by land, sea and air. It is of great significance to realize the effective connection and benign interaction among the links of commodity inspection, transportation, transportation, storage, circulation processing and information in the international logistics system.

2 LITERATURE REVIEW

As human society has increasingly become an organic whole and countries have become increasingly connected, public health incidents in individual regions are very likely to develop into complex crises involving the world. A public health crisis refers to a crisis event that occurs suddenly, causing or may cause serious damage to public health. It generally has the characteristics of suddenness, urgency, diverse causes, concealment, wide range of influence, and serious harm (Chu et al. 2014). The research on public health crises roughly covers two themes of problems exposed under the crisis and how to manage them. During the public health crisis, scholars analyzed the role of the government, financial security, management communication, and international cooperation in the crisis (Zhang & Shen 2015) and discussed the countermeasures and suggestions for solving such problems in the crisis (Xue & Zhu 2003). During the continued spread of COVID-19, the academic community has begun to pay attention to the impact of the epidemic on all fields of the world. In summary, studies on the impact of the COVID-19 epidemic are involved at the macro, meso and micro levels, specifically focusing on the impact of the COVID-19 epidemic on the world economy and the impact on the Chinese economy (Liang 2020, Zhong & Chen 2020), the impact on various industries and enterprises in China (Liu 2020, Zhang 2020). In terms of logistics, scholar Shi Xianliang explained the impact of the COVID-19 epidemic on the logistics industry from a macroeconomic perspective, which is reflected in two main ways: the sharp drop in logistics and the serious interruption of logistics (Shi 2020). Other scholars analyzed how the operation of the international logistics service supply chain was affected by the COVID-19 epidemic from the supply chain perspective. However, none of the existing studies has systematically studied the mechanism of the international public health crisis affecting the construction of the international logistics system, and proposed targeted optimization strategies for each link of the mechanism. This is also the innovation of this article.

Based on the principle of the international division of labor and cooperation, international logistics uses international logistics resources to realize the cross-border flow of goods in accordance with the conventions, so as to promote the optimal allocation of resources and the development of the economy in the world (Sun 2004). Under the trade orientation, the international logistics system is composed of a commodity inspection subsystem, transshipment and unloading subsystem, transportation subsystem, warehousing subsystem, circulation processing subsystem and information subsystem. Logistics practitioners, logistics facilities and equipment and related technologies implement to support the entire coordinated operation of the system (Song 2012). The research on the international logistics system is mainly carried out from the aspects of operation mode, current situation, future trends and how to construct, but the research perspectives are different.

Based on different backgrounds, such as 5G technology (Wang 2020), the Internet, economic globalization, and the "One Belt and One Road" (Huang 2017), scholars discussed the status quo and development strategy of China's international logistics system, and believed that the construction of an international logistics system should be In line with the development trend of today's world, it will transform to intelligent, globalized and informatized. With the booming progress of international logistics in recent years, there has been more and more research on the development and perfection of the international logistics system in academia. Judging from the existing research results, domestic scholars focus on infrastructure construction in the process of international logistics, international logistics personnel training, Information platform construction, and international logistics management to explore strategies to promote the development and improvement of China's international logistics (Xu 2015). Starting from the influence mechanism of the COVID-19 epidemic on the international logistics system, this article proposes systematic policy recommendations for the stable and efficient operation of various subsystems in the international logistics system, which not only provides a certain theoretical basis for China to accelerate the construction of an international logistics system, but also provides a certain direction for China to speed up the construction of an international logistics system.

3 MATERIALS AND METHODS

This article mainly starts with three parts. Firstly, this study elaborates on the necessity of accelerating the construction of an international logistics system under the international public health crisis from three aspects: international trade, digital transformation of enterprises and value chain. Secondly, starting from each link of the international logistics system, this paper discusses how the international public health crisis boosts the construction of China's international logistics system. Finally, in view of the various links affected by the crisis, it puts forward the strategies for China to accelerate the construction of an international logistics system under the international public health crisis. This research mainly adopts the literature research method and qualitative research method.

4 RESULTS & DISCUSSION

4.1 *The necessity for China to speed up the construction of an international logistics system under the international public health crisis*

4.1.1 *The international public health crisis restricts the development of international trade*

Once the international public health crisis breaks out, various countries and regions worldwide will take different approaches to the possible spread risk degree of immigration controls and isolation measures. Airlines will also gradually reduce or suspend relevant airline operations in our country. This will directly impact the short-term financing, logistics transport, and personnel exchanges in foreign trade. Taking the COVID-19 epidemic as an example, as of February 16, 2020, 133 countries and regions have taken entry control measures against China, and 45 countries and regions have suspended or partially suspended air, sea, road and rail passenger and freight traffic with China, including the United States, France, the United Kingdom, Japan and the Republic of Korea. A total of seven countries and regions temporarily closed or restricted customs clearance ports (Hong Kong, Mongolia, Russia, etc.). Surveys show that since the outbreak of the epidemic, more than 40% of Chinese enterprises' logistics, delivery and transportation time have been affected, many important international medical supplies logistics channels have been disrupted, and delivery time has been extended. During the outbreak of the event, the epidemic prevention materials must be sorted and repackaged in specific warehouses to complete the punctual distribution of the materials finally. The surge of epidemic prevention materials puts forward higher requirements for storage capacity, so the epidemic area warehouse is facing great pressure (Li 2020). All these will lead to a large number of goods in China's backlog at foreign ports, and the domestic demand for imported raw materials will be squeezed, which will have an important impact on China's import and export trade. As shown in Table 1, after the outbreak of COVID-19, China's total import and export value, as well as cargo turnover by air, waterway, road and rail, decreased significantly in February 2020, and then increased slowly. Therefore, the great impact on international trade has put forward urgent requirements for China to accelerate the construction of an international logistics system.

Table 1. Logistics and international trade performance in the first half of 2020.

	2020.01	2020.02	2020.03	2020.04	2020.05	2020.06
Total value of imports and exports (100 million yuan)	25790	15389	24370	24875	24693	26934
Air cargo turnover growth rate	–	–45.10%	54.46%	0.00%	20.81%	1.91%
The growth rate of waterway cargo turnover	–	–14.20%	24.77%	2.73%	6.35%	–0.65%
The growth rate of highway cargo turnover	–	–62.16%	196.57%	24.52%	1.18%	7.02%
Railway freight turnover growth rate	–	–14.61%	10.05%	–8.06%	15.46%	4.48%

4.1.2 *The international public health crisis makes the automation and digital transformation of enterprises urgent*

During the international public health crisis, on the one hand, the fierce competition in the domestic labor market, the rapid spread of the virus make it difficult for enterprises to supplement the labor force in time, resulting in a shortage of personnel within enterprises; On the other hand, in order to minimize the possibility of virus transmission after the occurrence of crisis events, enterprises will take a series of epidemic prevention measures, such as keeping a safe distance between production lines and employees, and implementing necessary epidemic prevention procedures, which will delay the normal production plan, thus affecting the output and productivity of enterprises. Due to the fear of the virus, consumers seldom go offline during the crisis, prompting many enterprises to turn to online sales, which puts forward new requirements for logistics services. Once the crisis event with high uncertainty occurs, the enterprises with strong sales channels and good digital operation ability will be less affected, while the enterprises with a slow digital process will sound the alarm. Automatic production and digital operation will become the mainstream trend in the future, and logistics capacity will inevitably occupy an indispensable proportion. Therefore, as a catalyst, the international public health crisis can accelerate the original orderly progress of the digitization process and accelerate the improvement of the logistics system on the agenda.

4.1.3 *The global value chain is facing a huge impact under the international public health crisis*

International public health crisis, domestic economic activity was almost at a standstill. International trade, especially the raw materials, intermediates, and trade parts, was interrupted by the outbreak or delayed, causing damage to the global value chain and even rupture. Because of the expected profit decline, foreign investment also faces a great reduction in profits and atrophy in the global value chain. As shown in Figure 1, compared with 2019, the inflow of foreign direct investment in all economies in the world in 2020 decreased significantly, among which developed economies decreased by 68.63%, developing economies by 12.25%, transitional economies by 77.59%, and the world as a whole decreased by 42.42%. The COVID-19 epidemic indicates that China's status as "the workshop of the world" has been established globally, and "Made in China" has been adopted by the world. However, China's manufacturing halted abruptly after the outbreak, and countries around the world began to feel the dangers of too much interdependence. As a result, the impact of the external environment may lead multinational companies to reconsider the distribution of their manufacturing industry. In the mode of division of labor, they prefer the division of labor within the enterprise; that is, the whole production process is controlled within the enterprise (Tong et al. 2020). Therefore, in the context of the international public health crisis, the international logistics system, global supply chain and value chain system must be readjusted and repaired. It becomes more urgent to accelerate the improvement of China's international logistics system and enhance the status of the global value chain at this node.

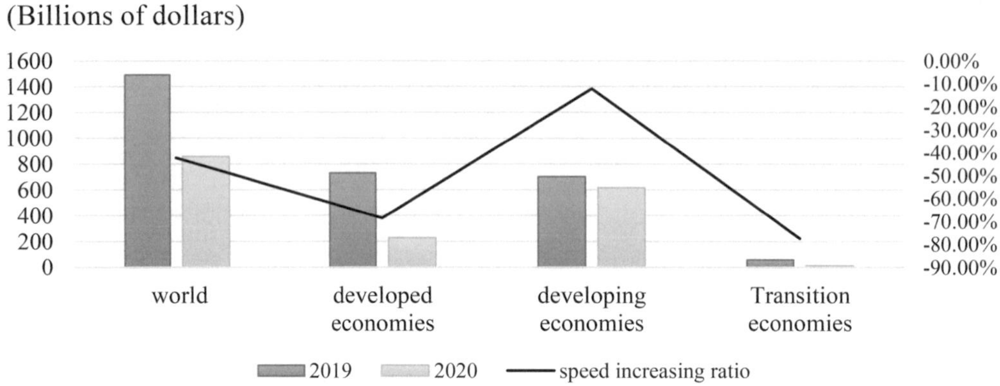

Figure 1. FDI inflows to economies.

4.2 *The mechanism of China's accelerating construction of an international logistics system under the international public health crisis*

The IHR defines an international public health event as a special event in which the spread of a disease across borders poses a public health risk to other countries and requires concerted international action. The COVID-19 outbreak swept over 200 countries and regions in a short period of time, with a huge impact on China and the world. On January 30, 2020, World Health Organization confirmed the COVID-19 epidemic as an international public health emergency. First of all, under the difficult situation that COVID-19 constitutes a global pandemic, all countries have actively taken epidemic prevention measures and basically stopped all non-essential domestic production and business activities. Second, international trade has been sharply reduced, and the global economy has slowed down during the epidemic. On the one hand, the upgrading of various epidemic prevention measures, such as restricting travel, isolating defense and strengthening virus detection, is of great significance for shortening the spread time of the virus and reducing economic losses. On the other hand, the continuous spread of the epidemic has basically brought the global economy to a standstill, and the international logistics service supply chain presents many problems such as poor cross-regional flow, sharp fluctuations in service prices, and unguaranteed timeliness of logistics (Liu 2020).

The modern international logistics system is composed of six subsystems, namely commodity inspection subsystem, transportation subsystem, storage subsystem, transport and unloading subsystem, circulation processing subsystem and information subsystem (Song et al. 2012). Therefore, this paper will discuss the mechanism of the international public health crisis promoting the construction of an international logistics system from inspection departments, logistics hubs, warehousing departments, transit centers, circulation processing enterprises and other departments, as shown in Figure 2. In order to effectively reduce the risk of the spread of the virus, the state has formulated strict epidemic prevention procedures during the period of the epidemic, and strictly checked the virus test results of import and export goods and people entering or leaving the country. In this way, in addition to testing the quality and specifications of products in the conventional procedures, virus detection takes more time in the whole process and greatly reduces the inspection efficiency. The transport links mainly include air, sea, land and other modes of transport. During the epidemic period, the number of medical supplies needed in the world has increased rapidly. The shortage of air cargo capacity and the low operational capacity have been caused by the reduction and suspension of many international routes. In terms of shipping, a number of shipping routes are forced to reduce or stop, coupled with the limited number of containers and the lack of effective scheduling, resulting in a significant rise in shipping costs; Inspection and disinfection stations for epidemic prevention and control have been set up across the country, which has lengthened the travel time of goods, impeded or even interrupted land transport, and delayed arrival of most logistics. In addition, international public health crisis events usually greatly impact international aviation and sea transportation, so land transportation will also face the problem of insufficient capacity. Due to the work efficiency of each link being significantly lower than usual, coupled with the limitation of the number of containers, logistics and warehousing links generally exist wharf warehouse bursting, lack of cabinets, the storage cost of goods greatly increased. During the outbreak of an international public health event, it is bound to lead to the shutdown of a large number of enterprises and the delay in the resumption of work and production. In addition, some workers are infected with the virus, resulting in the shortage of personnel in the logistics, transportation, unloading and distribution industries and the slow progress of goods inventory, inbound and outbound, and transshipment. During the epidemic, people's fear of the virus and unwillingness to travel greatly stimulated the development of e-commerce and online transactions. In this way, consumers will have higher requirements for product packaging design and logistics services. The packaging design of products should not only meet consumers' requirements of beauty, edible and safety, but also fully reflect the enterprise's trademark and brand value. This trend puts forward higher requirements for the circulation processing subsystem. Given the impact of the international public health crisis on all sectors, it is necessary for China to take prompt measures to put the construction of an international logistics system on the agenda.

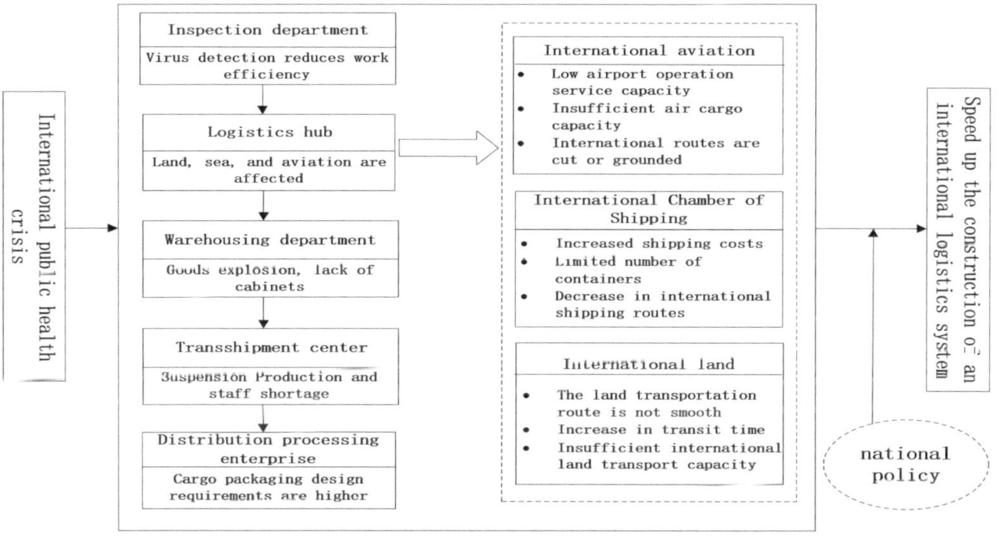

Figure 2. The mechanism of the international public health crisis boosting the construction of an international logistics system.

5 CONCLUSIONS

This paper reveals that it is imperative and urgent for China to accelerate the construction of an international logistics system since the international public health crisis restricts the development of international trade, accelerates the digital transformation of enterprises and has a huge impact on the global value chain. The occurrence of an international public health crisis will affect the good operation of all links in the international logistics system, from the inspection department, logistics hub, and warehousing department to the transfer center. Circulation processing enterprises are affected to varying degrees, resulting in the whole chain is not smooth or even interrupted. The transport link is the core link in the international logistics system. The international public health crisis mainly hinders the smoothness of the transport channel by affecting the air, sea and land transport modes. Based on this, the international community should actively introduce policies to respond and promote the implementation of the construction of an international logistics system. This paper puts forward some strategies to accelerate the construction of an international logistics system during the international public health crisis.

5.1 *Speed up the construction of international logistics infrastructure and build a comprehensive international aviation, maritime and land transportation network*

Based on the international logistics infrastructure, China will develop international air, sea and land multimodal transport, strive to reduce logistics and transport costs, and improve the reliability and efficiency of transport. Under normal circumstances, international public health crises will bring about huge casualties, and the number of medical supplies needed in epidemic areas is far from enough only by relying on local forces. At this time, various means of transportation need to be used to dispatch various medical supplies and daily necessities from all over the country and abroad. In terms of international aviation, there is still a gap between China and advanced countries and between Chinese logistics enterprises and international logistics giants, no matter from the perspective of transport capacity (number of cargo planes and logistics routes) or logistics nodes (aviation hubs) (Xi 2020). Sea transport and land transport are also the main modes of transport for China's import and export trade. The international public health crisis will severely impact sea transport and land transport through the rapid reduction of routes and trunk lines and the increase in costs. Therefore, to cope with the international public health crisis, it is essential to accelerate the construction of international logistics infrastructure and improve the international freight capacity. We will coordinate the investment orientation of the Asian infrastructure investment bank, BRICs New Development Bank and other financial institutions, so that they will give priority to the construction of international logistics facilities such as roads, railways, airports and ports. We should increase manpower, material resources and capital investment internally, promote the construction of freight hubs and stations such as airports, ports, stations, transshipment centers and storage centers, expand air routes to expand the scope of international logistics, and attach importance to the organic connection of various transportation modes, so as to create a comprehensive international air, sea and land transportation network.

5.2 *Develop an emergency strategic resource reserve plan to improve the flexibility of responding to crises*

Emergency strategic resource reserve usually includes manpower support, financial support, material support, facility support, technical support and special support. The establishment of a strategic resource reserve system for emergency response is an important guarantee for dealing with emergencies. In order to cope with emergencies such as international public health crises, China should establish and improve the emergency support mechanism for dealing with emergencies to ensure adequate preparation of personnel, materials and facilities. When an emergency occurs, coordinate and share resources to minimize losses. For China to construct the international logistics system, the role of the emergency logistics system cannot be ignored. The establishment of an emergency logistics system can effectively reduce the transportation time of relief materials and provide strong support for emergency rescue work (Xiong et al. 2015). To construct an emergency logistics system that can respond flexibly to crisis events, the government must first issue relevant laws and regulations as the guarantee for the construction and implementation of emergency logistics. Secondly, emergency logistics is not to establish a completely independent logistics system, but to fully coordinate and integrate the normal logistics system to deploy emergency personnel and materials effectively. In addition, it is also crucial to build an emergency logistics data platform by using modern information technology, which can efficiently integrate information from all aspects of society and quickly achieve the precise matching between logistics emergency demand and emergency supply.

5.3 *Develop logistics technology to form an efficient automated logistics system*

Logistics technology is a necessary means to realize logistics modernization, and the most important development trend of logistics technology is informatization. Large foreign logistics companies have achieved rapid development in logistics informatization. Information technology runs through all aspects of logistics activities, such as supply chain management systems and supplier inventory management systems, which are widely used in foreign logistics companies. The World Port of UPS, an American logistics giant, is the largest logistics automation hub in the world. Its high working efficiency enables it to carry a huge number of flights and packages every day. In contrast, China's network information technology has only developed rapidly in recent years, and the impact of various factors leads to the low level of information on international logistics. The degree of logistics informatization determines the ability to integrate logistics information resources, long-distance communication and decision-making, which is an important factor for China to develop the international logistics industry. With the continuous expansion of cross-border e-commerce transactions in China, China should accelerate the improvement of Internet infrastructure and the construction of a global logistics network, and improve the intelligent level of cross-border e-commerce platforms in logistics information traceability. In recent years, the extensive use of artificial intelligence in various industries is also urging China to accelerate the development of logistics technology and form an automated logistics system with unmanned driving and unmanned distribution as soon as possible.

5.4 *Actively promote international logistics cooperation and promote the establishment of unified international logistics standards*

Unified international logistics standardization is to regard international logistics as a large system, and realize the unification of technical standards of technical equipment, internal facilities, special tools and work standards of packaging, storage, transportation and transshipment in each subsystem of the system. With the deepening of economic globalization, international cooperation and international trade are playing a more and more important role in national economic development, and international logistics is the basic condition for the smooth development of international trade. Therefore, in order to better develop international business, all countries attach much significance to the standardization of domestic logistics and international logistics. The establishment of a unified international logistics standard can reduce logistics costs, break the trade barriers caused by inconsistent standards between countries and regions, and accelerate the entire international trade logistics process. Therefore, while promoting international cooperation with other countries, the government should also timely promote the research, formulation, and promotion of international logistics standardization to form a standard system that can accelerate the process of domestic logistics internationalization as soon as possible. Relying on the cooperation between China and other countries and regions in the world, China will comprehensively layout international logistics channels such as the "One Belt and One Road" international logistics channel, the RCEP East Asia core logistics network, and the logistics channels among G20 countries, so that a smooth flow of international air, sea, land and other all-round logistics can be realized.

5.5 *Cultivate professional talents in logistics and improve the operation and management level of the international logistics system*

Modern international logistics professionals should constantly learn the cutting-edge service and management ideas of the industry, lay a good foundation in professional knowledge and skills, master international settlement, transportation, insurance, inspection, customs declaration and other related knowledge and business operations, so as to meet the requirements of talent quality in the rapid development of the international logistics industry. The government departments should study the training plan of international logistics professionals as soon as possible and form a systematic training model for international logistics talents as soon as possible. Logistics enterprises should actively explore their talent training pattern in practice, enhance the induction training of relevant employees, and regularly organize professional logistics knowledge learning to enhance the theoretical knowledge base of logistics employees. Schools should set up specialized courses on logistics, understand the demand for professional logistics talents in today's society, formulate corresponding teaching plans and establish a perfect logistics discipline system, so as to improve the operation and management level of China's international logistics system.

ACKNOWLEDGMENT

This research was supported by the Zhejiang Province Soft Science Research Project (No. 2020C35067) and a grant from Hangzhou Philosophy and Social Science Foundation Project (Hangzhou Philosophy and Social Science Key Research Base: Hangzhou Wanxiang Polytechnic "Zhejiang Manufacturing Internationalization Research Center") (No. 2019JD64).

REFERENCES

Chu, A.N., Xu, Y.X., Lu, Q.J. (2014). Research on my country's Public Health Crisis Management Response Mechanism. China Health Policy Research, 07: 50–55.

Huang, H. (2017). Analysis of the impact of the "Belt and Road" initiative on international logistics. Business Economics Research, 23: 90–92.

Li, L. (2020). China's logistics industry under the new crown pneumonia epidemic: challenges and opportunities. Logistics Technology, 05: 126–128.

Liang, Y.F. (2020). Analysis of the impact of the new crown pneumonia epidemic on the world economy. International Economic Cooperation, 02: 4–11.

Liu, L.H. (2020). Research on the impact of the new crown pneumonia epidemic on the operation of the international logistics service supply chain. Supply chain management, 09: 5–13.

Liu, Z.B. (2020). The impact of the new crown pneumonia epidemic on Chinese industries: characteristics, risks and policy recommendations. Southeast Academic, 03: 42–47.

Shi, X.L. (2020). The impact of the new crown pneumonia epidemic on China's logistics and global supply chain and its countermeasures. Logistics Research, 01: 11–16.

Song, X.Y., Xiao, F.L., Zhang, R.S. (2012). Research on the Planning of Modern International Logistics System under the Trade Orientation. Logistics Technology, 15: 158–159+162.

Sun, Y. H. (2004). Construction of China's International Logistics System. Logistics Technology, 02: 14–15+27.

Tong, J.D., Sheng, B., Jiang, D.C., Yan, B., Dai, J.P., Liu, C. (2020). The global economy under the impact of the new crown pneumonia epidemic and the challenges to China. International Economic Review, 03: 9–28+4.

Wang, Q. (2020). Intelligent transformation of logistics industry based on 5G technology. Business Economics Research, 08: 136–139.

Xi, C.B. (2020). Challenges and countermeasures faced by international logistics under the epidemic. Logistics Technology and Application, 05: 58–62.

Xiong, X.K., Kang, G., Wang, Y. (2015). Construction of emergency logistics system based on collaborative operation mechanism. Logistics Technology, 06: 152–154.

Xu, D. (2015). Problems and countermeasures in the development of my country's international logistics. Price Monthly, 04: 52–54.

Xue, L., Zhu, Q. (2003). International References for Crisis Management: Taking the U.S. Public Health Emergency Response System as an Example. Chinese Administration, 08: 51–56.

Zhang, L.N., Shen, X.L. (2015). Research on International Cooperation in Cross-Regional Public Crisis Response: Taking Ebola Crisis Response as an Example. Administrative Forum, 06: 89–93.

Zhang, X.H. (2020). The impact of the new crown pneumonia epidemic on my country's small, medium and micro enterprises and its response. China Circulation Economy, 03: 26–34.

Zhong, Y., Chen, P. (2020). The impact of the new crown pneumonia epidemic on China's macroeconomy and the countermeasures. Theoretical Discussion, 03: 85–90.

Advances in Urban Engineering and Management Science – Khalil & Yang (Eds)

The feasibility research of post-occupancy evaluation of mega-event sports venue

Xianfeng Wu
China Architecture Design & Research Group Co. LTD, Beijing, China
Department of Architecture, Tsinghua University, Beijing, China

Haoyu Liu*
Nottingham University Business School, Jubilee Campus, the University of Nottingham, Nottingham, UK

Chang Liu & Xin Wan
China Architecture Design & Research Group Co. LTD, Beijing, China

ABSTRACT: Post-Occupancy Evaluation (POE) is an important theoretical tool for measuring the spatial effectiveness of built environments and is one of the important research areas of architecture. Due to a large amount of initial capital investment, high maintenance cost, complex management and negative impact on urban traffic, the construction of Mega-Event Sports Venue (MESV) is often considered a money-losing business. This paper describes the definition, characteristics, and existing problems of MESVs, systematically introduces MESVs from the perspective of sports development management, and presents the development, research methods of POE, as well as space efficiency evaluation of the built environment. The significance of POE for MESV is proposed. Finally, taking the Yanqing Competition Zone of the Winter Olympic Games as an example, the feasibility research of POE of MESV is discussed from the aspects of management and operation mechanism.

1 INTRODUCTION

Objectively speaking, from the planning and design of Winter Olympic venues to the mass sports orientation of promoting "300 million people participating in ice-snow sports", every step of innovation means success and every breakthrough represents progress. However, are these successes and advances in line with the concept of sustainable development in China's urban construction? Are their design, construction, operation, and maintenance at the same time taking into account the construction investment, environmental coordination, and sustainable development, rather than just seeking to fill up the domestic blank? This research introduces the definition, characteristics, and existing problems of MESVs from the perspective of sports development management, and puts forward the significance of POE for MESVs by introducing the development origin, significance, and research methods of POE theory as well as space efficiency evaluation of built environment. Finally, the feasibility of a periodical POE of the Yanqing Competition Zone of the Winter Olympic Games is elaborated on the aspects of management and operation mechanism.

2 THE SIGNIFICANCE OF MESV

Since the reform and opening-up of China, the development of sports has been on the fast track. The increasingly mega sports events were held in China (see Table **1**). The number of mega sports events increased in the recent ten years are, nearly twice of the last twenty years (see Figure 1). Behind the

*Corresponding Author: haoyu.liu@nottingham.ac.uk

DOI 10.1201/9781003305026-98

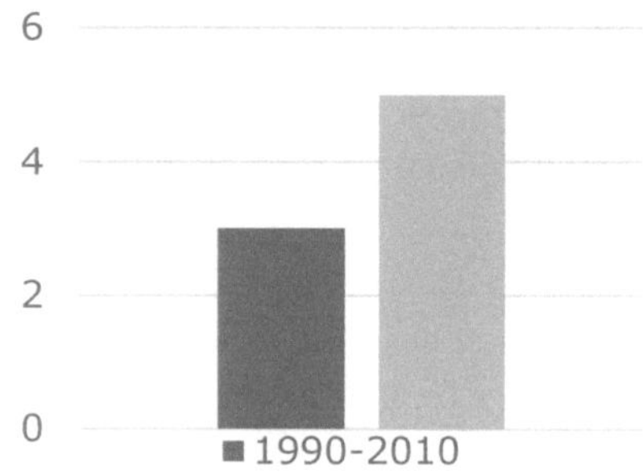

Figure 1. The number of mega sports events in China.

holding of this series of top-level events, the operation of capital plays a significant role in realizing political purposes, including stimulating city competition and marketing under the background of globalization as well as deepening rapid urbanization and sustainable development of cities in China. With the sustainable and rapid development of China's economy and the increasing improvement of people's living standards, cities have increasingly begun to actively hold or introduce sports events to meet the needs of people's diversified cultural lives. The advantage of MESV results from public welfare and its profitable nature. When public sports activities meet the needs of sports consumption of all members of the society, the quality and health of all members of the society are improved. Therefore, the construction and investment of mega-event venues are usually realized by the government macro-control and financial distribution. However, with the completion of the mega-event, how to transfer the venue from for mega-event to places for all remains an unsettled issue. Furthermore, the high consumption of the operation and maintenance, such as equipment renewal and manpower, material resources and financial resources, would be paid by the owner, which always becomes a great social burden. It is necessary to fully consider the comprehensive benefits of mega-event venues from the perspective of management to maintain normal operation and form a healthy and sustainable development path with the government's support.

Table 1. Mega sports events held in China.

Mega-event	1990	2008	2010	2011	2015	2019	2022
Asian Games	Beijing		Guangzhou				Hangzhou
Beijing Olympic Games		Beijing					
World Swimming Championships				Shanghai			
World Athletics Championships					Beijing		
FIBA Basketball World Cup						Multiple	
Beijing Winter Olympic Games							Beijing

3 THE EXISTING PROBLEM OF MEGA-EVENT VENUES

3.1 *Functional alienation and ambiguous standard*

At present, China's sports venue management has entered a new stage of market-oriented, making full use of various channels and ways to raise funds and establishing a market-oriented management mode. But this market-oriented benign operation mode is far from mature. First of all, according to the opinions, it should be clear that venues and facilities serving society must be open to the public to ensure that the nature of public welfare remains unchanged. The problem is that if the venues and facilities are operated in accordance with the public welfare, they will not survive. Therefore, in order to survive, the commercialization of venue facilities is becoming increasingly prominent. Therefore, the operational and functional mode of alienating venues and facilities for profit is prominent, which is obviously contrary to China's sports industry development. Besides, government public facilities have more distinctive functions compared with other competitive entities, venues, and facilities. If we only measure the operating benefits of venues and facilities from the perspective of revenue, and ignore, weaken or even lose or alienate the functions of venues and facilities in serving the society, this measurement standard should be worthy of review. It's better not to use the simple income to evaluate the operation of venues and facilities because when venues and facilities serve the society, it can't see economic benefits, but after the government purchases services, it provides service products to the society and the venues and facilities meet the social needs. Therefore, it is more realistic to establish a comprehensive evaluation system to evaluate the operation of a venue.

3.2 *The uncertain supporting policy of the government*

Although China's sports industry has been booming in recent years, the government's corresponding support policies on venue operation still remains uncertain. *"Public sports facilities should be open to the public and properly handle the relationship between public welfare and operation... We should build sports venues into healthy, scientific and civilized positions.* (CPC Central Committee and the State Council 2002)" It presents that the venues and facilities serving the society must be open to the public to ensure that the nature of public welfare remains unchanged. As the policy stresses, *"...we should improve the subsidy policy for the free or low fee opening of MESV and support the free or low fee opening of stadiums and gymnasiums to the society."* (General Office of the State Council 2020) That is to say, after nearly 20 years of policy research and feedback, the government has begun to try the necessary incentive mechanism to encourage the sustainable follow-up operation of MESV. This also means that the policy support and incentive policy of venue operation in China is still in the initial exploration stage. On the 14th five-year plan of China's development, we began to formulate a more detailed action plan for the construction of fitness facilities. But what is clearer is that after the construction and operation of a series of mega-event sports venues, China has realized the potential "white elephant" issue, and it has clearly put forward the construction policy of strict examination and approval, reasonable layout, and taking into account the community use of MESVs.

3.3 *Lack of professionals in sports management*

Due to the obvious lack of sports management and marketing knowledge in the knowledge structure of operators, and their ability to conduct in-depth research on sports market economic policies and theories, the operation and management of venues and facilities lack strength, the operation and management methods lag behind, the management level and effect are low, and the operation and management and operation and development cannot develop harmoniously (Lu & Zuo 2006). On the whole, the lack of professional management talents in the operation of stadium facilities has become one of the problems to be solved urgently. The operation of venues and facilities is a highly professional work, which needs compound talents who understand both sports and management to enter the work field. These talents are proficient in laws and policies, operation and development, blend competitive sports, mass sports, and sports industry operation mode, and carry out professional management and operation of venues and facilities in combination with the actual local situation. In addition, it is suggested that the professional management personnel of venues and facilities should implement special professional standards and the post should be certified and assessed, and the management of venues and facilities should be formally regarded as a special profession to recruit talents, so as to ensure the management level of venues and facilities.

4 MESV AND POE

4.1 *The development of POE*

In the 1980s, Preiser proposed a relatively complete theoretical basis for the POE of buildings, focusing on the needs of users, the advantages and disadvantages of design, and the performance of the built environment (Preiser & Vischer 2005). After more than 50 years of development, POE has been recognized and widely used by western countries. Its theory and method are further deepened and improved. It tends to be a complex subjective evaluation and comprehensive evaluation theory that combines soft and hard indicators. In the process of rapid development, a POE system that conforms to China's own national conditions is urgently needed. Meanwhile, it is necessary to maximize the practical value of theoretical research by combining the most cutting-edge research methods and technologies (Zhuang & Han 2019).

4.2 *The characteristic of POE on MESV*

First of all, the POE characteristics of MESVs should pay attention to their timeliness. The purpose and participants of mega-events in different stages are different. Therefore, the POE on MESV should be carried out in different periods. Secondly, the purpose of POE on MESV should be fully considered, which is to promote and improve the built projects, improve the projects under construction or guide the projects to be built. The content and method of POE should be set according to different purposes. In addition, we should keep pace with the times. With the development of society, the changes of the times, and the progress of science and technology, the corresponding standards should not remain unchanged, and the corresponding new research methods should also be boldly tried.

4.3 *The significance of POE on MESV*

The experience and quantity of the empirical research help create a systematic and holistic standard for MESV, from decision making and construction to management and operation. It is also helpful to change the situation of POE on MESV, which is normally under the control of each unit rather than authority with a general standard. The implementation of POE on MESV is indirectly beneficial to guiding the design, construction, and supervision. Other parties of MESV projects to pay attention to the quality of each phrase of MESV construction, stress the impact of the built environment of MESV on the occupier's behaviors and preferences and improve the quality of MESV. It contributes a benign circle to ensure a safe, comfortable, and stable built environment for people who use and enjoy the functions of MESV.

5 MANAGEMENT AND OPERATION MECHANISM OF POE ON MESV

5.1 *Implementing organization*

Implementation organization refers to who organizes POE on MESV, which is the first problem required to be solved in the specific implementation of POE. For MESV, POE of national planned investment projects can be organized by POE of national planning departments, and POE on MESV of provinces and municipalities can be organized by POE organizations of planning departments of provinces and municipalities. POE can also be organized by external organizations, such as consulting companies, design institutes, and university research institutions. At present, the research on POE on MESV in China is only carried out by designers or university research institutions that are related to the design of MESV. It has started to establish public building standards from industry associations in China. However, there is an urgent need and also a long way to go in terms of a top-down POE mechanism in the industry.

5.2 *Resources requirement*

Research of POE on MESV requires a series of resources, including personnel, investment, time as well as evaluation contents. POE is an extremely complex work that requires high-quality evaluation personnel. The POE group of MESV generally should consist of experts in economics, technicians, project managers, financial and statistical analysts, sociology or sports experts, etc. The participants involved in the evaluation are related to decision-makers, design consultors, construction teams, and users. Mega-event sports have generally required the government's support and a lot of public resources. In order to meet the expectations of the urban political elites, the evaluators generally fall into the dilemma. Even if the performance of sports events does not achieve a surplus in the short term, it will have a positive effect in the long term. For example, many scholars' research on the Olympic Games mostly concludes that sports events have a significant positive effect on urban tourism and can promote economic growth (Hiller 2010). To some extent, this kind of research exaggerates social value, misses some objective evaluations, and further impacts decision-makers' choices. For each specific project, due to the size of the project, complexity, the amount of manpower invested, and the specific requirements of the organization for POE content, the time requirements of POE on MESV are not completely consistent. Generally, it takes about three months from the proposal of the POE plan to the submission of the POE report. Taking too less time probably lacks depth and further affects the quality of the research. Getz (1997) believes that the POE of sports events can be divided into three aspects from the perspective of time: pre-evaluation, process evaluation, and post-evaluation (Getz 1998). For MESV, it can also be divided into periods. Taking Yanqing Competition Zone as an example, the 2022 Beijing Winter Olympic Games can be seen as a point on the axis, and the evaluation can be done in different periods for different purposes (see Figure 2).

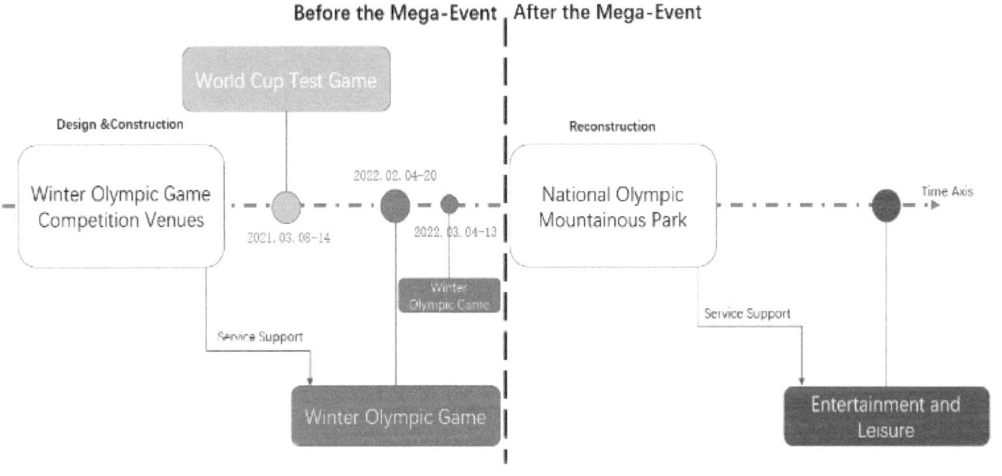

Figure 2. POE on Yanqing Competition Zone.

5.3 *The spatial configuration and performance perspective*

For MESV, the personnel is complex, and the streamlining is hard to be separated. It is necessary to consider how to classify the spatial configuration of the MESV for POE research. According to the management and operation logic, Yanqing Competition Zone can be divided into Front of House, Back of House, Field of Play, etc., which are mainly specific for its main participants, including spectators, staff, volunteers, players, etc. (see Figure 3). Furthermore, Building performance evaluation is also a key evaluation factor to measure whether a sports venue is sustainable operation and maintenance. It seeks to evaluate the MESV from six phases: planning, architectural programming, architectural design, construction, post-occupancy and reconstruction. However, the evaluation methods of the six phrases are different, and the operators and occupiers are also different. Therefore, the analysis for phrases is still carried out separately, and the feedback and optimization mechanism between each phrase is still relatively lacking.

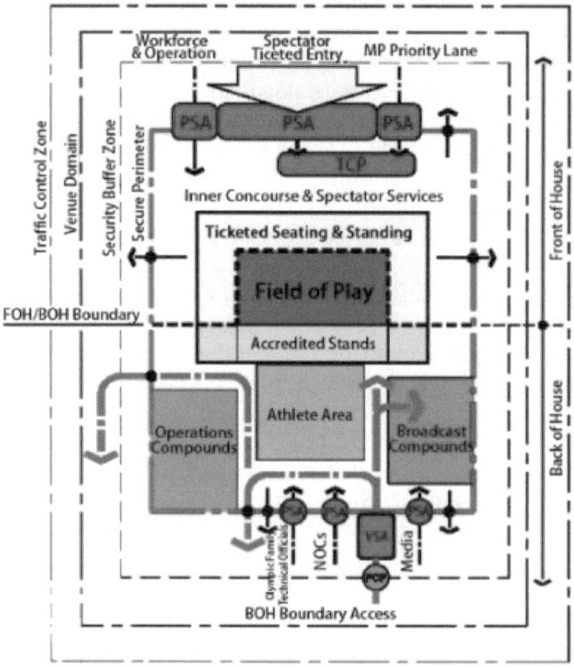

Figure 3. Venue planning principles.

6 DISCUSSIONS AND CONCLUSIONS

There are three aspects required to be stressed to solve the existing problems of POE on MESV, including confirming the functional and operational position of POE on MESV, reforming the governmental function of MESV management and operation and cultivating talents in operating management. A modified and revised national standard for POE on MESV would be very helpful to stimulate the development of POE on MESV from top to down. The research of POE on MESV can provide a comprehensive perspective to make evaluations and benefit life-cycle management. However, more related empirical studies on representative projects are imperative to prove the necessity of POE on MESV.

ACKNOWLEDGMENT

Project funded by Science and Technology Innovation Project of China Architecture Design & Research Group "Research on Key Technology and Post-Occupancy Evaluation of Winter Olympic Games Snow Venues" (Y2020199) & China Postdoctoral Science Foundation (2021M693038).

REFERENCES

CPC Central Committee and the State Council (2002). Opinions of the CPC Central Committee and the State Council on Further Strengthening and Improving Physical Education Work.
http://www.scio.gov.cn/xwfbh/xwbfbh/wqfbh/2011/0722/xgzc/Document/963599/963599.htm
General Office of the State Council (2020). Opinions of the general office of the State Council on strengthening the construction of national fitness venues and facilities and developing mass sports. http://www.gov.cn/zhengce/content/2020-10/10/content_5550053.htm

Getz D. (1998). Event Management and Event Tourism. *Annals of Tourism Research*, 25(1).
Hiller H. (2010). Mega-events, Urban Boosterism and Growth Strategies: An Analysis of the Objectives and Legitimations of the Cape Town 2004 Olympic Bid. *International Journal of Urban and Regional Research*, 24(2):449-458.
Lu J. & Zuo X. (2006). The management innovation of large sports venues in China. *Journal of Market Modernization*. 000(11Z), 52-53.
Preiser, W. & Vischer J. (2005) Assessing Building Performance. NY: Routledge
Zhuang W. & Han M. (2019). Post-Occupancy Evaluation: A review of methods and state-of-the-art techniques. *Time Architecture* (4).

Advances in Urban Engineering and Management Science – Khalil & Yang (Eds)
© 2023 The Author(s), ISBN 978-1-032-30426-7

Study on landscape renewal of old residential area based on microclimatic improvement

L.S. Cao*
School of Geography, Geomatics and Planning, Jiangsu Normal University, Xuzhou, Jiangsu, China

ABSTRACT: Urban microclimate research focuses on thermal comfort and microclimate effect, while empirical research on microclimate design considering summer and winter is limited. Taking the Kuiyuan community in Xuzhou as an example, ENVI-met was used to analyze the microclimate elements (air temperature, wind speed and humidity) and the characteristics of human thermal comfort PMV. Based on this, the utilization value of the site was judged, and the means of landscape renewal was proposed. Finally, the microclimate changes before and after landscape renewal were compared. It was found that the microclimate elements and thermal comfort were improved to varying degrees, which verified the feasibility and practical value of the landscape renewal strategy. This research provides a way to guide landscape design strategies through intuitive simulation data analysis.

1 INTRODUCTION

With the process of urbanization in China, the urban heat island effect has become a more and more serious environmental problem (Cheng et al. 2012). Landscape elements such as plants, water and topography play an important role in urban microclimate regulation, for their influence on solar radiation, humidity and wind, etc. (Altunkasa & Uslu 2020). At present, the research on urban microclimate is mainly on thermal comfort and microclimate effect, and taking into account the improvement of microclimate in summer and winter is relatively lacking (Cao et al. 2020). In summer and winter, people's needs for microclimate are completely opposite. The microclimate design research considers the two seasons to play an important role in improving the outdoor thermal comfort experience throughout the year (Liu et al. 2016).

Due to the early completion time, old residential areas often face problems such as high building density, few supporting facilities, and lack of outdoor activity space (Perini & Chokhachian 2017). At the same time, the aging of the residents in the old residential areas is serious, and the residents are sensitive to the climate. It is pointed out that the activities of the elderly are closely related to the outdoor climate conditions, which are more obvious in summer and winter (Liu et al. 2016). Taking the Kuiyuan community in Xuzhou City as an example, this study judged the design value of each public space by analyzing the microclimate factors and thermal comfort characteristics in summer and winter, then put forward landscape design methods based on microclimate improvement. Finally, the microclimate changes before and after landscape renewal in summer and winter were compared to verify the practicability of the design strategy.

2 COMMUNITY STATUS QUO

2.1 *Basic status*

Xuzhou City (34°15″W, 117°11″E) is located in the northwest Jiangsu Province, and it has a temperate monsoon climate with four distinct seasons, high temperature and rainy in summer,

*Corresponding Author: 6020180133@jsnu.edu.cn

 DOI 10.1201/9781003305026-99

frequent cold waves in winter, and an average annual temperature of 14°C. Kuiyuan community was built in 2000 and located in the south of Xuzhou City, with an area of 14.5 hm^2. The 6-story residential building is the main part of the community. It is a typical old community with a high aging population, lack of public space and greening. Excluding negative facility space (distribution station, garbage station, etc.), this study selected the midpoints of 10 public spaces named A-G as the coordinate reference points for the public space microclimate analysis. Because the areas of A and C are large, they are divided into 1 and 2 parts (Figure 1).

Figure 1. Current layout of the research object.

2.2 *Microclimate simulation*

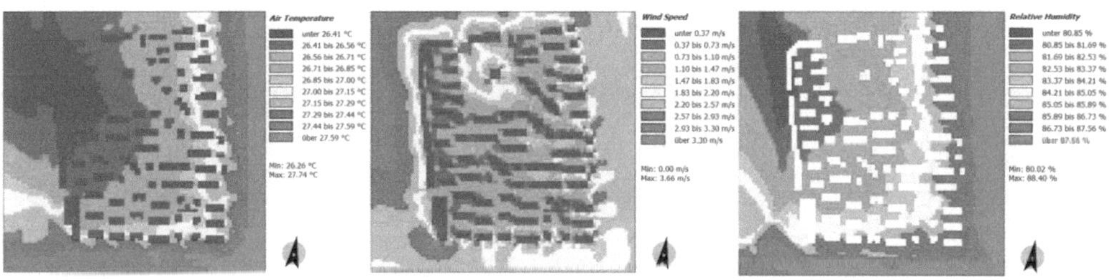

Figure 2. Analysis charts of microclimate simulation in summer (From left to right: air temperature, wind speed, humidity, The same below).

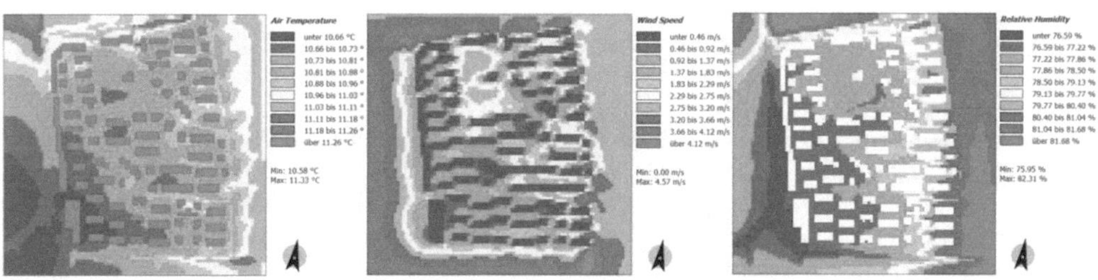

Figure 3. Analysis charts of microclimate simulation in winter.

In this study, ENVI-met software was used to simulate the microclimate effects such as air temperature, wind speed and humidity (Bouden & Ghrab 2005). According to the field investigation and design drawings, the site model was established in the ENVI-met spaces module, and the grid number was set to 100 ×100×30, the resolution was set to DX = 5 m, Dy = 5 m, DZ = 5 m. According to the meteorological data of Xuzhou over the years, the simulation date in summer was set to July 15, the wind speed was 2.8 m/s, and the wind direction was southeast 126°C. The temperature was 26.8°C, and the humidity was 74%. The winter simulation date was set to January 15, the wind speed was 3.6 m/s, and the wind direction was northwest 57°C. The temperature was 6.5°C and the humidity was 63%. The building facade material of the model was the system default material, and the underlying surface material was asphalt, hard pavement. The data output time was set at 15:00 when the residents' activities were frequent, and the microclimate effect was obvious, and the output height was set to 1.5m. The output result is shown in Figures 2 and 3.

In the summer, the air temperature was higher in the southeast and lower in the northwest, with a temperature difference of 1.18°C. The air temperatures in the B1, C, D, E and F five areas were relatively low. The wind speed was mainly between 0.73–1.83 m/s, and that in the B1 area was relatively high, in the D and E areas was relatively low. The narrow pipe effect had an obvious influence on the site wind speed. The humidity was high in the southeast and low in the northwest, and the humidity difference was about 7%. The humidity of D, E and F areas was relatively low, and that of the G area was the highest.

In the winter, the air temperature was higher in the northeast and lower in the southwest, with a temperature difference of 0.6°C. The air temperatures in the A1, and B1 areas were relatively high, while those in the C1, C2, E and F areas were relatively low. The wind speed was mainly between 0.9-3.2 m/s. Due to the narrow pipe effect, the wind speed in some areas was greater than 3.66 m/s. The humidity of the site was high in the east and low in the west, with a difference of about 5.1%, The humidity of the A1, B2 and G areas were relatively high, and that of the C1, D and E areas were relatively low.

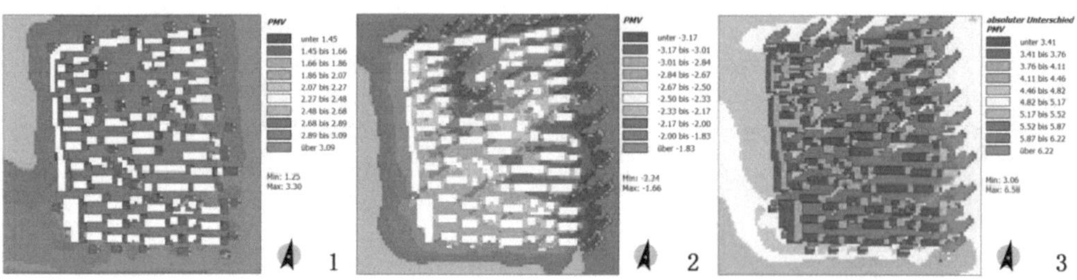

Figure 4. PMV analysis chart and superimposed chart of summer and winter.

2.3 *Thermal comfort simulation*

PMV (Predicted mean vote) was selected as the evaluation index of thermal comfort in our study (Djongyang et al. 2010). Using the Biomet module of ENVI-met, PMV at 1.5m height at 15:00 on the summer solstice and winter solstice was calculated, and then PMV values of the midpoint of 10 common spaces were extracted.

In summer (Figure 4-1), the PMV values were mainly between 2.68–3.09, which are generally high. The PMV values in the shadow of the building were significantly lower, showing that shading has a significant impact on PMV. The PMV values of D, F and G areas were lower, and these areas are cooler in summer. On the contrary, the thermal comfort experience in A, B and C areas was poor and needed improvement. In winter (Figure 4-2), the range of PMV values was −1.83 to −3.17. Large-scale building shadow has a significant effect on PMV. The PMV values of A1, C2, D and E were greater than – 2.00, showing good thermal comfort. The PMV values of other areas were lower than −2.50, and the thermal comfort needs improvement.

The PMV superposition analysis of summer and winter is shown in Figures 4-3. The values in the chart were the sum of the absolute values of PMV of the two seasons. The smaller the value, the better the comprehensive thermal comfort evaluation results of the two seasons. It can be seen that the sum of the absolute values of PMV in the A1, C2, D, E, F, and G areas was lower than 4.82, which is significantly lower than in other areas. These areas had a better thermal comfort experienced throughout the year.

3 LANDSCAPE RENEWAL STRATEGY

3.1 *Overall design*

The site is divided into three areas, A, B and C, by the roadway, and the three public spaces are located in the core of the three areas, which are suitable to be set as the main activity space. But their microclimate conditions are poor, such as high wind speed in winter in areas A1 and C2 and lack of shading in summer in areas A2, B and C1. Although the areas of D, E and G are small, the microclimate condition is relatively good, which can be considered as rest space. In summary, the analysis of the spatial utilization value of the research area is shown in Figure 5.

The circular footpath system was set on the site, and the results of thermal comfort evaluation are considered in the design of the footpath. The footpath was more tortuous in the area with a suitable microclimate and long sunshine time in winter, and the rest facilities were added in D, E and G areas to extend the resident's stay time. The footpath was relatively straight in the area with poor microclimate conditions to meet the needs of residents' rapid traffic. In order to cope with the decline of cognitive ability of the elderly, the striking red pavement was used on the footpath. The length of the slow footpath is about 1.3 km, connecting three main activity spaces, A, B and C, in series.

Figure 5. Analysis of spatial utilization value.

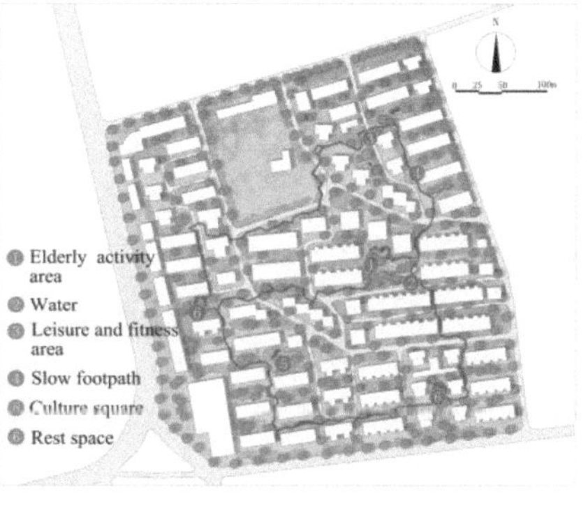

Figure 6. Landscape renewal design plan.

3.2 *Landscape elements design*

The ART (Aspect Ratio of Trees) was used to quantitatively describe the vegetation layout (Zhang et al. 2019). Koelreuteria paniculata, Liquidambar formosana and other local deciduous trees with large leaf area index were selected to form an effective shading effect and provide space for activities under the forest ($1 \leq ART < 2$). Deciduous trees can also ensure the sunshine demand of residents in winter. "Windproof in winter and ventilation in summer" was the principle of space design. Based on the vector map of the winter monsoon, the vegetation landscape was designed in the form of "arbor + shrub + grass" in the northeast direction of each activity space. Evergreen shrubs such as Euonymus ammodendron and Photinia serrulata with a height of more than 1.8m are densely planted to form an effective windbreak. A 6m wide ventilation channel was designed in the upwind direction of summer prevailing wind, and trees with a smaller leaf area index were sparsely planted ($ART \geq 2$). The fountain was designed in the core landscape area to increase the humidity of the site. The antiskid and permeable pavement was selected, which can not only enhance the water permeability of the site, but also affect the site temperature through transpiration in summer. The plan after landscape renewal is shown in Figure 6.

4 COMPARATIVE ANALYSIS

According to the landscape design strategy, plant, water, terrain, pavement and other design elements are added to the site model to build a new community ENVI-met model in the spaces module. Keeping the meteorological parameters unchanged, the microclimate of the site in summer and winter was simulated again, and the results are shown in Figures 7–8.

4.1 *Comparison of microclimate elements*

In terms of air temperature, the area of the low-temperature area in summer has increased significantly, and it can be seen that the air temperature has decreased significantly. The air temperature in the main activity space dropped by an average of 0.7–0.9°C. In winter, the average air temperature went up about 0.16 degrees, and the main activity space rose by an average of 0.3–0.5°C. Compared with summer, the improvement of air temperature in winter is not obvious.

In terms of wind speed, although we have adopted a ventilation design, the wind speed in summer was still reduced. The wind speed in the A2, B2 and C2 areas decreased by about 0.3 m/s. In winter, the effect of wind speed reduction was obvious, and the average wind speed in public space was reduced by 0.5 m/s −1 m/s. The high wind speed areas caused by the narrow pipe effect basically disappeared.

In terms of humidity, the humidity of the site overall increased by about 3% in summer, and the increase in area B was the most obvious. In contrast, landscape renewal has less impact on the humidity of the site in winter; the humidity only increased by about 0.7% in some areas.

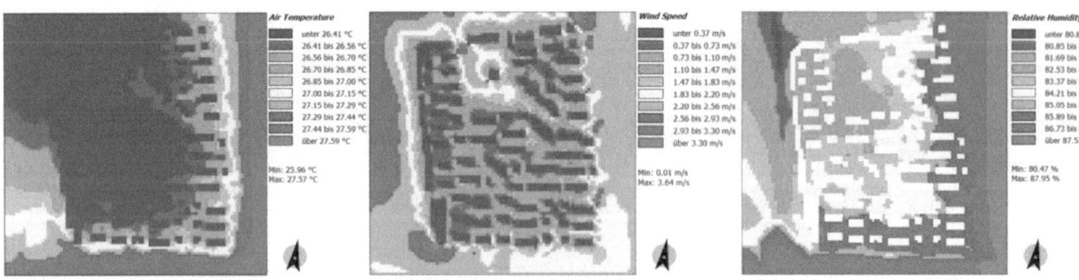

Figure 7. Analysis charts of microclimate simulation after landscape renewal in summer.

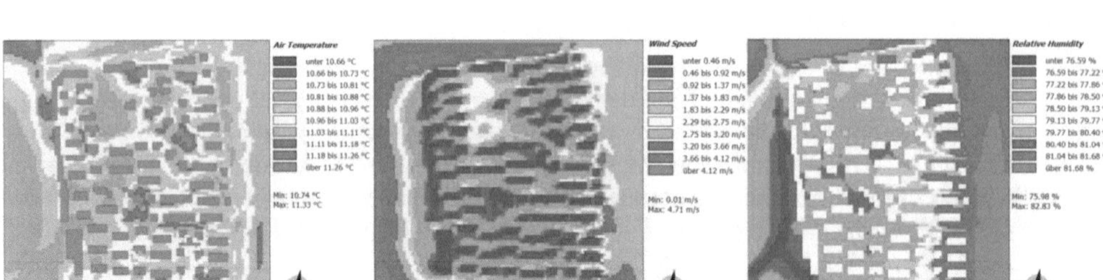

Figure 8. Analysis charts of microclimate simulation after landscape renewal in winter.

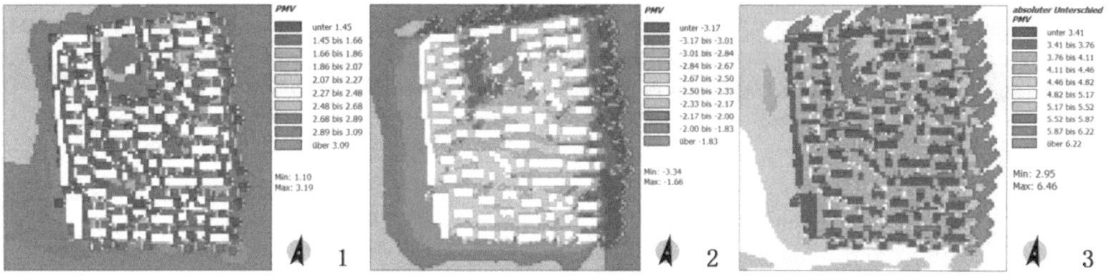

Figure 9. PMV analysis chart and superimposed chart of summer and winter after landscape renewal.

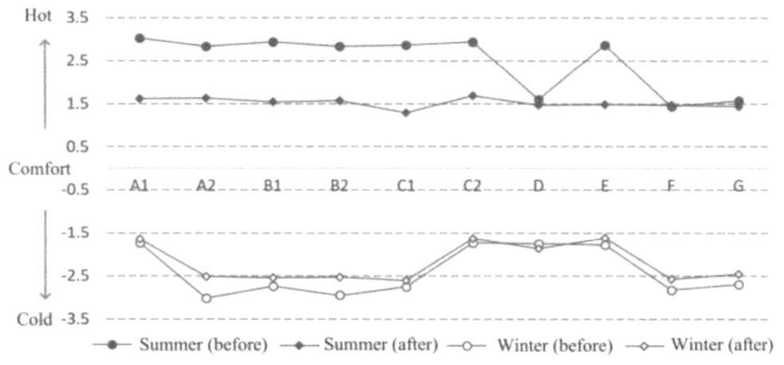

Figure 10. Comparison of PMV values.

4.2 *Comparison of PMV*

It can be found that the area of the blue region with lower PMV increased significantly in summer (Figure 9-1). The mean PMV value of the main public space decreased from 2.68 to 1.86, which indicated that the improvement effect of thermal comfort in summer was significant. In winter (Figure 9-2), the green area with a higher PMV value increased. The change range of PMV was improved from below −2.84 to between −2.84 and −2.50. Compared with summer, the improvement of thermal comfort in winter is weaker.

The change in PMV value of 10 public spaces is shown in figure 10. In summer, the PMV values of D, E and G three areas were lower before landscape renewal, and the change in PMV values was not obvious after landscape renewal. The PMV values of other public spaces are significantly reduced. The PMV values of other public spaces changed significantly, with an average decrease of 1.4. After landscape renewal, the difference in PMV value is small, and the thermal feeling tends to be the same. In winter, the value of PMV increased after landscape renewal, but it was not obvious. Except for the D area, the PMV values of other areas increased, ranging from 0.1 to 0.5. The PMV superposition analysis of summer and winter is shown in Figures 9-3. Compared with the previous, the areas of blue and green, which are more suitable for thermal comfort, were significantly larger. It can be seen that from the change in thermal comfort throughout the year, landscape renewal measures can effectively improve the overall microclimate conditions of the site.

5 CONCLUSIONS

Based on the analysis of microclimate elements and thermal comfort characteristics, this study judged the utilization value of each space and put forward the landscape renewal plan. Through the comparison of the landscape update before and after, it is found that the air temperature, wind speed and humidity in summer and winter have improved to different degrees. Among them, the improvement effect of summer air temperature and winter wind speed is the most obvious. The influence of landscape renewal strategy on microclimate in summer is stronger than that in winter.

Through the intuitive analysis of microclimate data to guide the next step of landscape renewal strategy, this study realized the docking between landscape design and microclimate regulation and verified the feasibility and practical value of landscape renewal strategy.

Comparing the improvement effect of climate elements and thermal comfort before and after landscape renewal, this study found that the shading effect of trees in summer and the windproof effect of shrubs in winter were the main reasons for the improvement of microclimate, which should be paid more attention in the future design. Although summer ventilation was considered in the design, the landscape renewal strategy still reduced the wind speed, which is also the main reason for the deterioration of local thermal comfort in summer. In addition, only 15:00 simulation data was used in this study. In future research, the simulation time can be extended to grasp more comprehensive microclimate change laws.

REFERENCES

Altunkasa, C. & C. Uslu (2020) Use of outdoor microclimate simulation maps for a planting design to improve thermal comfort. Sustain. Cities Soc. 57,102–137.

Bouden, C. & N. Ghrab (2005) An adaptive thermal comfort model for the Tunisian context: a field study results. Energy Build. 37(9), 952–963.

Cao, L. S. & H. Xu & H. Li (2020) Research on residential landscape design based on microclimate effect. Construction Science and Technology. 22, 84–88.

Cheng, V. & E. Ng & C. Chan & B. Givoni (2012) Outdoor thermal comfort study in a sub-tropical climate: a longitudinal study based in Hong Kong. Int. J. Biometeorol. 56(1), 43–56.

Djongyang, N. R. Tchinda & D. Njomo (2010) Thermal comfort: a review paper. Renew. Sust. Energ. Rev. 14(9), 2626–2640.

Liu, B. Y. & Y. Mei & W. Kuang, (2016) Experimental research on correlation between microclimate element and human behavior and perception of residential landscape space in Shanghai., Chinese Landscape Architecture. 32(01), 5–9.

Perini, K. & A. Chokhachian & S. Dong & T. Auer (2017) Modeling and simulating urban outdoor comfort: coupling ENVI-met and TRNSYS by grasshopper. Energy Build. 152, 373–384.

Zhang, L. & Q. Zhang & Y. Lan (2019) The cooling and ventilation effects of residential vegetation in Wuhan based on micro-climate simulation. Chinese Landscape Architecture. 35(03),92–95

Advances in Urban Engineering and Management Science – Khalil & Yang (Eds)
© 2023 The Author(s), ISBN 978-1-032-30426-7

Obstacle detection based on semantic segmentation for autonomous vehicles

Yunxiang Liu
Yunxiang Liu, Shanghai Institute of Technology, SIT, Shanghai, China

QianXun Guan
QianXun Guan, School of Computer Science and Information Engineering, Shanghai Institute of Technology, Shanghai, China

Xinxin Yuan*
Xinxin Yuan, Shanghai Institute of Technology, SIT, Shanghai, China

Yue Shi
Yue Shi, School of Computer Science and Information Engineering, Shanghai Institute of Technology, Shanghai, China

ABSTRACT: A convolutional neural network segmentation method is proposed to meet the real-time requirements of automated driving systems and the problem of fuzzy target boundary segmentation and inaccurate segmentation of similar objects caused by pooling and convolution operations in current semantic segmentation methods of traffic scenes. A model with an attention mechanism has been proposed to detect obstacles in traffic scenes. The model constructs a feature extraction block and introduces various extended convolutions in the feature extraction process to extract more detailed image feature information. After the information is decoded, dual-channel and spatial attention modules are added, and functional information of different scales is combined during up-sampling, making the semantic information richer. The test results show that the proposed lightweight semantic segmentation model has a mean intersection ratio (MIoU) of 71.5% on the Cityscapes dataset, achieving good obstacle detection results, and meeting the real-time requirements of autonomous vehicles.

1 INTRODUCTION

In recent years, with the development of computer vision, autonomous driving has also attracted much attention. In the case of self-driving cars, pedestrians, vehicles, and bicycles on the road can all be regarded as obstacles, which have a significant impact on the safe driving of the self-driving car. Semantic segmentation technology based on convolutional neural networks can detect and recognize these obstacles at the pixel level, and provide environmental information for decision-making, planning, and control of autonomous vehicles. In the field of autonomous driving, more and more researchers have begun to pay attention to the segmentation of road scenes. The image data obtained by autonomous driving is irregular, and statistics and analysis of these irregular data are required. It is almost impossible to process such a large amount of data in real time by relying solely on traditional algorithms. Therefore, the real-time segmentation of road scenes in the automatic driving process needs to be improved. Efficient and reliable semantic segmentation of traffic scenes is an urgent problem to be solved.

*Corresponding Author: xxyuan@sit.edu.cn

DOI 10.1201/9781003305026-100

Image semantic segmentation technology, as an important research direction of visual intelligent systems, is one of the key ways for intelligent systems to understand natural scenes (Rodriguez et al. 2017; Zhao et al. 2017). Semantic segmentation technology provides pixel-level image classification according to different semantic information represented in the image, and different colors represent different categories. If deep learning has not been applied to the field of computer vision, the classifier used by researchers to build semantic segmentation is usually the random forest method. Image block classification is the first commonly used deep learning method. Since the classification network usually has a complete connection layer, the input is a fixed-size image block, and the range of the sensing area is limited by the size of the image block. After that, he proposed a complete convolutional network (Long et al. 2015). It allows you to perform dense predictions without the need for a complete connection layer. Compared with the previous method, this method can generate segmentation maps of any size image, and the calculation speed is much faster. The main frameworks of semantic segmentation based on convolutional neural networks mainly include VGGNet and ResNet. Semantic segmentation models based on VGGNet include FCN, SegNet (Badrinarayanan et al. 2017), U-Net (Ronneberger et al. 2015) and DeepLab (Chen et al. 2018). Semantic segmentation models based on ResNet include PSPNet (Fang & Lafarge 2019), ICNet (Zhao et al. 2017) and DeepLab V3++ (Chen et al. 2018). In view of the structural characteristics of the ResNet network, the semantic segmentation network based on the ResNet framework has many layers. And the complex network structure requires high-performance hardware equipment. The semantic segmentation network based on the VGGNet framework is much simpler than the semantic segmentation network based on the ResNet framework, but it often cannot meet the real-time requirements. For self-driving cars, the real-time nature of the system is very important, and the accuracy of the segmentation results must also be taken into consideration.

Semantic segmentation methods require precise alignment of classification maps, so location information discarded by the pooling layer must be preserved. The pooling layer increases the receptive field of the upper convolutional kernel while converging the background, causing some of the positional information to be lost. In order to solve this problem, the researchers proposed two different forms of structure. The first is the encoder-decoder architecture. The encoder gradually reduces the spatial size through the pooling layer, and the decoder gradually restores the target details and spatial size. There is usually a gap between the encoder and the decoder, which allows the decoder to retrieve the details of the target better. The second is to use a structure called extended convolution (Koltun 2016). The goal is to extend the receiving field through deep convolution, reduce the use of pooling layers, increase the resolution of deep feature maps, and store more spatial information. With the development of research on semantic segmentation network models, DeepLab v1 (Papandreou et al. 2018) and DeepLab v2 (Papandreou et al. 2018) use conditional random fields to predict more precise information about target boundaries. Recently, DeepLabv3 + (Zhu et al. 2018) combines the encoder-decoder structure with the Atrous Spatial Pyramid Pooling module, which allows the encoder-decoder structure to obtain more information about the limits. ASPP can get more functional information, which is an area of semantic segmentation.

2 MATERIALS AND METHODS

Based on the DeepLabv3 + network structure, this article builds ResNet50 with extended convolution in the backbone network. Due to the model DeepLabv3 + also has shortcomings: In the process of segmentation of the model DeepLabv3 +, the target area and features do not focus on learning, cannot distinguish which features of the image and which area is an important feature and not an important feature, resulting in easy missing the region of interest. Therefore, under the principle of not increasing the number of parameters, the introduction of extended convolution allows the subsequent convolution layer to keep a large feature size, in order to achieve the goal of protecting the spatial resolution of the network, which is conducive to the detection of small targets and improve the overall performance of the model. By adding the attention module behind the convolution layer in the 3×3 convolution block of the decoder module and learning together with the main branch, the network can learn important types of features and regions of interest on the image without increasing the system time and drive parameters. The overall structure of the network is shown in Figure 1.

In the encoding part, DeepLabv3 is used as an encoder, and the ResNet50 structure with extended convolution is applied to gradually reduce the resolution of the feature map and capture high-level semantic information. In order to solve the problem of the scale diversity of segmentation objects, an ASPP module is added to the model, which uses empty convolution and average global pooling, 1 × 1 convolution, and a combination of expansion rates. 6 to encode the picture context information, respectively It is 12 and 18.

In the decoding process, a simple but efficient module is used to retrieve the details of the target boundary: 3 × 3 convolution is used to refine the features, then a simple bilinear 4-fold oversampling; in ASPP and Decoder modules, use Deep Separable Convolution to reduce the number of parameters and resolution of features extracted by the deep convolution neural network by controlling the convolution of the cavity to balance precision and execution time.

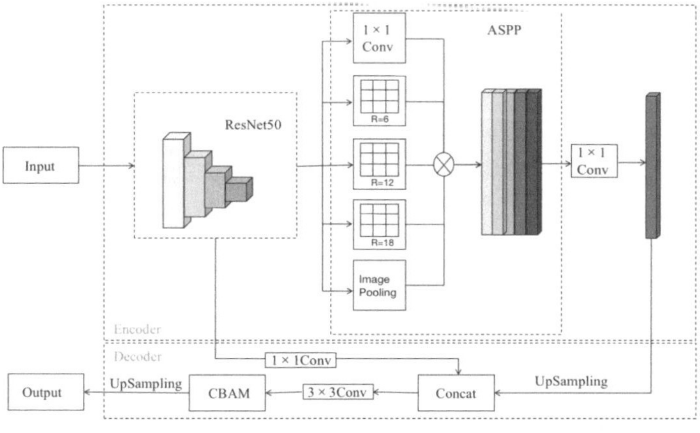

Figure 1. Improved traffic scene image segmentation structure.

Since the background signal of the traffic scene is complex, the problem of extracting the signal of interest from the interference network may arise. And many target details lost due to downsampling can hardly be recovered during the training process. According to the characteristics of traffic scene image segmentation tasks, extended convolution is introduced to avoid these problems. It can directly control the resolution of the feature map calculated by the deep convolutional neural network, and maintain the spatial resolution of the feature map in the convolutional network for image classification. Compared with the traditional convolution operation, it does not reduce the feature map, but increases the receptive field. As a result, the feature map loses as little information as possible when segmenting the target boundary and does not introduce high computational complexity. For a two-dimensional signal, the extended convolution formula applied to the input feature map is as follows:

$$y[i] = \sum_{k} x[i + r \cdot k]\omega[k] \tag{1}$$

where ω[k] denotes the convolution kernel with length k, y[i] denotes the output of the extended convolution, and r is the sampling ratio.

In semantic segmentation of traffic scenes, the main goal of the attention mechanism is to select more critical information from a lot of information for the purpose of the current segmentation task. Attention should not only indicate to the semantic segmentation model traffic scenes where it is concentrated, but also improve the expressiveness of the points of attention. The Attention Mechanism is used to increase model performance, focus on important features, and remove unnecessary features. The dual attention mechanism module used in this article is a lightweight attention module, as shown in Figure 2. It uses the attention module in space and channel simultaneously and can learn the characteristics, key targets, and regions of interest in channel and space dimensions. In addition, the efficient flow of information within the network benefits from learning what information this module should be underlined or removed. With the channel attention module and the spatial attention module, better results can be obtained than focusing only on the channel attention mechanism or the spatial attention mechanism.

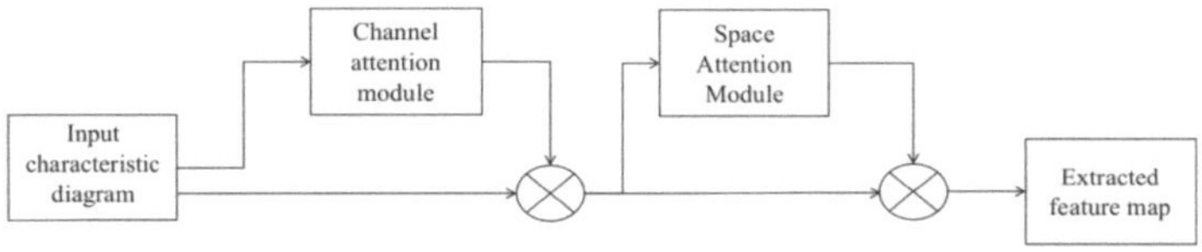

Figure 2. Attention model.

3 RESULTS & DISCUSSION

This experiment uses the deep learning framework Pytorch to complete training and testing in the Pytorch environment. The programming language is python 3.6. Based on the Linux system, the running platform is ubuntu 16.04. The experimental hardware is Intel Xeon E5-1650 v4 clocked at 3.6 GHz, with a memory of 32 GB, GPU Nvidia Geforce GTX 1080 Ti, with a memory of 11 GB. Pytorch has some framework features specifically designed for deep learning. Using the simple and easy-to-understand model algorithm interface provided by Pytorch, you can quickly build a neural network. After declaring the tensor matrix, the gradient can be calculated automatically. A lot of calculation codes are omitted, which improves learning efficiency. In the network training, the transfer learning method is adopted to transfer the features learned by the ResNet50 network trained on the image classification data set to the optimized traffic scene semantic segmentation model based on DeepLabv3+ with attention mechanism, and the convergence speed is faster.

In this test, the Cityscapes dataset contained 5,000 images of urban driving scenes, including street scenes from 50 different cities. Among them are 2,975 training images, 500 verification images, and 1,525 test images. This document selects 19 commonly used categories for classification. The neural network is trained using the training image and the test set labels that are not public. In this paper, only the validation set is used to evaluate the performance of the improved model. To meet the input requirements of the model, the resolution of the input image is adjusted from 2048 × 1024 to 513 × 513, and the image data is enhanced by random cropping, color shake, and rotation.

For semantic segmentation, there are generally four indicators to evaluate its performance: Pixel Accuracy, Average Pixel Accuracy, Mean Intersection over Union and Frequency Weighted Intersection over Union. Among them, the average intersection-and-merge ratio is simple and intuitive, which is one of the most-commonly-used important indicators to measure the accuracy of image segmentation. It is the ratio of an intersection-and-merge set of the model's prediction results for each type and the real value and the result of summation and re-average. In each group of experiments in this paper, MIoU is used as the evaluation standard of model performance. The specific formula is as follows:

$$MIoU = \frac{1}{k+1}\sum_{i}^{k}\frac{p_{ii}}{\sum_{j}^{k}p_{ij}+\sum_{j}^{k}p_{ji}-p_{ii}} \tag{2}$$

where P_{ij} denotes the false positive, P_{ji} denotes the false negative, and P_{ii} denotes the real number.

Table 1 compares the performance of DeepLabv3+ in terms of segmentation accuracy, storage space occupation and time performance before and after enhancement, and verifies the effect of CBAM and ResNet50 on the segmentation effect. When the original backbone network is upgraded to ResNet50, MioU is 8.5 percentage points higher than the original model; using this method, MioU is nine percentage points higher than the original model. Therefore, the CBAM structure and ResNet50 improve the segmentation accuracy of the model to varying degrees. Compared with the original model, adding CBAM has little effect on storage space occupation and can be ignored. Adding a lightweight CBAM module hardly increases the time consumption of model prediction results. It does not bring about a large amount of storage space and time consumption while improving the segmentation accuracy, which verifies the efficiency of the attention module. However, the use of ResNet50 in the backbone network will greatly increase the time consumption, which is a problem that needs to be further solved. For example, the head margin in this model is larger than usual. This and other measures are well thought out, and the norms use your article as part of the overall process rather than as a stand-alone document. Please do not modify any current names.

Table 1. Comparison of performance indexes of three models.

Model	MIoU/%	memory space/MB	mean time/ms
DeepLabv3+	62.6	4671	250
DeepLabv3++ResNet50	71.3	7552	576
DeepLabv3++ResNet50+CBAM	71.5	7552	578

4 CONCLUSIONS

This paper designs an optimized traffic scene image segmentation model based on DeepLabv3+, with an attention module. Higher-level semantic information is retained without increasing the number of parameters and changing the receptive field of the convolutional layer at each stage. By learning channel information and location information, a convolutional neural network with higher accuracy and richer features can be obtained when recovering boundary information. Adding CBAM to the DeepLabv3+ network enables the network to learn important feature types and regions of interest in the image, almost without increasing time and memory consumption. The convolutional layer behind the ResNet50 network introduces extended convolution to maintain a larger feature size while retaining more high-level semantic features, so that the network can better understand the content of the traffic scene image. The experiments in this paper show that the introduction of dilated convolution and attention modules improves the semantic segmentation accuracy of the network to varying degrees. Compared with the DeepLabv3+ network, the network proposed in this paper has higher accuracy, especially in the accuracy of small object segmentation and approximate object segmentation. The MIoU of the network reached 71.5%, which is 9% higher than the original DeepLabv3+ network. Subsequent work will consider model compression, optimize the improved DeepLabv3+ network, and improve the segmentation accuracy while considering time and memory consumption to achieve the lightweight of the convolutional neural network.

REFERENCES

Badrinarayanan V, Kendall A, Cipolla R (2017). SegNet: A Deep Convolutional Encoder-Decoder Architecture for Image Segmentation[J]. IEEE Transactions on Pattern Analysis & Machine Intelligence, 2017:1–1.

Chen L C, Papandreou G, Kokkinos I (2018), et al. DeepLab: Semantic Image Segmentation with Deep Convolutional Nets, Atrous Convolution, and Fully Connected CRFs[J]. IEEE Transactions on Pattern Analysis and Machine Intelligence, 40(4):834–848.

Chen L C, Zhu Y, Papandreou G (2018), et al. Encoder-Decoder with Atrous Separable Convolution for Semantic Image Segmentation[J]. Springer, Cham.
Chen L-C, Zhu Y, Papandreou G (2018), et al. Encoder-decoder with atrous separable convolution for semantic image segmentation [C] // Proceedings of the 2018 European Conference on Computer Vision. Cham: Springer, 2018: 833–851.
Chen L C, Papandreou G, Kokkinos I (2018), et al. DeepLab: Semantic Image Segmentation with Deep Convolutional Nets, Atrous Convolution, and Fully Connected CRFs[J]. IEEE Transactions on Pattern Analysis and Machine Intelligence, 40(4):834–848.
Fang, H., & Lafarge, F. (2019). Pyramid scene parsing network in 3D: Improving semantic segmentation of point clouds with multi-scale contextual information. ISPRS Journal of Photogrammetry and Remote Sensing, 154, 246–258.
F Yu, Koltun V (2016). Multi-Scale Context Aggregation by Dilated Convolutions[J].
Garcia-Garcia, A., Orts-Escolano, S., Oprea, S., Villena-Martinez, V., & Garcia-Rodriguez, J. (2017). A review on deep learning techniques applied to semantic segmentation. arXiv preprint arXiv:1704.06857.
Long J, Shelhamer E, Darrell T (2015). Fully Convolutional Networks for Semantic Segmentation[J]. IEEE Transactions on Pattern Analysis and Machine Intelligence, 39(4):640–651.
Ronneberger O, Fischer P, Brox T (2015). U-Net: Convolutional Networks for Biomedical Image Segmentation[J]. International Conference on Medical Image Computing and Computer-Assisted Intervention.
Zhao H, Qi X, Shen X (2017), et al. ICNet for Real-Time Semantic Segmentation on High-Resolution Images[J].

Author index

Bai, L. 546
Bai, Y. 584
Bao, L.X. 165
Bo, T. 312

Cao, L.S. 744
Chan, H. 15
Chang, G.X. 577
Chen, C. 171
Chen, E.Y. 490
Chen, F. 66
Chen, G. 338
Chen, J. 91, 300, 332, 562
Chen, K. 656
Chen, L. 149, 350, 463
Chen, R. 357
Chen, S.Q. 528
Chen, Y. 3, 66, 615
Cheng, J. 546
Cheng, L. 206
Cheng, M. 344
Cheng, Y. 57
Cheng, Z. 719
Chuan, S. 428, 433
Chul-Soo, K. 243
Cui, Y. 216

Danzhen, W.J. 200
Ding, J. 656
Dong, C. 357
Du, X. 369
Du, Y. 237
Duan, C. 615
Duan, H. 171

Fan, J. 678
Fan, X. 411
Fan, X.-F. 157
Fu, P. 463

Gao, Y. 546
Gong, J.-Y. 157
Guan, Q.X. 751
Guo, F. 671
Guo, Q. 703
Guo, X. 258

Han, S. 562
Han, X. 99, 116, 664
Hao, K. 363
He, T. 15
He, Y. 711
He, Y.X. 490
Hou, Y. 499
Hu, C.-M. 157
Hu, F. 149
Hu, H. 728
Hu, Y. 75
Huang, C. 584
Huang, J. 37, 42
Huang, W. 132
Huang, Y. 206, 631
Huangfu, T. 99

Ji, N. 165
Ji, X. 438
Jia, H. 447
Jia, N. 269
Jiang, J. 15, 81
Jiang, L. 615
Jiang, Y. 165
Jiao, Y. 615
Jin, H. 584

Khan, M.F. 263

Lan, M. 553
Lei, Y. 216
Li, G.H. 200
Li, J. 300
Li, J.F. 269
Li, M. 15, 288
Li, P. 66, 81
Li, Q. 139, 719
Li, S. 637
Li, W. 528
Li, X. 9, 281
Li, Y. 237
Li, Z. 269, 615
Liang, J. 622
Liao, F. 75
Lin, C. 631
Lin, W.Y. 32
Liu, B. 52
Liu, C. 226, 737
Liu, C.B. 398
Liu, D. 463
Liu, H. 226, 649, 737
Liu, J. 99, 116, 332, 389, 664
Liu, K. 139
Liu, L. 185, 649
Liu, R. 539
Liu, S. 546
Liu, T. 482
Liu, X. 319
Liu, Y. 251, 482, 507, 751
Long, C. 631
Lou, S. 600
Lu, J.T. 528
Lu, L. 584
Lu, X. 369
Luo, N. 226
Luo, Q. 490, 528
Luo, Z.Y. 398
Lv, J. 75
Lv, X. 251
Lv, Y. 216

Ma, B.B. 269
Ma, C. 52, 206
Ma, H. 281
Ma, J. 482
Ma, Y. 649, 678
Mei, D. 344

Nie, C. 263

Pan, L. 473
Piao, C. 258

Qi, C. 81
Qi, Z. 52

Qian, H. 86
Qiang, W. 312
Qin, L. 139
Qin, S. 572
Qiu, C. 463
Qiu, L. 622
Qiu, M. 216

Ren, J. 389
Ren, L. 678

Shao, P. 553
Shi, B. 363
Shi, J. 703
Shi, M. 608
Shi, Y. 312, 751
Song, Z. 251, 428, 433
Su, F. 376
Su, G. 631
Sun, G. 453
Sun, J. 319
Sun, L. 698
Sun, Z. 46
Syamsunur, D. 422

Tang, B. 32, 463
Tang, F. 562
Tang, L. 108
Tian, Z. 326, 382, 644
Tong, H. 132
Tong, J.X. 269
Tong, L. 376

Wan, X. 737
Wang, Q.L. 165
Wang, X. 422, 608
Wang, Y. 369, 473, 671, 719
Wang, Z. 350
Wei, C. 81
Wei, L. 276
Wei, W. 577, 608
Wei, Z. 185
Wen, Q. 99, 116, 389
Wu, D. 482, 572
Wu, S. 157
Wu, T. 192
Wu, X. 615, 737
Wu, Z. 243, 515

Xiao, Z. 515
Xie, D. 350
Xie, P. 281, 312, 319
Xie, Z. 631
Xin, Y. 691
Xiong, L. 438
Xu, B. 37, 42, 711
Xu, C. 369
Xu, G. 338
Xu, L. 86
Xu, P. 649
Xu, S. 192
Xu, Y. 691

Yan, J.J. 398
Yan, L. 728
Yan, X. 539, 562, 631
Yang, G. 57
Yang, J. 206, 515, 515
Yang, L. 149
Yang, P. 507
Yang, X. 15, 546
Yao, Y.M. 398
Yin, Z. 237
Ying, H.W. 398
Ying, S. 591
You, F. 515
Yu, J. 572
Yu, X.H. 577
Yuan, X. 751

Zang, Y.W. 398
Zeng, Z. 206, 608
Zeng, Z.Q. 577
Zhang, D. 350
Zhang, G. 411
Zhang, J. 294
Zhang, J.W. 125
Zhang, K. 350, 719
Zhang, Q. 405
Zhang, S. 108, 171
Zhang, T. 490
Zhang, Y. 165
Zhang, Y.-J. 405
Zhang, Z. 237
Zhao, K. 463
Zhao, T. 206
Zhao, W. 3
Zhao, X. 37, 42, 319
Zhao, Y. 344
Zhen, Q. 332
Zheng, J. 300
Zheng, Y. 664
Zhong, Y. 57, 376
Zhou, C. 86
Zhou, H. 32
Zhou, J. 281, 312, 319
Zhou, T.C. 200
Zhou, X. 664
Zhou, Z. 422
Zhu, G. 546
Zhu, H. 411
Zhu, J. 369, 562
Zhu, X. 376
Zhu, Y. 438